PREFACE

THIS BOOK has developed from my more than ten years of involvement in introductory courses in geography. Over this period I have had some experience with several types of beginning courses and approaches, including the "basic physical" approach, the "introductory cultural" approach, and the "world regional" approach. I imagine that everyone who teaches introductory geography at the college level has wondered about the relative effectiveness of each of these approaches. Classes continue to grow larger, and are made up not only of students who will choose geography as their professional field (or have already done so) but also of many students for whom this one introductory course will be the only contact with geography. Is it possible to create a course that stimulates the student who majors in geography by giving him a glimpse of the exciting things that lie ahead, but also fires the enthusiasm of other students for whom this course may make a lifetime impression of what geography can be?

The Regional Method: Is It Sound?

This book represents the objectives that I have tried to attain in my own efforts to develop a responsible introductory course by using the world regional approach in its basic sequence. And, after some years of experimentation, I am less prepared than I once was to argue that this approach is unsound as a vehicle for fundamental geography. When an introductory course is based solely on "theory" (even using this much-abused term in its broadest possible application), it can all too easily become an abstract, clinical exercise in spatial arithmetic. I used to give a three-lecture sequence to introduce students to the work of Von Thünen and Christaller, and one or two more modern exponents of location theory, with, I thought, very limited success. Then I began to wonder what would happen if I tried to place Von Thünen in the temporal and spatial context of Europe's nineteenth century search for order. Until that time, in my lecture sequence I discussed "concepts" in the first half of the semester, and then in the second half I applied the learned concepts to world regions. I decided to reorganize, and discussed Von Thünen and Christaller in the context of Europe's spatial transformation in response to the industrial and agricultural revolutions, the nation-state in relation to the emergence of the Soviet Union out of the Russian core, culture hearths in connection with the rise of the great Meso-american civilizations, urban hierarchy and function in a consideration of the cities of South America, and continental drift and peneplanation in a view of Africa. In terms of both the level of interest and the performance of the students there was no doubt that the reorganization was an improvement.

Inventory: Is It Inevitable?

It is sometimes said that discussions of real-world regions or countries tend to degenerate into inventories of physical and human content, and that these discussions deal mostly with current affairs, so that the student is left with something that is soon outdated. Perhaps this is true, but geography is primarily (although, of course, not exclusively) a social science, and I want to demonstrate to students who have had only one brief contact with geography that we *are* concerned with human issues as are other social scientists, and that we *do* have a way of approaching actual problems, both in our own society and in others. Geographers who ten years ago outlined to students the inequities of land ownership in certain countries of Latin America, the enforced territorial separation of the races in South Africa, or the locational characteristics of depressed urban areas in the United States said nothing then that is out of date today. An introduction to our field, with its emphasis on *world* regional geography, should also demonstrate a respect for cultures, societies, and political and economic systems other than our own (present and past); it should counter ethnocentrism and foster a lasting concern about social injustice, which so often has spatial overtones. And as for the inventory problem, this indeed can be an undesirable side effect of the world regional approach—but just as the language student must learn a small beginning vocabulary, it is surely not too much to ask the geography student to have some idea of major world distributions, of the location (relative to each other) of the major states of the world, and, yes, even some specific places. Obviously the inventory should not be the mainstay of an introductory course, but let us not carry the undesirability of a few real-world details too far.

In this book I have attempted to place approximately one hundred concepts and ideas in a regional perspective. Many of these concepts are truly geographical; others are ideas about which I believe students of geography should have some knowledge. Of course, I have not listed at the chapter openings every concept used in that chapter; nor are every listed concept and idea specifically identified within the chapter, although most of them are. Most teachers, I suspect, will want to make their own region-concept associations and, as readers of this book will readily perceive, the arrangement is quite flexible. For example, I have discussed problems of city classification and theories of urban structure in the context of South America; there is no reason why this discussion cannot be made relevant, say, to North America, and why the discussion of demography, which is part of a chapter on Europe, cannot be related to another region of the world. I do suggest that certain concepts be kept in their regional setting because they could hardly serve better than they do where they are: obviously the Mainland-Rimland concept of Professor Augelli has specific relevance to Middle America, and the concept of Continental Drift nowhere makes more sense than it does in Africa. These are just two examples; others will emerge in the reading. And undoubtedly there will be cases where the concept does not appear to "fit" comfortably in the regional context in which I have placed it. Surprisingly, this nevertheless has pedagogical advantages. When I ask my students why they sense a lack of ready association in some instances, they quickly respond that they simply don't see how so "Western" an idea can be of importance or relevance in a non-Western part of the world. How, for example, can maritime questions—boundaries, historic waters, and the like—mean much to peoples of Southeast Asia? Well, it does not take a very difficult exercise in research (a content analysis of some major United States newspapers is enough) to prove that this has been a vital matter to the Java-based government of water-fragmented Indonesia. Thus an association of apparently limited relevance can be turned into a major fact-finding adventure and can constitute a real teaching asset.

GEOGRAPHY REGIONS & CONCEPTS

HARM J. de BLIJ
UNIVERSITY OF MIAMI

with a chapter by

STEPHEN S. BIRDSALL
UNIVERSITY OF NORTH CAROLINA

and cartography by

HANS J. STOLLE
WESTERN MICHIGAN UNIVERSITY

John Wiley & Sons, Inc.

New York   London   Sydney   Toronto

Library of Congress Catalogue Card Number: 73-127662

ISBN 0-471-20060-3

Printed in the United States of America

10 9 8 7 6 5 4 3 2

To Katherine, Tanya, and Hugh James

The Method: How Flexible Is It?

In a thoughtful review of an earlier book, a reviewer castigated me for not leaving anything for the instructor to do. "A thematic approach is all very well," he said, "but what is the lecturer going to say?" I do not believe that such a problem will arise with this book. In fact, the instructor has a great range of opportunities to mold a course of his own liking, to transfer concepts, and to focus his presentations mainly on conceptual matters or on regionally oriented materials. Concepts are sometimes raised but not pursued in depth (this is true of one or two concepts in every chapter), so that the lecturer may choose to penetrate these concepts or ideas to greater detail. Neither does every paragraph on any region deal specifically with that particular region. This opens the possibilities of extrapolation to other regions: there are buffer zones in several continents, the insurgent state is a phenomenon in the Americas, Africa, as well as Asia. As the "bank" of known concepts grows, the opportunities to return to regions discussed earlier and to establish their relevance there also expand. For example, I have found that the idea of population/resource regions, which I do not reach until I discuss North Africa and Southwest Asia, excites a great deal of interest and quickly leads to comparisons to other world areas already discussed.

The number of ideas and concepts actually identified in this book, of course, is not limited to the one hundred I have listed by name. Many others are given stress in the text without any headlining. Still others appear not once but several times: complementarity is discussed in the context of evolving West Africa and also (as regional interdependence) in connection with Korea before partition. Again, this enhances the flexibility of the material presented, since the instructor can decide where he wishes to emphasize this aspect, and where he wants to play it down. No doubt there will be many better ideas for linking concepts and regions than I have had, concepts I have omitted that should be there, and association-opportunities that I have missed. I shall be grateful for any suggestions that readers may be willing to send me.

Since this book is written primarily for readers in North America, and since most of its audience will be better (or, at least, more directly) acquainted with North America than with any other part of the world, I included a somewhat more challenging chapter on this region. I was indeed fortunate that Professor S. S. Birdsall of the Department of Geography of the University of North Carolina was willing to undertake this work. Dr. Birdsall's chapter illustrates very well what has been attempted throughout this book: a balanced look at several sides of America, a respect for nonwhite and non-Western cultures. And also there is a search for balance in other respects. If we are going to read parts of Hoebel's *Anthropology*, then let us not be afraid to let students be impressed by Ardrey's *Territorial Imperative*. If it is suitable to read Gunther's *Inside Africa*, then surely there is reason to let students have a look at Snow's *Other Side of the River*. I have suggested some of these works in the bibliographic appendages to each chapter, and the lecturer who manages to get his students to take a look at sections of these volumes will add flavor and excitement to his classroom discussions. To some extent, the course of these discussions and debates in my own freshman classes has affected the writing of this book; invariably I find that students are extraordinarily interested in such issues as the old cultures of America and Africa, in the Ainu of Japan and the Maori of New Zealand, and in the difficult questions relating to the dispersal of ancient peoples.

If I speak with enthusiasm about this effort, I do not do so without an awareness of the inevitable shortcomings of a work as wide-ranging and vast as this. In my acknowledgments, I can reflect only a small measure of my truly deep gratitude to all those who willingly helped me over the several years in which this book was in progress. Colleagues wrote lengthy and enormously useful commentaries, students read the mimeographed second and third drafts and made many suggestions that have been incorporated. I hope that the innovation of concept-region associations will be a productive one but, whatever the merits of this book may be, they are largely attributable to all those who took time to advise and assist; for the demerits, of course, I am solely responsible.

A final word on the bibliographic entries at the end of each chapter. I have tried to confine my suggestions (1) to standard works on the region and/or concept under discussion and (2) to recently published books, still available, and preferably in a relatively low-priced paperback edition. With the large student numbers, and sometimes restricted library privileges, there are obvious advantages in this orientation to the bibliographic sections. In general I have not repeated in the discursive readings sections the works already footnoted in the text, unless they have special importance.

Coral Gables, Florida, 1970   *Harm J. deBlij*

ACKNOWLEDGMENTS

I owe a special debt of gratitude to Dr. S. S. Birdsall of the University of North Carolina, not only for his willingness to undertake the task of writing the chapter on North America, but also for his constructive interest in this volume and its objectives. I am also very grateful to our cartographer, Mr. Hans Stolle of Western Michigan University, whose work speaks for itself and constitutes a major asset to this book. I also record my appreciation for the meticulous and productive review of two of the drafts of this manuscript done by Professor Theodore W. Kury of the State University College at Buffalo, New York. A similarly intensive review was done by three members of the faculty of the Department of Geography of the University of Georgia, Dr. Merle C. Prunty, Jr., Dr. Louis DeVorsey, and Dr. John Tuck. To these colleagues I owe an enormous debt of gratitude, for their reviews went far beyond what an author might normally expect. Practically every page and paragraph of this volume were in some way improved by their suggestions; indeed, its whole structure bears the imprint of their ideas.

For a working paper on periodic markets in Africa I am grateful to Dr. John M. Hunter of Michigan State University, and Dr. Ronald J. Horvath of Michigan State University provided me with useful materials relating to the Von Thünen sections. For incidental details in connection with the Africa chapter, I am grateful to Dr. Derrick J. Thom of Utah State University; Dr. James W. King of the University of Utah reviewed in great detail the chapters on Africa south of the Sahara as well as North Africa and Southwest Asia. Dr. Robert C. Eidt of the University of Wisconsin, Milwaukee, reviewed the chapters on Middle and South America, and Dr. Dieter H. Brunnschweiler of Michigan State University considered these chapters also. For a detailed review of the chapters relating to Europe, I am indebted to Dr. Guido G. Weigend of Rutgers University, and Dr. John D. Eyre of the University of North Carolina commented on the chapters covering India and Southeast Asia. Dr. Paul E. Lydolph of the University of Wisconsin, Milwaukee, reviewed the sections dealing with the Soviet Union, and many of his suggestions are embodied in the chapter as it now appears.

Dr. Gordon J. Fielding of the University of California, Irvine, reacted to a draft of the chapter on Australia, and Professor Edward Myles of the University of Oklahoma commented on this chapter also. Dr. Paul Sanford Salter of the University of Miami gave selflessly of his time in his consideration of the chapters on China and Japan, and in other connections as well. Dr. Robert A. Harper of the University of Maryland similarly commented critically on parts of the manuscript. I emphasize again that I alone remain responsible for the contents of this book; indeed, it was not always possible to adopt all suggestions made by those who were kind enough to read drafts of the manuscript.

A number of persons have assisted me in other connections. Mr. J. W. Wright, a graduate student at the University of Miami, provided me with some important materials on China; Mr. R. Willich, also of the Department of Geography at the University of Miami, worked on recent census data. At Michigan State University, where I wrote a sizable part of this book, I had the expert help of Mrs. B. Naedele and Mrs. J. Lawson of the Department of Geography and Mrs. J. McKulsky, Mrs. K. Bishop, and Mrs. L. Ernsberger of the African Studies Center. Again at the University of Miami, Mrs. M. Pesetsky worked on parts of the second draft, Mrs. A. K. Brennan, Secretary of the Department of Geography, assisted me in several ways, and Miss Florence Dawson, a geography major, did a complete typing of the final draft. Miss Sue Haefley completed the work on the index. In the final stages she was assisted by Mr. Richard Vermeer and Miss Debbie Patterson. I am indeed grateful for the excellent work these people produced.

Finally, I thank the members of the staff of my publishers; no one could have asked for more and all too often these efforts go unreported. My persistent editor, Mr. Paul A. Lee, proved that he can also be patient. Miss Cherrie Haynes and Mrs. Harriet Sappe performed innumerable and vital operations in the receipt and dispatch of manuscripts, the organizing of conferences, and really countless other matters. Mrs. Audrey Labaton did the picture research on much of this volume and, as the photographs prove, she did truly excellent work. The picture research for the first two chapters was done by Mrs. Olivia Buehl. The design of this volume which, like the cartography, speaks best for itself, was done by Robert Goff, Design Director at John Wiley and Sons, Inc.; Production Manager Dennis Hudson supervised the transition from manuscript to publication. The editorial work was guided by Malcolm Easterlin, Editorial Supervisor, and distribution was organized by Frederic H. Hahn, Manager of Marketing Services. I am also appreciative of the work of Mrs. Joan Rosenberg, and Miss Elaine Miller, who did the copy editing. In the early stages of the preparation of this volume, much benefit was derived from the expert attention of Mr. Ken Burke.

To all these people and to many others who directly and indirectly contributed to the completion and publication of this volume, I again express my lasting gratitude.

H. J. deB.

CONTENTS

1

REGIONS OF THE WORLD

Psychologists tell us that the processes of human thought and communication are marked by the use of abstractions. With this device we are able to categorize and classify mentally a wide variety of objects and events, and thus formulate general ideas. For thousands of years the relationships between observed facts and prevailing ideas have been discussed and debated; often newly discovered facts forced a revision of long accepted ideas. In Europe and the Western world a fairly good record of these debates has accumulated since the time of Plato, and this record reflects the gradual but accelerating pace of progress in science and technology. In geography, as in other sciences, general ideas have emerged, some of which have had to be discarded, while others were strengthened as time went along. Such general ideas are called *concepts*.

Modern concepts often are complicated constructions that require mathematical development, but we use others, almost without realizing it, as a part of our everyday conversation. For example, if we were asked to identify the sort of climate that exists, say, halfway up the Amazon River, our response would no doubt be "tropical." By this we represent a set of prevailing environmental conditions: high temperature, much humidity, considerable annual precipitation, among others. In a very general way this is a conceptualization, although a rather imprecise one. It might be quite difficult to place a line on the map of South America showing just where this "tropical" climate does prevail, and if asked to be specific about those prevailing conditions, we would probably have to refer to local weather bureau statistics. But none of this would alter our impression, our mental generalization, of what we intend to convey when referring to a tropical clime.

Then again, in response to the question "What is Europe?" we might unconsciously answer in a conceptual way by saying "Europe is a region." In so doing we have presented one of the central generalizing concepts of the field of geography.[1] We expressed the opinion that Europe possesses a certain distinctiveness, a degree of homogeneity, a kind of functional cohesion that justifies its identification as a region. The trouble with this response is that it is a rather subjective, intuitive view; we have allotted to Europe a regional uniqueness which it may, upon closer analysis, not actually possess. Just as our facile generalizations regarding environment in the Amazon Basin would require modification upon more direct scrutiny, so our distant and rather

[1]For a detailed discussion of this idea see Derwent Whittlesey, "The Regional Concept and the Regional Method," in Preston E. James and Clarence F. Jones (Eds.), *American Geography—Inventory and Prospect*, Syracuse University Press, Syracuse, N.Y., 1954, pp. 21–68.

vague identification of Europe as a regional unit would collapse in the face of detailed examination. In fact, it would soon seem incredible that peoples and countries as diverse as the Portuguese and the Finns, or Ireland and Italy could, under any circumstances, be placed in the same class or category.

Thus a region, a term so easily and frequently used and a concept so deceptively simple, is actually very difficult to define—and it is even more difficult to put the concept to really productive use. Of course if it is assumed that every region is unique and independent, then geographical studies of regions become inventories of contents, without a great deal of general or universal relevance. For many decades geographers produced meticulously researched descriptions that attested to the powers of observation and the literary skills of their authors. But few

Map 1-1 EUROPE IN THE LAND HEMISPHERE.

contributions were made to the refinement of the regional concept. Whenever excessive attention is paid to things that appear to be unique, opportunities to generalize are lost.

The Region and Its Characteristics

Partly because of this preoccupation with uniqueness, it has not proved easy to make the regional concept "operational." It may be easy to visualize a region, but it can be very difficult to define and delimit it accurately and incontrovertibly.

What are the properties of regions—and are any of them really unique? Let us take one aspect of Europe as a region that no other region could possibly possess: its location. All regions have a position or location on the globe. Europe occupies the western extremities of the vast landmass that is known as Eurasia; in fact, it might be described as a peninsular extension of the huge continent. No other region can lie where Europe does, and in this respect—location—the European region and all other regions are surely unique. This argument is sound only if we identify the type of location to which we are referring. Europe lies approximately between 35 and 75 degrees north latitude, and between 10 degrees west and 30 degrees east longitude. These numbers represent its *absolute* location, and in these terms it is indeed possible to sustain the argument that Europe has an element of uniqueness.

But is location really relevant without reference to *other* locations? What would be a more meaningful response to the question "Where is Europe?"—a listing of the latitudinal and longitudinal figures stated above, or the point that Europe may be said to lie at the center of the Land Hemisphere (Map 1-1). Now we are referring to a different type of location, namely *relative* location. Map 1-1 shows the globe's continental landmasses as they are spread about Europe; the concentric circles, of course, indicate the straight-line distances from Europe (Hamburg was used as the focus) to other parts of the world. Since Europe's central location was never more significant than it was during the period of seaborne expansion, it would be appropriate to distort this map to show the proximity of the world's coasts to the European region. Such a map would represent the same principle: Europe's position is one of maximum efficiency for contact with the rest of the world.

When we examine regions from the standpoint of relative location, all sorts of conclusions can be drawn. The farther east one goes, from Western into Central and Eastern Europe and eventually Asia, the greater is the land area's remoteness from that essential ingredient of European growth: the sea. Not only is Europe a western peninsula of Eurasia but the water interdigitates with the land as it does nowhere else. Southern and Western Europe are practically made of peninsulas and islands, from Greece, Italy, France, and the Iberian Peninsula (Spain and Portugal), to the British Isles, Denmark, Norway, and Sweden. Southern Europe faces the Mediterranean, and Western Europe virtually surrounds the North Sea and looks out over the Atlantic Ocean. Beyond the Mediterranean lies Africa, and across the Atlantic, the Americas. Europe has long been a point of contact between peoples and cultures, and of circulation of goods and ideas. The hundreds of miles of navigable rivers, the bays, straits, and channels between the islands and peninsulas and the mainland, the Mediterranean and the North Sea, and later the oceans provided the avenues for these exchanges. Even on the landward side, Europe is open to contact. West and north around the arc of the Alps, a vast and virtually unbroken lowland leads from the coasts of the North Sea through Germany and Poland into Russia and beyond (Map 1-2). The first prominent north–south physical feature to break it is the Ural Mountains. There is considerably more relief in Southern and Southeastern Europe, but not enough to present effective obstacles to contact with the Arab–Islamic world—as is proved by the history of the Balkans.

A region has location; it also has *area*. This would seem so obvious as to hardly deserve mention, but this apparently simple statement leads us to several of the major problems related to the whole regional concept. We are now faced with the need to examine the content of the European region, to determine what Europe is and what it is *not,* to provide an adequate justification for the factors or criteria we have employed, and to present our conclusions by drawing the European regional boundary on the map. A region has area and therefore has limits. But before turning specifically to the European case, let us look in these terms at regions in general.

Clearly, the world is so complex that it cannot be studied fruitfully at once and as a whole. This is true of the human as well as the physical world. So vast is the range of human activities—economic, political, social—and so immense the variety of distributions and patterns of man and man's works over the globe that areal subdivision is a matter of necessity. The physical–biotic world also presents enormous contrasts calling for a systematic breakdown. But the criteria on the basis of which such areal subdivisions are to be established are a matter for unending debate. In the physical world, it is sometimes possible to identify regions quite easily: there is little doubt that leaving the Rocky Mountains, say, in the vicinity of Denver and entering the Great

Map 1-2 RELIEF OF EUROPE.

Plains involves leaving one physical region and entering another. But the human world is mobile, interconnected, and interacting; clear-cut regional boundaries are virtually absent. Indeed, let us not pretend that Europe's western and southern regional boundaries are absolute because they constitute the land limits of the continent. It would be easy to support the argument that Europe is but a part of a greater Western world for which the Atlantic has become an interior sea just as the Mediterranean once was to Rome.

Regions, then, are always artificial constructs, intellectual devices if you like, designed to function as organizing concepts in geography. The broadest possible definition of a region is probably that it "is an area of specific location which is in some way distinctive from other areas and which extends as far as that distinction extends."[2] "Distinctiveness" can mean almost anything, and for the regional concept to have some utility, something more specific is clearly needed. Therefore, it is always best to identify the *type* of region that is being defined in order to give an indication of the major criteria that are used when the region is established. The French language, for example, is spoken not only in France itself but in most of southern Belgium (the variety known as French Walloon), in part of Switzerland, and in a few sections of northwestern Italy. On the basis of a certain index of prevalence of the French language, a *linguistic* region could thus be defined (Map 1-3). From another point of view it can be argued that since a state is a political unit within which a single governmental system operates, it constitutes a *political* region. Again, on the basis of certain criteria related to temperatures and precipitation, *climatic* regions can be delimited. These, of course, are still very general examples, and they are based on relatively static features in the landscape: languages change, but only slowly; the political boundaries are there for us simply to record; climatic regions are established by manipulating long-term weather records. Things become much more complicated when we get involved in the functional aspects of regions and the processes that have shaped them.

Man's world consists of a multitude of interrelated parts. Not a single element of human

activity is without its relationships to other elements, and all form part of a great, worldwide complex of systems. Culture, a property of all peoples, consists (as we shall see later in this chapter) of a set of systems of human endeavor. Attempts can thus be made to identify regions on the basis of their functional organization. This involves a great deal more than merely mapping an area of linguistic, religious, or ethnic distinctiveness or "homogeneity." Many interrelationships must be examined—not only between human features, but between human and physical features as well. The determination of an agricultural region, for example, will require consideration of such features as slope incidence, soil quality, and climatic regime, as well as the nature of the available farming technology, manpower, the crops grown, the transport services, markets—all the inputs and outputs of the system.

Regions, then, may be seen in two ways: they may be areas within which homogeneity prevails in terms of one or more categories of phenomena (such as language or religious preference), or they may have functional coherence in terms of several interrelated, interacting categories of phenomena. The terms "formal" and "functional" are also used to identify the two types. Clearly an area that is part of several overlapping "formal" regions is likely to have a great deal of that elusive "distinctiveness" mentioned earlier in this discussion. A "functional" region, on the other hand, may not actually be visibly "homogeneous" at all. A city or town and its surrounding trade area form a functional regional system, but its limits may be indiscernable in the landscape. Only elaborate analyses will determine the location of the regional boundaries, and only maps will show them clearly.

How then does all this help us in establishing an eastern regional boundary for the European region? Let us first select a suitable approach. As we have seen, both formal and functional regions can be based on single or multiple categories of phenomena, although a functional region is only rarely sustained by a single feature (for example, a farm or a river system). But any attempt to identify functional coherence on a continental scale is unlikely to succeed; Europe is too complex to make this feasible. We can, however, establish a number of single-category formal regions, employing every phenomenon that should be considered as a measure of re-

[2] R. Hartshorne, *Perspective on the Nature of Geography,* Rand McNally, Chicago, 1959, p. 130.

Map 1–3 LANGUAGES OF EUROPE.

gional identity: ethnic characteristics, language, religious adherence, physiography (including natural landscapes, climate, and vegetation), social, political, and economic systems (and population distribution and productivity), degree of mobility (as reflected by the intensity of communication lines), urbanization, and cultural-historical evolution. Possibly there are others.

One way to carry out the recording procedure would be to place a grid of small squares over as large-scale a map of Western Eurasia as can be managed. Each of the squares is then given a numerical value for every category employed, and in the end a final map is drawn representing the total record. What may happen is that over much of what we intuitively "know" to be Europe, the categories we have chosen remain rather constant in their occurrence (in other words, they display low areal variance, and there is homogeneity here). But eventually there will be change; the values will rapidly decline in some categories and perhaps increase in others. The transition zone from what is Europe to what is not Europe emerges. Through this transition zone we must draw the regional boundary of Europe.

However precise our attempts to define the European regional boundary, it can never be absolute in its validity. First, our choice of criteria may be open to argument; second, our allotment of measured values to such phenomena as language or mobility may require revision. Some geographers use the method of *superimposition* to reach their solution to the problem. They draw a series of maps, to the same scale, on which they represent sharp breaks in the landscape. They use physical as well as human criteria: mountain ranges and major rivers may figure in their evaluation just as (in the case of Europe) adherence to Christian religions and the use of Teutonic, Romanic (Romance), and Slavic languages. Finally all these maps are placed on top of one another, and again a transition zone is recognized; through this transition zone a boundary is drawn, either by the use of some mathematical method (the determination of a median line) or by arbitrary means.

Whatever the method employed, the inevitable conclusion is that boundary lines between world regions are normally zones of transition and interdigitation rather than sharply defined breaks. The problem is that our carefully determined regional boundary may cut through several countries; it may even tell us that parts of Spain, Finland, and Yugoslavia are not "Europe." Furthermore, our analysis may produce a "perforated" region, in that pockets of territory *within* the regional boundary fail to meet the regional standards initially established. This may be a credit to the precision of the regional analysis, but it must lead to compromise: the cultural region on a world scale should have utility as well as validity.

Until now we have talked of Europe as a multiple-feature region, that is, a region whose limits are determined with the use of several categories of phenomena. One way to avoid the problems related to the transitional character of the resulting boundaries is to select only one category of phenomena, and to use *not* a grid of squares, but the grid of sorts already provided by the world map: the political framework. Take, for example, the single category of *size*—the total area of "European" states. On this basis Russia is in a class by itself (Table 1-1); the interval between it and the second largest state (France) is about fifty times as large as is the next largest intra-European interval (between Sweden and Finland). Indeed, Russia west of the Urals constitutes a larger area than all countries of Europe together.

Table 1-1

Rank	State	Area (square miles)
1	"European" Russia	2,079,565
2	France	212,822
3	Spain	194,345
4	Sweden	173,666
5	Finland	130,119
6	Norway	125,064
7	Poland	120,348
8	Italy	116,303
9	Yugoslavia	98,766
10	West Germany	95,928
11	United Kingdom	94,214
12	Romania	91,698
13	Greece	50,547
14	Czechoslovakia	49,367
15	Bulgaria	42,729
16	East Germany	41,815
17	Iceland	39,777
18	Portugal	35,340
19	Austria	32,374
20	Ireland	27,136
21	Denmark	16,619
22	Switzerland	15,941
23	Netherlands	12,978
24	Belgium	11,778
25	Albania	11,099

Table 1-2

Rank	State	Population (in thousands)
1	"European" Russia	183,642
2	West Germany	61,503
3	United Kingdom	56,719
4	Italy	55,490
5	France	50,993
6	Spain	33,209
7	Poland	33,117
8	Yugoslavia	20,912
9	Romania	20,206
10	East Germany	17,684
11	Czechoslovakia	15,018
12	Netherlands	13,198
13	Belgium	10,004
14	Portugal	9,796
15	Greece	9,011
16	Bulgaria	8,602
17	Sweden	8,017
18	Austria	7,509
19	Switzerland	6,304
20	Denmark	4,797
21	Finland	4,749
22	Norway	3,999
23	Ireland	2,956
24	Albania	2,003
25	Iceland	202

A similar, though less dramatic, picture emerges from a ranking of these countries by their total populations (Table 1-2). The interval between Russia and the most populous European state (West Germany, with nearly 60 million people) is almost seven times as large as that between France and Spain, Europe's next largest interval (over 17 million).[3] On these bases, European Russia constitutes a discrete region; this is so despite the fact that the figures used here are those for only the western part of that giant state. True, it would be possible to counter these conclusions with categories of phenomena pointing to a close regional association between the western Soviet Union and Europe, especially Eastern Europe. The choice of solution depends on the objectives involved, and the use to which we want to put the region we have defined.

This brings us to an issue which always presents itself when regional geography on a world scale is discussed. Having taken stock of all the many variables involved, we must decide where the regional boundaries lie—and most of the time we use the existing political boundaries to define world regions. This is done for a very simple reason: the world's political boundaries actually constitute the best available "grid" we have, a framework that is familiar and forms a

[3]See Table 1-2.

good frame of reference. We know very well that there are strong "Latin" influences in the southwestern United States and that the United States in turn has a visible impact in Mexico, but the best regional boundary between what has been called "Anglo" America and "Latin" America still lies along the political boundary between the United States and Mexico. If we agree that Poland lies in Eastern Europe and that Europe is a different culture region from the U.S.S.R., there is really no point in arguing that, because the Soviet impact in eastern Poland has been so great (the boundary there has even shifted back and forth), the regional boundary should lie somewhere inside Poland. Rather we should identify these areas of transition for what they are and retain the political framework for our reference. This is what we shall do in this book.

Looking again at the boundary of Europe, we observe it to begin in the north along the border between Finland and the Soviet Union. It follows this line into the Gulf of Finland, and traverses the Baltic Sea to the northern end of the Poland–U.S.S.R. boundary. From this point it coincides with the western boundary of Russia, entering the Black Sea at the mouth of the Danube River. It then turns southwestward into the Aegean Sea, separating Bulgaria and Greece from Turkey. This boundary defines a European realm, a "total" region so complex that it would be impossible to analyze every aspect of its human content. For this kind of region the late Professor D. Whittlesey introduced the term *compage*, but this concept never really took root.

The World Region as a Culture Realm

Thus, when we argue that Europe constitutes a region, we answer in a conceptual way—and thereby raise a lot of related problems. One of these is easy to understand: if Europe is a region, and parts of Europe themselves form regions (Scandinavia, for example), and other regions of ever smaller dimensions can be recognized (the Paris Basin, the Frisian language region), then we are soon tempted to try to establish a sort of hierarchy of regions. If Europe is a region, then Scandinavia is a subregion, and the Paris Basin is a sub-subregion, and so forth. Obviously this is an unworkable solution. We are trying to categorize changes in scale, and the variety of possibilities is infinite: there are not enough words to apply to all the possibilities. And even if we did produce the necessary words, their usefulness would be limited. What constitutes a "subregion" in Europe may not be the same in, say, South America. The categories of phenomena selected to determine subregions in Europe may not be usable in a similar effort in another part of the world. Thus there is the danger that we might succumb again to the view that regions are unique, ultimately not capable of classification and universal applicability.

This is one of the major problems in the science of geography. But to abandon regional geography because no satisfactory solutions have been found is surely premature. Every science has problems of classification and categorization,

MODIFIED AFTER: E. IMHOF, <u>SCHWEIZERISCHER MITTELSCHULATLAS</u>

Map 1–4 EUROPE: LOCATION MAP.

and regional geography is beset by an immense range of relevant phenomena. Besides, we would throw overboard a whole area of communication and exchange of ideas. The world is full of obvious, large-scale realities; ask anyone who is at all aware of the world around him to make a list of the major world areas he recognizes. Undoubtedly Europe will be one of them, and so will the Middle East, Southeast Asia, and South America. Such regional names as "the Arab World," "Black Africa," and "the Chinese Sphere" will often occur on the list. Are these broad "regions of intuitive consensus" useless because they are indeed subjective and imprecise in their definition? Of course not. Our respondent—consciously or not—is using categories of phenomena that appear to him to be of overriding importance in his impressionist regionalization of the world. In some cases the list of "regions" may turn out simply to be an enumeration of the continents, and here the observer merely recorded the major outlines of his mental map of the world's land and water areas. But more often there will be evidence of a realization of differences in ways of life, traditions, ideas, and values between various parts of the world—a recognition of world *culture* areas.

Geographers, too, have sought to divide the world "into a modest number of major cultural regions, realms, or worlds . . . [this has] . . . become a fairly popular organizing principle . . . though differing considerably in detail such broad regional breakdowns show such basic similarity as to indicate that geographers find themselves in agreement as to the existence of and general outlines of such realms."[4] Obviously we are dealing here with the most general of regions—or *realms,* to get away from a possible misuse of the regional concept, and to emphasize the scale involved—and the culture realms represented on the map are bounded not by lines, but by deep transition zones.

In the context of the culture realm, how is culture defined? Like region, culture is a concept. Anthropologists and cultural geographers have devised numerous definitions: Hoebel states that culture is "the integrated system of learned be-havior patterns which are characteristic of the members of a society and which are not the result of biological inheritance. Culture is not genetically predetermined; it is noninstinctive. It is wholly the result of social invention and is transmitted and maintained solely through communication and learning."[5] A more complex and oft-quoted definition is that of Kroeber and Kluckhohn: "Culture consists of patterns, explicit and implicit, of and for behavior and transmitted by symbols, constituting the distinctive achievements of human groups, including their embodiments in artifacts. The essential core of culture consists of traditional (i.e., historically derived and selected) ideas and especially their attached values; culture systems may, on the other hand, be considered as products of action, and on the other as conditioning elements of further action."[6] A culture, then, is a more or less cohesive functional unit of prevailing ways of life, ideas, and values, having a common heritage.

If these definitions, in their diffuseness, are reminiscent of those given for the regional concept, the similarity is more than accidental. As in the case of the region, we are faced here with problems of scale; the term culture may be applied to immense sections of the human world and it may be applicable as well to relatively small units. And although the word game is played again to identify discrete classes ("world civilizations" and "macrocultures" for the largest ones) the result is no more satisfactory than it was in the case of the region. Obviously the concept of a "culture-region" or "culture-realm" is the ultimate in ambiguity. Nevertheless, it is a useful device. Note that such terms as "core," "system," and "pattern" occur in the conceptual definitions for the region as well as culture. Learning more about the contents of culture and its expression on the surface of the Earth helps geographers in their efforts to refine the regional concept.

As the preceding definitions suggest, cultures incorporate *systems* of various kinds. Perhaps the most important of these is language; without language there could not be the communication

[4]H. H. Aschmann, "Can Cultural Geography Be Taught?" in *Introductory Geography: Viewpoints and Themes,* Association of American Geographers, Washington, 1967, p. 70. For an example of a world regional framework see P. E. James, "Geography in an Age of Revolution," *The Journal of Geography,* Vol. LXII, 1963, pp. 97–103.

[5]E. A. Hoebel, *Anthropology: the Study of Man* (3rd ed.), McGraw-Hill, New York, 1966, p. 5.

[6]A. L. Kroeber and C. Kluckhohn, "Culture: A Critical Review of Concepts and Definition," *Papers of the Peabody Museum of American Archaeology and Ethnology,* Vol. 47, 1952, p. 181.

necessary for the propagation of a culture. The belief system—religion, philosophy—is another. There are political systems and economic systems. In these terms a culture is a vast mechanism of many interacting parts.

We in the Western world tend to have a distorted view of our own as well as other cultures. There are many reasons for this. All too often, we view our own culture (as we perceive it) to be the norm, from which other cultures are deviations. At times we try to compare incomparables, and then identify other cultures as "primitive" or "underdeveloped." Perhaps this is a trait we share with peoples of other cultures, although we know very little as yet about the way other peoples view their external world—including Western culture. This attitude has even been reflected in academic research. In urban geography, for example, the Western urban center has long been the standard of comparison; concepts and generalizations have been derived from Western examples, so that the non-Western city until recently received scant and rather unrewarding attention. Another reason lies in the power position so long held by Europe and the Western world, and what we might call the Westernization of the world. Many aspects of European culture (or Western culture) have been superimposed upon other cultures, and we have had a preoccupation with these Western elements in the non-Western world. Who thinks first of the Indian of South America when the very culture realm is identified as "Latin" America? What comes to mind in "Black Africa"—villages, homesteads, and periodic markets or the shining skyscrapers of Johannesburg and Nairobi? What is the "Arab world?" Villagers, peasants, and nomads—or Western oil companies and Western pipelines?

A culture is a continuum. What has just been said reflects a worldwide process that affects all cultures that are in contact with other cultures: the process of *acculturation.* Cultures are constantly changing, but no culture has ever stimulated change in other cultures as European culture has done in most of the rest of the world. Europeans explored, conquered, colonized, and proceeded to introduce their cultural attributes (language, religion, politico-territorial organization, technology) into non-European areas. The degree of impact in some of these areas has been astonishing, in view of the comparative brevity of the European invasion—in much of

Black Africa, for example—and it attests to the strength of Europe's various cultural systems. In fact it also reveals one or two Western cultural traits we might not readily have attributed to ourselves: aggressiveness and arrogance.

Europe, then, is a culture realm. But cultural features of European origin dominate many other parts of the world, and the European culture realm should perhaps be viewed as part of the totality of Western culture, which includes much of the Americas as well as Australia and New Zealand. Later we shall have occasion to refer to the European sources of many of the cultural traits that prevail in these other regions dominated by Western culture; for the moment Europe is our specific concern. Is it possible, out of the immense complexity of European culture, to identify major, overriding elements? Within what we have called the European culture realm, a number of individual cultures have been in contact with each other for many centuries. They have competed with one another, and the stimulus of interaction and competition produced continuing change and progress. To a greater or lesser degree, industrialization is a characteristic of Europe, though more so in the west and northwest than in the east and south. Great cities developed, and an ever larger part of the population became concentrated there: a high level of urbanization is another element in the European culture complex. Thus the United Kingdom was the first country in the world where more people lived in cities and towns than in the rural areas, and today it remains among the most highly urbanized countries—with over 90 percent of its population of about 55 million residing in urban centers.

The focus of industrializing, urbanizing Europe—the culture *hearth,* as such a source area of innovation and dispersal has been called—has for centuries been positioned on the shores of the North Sea. The core of modern Europe still remains here in the west, and in very broad terms the impact of Europe's own economic and political revolutions is recorded with decreasing intensity toward Europe's eastern and southern margins. One way to examine this point is by reference to a map showing the distribution of the region's major urban centers, industrial areas, and their interconnections. This, indeed, is another element of European culture: the high intensity of circulation of goods and people—its mobility. In Western and Northwestern Europe literally hundreds of cities and towns are linked by rail, road, and water, as well as by air; a myriad of transport lines crisscross the countryside. But eastward this intensity decreases, and mobility is less. It is not surprising; not only has Western Europe been more open to the outside world, with its peninsulas and seas, but also its rivers and canals have facilitated cheap internal transportation from an early stage, and its lowlands have favored circulation and contact between cultural units. Farther to the east the effects of comparative isolation, greater obstacles to communication, and a generally lesser resource content have all played their role in resisting the impact of the industrial, commercial, and agricultural revolutions that had their source in the west.

Our preoccupation with Europe's innovations in science and technology and their contributions to manufacturing and industry—and the attendant urbanization process—should not lead us to forget another major element in European culture: field agriculture and the individual ownership of land. To us it may seem perfectly normal that land is owned by individuals, from a small garden to a ranch of several thousand acres. In other cultures such personal ownership may be inconceivable. In fact, this is an entire area of conflict that did much to strain relationships between European colonizers and peoples of other cultures. Associated with this aspect of ownership is the mechanization of agriculture. The use of increasingly sophisticated farm machinery enables fewer farmers to fill the needs of the growing urbanized population. In non-Western cultures the efforts of the individual farmer are generally still the key to yields.

A Cultural Landscape

A culture gives character to an area. In all cultures, aesthetics play an important role, and often it is possible from a single photograph of a village or a street scene in a town to identify that general part of the world where the picture was taken. The architecture, modes of dress, and other elements of the scene allow us to make a pretty good guess. Perhaps we recognize a certain "atmosphere" that we associate with a particular culture area of which we have some knowledge.

These are obviously very subjective observations, and they are of little or no use in any scientific investigation. Still, there is the point that the people of a given culture, living in a given area (perhaps a distinct region) of the Earth's surface, proceed to transform that area by building their structures upon it, covering part of it with concrete and asphalt roadways, diverting natural waters into irrigation canals, cultivating fields, and so forth. This composite of man-made features is conceptually identified as a *cultural landscape,* a term that came into use in the 1920's. The geographer whose name is most closely associated with this concept is Carl O. Sauer, whose own definition was deceptively simple: the cultural landscape constitutes "the forms superimposed on the physical landscape by the activities of man."[7] But the physical or natural landscape may itself undergo some changes; cultivation may lead to gullying if the fields are abandoned, dams in rivers can affect the river regimes hundreds of miles downstream, where there is no cultural feature in sight. The debates surrounding the concept of the cultural landscape can be followed in the geographical literature; perhaps the best view of it is the broadest, namely that it includes all identifiably man-induced changes in the "natural" landscape (physical as well as biotic).

What was the purpose of the concept in the first place? Primarily it was intended to serve

A Viking ship. Uncounted numbers of Norsemen made the journey across North Sea waters in boats such as this. (Culver Pictures, Inc.)

[7] C. O. Sauer, "Recent Developments in Cultural Geography," in E. C. Hayes (Ed.), *Recent Developments in the Social Sciences,* New York, Lippincott, 1927, p. 186.

as an organizing construct for the so-called cultural-historical school of geography, in which interest focuses upon the origins of cultures, their former distributions, diffusion, and present character. By identifying as exactly as possible the elements of the cultural landscape of a certain area at several significant stages in its historical development, answers are sought to the problems thus posed. To take a European case: the last successful invaders of England were the Normans under William the Conqueror (1066). These Normans ("Northmen"), although they came from a part of France called Normandy and brought a Latin culture to England, were themselves descendants of Scandinavian Viking raiders who a century and a half earlier had invaded and occupied the peninsula they named after themselves. In other words, these Vikings were acculturated to the prevailing (and superior) French culture, and then they crossed the Channel and superimposed their civilization upon England. By piecing together evidence from fields such as archaeology, linguistics, and oral history, and by uncovering ancient settlement patterns and old traditions, geographers would like to reconstruct the cultural landscape that prevailed in Normandy prior to the invasion of England—and they would like to do the same in England after the episode, then to compare the results and identify those elements of the cultural landscape of Normandy that were directly introduced in England, and survived there.

If such attempts to reconstruct the cultural landscapes of history seem to be rather static in character, there is also a so-called functional approach in cultural geography. Geographers who pursue this line of research also recognize the existence of a cultural landscape, but they are more interested in the functioning systems and the processes that produced it. Obviously, present-day, existing culture areas provide the best laboratories for such studies, and so the historical aspect diminishes in importance. But this is not to suggest that the lessons learned do not have application to the past. Take, for example, the distribution of settlements—villages, towns, and cities—in any given culture area. These names themselves suggest a sort of heirarchy or rank order, from the least to the most important. The distances between them are related to the prevailing modes of transportation, the size of the settlements, and the services they have to offer. Great cities are farther apart than

Old ways of contact: European integration began in the days of the Roman Empire, and many old roads still serve. Hedge-lined roads in England and tree-lined arteries such as this on the Loire (at Pouilly) remind us of the old heritage of Europe. (George Martin/Rapho Guillumette)

towns, and the average distances between towns are greater than between villages. Thus mobility and circulation help fashion this major aspect of the cultural landscape, and transport innovations—faster automobiles, superhighways—may signal the death knell for a small rural village now bypassed, but, on the other hand, cause a boom for a nearby, growing medium-sized town. In the Western world, and especially in America, villages are generally sufficiently far removed from each other that the automobile is required for travel between them. But in many parts of Africa, walking (and, increasingly, bicycle riding) remains the main mode of transportation. As in America, people in Africa trade, and they must carry their goods to market. If they must do so on foot, then the distance they can walk with their load in a single day will be a major factor in determining the distances their settlements lie apart. What often happens is that people at different times carry their goods to different surrounding markets, so that a system of "periodic markets" develops. Here is a problem geographers of the functional cultural school would be interested in: what happens when a

The Romans sought to organize what was within their sphere, and to exclude and repel what lay outside of it. Thus the Romans were early practitioners of the *demarcation* (the marking on the ground) of political boundaries. This old Roman wall still stands in Northumberland, Britain. (J. Allan Cash/Rapho Guillumette Pictures)

whole new system of transportation is superimposed on such a landscape? Some villagers will find themselves on or near the new railroad and/or highway, while others may lie outside the zone of immediate impact. Obviously there will be a major readjustment of circulation patterns. How does this adjustment take place?

The answers to such questions do indeed have historical relevance. Let us remind ourselves of an early stage in the evolution of the European realm: the centuries of the Roman Empire. Among the many innovations Rome brought to large parts of Europe was a true network of communication lines—a system of roads of quite amazing intensity and unrivalled extent. Conscious of the need for organization and territorial integration in order to retain control over their far-flung dominions, the Romans laid thousands of miles of roads, even in distant England, Wales, and Scotland. True, these roads were built to establish effective connections between Rome and its possessions, and to move armies and supplies as rapidly and efficiently as possible. But at the time, just two thousand years ago, most of Europe was a backward area of subsistence and tribalism. The processes of adjust-

ment that must have occurred in response to the penetration of the Roman roads no doubt resemble in some ways the processes of change that accompany the superimposition of modern transport lines upon an area of subsistence and barter economy. In any event, cultural geographers of the functional school are more interested in determining the nature of these processes than they are in the piecing together of all the identifiable elements, say, of the Roman cultural landscape.

Present-day Europe is a region not of one cultural landscape, but several, each with its particular components and impressions. Geographers have grappled with the problem of the less tangible characteristics of the culture area, which are often so significant in producing a regional personality—a *Gestalt,* as a school of German psychologists in the early part of this century came to call a mentally perceived configuration involving many different parts. Gottmann, a European geographer, put it as follows:

"To be distinct from its surroundings, a region needs much more than a mountain or a valley, a given language or certain skills; it needs essentially a strong belief based on some religious creed, some social viewpoint, or some pattern of political memories, and often a combination of all three. Thus regionalism has what might be called an *iconography* as its foundation: each community has found for itself or was given an icon, a symbol slightly different from those cherished by its neighbors. For centuries the icon was cared for, adorned with whatever riches and jewels the community could supply. In many cases such an amount of labor and capital was invested that what started as a belief, or as a cult of or even the memory of a military feat, grew into a considerable economic investment around which the interest of an economic region united."[8]

Nowhere is this concept as applicable as it is in the Old World, where many centuries have gone into the evolution of cultural landscapes and into the molding of those systems of beliefs and values that promote the development of strong regional iconographies.,

Regions of the World

Twelve world regions, then, form the structural basis for our study. These twelve regions are delimited on several bases, some of which have been mentioned in the preceding discussion and still others which will emerge in the individual chapters. The regions differ enormously from each other—in size, population, degree of development, and potentials—but some of their characteristics are already familiar to us. The Soviet influence in Eastern Europe, the political crises in the Middle East, the instability of Southeast Asia, and the remote isolation of Australia are all common-knowledge features of the world around us. Now we shall try to fit these and other prominent regional properties into a productive intellectual framework. The twelve regions are summarized below (Map 1-5).

1. *Europe.* The European realm is defined as lying west of the U.S.S.R., with its boundary extending from Finland to the Aegean Sea. Europe is a relatively small but populous region, and very complex. It is made up of about two dozen countries, large and small. Much of our

[8]J. Gottmann, *A Geography of Europe* (3rd ed.), Holt, Rinehart and Winston, New York, 1962, p. 69. For a more complete statement of the concept see the same author's article, "Geography and International Relations," *World Politics,* Vol. III, No. 2, 1951, pp. 153–173.

study will deal with the impact of this European realm upon the rest of the world, the dominant theme of world history over the past several centuries—but today Europe is sustaining external impacts itself. Its eastern part lies under the Soviet sphere of influence, and Western Europe is strongly affected by the United States.

2. *The Soviet Union.* Although a single state, the U.S.S.R. (Union of Soviet Socialist Republics) is the largest world region territorially, extending from its European borders to the Pacific Ocean and from the Arctic southward deep into central Asia. Its ideological fervor and political power justify its consideration as a world region, but the Soviet Union is not homogeneous ethnically or culturally. Just two examples are its minorities of Mongoloid stock and its sizable Moslem population—both of which have produced serious internal and external problems.

3. *North America.* The United States and Canada are here identified collectively as North America, in preference to the term "Anglo" America which is sometimes used. The disadvantage of "Anglo" America is that many North Americans do not like it—America's Indians, the United States' black people, and Canada's French-speaking Quebeckers, for example. For these people, "Anglo" refers to an essentially alien heritage. But there is a disadvantage to *North* America as well, since this might be taken to include also the countries to the south of the U.S., from Mexico to Panama. Our definition of North America, nevertheless, confines itself to Canada and the United States.

4. *Middle America.* The countries between the Rio Grande and Colombia in South America, along with the islands of the Caribbean, constitute a quite distinct culture realm. From pre-European times this was a significant hearth of development, where cities were founded a thousand years ago, crops were domesticated, and empires built. This, too, was the scene of the first European arrivals, and for some time it remained the focus from which the white man's influences radiated outward. Middle America is therefore treated separately as one of our twelve world regions.

5. *South America.* The triangular continent also was the scene of impressive Indian civilizations and it, too, was overrun by the white man. And the overwhelming European influence was Latin, Iberian in its origin. Spain and Portugal between them occupied almost all of the continent, and

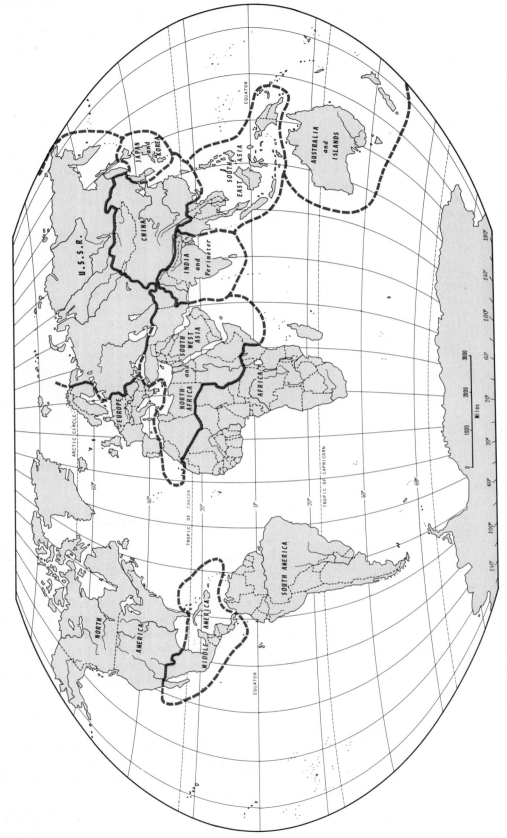

Map 1-5 REALMS OF THE WORLD.

16

from the colonial period emerged the modern political situation, with the former Portuguese sphere unified in the state of Brazil and the Spanish domain divided into nine countries. History and culture give "Latin" America a clear identity.

6. *Africa.* Africa south of the Sahara is the black man's Africa, but it is the Africa of white men, Asians, and Arabs as well. The transition zone between Black Africa and North Africa presents one of the most difficult boundary problems to be found anywhere. We define "Africa" to mean all those countries that lie to the *south* of the Mediterranean North African states, the Sudan, and Ethiopia. Thus we make use of existing political boundaries when the closest approximation of the cultural break between Black and Arab Africa would cut several countries in half, including the Sudan and perhaps Nigeria.

7. *North Africa and Southwest Asia.* Mediterranean North Africa, the "Middle East," and Iran and Afghanistan all form part of a tenuous and heterogeneous region whose common features are reflected in the terms often used to identify it. This is the so-called Arab world (although millions of its inhabitants are not Arabs), the "dry" world (although the exceptions to its dryness have made it what it is), the "world of Islam" (although millions belong to other faiths). Its widest definition is adopted here, so that it includes Ethiopia in the south, and Turkey in the north.

8. *India and the Indian Perimeter.* East of southwest Asia lies India, though not India alone. Our chapter heading makes reference to the set of countries and territories that surrounds Mother India, including East and West Pakistan, Nepal, Sikkim, and Bhutan, Kashmir, and the island state of Ceylon.

9. *China.* Certainly China alone constitutes a world region; with a population between one-quarter and one-third of all mankind, a great

cultural history, and an impressive political and economic resurgence under its distinctive version of Communist doctrine, China is almost a world in itself. China is a giant, but for all its greatness it lies barricaded to the rest of the world. There was a time when it was relevant to speak of darkest Africa, for Africa was indeed dark in the minds of people who otherwise knew the world; today it is darkest China, for our ignorance of it is great.

10. *Southeast Asia.* Between the giants, China and India (and, in another context, between China and the United States) lie the lands and islands of Southeast Asia. This is a varied and interesting region, which carries a record of very ancient occupation, several religious infusions, Chinese and European settlement, and political fragmentation.

11. *Japan.* Japan is but one country, and on the map it hardly looks like a world region. But in the non-Western world there is nothing else like this industrialized, urbanized, productive island nation of a hundred million people. Japan's search for raw materials for its industries encircles the world, and its industrial production floods the markets of many countries. The Japanese achievement is unparalleled; Japan's modern history is unique politically as well as economically. As we shall see, it is not difficult to support the contention that Japan is one of the major regions of the world.

12. *Australia.* Australia and the islands of the Pacific form a world region despite their small population—by far the smallest of all the regions listed here. Australia and New Zealand are western outposts off the southeast corner of Asia, as unlike neighboring Indonesia as Britain and America are unlike India.

These, then, are the twelve world regions that will form the focus of our study. As we discuss them individually we shall outline their boundaries and distinctive characteristics in detail.

ADDITIONAL READING

Several books give an insight into what geography is all about. Included are S. W. Wooldridge and W. G. East, *The Spirit and Purpose of Geography,* published in London by Hutchinson University Library in 1951; J. O. M. Broek, *Geography; its Scope and Spirit,* pub-

lished in Columbus by Charles E. Merrill Books in 1965; and, on a more difficult level, R. Hartshorne's *Nature of Geography,* published by the Association of American Geographers in 1946. In 1959, Professor Hartshorne wrote a *Perspective on the Nature of Geogra-*

phy, and this smaller volume was published by Rand McNally in Chicago in 1959. Concepts and scientific inquiry in geography are discussed by W. W. Bunge in *Theoretical Geography,* published by the Department of Geography of the University of Lund, Sweden, in 1962. More recently P. Haggett published a specialized work, *Locational Analysis in Human Geography,* available in New York from St. Martin's Press, 1966. In 1957, T. G. Taylor edited a collection of essays on various geographic fields of research, *Geography in the Twentieth Century,* and this volume was published by The Philosophical Library in New York in 1957. An earlier compendium is by P. E. James and C. F. Jones, editors, *American Geography: Inventory and Prospect,* published in 1954 by Syracuse University Press.

It is a good idea to acquaint yourself, on your next visit to the library, with the professional journals of geography. Many of the topics that are discussed in this book appear in articles in *The Annals* of the Association of American Geographers, *The Professional Geographer,* also published by the A.A.G., *The Geographical Review,* published by the American Geographical Society, *Economic Geography,* published at Clark University, and *The Journal of Geography,* the journal of the National Council for Geographic Education, published at the University of Miami. Various state and regional geographic organizations also publish journals: check the listing of journals under Geography.

2

THE GROWTH OF EUROPE

GEOGRAPHERS divide the Earth's surface into regions, anthropologists differentiate between cultures of man, and historians fragment time into *periods*. The regional concept, the concept of culture, and the concept of the time period have a great deal in common. It is no easier to identify discrete periods of human history than it is to define geographic regions. But if no region and no culture can adequately be studied in spatial isolation, neither should the temporal factor be forgotten.

The more distant periods of human history are the least well known. During the era of the dawn of man we refer vaguely to periods such as the Paleolithic, Mesolithic, and Neolithic (early, middle, and late *rock* periods, during which man used implements fashioned of flint as well as bone), and to succeeding *ages* such as the Bronze Age and Iron Age. The closer we come to the present, the more detailed is our knowledge of the historical sequence of events, their interrelationships, and their function in shaping existing world patterns. Hence, time periods tend to become shorter—and generalizing about the mass of available information becomes more difficult—as we proceed through history to the twentieth century.

Although man has lived in parts of Europe for many thousands of years, his beginnings on that continent are shrouded in mystery. He made his appearance during the last phase of the great Ice Age that covered much of Europe with glaciers (Map 2-1). The glaciers of this Pleistocene Ice Age advanced and retreated repeatedly, and during the warmer intervals European man emerged. He led a primitive existence; hunting and fishing (and some gathering of plant foods) were his modes of survival. The final advance of the great ice sheet took place quite recently— its retreat occurred within the past twelve thousand years—and during this time, the only

19

AFTER E. RAISZ, GOODE'S WORLD ATLAS

| | Tundra | | Steppe | | Forest |

Map 2-1 EUROPE DURING THE LAST ICE AGE.

areas in Europe suitable for habitation were in the south. Of all parts of Europe, these areas lie nearest to other regions where important developments were taking place (North Africa and the Middle East), and a stream of influences reached Europe from there.

This was true from the very beginning. During the Paleolithic period, the technique of chipping pieces of flint into hand axes seems to have been introduced from Africa. Archaeologists tell us that there were also migrations from North Africa into Mediterranean Europe, as excavations of ancient gravesites show. During the Mesolithic period an important local innovation occurred: the canoe came into use. But soon another external contribution was absorbed: agriculture, practiced in the Middle East since about 9000 B.C., began to supplement European hunting-fishing subsistence economies. Contact between Southern Europe and the Middle East and North Africa was probably made easier by several land bridges which may have crossed the Mediterranean Sea in those times. Such bridges may have existed between North Africa and the Iberian Peninsula at Gibraltar, across Sicily to Italy, and across the Turkish straits.

As the glacial ice made its last retreat, the great northern lowland of Europe was opened to settlement. And again Europe received an outside incursion—this time from Asia. During the Neolithic period and approaching the Bronze Age (which began about 2000 B.C.), strands of per-

manence in the European landscape could be detected. Europe's languages, for example, are related to those of various parts of Asia, not to those of Africa. What is less certain is the number of people involved in these westward migrations, and how many local residents of Europe were absorbed in the wave of invasions. Although organized along tribal lines and living a life of subsistence, the peoples of Central and Western Europe undoubtedly were sufficiently strong to offer more than token resistance to the invaders. Ultimately they were subjugated, for the westward flow continued relentlessly for many centuries.

It is important not to lose sight of the differences, in terms of the development of culture and civilization, between Europe and the lands across the Mediterranean, even at a time when Europe was receiving its crucial impact from the steppelands to the north of the Black and Caspian Seas. While fortified villages constituted the most Europe had to offer as sites for population concentrations, urbanization on a major scale already prevailed in the Middle East and North Africa. Trade was a vital part of the life of these advanced societies, and raw materials of ever greater variety were in demand. One obvious source for these raw materials (metals such as copper, silver, and tin; also oils and textiles) was in nearby Southeast Europe, and trade connections between Egypt and other parts of the Middle East developed with the lands of the Aegean Sea. It stimulated the bronze industry of Europe, and the effects of the trade were felt far in the European interior. For the first time a trade route across the Alps connected Central and Southeastern Europe. And again an outside stimulus provided the impetus for an important European development.

It is not surprising, then, that Mediterranean Europe was to become the scene of Europe's first stable, large-scale civilizations. During the first millenium B.C., Southeast Europe saw the Greeks lay the foundations of European civilization, while to the north of the Alps another significant event occurred: the wide dispersal of the Celtic peoples (or, more properly, the Celtic-speaking peoples). The Celts had been concentrated in the area of present-day southwestern Germany, eastern France, and northern Switzerland; they now spread as far afield as northern Spain, Czechoslovakia and the lower Danube area, and, of course, Britain. Little of the Celtic languages remains in the linguistic pattern of Europe today (Gaelic and Welsh are modified forms), and this attests to the comparative brevity of Celtic dominance across these parts of Europe. The Mediterranean challenge was soon to come.

The Beginnings. No one knows exactly where the Greeks came from, but among the many migrations that went on in Europe for thousands of years one route led from the general area of the lower Danube into the peninsula that is today Greece. During the second millenium B.C. these earliest Greeks subjugated the existing population there, mixed with it, and produced a civilization known as the Mycenaean. Although its representatives ranged far and wide, even beyond the shores of their territory to Crete and Cyprus and beyond, this civilization was short-lived. It was terminated as a series of population movements that lasted into the first millenium B.C. threw the peninsula into a state of disarray.

The decisive period for ancient Greece began with the expansion of the Hellenic peoples over a vast area of Southeastern Europe. From the Black Sea to Sicily and France, where Massilia (Marseilles) was a Hellenic port, and from Greece to the Nile Delta, the Hellenes produced a civilization that contributed much to Europe. But let us not forget the wide dispersal of these people, and the physical landscape that formed their homeland: settlements were positioned on river valleys and estuaries often separated by rugged mountains, on islands, and on peninsulas jutting out into the Mediterranean. There was contact between these various places but the water usually provided easier avenues than did the land. The Hellenic civilization was widespread—so much so that it was in touch with cultures of the Middle East and North Africa—but it was not contiguous.

While there was no Greek nation, then, there were a number of so-called *city-states,* each with a great deal of individuality, its own power and economic objectives, and its own political character. On land the influence of such places often declined rapidly with increasing distance, but on the sea their power was felt many hundreds of miles away. At times the objectives of several city-states coincided, and they would join in intercity *leagues* to promote their aims in combination and thus achieve greater strength, but fierce and intense competition between city-states also occurred. Such leading cities as Athens,

Sparta, and Thebes headed powerful rival leagues. But the Greeks of old were people of independent spirit, and they did not like control from above or afar. Time and again cities that had for a period been members of a league decided to go their own way despite the economic advantages of membership. No intercity league was without its defections and none lasted very long.

This independent spirit was also reflected in the variety of political institutions forged by the Greeks. In a way Greece was a microcosm of the Europe that was to follow it: kings ruled during the Mycenaean period, which was followed by a lengthy interval of feudal disorganization; with the rise of the Hellenes the final push to democratic forms of government began. An oligarchy of landowners, a wealthy and privileged nobility, delayed the emergence of democratic types of government. But eventually (though not everywhere to the same degree) the general population—excluding the slaves, of course—became involved in the running of the state. While such mass participation of the community was achieved in a number of states, others remained prone to dictatorial control, and fell victim to autocratic rule.

Political philosophy and political science became important products of Greek culture—and indeed, here was a culture of world significance. Plato (428–347 B.C.) and Aristotle (384–322 B.C.) are among the most famous of the Greek philosophers of this period, but they stand out among a host of other contributors to the lasting greatness of ancient Greece. The writings they left behind have influenced politics and government ever since, and constitutional concepts that emerged at that time are still being applied in modified form today. But there was more to Greece than politics. In such fields as architecture, sculpture, literature, and education the Greeks displayed their unequalled abilities as well. And because of the fragmentation of their habitat there was local experimentation and success—followed by exchanges of ideas and innovations.

Individualism and localism were elements the Greeks turned to great advantage, but internal discord was always present, actually or potentially. Eventually it got the better of them, especially because of the endless struggle between the two major cities, Athens and Sparta. After the fourth century B.C., there was a period of ups

A reconstruction of ancient Rome in the time of Emperor Aurelius. (The Bettmann Archive)

and downs, and finally the Romans defeated the last sovereign Greek intercity league, the Achaean, in 147 B.C. But what the Greeks had accomplished was not undone. True, they had borrowed from their Middle Eastern neighbors in such fields as astronomy and mathematics, but they had by their own energy transformed the eastern Mediterranean into one of the culture cores of the world.

The Successors. The Romans succeeded where the Greeks had failed. Theirs was not a nation dispersed through a string of city-states but a contiguous and unified empire in the true sense of the word, strongly centralized in one unrivalled city, Rome. But Rome itself had not achieved supremacy overnight. Italy around 400 B.C. (about the time of Greece's greatest glory) was a territory of many tribes and nations. Well-fortified hilltop towns dotted the area; these were the centers of power and protection for local groups. As in Greece, but on a smaller scale, towns entered leagues of mutual defense, and as time went on places whose natural protections were less than adequate were abandoned as their inhabitants moved to larger and safer centers. Gradually the regional influence of Rome in-

creased. Its original populations had been largely Italic (the name given to a people whose first representatives entered Italy across the Eastern Alps and who settled in Latium before 1000 B.C.), but an ever-wider variety of nationalities entered the place. The great achievements of Rome lay in its successful absorption of this influx, the stability of its political institutions, and an effective campaign of Roman colonization in outlying districts. The great empire that was to come had its local predecessor: by the middle of the third century B.C. there was a true Roman Federation extending from near the Po Valley in the north to the Greek-occupied south.

Having achieved unification on its peninsula, Rome now was a major force in the entire Mediterranean area. Its commercial interests and trade connections widened and Rome made friends—and enemies. Naturally there was competition with the Greeks, but there were challenges elsewhere as well. Carthage on the North African coast managed to resist the Roman thrust for decades. Nevertheless, the Roman Empire continued to spread. The defeat of Carthage removed the major obstacle to westward expansion, and to the east the Greeks were forced to recognize Roman superiority. But Greek culture had a major impact on the Roman civilization, and Greek identity survived despite the strength of the Roman invasion. In the eastern Mediterranean borderlands, for example, Greek remained the dominant language, while almost everywhere else in the Roman Empire the language of common use became Latin. It is more accurate to speak of a Roman-Hellenic civilization than of a Roman civilization alone, for the Romans provided the vehicle to disseminate Greek cultural achievements throughout Mediterranean and Western Europe.

Greek culture was a major component of Roman civilization; but the Romans made their own essential contributions. The Greeks never achieved politico-territorial organization on a scale accomplished by the Romans. In such fields as land communications, military organization, law, and public administration the Romans made unprecedented progress. Comparative stability and peace marked the vast realm under their domination, and for centuries these conditions promoted social and economic advances. The Roman Empire during its greatest expansion (which came during the second century A.D.) extended from Britain to the Persian Gulf and from the Black Sea to Egypt (Map 2-2). In North Africa the desert and the mountains formed its boundaries; in the west it was the sea, and to the north tribal peoples lived beyond the Roman domain. Only in Asia was there contact with a state of any significance. Thus the empire could organize internally without interference. But its internal diversity was such that isolation of this kind did not lead to stagnation. The Roman Empire was the first truly interregional political unit in Europe. Apart from the variety of cultures that had been brought under its control, and the consequent diffusion of ideas and innovations, there were a multitude of possibilities for economic exchange.

This process of economic development (for such it really was) made a profound impact on the whole structure of Southern and Western Europe. Areas that had hitherto supported only subsistence modes of life were drawn into the greater economic framework of the state, and suddenly there were distant markets for products that had never found even local markets before. In turn these areas received the farming know-how of the heart of the Roman state, so that they could increase their yields and benefit even

further. Foodstuffs came to Rome from across the Mediterranean as well as from Southeastern and Southwestern Europe; with its ever more complex population of perhaps a quarter of a million, the city was the greatest individual market of the empire, and the first real metropolitan urban center of Europe. But in addition to these agricultural advances the Romans also undertook a social-political reorganization of their dominions which served economic growth—and was to have a lasting impact on Europe in general. This was their decision to support the development of a number of secondary Roman capital cities throughout the empire,

The might of ancient Rome united Europe and encircled the Mediterranean. This is the forum of Leptis Magna, once a flourishing Roman city in Libya, North Africa, about 80 miles from Tripoli. (Arab Information Center)

Caspian Sea
(Mare Hyrcanium)

Parthian Empire

Tigris
Ctesiphon

Arabia

60°

ALBANIA
IBERIA Cyrus
ARMENIA
ASSYRIA
MESOPOTAMIA
COLCHIS
Euphrates
Aorsi
Huns
Alans
Siraces

SYRIA
Antioch
Damascus
Red Sea
(Sinus Arabicus)

CAPPADOCIA
CILICIA
Tarsus
Tyre
Hierosolyma
Jerusalem
PALESTINE
ARABIA

S a r m a t i a

Anadoci
Navari

Bastarnae
Carpi

Tanais
Palus
Maeotis

BOSPORUS Panticapaeum
Tauri
Chersonesus
Heraclea

PONTUS
Sinope
Trapezus
Paphlagonia
BITHYNIA
Nicomedia
GALATIA
ASIA
Pergamum
Sardis
Ephesus
LYCAONIA
PISIDIA
PAMPHYLIA
LYCIA
RHODES

Byzantium

Black Sea
(Pontus Euxinus)

Alexandria
Sais
Memphis
Tanis

LIBYA
EGYPT
Nile
Thebes

30°

Longitude East of Greenwich

Venedae
Mezzii

Guttones
Burgundians
Marcomanni
Quadi
Jazyges

LOWER
Ister
UPPER
DACIA
LOWER
UPPER
Save
Danube
MOESIA
THRACE
Adrianopolis
MACE
DONIA
Thessalonica
EPIRUS
ACHAIA
Nicopolis
Aegean Sea
Thasos
CRETE

Cyrene
Arsinoe
CYRENAICA
Berenice
Syrtis Major
Marmarica

Germania
Sueoi
Chauci
Lombards

RAETIA
NORICUM
PANNONIA
Aquileia
Augusta Vindelicorum
DALMATIA
Salonae
Scodra

Adriatic Sea
Ionian Sea

MEDITERRANEAN SEA
(Mare Internum)

Colonia Agrippina
Rhine
Augusta Treverorum
Belgica
Cisalpine
Gaul
Genua
Ancona
ITALY
Rome
Capua
Neapolis
Tarentum
CORSICA
SARDINIA
SICILY
Syracuse
Agrigentum
Utica
Carthage
Hadrumetum
Syrtis Minor

NORTH SEA
(Oceanus Germanicus)

G A U L
Noviodunum
Lutetia
Augustodunum
Lugdunensis
Liger (Loire)
Aquitania
Burdigala
Arelate
Narbonensis
Narbo
Massilia
Tyrrhenian Sea

BALEARIC ISL.

AFRICA
NUMIDIA
Hippo Regius
Cirta
Gaetulia

Eburacum
BRITAIN
Venta
Londinium
Camulodunum
Hibernia

Iberus (Ebro)
Tarraco
Valentia
S P A I N
Gallaecia
Asturica Aug.
Bracara Aug.
Durius
Salamantica
Lusitania
Emerita Aug.
Tagus
Tarraconensis
Baetica
Corduba
Carthago Nova
Gades

CAESARIENSIS
RETIA
Caesarea
MAURETANIA
TINGITANA
Tingis

50°
40°
30°

Miles
0   200   400   600

10°
20°
0°

MODIFIED AFTER: HAMMOND, RANDOM HOUSE DICTIONARY

Map 2-2 THE ROMAN EMPIRE.

cities that would have authority over surrounding rural areas and that would be the chief organizing centers for such areas. It is not difficult to see the Greek city-state concept in this, modified to serve Roman needs. But unlike the Greek cities, the Roman towns were connected by that unparalleled network of highways referred to earlier. European integration was proceeding at an unprecedented pace. On current maps Roman towns (and, here and there, Roman routes as well) are still represented: Lugdunum (Lyons in France), Singidunum (Belgrade), Eburacum (York), and Colonia Agrippensis (Köln), among others.

Under the unifying influence of the Roman Empire, Europe made more progress in a little over one century than it was to make in the millenium that followed the fall of the Romans from power. The centuries of Roman hegemony in Southern Europe saw a change over much of the continent, a change from subsistence and comparative isolation of peoples and communities to a situation of contact and integration. More and more, certain parts of the empire depended on the products of other parts; Rome, for example, could not have thrived without the constant importing of foodstuffs grown elsewhere, but the empire itself could not have been sustained without the political and ideological productivity of this great center. Thus there was *areal interdependence* among various parts of the empire—an areal interpendence that reflected the degree of progress that had been made during Roman times. When Europe entered the Dark Ages such interdependence was at a minimum.

When parts of a continent or an empire are interdependent, it means that there is economic specialization, a concentration of activity in certain places to produce certain particular goods or services. Perhaps it is better to refer to such a situation as one involving *functional specialization,* since the activity may also be other than economic. In the Roman Empire there was much of this, on the part of individual people (from slaves to the nobility and the senate), places (such as the port of Ostia, where the fleets carrying grain destined for Rome from across the Mediterranean unloaded their cargoes), and, indeed, whole regions (like Egypt, the farmland granary of Rome). There were many other cases of specialization in Roman Europe. Elba, the Mediterranean island, produced iron ore that

was shipped to Puteoli, near present-day Naples, for manufacture. Inland from present-day Cartagena in Spain were important silver and lead mines whose yield was exported through Cartagena's predecessor—the Roman port of Nova Carthago. In the economic as well as the political sphere, Rome accelerated Europe's evolution to an unprecedented degree.

The Contributions. The Roman Empire began its decline late during the second century A.D., but for centuries beyond that time it survived, as a whole or in part—though without its former greatness. Toward the end of the fourth century the military situation and administrative difficulties combined to lead to a fragmentation into two units, the Western Empire which continued to focus on Rome and an Eastern Empire centered on Byzantium (later renamed Constantinople and subsequently called Istanbul). The Western Empire collapsed finally during the fifth century, but the Eastern Empire, known as the Byzantine Empire, lasted until the middle of the fifteenth century, when it was overthrown by the Ottoman Turks.

The breakdown of the Roman Empire as far as Western and Southwestern Europe was concerned, then, took place fully 1500 years ago. And yet, much remains in Europe today to bear witness to the days of Roman hegemony. Perhaps the most obvious contribution lies in the languages of present-day Europe. Latin was the language of Rome's upper classes; it was first the commercial language and, increasingly, the common language throughout the empire. Latin did not displace the original Indo-European languages already spoken in Roman dominions and provinces, but in administration and education it was the medium of communication. People aspiring to full Roman citizenship sought to learn it. Eventually several regional dialects of Latin developed into the Romanic (Romance) languages of today, including Italian, French, Spanish, Portuguese, and Romanian.

A second Roman contribution to Europe lies in its functioning as a bridge for Christianity. The Christian religion, after all, had its roots in the Middle East. Its influences reached Rome via the Hellenic sector of the empire, and eventually it became the religion of the state. It is not necessary to emphasize the role of Christianity in the centuries that followed the Roman Empire; in many ways it was the sole strand of permanence in Europe during the Dark Ages.

Later it was cause for intense intra-European conflicts, and ultimately modifications of it were exported to other parts of the world.

Roman culture was above all else an urban culture, and many urban centers founded or developed by the Romans continue to function and grow today. While the economic base of many of these places has changed, others, notably the ports, still perform as they did in Roman times. And while the original surface of the old Roman roads has mostly disappeared, there are highways in Europe that follow exactly the routes laid out by Roman engineers. But more than anything else, the Roman Empire left Europe a legacy of ideas, of concepts that long lay dormant but eventually played their part when Europe again found the path of progress. In terms of political and military organization, effective administration, and long-term stability, the Roman Empire was centuries ahead of its time. Concepts of federal and confederal relationships as applied by the Romans had great currency more than a thousand years later. At no time was a larger part of Europe unified by acquiescence than was the case under the Romans. Never did Europe come closer to obtaining a *lingua franca* than it did during Roman times.

The Völkerwanderung. The breakdown of the Roman Empire was related to a great movement of European populations involving Germanic and Slavic peoples. This was the formative period of the final distribution of European peoples, as the Anglo-Saxons invaded Britain from Danish shores, the Franks entered France, the Allemanni reached Germany, and Burgundians, Vandals, and Visigoths penetrated the Western Empire. Essentially what happened was that the empire was replaced by a number of smaller states, mostly kingdoms, ruled by kings who abandoned whatever remained of the political institutions of the empire and who ruled by personal decree. These kings viewed their territories practically as huge estates, a part of their private wealth. Their subjects did not enjoy the safeguards of law and justice that the Roman Empire had once provided. Of course, it must be said that Roman ideals were already failing long before the Germanic invasions put an end to the Western Empire, but there was nevertheless a general retrogression. The cities, for example, were fortified to become headquarters of a military force, and contained a bishop and his clergy, and perhaps some remaining traders and craftsmen. But the money economy, which had so importantly affected economic life throughout Roman Europe (and far beyond—Roman coins have been found as far away as Sweden), was on its way out, to be replaced by barter in agricultural produce. As the urban centers declined so did their markets. Peasants had little reason to produce a surplus; areas the Romans had wrested from a subsistence mode of life quickly returned to it. Europe was entering its Dark Age.

In the east, also, population movements on a major scale were in progress. The Slavs were the last peoples speaking Indo-European languages to reach their present habitat, and they penetrated the Eastern Empire along the Danube River. Again their source is uncertain, but they spread westward to create frontiers with the Germanic peoples and southward through the Balkans to challenge Byzantine supremacy. The Eastern Empire, however, was not as ready to fall as was the west, and the tribal Slavs failed to overthrow Constantinople. Periodic incursions from the east occurred, but the most important of these was that of an Asian nomadic people, the Magyars, which started in the ninth century. The Magyars were not Slavs, but they penetrated to Central East Europe where they established a large enclave roughly coterminous with modern Hungary. Thus they separated the southern Slavs (Serbs, Croats) from those in the west (Poles, Czechs) and east (Russians, Ukrainians). They continue to occupy this position today.

All this was part of the great *Völkerwanderung* of Europe—a momentous stirring of European peoples and their migration to other parts of the continent. Just what motivated these movements is not always clear. No doubt some were caused by the attractiveness of richer lands not far away and now attainable because Roman power was on the wane. Other migrations were forced; that is, peoples were pushed forward by a powerful invasion of their domain by alien conquerors. The Huns, from the steppes of Central Asia, had precisely such an impact upon the east Germans during the fourth century A.D., and they kept a vast area in turmoil for many decades. Southwestern Europe during the eighth century was invaded by Arab and Berber warriors (the *Moors*), whose strength lay in the rise of Islam and the power of the Moslem caliphs who backed them. These Moors left strong imprints on the culture and the cultural landscape of southern Spain, carried to this day in the architecture of such cities as Cordoba, Granada, and Seville. But Moorish rule in Iberia was still less disruptive than the later occupation of the Balkans in Southeastern Europe by the Moslem Ottoman Turks. Oppressive and destructive, the Ottoman period set Eastern Europe back severely with ruthless subjugation and forced conversions to Islam. In any case, with the onset of the Völkerwanderung, Mediterranean unity and stability were forever broken, and all the performers on the European stage seemed to be seeking new positions for the next act.

The results of these events can be read in the modern map of European languages, but other features of modern Europe began to emerge as well. During the millenium that followed the fall of the Roman Empire, European states and nations began to consolidate. In the second half of the eighth century there was a revival of sorts under the leadership of Charlemagne, who forged an empire that seemed destined to become a successor to the Western Roman Empire and perhaps a recreation of it. But it was too extensive, too ineffectively administered, too backward economically, and its parts were inadequately interconnected. In 843 it broke up—and the breakup had momentous consequences, for one part came to form the roots of the state of France and another, across the Rhine and to the north of the Alps, the beginnings of Germany. Actually, a map of Europe toward the end of the twelfth century (Map 2-3) shows several other countries and peoples in their present position, including Denmark, England, Poland, and Hungary. By then the German fragment of Charlemagne's empire had evolved into the Holy Roman Empire, so named to emphasize the Germanic state's political, religious, and cultural inheritance acquired through the conquest and absorption of papal Rome itself.

Feudalism. Europe during the period from about the middle of the eighth century to the end of the twelfth was dominated by feudal institutions. Roman judicial processes and other institutions that protected individual rights had weakened. The new kings ruled by private law, personal power, influence, and wealth, and through the allegiance of counts, barons, dukes, and other representatives. Although some kings were successful in forging and maintaining states of considerable size (Charlemagne was foremost in this respect), the framework of government was not sufficiently strong, and there was not enough contact and communication within the country to withstand disorder and disorganization. Great as it was, Charlemagne's empire was inherently weak because it required a person of the power and stature of Charlemagne to maintain it. And although he managed to hold his dominions together for a long time, even Charlemagne was compelled to grant increasing powers to local lords.

In this way feudalism served the kings. In order to secure their allegiance, feudal rights were granted to counts and dukes in outlying parts of the state. There the royal agent was in charge of practically every aspect of life, from the peasantry to the military. These old units can still be seen in the locations and names of the political subdivisions of certain countries of the Old World, such as the *county,* the *duchy,* and the march or *mark.* And when the larger political unit disintegrated through the death of a king, a weak succession, or an invasion, the feudal counts and barons were quick to assert complete hegemony over their local domains, including all land and people within them.

Thus, during the feudal period much of Europe was organized into dozens of such jigsawlike feudal units, each with its nobility and vassals, and its more or less tenuous connections with higher authority. The connectivity that Rome had brought to Europe was now at a minimum, and reforms were needed to rekindle economic as well as political progress. These began during the second half of the eleventh century and continued throughout the twelfth. In England the Norman invasion (1066) destroyed the Anglo-Saxon nobility and replaced it with one drawn from the new immigrants. William the Conqueror strengthened the feudal organization of England to such a degree that he became the most powerful ruler in Western Europe; at the same time he introduced changes in the system of governmental organization that were to outlast the feudal period. A similar revival took place in the Germanic empire to the east under Henry III. And accompanying these developments was a new upward trend in the economy. The growing population required more food, and farm output had to be increased. But bad years or crop disease would afflict various areas with resultant famines or the threat thereof. Thus when large scale trade began once again in Europe, it was initially stimulated by the need for the distribution of food to areas where it was required.

One reflection of this economic revival could be seen in the rising prosperity of favorably positioned European towns, especially the ports. Along what is today the north German coast alone, the towns of Hamburg, Lübeck, Wismar, and Rostock thrived as Scandinavian trade in-creased and the Baltic Sea became an ever busier trade route. In Flanders, England, and northern Italy the ports prospered. Wealthy merchants built splendid homes and helped finance fortifications to protect the town against competition from land and sea. Soon several of these ports joined in an association to protect their common interests. In the mid-thirteenth century, for ex-

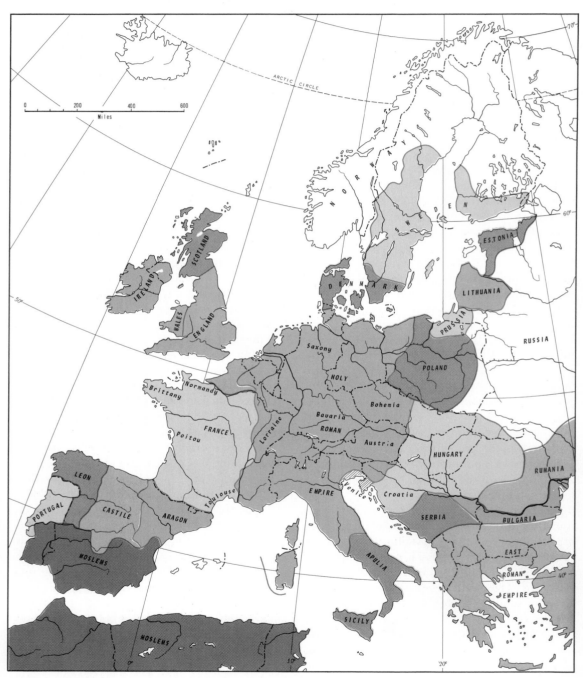

Map 2-3 EUROPE AT THE END OF THE TWELFTH CENTURY.

ample, Hamburg and Lübeck agreed to share responsibility for the safeguarding of the seaway that links the Baltic and North Seas; soon other towns found reasons to attach themselves to this union. Thus the famous *Hanseatic League* was born, a federation that eventually counted more than 70 member cities in Europe. The Hanseatic League reached its zenith of influence during the second half of the fourteenth century, when towns from the Baltic to Flanders were members of it. However, the Hanseatic League was not the only association of its kind; other unions of this kind included one on the other side of the Alps, the Northern Italian Merchant League.

The Crusades. Before turning to the rise of the monarchies and the emergence of modern Europe, we should note an initial period of European expansion (after the days of Rome) which occurred during feudal times. Over the period during which we have been viewing Western Europe, the East Roman Empire (the Byzantine Empire) continued to flourish, and the Moslem world was gaining ever greater strength. Islam now dominated the entire Middle East, including the spiritual source of Christianity. Arabs also held control over southern Spain. But when feudal Europe began to emerge from its darkest period, both the Moslem caliphates and the Byzantine Empire were in decline. In the Iberian Peninsula, the Spanish kings pushed back the Moors; from this struggle arose the kingdoms of Castile and Aragon. Late in the eleventh century the pope, influenced by Byzantine demands and promises, called for a full-scale crusade against the Moslems in the Middle East itself. The first such crusade was mounted with the support of the nobility over large parts of Europe, especially France and Germany, and it succeeded: among its prizes was the town of Jerusalem. But despite such early successes, reversals were soon to come. Gaining strength once again, the Moslems recaptured Jerusalem (1291), and from another source came a new threat: the Ottoman Turks, who in the fourteenth century invaded the Balkans and then turned on Constantinople itself. The city fell in 1453, and with this event the last direct connection which the Roman Empire was broken.

The Oscillation Principle

The preceding paragraphs have summarized the ebb and flow of European fortunes. Great successes and momentous progress were followed by decay and stagnation—and then, for reasons only partially understood, Europe started moving forward again. This story is not Europe's alone. Human progress is not continuous and uninterrupted. Great accomplishments inevitably seem to be followed by periods of disarray during which it is doubtful that what has been learned can even be adequately transmitted to future generations. Certainly Europe could not pick up where the Romans left off when the darkness of the first millenium A.D. came to an end. The world is full of examples of the same kind: the disorganization that followed Han China happens to be roughly contemporaneous with that of Roman Europe, but there are others at different times in the Indus Valley, in the forests of Yucatan, in the deserts of the Middle East, on the savannas of West Africa, on the plateaus of Peru. Wherever man has made progress on this planet, he has also seen failure and retrogression. But—usually elsewhere—the strands have been picked up again and another forward thrust has been made. What followed has usually been an improvement on what had been before. Seen in the large-scale perspective of human history, such improvement has almost never come in the same location. Greece took center stage from competitors in the Middle and Near East, and then yielded to Rome. Rome yielded to Western Europe, and Western Europe yielded to competitors in the Slavic east and, across the ocean, to North America in the west. It has happened elsewhere: from Yucatan the center of human organization was transmitted by the Toltec communities to the Aztec state of the Valley of Mexico; in West Africa ancient Ghana gave way eastward to Mali, then Mali, in turn, yielded eastward to Songhai.

Might we argue that human progress is an oscillatory process, not only in time but also in space? Could it be that human accomplishments, when disseminated in various directions from their source, fall on barren ground except in locales where human energies and particular circumstances favor their adoption and further development? We shall have occasion to return to this argument later.

The Rebirth of Europe

The emergence of modern Europe dates from the second half of the fifteenth century—some put these beginnings at 1492, the year of Columbus' first arrival in the New World. In Western

Europe strong monarchies gave rise to nation-states. The monarchies began to represent something more than mere authority; increasingly they became centers of an emerging national consciousness and pride. At the same time the long-prevalent trend toward fragmentation was reversed as various monarchies—through marriage or alliance—combined to promote territorial unity. Feudal privileges were regained by the central authority, the aristocracies (which tenaciously opposed monarchical rule) were put down, and parliamentary representation of the general population grew. There was renewed interest in Greek and Roman achievements and theories in science and government, an interest that coincided with a new infusion of Roman (Byzantine) learning. That infusion, of course, was related to the fall of the Byzantine capital, Constantinople, to the Ottoman Turks in 1453. In Western Europe the pace of progress was decidedly picking up, and the period is appropriately called the *Renaissance*.

Western Europe was on the threshold of discovery—the discovery of continents and riches across the oceans. But before we consider the momentous consequences of the events that followed Columbus' first sighting of land in the West Indies, it is well to look eastward for a moment, for in the east loomed an enormous threat. Having taken Constantinople, the Ottomans, with armies superior to all others on the continent, were forging an empire greater than any in the west. They pushed into the Balkans, took Hungary and penetrated as far as present-day Austria. They were masters of the Adriatic coasts and extended their power along the North African littoral from Egypt to Algiers. Certainly the Ottomans had the upper hand throughout the Mediterranean; had they not been preoccupied elsewhere they might, like the Romans, have made the Mediterranean Sea an interior lake to their empire. And in that case Western Europe would have been concerned with self-defense rather than with a scramble for the wealth of the distant lands of which it had become newly aware.

As it happened the Ottoman Turks were in conflict in Asia and the Near East as well as in Europe, and these eastern struggles siphoned off enough attention and energies to restrain their westward push. They had hostilities to contend with in Persia, Anatolia, Syria, and Egypt. Nevertheless, the Ottomans did enter into competition with one West European power, though not in Europe itself: from their bases in the Middle East they contested Portuguese eminence in the Indian Ocean.

If the Ottoman Empire was Europe's most powerful state, it was also under its most totalitarian regime. While political institutions in the west were becoming more and more representative of the people in general, no such change marked Eastern Europe. Indeed, to the east of a line drawn roughly from the lower Baltic Sea to the head of the Adriatic, Europe was in general decline. The peasants' lot deteriorated still further as monarchies weakened and feudal landowners took control. Towns were mere shells of their former existence, and communications were at a minimum. No rebirth could be witnessed here.

Europe's Renaissance, then, focused on the west, and the competing monarchies of Western Europe were fortunate that they could make full use of their advantages to engage in economic rivalry without interference from the east. Indeed, to the political nationalism that focused upon the monarchies we must now add economic nationalism in the form of *mercantilism*. The objectives of the policy of mercantilism were the accumulation of as large a quantity of gold and silver as possible and the use of foreign trade and colonial acquisition to achieve this. Mercantilism was a policy promoted and sustained by the state; it was recognized to be in the interests of the people of the state in general. Wealth lay in precious metals, and precious metals could be obtained either directly by the conquest of peoples in possession of them (as Spain did in Middle and South America) or indirectly by achieving a favorable balance of international trade. Thus there was stimulus not only to seek new lands where such metals might lie, but also to produce goods at home that could be sold profitably abroad. The government of the state sought to protect home industries by imposing high duties on imported goods and by entering into favorable commercial treaties. The network of nation-states in Western Europe was taking shape more and more rapidly; Europe had entered the spiral that was to lead to great empires and temporary world domination.

But the spiral did not always lead upward; it was not always one of progress. Superimposed on Western Europe's dynastic, national, and mercantile struggles was a series of religious conflicts that was a reaction to the secular values of the Renaissance itself. The sixteenth century was one of wars—wars in which Catholicism and Protestantism, Reformation and Counter-Reformation, and even various forms of nationalism played their roles. New alignments emerged, and in the aftermath the monarchies reached their zenith. But much was sacrificed. Absolutism returned to the kingdoms of the west, and most of the parliaments and assemblies that had become part of political life were destroyed. Louis XIV of France was the personification of the despotism that had returned to Europe. The nobility once again enjoyed privileged status. And a continuing struggle involving organized religion—the struggle between church and state—eroded society still further.

The Revolutions

While political power in Europe's states thus became increasingly centralized in the monarch and his circle of nobility, the economic spiral was not slowed down. The population was growing, and the kings realized that more people, so long as they could be effectively controlled, meant greater strength for the state. Europe's population had had its ups and downs: between about 1000 and 1300 A.D. it had nearly doubled, with marked increases in France, Germany, Italy, the Netherlands, and England. Then came a century and a half of decline, with great epidemics ravaging the countryside, but late in the fourteenth century the upward trend began again. In the mid-sixteenth century Europe's population probably was of the order of 100 million, two-thirds of it located west of the Balkans. Then, after a period of comparative stability, came the great increases of the eighteenth century. More people were now available for manufacturing and industry, for the mines at home and for colonization abroad. The mercantilist doctrine was still very much alive.

Ultimately, however, economic developments in Western Europe proved to be the undoing of monarchical absolutism and its system of patronage. It was the urban-based merchant who gained wealth and prestige, not the nobleman. Money and influence became concentrated more and more in the cities, and the traditional measure of affluence, land, began to lose its relevance in the changing situation. Not surprisingly, the merchants and businessmen of Europe wanted political recognition and began to exert pressure

to get it; indeed, there were those who had enough of state control and mercantilist protectionism and who opted for the so-called *laissez-faire* doctrine, whereby a minimum of state involvement in commerical affairs was demanded.

Europe was about to enter its age of revolution—an age that was to have an enormous impact not only on the continent itself but on the world as a whole. There were revolutions in industry, in agriculture, in politics. But in assessing these events it is important to recognize the base that already existed there. For example, a great deal of industry existed in Europe before the eighteenth-century industrial revolution began. In the Low Countries, especially Flanders, specialization had been achieved in the manufacture of woolen and linen textiles. In several parts of present-day Germany (especially Thüringen, now in southwestern East Germany) iron was mined and smelters refined the ores. In England water power was used in the manufacture of worsted textiles. Again, there were practices of crop rotation, horsedrawn plowing, and soil fertilizing before the revolution in agriculture took place. And long before the political (or democratic, as it has been called) revolution transformed the monarchies, Europe had had experience with representative forms of government.

Nevertheless, the changes that came to eighteenth-century Western Europe truly were revolutionary. In agriculture, a stimulus was provided by the ever more rapidly growing cities, which in the Low Countries and northern Italy contributed to improved organization of land ownership and cultivation. Some of the ideas prevalent there spread to England and parts of France, where traditional practices of communal landholding were replaced by individual ownership. Land parcels were marked off by fences and hedges; the owners readily adopted new innovations to improve yields and enlarge profits. The pastoral industry also benefited. New breeding techniques and the supply of winter fodder for cattle, sheep, and other domestic livestock greatly increased the weight of these animals. New crops were introduced, notably from the Americas: the potato became a European staple.

The Isolated State. Agriculture was becoming a profitable industry, and its market was guaranteed. Not only was the population growing, but an increasing number of people were moving to the towns and cities. Around these urban centers farmers produced needed agricultural commodities, and transported them to the towns. It is appropriate in this connection to tell the story of a geographer-farmer who did pioneering work on a subject that must have concerned the farmers in those days, a concern that is shared permanently by all those involved with market-oriented production of any kind: the effect of distance and transportation costs on the location of productive activity. The geographer, J. H. von Thünen (1783–1850), was directly involved in this matter, for in 1810 he obtained a comparatively large farming estate not far from the town of Rostock in German Mecklenburg. For four decades he ran this estate, meticulously keeping records of all the transactions involved. He published a series of volumes under the title *Der Isolierte Staat,* the first of which, dated 1826, contained an initial contribution to theory regarding the spatial structure of agriculture. In later volumes he used the information collected while carrying on his farm operations to support his theory. It has been a century and a half since Von Thünen wrote, but in a way he is the founder of spatial theory, and geographers remain very much interested in what he had to say.

He entitled his work *The Isolated State* because he wanted to establish, for purposes of analysis, a self-contained country, without outside influences to disturb the situation—a sort of regional laboratory within which he could clearly identify the factors that affect the location of farms. But he went even farther than that. Not only did he postulate an isolated state surrounded by wasteland, but *within* this "state" he made a few other assumptions, hoping thereby to isolate one remaining operative variable: distance. For example, Von Thünen's theoretical state had a completely homogeneous climate (perhaps it is better to say *uniform* soils and a *uniform* climate), and a flat, plainlike surface without such interruptions as a river valley or a mountain range. Furthermore, the state was circular in shape and had only one city and hence only one urban market, located at its center. Von Thünen had the farmers do their own transporting; to get to this market they had to travel directly overland by oxcart. This is the same as assuming a network of radially converging roads of equal quality everywhere, and in such a situation, of course,

costs of transportation would be directly proportional to distance.

In such a state, Von Thünen then asked himself, what would the spatial arrangement of agricultural activities be? For example, would it be reasonable to expect high-priced goods such as perishable vegetables and dairy products to be produced farther from the city than, say, potatoes and grains? Of course not. But what would it depend on? On a principle of economic rent (or *land* rent, as he himself identified it). What would happen is this: agricultural products would be grown in a series of zones, a set of concentric belts outward from the market center. Nearest to the city those crops would be grown that yielded the highest returns, for example, vegetables; dairy products would also be found in this first belt. Farther outward agriculture would become more extensive, and more durable goods such as potatoes and grains would be cultivated. And eventually, since transport costs to the city increased with distance, there would come a point where economic or land rent would become zero. Beyond lay the wasteland.

Broadly on this basis Von Thünen concluded that five belts would develop in the isolated state: (1) a zone of intensive agriculture and dairying, (2) a belt of forest for timber and firewood, (3) a broad belt of increasingly extensive field crops, (4) a belt of ranching and animal products, and (5) the surrounding wasteland (Map 2-4). What is likely to strike anyone looking at this is the forest belt, apparently able to command a location near the city—a position of high accessibility. Obviously the forest held great importance at that time. And so it is important to remember the time as well as the place in which Von Thünen wrote. He was dealing with a commercial but preindustrial state. Building materials and the prime source of heat lay in the forest. But what was important to early nineteenth century Rostock is no longer so; an industrial revolution was soon to follow the one in agriculture. However, before we are tempted to abandon Von Thünen's ideas as out of date, let us remember that "preindustrial" is a connotation that still applies to a number of towns and even cities in the world. When Professor R. J. Horvath investigated the situation around Addis Ababa, the capital of Ethiopia, he found a wide and continuous belt of eucalyptus forest surrounding the city, more or less where Von Thünen would have suggested it to be, and serving functions similar to those attributed to the forest belt of the isolated state.[1]

Von Thünen developed what we call a *model* and, as we have just seen, geographers are still comparing that model to the real world. The trouble with a model is that the more assumptions you make, the less it may be able to help in solving critical problems. So it is with Von Thünen, whose list of assumptions in developing this particular model is a long one. But his represents a particular type of thinking in geography, and a very productive one—and he made it clear that he had thought about this aspect of his work:

"... assumptions are necessary in order to clearly understand the effect which a given variable has. In actual life we have only a vague idea of the effect and operation of any single variable because it appears always in conflict with other variables operating at the same time. This procedure has thrown light on so many problems in my life and seems to me so generally applicable that I consider it the most important feature of my work . . ."[2]

During Von Thünen's lifetime Europe was changing explosively. He was among the first geographers in modern days to seek order in this one aspect of life, but as the title of his work implies, he avoided interrelationships between

[1]R. J. Horvath, "Von Thünen's Isolated State and the Area Around Addis Ababa, Ethiopia," *Annals* of the Association of American Geographers, Vol. 59, No. 2, June 1969, pp. 308-323.

[2]J. H. Von Thünen, *Der Isolierte Staat in Beziehung auf Landwirtschaft und Nationalökonomie,* in K. W. Kapp and L. L. Kapp (Eds.), *Readings in Economics,* Barnes and Noble, New York, 1949, p. 299.

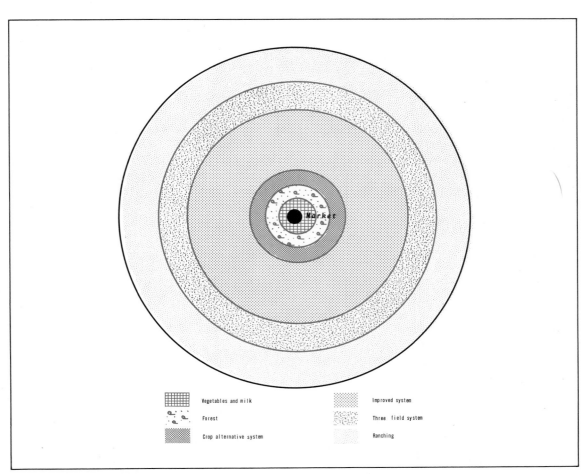

Legend:
- Vegetables and milk
- Forest
- Crop alternative system
- Improved system
- Three field system
- Ranching

Map 2-4 THE ISOLATED STATE.

several urban market centers and "states." One major consequence of Europe's revolutions, however, was in exactly the opposite direction: communications improved, markets grew and competed, and few parts of the continent remained untouched by the new age.

With the *industrial revolution* we witness the return of Europe to a situation of growing functional specialization and areal interdependence, something we last had occasion to refer to when discussing the Roman Empire and its trans-Mediterranean connections. But what developed in the eighteenth century was something completely new. Interdependence was no longer a matter of exchanging agricultural goods and transporting mineral ores to places of treatment. What lay ahead was an unprecedented intensification of industrial productivity, the transformation of virtually every existing form of manufacturing, and the zenith of Western European economic and political power.

At the center of it all was England, with its long-time, strong overseas commercial connections, its existing wealth in accumulated capital, its internal free trade, its favorable political institution—and its plentiful supply of coal. Coal became the essence of the industrial revolution, and unlike some of its potential continental competitors, England possessed what was needed. But England had many other advantages as well. With Western Europe it shared a good climate and soils capable of sustaining the rising population, but only England stood alone, in insular security, open to the ocean yet adjacent to a productive continent.

The beginning of the industrial revolution gave little indication of how far-reaching it was to be. The first innovations were mechanical in nature; that is, they were improvements in machinery that did not yet depend on new sources of power. The beneficiary was the textile industry, and the prime objective was to save on labor: the industry simply had not been able to keep pace with the growing trade. Better spinning and weaving equipment was invented, but as before the machines were driven by water power. Then Watt invented his first steam engine, and quickly learned how to refine it for specific uses. Pumps became available to keep water out of mines. The power loom revolutionized the weaving industry. Meanwhile, the manufacture of iron also benefited from discovery: coal rather than charcoal could now be used in blast furnaces, and

the pressure on fast-dwindling wood supplies was relieved. Time and again the inventions made for one particular industry served others as well, and the whole English industrial complex benefited.

The revolution also had its effect in transportation and communications. The harnessing of steam led to the development of the locomotive, and the first railroad in England was opened in 1825. In 1830 Manchester was connected to the port of Liverpool, and in the following decades several thousand miles of rail were laid. Ocean shipping was also entering a new age; the first steam-powered vessel crossed the Atlantic in 1819. Now England enjoyed even greater advantages than those with which it entered the period of the industrial revolution. Not only did England hold a monopoly over products that were in world demand, but Britain alone possessed the skills necessary to make the machines that manufactured them. Europe and America wanted railroads and locomotives, steamships and docks; England had the know-how, the experience, and the capital. Soon the fruits of the industrial revolution were being exported, and British influence around the world reached a peak.

Meanwhile, the spatial pattern of modern, industrial Europe began to take shape. In Britain, industrial regions, densely populated and heavily urbanized, developed in the "Black Belt"—near the coalfields. The largest complex was (and remains) positioned in north-central England (the Midlands), and includes Birmingham (where old gunmaking and jewelry industries gave way to the manufacture of silver and brassware, automobiles and cycles), Sheffield (cutlery), Leeds (where the availability of soft water promoted the wool industry; today the clothing industry ranks along with the manufacture of industrial and agricultural equipment and airplane parts), Manchester (the textile center of the region, with important financial and transport functions), and the port city of Liverpool. Secondary coal-industrial areas developed near Newcastle, in southern Wales, and along the Clyde River in Scotland.

In mainland Europe, a belt of major coalfields extends from west to east, roughly along the southern margins of the North European Lowland—due eastward from southern England across northern France and southern Belgium, the Netherlands, Germany (the Ruhr), western

Map 2-5   EUROPE: MAJOR INDUSTRIAL REGIONS.

Focused on Major Urban Concentrations   Situated on and near Major Raw Materials   Based on Long-term Regional Growth

Bohemia in Czechoslovakia, and Silesia in Po-land. Iron ore is found in a broadly similar belt, and the industrial map of Europe reflects the resulting concentrations of economic activity (Map 2-5). But nowhere are the coasts, the coal-fields, and iron ores located in such close prox-imity as they are in Britain.

We shall return to the consequences of the industrial revolution in the next chapter, but it remains now to consider a third field of revolu-tion in eighteenth-century Europe. In the 1780's a series of upheavals began that were to change the sociopolitical face of the continent. Over-shadowing these events was the French Revolution (1789–1795), but it was much more than that. The consequences of the *political* (democratic) *revolution* were felt into the twentieth century, and they were felt by every monarchy in Europe.

In France the revolution brought to the surface the dissatisfactions of not just one, but several sections of the population, and probably nobody envisaged at the beginning of it the depths of destruction to which the country would plummet. At the top of the social pyramid there was displeasure among the nobility and clergy with the rule of Louis XVI. In the middle layers there was a search for a constitutional, parliamentary monarchy along English lines. And the peasants were restive because of a depression, rising prices, and a general decline in their economic position. Ultimately the most extreme elements came to control the movement, and in 1792 a republic was proclaimed. The king was executed early in 1793, and revolutionary armies began to advance into neighboring countries, calling on all peoples to dispose of their rulers and to join in a liberating war of revolution. The war took on the earmarks of a nationalist struggle against Europe's dynastic monarchies.

But in France itself, while the armies penetrated farther into Europe, the pendulum of revolution swung the other way, and a class of people who had benefited from the crisis—through land speculation, hoarding of money, or other means—gained the upper hand. Now many of the revolutionaries met a violent death; nothing seemed capable of stopping the cycle of class conflicts. But the successful campaign of war thrust into national prominence a man who was to salvage the country, the general who had led French forces into Italy: Napoleon Bonaparte. When Napoleon took control via a military coup in 1799, he found himself in charge of a republic that controlled Europe from Prussia and Austria to Spain and from the Netherlands to Italy. It was an empire; but his first task was the consolidation of the situation in France itself.

Napoleon's efforts in this area laid the foundations for modern France. There were over three dozen "provinces" in France at the beginning of the revolution, each with its own status, rights, and privileges. These were replaced by 83 "departments," roughly of equal size, each defined by a new boundary, each with a local

headquarters. Every "department" had a position with respect to Paris that was equal to that of all the others. Napoleon sought also to construct a road network that would link each of these units as directly as possible with his capital. The system had administrative advantages and there were other assets too; in this way Napoleon could forever destroy the feudal rights that had been held over much of France's land, and could guarantee the peasants permanent ownership over the land they had acquired during the revolutionary chaos.

Having initiated this sweeping reorganization of France, Napoleon in 1804 was crowned emperor of his state, and now he undertook to consolidate the empire. Although the British successfully asserted their hegemony of the seas, Napoleon by 1810 either controlled or dominated through alliances all of mainland Europe except Portugal and the Ottoman Empire. Britain's only foothold was an alliance with Portugal, and in 1810 the British repelled Napoleon's invasion of that country. A more costly repulsion came in 1812, when the French leader sought to extend more effective control over Russia: Napoleon's international army was beaten back from Moscow and his prestige and power waned rapidly. Faced with an effective military challenge from Prussia and Austria and serious uprisings in Spain, Napoleon abdicated in 1814. While the Paris monarchy was restored by allied European forces, he was exiled to Elba. In 1815 he managed to escape and returned to France, only to be defeated for the last time at Waterloo.

But Europe had not seen the end of revolution. The reverberations of the French Revolution and its aftermath of empire were to be felt throughout most of the nineteenth century, and, indeed, beyond. Spain, Portugal, Italy, and Greece were shaken by revolts in the 1820's. At mid-century socialist uprisings were crushed in France, Italy, and Belgium. And in 1870 France and Prussia, the center of German power under the leadership of Bismarck, came to war. This war followed several years of increasing tension between the two powers, during which the French sought an anti-Prussian alliance with Austria and the Prussians tried to place a German on the Spanish throne. The French were quite confident that they could win the war, but Bismarck proved them wrong; early in 1871 France was forced to sign a peace treaty whereby it lost its territories of Alsace and most of Lorraine, and the victorious German army marched through the streets of Paris.

The Franco-German (or Franco-Prussian) War had another effect. Bismarck had long had his troubles in his effort to unite the German states of Prussia, Bavaria, and Württemberg; some historians argue that he provoked the hostilities against the French in order to submerge anti-Prussian feeling that was on the rise in southern Germany. Whether or not this was the case, certainly the war with France helped cement the unity of the German-speaking states. Even prior to the end of the war, conferences were in progress to create a greater Germany once and for all, and despite Bavarian demands for a certain measure of autonomy the German *Reich* was consolidated before the war with the French had been successfully concluded. It was a fateful development: in the century that followed the German state twice plunged much of the world into costly wars.

A Search for Order

When Von Thünen developed his "model" isolated state, he was trying to identify the impact of one particular circumstance on the development of the agricultural landscape in his part of Europe. He was an economic geographer, and he was fascinated by the emerging relationships between those who produced agricultural goods and those who required and consumed them. In the next chapter we shall see that other geographers became interested in the growing towns and cities of Europe, whose significance increased as the industrial revolution proceeded. Geographers felt that there must be laws that determine the number, size, and *distribution* of those cities—that is, their arrangement with respect to one another. Note that the focus is once again upon *spatial* organization. And there were also geographers who became interested in the political processes that were at work in Europe. Was it possible to seek in spatial relationships the answers to the problems of the rise and fall of states and empires?

Questions of this kind especially interested a German geographer, Friedrich Ratzel (1844–1904). Ratzel was concerned with the effects of what he called space (total area, in fact) and location on the fortunes of the state. He was the first to write about these matters in a systematic

way, and in so doing he laid the foundations of the field of political geography.

Like several other geographers with whom we shall become acquainted, Ratzel based his work in human geography on accepted principles of physical and biological science. He once described the state as "a piece of soil and humanity," but he meant a great deal more than that. He saw the state as a complex structure of innumerable interrelationships—of men among one another, and of men with the soil that afforded them space and life. In a word, he viewed the state as an organism, a vast organic being that would grow and die according to natural laws. And so he proposed the *organic theory* of state evolution, in which the critical element was territory or space. Just as an organism needs food to survive, so a state, to thrive and achieve strength and greatness, requires more and more space. Boundaries, Ratzel felt, signified the death knell for the state. What it needed were open frontiers into which it could spread its influence, whose assets it could absorb, and whose territory would ultimately be incorporated.

Ratzel did not come to this conclusion without careful study. This was not an incidental opinion; it was supported by painstaking comparative studies of European states. Much of Ratzel's writing was done late in the second half of the nineteenth century, and the great convulsions generated by the political revolutions did not lie far behind. From the time of Rome to that of France, space and greatness, greatness and empire seemed synonymous. From the consolidation and alliance of monarchies to the subjugation of weaker states by stronger ones, states in ascendancy were gaining territory, those in decline were losing it. The organic theory seemed an appropriate formula to explain all this. But note the differences between Von Thünen's isolated state and Ratzel's organic theory. Von Thünen saw increasing complexity and sought to understand one element in it as completely as possible. He tried to isolate one particular factor. Ratzel saw another kind of complexity, and although he too tried to understand the role of one or two elements (space and location), he ultimately sought to identify a "law" that would govern the whole range of phenomena with which he had concerned himself. It represents a difference in approaches and viewpoints in geography which we shall encounter again.

ne pursuit more fruitful than the other? ps we should consider other representative before we render judgment. But neither Thünen nor Ratzel have been forgotten. Modifications and applications of the Von Thünen model still keep geographers occupied today. And Ratzel's school of political geography contributed to the emergence of a philosophy of statecraft that shook the twentieth century world.

THE CONTENTS OF EUROPE

Europe merits identification as a continent despite the fact that it occupies a mere fraction of the total area of Eurasia—a fraction which, furthermore, is largely made up of the peninsular western extremities of that vast landmass. Certainly Europe's size is no reflection of its greatness, but it would be difficult to argue that Europe does not deserve consideration as a discrete unit of world significance. Probably no other part of the world is or ever has been so packed full of the products of human achievement. Innovations and revolutions that transformed the world originated in Europe. Over centuries of modern times the evolution of world interaction focused on European capitals and European states. Time and again Europe proved to contain the resources needed for continued progress, in physical as well as human terms.

Among Europe's greatest assets is its internal diversity. From the warm shores of the Mediterranean to the frigid Scandinavian Arctic and from the flat coastlands of the North Sea to the grandeur of the Alps, Europe presents an almost infinite range of natural environments. The insular and peninsular west is contrasted against the more continental east. A resource-laden backbone extends across Europe from England eastward, but there are poor areas too: even Europe has countries with "underdeveloped" economies. And neither is this diversity confined to the physical makeup of the continent. The European realm includes peoples of many different stocks. In the preceding chapter the broadest possible grouping was indicated when we identified the Latin, Germanic, and Slavic elements in the peopling of Europe. This generalization, however, fails to suggest the enormous variety of peoples within each of these units, just as it fails to account for important minorities (such as the Finns and the Hungarians) that also form a part of Europe's population complex.[3]

It is not this diversity of physical and human content alone, of course, that is such a great asset to Europe. If differences in habitat and race or culture were automatically to lead to rapid human progress, Europe would have had many more competitors for its world position than it did. But in Europe there were virtually unmatched advantages of scale and proximity, and coupled with frequent innovations and infusions of new knowledge and ideas, a sort of internal *complementarity* developed—first at a local, then at a regional, and finally at a continental level. When two areas or places find themselves in a position of complementarity, it means that area A requires products supplied by area B, and at the same time area B needs products supplied by area A. We had occasion to refer to such areal interdependence in the preceding section; let us here consider a more recent case. Among Mediterranean countries other than France, Italy is foremost in terms of economic development; it is a member of the core group of "Inner Six" Common Market states. But Italy is a coal-poor country, and for a long time its industries depended on coal supplies from the rich resources of Western Europe.[4] On the other hand, Italian farmers grow agricultural products that cannot be grown in the north: citrus fruits, olives, grapes. Such products are desired by northern markets. Hence Italy needs Western Europe's coal; Western Europe imports Italian fruits and wines. There are thousands of examples of this

[3] Modern, living races of Man in Europe are now classified by anthropologists into four units, identified by locale: (1) *Northwest European,* including the peoples of the British Isles, Scandinavia, and northern France and northern Germany; (2) *Alpine,* a belt extending from central France across Switzerland and northern Italy eastward; (3) *Mediterranean,* along the shores of southern Europe, and (4) *Northeast European,* of the Baltic-Balkans area, and into Russia. See S. C. Coon, S. M. Garn, and J. B. Berdsell, *Races: A Study of Race Formation in Man* (1950) p. 127; quoted from E. A. Hoebel, *Anthropology: The Study of Man,* McGraw-Hill, New York, 1966, p. 217.

[4] This dependence has been lessened by increasing coal imports from the United States; Italy has also made great use of its hydroelectric power potential to reduce coal requirements, while the Netherlands and West Germany are exploiting extensive fields of natural gas. Petroleum from North Africa also contributes to Europe's power supply.

The North European Lowland is flat or slightly undulating. This picture shows the Lowland in northern Germany, south of the town of Emden. Fertile fields surround the town. (Charles Rotkin)

kind in Europe alone, and if we possessed all trade information for the past several thousand years and drew a series of maps to represent these transactions, we would see a growing web of ever-increasing complexity, involving more and more places, eventually overflowing the continent and reaching out across the oceans to other parts of the world.

Opportunities for Exchange

Let us now turn briefly to the physical stage upon which all these human activities have been carried on. The first impression of Europe's topography is one of immense confusion, with large mountain ranges such as the Pyrenees and the Alps obstructing land communications, and islands and peninsulas suggesting even greater disconnectivity. But in fact there are more opportunities than obstacles. The great Alpine

backbone of Europe has been broken by erosion and glaciation, so that numerous passes exist to permit transportation routes to cross it. The waters separating the peninsulas from the mainland have afforded marvelous opportunities for sea traffic. And from the central mountains flow great rivers in all directions to the sea—rivers which, over long stretches, are navigable to large vessels.

Any first attempt to identify the topographic regions of Europe is likely to produce four such units. Of these, the *North European Lowland* presents fewest barriers to communications. This regions extends from southwestern France (where it is known as the Aquitanian Plain) through the Paris Basin and the Low Countries across northern Germany and eastward into Poland. Southeast England and southern Sweden may be considered part of this region, which forms

E. RAISZ, GOODE'S WORLD ATLAS

| | Western Uplands (mostly old rocks) | | Central Uplands |
| | Great European Plain | | Alpine System |

Map 2-6 PHYSICAL REGIONS OF EUROPE.

a continuous belt on the mainland from southern France to the Gulf of Finland (Map 2-6). Most of the North European Lowland lies below 500 feet in elevation, and local relief rarely exceeds 100 feet. Make no mistake, however—this region may topographically be low lying and flat or gently rolling, but there its uniformity ends. Beyond this single topographic factor there is much to differentiate it internally. In France it includes the basins of three major rivers, the Garonne, Loire, and Seine. In the Netherlands, for a good part, it is made up of land reclaimed from the sea, enclosed by dykes and lying below sea level. In southeast England, the higher areas of the Netherlands, northern Germany, southern Sweden, and farther eastward, it bears the marks of the great glaciation that withdrew only a few thousand years ago. In the Soviet Union local relief increases somewhat and the lowland is studded with lakes, again a legacy of glacial times.

Each of these particular areas affords its own opportunities for human endeavor, as soils and climates vary. Along with wheat and corn, there are areas of viticulture in the Garonne lowland. There are fine pastures in the Netherlands and England, and here dairying is an important industry. Fields of rye cover the north German countryside. Toward the northeast the glaciated

country is clothed by stands of mixed forest (including areas of conifers familiar to anyone who has seen the woodlands of Canada and northern Michigan), with pastures and cropland in intervening belts.

The North European Lowland has been one of Europe's major avenues of human contact. Entire peoples migrated across it; armies marched through it. At an early stage there were trade ties with southern parts of the continent. As stock raising gave way to agriculture, the forests which once covered most of this region were cut down. And as time went on new techniques of cultivation and new crops led to ever-greater diversification. It is hardly possible to speak of a "wheat belt" or "ranching area" in Europe, such as exist in North America. Even where one particular activity or crop predominates in the farming scene, some other pursuit takes place only a few hundred yards away. A short train ride almost anywhere will lead past fields of carefully demarcated pasture, then suddenly along fields of well-tended potatoes, back to pasture, perhaps through fields of bulb flowers, a stretch of rye, and then—we must be nearing a city—neat rows of vegetable gardens. Trimmed hedges, narrow canals, or fence limit the various fields; every bit of available land seems somehow in productive use. Small wonder that trade in agricultural produce long was Europe's mainstay.

It is not just the low relief that has favored contact and communication in Europe's lowland. The region has another crucial advantage: its multitude of navigable rivers, emerging from higher, adjacent areas, which wend their way to the sea. We have mentioned several in France; in addition, the Rhine-Maas (Meuse) system serves one of the world's most productive industrial and agricultural areas, to reach the sea in the Netherlands; the Weser, Elbe, and Oder penetrate northern Germany; the Vistula traverses Poland. Where the lowland broadens eastward, the Dvina Rivers, the Dnieper, Don, and Volga serve the Soviet Union, while in the southeast the Danube rivals the Rhine in regularity of flow and navigability. It is a radial pattern outward from Europe's higher central areas, and only the Mediterranean south fails to share proportionately in this wealth of natural landscape.

Thus the natural waterways as well as the land surface of the North European Lowland favor traffic and trade, but man improved the situation still more. Navigable stretches of rivers were connected by artificial canals, and through these and a system of locks a veritable network of water transport routes evolved. Waterways, roads, and later railroads combined to bring tens of thousands of localities in contact with each other. New techniques and other cultural innovations spread rapidly; trade connections and activities developed continuously. No region in Europe provided greater opportunities for all this than the lowland.

The North European Lowland forms a triangle, with its apex in the west and opening eastward. Along both its northwestern and its southern sides it is flanked by higher and more rugged land. From Finland, through Scandinavia, northern England, Scotland, and Ireland and southward into the Iberian Peninsula lie the *Western Uplands*.[5] France's Bretagne (Brittany) Peninsula also forms part of this region, which largely consists of plateau country between 1000 and 5000 feet in elevation, although some parts are higher. In Scandinavia and Spain several areas exceed 7000 feet; in Scotland elevations are lower, rising aboue 4000 feet in only a few places. Ireland and western France are lowest.

This western arc of highlands presents a situation very different from the North European Lowland. Whereas sedimentary rocks of various kinds sustain the low relief of the European Lowland, the Western Uplands are supported by crystalline rocks, mostly of great age and of considerably higher resistance to the forces of weathering and erosion. Every continent has such masses of crystalline rocks, which have withstood forces of destruction over hundreds of millions of years and remain today as geologic core areas. In North America, most of Canada is made up of such crystalline rocks; in South America that part of Brazil not covered by Amazon Basin sediments is of the same quality. Much of Africa south of the Sahara is únderlain by crystalline rocks of this kind, as is western Australia. Such large crystalline bodies are referred to as *shields,* and so it is appropriate to call the northern part of the Western Uplands the Scandinavian *Shield.* In North America the Canadian Shield is sometimes named after one

[5] It is important to remember that we are looking at the *topographic* regions of Europe, not the *physiographic* regions of the continent. The latter would involve many more considerations than elevation and relief!

A **view** over the glaciated Scandinavian shield. Note the truncated spurs and U-shape of the valley once occupied by a glacier. (Aerofilms Ltd.)

particular area where it is especially well expressed: the Laurentian Shield. And like the Canadian Shield, the Scandinavian Shield during its long history repeatedly underwent severe glaciation. The ice scoured bare the rocks buried deep beneath it, and millions of tons of ground-up waste were carried away to be deposited elsewhere. Where rivers formerly flowed the ice carved wider and deeper valleys, and when the ice age came to an end, barren wastes were exposed to water and wind. Where glaciers had emerged from their valleys to push out into the ocean, great fjords remained to be filled by the invading sea water.

But while the ice gouged and scoured, it could not destroy the Scandinavian Shield completely. The rocks prevailed, though they were doubtlessly lowered during the great frozen siege. When it comes to Man's occupance of these areas it is soon clear that the glaciers were both friend and enemy: enemy in the destruction of river gradients and navigability, the removal of soils, the scattering of boulders and other rock debris; friend in the flattening of topographic barriers and the carving out of useful harbors. The relief presents serious obstacles to transportation only in northern Scandinavia, where human activity is also obstructed by other factors; elsewhere there are hindrances, but no barriers.

All this applies to Scandinavia and the northern British Isles, but not to the Iberian Peninsula. Spain and Portugal were not glaciated during the Pleistocene—it was cold, and there was tundra in the north while conifers spread across the remainder of the area—but there was no permanent ice sheet. But in terms of topography and relief the Iberian Peninsula, especially in the central and western areas, is very much a part of the Western Upland arc. West and northwest of Madrid elevations exceed 5000 feet and in some places approach 8000 feet; near the Bay of Biscay in the far northwest the Cantabrian Mountains include elevations well over 8000 feet. Here again are old, hard rocks thrust upward during ancient periods of uplift and mountain building, and they continue to sustain a comparatively rugged landscape. In the older *meseta* country, the most common vista is plateaulike rather than truly mountainous. As in Scandinavia there is little opportunity for inland water transport, but the relief is not such that it inhibits effective land connections.

A third major topographic region of Europe lies to the south of the North European Lowland and north of the Alpine mountain chain. It extends from southern France through southern and central Germany into western Czechoslovakia and consists of terrain that is generally hilly and often plateaulike, with some prominent, forest-clad mountainous areas. The relief is not unlike that of the Western Uplands, although the period of mountain-building that pushed the rocks upward came later than that of Scandinavia and Scotland. This region, then, is called the *Central Uplands,* and in Western and Central Europe it marks the southern limit of the North European Lowland. Usually that limit is quite well defined as the flatness of the European Lowland abruptly gives way to higher elevations and much more varied relief. But what is seen today is but a shadow of the great mountains that once existed there. Among the layers of rock that were folded and twisted into great mountains were beds of very soft rock such as sandstone and shale, and these weaknesses of the mountainous structures were soon exposed by erosion. As these were attacked, harder rocks were left unsupported and collapsed. And so the mountains wore down, literally to their roots— but in the process something very important happened. Layers of coal, improved in quality through having been buried deeply and under great pressure, became exposed or were brought

The Spanish Meseta: a shepherd tends his flock in Guadelajara Province, 30 miles from Madrid. (Michael Kuh/Photo Researchers)

The Bavarian Mountains are less high, less impressive, and less obstructive to transportation than their southern counterparts. This is a typical picture of the relief in these areas. (Photo Researchers Inc.)

close to the surface. From Wales (the effects of the mountain-building were felt even there) through the German Ruhr, Upper Silesia, and the Donetz Basin, coal deposits were preserved within or on the margins of the mountain system. Ultimately, as we know, their importance to Man as a resource helped thrust Europe into the forefront.

Traveling southward from Denmark to Italy we get a good cross-section of European topography: first the flat or gently rolling plains of Denmark and northern Germany, then the increasing relief of the Central Uplands, and ultimately the formidable Alps. The Central Uplands, on this scale, seem no more than a series of foothills separating the coastal plainlands from the southern mountains, and although there is rough terrain in places, it is never sufficiently extensive to impede cross-continental transportation. The higher parts, which sometimes exceed 4000 feet, are tree-covered and identified as such by name: if our journey carried us somewhat to the west, we would see the Black Forest; to the east lies the Bohemian Forest. But the terrain of the remainder of the region is such that it can support intensive cultivation. And

with a variety of well-drained soils, alluvial valleys, and a good climate, there are very varied opportunities for agriculture. From the pastureland of southeastern France and the wine country of southwestern Germany these range to the wheat, oats, rye, and other grain crops grown elsewhere.

With the *Alpine Mountains* the land surface changes abruptly. In this southernmost topographic region of Europe transportation faces its greatest obstacles and, over large areas, agriculture and industry as well. This region is referred to as the Alpine Mountains rather than just the Alps because it includes mountain ranges sprawled across Southern Europe from Spain to beyond the eastern boundary into Turkey and the southern U.S.S.R. Indeed, even North Africa has its share of Alpine mountains in the Atlas ranges. The whole system constitutes the most recent manifestation in Europe of mountain-building; from the Pyrenees to the Caucasus and onward into Asia the rocks were folded and crushed into giant, mountainous arcs of which the most famous, the Alps, is by no means the only one. Geologists are not yet certain just why and how mountain building starts and stops, but

This photograph illustrates the relief of the Swiss Alps in the Berner Oberland. The cloud cover lies well below (right center). (Swiss National Tourist Office)

the Alpine *orogeny* (as such a revolutionary event in Earth history is called) is clearly not completely finished. Mediterranean Europe continues to be plagued by associated events such as volcanic activity and earthquakes; in 1968 calamitous earthquakes struck Sicily with a high loss of life.

The Alps and its related mountain chains (such as the Carpathians and the backbone of the Italian Peninsula, the Apennines) will no doubt undergo the same fate as the mountains that earlier stood where the Central Uplands now lie: erosion will do its destructive work and gradually wear them down. But the Alpine Mountains are still young, and notwithstanding the attacks of Pleistocene glaciers their height and prominence bear witness to their youth. When it comes to transportation, time has not given Man the help it has given him elsewhere in Europe. The Pyrenees, the Alps, and the Carpathians—and, as will be seen later, the Caucasus—are formidable obstacles to contact. What lower areas there are, such as the Po Valley

in northern Italy and the Hungarian Basin, are surrounded by difficult terrain.

Still, it could have been worse. Although the Pyrenees are a rather effective barrier between the Iberian Peninsula and France (a barrier which is reflected in the cultural landscape of Southwestern Europe), it never stood in the way of major transcontinental trade. There is a way around the Alps on both sides—through southeastern France and along the Danube Valley separating the Alps from the Carpathians—and the ranges of the Alps are themselves breached by deep valleys and interconnecting passes. Goods found their way across and around the Alps at a very early stage, for the potential complementarity existing between Mediterranean and Northern Europe soon was recognized. But the traffic was always impeded by the Alpine bottlenecks, and Europeans have labored for centuries to improve communications across the mountains.

Within Mediterranean Europe land communications were always difficult. Settlements located in valleys or basins remain separated by steep-sided mountain ranges, and more often than not access to the sea is easier than to another place on land. Apart from hundreds of larger and smaller islands that lie scattered through the Mediterranean from Cyprus to the Balearics, many mainland valleys open directly to the sea. Small wonder that navigation skills were quickly adopted, learned, and developed in these parts. And this is a good time to note an advantage enjoyed not only in Mediterranean Europe but more or less throughout the continent: the advantage of moderate distances. With the exception of Iceland, the islands of Europe are all quite near the mainland. There is a remarkably low variance in the width of such water bodies as the Aegean Sea, the Adriatic, the English Channel, the Skagerrak-Kattegat, the Baltic Sea, and the Gulf of Bothnia. Again, the coasts of the North Sea are about as far from one another as are those of the western Mediterranean, the Tyrrhenian, and Ionian Seas. Europe's offshore waters are not the almost endless expanses that are those of the Americas, Africa, South Asia, and Australia. They invited exploration and utilization; what lay on the other side was not some dark unknown but, more often than not, a visible coast of land of known qualities.

This applies on land as well. The Alps form

The rugged, divisive country of the Pyrenees, on the French-Spanish border. (Aerofilms Ltd.)

a divider, but what they divide still lies in close juxtaposition. Consider Rome and Paris: the distance between these long-time headquarters of Mediterranean and Northwestern Europe is less than that between New York and Chicago or Miami and Atlanta. No place in Europe is very far from anyplace else on the continent, and nearby places are often sharply different from each other in terms of economy and outlook. Short distances and large differences make for much interaction—and that is what Europe has had for many centuries.

The Alpine-Mediterranean region has undergone some major changes since Man began to organize states there. Arriving on the scene today we might wonder how such comparatively barren country could have supported such thriving empires; the Iberian plateaus are bare and treeless, much of Italy's landscape is subject to severe erosion, and in Greece the country seems denuded, almost crying out for a protective vegetative cover to reduce the glare and brightness of the Mediterranean sun. What has happened is that Man and his domestic animals over cen-

turies of occupance attacked and destroyed the natural vegetation, reducing it from forest to grassland and eventually to the most hardy scrub—and ultimately even the scrub gave way. The roots that had held the soil together now ceased to do their work, and the full force of erosion laid bare vast sections of the Mediterranean countryside. We get a glimpse of what this region must have looked like before this onslaught by viewing the now-famous patch of forest called the cedars of Lebanon, a sad remnant of a once-extensive cloth of forest.

The most serious destruction of the original vegetation of Southern Europe occurred after the end of Greek-Roman power in the Mediterranean. Of course the forests constituted a major resource to the Greeks and Romans: they needed the wood for the construction of dwellings and the building of ships. But although they did destroy the forest to a considerable extent in some areas, especially around major cities such as Athens and Rome, there was not, by 500 A.D., the wholesale destruction that was to mark the region later. This came with population increases, the clearing of land for cultivation, the growing need for timber to build and maintain ever-larger fleets, and the onslaught of grazing animals, especially sheep and goats. In Spain once stood a dense mixed forest of oaks and other tall trees; today the plateaus of Iberia are virtually treeless. The pine forests of Italy and the woods of Greece and the larger islands suffered a similar fate. In Southern Europe this was more disastrous than it was elsewhere, for the forest performed important functions in the retention of soil and moisture. What was possible in the North European Lowland, where successful cultivation followed forest clearance, was not possible along the slopes of the Mediterranean. For centuries an increasingly heavy price has been paid in lost farmland and lost incomes.

The dark picture just painted does not, of course, apply to all areas of Southern Europe without exception. The region still includes enormously fertile plains and valleys and, with its salubrious climate, its specialty crops find ready markets. On the good soils of the Po Valley winter wheat, the Mediterranean staple, thrives and it does well also in north-central Spain. Barley is grown in Spain and central Italy. Vineyards lie along the eastern coast of Spain, in the "heel" of Italy, along the western coast of the Greek Peloponnisos, on Crete and, on a smaller scale, in many other areas also. A wide variety of Mediterranean fruits, including citrus fruits, figs, and olives, as well as vegetables and even flowers are grown from one end of the region to the other. But the good land of Mediterranean Europe is patchy. Good soils are at a premium and cultivation on elaborate terraces creeps precariously up steep hillsides. It is quite a contrast to the total utilization of space that marks the farmlands of the western European Lowland.

Bases for Productivity

Europe is a high-latitude continent; however often one is reminded of it, it is still surprising to realize that the British Isles lie farther to the north than Newfoundland, that Paris lies north of Montreal, that Rome and Madrid lie north of Washington, D.C. But if we are going to be consistent about comparisons, let us not compare the western side of one continent with the eastern side of another. Let us view Western Europe and *western* North America in this context. Now we find that London lies only slightly farther north than Vancouver, B.C., that Paris is at about the same latitude as Seattle, and that Athens and San Francisco also lie at virtually identical latitudes. Certainly Europe has no monopoly on moderate climatic conditions at relatively high latitudes. But there is something about Europe on Plate 2 (see foldout map in rear of this volume) that still demands an explanation. The largest part of the continent lies under a climatic area colored turquoise and marked *Cfb* which, from the legend, we gather to mean a warm temperate climate with sufficient precipitation in all months and a rather cool summer. In North America, too, there is a small area so mapped, but it is a mere ribbon along the north Pacific coast. In fact, a glance at the map shows that Europe has by far the largest area of *Cfb*-type climate in the world. Why should this be so—and will the answer to this question provide answers to other problems related to the European environment?

Such answers are sought through a greater understanding of the behavior of air—the air of the Earth's atmosphere. The atmosphere is the key. Over the oceans air may become moisture-laden; such air may then travel many hundreds of miles into a continental interior before dropping its humid content in a series of rainstorms. A mass of air may lie over the Arctic, cooling and chilling, and then, when conditions are ripe, flood down into the midsection of a hitherto comfortably warm continent. In the tropics surface air may become so warm that it begins to rise, vacating lower altitudes so that air from elsewhere rapidly flows to the area of strong convection. The ground observer will feel a mild wind, and looking upward he will see the rising air cauliflower into giant, billowing clouds which soon develop dark and rainy undersides. Everywhere the atmosphere is almost constantly in motion, driven by the radiating power of the sun and fanned by the rotation of the Earth itself.

Topography has a great deal to do with climate. Mountain barriers can impede the flow of air in the atmosphere just as they can obstruct land transportation. Consider what has been said on previous pages about the topographic layout of Europe, and compare this with what we know of North America. In Europe there is not a single major mountain system that runs north-south; the nearest thing to it is the backbone of the Scandinavian Shield. Elsewhere, from the North Sea to the Urals the North European Lowland opens the whole of Europe's midsection to a flow of air that might be coming from the west. The southern limit of this area is marked by the Alpine Mountains, but they trend from west to east. These mountains will get in the way of air flowing southward from northern (Arctic) latitudes, but do not form an obstacle to eastward-flowing air. Now compare this to North America, where it is quite difficult to think of any really large mountain chain trending east-west. North America's major mountains run north-south—the Rockies in the west central area and the Appalachians in the east. Our North American mountains form an obstacle to any eastward flow of air coming from the Pacific, but they form a virtual funnel for air moving southward from northern Canada—or northward from the warm Caribbean. To look again at the European and North American areas of *Cfb* climate: the mountains may not be the only explanation, but certainly that ribbon in North America conforms suspiciously to the trend and position of the northern part of the Rocky Mountain system. In Europe the southern limit of *Cfb* climate conforms quite closely to the Alpine ranges—not exactly, of course, as in northern Spain where it extends south of the latitude of the Pyrenees, but the rule generally applies. In Scandinavia we see a situation resembling that of Alaska: a narrow band of *Cfb* climate limited, apparently, by the rise in elevation toward the interior.

Europe, then, is open to mid-latitude maritime influences in a way that no other continent even approaches. In the Southern Hemisphere only South America comes close to matching Europe's latitudinal position (the southern tip of South America lies about as far from the equator as does England) but the narrowness of the land-mass, the height of the southern Andes Mountains, and the rough terrain elsewhere greatly limit the area of *Cfb* climate. In Southern Africa it forms a narrow band below the great escarpment that leads to the high plateau interior. The Australian area is similarly quite small, but New Zealand is in position to receive the full impact of mid-latitude maritime conditions.

A word of warning regarding the delimitation of climatic "regions" such as we are employing here is in order. Although this is an easy and convenient way to identify climatic conditions with similar characteristics in various parts of the world, such regional classification is, of course, based on wholly artificial criteria. Climatic conditions change from place to place on the globe, but the changes are normally in the manner of transitions rather than sharp breaks. On our world map (Plate 2), for example, we would be hard pressed in the field to find evidence to support the line separating Europe's *Cfb* climate from that which lies immediately to the east (*Dfb,* a "snow" climate also referred to as "humid continental"). The basis of this line is found in the legend, but although the figures listed there were of course selected after careful consideration of indicators of climatic change such as vegetation and soils, they are nevertheless quite arbitrary and, according to some authorities, not really defensible. For example, while the basis of the classification and regionalization of the world's climates on Plate 2 lies in monthly temperature averages and monthly precipitation figures, no account is taken of the amount of the incoming moisture that is lost to evaporation and transpiration. Thus an area that is marked as "humid" on the basis of rainfall alone may actually be deficient in moisture: more moisture may be lost annually through evapo-transpiration that actually arrives, and the water budget of the area is "in the red," as it were. Nevertheless our classification would show it as "humid."

With these shortcomings in mind, let us consider Europe's climatic regions, and their opportunities and limitations for agriculture. Three major types are indicated on Plate 2, namely the temperate *Cfb* in the heart of Europe, the Mediterranean type *Csa* (and *Csb*) with its comparatively dry summers in the south and southwest, and the "snow" continental type *Dfb* in the east and northeast. In the far north the blue color denotes the icy conditions of the Arctic, and over the Alps and other high mountains an *H* refers to highland climates, where the dominating influence is the elevation rather than latitudinal position. This regionalization of Europe's climates suggests that four influences are at work in fashioning the prevailing conditions. The north-south influences are those of the Arctic and the subtropical Mediterranean, and the Alpine mountain wall is the effective boundary between the areas of dominance of the air masses representing these extremes. The east-west influences are respectively continental and maritime in nature, but the maritime conditions have in their favor the fact that Europe lies in a belt of air movement that is generally from west to east—the so-called *westerlies.* Thus it would be reasonable to divide the area of *Cfb* on Plate 2 into two zones: a truly *maritime* zone, positioned along the coastline of France, the Low Countries, Denmark, and Scandinavia, and secondly a *transitional* zone, where the impact of the ocean begins to decline and continental influences begin to be felt more strongly.

While it is true that no mountain barrier interferes with the eastward flow of maritime air across Europe, there is another important reason why so much of Europe is affected by marine influences. Water bodies such as the Baltic Sea, the Gulf of Bothnia, the North Sea, and the Bay of Biscay (as well as the arms of the Mediterranean) penetrate deeply into the land area and carry such influences far into the interior. Europe has more coastline per square mile of land than any other continent, and that means that more land areas are near water than anywhere else. Hence maritime air masses can reach far inland without losing their effectiveness.

What is this effectiveness—how is a land area made "marine" by lying near water? This question is likely to arise when Plate 2 is considered again, now from the point of view of other North Atlantic coasts. Take, for example, the huge area marked *BWh*—the Sahara Desert—extending from Arabia in the east to the very shores of the Atlantic in the west, and not far from Europe at that. Why are there no "marine" conditions visible here? We have part of the answer: Europe's latitudinal location is such that atmospheric conditions tend to move from west to east, from the sea toward land. In other parts of the world the prevailing air circulation may well be offshore, that is, from land to sea, so that maritime conditions have little chance to penetrate the coastal areas. But there is more to it than that. It does not take much research to come up with a map showing parts of the world with prevailing onshore winds and *still* no maritime effect onshore. Obviously it is not just the prevailing movement of air, but also the conditions in that part of the ocean over which the air moves that plays a role. When water is very cold and the air over the water is cold as well, that air cannot pick up much moisture. Even if such air should then be swept inland the precipitation it could produce would be minimal. But if the adjacent water is warm and the overlying air mild, then it can pick up thousands of tons of

moisture, to be carried inland as potential precipitation.

In this respect Europe is singularly fortunate, although climatologists are still not quite sure just how Europe benefits so much. So warm is the water of the North Atlantic along and near the European coasts that even ports in the latitude of Greenland never freeze. The air that comes off these waters helps make Western Europe's winters mild, and in summer, when land areas tend to heat up rapidly, the same air again has an ameliorating, cooling effect. Thus Europe's adjacent waters are the great modifying influence, being warmer than the land in winter, cooler in summer. For a long time climatologists and oceanographers attributed this warmth to an extension of the Gulf Stream. This great current is part of the North Atlantic Ocean's clockwise circulation pattern (a motion induced by the rotation of the Earth), and carries tropical warmth with it past Florida and the eastern United States—and then across the North Atlantic toward Europe. Then most of the water turns south, but, it was thought, an extension of the Gulf Stream, the so-called North Atlantic Drift, moved slowly northeastward along the European coast to Scandinavia. However, it was recently realized that it is indeed a long way from tropical areas via the coasts of North America to Scandinavia—that so much heat would be lost during the water's movement that the North Atlantic Drift surely could have all the influence long attributed to it. One new hypothesis is that the tropical circulation of the North Atlantic Ocean expands at certain (apparently irregular) intervals, and that the force of the Earth's rotation sweeps most of the expanding volume toward Europe, while a compensating flow of cold water from polar areas moves along North American coasts. Whatever the answer, Western Europe owes its temperate character to this maritime source.

The major characteristics of the area identified as *Cbf* on Plate 2, then, are mildness and moistness. Annual rainfall over the North European Lowland is mostly between 20 and 40 inches (see Plate 1), but this precipitation normally comes in slow, steady, soft rains or drizzles often lasting for days. For other parts of the world 20 to 40 inches of rain may be insufficient for cultivation, but in this part of Europe the accompanying cloud cover and the cool weather greatly reduce losses through evaporation and transpiration.

This rain, also, is rather evenly distributed throughout the year; it does not come in heavy downpours during a single season. On the average, places in Western Europe have rain on half the days of the year; many have even more rainy days than that. Hence there is sufficient moisture for plant growth all the year round, and only the growing season, which varies from 6 to 8 months, limits such cultivation. From the point of view of the farmer, few parts of the world can rival the security existing here that there will be good yields; sometimes a particularly cool or wet summer can have a negative effect on some grain crops, but crop failure through natural causes is rare indeed.

On both the soil map (Plate 3) and the map of natural vegetation regions (Plate 4) the European Lowland can substantially be recognized. Most of this region is blessed with representatives of the gray-brown podzolic soil group in a tight mosaic depending upon local conditions of bedrock, drainage, and age. These soils developed under the deciduous forest (7 on Plate 4) that once prevailed across most of this region, and they contain a great deal of organic material and plant food. Their texture is good and they are well drained; for centuries they have produced good yields for Europe's farmers. Of course, practices of conservation, fertilization, and reclamation have helped maintain the productivity of the soils of the Northern European Lowland, and the climate (especially the winter rest period) also plays its preserving role.

Toward the eastern and northeastern interior of Europe, the maritime influences of moisture and mildness begin to diminish. Rainfall becomes gradually less, and while summers become hotter, winters are colder. The growing season is shorter and the quality of the soils is somewhat lessened. We are now traversing the transitional zone leading to the region identified on Plate 2 as under *Dfb* climate. This climatic zone extends around upper Baltic and around the Gulf of Bothnia into Scandinavia. The conditions prevailing here are familiar to many North Americans, especially those living in the Great Lakes area and in south-central and southeastern Canada. Here the winter months bring cold, polar air which sweeps into the Balkans, into the Black Sea area, and through the Bothnia-Baltic corridor to southern Scandinavia. For farmers, the risks of cultivation increase: the precipitation becomes more variable and less reliable. The

natural vegetation of the northern two-thirds of this area, which extends far into Siberia, is the needleleaf forest (8 on Plate 4) similar to that of Canada; the soils (4 on Plate 3) are more weakly podzolized.

The "ice" climate marked *E* and shaded blue is the poleward continuation of the *D* climate just discussed. In Europe its extent is not great; northern Scandinavia and northern Iceland have areas under tundra climate, as the milder version of this type is often called. In North America this climatic type is far more extensive, with a belt running from Alaska to Labrador and northward across the islands to Greenland, where it becomes permanent ice. This is the least populated of Europe's climatic regions as defined on Plate 4 which, with its long, severe winters and brief, very cool summers is not surprising. Soils are barely developed (such soils are called *lithosols, rock* soils) and in the tundra areas only the hardiest vegetation survives.

The Mediterranean climatic type (*Csa* on Plate 2) presents quite a contrast to these northern, Arctic conditions. Here is another region of mild climate—not, this time, by courtesy of the warm Atlantic waters, but as a direct result of the sun itself. Indeed, in almost every way the Mediterranean climatic type presents a contrast also to the *Cfb* climate of the Northern European Lowland. There is vastly more sunshine, generally a little less precipitation, a strong seasonal imbalance to the arrival of this rainfall, a hotter summer, and an absence of the constant changeability marking the *Cfb* clime. In some ways the Mediterranean is an area of greater extremes than is Western Europe: the southern slopes of the Alps, the Po Valley, and the Adriatic coast receive well over 40 inches of precipitation (over 60 inches are recorded in the Alps themselves), while the southeast of Spain, southern Sicily, and parts of Greece get well under 20 inches (Plate 1). Most of Europe's rain arrives along cyclonic fronts, that is, along zones of contact between air masses of different qualities within large areas of low pressure. But where the mountains help to force moist air upward, the rainfall total increases: note the high figures for southwestern Scandinavia and western Scotland, which average over 80 inches of precipitation annually. High areas in the Mediterranean region also display these higher totals.

The outstanding feature of the Mediterranean climatic regime is its failure to follow one of the meteorology: that "rain follows the ...ost warm parts of the world rainfall ....hen the sun is highest in the sky, that ...he summer. But not in Mediterranean ... summers are sunny, hot, and dry, ...t of the rainfall comes in the low-sun season, the winter. The only exception on our map lies at the heat of the Adriatic Sea and in the Po Valley, where the *f* in *Cfa* indicates that rainfall is distributed fairly evenly throughout the year. According to one theory the rainbringing low pressure areas manage to penetrate this area from central Europe, even during the summer months; a different view holds that this precipitation is convectional. Elsewhere over the Mediterranean coasts a summer high pressure system prevails, and high pressure means little or no precipitation.

The exact nature of the climatic regime of the Mediterranean is not yet certain. One explanation involves the seasonal northward and southward migration of the belt of westerly air movement that affects maritime Western Europe. With the arrival of winter, it is argued, this belt shifts somewhat to the south, and the cyclonic systems that moved into the European Lowland during the summer now move south of the Alpine Mountains, through various corridors into the Mediterranean. Then when summer returns and the sun gathers strength, the westerlies move northward again, and most of the Mediterranean comes under the influence of a high pressure system with its resulting dryness. Whether or not this is actually the case, the Mediterranean environment in many ways is a difficult one for agriculture. Apart from the high incidence of steep slopes, the soils even on more moderate topography are often poorly developed; they may be *terra rossa* (Italian for red earth), with some characteristics of higher-latitude podzols and some aspects of tropical latosols, or even lithosols. The incidence of these kinds of soils is quite high (Plate 3), although in some places, where the *terra rossa* has had a chance to develop over a long period of time without the early destruction of the natural vegetation or excessive agricultural demand, it turns out to be quite a productive soil. Still, the best soils in the Mediterranean region are those of the alluvial valleys and basins such as the Po Valley. These soils are as productive as any in Europe, but one problem remains: irrigation must supplement the annual rainfall, even in the comparatively well-watered basin of the Po. In northern Portugal, corn is grown under irrigation; in coastal areas of Spain, citrus orchards (oranges) are artificially watered, as are those of northern Sicily, where lemons are the main product. Around major cities, especially Genoa and Naples, vegetables and fruits are grown under similar conditions. But the areas with irrigation are far exceeded in size and number by those without, and the lot of the peasant in this part of Europe is not an easy one.

European Regional Organization

Having now surveyed some of Europe's environmental attributes, let us return now to our theme of European growth—the growth of European regional organization. We have seen that Europe has an enormously varied resource base. Its soils and climates permit the cultivation of a wide range of crops. Its bedrock holds reserves of coal, iron ore, and other metals. In France, Austria, and northern Germany there are minor petroleum fields, the last with very important associations of natural gas, which extend into the Netherlands province of Groningen. The adjacent seas are rich in fish fauna, and fishing makes a major contribution to the economies of such countries as Norway, Britain, and especially Iceland. Thus Europe has been an area of almost unlimited opportunity. In the days of the Roman Empire some of those opportunities were realized, and for a time that continent seemed destined to achieve a lasting regional economic and political organization at a very early stage.

European regional organization as it has been emerging during the last century, however, has not been the product of continuous, progressive development from the days of Rome. In medieval Europe the agricultural opportunities were there, but there was not much progress in agricultural methods and yields. Like so many parts of the world which the industrialized states consider "underdeveloped," Europe in the Middle Ages was a land of villages surrounded by areas of subsistence agriculture. There was almost complete local isolation; interconnections were few, and the only vestiges of any regional authority were those of the feudal lord and the church. The Roman roads lay forgotten and in disuse. Places which had at one time exchanged goods were in decline.

As was indicated earlier, the first signs of change came during the twelfth and thirteenth centuries. In the Low Countries there was the development of the manufacture of wool goods—a textile industry requiring imports of wool from as far away as Spain, Germany, and even across the Channel from England. In Mediterranean Europe the Po Valley was the scene of similar developments: Venice was on the threshold of its great age of power on eastern Mediterranean shores. The trade across and around the Alps had begun in earnest, and soon Mediterranean ports such as Genoa and Marseille (in addition to Venice) were in profitable contact with Flanders. And as the manufacturing and trade functions of long-dormant places began to grow, so did their populations—and the requirements of these populations. Increasing quantities of food became necessary; farmers who had long engaged in subsistence or a minimum trade with a nearby village found new markets for their produce. Especially in Venice and Genoa the arts thrived. The architecture of the period still stands as a monument to the newfound opulence and confidence of the rising centers of power.

Without denigrating the importance of these developments, however, they still left most of Europe outside of any larger sphere of regional organization—economic, political, or cultural. The impact of the new economic revolution was slow to reach areas that lay any distance from the major trade routes. But with time—more than a century of it—changes came even to these places. Some of them were located at crossroads on new European trade routes. Others were positioned at locations on rivers where waterborne merchandise had to be transferred to land carriage. All of them attracted traders and craftsmen; many of them have been important European centers ever since. Perhaps the most significant indicator of the changing economic situation in Europe was the trade *fair,* held in important towns to market commodities contributed from surrounding areas. These fairs were first held in the textile towns of northern France and Flanders, and they attracted commodities from all over Europe and even beyond. Here they were organized in a sort of annual rotation among the towns reminiscent of the system of periodic markets that exists in many parts of Africa, except of course on a larger scale. Soon many other European towns, eager to attract goods and people from far afield to add to their luster and importance, were organizing their own fairs. Many had great success with these and rose to regional stature, including two major competitors in the interior, Lyon and Geneva. Other towns with important fairs were Milan (on the Venice–Flanders route across the Alps), Frankfurt, and Leipzig.

Thus we begin to recognize the modern heartland of Europe—the European *core area.* The old commercial-manufacturing area of northern Italy, as well as the area from northern France into the Netherlands, today form an integral part of this region; many of the towns that rose to prominence at the close of the Middle Ages are urban centers of prime importance in modern Europe. But despite the developments in manufacturing of Flanders and northern Italy, and despite the increased *focality* of the towns that organized fairs (i.e., their quality of acting as foci of commercial activity for large surrounding areas), much of Europe still awaited integration into a true regional framework, political or economic. Most parts of the continent still remained under subsistence forms of economy; most Europeans still lived and worked in comparative isolation. And it was to remain so, more or less, until the impact of the industrial revolution and the resultant changes in the resource situation were felt on the mainland. This whole period began only about two centuries ago; the lag between the momentous discoveries and innovations made in Britain and their application on the mainland was considerable. Europe's population prior to the events of the eighteenth century hovered substantially below 100 million; after 1700 it began to increase at a growing rate. At first the increase was relatively slow, largely reflecting changes in England. But late in the second half of the eighteenth century the rate of increase became higher, and shortly after the turn of the century the total had doubled from the 1700 figure. Now Europe was truly experiencing a population explosion, for within a century the population was to double again. The chain of events leading to present-day Europe, highly industrialized, highly urbanized, highly organized, had begun.

Of course, not all of Europe shared equally or simultaneously in the benefits of the industrial revolution. First, those countries which might have been in the best position to receive and use the innovations were not always ready politically. Across the Channel, France during the eighteenth century was beset by feudalism and localism and was headed for the great revolution of 1789. Disunity also prevailed in Germany. Then the period of Napoleonic imperialism brought chaos to much of the continent. And when a semblance of order had been established, the adoption of the newly discovered techniques turned out to be no easy matter. The charcoal furnaces that had long been in use were not fit to operate with coking coal, and after a first unsuccessful try, many ironmakers went back to charcoal. Also, the coal that was used at first was not of sufficient quality to serve adequately as coking coal. The better coals were buried deeper; they had to be mined to be brought to the surface. Finally, the iron industry began to relocate on and near the coalfields of Europe, and with this last phase of adjustment to the new age of industry the details of modern Europe's core take shape—fully a century after this impact had first been registered in England.

The major coalfields of Europe lie in a belt across the French–Belgian boundary, through northeast Belgium and the Netherlands province of Limburg into the Ruhr, and then along the southern margins of the North European Lowland to Polish Silesia. Several important deposits lie in the vicinity of the Germany–Czechoslovakia boundary. This latitudinal zone of coal deposits, related to the older rocks of the Central Uplands and its outcrops through the European Lowland, continues in the Donetz Basin of the Soviet Union, about which more will be said in Chapter 5. In Western Europe, the really significant deposits are shared by three countries: France, Belgium, and Germany. Germany is best endowed, with good deposits in the Saar in addition to the Westphalia deposits that serve the Ruhr. In Western Europe the chief iron ore deposits lie just across the German boundary in the French Lorraine—an area Germany once held, but lost with World War I. A second major iron ore deposit lies well outside the east-west belt just identified: this is the Kiruna field in Sweden, whose ores are largely exported to (West) Germany, Britain, and Belgium.

The principle generally is, then, that the iron ore goes to where the coal is—and so do the industrial centers whose main business is iron and steel production.[6] With the description of the west-to-east belt of coal and iron fields we have at the same time identified one of the two main axes of the European core, an axis that begins in England and extends through northern France and the Benelux countries into Germany and on into Czechoslovakia. Some of the English cities so oriented will be enumerated later (see p. 64); France's second-ranking industrial-urban area focuses on the northern coalfields near the textile center of Lille; in Belgium the zone extends from Mons near the French boundary through Charleroi to Liege, and in the German Ruhr-Rhineland a megalopolis is developing that already constitutes Western Europe's prime industrial region. Essen, Dortmund, Duisburg, Düsseldorf, and Gelsenkirchen are all part of this giant conurbation. Saarbrücken is the center for the Saar, while in the French Lorraine the cities of Metz and Nancy are the principal foci.

The Core. It is difficult to give an exact definition to the concept of a core area. Everyone can see that it exists: most countries, and especially the developed ones, have what might be called "heartlands," where the center of gravity of the state and nation are located. It would be a difficult argument to sustain that Paris and the Paris Basin do not constitute the core area of France. London obviously is the heart of England. The trouble begins when attempts are made to give more exact dimensions to this idea of a core area, and to draw lines on maps showing exactly where core areas of states and even continents are—what is within the core, and what is not. The preceding pages have dealt with areas of productivity and potential, and from this discussion one might conclude that core areas are an essentially economic phenomenon, and that a core area can be defined by recognizing within a certain relatively small percentage of a country's territory a very high proportion of its productive capacity, cities with large populations and strong focality, and a high concentration of human activity. With a few indices—that is, *how much* of the productive capacity, *how many* of the cities—that definition might satisfy an economic geographer, but quite possibly not a political geographer or a historical geographer. These geographers might argue that since the unit of observation is the state with its political boundaries, the true core area should include the capital city with its many organizing functions and its emotional, psychological significance to the nation. In most cases the capital city does lie within the core of the state, however, defined—but there are exceptions. In some states

---

[6]This does not always hold true: for example, coal is shipped in great quantities to the comparatively coal-poor French Lorraine. Sometimes coal and iron ore are both shipped to an intermediate location. But where long distances are involved the iron ore usually travels farthest.

European Core Region

Map 2-7 EUROPE: CORE AND REGIONS.

the historic capital remains only a monument to a possibly glorious past, now in the shadow of a new industrial area elsewhere, forging ahead aggressively under the leadership of cities rivalling the capital in all but their ties with the past. In Italy, for example, Milan and the Po Valley industrial-agricultural area are in this position.

But when thinking of the heart and spirit of Italy, who thinks of Milan first and Rome second?

Nevertheless, the concept is a useful one. A European coreland clearly exists, just as clearly as such a single core area does *not* exist in South America or Africa (Map 2-7). The major reason for the contiguous nature of the European core area, apart from the relative proximity of the various productive areas, is the intensity of the communications networks that were developed. It is correct to use the plural in this connection: the waterways, roads, and railroads of Europe each form true networks serving the core; together they interconnect its various parts like no other region in the world is interconnected. When it became necessary to haul large tonnages of coal and iron ore, the waterways went through a new phase of improvement. With the development of the German Ruhr-Rhineland the Rhine River became a veritable highway, as busy with freight traffic as any land highway filled with trucks. A railroad network evolved during the second half of the nineteenth century, and the costs of

The Rhine is Europe's most important waterway; more than a dozen ships can be seen using it on this small stretch between Bad Godesberg and Bingen. And, just like any other highway, the Rhine has side "streets," typified by the canal intersection photographed south of Amsterdam in the Netherlands. (Charles E. Rotkin, *both*)

land transportation went down. Within the European core and even well beyond into Eastern Europe, no location is more than 10 miles from a railroad except in a few small areas in the Alps.[7] Only eastern North America has a comparable density of rail facilities. The road network was also intensified, although without the sort of planning that has led to the North American system of transcontinental highways. Many European roads, though important in the general intracore circulation, were not fit to cope with the higher speed, high-volume requirements of recent decades. The era of the superhighway only came to Europe well after World War II.

A second basis for recognition of a European core area, in addition to the region's tight integration through its various transport networks, lies in the high degree of urbanization prevailing here—in contrast to Scandinavian Europe, Eastern Europe, and Mediterranean Europe. Well over half the people of this entire region, which includes England, the Low Countries, France, northern Italy, and West Germany live in cities and towns of over 20,000 population; over one-third reside in cities counting over 100,000 inhabitants. By official statistics, which vary somewhat from country to country, between two-thirds and three-quarters of the population of the European core is classified as urbanized. In Britain, which still has the highest percentage of urbanized inhabitants of any European country (or the world), this includes the cities of Glasgow and Edinburgh, the Midlands, and of course the great urban complex of London. In France, Paris with its 7 million people is the largest urban center on mainland Europe. Notwithstanding the fact that there are no important resources for industry near Paris, this is France's greatest industrial center. For Paris, communications and position did what was needed: they brought the city first a local, then a national, and then a regional focality that has ensured it a permanent place of prominence. The same elements contributed to London's greatness; from a small settlement in early Roman days this strategically located city evolved into the world's most important center during the days of European imperial expansion. Belgium's cap-

[7] The railroad network is a phenomenon of Western culture; the impact of Western (or European) culture on various parts of the world is quite effectively measured by the introduction of railroads. See M. Jefferson, "The Civilizing Rails," *Economic Geography,* Vol. IV, 1928, pp. 217–231.

ital, Brussels, and the chief port, Antwerp, constitute Belgium's core area, not far across the border from France's industrialized northwest. And to the north this zone of urban-industrial development continues with the evolving conurbation of the Netherlands, the Amsterdam–Rotterdam–The Hague triangle. Almost immediately across the Netherlands–Belgian boundary in Germany are the several major cities of the Ruhr-Rhineland, West Germany's core area and Europe's major industrial complex. Near the French Saar-Lorraine boundary the triangle Frankfurt–Saarbrücken–Stuttgart includes some 20 cities with populations exceeding 50,000. In the industrial-agricultural area of northern Italy's Po Valley there are no less than two dozen such urban centers—including Milan (over 2 million) and Turin (1 million).

Everywhere in the European core area, cities are thus growing toward one another, until it is easy to believe the predictions of some observers who say that before long Western Europe will be in large part one vast urban area. Take the train from Rotterdam to The Hague: not long ago such a train ride took one past fields of pasture and vegetable gardens. Today the outskirts of Rotterdam have barely been left behind when the modern apartment buildings of suburban The Hague come into view. There was a time in Europe when it was reasonable to talk of the *hinterlands* of such cities—when most cities each had a surrounding, tributary area from which it drew products and to which it contributed its own goods and services, without much competition from any other town of similar size and capacity. In modern Europe that aspect of isolation is among the many that have disappeared: there are areas where perhaps a dozen cities share overlapping hinterlands. Clearly they could not all survive if their sole means of existence depended on exchange with immediate tributary areas. This, then, brings us to a third aspect of the European core: specialization.

Specialization is the hallmark of the European economic effort, not only in industry but in agriculture as well. It is, of course, a means of survival in the face of growing competition for resources and markets. To simply produce steel cutlery may not be good enough when the competition is Swedish stainless; to produce table silver means engaging in competition with world-renowned products marked Sheffield. Hence it is profitable either to manufacture a product of

undisputed uniqueness or excellence, or to produce something that cannot be produced competitively elsewhere. It is easy to think of a host of European products that are associated with particular places or areas, from wines and beer to watches and furniture. Belgian (Flanders) lace, Delft "blue," Manchester textiles, and Portuguese sardines are among many products that have a long history of association with a certain place or area, an association that began even before the industrial revolution. We have indicated previously that such places at an early stage began to perform certain functions in the evolving European economic framework; they attained *functional specialization.* The decades after the revolution have witnessed an ever-increasing degree amount of functional specialization

Superhighways are penetrating more of Europe each year, and this cloverleaf, north of Cologne (Köln), symbolizes the modernization and integration of the European road transport network. (German Information Center)

The crowded urban landscape of London, England. London grew from a local river port to the capital of a world empire, and many of the buildings shown here date from the days of British world supremacy. (Charles E. Rotkin)

throughout Europe, especially in the European core. When a place specializes heavily in one particular activity, say the manufacture of textiles, it must import other kinds of products from elsewhere. British textile manufacturers eat Danish meat and Dutch eggs. Swiss watchmakers use Swedish steel cutlery and German crockery. French steelmakers use German coal; German steelmakers use Swedish iron ore. Thus there develops a greater and greater degree of interdependence among places and areas—a situation of areal interdependence. And the functional specialization that accompanies these developments knows no national or regional bounds. "Dutch" chocolate is about as Dutch as Brazilian coffee: what is Dutch about it is the specialized treatment, not the cocoa. British cotton goods originate in the cotton fields of Britain's former dependencies such as Uganda and Nigeria; the economic relationships are perpetuated in the Commonwealth. In turn, these countries may buy the products of British manufacture created out of their own raw materials. British specialization is what they pay the price for.

Specialization requires skills, and of these Europe has legion. Ever since the impact of the industrial revolution began to change European manufacture, new skills have been adopted, learned, and modified. Of all the world, Europe probably still today has the largest population of skilled manpower, capable of adjusting rapidly to innovations. Much of this skilled population is concentrated in the populous core, and it has been one of the major reasons why Europe managed to recover from the disaster of World War II.

In agriculture and even fishing, there is specialization. Spanish oranges, Sicilian lemons, and Dutch vegetables are sold in other countries—the last in Germany, Belgium, and France. Norwegian herring has ready international markets. But as might be expected, with so large and highly urbanized a population on a comparatively small land area, Europe does require food imports, despite its high yields and despite the intensity of agricultural land use. Wheat and corn, for example, must be imported, as well as meat (Argentina has long been a major source), and cane sugar. There is real self-sufficiency only in dairy products, potatoes (which are a staple throughout the Northern European Lowland) and rye, the bread grain of Germany and Poland.

These, then, are the features of the European core area: it consists in physical terms of the western part of the European Lowland, a section of the Central Uplands, and the Alps and Po Valley; its climate is mainly of the *Cfb* maritime-influenced west coast variety, with moistness and mildness as the main characteristics. Apart from the Alps, slopes are gentle to moderate, soils are varied but of good quality. Communications are enhanced by natural waterways which, through the annual distribution of precipitation, do not fluctuate enough to interrupt their usefulness in transportation. The region is densely populated, highly urbanized, heavily industrialized, and strongly specialized. Intense transport networks promote the exchange of goods; a high degree of areal interdependence has developed. Politically, the region is fragmented into four major units—Britain, France, Germany, Italy—and five minor ones (the Netherlands, Belgium, Luxemburg, Switzerland, and Austria). This fragmentation has inhibited the free flow of resources and commodities and has had a negative effect on European development.

Greater Unity—A Supranational Europe. The states of Europe have much in common despite their cultural differences. Increasingly they are being pulled into joint economic frameworks; they are more interdependent today than ever. The impact of the industrial revolution was shared by all of them; Europe took on a dynamic character as the interconnections and mobility between the states intensified and people and goods were exchanged at ever greater rates. But European states have a long history of competition and rivalry: in their colonizing campaigns they obstructed each other, and in Europe itself they struggled for military supremacy. France under Napoleon in the nineteenth century and Germany under Hitler in the twentieth tried to establish hegemony over the continent; each time Britain, in whose interest it was to keep Europe from being dominated by a single power, took to war to prevent it. Thus while economic realities pulled Europe ever closer together, political divisions remained sharp. Nationalism is still a potent force in European affairs.

Obviously Europe, by its division, has lost great opportunities for a lasting position of world leadership. Not since the time of the Roman Empire—itself the product of imposition by force—have parts of the continent as distant as Britain, Iberia, and Italy been unified under a single government. This is not to suggest that Europeans over the centuries have not desired or sought unity: Erasmus in the early 1500's, and

Rousseau, Kant, Victor Hugo, and Condenhove in later times expressed this ideal repeatedly. In the aftermath of World War I there was a movement in this direction, and a congress was held in Vienna to consider the possibility of Pan-European unification, but the events of the 1930's soon took their course and terminated whatever hopes there were. Since the end of World War II, however, unifying movements have again arisen, most of them based—as perhaps they should be—on economic needs and benefits. One lesson of *supranationalism* (international cooperation involving the voluntary participation of three nations or more in an economic, political, or cultural unit) is that political integration, if it is ever to occur, is likely to follow rather than precede economic cooperation. When people can see a better life, they are more willing to sacrifice a little bit of their autonomy to secure it than when they are asked simply to give up their political identity for an intangible "greater" state.

Even before the end of World War II, papers were signed by the exiled representatives of the Netherlands, Belgium, and Luxemburg to initiate *Benelux,* the first of several organizations of international cooperation to arise in postwar Europe. But a stimulus was provided also by the United States, since there had to be some sort of multinational cooperation to disseminate the massive aid provided to Europe under the terms of the Marshall Plan. This led to the formation of the *Organization for European Economic Cooperation,* initiated in 1948. Not long afterward the foreign minister of France, Robert Schuman, proposed a *European Coal and Steel Community,* with the principal objective of lifting the restrictions and obstacles in the way of the flow of coal, iron ore, and steel among the mainland's six prime producers: France, West Germany, Italy, and the three Benelux countries. The mutual advantages of this arrangement are obvious, even from a map showing simply the distribution of coal and iron ore and the position of the political boundaries of Western Europe; but the six participants did not stop here. Gradually, through negotiation and agreement, they enlarged their sphere of cooperation to include reductions and eliminations of tariffs, and a freer flow of labor, capital, and nonsteel commodities, and ultimately, in 1958, they joined in the "Common Market," the *European Economic Community.* This organization incorporates all of the European core area of the mainland, and its total assets in terms of resources, skilled labor, and market are enormous. Its jurisdiction is strengthened by various commissions and by legislative and judicial authorities.

One very significant development related to the creation of the Common Market was the decision of the United Kingdom not to join it. This was itself a move based on supranational considerations: there was fear in Britain that participation would damage evolving relationships with Commonwealth countries; for many of these countries Britain is the chief trade partner, and there were pressures on the United Kingdom not to endanger these ties—pressures both within the country and from the far-flung Commonwealth. Thus Britain stayed out of the Common Market, but it made its own effort to create closer economic bonds in Europe: in 1959 it took the lead in establishing the so-called *European Free Trade Association,* comprising, in addition to the United Kingdom, three Scandinavian countries (Sweden, Norway, and Denmark), the two Alpine states (Switzerland and Austria), and Portugal. This scattered group of countries, with their relatively small populations, generally limited resources, and restricted purchasing power, add up to something much less than the Common Market; they became known as the "Outer Seven," while the contiguous states of the core came to be known as the "Inner Six."

Within a few years of the creation of the Outer Seven, the United Kingdom changed its position on E.E.C. membership, and decided to seek entry. Now, however, the political attitude on the continent had changed; France took an inflexible position under De Gaulle and obstructed British participation, despite a desire on the part of other Common Market members to ratify admission. Thus Britain, long a beneficiary and indeed a promoter of division on the continent, had again taken a hand in dividing its mainland

competitors—this time unintentionally. Here the matter stood in 1968, without much indication that the impasse was soon to end. In the meantime, however, another step had been taken toward greater unity on the O.E.E.C. model, but with the United States and Canada as full members in a nonmilitary, Atlantic association. The objective was to reduce the divisive nature of the Common Market–Outer Seven split, and at the same time to broaden the basis for Western economic cooperation. Ratified in 1961, the *Organization for Economic Cooperation and Development* counted 20 members, including all those who signed the O.E.E.C. papers in 1948. The O.E.C.D., unlike the Common Market, has little power and no binding authority, and some participants argued that it represents nothing very constructive in the European drive toward unification. But it may yet play its role in finding a way to overcome some of the obstacles facing this objective.

Thus Europe may finally have reached the stage in history that will witness the development of the greater European unity for which it has long waited. Supplementing the functional integration of much of the continent in the economic sphere, supranational organizations also have arisen in the military arena—the North Atlantic Treaty Organization and the Western European Union—and even a fledgling political community called the *Council of Europe.* The Council, which has met regularly since 1949 in Strasbourg, is little more today than a forum for the exchange of ideas, but as such it is nevertheless important. It consists of two organs: a Council of Ministers and a Consultative Assembly. Its deliberations are a guide to European opinion, and while the Council has no executive authority, many of its stated views are known to have affected the decisions of European governments. The future may yet come to recognize this as the beginning of a European government.

ADDITIONAL READING

Many geographers and historians have discussed the idea we have defined as the oscillation principle. One of the most interesting and speculative books is J. Fairgrieve, *Geography and World Power,* published by the University of London Press in 1932 and still worth reading. Another book from the same period, but taking a different line to some extent, is M. I. Newbigin's *New Regional Geography of the World,* published in New York by Harcourt, Brace & Co. in 1930. A famous work is by E. Huntington, *Mainsprings of Civilization,* published by John Wiley & Sons in New York, in 1945. Huntington ascribed man's successes and failures to his physical environment, especially the climate; in this context see a work by a famed woman geographer in the United States, E. Churchill Semple, *Influences of Geographic Environment,* published by Holt in New York as long ago as 1911. As for the discussion of climate itself, the most useful source is probably A. N. Strahler's *Physical Geography* (Part I), published by Wiley in a third edition in 1969. Another good basic physical geography book to consider is J. E. Van Riper, *Man's Physical World,* published by McGraw-Hill, New York, in 1962.

A look at the current indexes of the *Annals* of the Association of American Geographers and other geographic publications will prove that Von Thünen's *Isolated State* still excites discussion and debate. References to Ratzel can be found in most histories of geography, and one of his more famous articles was reprinted recently by R. E. Kasperson and J. V. Minghi in *The Structure of Political Geography,* published by Aldine Company, Chicago, in 1969. On the subject of core areas, consider R. S. Platt's *Latin America: Countrysides and United Regions,* a McGraw-Hill publication dating from 1942, and A. K. Philbrick, *This Human World,* published by Wiley in 1963. For an extended discussion of supranationalism, consult Chapter 18 in H. J. de Blij, *Systematic Political Geography,* published by Wiley in 1967. On the race issue, it is good to look through a basic anthropology book such as E. Adamson Hoebel's *Anthropology: the Study of Man,* published by McGraw-Hill (third edition) in 1966. And for something controversial, see C. Coon, *The Origin of Races,* published in New York by Macmillan in 1963.

REGIONS OF EUROPE

CONCEPTS AND IDEAS

The Population "Explosion"
The Demographic Cycle
Urban Site and Situation
Law of the Primate City
The Forward Capital
Central Place Theory
Regionalism
Transhumance

EUROPE's core area lies in the west, and it is adjoined by other European regions which do not have the dynamic qualities of the core—its dense population, dynamic growth, enormous mobility, great productivity, and its high degree of urbanization. When we considered the European core area as a functional region we drew its boundaries according to criteria *other* than political. For example, we recognized that northern Italy is a part of this European heart, while central and southern Italy are not. We recorded that England lies within the core, while Scotland lies largely outside of it. Having looked into some of the properties of this European regional core, we now turn to another way of establishing European regions. We attempt to group the countries of Europe together in such a way that the several groups reflect their proximity, historical association, cultural similarities, common habitats, and economic pursuits.

On these bases five European regions are recognized: (1) the British Isles, (2) Western Europe, (3) Nordic or Northern Europe, (4) Mediterranean Europe, and (5) Eastern Europe. The present chapter deals with the first four of these regions, while Eastern Europe for several reasons will be discussed separately in Chapter 4.

The British Isles

Off the west coast of Europe lie two islands called the British Isles (Map 3-1). The largest of the two, which also lies nearest to the mainland (a mere 21 miles at the closest point) is the Island of Britain, and the smaller island is known as Ireland. Surrounding Britain and Ireland are several thousand small islands and islets, none of them of great significance.

Although Britain and Ireland continue to be known as the British Isles, the British government no longer rules over the state that occupies most of Ireland, the Republic of Ireland or *Eire*.

After a very unhappy period of domination, the London government gave up its control over that part of the island which was overwhelmingly Catholic; in 1921 it became the Irish Free State. The northeastern corner of Ireland, where English and Scottish Protestants had settled, was retained under British control and called *Ulster.* Thus the isle of Ireland is partitioned between a dominantly Catholic south and a dominantly Protestant north, a situation that has repeatedly produced tension and even violence.

The island of Britain also has political divi-

sions. Sometimes "Britain" or "Great Britain" is referred to as *England,* but this is a mistake. England is only the largest of the political areas on the island of Britain, and it was the center of power from where the rest of the Isles were originally brought under unified control. The English conquered *Wales* in the Middle Ages, and *Scotland* was first tied to England early in the seventeenth century, when a Scottish king ascended the English throne. Thus England, Wales, Scotland, and Ulster became the *United Kingdom* (U.K.).

Map 3-1 THE BRITISH ISLES.

Lowland Britain: England. Scotland and Wales lie mostly in the higher, more rugged country of Britain. England constitutes the generally lower south and southeast, and without doubt it constitutes a part of the great regional core of Europe. Indeed, it formed the very cornerstone of this European core, for new ideas and innovations constantly fanned out from England into Western Europe. But with equal certainty the United Kingdom gives the British Isles identity as a distinct and individual region. Its insular location off (but not too far off) mainland Europe helped ensure centuries of security; the last invasion of England took place more than nine centuries ago. Neither Napoleon nor Hitler attempted to cross the Channel. But the Normans and William the Conqueror, who in 1066 closed the book on invasions into Britain, had many predecessors. Britain's insularity did not mean that it did not receive infusions of various mainland peoples. Only archaeological evidence remains of the earliest of these immigrants, but the Celts, who once held sway over Britain, are still represented in small numbers in the modern population complex. Descendants of the Celts include the Gaelic people of Scotland's northern mountains, the Welsh, and, of course, the Irish. Subsequent to the Celts the Romans arrived about midway through the first century A.D., and they ruled most of Britain until after the beginning of the fifth century, leaving a permanent imprint of Latin civilization there. Then followed Germanic invaders: the Angles and the Saxons brought new elements of culture to Britain, derived from their source areas in German Schleswig-Holstein and Saxony. Even this immigration was not to be the last. From Scandinavian coasts the Norwegians and Danes arrived to plunder or settle. These Vikings penetrated inland and subjugated the local population, and before long Britain fell under Danish domination. The stage was now set for the Normans who, as we know, were themselves Latinized Scandinavians. Let no one say that cultural mixture cannot contribute to the development of a strong and progressive society.

Britain's physiographic makeup (that is, its topography, relief, climate, drainage, soils, natural vegetation—the total physical landscape) is also varied, and the relative location of the different parts is very important. The most simple, and at the same time the most meaningful, division of Britain is into a *lowland* region and a *highland* region. Lowland Britain happens to lie opposite the western end of the Northern European Lowland of the mainland, and it is essentially a continuation of it, with low relief, good soils, ample rainfall and great agricultural productivity. With the avenue of penetration provided by the Thames River (on which London lies), several good ports, and a large number of places of easy access, this part of Britain has always been open to foreign influences. The early invaders usually entered the island here, and when capable of it drove their predecessors farther north and westward. Even the Romans failed to gain complete control over all of Britain; they built walls to mark the limit of their sovereignty and left the mountains beyond to their adversaries. Thus over time lowland Britain became the center of power and of political and economic organization and integration. Except for some comparatively brief lapses, England—that part of Britain extending to the higher country of Wales and Scotland—steadily evolved into the core area of the British Isles. In its political development England was relatively free from the threat of external interference, unlike so many countries of mainland Europe; this freedom was ultimately translated into a system of parliamentary government that had no peer in the Western world.

The economic developments that to a considerable extent precipitated this political progress came in stages that could be accommodated without wrecking the whole society. As late as the fifteenth century England was an *exporter* of raw materials—wool—to northern France and Flanders. This in itself constituted a great advance over the period of subsistence agriculture, and it took much land out of communal ownership and put it under private use for profit by a method that is another characteristic of Western culture: enclosure. But what may in truth be called the commercial revolution of the sixteenth and seventeenth centuries came with the advent of mercantilism. Now there were new bases for wealth, and at this early stage perhaps as much as half the population was concentrated in cities, including the new class of merchants. Productive land was no longer the essential measure of wealth, but London and the other ports from which the profit-seeking fleets sailed were also located in England; nothing had happened to reduce England's primacy in Britain. London's focality was only strengthened.

The industrial revolution brought about a major transformation not only in British economic life, but in its regional structure as well. And once again England was the main beneficiary, for the majority of the coalfields on which it was all based were located in lowland Britain. As in mainland Europe, the coalfields tend to lie near a zone of contact between the European Lowland and uplands sustained by older, folded rocks. In Britain the southern part of the Pennines is surrounded by coalfields: to the east lies the most important of these (in more recent times, at least) on which such cities as Leeds-Bradford, Sheffield, Derby, and Nottingham are based; to the west the deposits formed the fuel source for the industries of Manchester, Salford, Bolton, Stockport, and Blackburn—among a number of other, equally important towns. To the south of the Pennine Chain lies a group of smaller coalfields, still of sufficient capacity to support the growth of Birmingham and its surrounding Coventry, Wolverhampton, and Walsall. Yet another coalfield within England lies in the northeast, centered upon Newcastle. Only two fields of significance in Britain lie outside England itself: those of south Wales (the Cardiff-Swansea area) and Scotland (in the corridor known as the Scottish Lowland).

Hence the impact of the industrial revolution was felt most strongly in England itself, but London, far from the nearest significant coalfield, was not directly involved. The center of activity shifted to the cities of the Midlands, of the northeast, of Wales and Scotland, many of them practically new and thrown together too quickly and often with little foresight. The population of these places skyrocketed as their industries prospered. Working conditions were bad and living conditions were frequently worse. As labor organized, it too became a major force in British politics, as the wealthy landowners and the merchants of earlier days had become. Once again there was a need for reform in government, and once again the situation, explosive as it was, could be accommodated without chaos. Meanwhile, Britain prospered. Nowhere is the principle of functional specialization better illustrated than here. Wool-producing cities lie east of the Pennines, centered on Leeds and Bradford. The textile industries of Manchester, on the other side of the range, are cotton-producing. Birmingham, now a city of 2.5 million people, and its nearby cities, concentrate on the production of steel and metal products, including automobiles, motorcycles and bicycles, and airplanes. The Nottingham area specializes in hosiery, as does Leicester; boots and shoes are made in Leicester and Northampton. In northeast England, shipbuilding and the manufacture of chemicals are the major large-scale industries, along with iron and steel production and, of course, coal mining. Coal became an important British export, and with its seaside locations in northeast England, Wales, and Scotland, it could be shipped directly to almost any part of the world—cheaper than it can be produced from domestic sources in those faraway countries. In Wales coal mining is supplemented by shipbuilding and the production of iron and steel. The same industries are based on the coalfields of Scotland, where Glasgow with nearly 2 million people has become one of Britain's major industrial cities. It is a measure of the impact of the industrial revolution that today four-fifths of the population of Scotland is concentrated in the Glasgow-Edinburgh corridor.

Today, the industrial specialization of British cities continues, but the resource situation has changed drastically. A one-time exporter of large quantities of coal, England has in recent years imported even this commodity. Iron ores have been depleted, and two-thirds of the ore required by British industries is bought elsewhere. Of course Britain's industries always did use raw materials from other countries: wool from Australia, New Zealand, and South Africa; cotton from the United States, India, Egypt and the Sudan, as well as Uganda. But with the breakdown of the Empire and greater competition from rising industrial powers, Britain is now hard-pressed to maintain its position; its connections are simply not as good as they once were. But there are other problems. Other countries are outdistancing Britain in the manufacture of products over which at one time Britain held a monopoly. West Germany, for example, is producing more steel, and the countries of the mainland sector of the European core area produce about twice as much steel as Britain now does. And Britain also has been overtaken in the field of technology. Plants which at one time were the epitome of industrial modernization still operate today—comparatively inefficient, expensive to run, slow, and wasteful. In other parts of the world, especially the United States and mainland Western Europe, newer equip-

The drab industrial scene shown here is representative of so many English manufacturing cities—this is Shaw, Lanchashire. (J. Allan Cash/Rapho Guillumette)

ment in various manufacturing fields does the job much better than British equipment does. Britain, at the vanguard of the industrial revolution, has not always kept pace with modern developments. One consequence of this situation has great geographical interest. With the relative decline of the industrial cities of the Midlands and north England, there is a tendency for industrial enterprises which can afford to do so, to relocate near the old focal point of Britain: London. Here, still, is the greatest domestic market of the British Isles, and increasingly that market is of importance to local manufacturers as competition for markets elsewhere intensifies. It reflects the decreasing importance of coal in the energy-supply picture (which is changing toward nuclear power), the desire to start afresh with up-to-date machinery, and a realization that London, in addition to forming a large domestic market, is also a good port through which to import whatever raw materials are required. London's old assets still mean much in modern days.

Lowland Britain, in addition to its industrial development, also has the vast majority of Britain's good agricultural land. Actually, the land is not nearly enough to feed the British people (55 million of them) at present standards of calorie intake. The good arable land is concentrated largely in the eastern part of England,

although there is a sizable belt of good soils behind the port of Liverpool, in the west. Where the soils are adequate there is very intensive cultivation, and grain crops (wheat, oats, barley) are grown on tightly packed, irregularly shaped fields covering virtually the entire countryside. Potatoes and sugar beets are also grown on a large scale. But much of lowland Britain is suitable only for pasture, and cannot for reasons of soil quality, coolness, excessive precipitation or other factors sustain field crops. Hence Britain requires large food imports every year; the country cannot adequately feed even half its people, and perhaps not even one-third. There are various ways to estimate this, but in all probability Britain imports nearly two-thirds of the food consumed by its population. This makes Britain the greatest importer of food in the world, and, for those countries that have food to sell, the world's top market. Meat, grains, dairy products, sugar, cocoa, tea, fruits—Britain is either the leading or one of the world's leading importers.

In a country where over 90 percent of the population live in urban centers of one kind or another, and where only 1 of every 25 gainfully employed persons makes a living out of farming, none of this is surprising. Most Britons work in commerce, manufacturing, transportation, or other urban-oriented economic activities; this is

The Scottish Highlands are shown in their characteristic relief, with their lakes and misty slopes. (Harm J. de Blij)

the most highly urbanized society in the world. The dependence on food imports just described developed during the nineteenth century, with Britain's population explosion and the competition from outside (especially Australia) of cheap farm produce—cheaper wheat came from America and Australia, for instance, than British farmers could profitably grow at home. Hence they turned in ever larger numbers to the raising of cattle and sheep, the main domestic livestock, and turned more acreage over to pasture and meadow. With their milk and meat they would have a better chance to beat foreign competitors to the local market than with grain crops. In recent decades the British government has attempted to turn this tide, and has subsidized farmers to encourage them to return to the cultivation of wheat and potatoes, among other crops. But it has been an expensive and not very fruitful exercise. Only a crisis involving a cutoff of overseas supply lines is likely to stimulate British agriculture to full calorie-producing capacity.

Highland Britain. England, as we have seen, is mainly lowland country, except in the Cornwall Peninsula in the southwest and in the Pennines and the Lake District of the north. Here the many advantages that made Britain great are concentrated. But the larger part of the British Isles is highland country, including Wales, Scotland, Ulster (the northern part of Ireland that is part of the United Kingdom) and, with the exception of the central plain, Ireland itself as

well. Nature has not been nearly as kind to these areas. Steep slopes, poor, thin soils, a rather cold climate, and excessive moisture generally prevail, and opportunities for livelihood are quite limited. On the other hand, these areas provided a more or less permanent refuge for the Celtic peoples who over the centuries continued to resist whoever was in power in England. Only Cornwall has lost its identity in this respect; Celtic languages are still important in the other territories. Welsh and Scottish nationalism (not to mention that of Ireland) share an antipathy to England. But the same hills and mountains that provided sanctuary also made life difficult for their inhabitants and tended to keep them divided into small and relatively isolated sections or clans. It took a hardy people to maintain itself in these parts, and while many did, others gave up the struggle and left for other parts of the world, in search of greater opportunity.

Still, even in these areas there is economic specialization, and not just mere survival. There is pasture on the hills, and sheep can be kept; their fine wool is used in the production of high-quality sweaters, coats, and suits, and anyone who has ever looked for one of these items in a store has probably used some Scottish or Welsh place names to ask for what was wanted: Tweed, Paisley, Cardigan. And in those favored places where grain crops could be grown that would survive in the rigorous climate, they were used not only for local consumption but also in

The mining town in Wales, with its row housing and slag heaps, has invaded the rural countryside. (Bruce Davidson/Magnum)

the production of products that are inextricably connected with their countries of origin: Scottish and Irish whiskey. And from their famous rugged landscape with its great mountains and wild coasts the Scots (and to a lesser extent, the Welsh as well) reap the benefit of a considerable tourist trade, as more Britons seek the emptiness of the north and west, to escape from their crowded environment.

Wales and Scotland did not escape the impact of the industrial revolution. The coalfields of south Wales performed the dual function of supplying parts of England with coal, while themselves forming the base for considerable industrial development. No field in Britain possessed quite the variety of coals that marked the Welsh fields; from lignite to coking coal to high-quality anthracite. Labor from the Welsh hill country flooded to Cardiff, Swansea, Rhondda, and Newport. But the industry was headed for trouble: as the coal seams were pursued deeper and deeper, production became more expensive, while other areas of supply in England and elsewhere in the world brought stronger competition. Wales suffered a creeping depression; there were too few opportunities for diversification and specialization. The industrial revolution was a mixed blessing to this part of the United Kingdom.

Scotland, on the other hand, was more fortunate. From the Atlantic to the North Sea, along the narrow "waist" of the Clyde and Firth of Forth, Scotland found itself with an extensive coalfield, an immediately adjacent deposit of iron ore, and, on the west side, an excellent natural harbor. In this particular area Scotland has its most extensive lowland, positioned between its Southern Upland and the Grampian Mountains. This lowland corridor, with Glasgow at one end as the industrial center, and Edinburgh at the other as the administrative and cultural focus, has become the core area of Scotland, and has forever broken Scotland's isolation. The Scottish coalfields did not simply function as a supply area for other parts of the world; they formed the basis for major manufacturing development, including shipbuilding and, with the nearby availability of raw wool, the production of textiles. When the local iron ores ran out the industries could afford to import that commodity from elsewhere. When diversification came to the textile industry it, too, could pay for the cotton it now required. And as in Wales, there was a great influx of people from adjacent areas to the newly productive region. But unlike Wales, Scotland has a good deal to show for its effort: two of Britain's major urban centers, Glasgow (2 million) and Edinburgh (750,000), an industrial-commercial core area of major significance, a variety of competitive and healthy industries, and one of Britain's major ports. And again unlike Wales, comparatively little of the countryside was ravaged by the mining operations; indeed, highland Scotland has been left amazingly untouched by the events along the Clyde.

Ulster. Across the Irish Sea, Northern Ireland, known also as Ulster, remains a part of the United Kingdom. Since 1922 this political entity, comprising about one-sixth of the island, has been separated from Ireland proper and has had representation in London rather than Dublin. It is a product of the ancient conflict between English and Irish, a conflict which is still not resolved and which occasionally produces trouble along the Ulster-Irish boundary.

With its own parliament handling local affairs, Ulster has a measure of local autonomy; but its relationship with the United Kingdom has clearly been an asset, at least from an economic point of view. This is the single political unit among the four comprising the United Kingdom that possesses practically no resources for primary industries. It is hardly better endowed with such raw materials than the Republic of Ireland itself—which is notoriously poor in this area. Yet Ulster has a considerable amount of industry, two major urban centers, and what may justly be called a diversified economy. All this is the result of its connections with United Kingdom interests: Scottish capital started both the shipbuilding industry of Belfast (over $\frac{1}{2}$ million) and the textile (linen) industry of the smaller town of Londonderry. For the former, the steel and coal are imported from Britain, a process which is facilitated by the absence of international boundaries, tariffs, etc. The linen industry, on the other hand, was initially based on local production of flax, but today it too requires large-scale imports of this commodity. In recent years aircraft manufacture has made its appearance at Belfast. But its lead over Ireland notwithstanding, Ulster is in a precarious position. About the only local contribution it can make to industrial requirements is labor, and of this there is usually too much. In 1968 there were major problems with unemployment here, and it was not the first time that problem had arisen.

Ireland. Ireland (Eire) is one of Europe's youngest independent states, and the country from which it fought itself free is its neighbor, Britain. In fact, it is more correct to say that the struggle was one between Irish and English, for certainly in Wales and Scotland there were pro-Irish sympathies. But to those who think that colonialism and its oppressive policies and prac-

tices, especially where land ownership is concerned, is something that can only be perpetrated by European peoples upon non-European peoples, Ireland's history is a good object lesson.

Without protective mountains and without the resources that might have swept their country up in the great forward rush of the industrial revolution, the Irish—a nation of peasants—faced their English adversaries across a narrow sea; and a sea was slim protection against England, whatever its width. Wales had given up the struggle against English authority centuries earlier (shortly before 1300); Scotland had been incorporated soon after 1600. But the Irish continued to resist, and their resistance led to devastating retaliation. During the seventeenth century the English conquered the Irish, alienated the good agricultural land, and turned it over to (mostly absentee) landlords; those peasants who remained faced exorbitant rents and excessive taxes. Many moved eastward, to the lesser farmlands, there to seek subsistence as far away from the British sphere as possible. The English placed restrictions on every facet of life—especially in Catholic southern Ireland, though rather less in the Protestant north. Irish opportunity and initiative were put down.

The island on which Eire and Ulster are located is shaped like a saucer with a wide rim, a rim that is broken toward the east, where lowlands open to the sea. Though less prominent or rugged than their Scottish and Welsh counterparts, the hills that form the margins of Ireland are made of the same kinds of rocks, and in places elevations exceed 3000 feet. A large lowland area is thus enclosed, and certainly Ireland would seem to have better agricultural opportunities than Scotland and Wales. But in fact the situation is little better. As Plate 1 (in rear of this volume) shows, Ireland is even wetter than England, and excessive moisture is the great inhibiting element in agriculture here. Practically all of Ireland gets between 40 and 80 inches of rain, and western Ireland receives over 60 inches annually. Hence pastoralism once again is the dominant agricultural pursuit, and a good deal of potential cropland is turned over to fodder. The country's major export is livestock and dairy products, and most of it goes to Britain.

How can 60 inches of rainfall be too much for cultivation, when there are parts of the world where 60 inches is barely enough? Again, the evapotranspiration factor comes into play, and here is good illustration of the deceptive nature of such climatic maps as Plate 2—deceptive unless more is known than the map tells us. Ireland's rain comes in almost endless, soft showers that drench the ground and keep the air cool and damp—in Scotland they are called "Scottish mists." Rarely is the atmosphere really dry enough to permit the ripening of grains in the fields. Without such a dry period the crops are damaged or destroyed, and so there are severe limitations on what a farmer can plant without too great a risk. In Ireland, the potato quickly took the place of other crops as a staple when it was introduced from America in the 1600's; it does well in such an environment. But even the potato could not withstand the effects of such wetness as prevailed in several successive years during the 1840's. Having long been the nutritive basis for Ireland's population growth, which by 1830 had reached 8 million, the potato crop, ravaged by blight and soaking, failed repeatedly; the potatoes simply rotted in the ground. Now the Irish faced famine in addition to British rule, and while over a million people died, nearly twice that number left the country within ten years of this new calamity. It set a pattern that has not yet been broken, as emigration continues to exceed natural increase. Together with Ulster's 1.5 million people, Ireland now counts slightly over 4 million—half of what it was when the famine of the 1840's struck. There are not many

places in the world that can point to a decline in population since the days of the industrial revolution.

In 1921, the Republic of Ireland and the United Kingdom cut their last political ties, other than those of Commonwealth; this was the conclusion of a process that had begun in the 1870's with Home Rule, a form of conditional autonomy. That more than a generation would be required to erase the memories of British oppression was indicated in 1949 when, after a period of studied neutrality during World War II, Ireland decided to withdraw from the British Commonwealth. The continued partition of Ireland into Ulster and the Republic is one source of continued friction, but voters in Ulster have themselves indicated a preference for the *status quo*. This preference is in no small way related to the old Catholic-Protestant division of the island, a division that remains as sharp as ever, although minorities exist on both sides of the border. Nevertheless, at the 50-year anniversary of Ireland's independence, relations between Britain and Ireland are improving. Britain is Ireland's first trade partner, and at present more Irish emigrees leave for Britain than for any other country. At the same time the Irish government is attempting to foster the spirit of Irish nationalism by encouraging the use of Gaelic and through vigorous support of Irish cultural activities. In addition, there is a drive to diversify the economy, and a wide variety of light indus-

An Irish landscape. (Courtesy Irish Tourist Board)

tries have sprung up, most of them in the area of the capital, Dublin. Not long ago a mere 15 percent of Ireland's exports were manufactured products (the rest being live animals, dairy products, and other farm-derived produce); today the figure is about 40 percent. There is new hope that Ireland is about to shed its rural-agrarian outlook for one that will move the country toward membership in the European core.

Population: the Crisis of Numbers

Throughout Europe and the world, countries affected by the industrial revolution experienced revolutionary changes in their rates of population growth as well. Prior to the period of industrialization, populations tended to fluctuate around an average which, over the centuries, showed only a very gradual increase. Often, times of rapid population increase were followed by crop failures and famines; disastrous epidemics also ravaged the population and decimated it. Systems of food distribution were yet inadequate to cope with famine conditions of a regional nature, and progress in medicine was not sufficient to forestall or even combat epidemic diseases.

All this began to change with the impact of the industrial revolution, which, as we know, led quickly to the agglomeration of people in places where resources were extracted from the earth and where products were manufactured: cities and towns. This process of urbanization brought with it a number of advances that had the effect of reducing existing death rates. For example, people's personal hygiene improved as soap became more generally used, running water was brought to living quarters, new sewerage systems were laid out, and homes were built of better materials than before—city dwellings could not be made of wood and thatch but required brick and slate. In addition, the products of the new industries themselves helped to raise standards of cleanliness. The smoother cotton textiles were an improvement over the much rougher cloth previously worn by the majority of the people. Bandages and dressings became more readily available and were discarded after use; medical and surgical techniques, aided by newly manufactured instruments, developed rapidly. Hospitals grew in size and number, and such disease-promoting habits as allowing waste to accumulate and burying the dead with insufficient care began to disappear. At the same time, farming methods improved and modern equipment made of iron and steel replaced the mostly wooden handtools that were in general use. New crop rotations, fertilizers, and higher seed quality all contributed to better yields and greater food production. Modernized transportation systems now made possible the more efficient distribution of food to the cities and areas where it was needed; people began to eat better than ever before.

The result of all this was a rapid growth of the population, which means a greater excess of births over deaths than was previously the case. These figures are normally given as births/deaths per thousand per year. Thus a country with a high birth rate, say 45 births per thousand inhabitants per year, and a high death rate as well, say 26 deaths per thousand per year, grows at a rate of 1.9 percent annually (an increase of 19 persons per thousand population), not counting immigration or emigration. Such figures may have characterized preindustrial times, though the gains often were wiped out by famine, epidemic, or other causes such as war. But then the industrial revolution occurred, and countries affected by it found that their birth rates remained quite high, but because of the new conditions, some of them described above, death rates fell. The imaginary country for which figures were just given, for example, might show a birth rate still as high as 37 per thousand, but a death rate as low as 11. Now the annual rate of increase would be 26 per thousand or 2.6 percent, and the population would be growing at a much greater rate. When such conditions affect not just one country (as England was first affected) but many countries, the result is an unprecedented, explosive expansion—indeed, a "population explosion."

Malthus. Long before the industrial revolution began, the question of population numbers and the supply of food was of concern to people in many parts of the world. Even people living in clans or small tribes realized that a larger number of mouths would mean a shortage of food, and they took steps to limit the size of their social unit. In Europe prior to the industrial revolution, there were also efforts to hold down the numbers—though mostly at the family level, and not through government inducement. But birth control is nothing new, and neither are Man's worries over how he will feed himself.

In 1798 the British economist Thomas R. Malthus published a study of population growth and food availability that connected his name unalterably with this whole issue. Malthus argued that population expands geometrically, that is, by an upward curve on graph paper, but that food supplies can increase only arithmetically, that is, by a straight line on the same paper. Hence, the curve would run farther and farther ahead of the straight line, and Man's increases would inevitably be halted by lack of food—by hunger. Viewing the situation in Britain, whose population was in the "explosive" phase brought on by the industrial revolution, Malthus predicted that a mere 50 years after his writing, this situation would come to pass.

As we know, Malthus was incorrect as far as Britain was concerned—that is, as far as Britain's ability to acquire its needed foodstuffs goes. But Malthus' warnings have not been forgotten, and there are many scientists today who feel that, in a world context, Malthus was fundamentally correct. Certainly Malthus did not consider it possible that as much food as is produced today could ever have been wrested from the earth. And his estimates of world population growth were widly exaggerated. Nevertheless, mankind has recently increased explosively. The total doubled (from under 600 to over 1200 million) between 1650 and 1850, then doubled again by 1950 (to 2500 million), and is now in the process of doubling again in less than 50 years. Modern adherents of Malthus' position argue that the major reason why most industrialized countries still feed themselves adequately lies in their ability to command the necessary imports from elsewhere; but as other countries industrialize, they will lose their sources of raw materials and hence their ability to produce and trade for their foodstuffs. These neo-Malthusians see a bleak future for a world in which some countries enjoy plentiful food supplies while others, especially agricultural states like India, face starvation diets, famines, and endless hunger.

For us in the Western world it is often difficult to perceive of this problem as one that could threaten our very peace and security; until recently even the hunger conditions prevailing in some parts of the United States were virtually unknown to the average citizen. And there are those who refuse to believe that Malthus' views have any relevance today. Anti-Malthusians say that the population "explosion" will eventually check itself, and that new sources of food have been found and will continue to be found on the earth—not only in newly opened areas such as the Soviet Union's Virgin Lands, or in areas with potential, such as parts of the Amazon Basin, but even in the oceans, seen as a vast supplier of Man's needs and awaiting only his technological skills to draw its riches. But these optimistic predictions notwithstanding, almost half of mankind goes inadequately fed today, and from the Americas to Asia people's lives are being shortened for exactly the reason Malthus predicted.

The Demographic Cycle

One reason for the optimism expressed by some demographers is based on what appears to have happened to population expansion in industrialized Europe. Stated most simply, in preindustrial times, birth and death rates were both high, and population increase was very gradual. With the onset of industrialization, urbanization, and the accompanying agricultural revolution, death rates went down while birth rates remained high. This is the second stage in the cycle of population growth—the *demographic cycle*—and it is the period of population "explosion." However, this period of rapid growth, in Europe at least, appears to have come to an end. A general leveling off marked a distinct third stage, when birth rates also declined, so that both birth rates and death rates became low and the net annual increase was reduced to a very small figure. In the United Kingdom, for example, the birth rate in recent years has been about 16 per thousand, and the death rate about 11, giving an increase of 0.5 percent—part of which is lost through an excess emigration over immigration. In the case of France, the population has actually showed signs of leveling off completely, with a very small net increase at times more than offset by emigration. The French government has offered inducements for families to have more children, since from many points of view this is an undesirable situation.

In any case, anti-Malthusians, pointing to this European phenomenon, argue that the same sequence of events will occur elsewhere in the world when industrialization runs its course, and that the "explosion" of population through which we are going today is but a temporary phenomenon that will indeed check itself as it did in Britain, France, and other industrialized countries. But this presupposes some things that may not actually be so:

"When today a state experiences the impact of modern change of 'development', are the changes that occur similar to those recorded in Europe after 1700? Can the lessons of geography and politics be applied in the presently emerging world? The answer may well be no. Those states that are today still in the first stage of the cycle (some African and Asian states) are in many ways unlike preindustrial Europe. Population densities are higher today than they were three centuries ago, so that these emergent states have a different point of departure. Europe's mortality rates declined relatively slowly as the industrial revolution took effect; the Western contribution to its colonial realms brought far more sudden declines in death rates. Thus the rates of population increase (the 'explosion') in developing, urbanizing, industrializing Western Europe in its second stage were probably less than they are and will be in many developing states now experiencing these conditions. If these arguments are valid (that conditions during Western Europe's first and second stages both differ importantly from the same stages currently prevailing elsewhere), then there is reason to believe that the third stage in, say, India may not be identical with that of France or Great Britain. In other words, the thesis that the ultimate fate of the population growth cycle is stability, based upon the European experience, may well be invalid."[1]

Indeed, there is even doubt that all industrializing countries will necessarily experience a lowering of the birth rate comparable to that of, say, the United Kingdom. Neither the U.S.S.R. nor the United States have reached this third stage—some reduction in the rate of increase in population has been recorded, but no leveling off by any means. In the case of the United States, the period just before World War II seemed to presage such a drop in the rate of increase, but since the war the United States' birth rate has remained about twice as high as that of several of Europe's industrialized-urbanized states. And while the Soviet Union certainly showed a decreasing birth rate (from 45 per thousand in 1920 to 25 at present), the death rate also declined considerably, from about 28 per thousand to less than 8. We shall

refer to the population question again in the regional discussions that follow Europe.

Western Europe

The essential criteria on the basis of which a Western European region can be recognized are those that give Western Europe the characteristics of a regional core, as discussed on pp. 52–57: this is the Europe of industry and commerce, of great cities and enormous mobility, of functional specialization and areal interdependence. This is dynamic Europe, whose countries founded empires while forging parliamentary, democratic governments. For some time, these countries lay at the center of gravity of the world.

But if it is possible to recognize some semblance of historical and cultural unity in British, Scandinavian, and Mediterranean Europe, that quality is absent here in the melting pot of Western Europe. Europe's western coreland may be a contiguous area, but in every other way it is as divided as any part of the continent is or ever was. Unlike the British Isles, there is no *lingua franca* here. Unlike Scandinavia, there is considerable religious division, even within individual countries. Unlike Mediterranean Europe, there is not a common cultural heritage. And indeed, the core region we think we can recognize today based on economic realities could be shattered at almost any time by political developments. It has happened before.

The leading states of Western Europe, of course, are West Germany and France. In addition, there are the Low Countries of Belgium, the Netherlands, and Luxemburg (also known collectively as Benelux), and the Alpine states of Switzerland and, rather peripherally, Austria. It will be recalled that northern Italy, by virtue of its industrialization and its interconnections with Western Europe, was identified as part of the European regional core; for the purposes of the present regionalization of Europe, however, all of Italy will be considered as part of Mediterranean Europe. Thus the region comprises 382,000 square miles (213,000 of them in France), and is inhabited by just under 150 million people, more than 40 percent of them in Germany. Attention turns first to these two states.

The contrasts between France and Germany are many. France, in the first place, is a very old state—by most measures the oldest country in Europe. Most of what is today France was

[1] H. J. de Blij, *Systematic Political Geography*, New York, John Wiley and Sons, 1967, pp. 68–69.

occupied by the Romans fully two thousand years ago, and the peoples of the area were endowed with Roman civilization. Out of the breakup of the Roman Empire arose the first discrete political entity in France, and ever since, through the days of Charlemagne, the kings, and Napoleon, there has been a France in Western Europe. For many centuries France was the cultural center of the continent, as scholarship and the arts thrived. Paris, the focus of France—even in history it is inconceivable to think of another—long has been the cultural headquarters of the region.

Not so in Germany. During and long after the Roman period the forested lowland was occupied by isolated groups of people, and political organization came slowly and haltingly. Midway during the nineteenth century, long after the French Revolution and Napoleon's effort to impose unity on Europe, Germany still consisted of a loose, ineffective association of autonomous states including Brandenburg in the east, Bavaria in the south, and Westphalia, Alsace, and Lorraine in the west. Then Prussian power, centered in the northeast, began to make itself felt. It achieved pacification of the eastern borderlands with the Slavs, and imposed unification by military authority on the remainder of Germany—not through conquest, but by a display of effective authority. The leading figure in all this was Bismarck, under whose aegis the German states fought a highly successful war against France (1870–1871). That was enough to cement German unity, and in 1871, just four generations ago, the German Empire was proclaimed.

From the map of Europe it would seem that Germany, smaller than France in area, also has a disadvantageous position on the continent when in comparison with its rival (Map 3-2). France has a window on the Mediterranean, where Marseilles, at the end of the cross-continental route of which the Rhône-Saône valley is the southernmost link, is its major port. In addition, France has coasts on the North Atlantic Ocean, the English Channel, and, at Calais, even a corner on the North Sea. Germany, on the other hand, has short coastlines only on the North Sea and the Baltic, and for the rest it seems heavily landlocked by the Netherlands and Belgium to the west, by the Alps to the south, and by Poland, to which Germany lost important territory after World War II, in the east. But such appearances can be deceptive. In effect it is France which is at a disadvantage when it comes to foreign trade, not Germany. None of France's harbors, including that of Marseille, is particularly good. Only very few of France's many miles of rivers and waterways are navigable to modern, large, ocean-going vessels. Indeed, France in an earlier day was better served by water transport than it is at present; larger ships and deeper drafts have caused a decline in the usefulness of France's waterways. Among the Atlantic ports that France does have, several have major disadvantages. Le Havre and Rouen serve as outlets for Paris and the Paris Basin; large ships can navigate up the Seine only as far as Rouen, now France's second port. Brest and Cherbourg lie at the end of peninsulas, far from the major centers of production and population. Nantes, La Rochelle, Saint Nazaire, and Bordeaux have only local significance. A good deal of trade destined for France goes not through French, but through other European ports, such as London, Rotterdam, and Antwerp.

Germany is much better off. Although the mouth of the great Rhine waterway lies in the Netherlands, Germany has most of its course—a course that runs through Europe's leading industrial complex, the Ruhr. For the western part of Germany, the Rhine is almost as effective a connection with the North Sea and the Atlantic as a domestic coast and harbor would be; Rotterdam, the world's largest port in terms of tonnage handled, is a more effective outlet for Germany than any French port is for France. But Germany has its own ports as well: Hamburg, 60 miles inland on the Elbe, is the major break-of-bulk point where cargoes from ocean-going vessels are transferred to barges, and vice versa. Hamburg has suffered from the consequences of World War II; through a series of canals its hinterland extended as far across the North European Lowland as Czechoslovakia, and the partition of Germany has reduced its trade volume with most of this area. A North-South Canal is under construction to connect Hamburg with the Mittelland Canal within West Germany. Not far to the west, on the Weser River, is the port of Bremen, a smaller city and located in a less productive area, but of historic and more recently of military importance, as it became the major American military port after World War II. Prior to the war, Hamburg and

Map 3-2 GERMANY AND FRANCE.

Bremen had benefited from the completion of the Kiel Canal across the German "neck" of Denmark, since they thereby attained new significance as gateways to the Baltic. And on the Baltic side, Germany (including East Germany) has several ports that are easily in a class with the French entries of Nantes, La Rochelle, and Bordeaux: such places as Kiel, Lubeck, and Wismar, each with well over a quarter of a million population, and each with some industrial as well as port development.

In aggregate, then, Germany's outlets are far superior to those of France. The same is true for the inland waterways that interconnect each country's productive areas. In France, although canals link the Garonne and Rhone Rivers and the Loire and Seine, the important waterways that carry the heavy traffic lie in the northeast and north, where the productive areas of the Lorraine, the Paris Basin, and the Belgian border areas are opened to the Meuse (Maas) and Rhine, and thus to Rotterdam and Antwerp. In Germany, the Mittelland Canal, with an east-west orientation, literally connects one end of the country to the other. In turn it intersects each of the northward-flowing rivers: the Oder-Neisse, the Elbe (to the port of Hamburg), the Weser (to Bremen) and the Dortmund-Ems Canal that links with the Ems River and with the port of Emden. The Dortmund-Ems Canal provides the Ruhr with a German North Sea outlet (and *in*let: most of Sweden's Kiruna iron ores arrive through the port of Emden); the whole system constitutes a countrywide network of great effectiveness.

Another strong contrast between France and Germany lies in the nature and degree of urbanization in the two countries. True, Paris is without rival in France and in mainland Europe, but there is a huge gap in France between it (population: over 7 million in the city proper, over 9 million with adjacent urbanized areas) and the next ranking city, Lille (near the coalfields on the Belgian border: population under 1 million). Two questions arise, one no less interesting than the other: why should Paris, without major raw materials in its immediate vicinity, be so large—and why should Lille, Lyons, and Marseilles (all near 1 million in population) be no larger than they are?

Paris owes its origins to advantages of *site,* and its later development to a fortuitous *situation.* Whenever an urban center is considered, these are two very important aspects to take into account. A city's *site* refers to the actual, physical attributes of the place it occupies: whether the land is flat or hilly, whether it lies on a river or lake, whether the port (if any) has shallow or deep water, whether there are any obstacles to future expansion such as ridges or marshes. By *situation* is meant the position of the city with reference to surrounding or nearby areas of productive capacity, the size of its hinterland, the location of competing towns—in other words, the greater regional framework within which the city finds itself. Paris was founded on an island in the Seine River, a place of easy defense and in all probability also a place where the Seine was often crossed. Exactly when settlement on this, the Ile de la Cité, actually began is not known, but probably it was in pre-Roman times, for it functioned as a Roman outpost some two thousand years ago. For many centuries this defense aspect of the site continued to be of great importance, for the authority that existed here was by no means always strong. Of course the island soon proved to be too small and the city began to sprawl along the banks of the river, but that did not diminish the importance of the security it continued to provide to the government.

The advantages of the situation of Paris were soon revealed. The city lies near the center of

One of Europe's many busy canals. Barge traffic carries the products of the countryside to the markets. (Charles Rotkin)

a large and prosperous agricultural area, and as a growing market its focality increased continuously. In addition, the Seine River, itself navigable to river traffic, is joined near Paris by several tributaries, all of them navigable as well, leading to various sections of the Paris Basin, and even to parts beyond (Map 3-3). To the west, the Seine runs into the English Channel, and although today's ocean-going ships cannot reach Paris, the smaller vessels of the past could—and today only one transshipment, at Le Havre or Rouen, is necessary. But more important than even this seaward connection are the connections afforded by the rivers that join the Seine (the Marne, Oise, Yonne) from the northeast, east, and southeast, and the canals that extend them even farther. Paris can be reached from the Loire, the Rhône-Saône, the Lorraine industrial area, and from the Franco-Belgian area of coal-based manufacturing. And, of course, there are land communications as well. As Paris gained greater political influence, contact with distant parts of France improved, often by design; government after government invested heavily to bring such areas into the capital's orbit. The political reorganization brought to France by Napoleon involved not only a reconstruction of the internal administrative divisions of the country, but a radial system of roads which, like Roman roads once led to Rome, all led to Paris.

When the industrial revolution came to France, Paris already was mainland Europe's greatest city—but Paris did not (as London did not) have coal or iron deposits in its immediate vicinity. Nevertheless, Paris was the greatest local market for the manufactures produced in Picardy on the Belgian border, in Lorraine on the German border, in Lyon at the Rhône-Saône confluence, and elsewhere. A railroad system was now added to the existing water and land connections, and once again the centrality of Paris was strengthened. As in the case of London, Paris itself began to attract major industries, and the city, long a center for the manufacture of luxury items (jewelry, perfumes, fashionable dresses), experienced great industrial expansion in such major areas as automobile manufacturing and assembly, the metal industries, and the production of chemicals. With a ready labor force, an ideal position for the distribution of finished products, the presence of governmental agencies, and France's largest domestic market, the continued growth of Paris is no surprise.

Paris is an excellent example of a historic capital that has become the *primate* city of the country. In 1939 M. Jefferson introduced an idea he called the Law of the Primate City, which he stated as follows: "A country's leading city

MODIFIED AFTER: J. DOLLFUS, ATLAS OF WESTERN EUROPE

Map 3-3 NAVIGABLE WATERWAYS OF WESTERN EUROPE.

is always disproportionately large and exceptionally expressive of national capacity and feeling."[2] As he suggested, there are cities that may have special advantages of location, perhaps within the national territory (as Paris) or perhaps with reference both to the national territory as well as to a larger area. In Denmark, for example, Copenhagen has had both: Denmark at one time was larger than it is today, and included much of southern Sweden; thus Copenhagen had a central position with reference to this national area. But Copenhagen also lies at the entry to the Baltic Sea, and as such enjoys major advantages as a break-of-bulk station. In any event, it appears that the asset of dominant size gives a city permanent primacy within the national area and at times an accelerating rate of growth—growth that often does not seem to be related to any factors other than that of size. It is a cyclic situation: as the largest city, the capital attracts business and industry and stimulates specialization, for it is the leading domestic market; it grows larger by its attraction. Thus the primate city, which on the basis of raw materials and location might justify a size somewhat greater than that of its nearest competitors, actually turns out to be several times as populous as the next urban center in the national territory. In Europe, Athens (a city of 2 million in an impoverished country of 9 million), Copenhagen, and Paris are cases in point.

If it is not difficult to account for the greatness of Paris, it is quite another matter to account for the relatively limited development of French industrial centers elsewhere. In France there is no Birmingham, no Glasgow—and certainly no Ruhr. And yet there is coal, as we have seen earlier, and there is good iron ore. There is also a long history of manufacturing: the linen industry of Lille existed in the Middle Ages, and the silk industry of Lyon also is no newcomer to the industrial scene. What is lacking? For one thing: large amounts of readily accessible high-quality coal. For another: the juxtaposition of such coal with cheap transport facilities and large existing population concentrations. Where France has a situation approaching these requirements (as in Picardy and the Lorraine) there has been considerable industrial development. But nowhere does France have anything to compare to the Pennines-Midlands fields or

[2] M. Jefferson, "The Law of the Primate City," *The Geographical Review,* Vol. 29, No. 2, 1939, p. 226.

the Westphalia fields. Coal is imported from Germany to supplement the local supply; the Lorraine iron ores, sent in return to the Ruhr, are not nearly as high-grade as those the Ruhr buys from Sweden. France simply does not have the advantages enjoyed by British and German industries. So French manufacturers have done what Europeans have done almost everywhere: they have specialized. French industry produces precision equipment of many kinds, high-quality china, luxury textiles, automobiles, and, of course, wines and cheese. Unable to compete in volume, they compete in quality and specialty.

From the above it might be deduced that while Britain and Germany hold a clear lead over France in terms of manufacturing, France, with its lower level of urbanization, should have greater strength in agriculture. That impression is correct. With a much larger area of arable land than Britain or Germany, and with the benefits of the temperate climate without excesses of moisture or temperature, France has great agricultural resources. Apart from the Paris Basin and adjacent parts of the North European Lowland in France, there are several other major agricultural areas, including the valleys of the Rhone-Saone, Loire, and Garonne. Elsewhere, as in the sandy and marshy parts of the Aquitaine, in the higher parts of the Massif Central, and in the Jura and Vosges Mountains, the soil is not suitable or the slopes are too steep for farming. Between these extremes there is a wide variety of conditions, each with its special opportunities and limitations: wheat is grown on the best soils, and it is France's leading crop; oats will grow under slightly less favorable conditions, and barley will tolerate even greater disadvantage. Southern and Mediterranean France (the *Midi*), of course, produce grapes and the usual association of fruits. French agriculture is marked by an enormous diversity of production, and apart from the predominance of wheat it is hardly possible to isolate regions of exclusive land use—except in the vineyards of the Rhône-Saône, Garonne, and Loire Valleys. A host of crop rotations are practiced; animal manure and commercial fertilizer are widely used (France is Europe's top beef and milk producer), and despite a rather low level of farming efficiency, especially in the south, France's annual agricultural output is second to none in Europe.

Perhaps there is no better way to measure the importance of French agriculture than to com-

The intensively used land of the wine country near Champagne, France. (Goursat/Rapho Guillumette)

pare France's almost complete self-sufficiency in this area to the dependence of Britain on imports for two-thirds of its annual calorie intake (see p. 65). France has long had an agricultural orientation; there was a time when the country's great strength in Europe stemmed from its agricultural productivity. In a way this agrarian tradition may have added to the impediments already placed in the way of French industrial development, for it is difficult to move people from land that is theirs—and productive. Nevertheless, people have migrated to the cities from Brittany, the Massif Central, and the mountain areas. In modern times agricultural self-sufficiency is no longer a measure of the health or strength of a state; industrial production and its associated features certainly are. France's urbanization is a reflection of its limitations in this field.

Germany has no Paris—and it is far from self-sufficient in farm produce. But despite the absence of a city with the primacy and centrality of Paris, Germany is a far more highly urbanized

country than France (by official definitions the comparison is 71 against 56 percent of the population). It has many more urban centers with over a half million population—17 in 1967—than France which has five. And as might be expected, Germany is also a much more strongly industrialized state than its southwestern neighbor.

The majority of Germany's cities lie in or near the zone of contact between the two major physiographic regions, the plain of the north (part of the North European Lowland) and the uplands of the south (Map 1-2). This, as we know, is also the chief zone of coal occurrences, and of course the development of several of these cities is bound up with the availability of this power base. Prewar Germany counted three major areas of industrial growth in this zone extending from west to east: the Ruhr, based on the Westphalia coalfield, the Saxony area near the Czechoslovakian boundary (now in East Germany), and Silesia (now in Poland). In the south, near the French border, there are coal deposits in the Saar, and in many other parts of Germany minor coal deposits lie scattered. The best fields, however, are those serving the Ruhr and Silesia; Germany has lost Silesia, but the Ruhr has become the greatest industrial complex of Europe. After Britain, Germany is Europe's top coal producer today; it is Europe's first iron and steel producer bar none.

The Ruhr, named after a small tributary of the Rhine, provides us with an illustration of the advantages of a combination of high-quality resources, good accessibility, and a position near large, high-capacity markets. The coal of the Ruhr area is of excellent quality for steelmaking purposes. Although the iron ore locally available was soon exhausted, ores can be shipped from overseas with only a single transshipment. Long before the Ruhr emerged as Germany's major complex of heavy industries, cities had been growing nearby along the Rhine, including Köln (Cologne) and the present capital of West Germany, Bonn. The surrounding agricultural area was productive and densely populated. Only the initial period of political uncertainty restrained development somewhat; after 1871 the Ruhr expanded rapidly and increasingly received the benefit of government support in the form of subsidies and tariff protection. The Germans were well aware of the power base the Ruhr constituted, and they did all they could to promote its growth. Chief among the manufactures made here are iron and steel (and the products created out of these, such as railroad equipment, vehicles, machinery etc.), textiles, and chemicals, but of course there is an almost endless variety of associated products.

While the Ruhr region specializes in heavy industries, Germany's second industrial complex, that of Saxony, is skill- and quality-oriented. Today this region, which includes such famous cities as Leipzig (printing and publishing), Dresden (ceramics), and Karl-Marx-Stadt (textiles), lies in East Germany. Again, this was a region of considerable urban and manufacturing development long before the industrial revolution occurred, and nearby coal supplies, though of lower quality than those of the Ruhr, stimulated new industries and accelerated growth.

West and East Germany together have a population of nearly 80 million, and a great deal of good agricultural land would be required to feed so large a population. Farming is limited by extensive areas of hilly country in the south, and there are problems with sandy and otherwise difficult soils in the north. Nevertheless, with staples including the ever-present potato, rye,

The Ruhr is Germany's industrial core area. Shown here is a sample of the concentration of activity that characterizes this region; the photograph was taken near Essen. (German Information Center)

and wheat, Germany manages to produce about three-quarters of the annual calorie intake of the population. This is not quite as good as France, but it is a great deal better than Britain manages to do; Germany, also, has greater industrial productivity than France. In view of the size of the population (even West Germany alone has a substantially more people than the United Kingdom or France), the German achievement in agriculture is remarkable.

As a look at the map of Germany quickly shows, there are some cities whose location cannot be readily explained in terms of any of the functions hitherto discussed (the ports, the industrial zone, major agricultural regional centers). One of these is Munich, a city that has political-administrative origins as the long-time seat of power for a succession of Bavarian kings. Although Munich lies in a rather productive agricultural plain in the southern, higher part of Germany, its initial growth resulted in large measure from a concentration of government here—and the wealth of the kings. In its location north of the Alps it had commercial advantages, especially after the demise of its rival, Augsburg, and manufacturing was generated. The city's domination in the region grew, and today it is Germany's third city. Like Bremen, it reaped particular benefits from the Allied occupation of Germany, for which it became an organizational focus. The other major German city in this category is the former capital, Berlin (4.5 million). Unlike Paris, Berlin is no ancient German focus. It was chosen deliberately, and arbitrarily, to serve as a binding agent for Germany in what was in the early period of the *Reich* the least developed (and, eastward, the least secure) part of the state. The concentration of governmental activities there, the centralization of authority, and the construction of a network of communication lines across the North German Plain focusing upon the city, were all designed to counterbalance the influence of the more developed west, and to solidify the German presence in the east. This capital, then, was chosen deliberately to fulfill functions in addition to the accommodation of the country's administration. It was to funnel German influence in a direction where it would be felt by those who might threaten it, and it was to stimulate development in an area less rich in raw materials than some other parts of Germany. In this respect Berlin is not unique. Other countries have used

their capital cities to direct national attention to new areas or to promote the national position in a disputed zone. In recent years Brazil created a wholly new capital in the interior at Brasilia, in an expression of increased interest in that country's huge but underdeveloped interior; the Brazilians seem to have hoped that the law of the primate city would soon render Brasilia a self-perpetuating, growing force on the Mato Grosso. Pakistan's choice of Islamabad to become the new capital, located in the deep interior and to replace coastal Karachi, again illustrates use of the capital city to foster national influence in distant areas. Islamabad happens to lie very near a much-disputed territory named Kashmir. Such capital cities are classified as *forward* or *pioneer* capitals. In the case of Berlin, the power and prestige of the German state were soon reflected in the city's growth and prosperity; Berlin indeed quickly became a primate city. It rose to become Germany's largest urban center, with a sizable industrial base along with its governmental, commercial, and other functions.

Central Place Theory

From France and Germany have come many important contributions to the field of geography. Unfortunately, German geography is all too often remembered first for the aberrant "science" of *geopolitik,* an outgrowth of the otherwise reputable area of political geography, which provided an academic justification for twentieth-century Germany's campaigns of war against "encirclement." But Von Thünen had his successors too, and one of them, Walter Christaller, is now recognized as one of the founders of a whole field of geographic study still full of excitement and opportunity today. It took almost a generation for Christaller's work to come to general attention. His major opus was published in 1933, just as the Nazi regime took power, and its title must have had little appeal: *The Central Places of Southern Germany.* Not until 1954 did an English translation (done by C. Baskin and published by the University of Virginia) appear in the United States.

What interested Christaller was the distribution of towns—their number, size, and where they were located with respect to one another. He wanted to discover the laws of economic geography that govern this distribution, just as Von Thünen a century earlier had wanted to discover the laws governing the evolution of the spatial arrangement of agricultural activities around a single urban center. In fact, Christaller carefully read Von Thünen's work, and Von Thünen himself had admitted that his postulation of a single city without the competition of other cities was a weakness in his work. Thus Christaller set about solving the problem, and, like Von Thünen, he first established a set of assumptions. These will sound familiar to readers of Von Thünen, for they included (1) an unbounded, flat plain with soils of equal fertility, (2) an even distribution of population and purchasing power, (3) a uniform transport system permitting direct travel toward each urban center, and (4) a constant maximum distance or range in all directions from the town for the sale of any item produced in that town.

Armed with these assumptions, Christaller proceeded to construct a model for the distribution of central places. Not all towns or settlements are central places: a central place, as its name implies, lies at the center of a region and is in functional contact with this region: it sells goods and provides services to the people in it. Farm settlements, mining stations, and customs collecting points, for example, are not necessarily central places. On the other hand, some places, Christaller realized, are more "central" than others. The central functions of larger towns will cover regions *within* which smaller places with lesser central functions nevertheless exist. What was needed was a means to calculate this degree of centrality of the various towns. In order to do this, Christaller identified *central goods and services* as those provided only at a central place and provided for and to the people of the surrounding region—as opposed to those services that might be available anywhere, without focus, and those products that might be manufactured for distant, even foreign markets. Next came the question of the range of sale of such central goods: the distance people were willing to travel to purchase them. There is a lower limit as well as a higher limit for this: the lower or threshold limit is marked by a line enclosing the minimum number of consumers required to permit the product to be manufactured economically—a circle, of course, since we are dealing with a constant range in all directions from the central place. The higher, or real, limit is reached halfway between one central place and the next town capable of producing the product at the same price; under the assumptions used, a person will

not go 51 miles to one place if he can go 49 miles to another to get the product he intends to purchase.

Thus each central place has a complementary region, within which the town has a monopoly on the sale of certain goods, since it alone can provide such goods at a given price. From what has just been said, it would seem that such complementary regions would be circular in shape, but in constructing the model on this basis, problems are encountered: either the circles are adjoining and leave "unserved" areas, or they overlap, and then there is no monopoly over the sale of goods. The solution is a system of complementary regions that are hexagonal in shape (Map 3-4a).

With these ideas, the hexagonal complementary region and the "range" of goods, Christaller derived his model for the spatial arrangement of towns. How many goods of different kinds are provided by a large central place? Let us say 100, and rank them in order of importance and range from 1 to 100. Thus the product ranked number 1 will have the greatest "real limit," and that ranked number 100, the smallest. Now if we arbitrarily placed three central places, each producing 100 central goods, say 100 miles away from each other, the following would happen. (These, incidentally, Christaller called G places.) The hexagonal region represents the "real limit" of the product ranked number 1. The product ranked number 2 has a real limit marked by a circle just inside the hexagon; by the time we draw the ever-smaller circles for the various lower-ranked goods, we begin to leave a fairly large population without service— theoretically, of course. Soon, by the time we reach product ranked about number 30, this intervening area is large enough to accommodate its own hexagon and central place; thus a network of B places, as Christaller identified them, is born (Map 3-4b). Now both G and B places provide goods ranked 31 and lower, and as we proceed and reach products ranked somewhere between 50 and 60, we find that the situation described above repeats itself, and once again there is room for a system of central places of a still lower order (K places) and their hexagonal complementary regions (Map 3-4c). These places now make available, cheaply, goods ranked down to 100.

Christaller himself realized that his assumptions were far-reaching, and that there is much

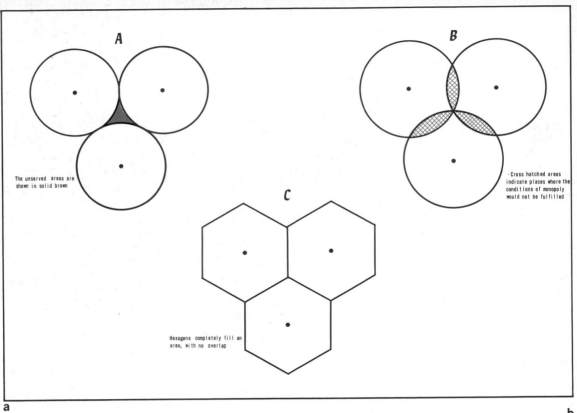

A

The unserved areas are shown in solid brown

B

·Cross hatched areas indicate places where the conditions of monopoly would not be fulfilled

C

Hexagons completely fill an area, with no overlap

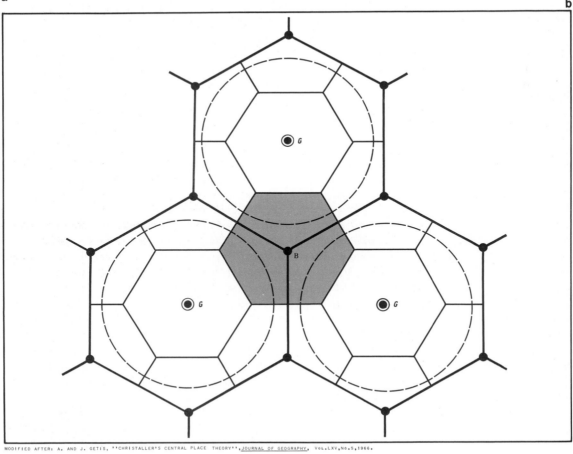

MODIFIED AFTER: A. AND J. GETIS, ''CHRISTALLER'S CENTRAL PLACE THEORY'', JOURNAL OF GEOGRAPHY, VOL. LXV, NO. 5, 1966.

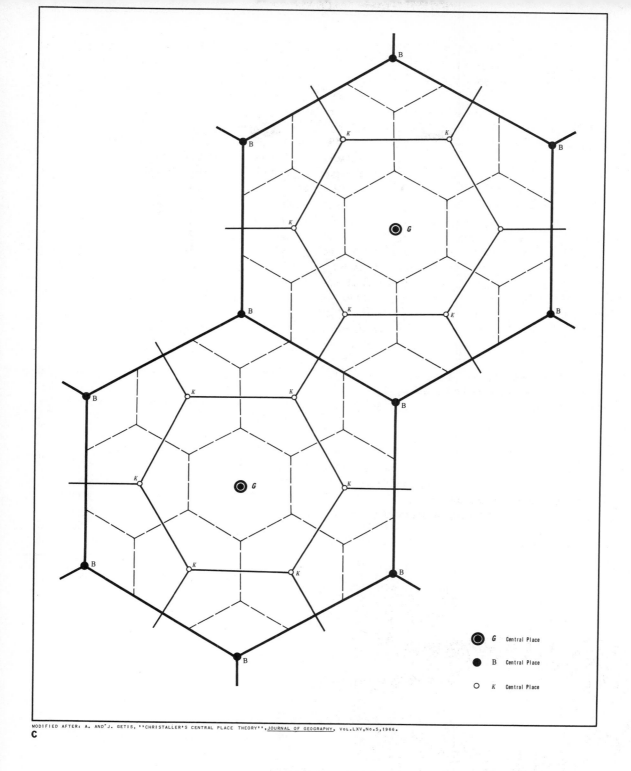

MODIFIED AFTER: A. AND J. GETIS, ''CHRISTALLER'S CENTRAL PLACE THEORY'', JOURNAL OF GEOGRAPHY, VOL.LXV,NO.5,1966.

c

Map 3–4a, b, c CHRISTALLER'S CENTRAL PLACE THEORY.

in the real world to distort his hexagonal systems. But like Von Thünen, he had opened a Pandora's box of debate and discussion. All sorts of support and criticism were voiced; some geographers saw modified hexagons everywhere, others saw none at all. Soon there were efforts to alter the model, to introduce different or additional elements, and to relate it to parts of the world other than Germany. Christaller himself participated; in 1950 he published a new volume entitled *The Foundations of Spatial Organization in Europe,* in which he enthusiastically endorsed his own view: "When we connect the metropolitan areas with each other through lines, and draw such a network of systems on the map of Europe, it indeed becomes eminently clear how the metropolitan areas everywhere lie in hexagonal arrangements." (pp. 17, 18). Whether or not his interpretation is correct, Christaller's major contribution perhaps lies less in the merits of the models he devised than in the stimulus his scientific approach to geography provided.[3]

Benelux

Three political entities are crowded into the northwestern corner of Western Europe: Belgium, the Netherlands, and tiny Luxemburg. In combination these countries are often called the Low Countries, which is a very appropriate name; most of the land is extremely flat and lies near (and, in the Netherlands, even below) sea level. Only toward the southeast, in Luxemburg and Belgium's Ardennes, is there a hill-and-plateau landscape with elevations in excess of 1000 feet. Another identification for these countries is *Benelux,* a combination of the first letters of the name of each. This applies to the economic union of the three, discussed and agreed upon even before the end of World War II, in London, and initiated shortly afterward. Some of the objectives are greater economic cooperation, the creation of common tariffs, and the elimination of import licenses within the Benelux area.

This attempt to foster economic unification is only the most recent event in a relationship between these countries that has experienced centuries of ups and downs. Toward the close of the Middle Ages, the wool-based textile industry of Flanders, which lies partly in present-day Belgium and partly in the Netherlands,

gave this area prominence and prestige in Europe. In Chapter 2 we noted how the consolidation of early European states often occurred through royal intermarriage; in this manner Flanders and most of the other Low Countries fell under Spanish domination in the fifteenth century. But the majority of the population, especially in the northern areas, was Protestant—and soon conflicts arose between Catholic rulers and Protestant subjects. From 1568 to 1648 the Low Countries fought the Eighty Years' War against Spanish overlordship, and with independence the Netherlands gained control over a considerable section of Flanders as well as other southern provinces where Catholicism prevailed. This was the origin of the internal, and continuing, split between Catholic and Protestant in the Netherlands.

Belgium, however, had not yet been born. What is today Belgium remained first under Spanish control, and later became French; with the defeat of Napoleon it was finally reunited with the Netherlands, after centuries of separation. But by then, divisions were too strong for such a union to last, and in 1830 Belgium was established as an independent state, itself still divided—not religiously so much as linguistically, with its Flemish-speaking north and northwest and Walloon (French)-speaking southeast and south. Luxemburg, a Grand Duchy formerly of considerable proportions and even functioning temporarily as a buffer between French and German interests in this part of Europe, was recognized as an independent state in 1867. With 999 square miles and a population under 350,000, this is Europe's smallest functioning state; it benefited greatly from its economic union with Belgium (1922), when Belgian markets were opened to its metal and leather products.

As between France and Germany, there are strong contrasts between Belgium and the Netherlands (Map 3-5). Indeed, there are such contrasts that the two countries find themselves in a position of complementarity. Belgium, with its coal base along an axis through Charleroi and Liege, where there are heavy industries, and with its considerable manufacturing of lighter and varied kinds along a zone extending from Charleroi through Brussels to Antwerp, produces a surplus of industrial products including metals, textiles, chemicals, and items such as pianos, soaps, cutlery, etc. The Netherlands, on the other

[3]This section owes much to Arthur and Judith Getis' article, "Christaller's Central Place Theory," *Journal of Geography,* Vol. LXV, No. 5, May, 1966, pp. 220–226.

Longitude East of Greenwich

NORTH SEA

WEST FRISIAN ISLANDS

AMELAND
TERSCHELLING
VLIELAND
Waddenzee
TEXEL

Den Helder

Ijssel Meer

Enkhuizen
Alkmaar

Haarlem
Amsterdam

Leiden
The Hague
Delft Gouda
Rotterdam
Utrecht
Old Rhine
Hilversum

Leeuwarden Groningen
Assen
Emmen

Kampen Zwolle Vechte
Almelo
Apeldoorn Deventer Hengelo
Enschede

Lek
Arnhem

NETHERLANDS

GOEREE
Dordrecht
Waal
Nijmegen
Rhine

SCHOUWEN
Ooster Schelde
NORTH
WALCHEREN BEVELAND
Flushing SOUTH
Wester Schelde

Hertogenbosch
Maas (Meuse)

Breda
Bergen op Zoom Tilburg
Eindhoven

Venlo

Ostend
Bruges
Roeselare

Antwerp
Sint-Niklaas
Turnhout

Roermond

GERMANY

Gent
Aalst
Mechelen

BELGIUM

Schelde

Brussels
Senne
Leuven

Hasselt

Maastricht

Tournai
Mons
Charleroi Sambre

Namur
Meuse

Liège
Seraing
Verviers

Ourthe

FRANCE

LUXEMBOURG

Luxembourg

Esch-sur-Alzette

○ 20 000 - 100 000
◉ 100 000 -500 000
◉ 500 000 - 1 000 000
● Over 1 000 000
—— Canals
:::: Tidal Flats
▒▒ Land below Sealevel

0 10 20 30 40 50
Miles

MODIFIED AFTER: LAUTENSACH, ATLAS ZUR ERDKUNDE

Map 3-5 BENELUX: LOCATION MAP.

hand, has a large agricultural base (along with its vitally important transport functions); it can export dairy products, meats, vegetables, and other foods. Hence there was mutal advantage to the Benelux union, for it facilitated the flow of needed imports to both countries and it doubled the domestic market.

The statistics given on pp. 7 and 8 in Chapter 1 suggest a common characteristic of the Netherlands and Belgium: a very high population on a very limited area. In general, such averages as are produced by dividing the number of people of a country into the number of square miles of its area do not mean much, since there is such great variation from place to place. But by any standard of measurement, these two countries are very densely populated: on an average there are 1000 people for *every* squre mile in the Netherlands, and 800 in Belgium. This despite the fact that, over the centuries, many hundreds of thousands of Hollanders and Belgians have emigrated to other countries, including Australia, Canada, South Africa, and the United States. In the Netherlands which, with its agricultural strength, faces the problem in its most acute form, preparations are being made for an even more difficult future through the reclamation of land from the sea. The greatest of these projects, so far, is the draining of the Ijssel Meer (Lake Ijssel), formerly known as the Zuider Zee (Southern Sea). The name change

is significant, the Dutch built a dyke across the open end of this sea, and subsequently parcelled the resulting lake off into reclamation *polders*. In the process they are adding well over a half million acres of land to their total area, and they are by no means finished. The islands in the southwest are being connected by a seawall for the same purpose, in the great Delta Plan designed to close the Rhine-Maas estuaries except for the approaches to Rotterdam and Antwerp. Ultimately the northern islands, curving around the Wadden Zee, may some day also be linked, and the Wadden Zee laid dry.

The three cities of the Netherlands' core triangle each have about 1 million inhabitants. Amsterdam, the constitutional capital, has canal connections to the North Sea as well as the Rhine, but it obviously does not have the locational advantages of Rotterdam. Nevertheless, Amsterdam remains very much the focus of the Netherlands, with a busy port, a bustling commercial center, and a variety of light manufactures. Rotterdam, a gateway to Western Europe, commands the entries to the Rhine and Maas (Meuse) Rivers, and its development, especially during the past century, mirrors that of the German Rhur-Rhineland. With its major shipbuilding industry, Rotterdam has importance also in fields other than transportation. The third city in the triangle, The Hague, lies near the North Sea coast but is without a port. It is the seat of the Dutch government, and in addition to its administrative functions it benefits from the tourist trade of the beaches of suburban Scheveningen.

These cities of the core area stand out among a dozen other urban centers with populations in excess of 100,000, but the Netherlands is still somewhat less strongly urbanized than Belgium. Hence the population density in the rural areas is very high, and practically every square foot of available soil is in some form of productive use, even the medians of highways and the embankments of railroad lines. In contrast to Belgium the local resource base has always been largely agricultural. Only in the southeast, the province of Limburg, wedged between Belgium and Germany, partakes of the belt of coal deposits that extends from France through Belgium. But other than ensuring self-sufficiency in coal for heat and fuel—and some minor industrial development in the province itself—this coal has not been as important to the Nether-

The industrial scene near Antwerp, Belgium. (Charles Rotkin)

lands as have adjacent deposits to its neighbors. Not long ago, newly discovered natural gas reserves were opened up in the northeast; these reserves, incidentally, extend across the border into West Germany. But all in all, what the Dutch need most is farmland—and this they are adding by their own hands.

Transport and Trade. In every respect the Low Countries display the characteristics of what we have defined as Europe's core area. Coal-based industrial development is considerable. Agriculture is intensive and technically advanced; produce is high-value. Urbanization is considerable. And in trade and transportation the Low Countries are in the forefront. With the existing limitations of space and raw materials, the Netherlands especially, but Belgium as well, have turned to the handling of international trade as a major and profitable activity. It is hard to think of any countries that could be better positioned for this than they are. Not only do the Rhine and Maas form highways that begin in the distant interior and end in the Low Countries, but Benelux itself is surrounded by the three leading productive countries in Europe—and in the world. Time has repeatedly added to the advantages inherent in this position. Canals have linked inland waterways and expanded the "catchment area" of trade to pass through Rotterdam and, to a lesser extent, Antwerp. Industrial development in Germany and

France has accelerated, production has increased, more raw materials have been required. And despite the Dortmund-Ems Canal and the functions of Hamburg, Bremen, and Le Havre, the Low Countries still are the threshold to Western Europe. At the present time a giant new port complex is developing not far from Rotterdam, whose name—Europoort, portal to Europe— suggests another aspect of the core area, and one which will be discussed later: the growing effort to create a unified Europe.

This is not to suggest that the Low Countries share equally in this great locational advantage. Holland has the better position, and through centuries of competition between Rotterdam and Antwerp, the latter has fared badly. Antwerp cannot be reached except through the Schelde River, which cuts through the Dutch province of Zeeland. After the Eighty Years' War, Holland closed the Schelde, cut off Antwerp, and did not permit the city full access to the oceans until nearly a century and a half later. The Netherlands subsequently turned down Belgian requests for a canal to link Antwerp with the Maas (in part this canal was to lie through Dutch territory). And repeatedly the maintenance of the Zeeland portion of the Schelde has been a issue of contention. Thus Antwerp, after an early period of greatness and promise, became mostly a Belgian outlet; Rotterdam forged ahead to a lead it has never yielded. In this respect Antwerp

A typical low-country rural scene: dairy cattle, lush pastures, canals traversing flat-lying polders. (Netherland Information Service)

rather resembles Amsterdam; Amsterdam, too, has canal and rail connections with the interior, but they are simply not efficient enough to affect Rotterdam's primacy. Hence Amsterdam, after its early period of greatness as the chief base for Dutch fleets and the center of a growing colonial empire, also gradually yielded its position to Rotterdam. Today Amsterdam's port handles mainly Netherlands trade, while Rotterdam specializes in the transit of international trade.

The Alpine Countries

Switzerland and Austria share a landlocked location and the mountainous topography of the Alps—and little else. On the face of it, Austria would seem to have the advantage over its western neighbor: it is twice as large in area and has a substantially larger population than Switzerland. A sizable section of the Danube Valley lies in Austria, and the Danube is to Eastern Europe what the Rhine is to Western Europe and the Volga is to the Soviet Union. In addition, no city in Switzerland can boast of a population even half as large as that of the famous capital of Austria, Vienna (under 2 million). The map shows, further, that Austria has considerably more land that is relatively flat and cultivable—a prize possession in these parts of Europe —and, as Plate 4 (in the rear of this volume) suggests, more of the remainder of it is under forest, a valuable resource. And that is not all.

The Swiss village of St. Moritz Bad, dominated (as are so many Swiss towns) by the country's mountainous scenery. (J. Allan Cash/Rapho Guillumette)

From what is known of the raw materials buried under the Alpine topography it appears that Austria is again the winner: in the east, where the Alps drop to the lower elevations, iron ores have been found, and elsewhere there are deposits of coal, bauxite, graphite, magnesite, and even some petroleum. Switzerland, for all practical purposes, does not have any usable mineral deposit.

From all this we might conclude that Austria should be the leading country in this part of Europe, and that impression would probably be strengthened by a look at some cultural maps. Take, for example, that of European languages (Map 1-3, Chapter 1). Austria is a unilingual state—that is, only one language is spoken throughout the country. In Switzerland, on the other hand, no less than four languages are in use. German is spoken in the largest part, French over the western quarter, and in the south there is an area where Italian is dominantly used. In the mountains of the southeast lies a small remnant of Romansh usage. This is hardly a picture of unity; and when we examine a map of religious preferences, we note that somewhat more than half of Switzerland's people are of Protestant orientation, and most of the remainder Catholic, a division similar to that prevailing in the Netherlands, while Austria is 90 percent Roman Catholic.

Yet in the final analysis, it is the Swiss people who have forged for themselves a superior standard of living, and it is the Swiss state, not Austria, that has achieved greater stability, security, and progress. Though poor in raw materials, Switzerland is an industrial country, and it exports on an average about four times as high a value of manufactured products annually as Austria. Its resources amount to the hydroelectric power supplied by mountain streams and the specialized skills of its population, and yet over half of all Swiss employment is in industry. While both Austria and Switzerland are clearly Western European countries, a good case can be made for the position that Switzerland is part and parcel of the European core, but Austria is not.

In some ways the origins of Switzerland are similar to those of the Netherlands. The Dutch, at the mouth of the Rhine, found themselves in a position to profit by the handling of trade making the transfer to and from the sea. The Swiss state had its origins late in the thirteenth century, at a time when trans-European connections were being rekindled by developments in the Po Valley and in Flanders. At that time, the valleys between the mountains were occupied by relatively isolated and more or less autonomous units or *cantons*. In 1291 three of these cantons (Uri, Schwyz, and Unterwalden), located in the central part of present-day Switzerland, joined in a league. The first objective of this merger was mutual defense against domination by the authority of the Holy Roman Empire, then in control of what is now Germany and Austria as well as northern Italy. But the inhabitants of the cantons also constructed a roadway and bridge through part of their territory, which had the effect of facilitating an Alpine crossing via the now-famous St. Gotthard Pass. More trade now funneled through this passage, and as the complementarity of the two parts of Europe grew, so the cantons prospered. Before long, other cantons joined the original three, including some, like Luzern and Zürich, that were located elsewhere along this trade route.

As a physical map of Switzerland shows, the easiest land communications in this country are from east to west, where the Swiss Central Plateau lies between the Jura Mountains in the north and the Western Alps in the south. This plateau, too, was an important route of history. It intersects the St. Gotthard Pass in the heart of the country, so that Switzerland developed around a major European crossroads. Cantons located on this plateau also joined the original league, as Bern did in 1353 and Fribourg in 1481. A very important merger was that of Basel, on the northern flank of the Jura and at the southern terminal of the navigable part of the Rhine River (1501). The Swiss, meanwhile, busied themselves improving other trans-Alpine roads, recognizing that both prosperity and security lay in the transit of a high volume of the trade of major European powers. And at this early stage the Swiss political position in Europe was determined: sovereignty and neutrality. As early as the sixteenth century the Swiss called for treaties with Austria and France guaranteeing both, and they got them. Wider recognition in Europe came in 1648, and in 1815, after the Napoleonic adventure, Swiss neutrality was reaffirmed. Meanwhile, cantons continued to join the confederation. French-speaking Vaud entered in 1803, Grisons (the stronghold of Romansh) in 1803, and Italian-speaking Ticino

in 1813. Thus the complexity of Switzerland's human geography grew.

When the industrial revolution stimulated the productivity mobility in Europe, Switzerland again was ready to handle the traffic. Swiss road building has a world reputation for excellence, and an unparalleled system of tunnels facilitates Alpine crossing; a considerable network of railroads connects Italy, France, and Germany. The system is continually being improved and the time required for people and goods to cross the Alps has been steadily reduced (Map 3-6).

The world map sometimes seems to suggest that mountain countries share a set of limitations on development that preclude them from joining the "developed" nations. It is very tempting to generalize about the impact of mountainous terrain (and the frequent corollary, landlocked location) as preventing productive agriculture, obstructing the flow of raw materials and the discovery of resources, hampering the dissemination of new ideas and the diffusion of innovations. Tibet, Afghanistan, Ethiopia, Lesotho, and the Andean portions of South American states seem to prove the point. That is why Switzerland is such an important lesson in human geography: all the tangible evidence suggests that here, at last, is a European area of stagnation and lack of internal cohesion—but the real situation is exactly the opposite. The Swiss, through their skills and abilities, have overcome a seemingly restrictive environment; they have made it into an asset that has permitted them to keep pace with industrializing Europe. First they used the passes to act as "middlemen" in interregional trade, then they used the water cascading from the mountains to reduce their dependence on power from imported coal by hydroelectric means, and finally they learned to accommodate those who came to visit the mountain country—the tourists—with professional excellence.

Farmers as well as manufacturers in Switzerland are skilled at getting the most (in terms of value) out of their efforts. The majority of the country's population is concentrated in the central plateau, where the land is at lower elevations. Here lie three of the major cities, Bern (the capital), Zürich (the largest city), and Geneva. Here, too, despite the fact that the land on this "plateau" is far from flat, lie most of the farms. The specialization is dairying, for several reasons: (1) the industry produces items that can command a high price; (2) little of the central

MODIFIED AFTER: RAND MCNALLY, *GOODE'S WORLD ATLAS*

Map 3-6 THE ALPINE STATES.

The Austrian Alps. (Aerofilms Ltd.)

plateau is suitable for the cultivation of grain crops; and (3) the industry affords an opportunity to use a mountain resource, namely the Alpine pastures that spring up at high elevations when the winter snows melt. In the summer, much of the herd is driven up the slopes to these high pastures. The herders—and sometimes the whole farm family—take up living in cottages, built specifically for this purpose near the snow line. With the arrival of autumn the cattle and goats and their keepers abandon pasture and cottages and descend to the plateau or to the intermontane valleys. Swiss farming is famous for this practice, known as *transhumance,* but it is not by any means unique to Switzerland.

Thus the Swiss farmer is engaged in substantially the same activity as his British and Dutch counterpart. Dairy products, cheeses, and chocolate are the chief products, and the last two are exported to a wider market. In manufacturing, also, the situation in Switzerland is quite like that in other countries of the European core—except that no other country has to import virtually *all* its raw materials. But again, specialization is the rule and high-value exports the

result. Swiss industries attempt to import as few bulky raw materials as possible, while manufacturing items whose price is determined more by the skills that have gone into them than the materials used in them. Precision machinery, instruments, tools, fine watches, and luxury textiles are among the major exports, and the reputation of Swiss manufactures guarantees them a place on the world market.

Nevertheless, Switzerland must import over one-third of its food requirements in addition to its industrial raw materials, and without further sources of income the unfavorable trade balance would have the country in trouble. Instead, there are no problems—thanks to the thriving tourist industry and the country's role as an international banker and insurer. The tourist industry, of course, makes use of Switzerland's magnificent scenery, but the Swiss have made tourism another field of specialization. In their hotels, lodges, rest homes, and other temporary abodes for visitors they have set a standard of excellence which alone is enough to draw thousands of visitors each year. And the banking-insurance industry is founded on centuries

Austria's primate city, Vienna. The view shows the Donau Canal and in the background, the Leopoldsberg. (Louis Goldman/Rapho Guillumette)

of confidence inspired by Switzerland's stability, sovereignty, and neutrality.

Unlike Switzerland, Austria as it is presently constituted is a relatively young state. This is not to suggest that human organization in what is today Austria is a recent phenomenon; Austria at one time lay at the center of the Austro-Hungarian Empire, one of the great centers of power in Europe and one of the casualties of World War I. Before that, Vienna was the center of the Holy Roman Empire and seat of the Habsburg Monarchy; Austria in truth was a core area on the eastern margins of Western Europe.

Modern Austria, however, is merely a remnant of its former greatness. The Austro-Hungarian Empire had linked the Germanic Austrians and the non-Slavic Hungarians in a mid-European bastion of power and influence, centered on the Danube River and focused on the great city of Vienna. The Empire dominated peoples in Czechoslovakia, Romania, Yugoslavia, Poland, and even Russia; but it always suffered from internal stresses related to the complex of nationalities it controlled. The Empire failed where Switzerland had succeeded: in the attainment of commitment from those groups, national, linguistic, religious and otherwise, within its borders. After its disastrous participation in

World War I the Empire was divided into several component units, of which Austria was one.

Thus while Switzerland could keep abreast of economic changes in Europe, Austria, far from neutral and stable, was constantly involved in political conflict and struggles for power. The Brenner Pass—Austria's equivalent of the St. Gotthard—constituted a gateway from Italy to north-central Europe, but it never served Austria as the Swiss transalpine routes served the confederation. When, in 1919, Austria emerged as a separate entity, it faced not a time of plenty, but a period of reconstruction and reorganization of the national economy. And before the country had the opportunity to recover from the economic disaster of the early 1930's, it lost its independence to Nazi Germany, which forced incorporation upon it in 1938. Most recently Austria regained a measure of independence in stages, first in 1945 with the end of World War II (but under allied occupation), and then in 1955, when foreign forces withdrew under condition of continued Austrian neutrality.

Since 1945 Austria's most difficult problem has been its reorientation to Western Europe. Even Austria's physical geography seems to demand that the country look eastward; it is at its widest, lowest, and most productive in the east, the Danube flows eastward, and even Vienna lies near the eastern perimeter. But the days of domination over Eastern European countries are gone, and the markets there are no longer available. And with the interruptions and setbacks of the twentieth century, Austria is poorly equipped to catch up with competitive Western Europe.

Still, there are resources and opportunities, and Austria may finally be on its way to solvency. Hydroelectric power plants are being built and the time is approaching when all power requirements will be met by "white coal." Iron and steel production, which began during the nineteenth century, was intensified during the war years and has been supported by American subsidies. But this industry is not marked by the high-quality products that are Switzerland's mainstay; it requires expensive coal imports (and iron ore as well). Moderate development marks the chemical and textile industries, but the greatest promise still appears to lie in (1) the forest industry and (2) the tourist industry. Austria has extensive forests and several of its neighbors have pulp and other wood requirements.

And though perhaps not as spectacular as Alpine Switzerland, and certainly not as well located with reference to Europe's major tourist sources, Austria's Alps have great potential as a source of revenue. Austria certainly has the essentials to join the European core; what is needed now is what Switzerland has long had: stability and time.

Nordic Europe (Norden)

In three directions from the core area, Europe changes quite drastically. To the south, Mediterranean Europe is dominated by Greek and Latin influences and by the special habitat produced by a combination of Alpine topography and a "Mediterranean" climatic regime; to the east lies "continental" Europe, the Balkans, less industrialized, urbanized, or mobile than the core; and to the north lies Nordic Europe, all of it except Denmark separated by water from what we have here defined as the core of Europe. It is interesting that when we began our discussion of Europe, we made a point of Europe's peninsulas and the waters separating them from the mainland: their importance and their role in the history of European development. And now those peninsulas, where once the power of Greece and Rome lay, and from which Vikings raided, conquered, and settled, and from where great fleets sailed to colonize lands across oceans—those peninsulas mean so much less. It is difficult to miss the significance, in this connection, of Map 2-7; apart from England, the core area has not spread to a single insular or peninsular part of Europe, and neither has it penetrated far into Eastern, continental Europe. This is not to suggest that Nordic or Mediterranean Europe is "underdeveloped" Europe. Quite the contrary, in general a great deal has been achieved in many places without much help from nature. But in terms of resources, and in a European context, the peninsular areas of Europe are not particularly rich.

This is especially true of Scandinavia and Finland, and it is reflected in many ways. Take the combined population of the five countries (Denmark, Norway, Sweden, Finland, and Iceland) for example: it barely exceeds 20 million, which is less than that of Benelux. The total area, on the other hand, is some 486,000 square miles—larger than the entire European core, including Britain, France, Germany, Benelux, Switzerland, and northern Italy. People go where there is a living to be made; the living in most of Scandinavia is not easy.

Several aspects of Norden's location have much to do with this. First, this is the world's northernmost group of states; while the Soviet Union, Canada, and the United States possess lands in similar latitudes, each of these much larger countries has a national core area in a more southerly position. The North Europeans themselves call their region *Norden*, an appropriate term indeed. Second, Norden, from Western Europe, is on the way to nowhere. How different would the relative location of the Norwegian coast be if important world steamship routes rounded the North Cape and paralleled the shoreline on their way to and from the European core. Norway's ports of Trondheim and Bergen, and the capital of Oslo as well, would be different places today. Third, all of Norden except Denmark is separated by water from the rest of Europe. As we know, water has often been an ally rather than an enemy in the development of Europe—but mostly where it could be used for the *interchange* of goods. Norden

The Scandinavian Shield is an area of old, rugged topography, tree-covered and incised by glacier streams. This photograph was taken in Norway. (Karl Gullers/Rapho Guillumette)

Trondheim, Norway. (J. Allan Cash/Rapho Guillumette)

lies separated from Europe *and* relatively isolated in the northwest corner. Greater proximity to and contact with Western Europe no doubt would have been a stimulus to Scandinavian development. Thus Denmark and southern Sweden are really extensions of the North European Lowland and an exception to the Scandinavian rule. After Sweden the most populous Scandinavian country, Denmark is also the smallest state in the region; its average population density is over six times that of Sweden. On Denmark's flat and partially reclaimed country an intensive dairy industry has developed—so intensive, in fact, that cattle feed has to be imported similar to the way an industrial country imports iron ore. The produce is sold to Europe's greatest food importer, Britain, and across the border to Germany; these two countries annually buy about half of Denmark's export production. For these transactions, Denmark is very favorably situated (Map 3-7).

Nordic Europe's relative isolation did have some positive consequences as well. The countries have a great deal in common; they were not repeatedly overrun or invaded by different European groups, as was so much of the rest of

Europe. At different times in history each of the Scandinavian states seems to have had an advantage: Denmark ruled Norway from 1397 to 1814, and Sweden held hegemony over Finland over about the same span of time, though southern Sweden itself was freed from Danish control as late as the seventeenth century. Then Sweden merged Norway under its crown in 1814; Norway, in fact, did not attain real independence until 1905. Finland, which went from Swedish to Russian control in 1809, attained independence from Russia in 1917. Through the centuries Denmark and Sweden were at an often uneasy standoff, and neither ever was able effectively to control the other. On the other hand, Iceland was under Danish rule from the 1300's until 1918, and even after independence it remained nominally under the Danish crown until 1944. All this had several important results. The three major languages, Danish, Swedish, and Norwegian, are mutually intelligible, and people can converse with each other without the need of an interpreter. Icelandic belongs to the same language family; only Finnish is of totally different origins—but the long period of contact with Sweden has left a sizable Swedish-speaking

ARCTIC CIRCLE

Isafjördhur
Siglufjördhur
Húsavik
Vopnafjördhur
Stadhur
Eskifjördhur
65°
Borgarnes
Akranes
Reykjavik Thjörsa HEKLA
Eyrarbakki △ 4747'
Vik

I C E L A N D

15°

20°

ARCTIC   OCEAN
BARENTS
SEA
70°

NORTH CAPE
Hammerfest
Vardö
Kirkenes
Vadsö
Varangerfjord

Tromsö
Inari
U. S. S. R.

VESTERÅLEN
Harstad
F
Muonio
Ounas
LOFOTEN ISL
Svolvaer
Narvik
I
Vestfjord
Kiruna
Torne
KEBNEKAISE
6965'
Gällivare
Kemijärvi
Kemi
N
Bodö
SULITJELMA
△ 6158'
Porjus
Lule
Boden
Rovaniemi
Ylikitka
65°
Mo i Rana
Uddjaur
Skellefte
Haparanda Tornio
Luleå
Kemi
Oulu
L
N
O
R
W
A
Y
Namsos
Vilhelmina
Ume
Skellefteå
Gulf of Bothnia
Raahe
Oulujärvi
Kajaani
Trondheimsfjord
Ångerman
Umeå
A
Kristiansund
Trondheim
HELAGSFJÄLLET
Åre
Indals
Härnösand
Kokkola
N
Molde
Störén
△ 5892'
Storsjön
Jakobstad
Kuopio
Röros
Bracke
Vaasa
D
SNÖHETTA
Tynset
Sundsvall
Jyväskylä
Mikkeli
△ 7500'
S
Hudiksvall
Päijänne
Saimaa
GLITTERTIND
Lillehammer
Ljusnan
Pori
Kymi
Tampere
△ 8104
W
Hamar
Söderhamn
Rauma
Hämeenlinna
Lahti
Kotka
Sognefjord
Gloma
E
Gävle
Voss
Mjösa
Klar
D
Falun
ÅLAND ISL
Turku
Helsinki
Gulf of Finland
60°
Bergen
Eidsvoll
Dal
Uppsala
Hango
N
E
Oslo
Västerås
Stockholm
Stavanger
Moss
Karlstad
Eskilstuna
Fredrikstad
Kristinehamn
Örebro
Gulf of
Riga
LINDESNES
Kristiansand
N
Vänern
Norrköping
Skövde
Linköping
NORTH
SEA
Skagerrak
Hjörring
Frederikshavn
Kattegat
Vänersborg
Vättern
GOTLAND
Borås
Jönköping
U. S. S. R.
Göteborg
Aalborg
JUTLAND
Randers
Kalmar
OLAND
BALTIC
Ringköbing
Arhus
Halmstad
Karlskrona
SEA
55°
Helsingör
Hälsingborg
Vejle
DENMARK
SJAELLAND
Lund
Esbjerg
FYN
Copenhagen
Malmö
Odense
Naestved
BORNHOLM
55°
Nyköbing
GERMANY
POLAND
10°   Longitude East of Greenwich   15°
20°

Under 50 000
50 000 — 500 000
500 000 — 1 000 000
Over 1 000 000
Canals

0   50   100   200
Miles

MODIFIED AFTER: RAND MCNALLY, GOODE'S WORLD ATLAS

Map 3–7   NORDEN: LOCATION.

resident population in Finland, where this language has official recognition. Furthermore, in each of the Scandinavian countries there is overwhelming adherence to the same Lutheran church, and in each it is the recognized state religion. There are no religious conflicts comparable to those that have affected and continue to trouble parts of Europe. Finally, there is considerable similarity in the political evolution of the Scandinavian states, and in their socioeconomic policies. Democratic, representative parliaments emerged early; individual rights have long been carefully protected. This progress was possible, as it was in England, because of the lack of immediate outside threat to Scandinavia through most of its modern history. The manner in which the common cultural heritage that marks Scandinavia evolved in many ways parallels that of the British Isles.

Resources and Responses. The five Nordic countries share more than cultural ties. In their northerly location they also—within a certain range of variation, of course—face common conditions of climate and habitat. The three largest countries, Sweden, Finland, and Norway, all have major concentrations of population in the southern part of their land area. Though the waters of the North Atlantic Ocean help temper the Arctic cold and keep Norway's Atlantic ports open, they rapidly become less effective both northward and landward. Northward, the frigid polar conditions reduce the temperature of the water and thus its effect on overlying air masses.

But more important is Scandinavia's high mountain backbone, which everywhere stands in the way of air moving eastward across Norway and into Sweden. Not only does this high upland limit the maritime belt to a narrow strip along the Norwegian coast, but by its elevation it brings Arctic conditions southward into the heart of the Scandinavian Peninsula. This is well illustrated on Plate 2, as is the "shadow" effect of the highland upon Sweden. Note that Denmark and southern Sweden lie in the temperate, marine-influenced *C* climatic region, and that the remainder of Sweden, despite its peninsular position, has a climate of continental character. One cannot but speculate on what Western Europe's environment might have been if that Scandinavian backbone had continued through the Netherlands, Belgium, France, and into Spain!

Only Denmark combines the advantages of temperate climate with land sufficiently level and soils good enough to sustain intensive agriculture. Norway, with its long Atlantic coast, is almost entirely mountainous; only in the southeast, around Oslo, in the southwest, south of Bergen, and on the west coast, near Trondheim, are there areas of agricultural land and reasonably good soils (apart from tiny patches of bottomland in fjorded valleys near the coast). Less then 4 percent of Norway's area can be cultivated, and even in the small areas that make up this 4 percent conditions are far from ideal. Norway is by far the wettest part of Norden and,

Denmark's agricultural landscape reflects its mechanized farming and intensive land use. (J. L. Stage/Magnum)

as we saw in Ireland, with cool temperatures the moisture is soon excessive. So it is in Norway, where potatoes and barley generally replace the crops the farmers would rather grow, namely wheat and rye. Thus much of the farmland lies under fodder, and pastoralism is the chief agricultural activity. Certainly Norway came off second best in the division of the Scandinavian Peninsula, and it turned to the sea to make up for this. And in the seas the Norwegians have found considerable profit: the Norwegian fishing industry is one of the world's largest, and fishing fleets from Norwegian coasts ply all the oceans of the world. One of the most productive fishing grounds of all happens to lie very close to Norway, in part in its own territorial waters. Here, in the North Atlantic, Norwegian fishermen take large catches of herring, cod, mackerel, and haddock. But the oceans have served the Norwegians in yet another manner. Over the years Norway has developed a merchant marine which, in tonnage, is the fourth largest in the world, and competes with such giants as the British, American, and Japanese fleets. This fleet carries little in the way of Norwegian products; rather it handles that of other countries, performing transfer functions Norway, through its location, is otherwise denied.

Sweden, Norway's eastern neighbor on the Scandinavian Peninsula, is more favored in almost every respect; it may not have Norway's access to the Atlantic, but it needs the Atlantic less. Sweden has two agricultural zones, of which the southernmost and leading one lies at the very southern end of the country, just across from Denmark. In fact, this area resembles Denmark in many ways, including its agricultural development, except that somewhat more grain crops, especially wheat, are grown here. Malmö, the area's chief urban center, is Sweden's third largest city with a quarter of a million people; its main function is that of an agricultural service center. Sweden's other agricultural zone lies astride a line drawn from the capital, Stockholm, to Göteborg. Here dairying is the main activity, but agriculture is overshadowed by manufacturing. Swedish manufacturing, unlike that of some of the Western European countries we have considered, is scattered through dozens of small and medium-sized towns; unlike Denmark and Norway, there are resources in Sweden to sustain such industries. For a long time Sweden served in large part as an exporter of raw or semi-

Fishing is very important to many European coastal peoples. In this early morning scene, Norwegian fishing boats leave the harbor at Henningsvaer. (J. Allan Cash/Rapho Guillumette)

finished materials to industrializing countries, but increasingly the Swedes are making finished products themselves, specializing much in the way the Swiss have done. Already the list of famous products is quite long: who has not heard of Swedish safety matches, Swedish furniture (two industries using the extensive forests as a resource base), Swedish stainless steel and ball-bearings (based on steel produced through hydroelectric refining processes), or Swedish automobiles with their claim of near-indestructibility? And there is a great deal more, much of it based on relatively small local ores: apart from iron, of which there is a great deal, there is also copper, lead, zinc, manganese, and even some silver and gold. There are small metallurgical industries at one end of the scale—and the huge shipbuilding works of Göteborg at the other. In electronics and engineering, glassware and textiles, Sweden shows once again that the skill and expertise of the people is as important as the resource base itself.

All three of the larger Nordic countries possess extensive forest resources, and all three have exported large quantities of pulp and paper. At first the mills used direct water power, but with the advent of steam power, they relocated at river mouths and used, in part, their own waste as fuel. No Nordic country depends to a greater extent upon its forest resources than does Finland, whose wood and wood product exports normally account for between two-thirds and three-quarters of all annual export revenues. In this respect the region long resembled the colonial dependencies of Western European powers, underscoring their "underdeveloped" position compared to raw-material-consuming Western Europe. But even logs and timber can be transformed into something specialized before export: the Norwegians make much high-quality paper; the Swedes, matches and furniture; and the Finns, plywood and multiplex, veneers, and even prefabricated cottages. Not all the forest resources are thus transformed, of course, and Finland, like Sweden and Norway, continues to export much pulp and low-grade paper. In Finland, however, the alternatives are fewer. Most of the country is too cold to sustain permanent agriculture, and where farming is possible, mostly along the warmer coasts of the south and southwest, the objective is self-sufficiency rather than export. Known mineral deposits are few: some copper lies at Outokumpu and iron ore at Otanmakiin. Nevertheless, the Finns have succeeded in translating their limited opportunities into a healthy economic situation, in which, once again, the skills of the population play a major role. The country is nearly self-sufficient in its farm products, and the domestic market sustains a textile industry (90 miles inland at Tampere) and metal industries for locally needed machinery and implements, at Turku and the capital,

Helsinki. For these, and for the shipyards at Helsinki and Turku, the raw materials must, of course, be imported.

The westernmost of Norden's countries—and the most westerly state of Europe—is Iceland, a hunk of basaltic rock that emerges above the surface of the frigid waters of the North Atlantic at the Arctic Circle. Three-quarters of Iceland's 40,000 square miles are barren and treeless; one eighth of the island's rough and mountainous terrain is covered by imposing glaciers. Below, the earth still rumbles with earthquakes and hot springs are numerous. Volcanic eruptions have left their mark on the country's past, and have claimed thousands of lives.

Iceland's population has the dimensions of a microstate (200,000), and nearly half this population is concentrated in the capital, Reykjavik. Iceland shares with Scandinavia its difficulties of terrain and climate, only in even greater degree, and it also has ethnic affinities with continental Norden. The majority of the earliest settlers came from Norway during the ninth century, and the Icelanders claim that theirs was Europe's first constitutional democracy; in 930 A.D. a parliament was elected. Later, Iceland fell under the sway of Norway and then Denmark, and it was not until 1918 that it regained its autonomy. In 1944 all remaining ties with the Danish Crown were renounced and the republic was finally resurrected.

With its severely restricted possibilities for agriculture, Iceland's twin assets are the sea and its location. From the sea comes a large harvest of fish, the country's leading export, and fish processing is the leading industry. Iceland's location has strategic implications which were proved first during World War II, when Reykjavik was a crucial intercontinental air relay station, and later through its involvement in the North Atlantic Treaty Organization. A sizable part of its income is derived from a lease of space for military installations under the treaty.

In summary, Nordic Europe in some respects resembles Western Europe, but in other ways it is radically different. In terms of the elements required for industrialization, Norden's base is inadequate—but the Swiss have taught us that this need not exclude these northern countries from participation in the general progress. Without appreciable coal, Scandinavia has much of the hydroelectric power potential of Europe—and a great deal of it already has been

The forests, the lakes, and the grain farming of Finland are all represented on this photograph, characteristic of southern Finland. (Finnish Tourist Association)

The Kiruna iron mines of Sweden. (Swedish National Travel Office)

employed as a substitute power supply. Sweden, with the largest area, largest population, and greatest array of raw materials, also has the most diversified economy; but even Sweden is still moving toward more domestic treatment and finishing of these raw materials and less exporting in this form. For many decades, Sweden has been sending some of the world's best, highest grade iron ore from its northern mines at Kiruna, via Luleå on the Gulf of Bothnia and Narvik on Norway's Atlantic coast, to smelters in Western Europe. Now a Swedish steel plant at Lulea, powered by a hydroelectric source, refines Swedish ores for Swedish use in the production of precision equipment and machinery. Norway, too, has built a hydro-powered iron ore smelter at Mo i Rana, on a fjord just south of the Arctic Circle. But in general an overdependence on one or two industries persists, a situation we shall encounter again in other parts of the world. Thriving Denmark, for example, has a dairy industry that is tied to Western European markets; if those markets were somehow closed, the country, without alternatives, could face disaster. Finland is seeking to diversify from its single dependence on its forest resource. Iceland has little opportunity for any such diversification: its economy is almost entirely dependent on the fishing industry; the country is virtually without mineral resources and agricultural possibilities are severely limited.

Norden's urbanization reflects the region's achievements and limitations. The largest city is Copenhagen (1.5 million), which long benefited from its strategic location at the narrow entry to the Baltic: not only did it handle much trade, but it levied tolls on ships passing through the bottleneck. The Kiel Canal notwithstanding, Copenhagen continues to prosper, even without the tolls. With its good natural harbor and port facilities, it acts as a place where goods are collected, stored, and transshipped—an *entrepôt,* a function which is greatly enhanced by a duty-free zone where goods can be kept prior to redistribution to Baltic or Atlantic ports. If it is possible for ships to sail through the Sound between Sweden and Denmark, why Copenhagen's break-of-bulk function? The reason is that large Atlantic Ocean vessels often cannot enter the Baltic, and smaller Baltic Sea tonnages need to be collected somewhere to make transporting them farther an economical proposition.

Copenhagen benefited, during an earlier period in Danish history, from its central position in Denmark as well as its location on the narrow Baltic entry. That central position was the result of Denmark's hegemony over what is today southern Sweden. With its early rise to prominence, the city has never lost its momentum; though no longer at the heart of a large state it still thrives as the seat of the Danish government, as a transport focus, and as a manufacturing center. Shipbuilding and food processing are, predictably, the major industries, but there are many others, from the manufacture of marine diesel engines (a Danish invention) to textiles to fine china and glassware.

So large a city in so agricultural a country can only be explained in terms of history and of external stimuli, and Copenhagen has much of both. Denmark's second city, Aarhus, by comparison, only has some 200,000 inhabitants. In Sweden, Stockholm, the capital, is also the largest city, but it has much more competition. With over a million people (almost double that, if the outlying suburbs are included), Stockholm displays its long-time position as Sweden's focus. Like Copenhagen, it too was at the center of a greater state, when Finland was under Swedish domination and Sweden was supreme over the Baltic. Then, when southern Sweden became the heartland of the state, Stockholm found itself rather on the periphery of the new core, but recently the new stimulus toward development of the north has swung the pendulum back toward Stockholm again. On the other hand, Stockholm has lost its position as chief Swedish port to Göteborg ($\frac{1}{2}$ million), and a westward movement of industry, toward this outlet, can be observed. But the functions of Swedish cities are well divided. Government, finance, and business reside in the cultural focus in Stockholm; foreign shipping, shipbuilding, and other industries center on Göteborg, and the third city, Malmö, is the agricultural service center for the south.

Like Copenhagen and Stockholm, Oslo, Norway's capital, is an old city. With over a half million people, it is the undisputed leader in this country, at the heart of the core area, and the focus of major communication lines. But Oslo does not have the advantage of a wealthy history, and although its attributes include not only good port facilities (this is Norway's leading port) but also a variety of industries, Oslo cannot compare to Copenhagen or Stockholm. Norway's second port is Bergen, center of the fishing industry. In Finland, Helsinki, with well over a half million inhabitants, is also a relatively young city, having emerged largely during the past century and a half from a small fishing port to international stature as the capital, leading port, and main commercial center of the young state.

A great deal of actual and potential complementarity and interdependence exists among the Nordic countries. Norway, with its merchant marine, buys tankers built in the shipyards of Göteborg, Sweden. Sweden and Norway import Danish meat and dairy products. Denmark, most treeless of the Nordic countries except Iceland, needs timber, paper, pulp, and associated products and can import from Sweden, Finland, and Norway. Norway, best suited of all for the cheap production of hydroelectric power, could supply this vital commodity to Denmark, which needs it most, and Sweden. Norway, in turn, can use equipment and machinery built in Denmark, Sweden, and Finland. The opportunities are many; regional associations are already strong, and they will doubtlessly grow stronger.

Mediterranean Europe

From Northern Europe we turn now to the four countries of the south: Greece, Italy, Spain, and Portugal. From near-polar Europe we now look into near-tropical Europe, and it is reasonable to expect strong contrasts. And there are many. But there are also some similarities, and some very telling ones. Once again we are dealing with peninsulas: three of them, this time, two occupied singly by Greece and Italy, and one jointly by Spain and Portugal (the Iberian Peninsula). Once again there is effective separation from the Western European core: Greece lies at the southern end of Eastern Europe and has the sea and the Balkans between it and the west; Iberia lies separated from France by the Pyrenees, which through history has proved to be quite a barrier; and then there is Italy. Southern Italy lies far removed from Western Europe, but the north is situated very close to it, and presses against France, Switzerland, and Austria. For many centuries northern Italy was in close contact with Western Europe, and developed less as a Mediterranean area than as a part of the core area of Western Europe. In a very general way northern Italy is as much an exception to Mediterranean Europe as Denmark is to Scandinavian Europe, and these two areas happen to lie opposite one another across Germany and Switzerland, in physical as well as functional contact with the European core (Map 3-8/9).

The Scandinavian countries share a common cultural heritage; so does Mediterranean Europe. Firm interconnections were established by the Greeks and Romans; under the Greeks, ports as far away as Marseilles and Alexandria formed part of an integrated trade system, and under the Romans virtually the whole Mediterranean region was endowed with similar cultural attributes. This unity, we know, did not last. New political arrangements replaced the old, and languages differentiated into Portuguese, Spanish, and Italian; in Greece, the Roman tide was to a considerable extent withstood. But the underlying, shared legacy remains strong to this day.

Like Northern Europe, Mediterranean Europe lies largely within a single climatic region, and from one end to the other the opportunities and problems created by this feature of the environment are similar. The opportunities lie in the warmth of the near-tropical location, and the problems are related largely to the moisture supply: its quantity and the seasonality of its arrival. And in the topography and relief, too, Mediterranean countries share similar conditions—conditions which would look quite familiar to a Norwegian or a Swede. Much of Mediterranean Europe is mountain or upland country, with excessively steep slopes and poor, thin, rocky soils; for its agricultural productivity the region largely depends on river basins and valleys and coastal lowlands. As with practically every rule, there are exceptions, as in northern Italy and northern and interior Spain, but generally the typical Mediterranean environment prevails.

Neither is Mediterranean Europe much better endowed with mineral resources than Scandinavia. Both Greece and Italy are deficient in coal and iron ore; for many years Italy, industrializing despite this shortcoming, has been one of the world's leading coal importers. Recently Italy's fuel position has improved somewhat through the exploitation of minor oil fields in Sicily and by the use of natural gas found beneath the Po Valley surface, but these are far from enough to satisfy domestic needs. Only in Spain are there really sizable fields of coal and iron ore positioned close to one another, in the Cantabrian Mountains of the north. But Spain, best endowed with raw materials of all Mediterranean countries, has not responded with primary industries on the scale of Western Europe. On the contrary, a good deal of the highest grade iron ore has been exported, as has the good coking coal, as though Spain were destined to be an underdeveloped, raw material-supplying country rather than an industrial power. In this respect Spain only mirrors the whole Mediterranean region; there are scattered mineral deposits, some

of them valuable and capable of sustaining local, skill-dependent metallurgical industries, chemical industries, and other enterprises—but instead, almost all these raw materials are exported to the core region. Scandinavia and Mediterranean Europe may share a resource poverty, but Scandinavia gets the most out of what it has—and Mediterranean Europe often the least.

This is to a lesser extent the case with both Scandinavia's and Mediterranean Europe's answer to the fuel shortage: hydroelectric power. Like Norway, Italy is one of the world's leading states in the development of its hydroelectric power potential; in the thirty years after 1939, the country has quadrupled its output, further reducing its dependence on imported sources of power. But Italy is most favorably endowed in this respect, much more so than either Greece or Iberia: its hydroelectric production is about double that of its three Mediterranean partners combined. The Mediterranean environment, for all its mountains and uplands, is not all that advantageous for hydroelectric power development: with its seasonal and rather low precipitation there are often water shortages, when streams fall dry and water levels in dams go down. In northern Italy, where the largest market is located, conditions are better: rainfall distribution is much less seasonal and water supply from the Alpine ranges is much more dependable.

Some centuries ago, Scandinavians and Southern Europeans would have recognized another area of similarity between their environments: the prevalence of an extensive forest cover, a valuable resource in pre-industrial as well as industrial times. The Scandinavians still have it, and their exports revenues include a sizable percentage derived from these forests—but Mediterranean Europe stands largely bare and denuded. Trees were cut down for housebuilding, for centuries of shipbuilding, for fuel, and to make way for agriculture. In the period of Spain's greatness the forests of the Meseta constituted one of Iberia's major assets. Today this area stands windswept and treeless.

Population. Each of the states of Mediterranean Europe has behind it an age of power, when it was the center of a greater empire with distant colonies. Greece really had two such periods, the first during the age of Greek hegemony over the Mediterranean Sea (before the rise of Rome), and the second during medieval times, when it was the core of the Byzantine Empire, which of course was a much more land-based state. Italy, after the heyday of Rome, went into a long period of division, during which various Roman places attained eminence and prosperity. There was a state in the Po Valley centered on Venice, which itself had much influence in other parts of the Mediterranean. Genoa thrived as an independent port, and in northern Italy cities such as Florence, Pisa, and Milan benefited from the growing trade across and around the Alps to Western Europe. But a unified Italy did not emerge until late in the nineteenth century (1871), and by then many of these places were in decline or of only local importance; despite an illfated colonial adventure in Africa, Italy never regained its former stature.

Spain's period of empire was preceded by an age during which it was not the colonizer, but the colonized: early in the eighth century it was invaded by Arabs and later by Berbers from North Africa who brought a superior civilization to Iberia. Over the next seven centuries much

progress was made in agriculture, irrigation, architecture, and some industries such as metal and leather working, but in the end the Moors were ousted—at just about the same time that an Italian mariner in the service of the Spanish court sailed westward and reached the New World. The double momentum of victory and discovery carried Spain to a century of unprecedented wealth, power, and prestige, and brought it possessions in the Americas, Africa, and the Pacific. Madrid became the capital of the world, and in Europe Spain had no equal. But Spain derived relatively little lasting benefit from its period of supremacy. Much of what the Moslems had contributed was lost with their ouster, and all the gold and silver from America failed to bring change to an economic system that needed—and continues today to need—transformation. Portugal alone still holds large parts of its colonial empire. It, too, had an age of splendor as it ruled supreme over the sea routes to the Orient. In South America, Africa, and Asia, Portugal acquired dependencies, and even

MODIFIED AFTER: J. BARTHOLOMEW, TIMES ATLAS

Map 3-8 WESTERN MEDITERRANEAN EUROPE.

though it was conquered by Spain late in the sixteenth century and held for 60 years, Portugal retained much of what it had taken. Though Portugal lost some colonies to the Dutch, who were then in the ascendancy, and later lost Brazil (and only a few years ago, its Indian colony of Goa), this is the colonial power today with the largest surviving overseas empire, including the huge territories of Angola and Moçambique in Africa.

The colonization and occupation of distant territories required manpower, and in this respect the Mediterranean countries seem always to have been better off than Scandinavia. Northern Europe's most populous state, Sweden, still ranks below Mediterranean Europe's smallest country, Greece, in these terms. Indeed, Southern Europe's four countries have a combined population that is fully five times as great as that of Scandinavia. So much of Scandinavia lies in climatically and topographically inhospitable areas that the effective, reasonably habitable territory is far smaller than the total size of the region. Much of the Mediterranean region also presents obstacles, but there are more opportunities as well. For example, there is nothing in Scandinavia to rival the Po Valley alone, where more than 40 percent of Italy's population is concentrated. Population densities as low as they are in most of northern Scandinavia are found in the Mediterranean only in the most rugged parts of the Pyrenees. Obviously this single aspect (of population numbers) alone does not explain the differences between Scandinavian and Mediterranean countries when it comes to their colonizing efforts, but no doubt it played a role. In Northern Europe, Denmark was the most adventurous colonizer, and Danish trading stations and footholds appeared on some tropical coasts. But no Scandinavian country has a history of colonial acquisition to match that of its Mediterranean counterparts, and thus none enjoyed the benefits: power, wealth, and growth.

From what we know of Mediterranean dependence on agriculture (in the absence of major mineral deposits), the historical development of

Map 3-9 EASTERN MEDITERRANEAN EUROPE.

important cities and trade routes, and the general topography of Mediterranean Europe, we can fairly well predict the distribution of population (Plate 6). In Italy, more than 20 million people are concentrated in and near the basin of the Po River, and elsewhere heavy concentrations also exist in coastal lowlands and riverine basins all the way from Genoa to Sicily. Thus the least heavily peopled part of the peninsula is the Appenine mountain chain (and the rim of Alpine areas in the far north), but the eastern and western flanks of the Appenines are very crowded. In Greece, today as in ancient times, heavily settled coastal lowlands are separated from one another by relatively empty and sometimes barren highlands. Especially dense concentrations occur in the lowland dominated by Athens and on the western side of the Peloponnesos. Both Spain and Portugal also have heavy settlement on coastal lowlands, though the Meseta is somewhat more hospitable than Greece's rocky uplands. The Mediterranean shorelands around Barcelona and Valencia have attracted a high population total; Barcelona stands at the center of Spain's leading industrial area (Catalonia). The Basque provinces from Bilbao to the Pyrenees are a major manufacturing area, especially for metal and machine industries. Spain's other population agglomerations have developed in the northwest, near the mining areas of the Cantabrian Mountains, in the south, where the broad lowland of the Rio Guadalquivir opens into the Atlantic, and where the Huelva–Cadiz–Cordoba triangle incorporates a sizable urban and rural population, and in the center of the country, in and near—especially southwest of—the capital, Madrid. In Portugal, which has Atlantic but no Mediterranean coasts, the majority of the population is nevertheless located on the coastal lowlands rather than on the Iberian plateau; Lisbon and Porto form centers for these coastal concentrations.

Thus Mediterranean population distribution is marked by a dominant peripheral location, by a heavy clustering of high concentrations and great densities in productive areas, usually coastal and riverine lowlands, and by a varying degree of isolation on the part of these clusters; contact between central and southern Greece, between the east and west coasts of Italy, and between Atlantic and Mediterranean Spain is not always effective. Terrain and distance are the obstacles. And although it is difficult to say exactly what constitutes overpopulation, there obviously is excessive population pressure on land and resources in many parts of Mediterranean Europe. Other than Ireland, no country in Europe has sent more of its people to overseas realms than has Portugal; standards of living here, and in Greece, and to a lesser extent in Spain and Italy, are quite low.

Rule and Exception. Perhaps the most stunning contrast between Scandinavian and Mediterranean Europe lies in the living standards of the people. While economic specialization and limited population growth, along with generally enlightened government policies and attempts to ensure a fairly equitable distribution of wealth, have produced standards of living in most of Scandinavia that are comparable and even superior to those prevailing in the European core, much of Mediterranean Europe seems hopelessly backward. Greece is perhaps the least favorably endowed of the four countries; less than a third of its area is presently capable of supporting some form of cultivation, and on this land the average density of population is around 800 persons per square mile—less than an acre per person. Thus many of Greece's farmers are engaged in sheer subsistence, their income is low, and their ability to buy improved farm equipment or fertilizer is minimal. Water supply is an ever-present problem, and the capital required to remedy it is very scarce. And yet agriculture is Greece's mainstay, for industrial opportunities are few. Thus Greek farmers, where conditions permit, turn to crops such as wheat and corn for the home market and tobacco, cotton, and typical Mediterranean produce such as olives, grapes, citrus fruits, and figs for exports. Greece ranks third among the world's leading exporters of olive oil (after Spain and Italy), and in some years nearly half its export revenues come from tobacco.

If this list of Greek agricultural products has an unusual ring to it, it should be remembered that the Peloponnesos is really Greece's Mediterranean zone, from which most Mediterranean crops are derived, while northward Greece takes on more continental characteristics—including more field and less garden agriculture. Eastward, the two areas meet in a point at Athens, whose urban area (including the port of Piraeus) counts some 2 million inhabitants.

Athens has an ancient history, but the present-day city began to emerge only about a century and a half ago, when Greece regained its independence (1830). After its days of greatness during Greek hegemony in the Mediterranean region, Athens went into decline, as did most of the other cities that had once been important centers of Greek power. When Greece became part of the Turkish Ottoman Empire, Athens was reduced to a shadow of its former stature, just a village around the hill on which the Acropolis stands. But then in 1830 it became the capital of sovereign Greece, and the headquarters once again of a state. Athens grew rapidly, and although in the decades that followed Greece was plunged time and again into ruinous wars and conflicts, the city continued to grow. It is the administrative, commercial, financial, cultural, and indeed the historic focus of Greece, and despite its limited industry it has grown to a size far beyond what would seem reasonable for such a relatively poor and agrarian country. With Piraeus it stands at the head of the Aegean Sea; Athens also has a large and busy airport. With its heritage of ancient structures still a direct reminder of past glory, Athens has become one of the Mediterranean's major tourist attractions, and thus a source of much-needed revenues. Another source of revenues is one resembling that developed by Norway: a large, worldwide merchant marine competing for cargoes wherever and whenever they need to be hauled.

At the other end of the Mediterranean, the Iberian Peninsula is less restrictive in the opportunities it presents for development. Iberia is much larger than Greece, and proportionately less of it is quite as barren and rocky as the Greek land. Also, raw materials are in far more plentiful supply. This is not to suggest that these opportunities have been put to maximum use. The rural areas are overpopulated; one price Spain paid for its slow industrial development was that its population "explosion" had to be accommodated largely in the rural areas, where pressure was already high. Land was divided and subdivided, farms grew smaller and smaller, and less and less efficient; poorer soils were turned over to farming even though they were marginal and their productivity was bound to be low. Most of northern Spain is fragmented into these tiny parcels; the situation is most serious in Galicia (the northwest corner, north of Portugal), where farms have been subdivided beyond the level of viability.

The dry and rocky countryside of Greece. (Fredrick Ayer III / Photo Researchers)

These, then, are some of the reasons why Southern Europe's per-acre yields are always shown to be so much lower than those of Western Europe—60 percent lower, on the average—and why so many farmers are caught up in a cycle of poverty. Another reason lies in the division of land, something that is less of a problem in Greece, where most of the land is already held in small, private holdings. But in both Spain and Portugal much land remains in the hands of absentee landowners whose huge estates are farmed by tenants. The system is reminiscent of another era in Europe—feudal, medieval days perhaps—but while land reform is badly needed, it has been successfully resisted by the conservative estate owners. In Portugal as much as 16 percent of the entire country is under such large estates, most of them in the hinterland of Lisbon, where the coastal lowland broadens into the country's largest agricultural area.

Geographers are often interested in matters relating to the attitudes of people, especially when such attitudes can somehow be isolated in research and can be proved to have a direct impact on the economic, cultural, or political relationships prevailing within or between countries. Professor R. E. Kasperson not long ago did such a study in the Dodecanese Islands, formerly

Fishing plays a large role in the Portuguese economy. Portuguese fishing boats are at anchor near Estoril. (Tom Hollyman/Photo Researchers)

Italian and today a part of the Greek state.[4] He found that the numerous changes in political control that have beset the islands have repeatedly altered the whole structure of Dodecanese political and economic geography. In Spain, too, attitude appears to have much to do with the economic situation both in agriculture and in industry, and even with regional development. Spain's major industrial area, we know, is located in Catalonia, and not on the coast of Biscay along the mineral-rich Cantabrian ranges. Thus Cantabrian coal would have to be shipped all the way around the Iberian Peninsula to provide power for Catalonian industries; that being the case, it might as well be imported from outside Spain. And indeed, much of it is imported. No, it is not the favorable location or the rich local resource base that has stimulated industrialization in Catalonia; it seems in the first instance to be the different attitude and outlook of the Catalans that has produced this development. Vigorous and progressive, the Catalans have forged ahead of the rest of Spain, aided by a strong regional identity in the form of a distinct language (Catalan differs from Castilian Spanish, the language used in most of Spain), a major urban-cultural focus (Barcelona, always a competitor of Madrid), and a certain local, Catalonian nationalism.

Catalonia does what many industrializing

though resource-poor areas do: it imports most of its raw materials and depends on a few local assets for success. Among the assets are the hydroelectric power available from the streams coming off the Pyrenees, the local labor force and its skills, and the local market, comparatively poor as it may be. But although it has had considerable success, it cannot be counted among Europe's leading industrialized areas. Unlike the Po Valley, Catalonia never stimulated effective trade contact across the mountains to the north. Compared to the Midlands of England, there is less diversification here although the last decade has seen considerable expansion: most of the industrial establishments still produce either textiles (mostly cotton goods) or chemicals. In large measure, of course, this reflects the very limited capacity of the Spanish market, with its millions of poor peasant families and its underpaid labor. But in going against the trend in reluctant Spain, the Catalonian achievement is a major one, and it reminds us how people, by their determination and skills, can transform the economic map.

Very slowly, Spain has recently been altering course—all the while falling farther behind the accelerating development of the European core. In the northwest, local iron and steel production is increasing, as is coal production. The hydroelectric power output has multiplied. But the general situation has changed little: this is a characteristically underdeveloped economy, exporting a number of untreated raw materials and

[4]R. E. Kasperson, *The Dodecanese: Diversity and Unity in Island Politics,* University of Chicago Press, Chicago, 1966.

importing a wide variety of foods and consumer goods. The exports reveal Spain's varied resources: in addition to iron ore and coal they include copper, zinc, lead, mercury (Spain is the world's leading producer of this element), and potash; the agricultural exports sound more familiar, including olive oil from Andalusia in the south, citrus fruits from the coastal zone around Valencia, wines from the Ebro Valley. At the center of it all, at the foot of the Guadarrama Range, stands Madrid, capital since the sixteenth century and still by far the dominant city of the whole Iberian plateau. Chosen because of its position of centrality on the Iberian Peninsula, Madrid (2.5 million) mirrors the problems of Spain: a facade of splendor hides large areas of severe urban blight, just as the tourist-admired beauty of Iberia conceals a great need for social reform.

Much of what has been said in this and the preceding chapter concerning Mediterranean habitats and economies applies to southern Italy and the islands of Sicily and Sardinia. But in a way Italy is not one country; it is two. While the north has had the opportunities and advantages (including those of proximity to the European core) to sustain development on the Western European model, the south has for centuries been stagnant and backward. While the north has developed a real urban complex counting several cities with over a half million people and many with over a hundred thousand, the south counts one major city—Naples (2 million), undoubtedly the poorest of Italy's large cities with staggering urban blight. Together, north and south count well over 50 million inhabitants (more than Spain, Portugal, and Greece combined), bound by Rome, situated, fortuitously from this point of view, in the transition zone between the two contrasting regions. In every way Italy is Mediterranean Europe's leading state: in the permanence of its contributions to Western culture, in the productivity of its agriculture and industries, in the percentage of its people engaged in manufacturing, in living standards. But the focus of Italy has

One of the source areas of Europe: Rome, capital of Italy, once the headquarters of a mighty empire. (Charles Rotkin)

A different view of Italy: the industrial north. This is Genoa, the port. (Charles Rotkin)

shifted from where it was during Roman times; Latium and Rome no longer form the peninsula's center of gravity. True, government and church still are headquartered in the historic capital and its adjunct, Vatican City, and in terms of population totals Milan, the northern industrial rival, has established no clear lead. And neither is Rome (2.5 million) ever likely to lose its special position in Italy and the world; it was chosen for psychological reasons to be the new Italy's capital, and no doubt would be chosen today if the choice had again to be made. But Italy's core area, certainly in economic terms, has moved into the area called Lombardy, centering on the valley of the Po River.

Northern Italy has a number of advantages. The Appenines, which form the backbone of peninsular Italy, bend westward, leaving the largest contiguous low-lying area in the Mediterranean between it and the Alps. This area, narrow in the west (where Alps and Appenines meet; Turin is located here), opens eastward to a wide and poorly drained coastal plain on the Adriatic Sea. Here lies one of the great centers of medieval Europe: Venice, Italy's third port, still carrying the imprint of the splendor brought by that early age. As the climatic map (Plate

2) shows, this area has almost wholly a non-Mediterranean regime, with more even rainfall distribution throughout the year. Certainly the Po Valley has great agricultural advantages, but what marks the region today is the greatest development of manufacturing in Mediterranean Europe. It is all a legacy of the early period of contact with Flanders and the development of trans-Alpine routes; when the stimulus of the industrial revolution came the old exchange was vigorously renewed. As we have seen, hydroelectric power from Alpine and Appenine slopes is the only local resource, other than a large, skilled labor force. But northern Italy imports large quantities of iron ore and coal, and today ranks as Europe's fourth largest steel producer, after France—though its production is less than half that of France. The iron and steel is put to a variety of uses. Competing for the first position among the industries are the metal industries and the textile industries, the latter enjoying the benefit of a much longer history, dating back, indeed, to the days of glory during the Middle Ages. The metal industries are led by the manufacture of automobiles, for which Turin is the chief center. Italy is famous for high quality automobiles; other metal products, such as typewriters, sewing machines, and bicycles are also produced. The Italian industry seeks to create precision equipment, in which a minimum of metal and a maximum of skill produce the desired revenues. Italy also has an impressive shipbuilding industry at Genoa on the Ligurian coast; with under 1 million people Genoa is Italy's leading port, located as it is on the Atlantic side of the peninsula.

The principal city in the region is Milan (3 million), leading industrial center in Mediterranean Europe. This is Italy's financial and manufacturing headquarters, and although it has seen unprecedented growth in recent years it has its roots in an earlier age of greatness. This period is still visible in the urban landscape with its impressive public buildings, palaces, and churches, but towering above these are the modern multistoried office buildings that house the offices of Italian industrial concerns. No city in Italy rivals the range of industries based here: from farm equipment to television sets, from fine silk (Milan competes with Lyon in this field) to medicines, from chinaware to shoes. The Milan–Turin–Genoa triangle is Italy's industrial heart, and it is the center of a larger, integrated region that forms part of the greater Western European core area.

The Weakest Link. These, then, are four of Europe's five regions. The four are interconnected in a multitude of ways, despite their individual identities. Scandinavian vacationers spend their holidays on Mediterranean beaches; Spanish fruits sell on Swedish markets. If we took a map and drew lines to represent all these interregional connections the map would soon turn black with clusters of lines—except in one direction, namely eastward. Certainly there are exchanges between Eastern Europe and the West, but the Iron Curtain still functions as a barrier. Between West and East Germany, between West Germany and Czechoslovakia, and between Austria and Hungary the international boundary takes on a far more divisive quality than is the case in, say, Western Europe. Here we are at an ideological boundary, a boundary that separates spheres of influence, a boundary that marks the limits of contrasting economic and political systems. Hence Eastern Europe today is not just another European region. Under Soviet sway, it has gone in directions altogether different from the postwar West.

ADDITIONAL READING

On population, one of the more useful volumes is *Population* by W. Petersen, published by Macmillan in New York in 1961. Still available and worthwhile is a paperback called *On Population,* edited by F. Osborn and published by The New American Library, New York, in 1960. Included in this little book is Malthus' own last statement, "A Summary View of the Principle of Population," and essays by J. Huxley and F. Osborn. Also consult the works of Kingsley Davis and Philip M. Hauser. Urban geography is discussed in H. M. Mayer and C. F. Kohn (editors), *Readings in Urban Geography,* published by the University of Chicago Press in 1959. R. E. Dickinson's *City and Region,* published in London by Routledge in 1964, contains a great deal of material on site and situation and related matters. The Law of the Primate City is best amplified by a reading of the article by M. Jefferson quoted in the text, and discussions of capital cities include V. Cornish, *The Great Capitals,* published by Methuen in London in 1923. Capital cities are also discussed in *Political Geography* by N. J. G. Pounds, published in New York by McGraw-Hill in 1963. Central place theory literature is scattered through the professional geographic journals; also see B. J. L. Berry, *Geography of Market Centers and Retail Distribution,* one of the volumes in the Foundations of Economic Geography Series published by Prentice-Hall, In Englewood Cliffs, New Jersey.

General works on the geography of Europe include *A Geography of Europe* by J. Gottmann, published by Holt, Rinehart and Winston in a fourth edition in 1969, and G. Hoffman's *Geography of Europe,* published in New York by Ronald Press in 1961 in a second edition. The Hoffman volume also includes discussions of Soviet Asia. For a view into the past, see W. G. East, *An Historical Geography of Europe,* published in London by Methuen in 1962. Regional works on the British Isles include G. H. Dury, *The British Isles: a Systematic and Regional Geography,* published in New York by Norton in 1964, and a series under the general editorship of W. G. East, published by Thomas Nelson and Sons in London, called *Regions of the British Isles.* On Nordic Europe see V. H. Malmström, *Norden: Crossroads of Destiny,* a Van Nostrand publication of 1965 in the Searchlight Book Series. On Western Europe and the core area, a volume entitled *Europe's Coal and Steel Community; an Experiment in Economic Union* by L. Lister was published in New York by the Twentieth Century Fund in 1960. Another interesting book is *A Regional Geography of Western Europe* by F. J. Monkhouse, published in London by Longmans in 1959. On the Mediterranean region, see D. S. Walker, *The Mediterranean Lands,* published by Wiley in 1962, and J. M. Houston's *Western Mediterranean World; an Introduction to its Regional Landscapes,* another Longmans publication put out in 1964.

There are numerous geographies of Europe's individual countries (or groups of countries, such as the Low Countries and the Alpine states). A good beginning source is a series of sketches of many of these countries published in *Focus,* which can be obtained from the American Geographical Society in New York. A number of individual European states have also been the subject of paperbacks in the Van Nostrand Searchlight Series, including *Divided Germany and Berlin* by N. J. G. Pounds (No. 1), *Spain in the World* by S. Bradford (No. 3), *The Common Market: the European Community in Action* by J. W. Nystrom and P. Malof (No. 5)—which deals with a group of countries, of course—and others will appear in the future. A geography of France by a French geographer is E. de Martonne, *Geographical Regions of France,* published in translation by Heinemann Educational Books, in London, and recently (1962) reprinted. On Germany, see R. E. Dickinson's *Germany: A General and Regional Geography,* published in London by Methuen in 1961.

4

EASTERN EUROPE—THE SHATTER ZONE

BETWEEN the might of the Soviet Union and the wealth of industrialized Western Europe lies a region of transition and fragmentation: Eastern Europe. Its position with reference to the major cultural influences in this part of the world at once explains a great deal: to the west lie Germanic and Latin cultures, politically represented by Germany and Italy, and to the east looms the culture realm that is Russia's. And to the south lie Greece and Turkey, whose impact also has been felt strongly in Eastern Europe; the Byzantine and Ottoman Empires took in sizable portions of this ever-unstable region.

The map of Europe shows a whole belt of countries, past and present, that lies between the Soviet power core and Europe's peninsular west: from Finland in the north through Estonia, Latvia, and Lithuania (now Soviet Socialist Republics) on the Baltic, all the way to Greece and its neighbors in the south. But the Baltic states have been absorbed by the U.S.S.R., and although Finland shares with Eastern Europe a history of Russian and Soviet pressure, its location, environment, cultural and economic development, and present political status tie it more closely to Scandinavian than to Eastern Europe. On similar grounds Greece may be viewed as a part of southern or Mediterranean Europe rather than Eastern Europe; it alone in the entire region between the Baltic and the Mediterranean has failed to adopt Soviet modes of resource utilization and political organization.

As here defined, then, Eastern Europe consists of seven states: Poland, the largest in every way, Czechoslovakia and Hungary, both landlocked, Romania and Bulgaria, which face the Black Sea, and Yugoslavia and Albania, political mavericks that have Adriatic coastlines. These countries form the easternmost and fifth regional unit of Europe (as outlined in Chapter 3, p. 61); their eastern boundaries also form the eastern limit of Europe itself, as was proposed in Chapter 1. This is Europe at its most continental, most agrarian, most static. In total it was not as well endowed as Western Europe or the Soviet Union with the essentials for industrialization, and it was the most remote of all parts of Europe from the sources of those innovations that brought about the industrial revolution. Since World War II Eastern Europe has looked eastward rather than westward for directions in its political and economic development. The Soviet Union gained control, and with the cooperation of local Communist parties, Communist forms of resource utilization and political organization were introduced. Although the countries of Eastern Europe were not remade into Soviet Socialist Republics (as Estonia, Latvia, and Lithuania were), they did become virtual satellites of the Soviet Union, and were drawn completely into the Soviet economic and military sphere.

But nothing has ever succeeded in unifying Eastern Europe, and it is doubtful even that Soviet power can do it. As early as 1948 one of the satellites, Yugoslavia, began to move away from the Soviet course; while an uprising was suppressed in Hungary in 1956 by Soviet military action, another rebellion, in Poland's industrial areas, brought modifications in the political and economic situation there. Meanwhile, Albania was choosing the Chinese side in the emerging Sino-Soviet ideological conflict, and in 1968 important political reforms took place in Czechoslovakia. The pendulum of power has swung across Eastern Europe many times, and it is likely to do so again.

Balkanization

The southern half of what has been defined here as Eastern Europe is sometimes referred to as the Balkans, or the Balkan Peninsula. This refers to the triangular landmass whose points are at the tip of the Greek Peloponnesos, the head of the Adriatic Sea, and the northwestern end of the Black Sea. The name comes from a mountain range in the southeasterly part of this triangle, but it has become more than merely a name for this area—it has become a concept, born of the reputation for division and fragmentation of this whole Eastern European region. Any good dictionary of the English language will carry a definition for the terms *Balkanize* and

[1] *Webster's Seventh New Collegiate Dictionary*, G. and C. Merriam Company, Springfield, Mass., 1963, p. 66.

Balkanization—for example, "to break up (as a region) into smaller and often hostile units."[1] Certainly division and hostility have been outstanding characteristics of Eastern Europe. Its peoples—Poles, Czechs, Slovaks, Magyars, Bulgars, Romanians, Slovenes, Croats, Serbs, and Albanians—have often fought each other; at other times they were in conflict with forces from outside the region. Various empires have tried to incorporate all or parts of Eastern Europe in their domains; none was completely successful, and the most recent effort, by the Soviet Union, is not likely to succeed either. Countless boundary shifts have taken place; time and again local minorities rose in revolt against their rulers. Culturally Eastern Europe is divided by strong regional contrasts, expressed in language and religion and sustained by intense nationalisms. Among these, none has emerged with sufficient strength to impose unification on the region; the nearest thing to an "indigenous" empire was that of the Austro-Hungarians.

Neither has the physical geography helped reduce Eastern Europe's divisiveness. An eastern arm of the Alpine mountain system, the Carpathians, cuts across eastern Czechoslovakia and north-central Romania, separating Poland on the North European Lowland from the countries to the south (Map 1-2). In the middle of the region lies the Hungarian Basin, drained by the middle course of the Danube River and surrounded by the Carpathians and Transylvanian Alps to the east and the Austrian Alps and the mountains of Yugoslavia to the west and south. The whole southern part of the region is quite mountainous. All this had the effect both of impeding communications and favoring the development of strong cultural regionalisms. There are even significant climatic contrasts: the countries of the Balkan Peninsula enjoy the warmth associated with Mediterranean conditions and receive somewhat more precipitation than the north, where growing seasons are shorter and agricultural possibilities are generally more limited.

Balkanized Eastern Europe, then, is a region of endless contrast and division, with Slavic and non-Slavic peoples, Roman Catholic, Eastern Orthodox, and even Moslem religions, and mutually nonintelligible languages with Slavic, Romanic, and Asian roots. History has seen peoples with common ties separated by political boundaries, and peoples with few similarities thrown together in a single state. Outside powers have capitalized on Eastern Europe's internal divisions by exploiting them. Empires of the past have left traces of their hegemony in the language of Romania (once the Roman province of Dacia, the Romanian language is a derivation of Latin) and in the religion of Albania (which is dominantly Moslem as a result of the Turkish Ottoman period), to identify but two instances; today the Soviet Empire is having an impact on the political and economic organization of Eastern Europe's states. In a world in which boundary disputes and struggles for territory have long been common-place, the case of Eastern Europe has been sufficiently unusual to have given its name to this phenomenon in political geography: Balkanization.

Shatter Belt and Buffer Zone

With such strong internal contrasts, how can Eastern Europe be called a region? The answer lies in this very characteristic—its function as a recipient of influences from west as well as east, its absorption of the elements of so many cultures, its chronic political instability, and its position as a zone of transition between highly developed and strong Western Europe and the powerful Slavic east. Eastern Europe is often referred to as a *shatter belt,* a term descriptive of the impact of the many internal and external pressures that have kept the region in nearly constant political change, sometimes crushed by the great power of adjacent empires, at other times free to quarrel internally and thus intensify the divisions already there.

Thus, Eastern Europe is truly a shattered area—shattered by divisive and opposing forces from without and within. But it is not the only area of this kind in the world. Shatter belts perform functions: they absorb the impact of expanding neighbors, buffer the blows of territorial aggrandizement (often by giving up territory themselves), and most of the time keep the major adversaries separated. In Eastern Europe, these adversaries have principally been the Germanic peoples to the west and the Russians to the east. While the Russian Empire grew, some of Eastern Europe's countries lost parts of their areas to the tsars; Poland disappeared altogether as a state for more than a century. At the same time, the growing power of the evolving German state led to the creation of "mark" or "outpost" states against the Slavs. Prussia and Austria were such

Eastern Europe has cold winters: its continental character stands in contrast to more maritime Western Europe. This snowy scene was photographed in Rumania by Jerry Cooke.

outposts at one time. During the nineteenth century the Austrians performed the miracle of welding an empire of sorts through an alliance with the Hungarians (non-Slavs themselves, of course). The Austro-Hungarian Empire incorporated parts of present-day Poland, Czechoslovakia, Romania, and Yugoslavia.

By World War I, Eastern Europe had just about been taken over by its designing neighbors, the Russians, Turks, Germans, and Austro-Hungarians; only the last among these could be viewed as anything like a "domestic" political entity. But the war saw three of these imperialist powers—the Germans, Austro-Hungarians, and Turks—allied and on the losing side. In 1919 Eastern Europe was reconstituted at the Paris Peace Conference, and the map that emerged from that meeting is the one that has become familiar to most of us—with the inevitable modifications of a half century, of course. The recreation of Eastern Europe at the same time resurrected its position as a buffer between East and West. Poland reemerged, interposed between the Soviets and the Germans. Czechoslovakia was newly created out of the Austro-Hungarian Empire; it combined the Czechs and Slovaks in one state. The Poles, Czechs, and Slovaks are Slavic peoples, but they are distinguished together as the West Slavs (as opposed to the Russians and Ukrainians, who are the East Slavs). The state of Hungary was separated from Austria, and to the south, the South Slavs—the

WINKEL'S PROJECTION

Map 4–1 WORLD BUFFER ZONES AND SHATTER BELTS.

Serbs, who had thrown off the Turkish yoke and had established a Kingdom of Serbia, the Croats, and the Slovenes—were united in a complex single state that was later to be called *Yugo*slavia, land of the South Slavs. Romania, Bulgaria, and Albania, which had all been under Turkish domination into the nineteenth century, had their boundaries somewhat modified and were each recognized as independent states by the Treaty of Versailles.

The interwar period witnessed a renewal of German expansionism. Poland, which was cut off from the sea by the German area of East Prussia, obtained from the 1919 Conference an outlet to the Baltic Sea, the now-famous Polish Corridor. This corridor had the effect of fragmenting and dividing German territory, and it was the object of German claims. Czechoslovakia, too, faced German pressure. As the map shows, the western boundary of this country runs along the Erzgebirge (Ore Mountains) and the Bohemian Forest, and forms a kind of V-shaped point into Germany's midsection. This was cause for German claims against the young country, and in 1938 boundary revisions were forced upon it. Shortly thereafter, we know, it was overrun by German forces. Austria, too faced the German might; in 1938 it was forced into a union with Nazi Germany. With World War II, the buffer function of Eastern Europe fell apart. German forces cut across the region and entered the Soviet Union, but the campaign was an unsuccessful one and the Red Army pushed the Germans back, across the region, and into East Germany itself. Eastern Europe's countries, briefly under the domination of (or allied with) Germany, now fell prize to the great power on the other flank, the Soviet Union. And this brought the latest phase in the complex historical geography of this region—that of the imposition of Soviet forms of economic and political organization, complete with Five-Year Plans, attempts at collectivization of agriculture, emphasis on heavy industry, and totalitarian rule.

What has happened in Eastern Europe, again, is not unusual when it comes to the fate of areas that form buffers between opposing powers. Before all countries of the world were delimited by well-defined boundaries, there were areas that remained sources of dispute, never effectively appropriated by any of the interested units. Sometimes these areas were not worth much, and consisted of desert or swamp or dense forest. But

The urban landscape of Skoplje in Yugoslavia reflects the area's varied history. While modern buildings rise, minarets and mosques still mark the scene; the one in the left foreground dates from the fifteenth century. (Ralph Gray/PIX)

at other times both sides might have coveted such an area without having the necessary power to occupy it and then maintain effective control there. For these areas, lying beyond the integrated part of the state, the term *frontier* came into use; thus the American frontier was competed for by the American white man and the American Indian, and it disappeared over time. The South African frontier lay between the African Zulu Empire and its sphere of influence and that of the Hollanders at Cape Town; for nearly two centuries these two power cores in Southern Africa grew without direct confrontation, separated as they were by this area. In Australia the frontier that lay inland from the coastal settlements was slowly absorbed. In South America, a frontier developed between Portuguese interest in Brazil and Spanish interests in Argentina. In Asia, British and Russian interests confronted each other across a mountain frontier (Map 4-1).

Today, such frontiers no longer exist. All the land of the world has been divided among the states, and while there are still disputes, such disputes revolve around demands for a relocation of boundaries and are boundary disputes, not frontier penetrations. In South America, Uruguay and Paraguay emerged as small political entities between the powers of Brazil and Argentina. In Asia, a tier of mountain states from Afghanistan through Nepal to Sikkim and Bhutan emerged, among which Afghanistan was

The **modernization** of Czechoslovakia is mirrored by the towers and stacks of the Skoda automobile factory as they rise over Pilzen. (Elliott Erwitt/Magnum)

deliberately recognized by both the Russians and the British as a divider between their respective spheres of interest in Asia. In Southeast Asia it was recently possible to witness a similar sequence of events as the country of Laos was provided with a neutral status to serve as a cushion between Communist and non-Communist spheres. Thus the function of separation is still performed by certain areas, but these areas are now parts of national states. Such states are called *buffer* states, and groups of them are referred to as buffer zones. Thus, the buffer zone is, in effect, a higher-order frontier. While the frontier was an ill-defined area lying open to penetration or positioned between two power cores of regional stature, the buffer state lies between powers of world significance (Map 4-1).

So it is with Eastern Europe. The boundary system that was devised in 1919 did not terminate this region's function as a buffer; it did reorganize it and to some extent strengthened the political units in it. But nothing could counteract Eastern Europe's in-between position. Much of Eastern Europe remained backward and weak; as late as the 1930's some two-thirds of the people still depended directly on agriculture for a living, and much of that agriculture was of a subsistence variety. The industrial revo-

lution made less impact here than in almost any other part of Europe. Few large cities existed, communications were poor, many people lived in isolation, not really in touch even with the core area of their own country. All this, of course, spells weakness—a fatal weakness that has afflicted this region for many generations. Only Czechoslovakia during the interwar period made substantial progress in the economic sphere—but then, the western part of that country, once the Kingdom of Bohemia, had long traditions of contact with the west, and the capital, Prague, was one of Eastern Europe's few major cities. But during the 1930's the German eastward push was being mounted, and it is not surprising that Eastern Europe's Slavic people looked eastward for support against the Nazi tide. As has happened in other buffer zones, they ultimately got more of it than they bargained for.

Irredentism

The boundary framework with which Eastern Europe was endowed in 1919 did not eliminate the internal territorial problems of the region. In fact, it was really impossible to arrive at any set of boundaries that would totally satisfy all the peoples involved; so intricate is the ethnic patchwork that different people simply had to

Map 4–2 BOUNDARY CHANGES IN EASTERN EUROPE.

MODIFIED AFTER : R.MURPHEY, AN INTRODUCTION TO GEOGRAPHY

be joined together. But people who had affinities with each other were also separated by the new international borders, and every country in Eastern Europe as constituted in 1919 found itself with minorities to govern. Frequently these minorities were located near the boundaries, and adjacent states began to call for a transfer of their authority, on ethnic, historical, or some other grounds. For example, Transylvania, a part of the Hungarian Basin, was severed from Hungary and attached to Romania—but Hungary openly laid claim to it. Macedonia had been divided between Yugoslavia and Greece; Bulgaria and Albania wanted parts of it. Between Yugoslavia and Italy the area known as the Julian March became the scene of territorial competition. These are just a few instances, and there were several more—not to mention German claims on Poland and Czechoslovakia and Soviet claims on Romania.

This situation, where a certain state, through appeals to a regionally concentrated minority in an adjacent state, seeks to effect the incorporation of the people and territory involved through a boundary shift, is referred to as *irredentism*. The term comes from the Italian, *Terra Irredenta* or Unredeemed Italy, relating to an Italian-speaking area claimed by the Italians in this manner. Irredentism is a worldwide phenomenon. War settlements, for example, have led to irredentism toward the "lost" territories: to mention a European case, Italy in 1919 received a large part of the Austrian province of Tyrol under the Treaty of St. Germain. It was a war prize, but the territory involved, according to the latest census, still was overwhelmingly German-speaking: over 200,000 people spoke German and less than 20,000, Italian. Austrian irredentism toward this part of Tyrol continues, and sporadic incidents of violence occur in the area. In the dispute between the Greek majority and the Turkish minority on the island of Cyprus, both Greece and Turkey have carried on irredentist policies, expressed by the call for *enosis*—union—by Greece. In Africa, the most famous case is that of the Somali Republic, which has long desired incorporation of eastern Ethiopia, a dominantly Somali-occupied area. When General De Gaulle spoke of a free Quebec in his 1967 Canadian visit, he was appealing to the kinds of emotions that usually sustain irredentist drives. Irish Catholics in Eire support Catholics in Ulster.

Eastern Europe was beset by situations of this kind, and only after the end of World War II were some of them eliminated by further boundary revision (Map 4-2). Poland was literally moved westward, the Soviet Union taking almost half of that country in the east, but Poland gaining a very large area from Germany in the west. The U.S.S.R. also took eastern portions of Czechoslovakia and Romania, and Bulgaria and Romania settled their dispute over the southern Dobruja, which was returned to Bulgaria. The problem of the Julian March was also settled, though not immediately: through United Nations' mediation the port of Trieste came under Italian administration, while most of the disputed territory involved went to Yugoslavia.

Another major factor in the simplification of the ethnic situation in Eastern Europe has been the dominance of the Soviet Union, not only over this region, but over East Germany as well. East Germany in a sense has become an Eastern European state, certainly as far as economic and political reorganization are concerned; but the new order has also involved a migration, voluntary as well as enforced, of Germans from several East European countries—where they formed significant minorities—to this part of their homeland. Thus the German minorities of Poland and Czechoslovakia have been drastically reduced (in the former, practically eliminated), and long-troublesome East Prussia no longer exists. Elsewhere, migrations have carried Hungarians from Czechoslovakia and Romania to Hungary, Ukrainians and Belorussians from Poland to the Soviet Union, and Bulgars from Romania to Bulgaria.

This is not to suggest that Eastern Europe's minority problems are solved. Many Hungarians still live in Romania, and Hungary has not forgotten its interests in Transylvania. Yugoslavia still has a large Albanian minority in the south, a Hungarian minority in the north, and some Romanians in the east. When the umbrella of Soviet overlordship is removed, Eastern Europe may still not be free of its irredentist problems.

The North: Poland and Czechoslovakia

The two northernmost countries of Eastern Europe, Poland on the Baltic Sea and landlocked Czechoslovakia, have exceptional qualities. Poland, in terms of territory and population, is the region's largest state. With over 120,000 square miles, most of them in the North European Lowland, and with well over 30 million people, it ranks sixth in Europe in both categories. Czechoslovakia, mountainous and surrounded, is less than half as large as Poland and has fewer than half as many people. Poland has lagged behind Czechoslovakia in its economic development. It is a reflection of conditions in Eastern Europe in general that the industrial revolution took so long to make an impact here; Poland has opportunities for industrialization that were not fully exploited until the innovation of national planning, introduced after World War II. In the 1930's some two-thirds of Poland's people still depended directly on agriculture for a living, and even today the country still retains a strong rural flavor. Czechoslovakia, on the other hand, has been more directly exposed to influences from Western Europe. It lay more directly in the

paths of commercial exchange and industrial development along Europe's east-west axis, and the Czechs had long-standing ties with the west. One of these ties is the Elbe River, which originates in the basin that is Bohemia, cuts through the Erzgebirge (Ore Mountains), and flows to the North Sea port of Hamburg. Before the Iron Curtain was lowered, this waterway was a major factor in Czechoslovakia's westward orientation.

Poland looks to the Baltic, where it presently has a lengthy coastline. No one knows what the future will bring, but at least Poland has the difficult days of the Corridor behind it, and has its own and undisputed outlets to the sea. Its major river is the Vistula, navigable to a point very near the capital of Warsaw. The other important river system marks the boundary with Germany, namely the Oder and Neisse Rivers (the boundary is called the Oder-Neisse Line). Today Poland reaches to within 40 miles of

This photograph of the industrial complex of southern Poland. (Elliott Erwitt/Magnum)

Berlin, and includes such formerly German provinces as Pomerania and part of Brandenburg. On the other side of the country, Russian territory comes within 90 miles of Warsaw (Map 4-3).

From the map it would seem that Poland has many assets. The country is compact; that is, it does not have extensions or proruptions, and encloses the largest possible area within the shortest possible boundary. From the point of view of national integration and the establishment of effective communications, this should be an advantage. Neither are there significant mountain obstacles to transportation; most of Poland is of low relief. In the south, along the Czechoslovak boundary, Poland shares the foothills of the Sudetes and Carpathians—foothills which we know to be productive of the raw materials necessary for industry. The capital, Warsaw (1.5 million), lies near the center of the country, in the middle of productive agricultural areas, and at the focus of a radiating network of transport lines that reach all parts of the state. Warsaw is Poland's primate city; it is the historic,

cultural, and political center of Polish life, and it has considerable industry, most of it geared to the domestic consumer market.

But there are things the map of modern Poland does not show us. Much of the industrial development of the state after World War II is based upon resources that fell to Poland only after the new boundaries were delimited; Silesia was first developed by the Germans and was taken over recently, to be integrated in the Polish economic framework. Now a major industrial complex is developing in southern and southwestern Poland, and the Krakow–Czestochowa–Wroclaw triangle is emerging as Poland's industrial core area, based upon excellent coal resources and lesser local iron ores, which are supplemented by imports from the Ukraine. The city of Nowa Huta was created, Magnitogorsk-style, as part of a Five-Year Plan; it is built around an iron and steel plant and gained over 100,000 residents in just ten years. Certainly centralized planning, plus the newly gained resources of Silesia, have added up to accelerated industrialization and urbanization. About half

Map 4–3 POLAND AND CZECHOSLOVAKIA.

the Polish population is now classed as urbanized—still a long way from Western European figures, but a major change in Poland, nevertheless. The Poles themselves would like to see greater industrial activity and urbanization in the central sections of the country as well, and in their planning have sought to stimulate the development of a Central Industrial Region, which includes Warsaw and the textile center of Lodz (with a population of 1 million, Poland's second city), the "Polish Manchester." But in almost every way, Poland's best opportunities lie in the southern half of the country, including good farmland. Southern Poland has a black soil belt—it broadens eastward into the Ukraine—which sustains intensive farming, with wheat as the major crop. To the north the poorer glacial soils support rye and potato cultivation, and farther north still, pastureland and moors predominate.

Poland has not accepted Soviet economic innovations and political dominance without reservations, which have at times been expressed in violence. It cannot, however, be doubted that Soviet "cooperation" has boosted Poland's industrial and urban progress. On the other hand, a measure of the failure of the new order can be gained from the attempts that were made to collectivize agriculture. In Poland, as in some other countries of Eastern Europe, this program had to be slowed down or temporarily shelved because of peasant resistance, and today less than a quarter of Poland's farmland is under collective management. The Communist pattern of great investment and support for industry, even at the cost of agricultural progress, is evident in Eastern Europe as well.

Czechoslovakia shares with Poland the industrial region of which the Krakow–Czestochowa–Wroclaw triangle is a part. The source of the raw materials lies astride the gap between the Ore Mountains and the Carpathians, and in Czechoslovakia the area is referred to as Moravia. Lying midway between Bohemia in the west and Slovakia in the east, this growing manufacturing area is of increasing importance to the country, with its coal supply and its heavy industry Ostrava (300,000) lies at the focus of a manufacturing complex that includes metallurgical and chemical industries.

Unlike Poland, Czechoslovakia is an elongated country, bounding Bavaria in the west and the Ukraine in the east. Once again this shape characteristic is quite significant. The western section of the country, mountain-enclosed Bohemia, has always been an important core area in Eastern Europe, most cosmopolitan in character, and Western in its exposure and development. With its Elbe River outlet the westward orientation of this part of the country was maintained for centuries. But in 1919 the Slovaks were attached to Bohemia, and the Slovak eastern part of Czechoslovakia, mountainous and rugged, lies in the drainage basin of Eastern Europe's greatest river, the Danube. And the Danube flows not westward, but eastward to the Black Sea. Slovakia is much more representative of Eastern Europe than is the Czech part of the state. It is mostly rural and neither as industrialized nor as urbanized as the Bohemian-Moravian west. In the days of the Austro-Hungarian Empire, this was a peripheral frontierland of comparatively little importance, while Bohemia was significant as a manufacturing area even then; during the early decades of the Czechoslovakian state, the region and its inhabitants took second place to the more advanced west. This situation has continued lately: a steady stream of manpower leaves Slovakia every year, in search of work in the factories of Bohemia-Moravia. Labor shortages occur on the farms at every harvest time.

Czechoslovakia's center of gravity, then, lies in the west, in Bohemia-Moravia. Unlike Warsaw, Prague is not a centrally positioned capital; it is located in the west, and it is very clearly a Bohemian city. It is more than four times as large as the country's next urban center (the Moravian headquarters of Brno with 350,000 people). Slovakia's capital, Bratislava on the Danube, is somewhat smaller still. In every way Prague is Czechoslovakia's primate city. It is the political and cultural focus of the country, and constitutes its major industrial region. Founded at a place where the Vltava River can be easily crossed, the city lies near the Elbe River, and in the middle of the country's greatest concentration of wealth. The mountains that surround Bohemia contain a variety of ores, and in many of the valleys stand small manufacturing towns that specialize, Swiss-style, in certain kinds of manufacture—for example, pencils are produced from local graphite at Budejovice, glass and crystal at Teplice-Sanoy and Yablonec, and so on. In Eastern Europe the Czechs have always led in the fields of technology and engineering skills, and the beginnings of the metal industries

Prague is the political, economic, and cultural headquarters of Czechoslovakia—the country's primate city. Here the National Theater on the Vtlava River stands among other ornate structures, all evincing the city's long and proud history. (George Novotny/Photo Researchers)

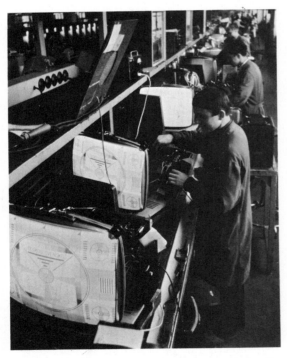

Assembly-line production is the hallmark of the Soviet-inspired push for industrialization in Eastern Europe. Here, Czech workers manufacture television sets. (Jerry Cooke)

were based on local coalfields and iron ores. When these became exhausted, the industries nevertheless kept growing as raw materials were imported and metal-finishing industries came to the fore, especially around Prague itself. Pilsen also grew into a major industrial center, with the famed Skoda steel works as the leading establishment, and the breweries perhaps as the most famous. Czech automobiles, shoes, tools, textiles, and a host of other goods find their way to capitalist as well as Communist markets.

The Bohemian basin is also well endowed with good farmland. The Elbe traverses a thriving agricultural area where barley, wheat, and oats are the grain crops and sugar beets are inserted in the rotation pattern. Warm-climate crops, such as grapes and tobacco, begin to make their appearance eastward, in the plains around Brno and in southern Slovakia.

The predominance of Bohemia in the economic life of the country may be gradually reduced as the heavy industries of Silesia-Moravia develop. The Czechs have given an indication of their desire to bring the Slovakian part of the country into the industrial sphere as well by incorporating, in a recent Five-Year Plan, a scheme for iron and steel production at Kosice—with coal from the Ostrava area and iron ore from the Krivoi Rog.

The South: The Balkan States

The five countries that make up the Balkan Peninsula (excluding Greece) seem almost to have been laid out at random, with little if any regard for the potential unifying features of this part of Eastern Europe. This apparent randomness is the result of centuries of territorial give and take, of migrations and invasions, of consolidation and shattering of states and empires. The comparative orderliness of Poland and Czechoslovakia, one a uniform plainland state and the other a rather well-defined mountain state, is lost here in the Balkans, where mountain ranges and mountain masses abound, large and small basins are sharply differentiated and separated, and where the ethnic situation is even more confused than elsewhere in Eastern Europe. At least the Slovaks had this in common with the Czechs: they were neither Poles nor Hungarians, and their attachment to the Czech state seemed a reasonable solution. But in the Balkans such solutions have been hard to come by.

The great unifier that might have been in the Balkans is the Danube River, which comes from southern Germany, traverses northern Austria, and then crosses Eastern Europe forming first the Czech–Hungarian boundary, then the Yugoslavia–Romania boundary, and finally the Romania–Bulgaria border. It is indeed anomalous that a great transport route such as this, which could form the focus for a large region, is instead a dividing line. After it emerges from the Austrian Alps, the Danube crosses the Hungarian Basin, which, although largely occupied by Hungary, is shared also by Yugoslavia, Romania, and Czechoslovakia. Then it flows through the Iron Gate (near Orsova) and into the basin that forms its lower course; this basin is shared by Romania and Bulgaria. No other river in the world touches so many countries, but the Danube has not been a regional bond. Only two Eastern European capitals—Budapest (Hungary) and Belgrade (Yugoslavia) lie directly on the river. Only Hungary is truly a Danubian state, as the river turns southward just north of Budapest and crosses the entire country, along with its tributary, the Tisza (Map 4-4).

In a very general way it can be argued that progress and development in Eastern Europe decline from west to east and also from north to south. In both Poland and Czechoslovakia the western parts of the country are the most productive. Hungary has the largest and most productive share of the Hungarian Basin; it is better off than Yugoslavia immediately to the south. In Yugoslavia itself, the core of the country lies in the Danubian lowland, and the southern mountainous areas lag by comparison. As a whole, however, Yugoslavia is far ahead of its small southern neighbor, Albania. On the other side of the peninsula, Romania has the advantage over Bulgaria to the south. Thus Hungary is the Balkans' leading state in several respects which, with its pivotal position in Eastern Europe and its role in the Austro-Hungarian Empire, is not surprising.

The Hungarians (or Magyars) themselves form a minority in the Balkans, since they are neither of Slavic nor of Germanic stock. They are a people of Asian origins, distantly related to the Finns, who arrived here (see Chapter 2) in the ninth century A.D. Ever since, they have held on to their fertile lowland, retaining their cultural identity (including their distinctive language), though at times losing their political sovereignty. The capital, Budapest, was a Turkish stronghold for more than a century during the heyday of the Ottoman Empire. Its recent growth (to over 2 million) was achieved during the period of the Austro-Hungarian Empire and the creation of the all-Magyar state of Hungary after World War I. Today Budapest is about 10 times as large as the next ranking Hungarian city, a reflection of the rural character of the country. With its Danube port and its extensive

The Iron Gate of the Danube. (Ron Nelson)

industrial development, its cultural distinctiveness and its nodal location within the state, Budapest epitomizes the general situation in Eastern Europe, where urbanization has been slow and where the capital city is normally the only urban center of any size.

Hungary's rural economy has not been without its problems. When the country was delimited in 1919, there was a great need for agrarian reform. Large estates were carved into small holdings, and productivity rose. Then World War II and its attendant destruction, especially of livestock, set the rural economy back; after the war, the Soviets saw fertile Hungary as a potential breadbasket for Eastern Europe. When farm production failed to rise, collectivization was encouraged, but the peasants who had become small landholders resisted the effort. But

Hungary had been on the German side during World War II, and the Soviets were conquerers here, not liberators; hence they pushed their reform program vigorously (as they did in East Germany). Today nearly three-quarters of Hungary's land is in collective operation, but farm yields are not expanding nearly as fast as the country's planners would like. Certainly the country is self-sufficient, with its harvests of wheat, corn, barley, oats, and rye, but there is less surplus for export than desired.

Industrially, also, Hungary has not yet been able to take full advantage of its potential. There is coal, notably near Pecs, not far from the Danube in the southern part of the country, and iron ore can be brought in via the Danube for iron and steel production. This has been done in quantity only recently; for a long time Hungary,

Map 4–4 THE BALKANS.

like so many countries in the developing world, has been exporting millions of tons of raw materials to other producing areas. This is especially true of the one mineral Hungary has in major quantities, bauxite, mined near Gant but refined for manufacture in Czechoslovakia and the Soviet Union rather than in Hungary itself.

As we have noted, the Hungarians did not gain control over all the Hungarian Basin, since they share this physiographic region with several of their neighbors. Budapest has a position much more central to the Basin as a whole than to Hungary itself; in fact the city lies a mere 20 miles from Hungary's northern boundary.

To the south, Yugoslavia holds a large section of this Danubian lowland. And this area has long been the heart of that state—as long, that is, as there has been a unified country within these boundaries. Its productivity increased with the agrarian reforms begun in the 1920's, and some other ethnic groups were pulled together under the royal house of Serbia. Real reform came after 1945, when the socialist regime of Marshal Tito, the country's wartime leader against the Germans, took over. Seeking a balanced approach to its economic problems, Yugoslavia went slow in its collectivization efforts (less than one-sixth of the farmland is under collectivized production today), while industry was guided by national planning. Before the war, Yugoslavia had been an exporter of some agricultural products, some livestock, and a few ores. After the war, with effective political control and economic centralization, its government sought to change this—not at the expense of agriculture, as was the case in so many Communist-influenced countries, but in addition to it.

One problem that faced Yugoslavia resembled one in Czechoslovakia: people kept moving from the mountainous, less-developed areas to the rich farmland of the Danube lowland, and to the industries of Belgrade (750,000), the capital, and Zagreb (½ million), the second city. It emphasized how mountainous, southern Yugoslavia was the country's backwater, with little economic progress or visible change of any kind. The postwar regime, through the method of planning began to bring this large area of the country into the national economic framework. This is not easy: pastoralism is the chief occupation there, and although a variety of minerals exists, these are scattered throughout the difficult terrain. For many years Yugoslavia has exported such ores

A **view** over another of Eastern Europe's primate cities: Budapest, capital of Hungary. (Jerry Cooke)

as copper, lead, zinc, and chrome, especially from deposits located conveniently near the Adriatic coast; one problem in the development of local industries lies in the provision of adequate power supplies. There are opportunities to remedy this: the north has been producing some petroleum, there is coal in the southeast, and in the mountainous topography many sites for hydroelectric dams exist. Already some of the mountain towns, like Sarajevo, Skopje, and Niksic, boast new and vigorous industries including metallurgy, machinery, and leather production.

If the generalization concerning declining development and southerly location in the Balkans

This photograph illustrates the rugged karst topography of Yugoslavia; the town is Dubrovnik. This is what a good part of Yugoslavia's coastline looks like. (Fritz Henle/ Photo Researchers)

A rural scene in Albania. This, according to the photographer, is typical country in Albania, near the town of Shkoder-Scutari. The beautiful mosque in the lower right is one of the best in the whole of Eastern Europe; it now serves as a cattle stall. (Gunter R. Reitz/PIX)

is to hold true, then Albania, Yugoslavia's Adriatic neighbor, should be less developed even than mountain and plateau Yugoslavia. And so it is. With less than 2 million people and just 11,000 square miles, Albania ranks last in Europe (excluding Iceland) in both territory and population. Its percentage of urbanized population (about 33%) is also Europe's lowest. Most Albanians eke a subsistence out of livestock herding and the one-seventh or so of this mountainous country that can be cultivated at all; the largest town is the capital, Tirane, which has about 100,000 inhabitants and a few factories.

Perhaps because of its abject poverty and the limited opportunities for progress (consisting of some petroleum exports, tobacco cultivation, and chrome ore extraction), Albania has turned to China for ideological as well as material support. Unlike Yugoslavia, which moved away from the Soviet orbit in the direction of moderation, Albania committed itself to the Chinese version, thus choosing the opposite way. Certainly Albania is worth more to China in this context than it is to the Soviet Union, which already dominates much of Eastern Europe; Albania clearly preferred a prominent place in China's priorities than a lowly one in the Soviet Union's. This country has little to bargain with except a somewhat strategic position in Eastern Europe, on the Mediterranean, and at the entry to the Adriatic Sea. As an ally its actual and potential value to China is considerable, and the current

association constitutes for Albania a maximization of its limited advantages.

The Balkans' two Black Sea states, Romania and Bulgaria, also confirm the southward lag of progress and development. Romania is both richer and larger—twice as large as its southern neighbor in terms of territory as well as population. Both countries' boundaries epitomize those of the Balkans in general: areas of considerable homogeneity are divided (for example, the Danube lowland, which these two countries share); areas that would seem to fit better with other countries are incorporated (such as the Transylvanian part of the Hungarian Basin, now part of Romania); national aspirations are denied (such as Bulgaria's longtime desire for a window on the Aegean Sea). But on the whole, Romania, despite some major territorial losses to the Soviet Union as a result of World War II, remains quite well endowed with raw materials.

The potential advantages of Romania's compact shape are to some extent negated by the giant arch of the Carpathian and Transylvanian Mountains. To the south and east of this arch lie the Danubian plains and the hills-and-valleys of the Siretul River (Moldavia) respectively, and to the west lies the Romanian share of the Hungarian Basin—with a sizable Hungarian population forming one of Eastern Europe's ubiquitous minorities here. As we have noted on previous occasions, a mountain–plain association often provides mineral wealth, and Romania is no exception. In the foothills of the Carpathians, near the westward turn of the Transylvanian Alps, lies Europe's major oil field. From Ploesti, the urban center of this area, pipelines lead to the Black Sea port of Constanta, to Odessa in the Soviet Union, to Bucharest and to Giurgiu on the Danube. While the Soviet Union is of course Romania's chief customer, both oil and natural gas are sold in Eastern Europe as well; the gas is piped not only to Bucharest, but to Budapest also. Pipelines for oil are being laid into several Eastern European countries.

Romania has a great deal more than this; here is one Eastern European country where energy supply is not a problem. In fact, in the nineteenth century, when Romania first appeared on the map of Europe as a minor kingdom located between the Carpathians and the Black Sea (a kingdom that had succeeded in wresting autonomy from the Turks), there already was some industry here. Oil and gas have been won from the Ploesti fields since the 1850's; an early iron industry existed at the western end of the Transylvanian Alps as early as the late eighteenth century, later to be modernized and centered on the town of Resita. When Romania arose as a greatly enlarged state after World War I, there seemed to be great hope for the future. Its industrial resources included not only good coal and large iron reserves, but copper, lead, manganese, and bauxite as well. Its farmlands were extensive and fertile and awaited only the necessary reforms to bring the promised yields. With a population of nearly 20 million, labor would be no problem.

But Romania lagged. Agrarian reforms came slowly, foreign capital was needed to exploit the oil fields. National consolidation hardly progressed; the ruling class in Bucharest and the landowners in the surrounding areas looked toward Paris and Western Europe for cultural sustenance, and the capital remained an island of style and wealth in a sea of stagnation. At an earlier stage in history the kings of Bohemia, too, had gone to France and brought French culture back with them; but Prague nevertheless became a functional center for the Czech people. Few improvements in communications between the core area and the mountainous interior were made, and urbanization was slow. The state was simply not well organized, and suffered in consequence. Health standards, literacy rates, and annual incomes increased little. And then World War II destroyed much of what Romania had built up, especially in the oil fields, which fell to the Germans and whose facilities were heavily damaged by bombing attacks.

After World War II, the Romanian state found itself with some boundary revisions costing it over 20,000 square miles, and with three million fewer people. It also found itself under Soviet domination, and thus began the familiar sequence of national planning, with emphasis on industrialization, further agrarian reform involving collectivization, and accelerated urbanization. But again, progress has not been spectacular—not as impressive, even, as in some of the other Soviet-dominated Eastern European countries. Agricultural mechanization and the provision of adequate fertilizers have been delayed, and farm yields never have reached the levels they should. Despite its considerable domestic resources industrial development has largely been geared to the local market. Bucharest at 1.5 million people is still several times as large as the next ranking town, the interior industrial center of Cluj with chemical, leather, and wood manufactures (240,000). The causes of all this are difficult to pinpoint, but there seems to have been a lack of national commitment to the cause of planned development, which may be a short-term factor; in the longer sense Romania has suffered from a chronic ailment of Eastern Europe, namely political instability.

The Romanians are a people of complex roots, whose distant Roman heritage is embodied in a distinctive language, but whose subsequent adventures have brought strong Slavic influence, especially to the people in the rural areas, as well as admixture with Hungarian, German, and Turkish elements. The Bulgars too, have been affected by Slavic as well as other influences. These people, numbering about 8.5 million today, also have a history of Turkish domination, and a Bulgarian state did not appear until 1878. Bulgaria in a way was the westernmost buffer state in the long zone that emerged in Eurasia between Russian and English spheres of influence; the Russians helped push the Turks from Bulgaria, and British fears of a Russian penetration to the Aegean led to the Treaty of Berlin, at which time Bulgaria's boundaries were delimited.

Bulgaria is a mountain state, except in the Danube lowland, which in this country is narrower than in Romania, and in the plains of the Maritsa River, which to the south forms the boundary between Greece and Turkey. Even the capital, Sofia, is located in the far interior, away from the plains, the rivers, and the Black Sea; the mountains were used as protection for the headquarters of the weak embryo state, and it has remained here. In any case, Bulgaria is not a country of towns—it would be more appropriate to see it as a country of peasants. Nearly 60 percent of the population is still classed as rural, and after Sofia with nearly one million people, the next ranking town has under 200,000 (Plovdiv, the center of the Maritsa agricultural area). The east-west trending mountain range that forms the country's backbone and separates the Danubian lowland area from the southern valleys carries the region's name, the Balkan Mountains. Much of the rest of Bulgaria, especially the west and southwest, is mountainous as well, and people live in clusters in basins and valleys and are separated by rough terrain. The

Turkish period destroyed the aristocracy and eliminated the wealthy landowners, and for many decades Bulgarian farms have been small gardens, carefully tended and productive. These garden plots survive today, although collectivization has been carried farther here than in any other Eastern European country except perhaps Albania. On the big *kolkhozy* the domestic staples of wheat, corn, barley, and rye are grown, but the smaller gardens produce vegetables and fruits; plums, grapes, and olives remind us that we are approaching the Mediterranean. From this point of view Bulgaria is a valuable economic partner to the Soviet Union, where fruits, vegetables, and tobacco are always needed and where the conditions to grow this particular assemblage are very limited.

At the time of transition to Soviet domination, Bulgaria was in a different position from both of its Eastern European neighbors, Romania and Yugoslavia. It was poorer than both, and still has less in terms of industrial and agricultural

The old and the new in Bulgaria: the Turkish mosque in Sofia's Dimitrov Street is a reminder of the days when Bulgaria was part of the Ottoman Empire; today 14 percent of the population are still Moslems. (Camera Press–PIX)

resources. But there was a lengthy history of association of one kind or another with the Russians and Soviets, and the Bulgars are more Slavicized than the Romanians. In the area of agriculture a groundwork of sorts was laid by the Turks, who left behind a country parcelled out into smallholdings. There was very little industrial development, and the Communist period brought vigorous searches for minerals—with some success—and industrial expansion. Despite Sofia's difficult location, a steel industry was established in the capital using imported ores and coal found near Dimitrovo nearby to the west. Certainly the post-1945 period brought advances to Bulgaria in the economic sphere.

Thus Eastern Europe is going through a phase of eastward orientation and Communist political and economic organization—an organization that is in most cases interrelated with that of the Soviet Union itself. But viewing the history and historical geography of Eastern Europe, it appears likely that this phase, too, will lead to something else. Stability has never been a quality of Eastern European life, and the superimposed stability of the postwar period already has been disturbed in some instances. And despite the repatriation of large numbers of people who formed minorities in Eastern Europe's countries, old national goals have not been forgotten. Bulgaria still considered the Macedonian question to be unsettled—and through Macedonia, of course, this country could obtain a way to the

Women work on the fields of a collective farm in Bulgaria. Perhaps more than any other Eastern European country, Bulgaria has adhered to the Soviets' ideological line. Bulgaria is the most faithful of the satellites; practically all farming is now collectivized. (David Holden/Camera Press–PIX)

open sea. Hungary is very well aware of the Magyar minority in Romania's Transylvania. And there are signs of dissatisfaction with the status quo as it relates to the Soviet Union—in Czechoslovakia, in Romania, in Hungary, and in Poland. Eastern European nationalism is a potent force, and national pride and Soviet planning suggestions are not always compatible. Hence it may be that Eastern Europe will loosen its political and economic ties with the Soviet Union—though hopefully not to rekindle old quarrels and bring on another period of destructive instability.

ADDITIONAL READING

The main concepts and ideas discussed in this chapter are from the field of political geography, and useful sources are L. M. Alexander, *World Political Patterns,* published in Chicago by Rand McNally in 1963 in a second edition, and *Political Geography* by N. J. G. Pounds, published by McGraw-Hill in New York, also in 1963. Several articles directly relevant to these issues are reprinted in H. J. de Blij, *Systematic Political Geography,* published in 1967 by Wiley. Very useful also is G. W. Hoffman's *Balkans in Transition* (No. 20) in the Van Nostrand Searchlight Series. Also see K. London (editor), *Eastern Europe in Transition,* published in Baltimore by the John Hopkins Press in 1966.

Individual countries also have been discussed by geographers. On Poland, see the paperback by N. J. G. Pounds, *Poland Between East and West* (No. 22) of the Van Nostrand Searchlight Series. On Czechoslovakia, the most useful source probably is H. G. Wanklyn's *Czechoslovakia,* published in New York by Praeger in 1954. Yugoslavia is the subject of a volume by G. W. Hoffman and F. W. Neal, *Yugoslavia and the New Communism,* published by the Twentieth Century Fund in New York in 1962. J. M. Montias in 1967 published *Economic Development in Communist Rumania* (M.I.T., 1967), and Hungary is the subject of a volume by M. Pěcsi and B. Sárfalvi, *The Geography of Hungary,* published by Collet's in London in 1964. On Greece, one of the best sources by a geographer remains the chapter focused on that country by J. Gottmann in his *Geography of Europe,* a Holt publication now in its fourth edition, 1969.

5

THE SOVIET UNION—REGION AND REALM

THE SOVIET UNION,[1] by virtue of its enormous size, dominates the politico-geographical map of the world. From the Baltic Sea to the Pacific Ocean, and from the shores of the Arctic Ocean to the borders of Iran, the U.S.S.R. (Union of Soviet Socialist Republics) occupies the heart of Eurasia—and almost half the area of that vast landmass. It is surrounded by a crowd of countries from Finland to Turkey to Afghanistan to Japan, and some of these countries have lost parts of their territory to what is truly a Russian empire. With more than 8.5 million square miles, the Soviet Union covers one-sixth of the earth's land surface and is indeed a state of continental proportions, larger than several of the world's culture realms (see Chapter 1), larger than all of South America, larger than Africa south of the Sahara, and more than twice as large as China or the United States of America.

In terms of total area, then, the Soviet Union among the states of the world is in a class by itself. But this is not the result of the recent emergence of the Soviet Union as a major world power (as we are sometimes led to believe); compared to the expansionism of Russia under the tsars, the U.S.S.R. since 1917 has absorbed relatively little territory. Russia for centuries has been the largest country in the world, though it was never the most powerful: mere size is not a guarantee of strength. With less than 250 million inhabitants, the U.S.S.R. has one-third as many people as China, half as many as India, and not very many millions more than the United States. And, as Plate 6 (in the rear of

[1] The Soviet Union is short for the U.S.S.R., or Union of Soviet Socialist Republics, which is the official name of the Soviet state. Sometimes the Soviet Union is still referred to by the name it had prior to the overthrow of the last of the tsars, *Russia*. However, Russia actually is only one (though by far the largest) of the 15 Union Republics that comprise the U.S.S.R.

this volume) indicates, the great majority of the Soviet Union's inhabitants still remain concentrated between the Urals and the boundary with Eastern Europe. Large parts of the vast eastern two-thirds of the country are very sparsely populated, and as such are perhaps more a liability than an asset to the U.S.S.R. In the division of the earth's mineral deposits among the countries of the world, the size of a country is no doubt a factor—the larger the country, the greater the chance that it will possess a larger volume and range of deposits than a smaller country. (Location with relation to known mineralized belts is, of course, another factor.) But possession alone does not always mean utility. If a mineral deposit is discovered to lie a thousand miles from the nearest populated area, the cost of exploiting it may be prohibitive; a railroad would be needed to haul the mineral all that distance to the place of use. With no towns along that railroad, no production, no other trade to handle, no function for that railroad other than to carry the ore, it may not be economically feasible to build it. Indeed, it may be cheaper to buy similar ore from another country and import it. Only objectives such as self-sufficiency or nondependence on foreign countries could lead to exploitation of remotely located domestic ores. And this, of course, is what can happen in the kind of economy the Soviet Union has: a *planned* economy.

A Harsh Environment

With such a huge area, why should the population of the Soviet Union be distributed as it is, with its western concentration and its ribbon-like eastward extension across south-central parts of the country to the Pacific coast? Again the world map provides a clue: the U.S.S.R., despite its bulk, is a high-latitude country, positioned very far north into Arctic lands and frozen wastes. More than 80 percent of the total area of the Soviet Union lies farther to the north than the North American Great Lakes! And, as Plate 2 indicates, the climate tends to increase in severity in an easterly direction. Most of the U.S.S.R. lies under Snow Climates (*D*), but in the western sector of the country, at least, there is a frost-free season of between five months (in the north, around Leningrad) and six months (south of Kiev). But eastward the frost-free period becomes smaller: at Novosibirsk it is just over 120 days, and around Irkutsk, near the southern end of Lake Baykal, it is down to 95

days. The map gives us an indication of this by the last letter of the climatic formula. Note that the *D* climate is a *Dfb* clime in the west, which means that it is rather like that of the upper Midwest in the United States. But the *b* changes to a *c* in the mid-U.S.S.R., and that means that there are *fewer* than four months with average temperatures warmer than 50 degrees Fahrenheit. And in the Soviet Far East the *c* even changes to a *d*, signifying the bitterest of winter cold. In this eastern area only the waters of the Pacific bring some moderation to these severe conditions—note that a *Dwb* appears near the coast. At Vladivostok, the Soviet Union's important Pacific port, the weather bureau records as many as 151 frost-free days in an average year.

The Snow Climate zone (which is also called the Humid Continental Climate) extends in a broad belt all the way across the U.S.S.R., as Plate 2 shows. When we compare this climatic zone with the vegetation map (Plate 4), we note that this is a zone of forests—needleleaf in the northern areas and deciduous forests in the south, with a belt of mixed trees marking the transition. Under such vegetation, especially under the deciduous and mixed forests, soils develop that are fairly fertile. On Plate 3, the blue area marking the podzols and weakly podzolized soils (area 4) closely overlaps the needleleaf forests, and southward, where the deciduous trees begin to make their appearance, the soils respond: gray-brown podzolic soils (area 5, Plate 3) form a belt that extends all along the southern margins of the needleleaf forests.

Toward the north, the Snow Climate turns into an Ice Climate. The northern parts of the Soviet Union, from the Finnish border in the west to the Bering Strait in the east, are bitterly cold and barren. Note that the so-called Ice Climate (*E* on Plate 2) coincides with a vegetative area called *tundra* (9 on Plate 4), and that both overlap closely with a soil region also called tundra (2 on Plate 3). In fact, just as the Snow Climate is sometimes referred to as the Humid Continental Climate, so the Ice Climate on our map is also named after its vegetative and soil characteristics: the Tundra Climate.

In the tundra, the environment is forbidding. No month of the year has an average temperature over 50°F, and above-freezing averages may occur only in one or two months—in the "warmer" parts, as many as four. Tundra climes have the shortest of short summers, and the soils

This photograph well conveys the bitter cold of the Siberian interior. Rising over the village of Tompo with its furiously smoking chimneys are the Verkhoyanskiy Mountains, bare except for a few straggly pines. (Camera Press—PIX)

and vegetation reflect this. There are no trees at all, and patches of grass are mixed with areas of mosses, lichens, and bushes. Marshes and rocky areas add to this picture of barrenness. Toward the south, things become a little better. Birches and willows appear, first in some of the stream valleys, and then in isolated stands. Gradually the vegetation becomes thicker, and then the needleleaf trees appear and the forest develops. On our map (Plate 2) the Ice or Tundra Climate gives way directly to the Snow or Humid Continental Clime, but some climatologists record the existence of a transition zone between the barren Tundra and the forested *D* climatic region. This transition zone has various names, and one of them is the *Subarctic* Climate. The coniferous forest in these high latitudes is called the *taiga*.

Permafrost. The cold conditions that prevail in these near-polar parts of the Soviet Union have produced a whole association of unusual features, not only in the treeless landscape, the severe winter blizzards, and the brief growing season, but also in the soil and below it. Here in northern Siberia, as in northern Canada and in Greenland, the ground below the surface over large areas is permanently frozen. This frozen ground is called the *permafrost,* and its effect on soil formation is great. Because the permafrost is impervious, the rainfall and meltwater that percolate downward during the short summer gets caught, and it saturates the lower soil layer. In fact, there are places where the soil is so thin that the whole soil layer is saturated and standing water appears at the surface—no wonder that this is a land of marshes and bogs. But good

drainage is an essential ingredient in good soil development, and thus the permafrost (which may begin a few inches or several feet below the surface) inhibits soil formation in an area where low temperatures already slow it down.

The alternate freezing and thawing of the soil between the permafrost and the surface produces some unusual phenomena at the surface. Just as mud cracks appear when water is dried out of

AFTER: KIMBEL AND GOOD

Map 5-1 DISTRIBUTION OF PERMAFROST.

The wild Soviet Arctic: fir forests stand along the Pechora River as it travels toward the Arctic Circle and the Barents Sea. (D. Donskoi and F. Grinberg/Camera Press—PIX)

clayey, muddy ground in a desert, so great cracks develop in the tundra soil when frost returns in winter, withdrawing water from the ground to form ice. These cracks have regular patterns, and over time they tend to fill up with stones; eventually *stone rings* appear. Flying over the tundra one can see these features extending over miles of the surface, as though man arranged them to mark off the land. But the permafrost is no friend of man. Not only does it inhibit soil formation: the permafrost also forms a major engineering headache when it comes to building roads and laying railroads. By stopping the downward percolation of water in the surface layers, the permafrost forces horizontal movement of this water. As a result, strong hydraulic pressures are built up. Many a road surface has been buckled by these forces.

Thus the northern margins of the Soviet Union are inhospitable, bedeviled by the severest of climates, unproductive soils, and transportation difficulties. And, as we are reminded by Plate 2, it is in Siberia, in the eastern U.S.S.R., where these conditions reach their greatest development: the *Dfb* zone, which in comparison seems mild and favorable, thins out

and disappears north of Lake Baykal. There the subarctic and tundra take over almost the whole width of the country. Small wonder that the population thins out too!

Thus the northern fringes of the *D* climatic region are excessively cold; what about the southern margins? Here lie some of the Soviet Union's best soils, but the climate is difficult once again. Now it is not coldness, but dryness that presents the greatest problem. Plate 2 shows that the southern parts of the U.S.S.R. lie to a great extent under *B* climatic conditions, which means that this is a steppe, with low precipitation. And the steppe merges southward into the great deserts of inner Asia.

But despite their marginal precipitation (Plate 1 reveals that most of the steppe area receives between 8 and 20 inches of rainfall annually) the Soviet steppes are very productive farming areas. Why this is so is suggested by Plate 3. Here, from near the Romanian border across the southern part of the country as far east as the Altay Mountains, lie the most fertile soils the U.S.S.R. possesses, in a broad belt which is widest in the west and narrows eastward. These, the map tells us, are the soils also found in some

The Soviet subtropics: the Black Sea port and resort of Yalta, seen from the air. (M. Redkin/Sovfoto)

of North America's most productive farming areas. This soil zone (8 on Plate 3) in the Soviet Union is called the black earth belt, and the soils themselves are identified as *chernozems*. Chernozem soils are rich in humus, thick, resilient, and very productive. Humus is that essential ingredient that makes soils fertile: the decomposed organic material (such as leaves, twigs, grass) that has been broken down by bacteria and has become part of the soil itself. This humus not only makes the necessary nutrients available to the newly growing vegetation, but it also helps the soil develop a good structure and it retains much of the sparse moisture that comes in the steppe. Steppe lands are grasslands, and it follows that grasses form better humus for the soil below than leaves and twigs. Hence these chernozem soils are more fertile than the soils under the deciduous and mixed forests that cover more northerly parts of the U.S.S.R.

Toward the desert margins, the chernozem soils give way to the so-called chestnut and brown soils (9 on Plate 3). These soils show the effects of the increasing dryness that prevails here. They are lighter in color than the dark chernozems, they become thinner, and they are less fertile. In the North American Great Plains, too, these soils appear on the western side of the chernozems; in the U.S.S.R. they lie to the south of the black earth belt. But these soils are still more fertile than most others, and the Soviets have tried very hard to overcome the problem of drought and to bring them into production. Toward the drier areas, rainfall reliability decreases (that is, variability increases), and means must be found to make water available through irrigation or to develop especially drought-resistant crops.

Finally, the Soviet Union has a small area of mild subtropical climate around the shores of the Black Sea and south of the Caucasian Mountains. There are actually several small areas, each favored by special topographic circumstances. The south coast of the Crimean Peninsula in the northern Black Sea is protected from cold northern winds by a range of mountains; the Caucasians shelter much of the eastern Black Sea and the area known as Transcaucasia. The climatic regime here is warm temperate, or *C* on our map. In appearance this area is a world apart from the rest of the U.S.S.R., with vineyards, fruit orchards, tobacco fields, and attractive resort towns including Yalta and Sochi. This,

indeed, is known as the Russian Riviera, but although it may be warmer than the rest of the Soviet Union, the winter is chilly even here. Yalta's coldest month, January, averages only 39°F, and Baku, on the Caspian Sea south of the Caucasian Mountains, has a January average of 38°. And the all-pervading dryness of the Soviet Union also makes its mark in the Soviet subtropics: Yalta gets 20 inches of precipitation annually, but Baku and the southern Caspian receive only 10 inches or less—and are consequently mapped as steppe (Plate 2). Only in the eastern Black Sea fringes are really moist conditions encountered, as at Batumi, which records over 90 inches. Plate 1 reflects these conditions clearly.

The Continentality of the U.S.S.R. Despite its great size, the Soviet Union is almost a landlocked country, and its continental position is emphasized by the prevailing climates. Most of the moisture the country gets appears to be derived from the North Atlantic, and by the time the air masses carrying this moisture have reached across Europe into the U.S.S.R., they have lost a good deal of it. Dryness increases eastward; as Plate 1 shows, there are only patches of blue beyond the Ural Mountains. The Soviet Union lies open to the frigid air from the Arctic, and certainly the frozen polar areas do not help ameliorate the country's continentality. And high mountains rim the giant state in the south, southeast, and east, cutting it off from any warming influences from the Indian or Pacific Oceans. Where there are good soils, as in the black earth belt, there are damaging, serious droughts and hot, destructive winds. Where there is somewhat more moisture, as in the more northerly areas west of the Urals, the soil is less fertile and the growing season declines. The Soviet Union is large, but its size is not matched by equally great environmental diversification. The severity of its climate, coldness, and drought constitute a major disadvantage.

A Vast Realm

Although the Soviet Union is nearly three times as large as the contiguous United States, its physiography has much less regional variety. Just a glance at the map of North America produces an image of numerous physiographic regions—the clear-cut Appalachians, the Great Plains, the coastal plain, the Rocky Mountains, the Colorado Plateau, the mountain ranges of the Pacific Coast, and several others. No such clear-cut division of the Soviet land surface emerges, although a look at any topographic map quickly shows a basic feature of the Soviet realm: the east and south are high and rugged, and the north and west are lower and more level. In the east the dividing line between the mountainous country and the plainlands is marked approximately by the Lena River, which originates near Lake Baykal and skirts the western slopes of the Verkhoyansk Mountains. From the vicinity of Lake Baykal the line runs along the northern slopes of the mountains of middle Asia and the Caucasus. Only the Ural Mountains, which trend north-south right through the middle of the Russian plains, constitute a major exception to this broad division of the U.S.S.R.'s physiography.

Regions. The simplest subdivision of the Soviet Union's physical geography produces eight regions (Map 5-2). West of the Ural Mountains lies (1) the *Russian Plain.* This is essentially the eastward continuation of the North European Lowland, and extends from the shores of the Barents Sea in the north to the Black Sea in the south. Approximately the northern half of this plain was glaciated during the most recent (Pleistocene) glaciation, and the region's appearance will be familiar to anyone who has seen the glaciated parts of North America. There are many evidences of this recent glaciation: marshes, lakes, disorganized drainage, and low ridges and mounds of rubble piled up by the glaciers as they advanced. Coniferous forests cover much of this northern part of the Russian Plain. The southern areas, which remained unglaciated, were nonetheless covered by clay and sand washed southward when the glaciers melted, and by loess carried still farther by the wind. There are also areas of bedrock, and in these rocks some of the Soviet Union's major mineral deposits have been found, including the iron ores of Krivoy Rog, the coal of the Donets Basin, and the manganese ores of Nikopol.

The Moscow Basin lies in the middle of the Russian Plain. There are low-grade coal fields in this basin, notably near Tula, south of Moscow. The basin was once invaded by the sea, and the rocks here lie nearly horizontal. In the west lie the famous Pripet Marshes, which in historic times constituted a barrier between the European sphere and the emerging Russian state.

The Russian Plain is bounded on the east by region (2), the *Ural Mountains.* Although the Urals form a prominent range, they are topographically rather unimpressive. They have undergone erosion for a long time and are not very high in elevation. Most of the Urals lie under 2000 feet above sea level, although the elevation rises somewhat toward the north. Mount Narodnaya, at just over 6200 feet, is the highest point in the entire Ural Mountains. As the map of Soviet communication lines shows, the Urals are hardly an obstacle to east-west transportation. Wide passes cross the range in several places, and railroad lines, roads, and pipelines connect east and west. The southern Urals are quite heavily populated, and the range is rather rich in mineral deposits. Iron and nickel ores, bauxite, platinum and chrome, and oil as well as coal are located in or near the Ural Mountains.

South of the Urals, east of the Caspian Sea, north of Iran and Afghanistan, and west of a line from Tashkent to Tselinograd lies (3) the *Caspian-Aral Basin.* This is mainly steppe and desert country, with a system of internal drainage centered on the Aral Sea. Salt flats, dunes, bare rock, and windswept plains mark this region. Its northern section is known as the Kirgiz Steppe, and in the south the Soviets are trying to conserve what little water there is in the basin through irrigation schemes on the Amu Darya and Syr Darya, both rivers draining into the Aral Sea.

East of the Ural Mountains lies Siberia. The western part of this enormous area is made up of the basin of the Ob River, a 3200-mile-long stream that begins its course near the point where the Soviet, Chinese, and Mongolian boundaries meet. The Ob River's basin is known as (4) the *West Siberian Plain,* a region that has been described as the world's largest unbroken, true lowland plain. Part of this plain is drained by an Ob tributary, the Irtysh, which has its source in China. We get an idea of the flatness of this area when we realize that the Ob, over the last thousand miles of its course, falls only 300 feet! What matters in this lowland lies concentrated in the south. The northern section is permafrost-ridden, the central areas are marshy, but in the south lies part of the eastward extension of Soviet settlement. The cities of Omsk and Novosibirsk lie near the southern margins of the West Siberian Plain.

East of the West Siberian Plain, roughly be-

Map 5-2 PHYSIOGRAPHY OF THE U.S.S.R.

The Soviet east: a scene in the Kirghiz Republic showing part of the Central Tien Shan in the background. Agriculture flourishes in the lowest areas, while the lower slopes afford rich pastures. (Camera Press—PIX)

tween the Yenisey and Lena Rivers, lies region (5), the *Central Siberian Plateau.* This is forbidding country: it is a dissected, low tableland, subject to severe winter cold, short summers, and almost completely afflicted by permafrost. Although some important minerals have been found in this region (including nickel, copper, and platinum around Norilsk in the northwest), this remains a sparsely settled part of the U.S.S.R.

The eastern, mountainous sector of the U.S.S.R., from the Verkhoyansk Mountains in the north to the Sikhote Alin Mountains in the south, forms region (6), the *Eastern Highlands.* These are high mountains, with some exceeding 10,000 feet. The topography is a jumbled mass of ranges and ridges, precipitous valleys and active volcanoes. Lake Baykal lies in a trough that is over 5000 feet deep; on the Kamchatka Peninsula, volcanic Mount Klyuchevskaya reaches close to 16,000 feet in elevation. The northern part of this region is the Soviet Union's most inhospitable section, with prevailing ice climates (see again Plate 2), permafrost, and practically no agricultural possibilities. Southward, along the Pacific coast, conditions ameliorate considerably. Here the Amur and the Ussuri valleys have alluvial soils, and the growing season is long enough to allow some crops to mature. This, however, is a true frontier region. The forests provide ample opportunities for lumbering; a fur trade exists, and gold and diamonds are won by hardy miners.

South of the three Siberian regions just discussed (regions 4, 5, and 6), lies an assemblage of mountains and mountain ranges which, for want of a better name, are combined as region (7), the *Central Asiatic Ranges* of the Soviet Union. They extend from Lake Baykal in the east, along the borders of Mongolia and China, to the vicinity of Lake Balkash. Included also is the more plateaulike Kazakh Upland. Some of these ranges are very high, and a few places exceed 20,000 feet; mostly they crest at well over 10,000 feet. Since the snow line is at about 13,000 feet, these ranges carry extensive glaciers and snow fields, and they provide important annual meltwaters to the rivers that rise on their slopes. Alluvium, carried down in these waters, contributes to the farmlands in the lower valleys of the rivers. And the lower mountain slopes, by virtue of their somewhat higher precipitation than the adjacent lowlands, have valuable pastures.

Finally, region (8), the *Caucasus and Transcaucasia,* occupies the area between the Black and Caspian Seas. These high ranges are related in origin to Europe's Alps, and in appearance they much resemble their European counterpart. Mount Elbrus, at 18,481 feet, is higher, in fact, than any point in the European Alps. The Caucasus has proved to be of economic importance: there are oil fields in this region, and iron, manganese, and bauxite have also been located.

Across this state of western plains and eastern mountains, northern forests and southern steppes, the Russian drama has been played. Still today, the vast majority of the Soviet Union's people remain clustered west of the Ural Mountains, in the region we have identified as the Russian Plain. We know, now, that there are environmental reasons for this distribution. But is the Soviet Union a European state? Does this western concentration of people also mean that there are and have been close and vital interrelationships between the old Russia and Europe? Is the Ural Mountain range, after all, a better boundary for Europe than the western boundary of the U.S.S.R.?

A European Heritage

Certainly a backward look at the consolidation and rise of the old Russian state provides many parallels with European development. As Professor Hooson wrote, "Russian national consciousness emerged from a melting-pot of immigrant peoples and took geographical root at about the same time as this was happening in England."[2] Later, we can see parallels in the impact of industrialization in Russia and in Britain and Western Europe. Undoubtedly Rus-

[2] D. Hooson, *The Soviet Union: People and Regions,* Wadsworth, Belmont, 1966, p. 49.

The glory of old Russia: one of the old headquarters was Novgorod, whose walled Kremlin is shown in the overview. The picture below shows the Sofia Cathedral, built in 1045 and still standing despite neglect. (Emil Brunner/Sovfoto, PIX)

sia from the Middle Ages onward was a major force in European affairs. Indeed, as we shall see, the major justification for the recognition of the Soviet Union as a separate political and cultural entity lies in the events of the twentieth century.

The Past. Much of what happened in Russia before the Middle Ages can only be pieced together from fragmentary bits of information. This part of Eurasia, too, experienced its *Völkerwanderung,* and for two thousand years wave upon wave of migrants crossed the plains on which Russia was eventually to emerge. The dominant direction seems to have been from east to west; many peoples came from Central Asia and left their marks in the makeup of the population. Scythians, Sarmathians, Goths, and Huns—they came, settled, fought, were absorbed, or driven off. Eventually the Slavs emerged as the dominant people in what is today the Ukraine; they were peasants, farming the good soils of the plain north of the Black Sea. Their first leadership came not from their own midst, but from Scandinavia. A Scandinavian people known as the Varangians had for some time played an important role in the fortified trading towns (*gorods*) of the area, and it was a Scandinavian prince, Rurik, who in the ninth century A.D. ruled over the first Slavic state.

This political organization came about for reasons that will be familiar to anyone who remembers the importance of the trans-Alpine routes across Western Europe. The objective was to render stable and secure an eastern crossing of the continent, from the Baltic Sea and Scandinavia in the northwest to Byzantine Europe and Constantinople in the southeast. This route followed the Volkhov River to Novgorod (positioned on Lake Ilmen), then led to Kiev, and southward to the Black Sea along the Dnieper. Novgorod, so near Scandinavia and close to the shores of the Baltic, was the "European" center, with its cosmopolitan population and its Hanseatic trade connections. Kiev, on the other hand, lay in the heart of the land of the Slavs, near an important confluence on the Dnieper, and not far from the zone of contact between the forests of central Russia and the grassland steppes of the south. Kiev had centrality; it served as a meeting place of Scandinavian and Mediterranean Europe, and for a time during the eleventh and twelfth centuries it was a truly European urban center as well as the capital of Kievan Russia. Kiev and Milan had something

in common: both were positioned near major breaks in the landscape, Milan at the entry to the difficult Alpine crossing, and Kiev at the edge of the dense and dangerous forests that covered middle Russia.

With these assets Kiev might have become the primate city of Russia (it is said to have overshadowed Paris and London in its medieval splendor), but about the middle of the thirteenth century Kievan Russia fell to yet another invasion from the east. Now the Mongol empire, which had been building under Genghis Khan, sent its Tatar hordes into Kiev (led, incidentally,

Kiev, in its days of glory, rivaled Paris in its splendor. Still today it is called the Paris of the Soviet Union. The top picture shows the city's main, tree-lined avenue. Kiev took a terrible beating during World War II and much of the city was reduced to rubble (below). (Tass/Sovfoto, Robert Capa/Magnum)

MODIFIED AFTER: LAUTENSACH, ATLAS ZUR ERDKUNDE

Map 5-3 U.S.S.R. LOCATION MAP.

142

by the grandson of Genghis, Batu), and the city as well as the state fell. Many Russians sought protection by fleeing into the forests that lay north of Kiev; the horsemen of the steppes were not very effective in that terrain. Thus Russia's position of complementarity was lost, and what remained of the country was cut off from Mediterranean Europe. Novgorod inherited from Kiev a temporary position of primacy, but Novgorod was not really a Russian city. What remained of Russia really lay in the forests between the Baltic and the steppes, and there, for a time, a number of weak, feudal states arose, many of them ruled by princes who paid tribute to the Tatars in order to retain their position. From among these feudal states, the one that centered on Moscow emerged as supreme. In the fifteenth century its ruler conquered Novgorod; then in the sixteenth, Ivan IV (the Terrible) took over authority and assumed the title of tsar (another Mediterranean contribution: czar or *caesar*). Ivan the Terrible made an empire out of Muscovy. He established control over the entire basin of the Volga River, succeeding in his campaigns against the Tatars, and extended Moscow's authority into western Siberia.

But Moscow's successes were attended by a great deal of oppression at home, and many of Muscovy's subjects moved away from the central authority, mainly southward into the steppelands. Some of them became farmers; others joined in seminomadic bands and came to be known as Cossacks. These Cossacks often roamed and plundered, or hired themselves out to do almost anything for anyone capable of paying the price—Cossacks even carried out warfare on behalf of the very authority they had escaped. The eastward penetration of Siberia was actually begun late in the sixteenth century by just such a band of Cossacks, led by the Yermak, hired by the rich, fur-trading Stroganov family to seek the riches of the unexplored interior. Their initiative stimulated a movement that has not yet stopped; throughout the seventeenth century these first Cossacks, who fought and defeated the Tatars in their way, were followed by other explorers and pioneers, most of them in search of the one resource known to exist there—fur-bearing animals. Before the middle of the seventeenth century they had reached the Pacific Ocean; in 1649 Okhotsk was established. Now there was trouble with the Chinese, and in 1869 a treaty divided the far eastern lands between China and Russia. But the Russians did not stop there. They entered Alaska and established settlements along its coasts, and continued southward along the west coast of North America into what is today California; the last station was founded just a matter of miles north of San Francisco.

The Rivers as Avenues. Russian expansionism, both west and east of the Urals, was much facilitated by the river systems in the Russian Plain and in Siberia (Map 5-3). We have already referred to the old patterns of trade and contact across western Russia, and Moscow's centrality is further emphasized by the radial drainage pattern that exists here—a drainage pattern that had much to do with the channeling of communications and the outward growth of Muscovy power. Rivers flow northward into the Arctic Ocean, northwestward into the Baltic Sea, southward into the Black Sea (the Dnieper) and the Sea of Azov (the Don), as well as the Caspian Sea (the Volga). Russia's rulers always saw to it that they had secure control over these natural avenues. This control was made possible through the construction of so-called *ostrogs,* stations built along the river courses at strategic points.

When the conquest of Siberia began, this technique of consolidating control through river *ostrogs* and way stations placed along overland portage routes proved a vital asset in Russia's eastward push. Siberia is crossed by several great rivers with numerous tributaries, including the Ob-Irtysh system, the Yenisey, the Lena, and, in the far east, the Kolyma Rivers. Tobol'sk, on the great bend of the Irtysh, Yeniseysk on the Yenisey about 150 miles north of Krasnoyarsk, Yakutsk on the Lena, and Irkutsk near Lake Baykal all began as *ostrogs,* and they played vital roles as centers of early Russian influence and power.

This eastward expansion, carried out by relatively few people, did not, of course, take place overnight. It began late in the sixteenth century and continued until 1812, the year of the founding of Fort Ross near San Francisco Bay. But by the time Peter the Great took over the leadership of Russia (he reigned from 1682 to 1725), Moscow lay at the center of a great empire—great, at least, in terms of the territories under its hegemony. And as such, Russia had many enemies. The Mongols were finished as a threat, but the Swedes, no longer allies in trade, certainly were a problem. In command of Finland,

The long-term capital of the Russia of the Czars was Leningrad; the famous Hermitage palace is in the left foreground. The Palace Square was the scene of some of the first major expressions of revolt; reprisals by the Czars' militia helped fire the revolution. (Tass/Sovfoto)

the Swedes threatened from the north; to the northwest there were the Lithuanians. To the west there was a continuing conflict with the Germans and the Poles. And to the south, the Ottoman Turks were heirs to the Byzantine Empire and they, too, posed a threat. Peter consolidated Russia's gains and did much to make a modern, European-type state out of the loosely knit country. He wanted to reorient the empire to the Baltic, to give it a window on the sea and to make it a maritime as well as a land power. In 1703, following his orders, the building of Petersburg (later St. Petersburg) began. This city, known today as Leningrad, is positioned at the head of the Gulf of Finland, which opens into the Baltic Sea. Not only did it thus provide Russia with an important maritime outlet, but St. Petersburg also was designed to function as a forward capital: it lay on the doorstep of Finland, which at that time was a Swedish possession. And in 1709 Peter's armies defeated the Swedes, confirming Russian power on the coast; in 1713 Peter took the momentous step of moving the Russian capital from Moscow to the new Baltic headquarters. It was to remain there until 1918.

Peter was an extraordinary leader, in many ways the founder of modern Russia. In his desire to remake Russia, to pull it from the forests of the interior to the western coast, to open it to outside influences, to end its comparative isolation, he left no stone unturned. Not only did he

move the capital; he himself, aware that the future of Russia as a major force lay in strength at sea as well as power on land, went to Holland to work as a laborer in the famed Dutch shipyards, to learn how ships were most efficiently built. Peter wanted a European Russia, a maritime Russia, a cosmopolitan Russia. He developed Petersburg into one of the most magnificent cities in the world, and employed Italian architects to make it so: with its wide avenues and elaborate palaces it remains to this day a city apart in all of the Soviet Union, reminiscent of Paris, London, or Milan. The new capital became and long remained one of Europe's principal cities, a seat of power with which the whole continent had to reckon.

During the eighteenth century the tsarina Catherine the Great, who ruled from 1762 to 1796, continued to build Russian power, but on another coast and in another area: the Black Sea in the south. Here the Russians confronted the Turks, who had taken the initiative from the Greeks: the Byzantine Empire had been succeeded by the Turkish Ottoman Empire. But the Turks were no match for the Russians. The Crimea soon fell, as did the old and important trading city of Odessa, and before long the whole northern coast of the Black Sea was in Russian hands. Soon afterward the Russians penetrated the area of the Caucasus, and early in the 19th century they took Tbilisi, Baku, and Yerevan. But as they pushed farther into the corridor between the Black and the Caspian Seas they faced growing opposition from the British (who held sway in Persia, now Iran) as well as the Turks, and their advance was halted short of its probable ultimate goal: a coast on the Indian Ocean.

But Russian expansionism was not yet satisfied. While extending the empire southward, the Russians also took on the Poles, old enemies to the west, and succeeded in taking most of what is today the Polish state including the capital of Warsaw; to the north, Russia took over Finland from the Swedes (1809). During most of the nineteenth century, however, Russian preoccupation was in Asia, where Tashkent and Samarkand came under St. Petersburg's control. The Russians were bothered still by raids of nomadic Mongol horsemen, and they therefore sought to establish their authority over the steppe country to the edges of the high mountains that lay to the south. Thus Russia gained a considerable number of Moslem subjects, for this was Moslem Asia they were penetrating, but under tsarist rule these people acquired a sort of ill-defined protectorate status and retained some autonomy. Farther to the east, a combination of Japanese expansionism and a decline of Chinese influence led Russia to annex from China several provinces along the Amur River. Soon afterward (1860) the port of Vladivostok was founded.

Now began the course of events that was to lead, after five centuries of almost uninterrupted expansion and consolidation, to the first setback to the Russian drive for territory. In 1892 the Russians began the building of the Trans-Siberian Railway, in an effort to connect the distant frontier more effectively to the western core. As the map shows, the most direct route to Vladivostok was across Chinese Manchuria. The Russians wanted China to permit the construction of the last link of the railway across Manchurian territory, but the Chinese resisted. Russia responded (taking advantage of the Boxer Rebellion in China) by annexing Manchuria and occupying it. This brought on the Russian-Japanese war of 1905, in which the Russians were disastrously defeated; Japan took possession of southern Sakhalin. Thus for the first time in nearly five centuries, Russia sustained a setback that resulted in a territorial loss.[3]

Thus, Russia, recipient of British and European innovations as were Germany and France

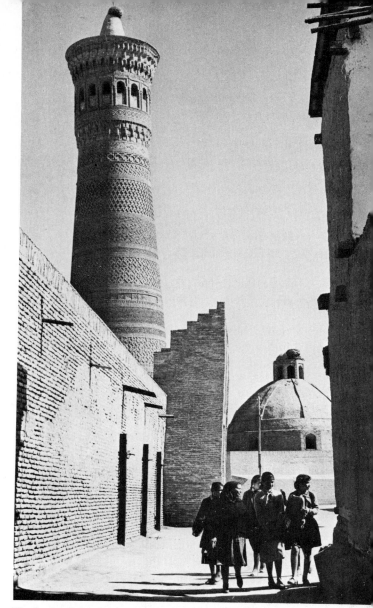

The Soviet Union is a land of cultural as well as physical variety. Here is a view in the town of Bukhara, in the Abu Ali-ibn Sini Oblast, with an impressive minaret. Until the 1870's, criminals were thrown to their death from this tower. (Camera Press—PIX)

and Italy, was just as much a colonizer too. But where other European powers traveled by sea, Russian influence traveled overland, into South Asia toward India, into China, and to Pacific coasts. What emerged was not the greatest empire, but the greatest *contiguous* empire in the world; it is tempting to speculate what would

[3] Earlier, a combination of war costs at home and British opposition in North America had led Russia in 1867 to relinquish its claims to Alaska through sale to the United States; fur-collecting stations along the west coast of Canada were also abandoned.

have happened to this sprawling realm had European Russia (for such it still was) developed politically and economically in the manner of other European power cores. At the time of the Japanese war the Russian tsar ruled over more than 8.5 million square miles, just a fraction less than the area of the Soviet Union today. The modern Communist empire, to a very large extent, is a legacy of St. Petersburg and European Russia—not the product of Moscow and the Revolution.

Revolution and Reorientation

While the empire was built, the Westernization of Russia continued—but at a very slow pace. Peter the Great had encouraged the establishment of industries that would help the state meet a larger part of its own needs; merchants and nobles were induced to invest money in private industries which, while subject to state regulation, also enjoyed benefits such as market guarantees, state subsidies, and even the right to purchase and hold peasant labor in serfdom. It is important not to overestimate the degree of Peter's success just as it is important to assess its significance in terms of his Western outlook; Peter could only hope to affect a small upper layer of his country's population, and the vast majority of Russians remained poor peasants, often without land and, as in Europe itself, in the obligatory service of the landowning nobility. This continued to be true into the second half of the nineteenth century, by which time, however, the impact of the industrial revolution came to be felt in Russia, and the old order changed. In the vastness of Russia the fortunes of men varied wildly: there were peasants who lived in abject serfdom, landless and hopeless; there were small landowners who managed to make a reasonable living off their farms; there were small, private (village) factories that did well for their owners. Then there were the estates and factories of the nobility, still run according to the old state regulations, and, of course, there was room even for free, capitalist-type industries. By the middle of the nineteenth century it was clear that the old system of serfdom was no longer an asset but, indeed, a liability; in 1861 emancipation was proclaimed.

The effect of emancipation was staggering: the iron industry, at that time based in the Urals and producing pig iron, was largely an industry owned by the nobility and run by serfs; it suffered a precipitous decline at a time when the innovations of the industrial revolution were sharply increasing iron production elsewhere. The textile industry, also largely noble-owned, was sharply depressed. But as the opportunities and possibilities afforded by the new technology reached Russia, this backward trend was reversed. With the coal deposits of the Donetz Basin and the iron ores of the Krivoi Rog, a new iron industry emerged which soon eclipsed anything that had been seen in the Urals. The new industrial period may be said to have begun about 1880; soon places like Kharkov and Rostov experienced growth comparable to that of Birmingham, Essen, or Pittsburgh. At St. Petersburg, meanwhile, developments occurred that were similar to those of other resource-poor but fortuitously located places: the raw materials came by water transport and the industries prospered.

But Russia was nevertheless a very poor country, and much of the capital for all this industrialization was foreign, especially British and French. This meant that the country had to repay by selling goods at low prices, all the time facing problems with rapid price changes on the world markets; government assistance was needed at home in the form of tariff protection and abroad in the search for loans. Thus, unlike England, for example, the industrialists in Russia did not have much political power, for they simply depended too much on government goodwill. At the same time, the negative social consequences of rapid industrialization, which had previously led to such poor working and housing conditions in the British manufacturing towns, were felt even more severely in Russia. But the relationships between authorities and the industrialists were not such that reforms came easily. Exploitation of the workers was so widespread that another age of serfdom seemed to have arrived. Ugly strikes occurred. The peasants, too, faced often miserable conditions. Still exploited by the nobility, many still landless, they were given to frequent outbursts of violence. But there was little organization to all this opposition: the autocracy remained firmly in control and suppression was very effective. And then, in 1905, the country faced defeat in the war with Japan. Comparing the situation in Russia with that in France in 1789, Professor Clarkson writes:

"On the very eve of the revolution, the situation somewhat resembled that of France in 1789. A monarchy which had largely lost the confidence of its subjects and was beginning to feel unsure even of itself, a nobility with little more than decorative functions, a rising bourgeousie, a restless lower class in the towns, and an oppressed peasantry were elements of instability that added up, not to a united revolutionary urge, but to several disparate and uncoordinated currents."

But there were important contrasts too:

"In fundamental ways, however, there was no parallelism between the France of 1789 and the Russia of 1905—or 1917. Among many differences were the existence in Russia of large-scale capitalist industry and of a peasantry with no ingrained sense of individual property rights in land."[4]

Another difference lay in the internal military situation; while the French autocracy had difficulties maintaining control over a comparatively small area, the Russian empire, its economic difficulties notwithstanding, continued to expand. Eastward, of course, there was little effective opposition and the conquest of large areas was achieved by relatively few men. But on the margins of European Russia, there was a ring of actual and potential enemies and large numbers of Russian soldiers were kept occupied throughout the nineteenth century. The tsars made a good deal of use of mercenary troops, men who traded several years of military service for some privilege, for instance a title to a piece of Ukrainian soil, and unquestionably one major reason for the unfavorable economic position of the country, just at the time capital was needed, was the enormous expenditure on military and police forces.

Significantly we are comparing the France of 1789 to the Russia of more than a century later; it is a good estimate of the lag involved in the arrival of the full force of the industrial revolution. And whatever progress Russia did make late in the nineteenth and early in the twentieth century was all but wiped out by the seemingly endless series of social upheavals and military defeats that followed. It is important to remember that the revolution that swept the Bolsheviks to power was not the first such uprising; throughout the period that led up to World War

[4] J. D. Clarkson, *A History of Russia*, Random House, New York, 1963, pp. 376 and 377.

I there were large and small scale rebellions against the tsars. A combination of the calamitous defeat suffered by the Russian forces during that war, the now-famous 1917 Revolution, and the ensuing civil war, brought about a new and different state, a rejection of much of what had gone before, and a reorientation of the whole country away from the Europe of which tsar Peter the Great had first wanted it to be a part.

Territorially, World War I led to considerable losses. Finland was freed, and independent republics were created along the Baltic in Estonia, Latvia, and Lithuania. Poland, long under Russian domination, was once again a sovereign state. And in the southwest, territory was lost to Romania. Seen on the European map, these are sizable setbacks, but in terms of the total area of the empire, the losses actually constituted a relatively small reduction, although the percentage of population involved was somewhat greater. In any case, World War II was to result in the recapture of the three Baltic states, a section of Poland, two small but significant areas from Finland, the areas lost to Romania, and—the Russians had not forgotten—southern Sakhalin and the Kuril Islands from Japan.

Much more important than these territorial consequences were the political and economic developments that arose out of the 1917 Revolution within Russia itself. The Revolution had not been a united front against the established autocracy; like so many previous attempts at rebellion it was torn by division and strife. But the means whereby the authority in Petrograd (as St. Petersburg had been renamed in 1914, to rid it of its German nomenclature) was maintained had been greatly weakened, and the Revolution's force spilled into the countryside, where the Bolsheviks' red armies battled the "whites." Even before the Bolsheviks' victory was complete, the signs of things to come were already evident. In 1918, the headquarters of the government were moved from Petrograd to Moscow, the focus of the old state of Muscovy, deep in the interior, not even on a major navigable waterway, amid the remnants of the same forests that centuries earlier had afforded the Russians protection from their enemies. It was a symbolic move, symbolic of a new period in Russian history, an expression of distrust toward a Europe that had contributed war and misery to Russia, but whose promises—of maritime power, of industrial modernization, of political advance-

RUSSIAN SOVIET FEDERATED SOCIALIST REPUBLIC

U.S.A.

U.S.S.R.

JAPAN

N KOREA

S KOREA

MONGOLIA

CHINA

Lake Baykal

NORWAY

SWEDEN

FINLAND

Tallinn

Riga

Kaunas

Minsk

Belorussian S.S.R.

Moscow

ARCTIC CIRCLE

U.K.

N

E

W

S

GERMANY

POLAND

CZECH.

RUMANIA

Kishinev

Moldav. S.S.R.

Black Sea

Georgian S.S.R.

Tbilisi

Armenian S.S.R.

Yerevan

Baku

Caspian Sea

TURKEY

SYRIA

IRAQ

SAUDI ARABIA

IRAN

AFGHANISTAN

PAKISTAN

INDIA

Kazakh S.S.R.

Lake Balkhash

Alma-Ata

Frunze

Kirghiz S.S.R.

Tadzhik S.S.R.

Tashkent

Uzbek S.S.R.

Aral Sea

Turkmen S.S.R.

Ashkhabad

Longitude East of Greenwich

Miles

0 200 400 600 800

20° 40° 60° 80° 100° 120° 140° 160°

60°

40° 60° 80° 100°

MODIFIED AFTER:LAUTENSACH, ATLAS ZUR ERDKUNDE

Map 5-4 REPUBLICS OF THE U.S.S.R.

ment—had never been fulfilled. The new Russia looked inward; it sought a means whereby it could achieve with its own resources and its own labor the goals that had for so long eluded it. The chief political and economic architect in this effort was also a revolutionary leader, V. I. Lenin. Among his solutions to the country's problems are the present-day political framework of the Soviet Union and the planned economy for which it is famous.

The Political Framework

Russia's great expansion had brought a large number of nationalities under the tsars' control, and now it was the turn of the revolutionary government to seek the best method of organizing all this variety into a smoothly functioning state. The tsars had conquered, but they had done little to bring Russian culture to the peoples they ruled; the Georgians, Armenians, Tatars, the Islamic Khanates of middle Asia (among literally dozens of individual cultural, linguistic, and religious groups) had not been "Russified." The Russians themselves, however, in 1917 constituted only about one-half of the population of the entire country (the proportion today remains the same), and thus it was impossible simply to create in an instant a Russian state over the whole of the political region. Account had to be taken of the nationalities.

This question of the nationalities became a major issue in the Soviet state after 1917. Lenin, we know, brought the philosophies of Marx to Russia; in his earliest proclamations he talked about the "right of self-determination for the nationalities." The first response on the part of a number of Russia's subject peoples was to proclaim independent republics, as was done in the Ukraine, in Georgia and Armenia, in Azerbaydzhan, and even in middle Asia. But Lenin had no intention of permitting the actual breakup of the state, and in 1923, when his blueprint for the new Soviet Union went into effect, the last of these briefly independent units was absorbed into the sphere of Moscow. Nevertheless, the whole framework of the Soviet Union remains essentially based on the cultural identities of the peoples it incorporates. Since the days of Lenin there has been modification, but the major elements of the original system are still there. The country is divided into 15 Soviet Socialist Republics (Map 5-4), each of which broadly corresponds to one of the major nation-

alities in the state. As Map 5-4 shows, by far the largest of these S.S.R.'s is the Russian one, the Russian Soviet Federated Socialist Republic, which extends from west of Moscow to the Pacific Ocean in the east. The remaining 14 republics lie in a belt which extends from the Baltic Sea (where Estonia, Latvia, and Lithuania retain their identity as S.S.R.'s) to the Black Sea (north of which lies the Ukrainian S.S.R.), and through Georgia and Armenia, both Union Republics, to the Caspian Sea, beyond which lie several Asian Soviet Socialist Republics.

Theoretically these 15 republics have equal standing in the Union, but in practice the Russian Republic has the leadership and the others look upward. With half the country's population, the capital and most of its major cities, and over three-quarters of its territory, Russia remains the nucleus of the Soviet Union. The Russian language is taught in the schools of the other republics; the languages of these other areas are not taught compulsorily in Russia. Although the country's constitution, again theoretically, would permit each republic to carry on its own foreign policy, to issue its own money, and even to secede from the Union, none of this has occurred in practice. On the other hand, even though it can be argued that the Soviet Union is highly centralized and a federal state only in theory, no republic has made use of wartime conditions to break away. The Germans, during their invasion of the Soviet Union in World War II, hoped that an anti-Russian nationalism would provide them with support in the Belorussian S.S.R. and the Ukrainian S.S.R., but they found only isolated instances of cooperation and no attempts were made to proclaim sovereign states outside Moscow's sphere. Undoubtedly there is a great deal of acquiescence to the *status quo* among Soviet peoples, and, in view of the racial, cultural, linguistic, and religious complexity of the population, this is no small achievement.

The 15 republics identified above do not themselves constitute homogeneous entities either. Most of them include smaller groups of people with some cultural identity; once again, the subdivisions of the republic are largely based on the recognition of these identities. In decreasing rank, there are the so-called Autonomous Soviet Socialist Republics (thus, a republic *within* a republic) or A.S.S.R., the Autonomous Oblast, and the National Okrug. The Autonomous Oblast (A.O.), as opposed to an *Oblast,* has within its borders a nationality group of some size and importance; the Oblast is simply an administrative creation without such considerations. The National Okrug or N.O. is usually very large, and exists for the administration of large, sparsely-populated areas such as northern Siberia. Sometimes the A.O. and the N.O. are combined into a *kray*. This political system is quite complicated, needless to say, but it has served the state well, in two important areas: (1) the integration of all peoples into a framework that would provide them with avenues for effective representation in and contact with Moscow and (2) the organization of territory and population in such a manner that state control over the resource base and economic production would be effective.

The Soviet political framework, then, serves the interests primarily of Russia; it is federal in name only, and the centralization and, indeed, totalitarianism it has promoted in many ways is reminiscent of the period of the tsars. Stalin, heir to Lenin's position of leadership, was as ruthless a ruler as Russia had known in the pre-1917 era. But if the system developed after 1917 proved to have some of the liabilities of the pre-1917 era, it also had some hitherto unknown assets. For the first time the entire country was brought under effective national control. For the first time the Russians, recipients of European innovations, now began to transmit these innovations to non-Russian and non-Slavic parts of the realm. For the first time there were elements of compromise in the relationships between Russians, other Slavic peoples, and the non-Slavic population of the country. Certainly there were conflicts, and individual groups at various time have held separatist feelings. But in general there has been an awareness of progress brought by the Soviet administration; after the terror of the Russian conquest itself the Soviet contributions in such areas as Armenia and Georgia are proof positive that membership in the Soviet Union does have its advantages. Standards of living have been raised, educational opportunities have improved, housing projects have taken place of city slums, and the economy has accelerated—all this without any enforced abandonment of local culture, including language. The Soviets, of course, like to claim that their half century of control has brought unanimous approval of their system throughout the country. While this is not so, their treatment of minority nationalities does appear to have achieved greater domestic approval than that of some colonial powers and, indeed, of some governments of sovereign states.

The Nation-State. The Soviets anticipate that time, and their continued program of Russification, will eventually submerge the cultures of the various minority nationalities and produce a unified and quite homogeneous state-culture. What the Soviets seek, then, is something desired by all governments: a country with as homogeneous a population as possible, with a single, universally spoken and used language—in short, with a single culture, without divisive regionalisms, racial friction, or religious division. Few states in the world today have achieved such unity; France is often taken as an example of a country that has come close. Political geographers, viewing the efforts of governments to erase internal differences in the state territory, have recognized that this is a manifestation of a search for an ideal, an end-product of the evolution of states that has gone on for many centuries. In the Soviet Union the program is one of slow superimposition of Russian culture over non-Russian culture; in the United States, efforts to legislate full equality for nonwhite Americans is similarly aimed at the eventual reduction and elimination of the divisions in our multiracial "nation." The concept involved here is that of the *nation-state*, which may be defined as a political unit comprising a clearly defined territory and inhabited by a body of people, both of sufficient size and quality, and sufficiently well organized, to possess a certain amount of power; the people considering themselves to be a nation with certain emotional and other ties which are expressed in their most tangible form in "law" and "government" (ideology).

The Planned Economy

If the political framework that emerged after the 1917 Revolution was of far-reaching significance, the economic system was altered even more profoundly. But economic organization and consolidation did not come as easily as political control. There was a headlong rush toward state ownership of industry and the confiscation of land from the nobility, but the state was not capable of administering the industries and the land turned over to the peasants failed either to commit them to the Revolution or to greatly increase agricultural productivity. It should be remembered that Russia in the early 1920's was still largely an agrarian country, and that the majority of the people still lived from agriculture—either in a subsistence way or through productive farming. As it happened, not only was land confiscated from the nobility; many ordinary but well-to-do farmers also lost their land. Throughout the 1920's there were peasant revolts in the country, just as there had been prior to the Bolshevik takeover.

Thus the Russian economy faced chaos, and in 1921 the government proclaimed a New Economic Policy (N.E.P.) which was designed to cope with all this and which was a long way from "true" communism. The peasants, for example, were given the right to exchange their goods once again; the market principle was restored, though with some limitations. In fact, some of the confiscated lands were returned to their former owners. And in industry, a private, as well as a socialist, sector was recognized, so that there was room for the smaller individually owned industrial establishments as well as the large, urban, heavy manufacturing plants. But even these state-owned enterprises were to some extent relieved of direct, central-government control; one aspect of the 1921 N.E.P. was a decentralization of the administration of such establishments. As we know, however, this was only a temporary relaxation of the objectives of Moscow, instituted to accommodate the pressures on the administration and economy. Sight was not lost of the original intent of those who had led the Revolution: the rapid industrialization of the Soviet Union in directions determined by the state, and the collectivization of agriculture. Slowly but surely, through transport restrictions, by long delays at milling and treatment facilities, in reduced loan availability, and in many other ways, the private entrepeneur was put at a disadvantage compared to his "collectivized" competitor.

Meanwhile, a State Planning Commission (*Gosplan*) was at work to seek a means whereby the objectives just listed could be more directly pursued, with less compromise toward capitalist ways. In 1928 the commission announced its first practical step—as it turned out, the first in a long series of such steps that have become the hallmark of the Soviet economy. This was the first *Five-Year Plan*. In the decades that have followed, a number of successive Five-Year Plans have been proclaimed and carried out, although with

This is a good-looking Soviet collective farm in the Black Sea steppe region. The tree belt across the top of the picture has been planted to cope with the dust storms that blow in periodically from the even drier areas beyond. (Sovfoto)

interruptions during World War II and in-term changes during the 1950's. These Five-Year Plans should be seen in two ways: first as guidelines and aims for the economy during the particular period (although the guidelines are often changed and the aims almost never exactly achieved), and second as exhortations for the Russian workers to labor as hard as they can toward the goals put before them by the state, which are invariably tied to promises of better days ahead.

The first Five-Year Plan was designed to accomplish the elusive collectivization of agriculture, delayed, by then, for fully a decade after the Revolution. This program during the period from 1928 to 1933 led to the incorporation of nearly 70 percent of all individual peasant holdings into larger units operated jointly and under government supervision. There was much opposition to the program, from the slaughter of livestock to avoid their being surrendered to the larger collective herds to outright revolt. On the other hand, the poorest peasants, those without land or livestock, voluntarily and gladly participated. But the initial impact on farm yields was negligible, and in 1929 it was agreed that the process would have to be speeded up. The Communist ideal was the huge state farm or *sovkhoz*, literally a grain and meat factory in which agricultural efficiency through mechanization and minimum labor requirements would be at its peak. But such a huge enterprise is not easily established, and so by way of an intermediate step the smaller, local collective farm or *kolkhoz* was recognized as a more easily attained replacement. Now the drive to force the peasants into *kolkhozy* began in earnest, even with the use of force; sabotage was the answer. But in the end the state won the struggle, and at the end of the second Five-Year Plan, in the late 1930's, well over 90 percent of all peasant farms were in collectives. In some cases, an example was set: fences and hedges were taken down, small farm parcels were consolidated into huge fields, and the peasants themselves were moved to live in the small village that forms the focus of every collective farm, clustered around the administrative headquarters of the collective. Here they have their small houses, usually with a parcel of privately owned land which they may cultivate for their own benefit. Here they have medical care, their children attend school, and recreational facilities of a sort are provided. But these are samples of Soviet ideals. Most Soviet farmers still live in the old villages of Empire days, transformed into collectives. Thus, when one travels through the Soviet countryside, one is struck by the absence of the house-and-barn units that dot the American rural scene.

However, since the Revolution, the Soviet Union is recognized less for its achievements in agriculture than for its massive industrialization and hence the goals of the first Five-Year Plan, even in the field of agriculture, were in considerable measure related to other objectives in industry. Thus the collectivization and mechanization of agriculture would free many thousands of peasants for industrial work in the cities and, with a larger and more dependable volume of farm production, the urban populations would be better fed. Major investments were made in transportation, electrification, and other heavy industry-oriented facilities, and certainly the results were spectacular. During the first Five-Year Plan, while the rest of the world experienced the effects of the Great Depression, Soviet industrial production climbed continuously and sharply. In terms of steel production, for example, the Soviet Union in 1932 was ahead of all countries but the United States; needless to say it heralded this achievement as proof of the effectiveness of the Plan, and it would be difficult to argue otherwise. From then on, Soviet manu-facturing has been in the forefront of the world.

In a way, then, the Soviet Union operates as a giant corporation, with the government and its agencies acting as a board of directors that determines the levels of productivity, the resources that will be exploited or opened up, the kinds of goods that will be produced and made available, the wages paid, and so forth. The government's economic planners set the country on a certain course, and although there are year-to-year modifications in all such programs, the major features of the Plans normally remain visible. But it would be a mistake to see the Soviet Union's economy (and that of other communist states) as the only economy in the world that involves "planning." Economic changes in the United States, too, are the result of planning activities, but here the plans are laid at different levels, from the giant corporations to small industries. Certainly the government takes a hand in shaping the progress of the United States economy, but it does not control as the Soviet government does. On the other hand, it is easy to exaggerate the effectiveness of Soviet national planning in bringing about the desired goals. The measures used to gauge the degree of success in meeting production schedules have varied, and the psychological impact that is so important a part of each Plan has not always had the desired result. Always agriculture seems to have been the problem area: the long haul toward adequate farm production was probably unnecessarily long because of the low prices paid, the lack of incentives, and the general secondary position of agriculture compared to that of industry. But the Soviets sought power, and the road to power, they realized, lay in heavy industry. And in this field there really has been revolutionary development, so much so that two generations after the Revolution the Soviet Union is in a position to make a challenge for world power. Today the productivity of agriculture is approaching satisfactory levels as investments that were until recently reserved for industry can now be diverted to farming; consumer goods, of which the Soviet citizen has long been virtually deprived, are increasing in volume on the Soviet market. Accompanying this there are signs of political and economic relaxation: the profit motive is helping to solve agricultural problems on collectives where farmers may be rewarded for individual output as well as time.

It is tempting, naturally, to speculate whether

The warmer south: harvesting tea on the State Tea Plantation at Ochkamur, Georgia. (Camera Press—PIX)

the Soviet economy could have developed so spectacularly without state control over all productive capacity, and without the economic planning for which it has become known. Many western economists, pointing to the rapid industrialization of Russia during the 1880's and 1890's, argue that if the trends set at that time had been continued, the country would be farther ahead industrially than it is today. But the Revolution and ensuing civil wars largely obliterated this early progress, and what the Soviets remember is the rise of their country out of a divided, agrarian, poor, unproductive, inefficient, and stagnant past. Hundreds of thousands of the country's citizens have seen this emergence of the U.S.S.R. from poverty and obscurity to greatness; millions have seen promises made by economic planners come true. Certainly most Soviet citizens would have little doubt.

Management and Crop Yields. In a prize-winning article published in 1968, Professor S. S. Birdsall used the method of *analogue area analysis* as a means of comparing the system of agricultural management of the Soviet Union with that prevailing in the United States.[5] He chose an area in the U.S.S.R. and another in the United States, with essentially the same physical base in terms of climate, soils, and general relief. The two areas were the Minskaya Oblast, which lies in the central part of Belorussia, and six farming counties in central Wisconsin. After considering their physical environment, Professor Birdsall concludes that they "appear nearly identical in

terms of their respective physical bases for agriculture. Average length of the frost-free period, and average temperatures and precipitation for (the) period are the same in both study areas. The relief in neither area is great; the soils in both areas are gray-brown podzol to sandy podzol with many swampy low spots. The areas so widely separated on the earth's surface could hardly be more similar physically."[6] Now he turns to these areas' respective agricultural productivity, and concludes that the Wisconsin counties are much more productive than the Minskaya Oblast. In grain yields, for example, the Wisconsin area has yields that are 1.5 to 2 times as large as those of Minskaya. Even in pastoral output the Wisconsin counties are far ahead—in milk yield per cow (1.5), in wool yield per sheep (1.8 to 3.6), and in egg production, also 2 to 3 times higher in Wisconsin. Only in root crops do the Soviet farmers approach American yields, and in one case, potatoes, Soviet yields are actually higher. But after considering the causes that might explain the differentials produced by his research, Professor Birdsall leaves no doubt about the chief factor: "It may be safely stated that the lower agricultural productivity of the Minskaya Oblast cannot be said to be caused by factors beyond the control of those ultimately in charge of farming operations,

[5] S. S. Birdsall, "The Effect of Management on Crop Yields in Soviet Agriculture," *The Journal of Geography,* Vol. LXVII, No. 2, February 1968, pp. 95–103.
[6] Ibid., p. 99.

Spykman
RIMLAND (1944)

Hooson
ORIGINAL AND NEW
SOVIET HEARTLAND (1962)
Original Heartland ▨ New Heartland

Mackinder
PIVOT AREA (1904)

Mackinder
HEARTLAND (1919)
Heartland ▨ East Europe

BASEMAP MODIFIED AFTER: MCGRAW HILL BOOK CO., MANS'S DOMAIN

Map 5-5 THE WORLD ISLAND.

154

i.e., the government . . . the Minskaya Oblast could ultimately have agricultural productivity at least 150 percent of its present level. This could be achieved by adopting the managerial methods (including decisions on technology utilization) used in the selected counties in Wisconsin."[7]

Here, then, is a relatively simple and direct study of the impact of Soviet decision-making upon progress in one sector of the economy. Indeed, there are those who argue that the Soviet Union has achieved power and greatness not so much because of, but *despite* the economic policies of its leaders. Of course it is impossible to determine what a capitalist state would have achieved in the immense realm of the Soviets, and certainly the rise of the modern U.S.S.R. was not foreseen by some of those same outsiders who now say that trends of the late nineteenth century foreshadowed a great Russia.

The Heartland

One geographer who did foresee the development of a powerful state in the interior of Eurasia was Sir Halford Mackinder (1861–1947). As a young man, Mackinder could observe the culmination of centuries of Russian expansion in Europe and Asia, and the maximum development of the British Empire as well. Russian power was land power, and British power was based on control over the seas. Not unnaturally Mackinder was led to speculate on the comparative strengths of land and sea power. Viewing the physical map of Eurasia, he pointed out that a large part of this vast landmass is not penetrated by navigable rivers; thus, it would be safe as a fortress from the warships of any sea power. This area, he argued in an article published in 1904, would be of crucial significance in world history.[8] He called it the *Pivot Area* (Map 5-5), and asserted that any power based in this region would be sheltered and could gain such strength as to eventually dominate the world.

Actually, a careful look at Mackinder's Pivot Area in terms of its power resources indicates that it is not as well endowed as Mackinder suggested. Note that it excludes a good part of the fertile Ukraine, the Baltic and Black Sea areas, and Belorussia; much of it suffers in extreme fashion from Russia's big maladies: cold and drought. It even misses some of the major coal and iron deposits of the present-day Soviet Union. These were some of the points made in the debate that followed the publication of Mackinder's article, and soon he modified his idea. Recognizing the importance of the mineral wealth and the agricultural possibilities of the area immediately to the west of the Pivot Area (thus, including all of western Russia, the Ukraine and Belorussia, Poland and other parts of Eastern Europe, and even eastern Germany), he now incorporated this zone into a hypothesis he stated as follows:

"Who rules East Europe commands the Heartland
Who rules the Heartland commands the World-Island
Who rules the World-Island commands the World"

In this hypothesis, the *Heartland* is the new name for the Pivot Area with its Eastern European appendage, and by the *World-Island* is meant not only Eurasia, but Africa as well (Map 5-5). The whole idea soon became known as the Heartland Theory, and the discussions continued. Mackinder himself was of the opinion that effective control of the Heartland could only be achieved by an alliance of two or more states, and he little expected any real political unification of the whole area. Critics continued to point out that even the new Heartland still failed to incorporate major productive areas from Western Europe to Japan. But as late as 1943 Mackinder nevertheless insisted that his concept was "more valid and useful . . . than it was either twenty or forty years ago."[9] As for the resources of the Heartland, he argued that they were "the equal—or thereabouts—of all that lies upon and beneath the United States and the Canadian Dominion."[10]

Mackinder, in other words, made a prediction which held that the political unification of the Heartland and the economic exploitation of its resources would produce a state so powerful that it was destined to "command the world." This feature of the theory—its inevitability, its so-called *determinist* character—came in for heavy criticism, for it failed to take into account, or so it seemed, any changes in technology and even a further discovery of raw materials either within the Heartland or outside of it. One geographer,

[7] Ibid., p. 102.
[8] H. J. Mackinder, "The Geographical Pivot of History," *Geographical Journal*, Vol. XXIII, 1904, pp. 421–444.

[9] H. J. Mackinder, "The Round World and the Winning of the Peace," *Foreign Affairs*, Vol. 21, July 1943, pp. 595–605.
[10] *Ibid.*, p. 603.

Map 5–6 THE GLOBAL VIEW.

POLAR EQUIDISTANT PROJECTION

156

N. J. Spykman, for example, felt that the potentialities of the area *between* the Heartland and the oceans were greater than those of the Heartland itself, and he made the following entry in the war of words:

"Who controls the Rimland rules Eurasia;
Who rules Eurasia controls the destinies of the world."[11]

By *Rimland,* Spykman meant that vast belt of countries that separates the Soviet Union from the sea, starting in Finland and extending through western and southern Europe, southwestern, southern, and eastern Asia.

As for the question of changing technology, the entry of air power seemed to doom Mackinder's hypothesis, which after all was based on land versus sea power. Here A. P. De Seversky entered the debate with the view that the land and sea forces of the world's major powers had been subordinated by air power; no part of the Heartland was now sheltered any more than any other. Indeed, as De Seversky looked at the globe (Map 5-6), all of the Heartland lay within a so-called *Area of Decision,* where Soviet and United States air power confronted each other at their fullest strength. And, as we know, air power has changed a great deal since De Seversky wrote (1950). Even Mackinder would agree that the impregnability of the Heartland is a thing of the past; his original hydrographic considerations, as D. W. Meinig has indicated, are today virtually meaningless.[12] And yet the Heartland Theory continues to excite comment; in modern political geography its qualities in terms of theory may not have much value (it is too broad, too general, too imprecise), but in the context of the Soviet rise to world power status it is frequently reappraised.[13] Two of Mackinder's assumptions became reality: (1) the political (or at least the ideological and eco-

nomic) unification of the Heartland area, and (2) the power-oriented exploitation of its resources through the innovation of Soviet economic planning. Whether Mackinder had the benefit of historical accident or special insight, the fact is that the Heartland today sustains one of the two most powerful states in the world.

Regions in the Soviet Realm

Earlier in this chapter we looked at the physical makeup of the U.S.S.R., and divided the huge land area into eight individual regions. Now we are about to try another regionalization of the Soviet state, but this time our primary interest is less the land surface itself than the mark man has made on it through his exploitation of soil and mineral resources. Neither shall we incorporate the whole country in our seven regions. What matters is the distinctiveness of the Soviet Union's clusters of human activity and achievement.

1. *Moscow and the cities of the West.* It is appropriate to begin with the Soviet Union's core area, for here lies the origins of the state, and much of what we shall discuss later is present now and still emerging. Moscow, we know, was heir to Kiev as the early center of Russian consolidation and expansion. Though located in the southern forests and in rather meager country, Moscow had centrality—a situation that proved very favorable during the decades of conquest and growth. On the map it is the river system west of the Urals that really emphasizes Moscow's favored position, with the Volga and its Oka tributary draining toward the southeast, the Dvina north to the Arctic, the West Dvina west to the Baltic, and Dnieper south to the Black Sea. These were the routes of expansion; in the lands between the rivers the tsars' control was secured through land grants to loyal citizens. Southward, the frontier of expansion penetrated the better soils of the black earth belt, and a Russian breadbasket developed there. *Ostrogs* on the Volga developed into towns, and Moscow began to feel some urban competition: as early as the seventeenth century these Volga sites were manufacturing products from locally grown flax and hemp, from wood, and from the furs trapped in the forest margins.

Despite its early rise and continued primacy as the country's largest city, Moscow does not lie in an especially well-endowed region of Russia—in terms of soils or minerals. True, there was

[11] N. J. Spykman, *The Geography of the Peace,* Harcourt, Brace, New York, 1944, p. 43.

[12] See D. W. Meinig, "Heartland and Rimland in Eurasian History," *Western Political Quarterly,* Vol. 9, 1956, pp. 553–569.

[13] For some evaluations see A. R. Hall, "Mackinder and the Course of Events," *Annals of the Association of American Geographers,* Vol. 45, No. 2, June 1955, pp. 109–126; W. G. East, "How Strong is the Heartland?" *Foreign Affairs,* Vol. 29, 1950, pp. 80 ff.; D. R. Mills, "The U.S.S.R.: A Reappraisal of Mackinder's Heartland Concept," *Scottish Geographical Magazine,* Vol. 72, December 1956, pp. 144–152; and W. G. East, "The Soviet Union and the 'Heartland'," in W. G. East and A. E. Moodie (Eds.), *The Changing World,* World Book Company, New York, 1957, pp. 432–449.

early industrialization there, including iron-working at Tula south of Moscow, but the iron ores were brought in from the Urals. The area around Moscow could provide only charcoal (from the local forests) and some low-grade coal (found near Tula). But when we studied Europe we noted the self-perpetuating quality of a large city which, while it may not be positioned on any appreciable resource base, still grows into the country's leading manufacturing complex. London and Paris were examples of this: old cities with locational assets and political advantages, with the greatest single concentration of population and therefore the largest domestic market and biggest supply of labor. Industries, especially those producing consumer goods, are attracted to these cities, and their factories employ a larger percentage of the labor force than is employed in any other industrial area. Paris and London have nodality, and in this respect, too, Moscow has advantages to match.

Moscow, then, lies at the heart of what is commonly called the *Central Industrial Region*. The definition of this region varies, as all regional definitions are subject to debate. Some geographers prefer to use the name *Moscow Region,* thereby emphasizing the point that for over 250 miles in all directions from the capital, regional orientations are toward the historic focus of the state. One advantage in this last name is that it avoids the suggestion that this region is the only industrial center of the U.S.S.R., or that it leads in all sectors of the industrial economy. The fact is that the Moscow Region has several competitors in the U.S.S.R.; the focus of heavy industry, for example, has long been to the south, in the Ukraine, and certainly Leningrad, to the north, has been an important center—there was a time when it seemed destined to take over the lead permanently from Moscow itself. Thus other geographers prefer a wider definition of the Central Industrial Region, to include not only the Moscow-Gorky industrial coreland but also most of southern Russia to the Ukrainian border, including such cities as Kursk, Voronesz, and Borisoglebsk.[14]

In this broader view of the Central Industrial

Region, the existence and recent exploitation of iron ores at Kursk—the Kursk Magnetic Anomaly—takes on great importance. Kursk lies south of Moscow, not far from the Ukrainian border. The existence of a huge iron ore deposit there had been known for a long time, but there were technological problems that made its exploitation impractical. In the 1950's, however, these obstacles were overcome, and mining began at the K.M.A. in 1959. Now, for the first time, the Central Industrial Region is benefiting from a major domestic resource: the iron ores southwest of Kursk are said to be among the largest in the world. They have already given Tula, the old iron-working center, a new lease on life, as the small steel industry there is being expanded. And, of course, the iron ores are being railed elsewhere too, especially to the steel mills of the Ukraine itself.

But this new development nowithstanding, the Moscow-Gorky area remains the leading urban-industrial cluster within the Central Industrial Region. Moscow (8 million) still remains the Soviet Union's primate city, exceptionally large and exceptionally expressive of national culture and ideology. Moscow survived Peter the Great's decision to move the country's political capital to St. Petersburg, although it temporarily lost its primacy. Among the assets that helped Moscow maintain its strength in the Russian state during the rise of St. Petersburg were its centrality, its greater proximity to the developing south, its function as the religious headquarters of the country, and its large landlord-aristocracy that controlled the agricultural economy.

Moscow thus held its position as a cultural and economic focus of Russia, and when the Communist period began it was only natural that the city should regain its former political status. With the present distribution of population, Moscow has great centrality: roads and railroads radiate in all directions to the Ukraine in the south, to Minsk, Belorussia, and Europe in the west, to Leningrad and the Baltic coast in the northwest, to Gorky and the Ural area in the east, and to the cities and waterways of the Volga in the southeast; a canal links the city to this, the U.S.S.R.'s most important navigable river. Moscow is the focus of an area that includes some 45 million inhabitants (about $\frac{1}{5}$ of the country's total population), many of them concentrated in such major cities as Gorky (over 1 million), the automobile-producing "Soviet

---

[14]For a comparison between these definitions, see the confined C.I.R. as mapped in P. E. Lydolph, *Geography of the U.S.S.R.,* Wiley, New York, 1964, p. 31. and the more extensive region on the map on p. 120 in D. M. Hooson, *The Soviet Union: People and Regions,* Wadsworth, Belmont, 1966.

Moscow by night. In the center is St. Basil's Cathedral, built by Czar Ivan the Terrible in 1552 to mark the fall of Kazan. In the left rear is Moscow University's Palace of Sciences, and in the right foreground is the Spassky (Savior's) Tower, one of the main entries to the Moscow Kremlin, first built in 491 A.D. and reconstructed in 1769. (Camera Press—PIX)

Detroit," Yaroslavl (½ million), the tire-producing center, Ivanovo, the "Soviet Manchester" with its textile industries, and to the south of Moscow, the mining and metallurgical center of Tula, where lignite deposits are worked.

The paucity of natural resources within the Central Industrial Region is counterbalanced by that excellent transport system which facilitates the inflow of a wide range of commodities, including energy sources such as gas from Stavropol, petroleum from the Volga fields, and electricity from hydroelectric dams as far away as Kuybyshev and even Volgograd. The greatest flow, in addition to food, is of materials that are processed, treated, or finished in the C.I.R. and then distributed to the consumer population. Timber, all kinds of metals, wool, cotton, flax, and leather are but a few of the many items that flood into the Moscow area, and the manufac-

tures of the C.I.R. are marked, as are those of London and Paris, by high values in relation to bulk. Moscow has always been the leading center of skilled labor in the Soviet Union, and the available skills are those which, as Russia found out after Western Europe, do not develop quickly and cannot be imparted overnight to peasant labor freed of farm work. Time and place have favored Moscow and the Moscow area, and the Central Industrial Region truly is the heart of the Heartland.

Leningrad, as the Soviets renamed Petrograd in 1924, after Lenin's death, remains the Soviet Union's second city, with under 4 million people. Leningrad has none of Moscow's locational advantages, at least not with reference to the domestic market; it lies well outside the Central Industrial Region and, in effect, in a northwestern corner of the country, 400 miles from

Moscow. Neither is it better off than Moscow in terms of resources. Fuels, metals, and foodstuffs all must be brought in, mostly from far away; the Soviet emphasis on self-sufficiency has even reduced Leningrad's asset of coastal location—raw materials might be imported much more cheaply across the Baltic than across Middle Asia. Only bauxite deposits lie nearby, at Tukhvin. But Leningrad was at the vanguard of the industrial revolution in Russia, and its specialization and skills have remained. In the late 1950's the city and its immediate environs contributed nearly 5 percent of the U.S.S.R.'s manufacturing, much of it through high-quality machine building.[15] In addition to the usual association of industries (metals, chemicals, textiles, and food processing), Leningrad has major shipbuilding plants and, of course, its port and the nearby naval station of Kronstadt. The aggregate, though not enough to maintain its temporary advantage over Moscow, has kept the city in the forefront of modern Soviet development.

2. *Povolzhye.* Povolzhye is the Russian name for an area that extends along the middle and lower Volga River. It would be appropriate to call this the Volga region, for the great river is its lifeline, and most of the cities that lie in the *Povolzhye* are situated on its banks. In the early 1950's a canal was completed to link the lower Volga with the lower Don River, extending this region's waterway system still further. On the map it can be seen to form a great Y-shaped route, with its base adjacent to the Ukraine, and its northern arms extending into the Central Industrial Region to the west and the Urals to the east.

The Volga River was an important historic route in old Russia, but for a long time neighboring regions overshadowed it. The Moscow area and the Ukraine were ahead in industry and agriculture; Plates 1 and 2 show the Volga River clearly and the one disadvantage emerges strongly—the lower Volga is almost desert-dry. To the north the forest takes over, and farming occurs in the cleared areas, but these forest zones are not nearly as fertile as the soils of the black earth belt. And the industrial progress that came late in the nineteenth century to the Moscow area left the *Povolzhye* little affected. Its major function remained the transit of foodstuffs and

[15] R. E. Lonsdale and J. H. Thompson, "A Map of the U.S.S.R.'s Manufacturing," *Economic Geography,* Vol. 36, January 1960, p. 42.

raw materials to and from other regions.

This transport function is still important, but things have changed in the *Povolzhye.* First, World War II brought a time of furious development, for the Volga region, located east of the Ukraine, was protected by distance from the invading German armies. Industries of the Ukraine were dismantled and reassembled in cities of the *Povolzhye;* with Moscow threatened, the Volga city of Kuybyshev became a seat of national government. Second, the Volga region proved to be the greatest source of petroleum and natural gas in the entire Soviet Union. From near Volgograd in the southwest to Perm on the Urals' flank in the northeast, an enormous oil field is estimated to contain over four-fifths of the country's total reserves. Production from this area exceeds two-thirds of that of the whole Soviet Union, and the cities of the *Povolzhye* became important refining centers, especially Kuybyshev and Syzran. The presence of this local power source stimulated other industries as well, notably the making of chemicals. And third, the transport system has been greatly expanded. The Volga-Don Canal links the Volga waterway to the Black Sea; the Moscow Canal extends the northern navigability of the system into the very heart of the Central Industrial Region, and via the Mariinsk Canals the Baltic Sea can be reached. Meanwhile, the course of the Volga River itself is being controlled by the construction of a series of great dams. A great reservoir was created by a dam upstream from Kuybyshev, which backed water up both the Volga and the Kama Rivers. Another dam was built at Volgograd, and further construction is in the process of transforming the remainder of the great Volga into a step-like series of artificial lakes, providing complete control over its waters. The by-products of such water control, hydroelectric power and irrigation possibilities, are adding still further to the development of the *Povolzhye.*

The major cities of the Volga region lie spaced at rather similar intervals along the river: Astrakhan, Volgograd, Saratov, Kuybyshev, and Kazan are all about 250 miles apart. The northernmost of the large *Povolzhye* cities, Kazan, lies just outside the Central Industrial Region, near the confluence of the Volga and the Kama. The largest city, Kuybyshev (1 million) is the focus of the whole region, the center of the oil fields that have sprung up, and the site of a growing industrial complex in which chemicals play an important part. The populations of the Volga cities are quite similar: smallest is the Caspian port of Astrakhan (300,000). Born of the Volga trade, the cities of the *Povolzhye* were not, in their subsequent industrial development, dominated by a particular association of local resources; hence a wide variety of industries were established, a variety that is characteristic still of the region, the recent oil-gas period notwithstanding. Manufactures range from Kazan's leathers and furs, Volgograd's metallurgy, and Saratov's chemicals to Astrakhan's fish and salt processing, Kirov's woodworking, and Kuybyshev's oil refining. With their fortuitous location for the import of raw materials and the domestic distribution of finished products, all have diversified manufacturing bases, and the *Povolzhye's* contribution to the national economy is rapidly rising.

3. *The Ukraine.* Soviet strength has been built on a mineral wealth of great variety and volume. With its immense area the country possesses an almost limitless range of raw materials for industry, although some deposits are located in remote areas and require heavy investments in transportation. Among these isolated resources are the nickel mines of the Kola Peninsula (between the White Sea and the Barents Sea) and at Norilsk near the mouth of the Yenisey River, and the copper mines in the Kazakhstan desert. The exploitation of these distant deposits, as we have mentioned, has to do with the long-standing Soviet desire for self-sufficiency. While it might be cheaper to import copper and nickel from foreign sources, Soviet planners could allocate the necessary funds for the mining and transportation of any domestic resources.

Many of the U.S.S.R.'s major sources of raw materials, actually, lie in quite favorable locations (as, for example, the Volga oil fields, which have been connected by pipelines directly to the markets in the Central Industrial Region). In the Ukraine the Soviet Union has one of those regions where major deposits of industrial resources lie in relatively close proximity. The Ukraine began to emerge as a major region of heavy industry toward the end of the nineteenth century, and one major reason for this was the Donets Basin, one of the world's greatest coal-fields. This area, known as *Donbas* for short, lies north of the city of Rostov; in the early decades this Donbas field produced over 90 percent of all the coal mined in the country. Of course other

fields have since been opened up, but today, nearly three-quarters of a century later, the Donets Basin alone still accounts for between one-quarter and one-third of the total Soviet output, which is still about double that of the second ranking producing area. And the Donets Basin fields have some other qualities: the coal is of high grade, some of it is anthracite, and most of it is suitable for coking purposes. Indeed, over half the anthracite and coking coal production of the Soviet Union in recent years have continued to be derived from the Donbas—notwithstanding the fact that the most accessible coals have been mined out and production is becoming more difficult and expensive.

What makes the Ukraine unique in the Soviet Union is the location, less than 200 miles from all this Donets coal, of the Krivoy Rog iron ores. Again the quality of the deposit is high, although the better ores are being worked out and the industry is now turning to the concentration of poorer grade deposits. Major metallurgical industries arose on both the Donets coal and the Krivoy Rog iron: Donetsk and its satellite Makeyevka dominate the Donets group (these constitute what might be called a Soviet Pittsburgh), while Dnepopetrovsk is the chief center of the Krivoy Rog cluster (Krivoy Rog remains the largest iron and steel producer). One way or another all the major cities located nearby have benefited from the fortuitous juxtaposition of minerals in the southern Ukraine: Rostov, Volgograd, and Kharkov near the cluster of Donbas cities and Odessa, and even Kiev not far from the Krivoy Rog agglomeration. And like the Ruhr, the Ukraine industrial region lies in an area of dense population (and hence, available labor), good agricultural productivity, and adequate transportation systems; it lies near large markets as well. In addition, it has provided alternatives when exhaustion of the better ores began to threaten; not only are large lower grade deposits capable of sustaining production in the foreseeable future, but iron ores also exist near Kerch, on the eastern point of the Crimea Peninsula, quite close enough for use in the established plants of the Ukraine. The biggest development in recent decades, of course, is the opening up of the Kursk Magnetic Anomaly, which we discussed previously; this area may technically lie outside the Ukraine, but its ores are of critical value in keeping the heavy industrial complexes of the Ukraine going.

And this is not all. The Ukraine and areas immediately adjacent to it have several other essential ingredients for heavy industry. Chief among these is manganese, which is needed in the manufacture of steel. About 12 pounds of manganese ore, on the average, goes into the making of a ton of steel. The Soviet Union has ample supplies of this vital commodity. Just south of Krivoy Rog, at Nikopol, lies one of the world's largest manganese ores. Another great manganese deposit lies between the Greater and Lesser ranges of the Caucasus, at Chiatura. In fact, these two manganese sources are generally considered to be the two leading deposits in the world. It is noteworthy that the United States, the world's greatest steel producer, is quite deficient in manganese and has to import this material from various sources; after World War II, but before the beginning of the Cold War, the Soviet Union exported a great deal of manganese to the United States. But these imports were cut off shortly after 1948.

Still there is more. To the southeast, along the margins of the Caucasus in the Russian S.F.S.R. and on the shores of the Caspian Sea in the Azerbaydzhan S.S.R., there are significant oil deposits—certainly not far from the Ukrainian industrial hubs by Soviet standards of distance. A pipeline connection runs from the Caspian coast along the Caucasus foothills to Rostov and Donetsk. The Volga-Urals field has overtaken these southern oil fields, which center on the old city of Baku, but production continues here.

The Ukraine provided the Soviets with their opportunity to gain rapidly in strength and power, and when World War II broke out, this was the center of heavy industry in the country. But fortuitous as its concentration of raw materials and manufacturing was, it also contributed to Soviet vulnerability. The industries of the Ukraine and the oil fields of the Caucasian area were prime objectives of Germany's invading armies. As the Soviets were forced to withdraw, they dismantled and even destroyed more than a thousand manufacturing plants so that they might not fall into the hands of the aggressors. As the Ukraine was cut off from that part of the country which remained free, mining and industrial development to the east were pressed to make up for the resulting deficiencies. And thus began a series of developments that has pulled the Soviet center of gravity steadily eastward; Soviet planners were impressed with the need for greater regional dispersal of industrial production, not just because the Ukrainian and other western mines might eventually be worked out, but also for strategic reasons.

The Ukraine (loosely translated, the name means "frontier"), for all the Soviets' encouragement of an eastward march of population and economic development, remains one of the cornerstones of the U.S.S.R. With nearly 45 million people it has between one-quarter and one-fifth of the entire Soviet population (though less than 3 percent of the land), and its industrial and agricultural production is enormous. With all that spectacular industrial growth, it is easy to forget that part of the Ukraine was pioneer country because of its agricultural possibilities, not its known minerals. Even today half the people live on the land rather than in those industrial cities; in the central and western parts of the republic rural densities may be as high as 400 per square mile. Wheat and sugar beets cover the landscape where the soils are best—in the heart of the Ukraine—and the moisture is optimum. Map 5-4 indicates the position of the Ukraine in the Soviet framework.

What has been said about the Ukraine and

A Soviet breadbasket: harvesting wheat on the fertile flatlands of the Ukraine. (Camera Press—PIX) the comparative recency of the major eastward push of the U.S.S.R. should not lead to the conclusion that there was no progress at all in mining and industry in eastern areas—the Urals and beyond—prior to World War II. Indeed there was—but not with the speed, intensity, and apparent urgency that have marked postwar development. Between the Volga River and Lake Baykal new mines, new industries, even new cities have arisen, whose contribution to the Soviet economy is increasing to such a degree that it is possible to recognize an eastward extension of the Soviet core area. As Professor Hooson put it in the title of a book relating these changes: are we seeing here the evolution of "A New Soviet Heartland?" [16]

This apparent eastward extension of the Soviet core, seen on the Eurasian map, is a continuation of that remarkably persistent west-east belt of available raw materials that begins in England, traverses mainland Europe from northern France through the Ruhr to Silesia, and then continues across the southern Ukraine (where Krivoy Rog's iron and the Donets coal are part of it) into Asia. At present the belt east of the Ukraine shows three major concentrations of resource-based industries: (1) the southern and eastern Urals, (2) the Karaganda area in the Kazakh S.S.R., and (3) the Kuznetsk Basin and its eastward projection toward Lake Baykal. In addition, important developments are taking place in the Soviet Far East, and a fourth region may be emerging between Komsomolsk and Vladivostok in the basin of the Amur River.

4. *The Urals.* The Ural Mountains, seen by some as the eastern boundary of Europe, form a north-south break in the vast Russian-Siberian plainland. In the north the Urals are rather narrow, but to the south the range broadens considerably; nowhere are the Urals particularly high, nor do they form any real obstacle to east-west transportation. Roads, railroads, and pipelines cross the range in many places. The middle and southern Urals lie amidst the Soviet Union's second-ranking heavy industrial region. (It ranks third overall.) At one time this was Russia's chief source of iron, and since the slopes of the mountains are clothed in timber charcoal was in plentiful supply. But when modern methods of steel production made their appear-

ance, the focus shifted to the Ukraine; the question of coal supply to the Urals has always been a problem. Not much good coking coal exists here, and it is railed in from hundreds of miles away—even from the Kutznetsk Basin, a thousand miles to the east. The range of metals located in and near the Urals, on the other hand, is enormous. The iron ores are still productive and famous for their high quality; there is copper, nickel, chromite, bauxite, and a host of other deposits. From one side of the Urals to the other, industrial cities and towns have developed, including Magnitogorsk (named after the rich iron ore with its quality of magnetism, magnetite), Chelyabinsk (on the eastern flank of the Urals), and Orsk; farther north lie Sverdlovsk, Nizhnil, Tagil, and Serov.

This area, more or less on a par with the Ukraine in terms of total production of manufactures, rose to prominence during World War II and in its aftermath. Heavy exploitation of the available iron ores soon led to exhaustion of the highest grades, and today the situation is such that even for this commodity long-distance transport hauls have become necessary. Not long ago the situation was greatly improved by the discovery of a large group of iron ore bodies in the valley of the Tobol River, some 200 miles east of Magnitogorsk. The center of the mining area is Kustanay. And other possibilities still remain: there is coal in the Pechora Basin in the far north, on the western flank of the Urals, and the northern metal manufacturing cities can use the huge (though low grade) iron deposit of Kachkanar. The whole Urals industrial region is so well developed, so specialized, and the product of such heavy investment that it is likely that Soviet planners will see to it that the flow of raw materials is not interrupted.

If important developments have taken place to the east of the Urals, in the direction of Lake Baykal, certainly there have also been significant changes in the other direction, toward the Volga River. It would perhaps be reasonable to speak of a Urals-Volga industrial region, for the problem of energy supply in the coal-poor Urals area has been considerably relieved by the oilfields found to lie west of the Urals; this fuel is pipelined into various parts of the Soviet core area, to Eastern Europe, and even as far into Soviet Asia as Irkutsk near Lake Baykal. Natural gas, too, is piped cheaply from distant sources to the Urals, for instance, from Bukhara in the Uzbek

[16] D. J. M. Hooson, *A New Soviet Heartland?* Princeton, Van Nostrand, 1964.

The problems of distance and access are reflected by this photography of the Turkestan-Siberian Railway in the Boomsky Gorge, in Kirghizia. To the right is the Chu River, and the road to the left is known as the Frunze-Przhevalsk Highway. (V. Baidalov/Sovfoto)

S.S.R. Thus the Urals industrial area, the Volga area (centered on Kuybyshev), and the Moscow-Gorky area are growing toward one another, and the core region of the Soviet Union is being consolidated. But eastward, too, there is growth; the core region by any measurement has long ago crossed the Urals and is expanding into the Asian interior.

5. *The Karaganda Area.* More than 600 miles east-southeast of Magnitogorsk lies one of those "islands" of mining and industrial development that form part of the ribbonlike eastward extension of the Soviet Heartland. This is the Karaganda-Tselinograd area, positioned in the northeastern part of the Khazakh S.S.R. Although it is much smaller than the Ukraine and the Urals, this is no longer merely a mining outpost, as it was before World War II when it sent coal to the Urals. Today it possesses a variety of industries, including iron and steel, chemicals, and related manufactures, and Karaganda, no longer a frontier town, has a population approaching three-quarters of a million. Rather than a supplier of raw materials, the Karaganda area now exchanges such commodities with other productive areas: in exchange for the coal it continues to send to the Urals, it receives iron ores from the Urals' own source area at Kustanay.

Among Karaganda's advantages is that of position in the path of the Soviet economy's eastward march. Not only are there rail connec-

tions and cost-reducing exchanges with the Urals, but eastward, too, opportunities for such arrangements exist. In addition, further mineral discoveries have been made in the vicinity, including a sizable deposit of medium-grade iron ore. Less than 200 miles to the northwest and on the railroad to the Urals, Tselinograd has emerged as the urban focus for the great Virgin Lands agricultural project, launched in the 1950's. And the problem of water supply, a serious one here in the dry Kazakh steppe, has been relieved by the construction of a 300-mile long canal from the Irtysh River. Nevertheless, Karaganda and its environs cannot compare to the Ukraine or the Urals, with their clusters of resources, urban centers, transport networks, agricultural productivity, and dense populations. Karaganda remains far from the present-day hub of the Soviet Union; it is still a marginally located, subsidiary developing area. Distances are great, markets are far, costs of production are high. But the future is likely to see the ever-greater integration of the Karaganda area into the spreading Soviet core. Soviet planners, by their attempts to bring vast eastern agricultural areas into production and by their willingness to have the state bear the costs involved in the industrial growth of the Karaganda region (again in large measure for strategic reasons born of World War II), have shown their hand to favor such developments.

6. *The Kutznetsk Basin.* Fully 1200 miles east of the Urals lies the third-ranking area of heavy manufacturing in the Soviet Union. In the 1930's this area was opened up as a supplier of raw materials to the Urals, especially coal, but this function of the *Kuzbas* has steadily diminished in importance while local industrialization has accelerated. The original plan was to move coal from the Kuzbas to the Urals and to let the returning trains carry iron ore to the coalfields, but subsequently good iron ores were located in the area of the Kutznetsk Basin itself. As the resource-based Kuzbas industries grew, so did the urban centers: Novosibirsk (with over a million inhabitants) stands on the Siberian Railroad as the symbol of Soviet enterprise in the vast eastern interior. To the northeast lies Tomsk, one of the oldest Russian Siberian towns in the whole region, founded three centuries before the Bolshevik takeover and caught up in the modern development of the Kutznetsk area. Southeast of Novosibirsk lies Novokutznetsk, a city of a half

Modern conveniences are still not available in many a Soviet frontier town. In this rather typical scene, the old quarter of Yakutsk is still served by a water cart, drawn through the muddy, unpaved streets by a horse. The homes in the picture have no plumbing. Life in winter is tough. (Camera Press—PIX)

million people specializing in the manufacture of heavy engineering products such as rolling stock for the railroads; aluminum products (of Urals bauxite) are also made here.

Impressive as the proximity of coal, iron, and other resources may be in the Kutznetsk Basin, the industrial and urban development that has taken place here must in large measure be attributed, once again, to the ability of the state and its planners to promote this kind of expansion notwithstanding what capitalists would see as excessive investments. When a railroad train full of coal leaves the Kuzbas bound for the Urals, it may arrive at its destination one-third empty—having consumed as fuel that much of its cargo. When a product is made in Novosibirsk to be sold on the markets of the Moscow area or even in the Ural cities, those markets are hundreds—many hundreds—of expensive miles away. But the Soviets have been willing to make those heavy investments; they long denied the citizenry the consumer goods that different economic tactics might have been available. In return they were able to push the country vigorously ahead on the road toward industrialization, with the hope that certain areas would successfully reach "jumping off" levels, after which they would require less and less direct investments and would to an ever greater extent

be self-perpetuating. The Kuzbas, for example, was expected to grow into one of the Soviet Union's major industrial regions, with its own important market, and with a location, if not favorable to the Urals and points west, then at least fortuitous with reference to the developing markets of the Soviet Far East.

Between the Kutznetsk Basin and Lake Baykal, east of the Kuzbas, lies one of those areas that has not yet reached the "jumping off" point. This area has impressive resources, including coal, but Soviet planners have been uncertain exactly where to make the necessary investments to make this a significant industrial district. There are possibilities of hydroelectric power from the Yenisey River, timber is in plentiful supply, brown coal is used for electricity production, and oil is piped all the way from the Volga-Urals fields to be refined at Irkutsk. Thus, at Krasnoyarsk and Irkutsk, both cities in the half-million class, a variety of industries including metal-processing, woodworking, and chemical manufacture have already been established. But the whole area lies several hundred miles farther again from the established core of the country, and while the distant future may look bright (especially if Soviet plans for the Pacific areas become reality), the present situation still presents difficulties.

7. *The Pacific Area.* The Soviet Union has about 5000 miles of Pacific coastline—more than the United States including Alaska. But most of this coastline lies north of the latitude of Washington State and, from the point of view of coldness, the Soviet coasts lie on the "wrong" side of the Pacific. The port of Vladivostok, the end-point of the Trans-Siberian Railway, lies at a latitude about midway between San Francisco and Seattle, but must be kept open throughout the winter by icebreakers—something unheard of in these American ports. The climate, to say the least, is severe. Winters are long and bitterly cold, and summers are cool. Nevertheless, the Soviets are determined to develop their Far Eastern Region as intensively as possible, and their resolve has recently been spurred by the growing ideological conflict between the U.S.S.R. and China. Farther to the west, high mountains and empty deserts mark the Chinese-Soviet boundary; south of the Kuznetsk Basin and Lake Baykal, the Republic of Mongolia functions as a sort of buffer between Chinese and Soviet interests. But in the Pacific area, the Soviet Far Eastern Region and Chinese Manchuria confront each other across an uneasy river boundary (along the Amur and its tributary, the Ussuri). Hence the Soviets want to consolidate their distant outpost, and are offering inducements to western residents to "go East."

Although it has a rigorous climate and difficult terrain, the Far East is not totally without assets. While most of it still consists of very sparsely populated wilderness, the endless expanses of forest (see Plate 4 in rear of volume) have slowly begun to be exploited by a timber industry whose major problems are the small size of the local market, the bulkiness of the product, and—as always in Soviet Asia—the enormous distances to major markets. Lumber stations and fishing villages break the emptiness of the countryside; fish is still the region's major product, and from Kamchatka to Vladivostok Soviet fishing fleets sail the Pacific in search of salmon, herring, cod, and mackerel, to be frozen, canned, and railed to the western markets.

In terms of minerals, too, the Far East has some possibilities. A deposit of coking-quality coal lies in the Bureya River valley (a northern tributary of the Amur), and a lignite field near Vladivostok is being exploited; on Sakhalin there are minor quantities of bituminous coal and oil. Not far from Komsomolsk (250,000) lies a body of poor-grade iron ore, and this city has become the first steel producer in the region. In addition to lead and zinc, the tin deposits north of Komsomolsk are important because they constitute the country's major source. Thus the major axis of industrial and urban development in the Pacific area has emerged along the Ussuri-lower Amur river system, from Vladivostok in the south, with its naval installations, shipbuilding, and fish processing plants, to Komsomolsk, the iron and steel center, in the north. Near the confluence of the Amur and Ussuri Rivers lies the city with the greatest advantages of centrality, Khabarovsk. Here metal manufacturing plants process the iron and steel of Komsomolsk, chemical industries benefit from Sakhalin oil, and timber is used in sizable furniture industries. Khabarovsk, a city of somewhat under a half million people, has become the region's leading urban center. Railroads lead south toward Vladivostok, north to Komsomolsk and beyond, and west to the Bureya coalfields, to Svobodny, and eventually to the Kuzbas and on to Moscow.

The farther eastward the Soviet Union's developing areas lie, the greater is their isolation and the larger their need for self-sufficiency, in view of the ever-increasing cost of transportation. Thus the Pacific area exchanges far fewer commodities with points westward than does the Kuzbas, although some raw materials from the Pacific area do travel to the west in return for relatively small tonnages of Kuzbas resources. In terms of food supply, the Pacific area cannot

An eastern Soviet outpost: Vladivostok. Shown here are the railroad station at the end of the trans-Siberian Railway and part of the port. (D. Kozlov/Sovfoto)

come close to feeding itself, despite its rather small population (the 1968 estimate is 15 million) and the farmlands of the Ussuri and lower Amur River valleys. But Soviet planners have not been sure just how to direct industrialization here. Large investments are needed in energy supply (and there are magnificent opportunities for hydroelectric development in the Amur and tributary valleys), but the choice is difficult: to spend here, in an area of long term isolation, modest resources, labor shortage, and harsh environment, or to spend money where conditions are less bleak. Once again the international situation comes into play. After World War II there seemed hope for productive and profitable exchanges with China and, increasingly, with Japan. Then the *detente* with China led to some stagnation, but subsequently Soviet strategists sought to bring the area more securely into the effective national territory. This desire must have been spurred by the clashes with China on the Sino-Soviet boundary, but it may be that the course of events will move too fast to await the consolidation of this region through population growth and economic development.

Population Problems

Little more than a century ago, Russia had more than twice as many inhabitants as the United States. As late as 1900 the Russian population was of the order of 125 million while that of the United States was not much more than half this figure. Yet today the Soviet population is just 245 million, only 35 million larger than that of the U.S.A. What has brought about this remarkable change in population ratio between the two countries?

The unhappy answer is that the twentieth century brought repeated disaster to the Soviet people, notwithstanding the fact that their country rose to a position of world power during that period. By comparison, the population of the United States, though suffering casualties in war as well, sustained far fewer losses. Part of the U.S.S.R.'s losses were self-inflicted; for example, during the collectivization period of the early 1930's, famines and killings probably cost almost 5 million lives. But the major destruction, staggering in its magnitude, was caused by the two World Wars. World War I, its aftermath of civil war, and the attendant famines resulted in some 17 million deaths and about 8 million deficit births (that is, births that would have

taken place had the population not been reduced in such numbers). World War II cost the Soviet Union approximately 27 million deaths and 13 million deficit births. Thus, during the twentieth century, the Soviet population lost over 70 million in destroyed lives and unborn children, while United States losses in all its twentieth century wars stands well under 1 million. In addition, the Soviet Union was marked by fairly heavy emigration, while the United States received millions of immigrants. Hence the U.S. population grew rapidly while the Soviet increase was severely limited: in the half century from the beginning of World War I to 1964 the U.S. gained some 90 million people (from a base of 100 million) while the U.S.S.R. gained about 65 million. Thus U.S. population nearly doubled in that period, while the Soviet population in a half century grew by only one third.

Population Policies. Naturally the results of these twentieth century calamities have concerned Soviet governments and planners. The deficit of births is today reflected in a shortage of young people entering the labor force; at higher ages females outnumber males to a greater extent perhaps than in any other country in the world. All this occurs just at a time when Soviet expansion into the Asian Heartland is gaining momentum, and thus official policy has been to encourage families to have several children—an attempt, thus, to raise the birth rate of the country. But this is not as simple as it sounds. Without religious obstacles to birth control, this practice has long been widespread in the Soviet Union; abortion and sterilization are considered matters of private concern only and are legal. And it is not much easier for an urban family in the Soviet Union to have a large number of children than it is for families in the cities of other parts of the world: in this respect the Soviet response to industrialization and urbanization is similar to that of the rest of the developed world. Thus the government devised a system of awards to mothers who have many children: when a wife bears her tenth child she receives the Order of Mother Heroine from the praesidium of the Supreme Soviet.

In this respect the Soviet Union is not alone. In recent years the French government, also concerned over the acute leveling off of the demographic curve for that country and the emigration of younger people, likewise, designed a system of rewards for productive families, chiefly in the form of tax relief. Such attempts to encourage national population increase are *expansionist* population policies, reflecting official desires to sustain population growth.

Certain countries, notably those whose populations are expanding so rapidly that food supplies are inadequate every year (such as those agricultural countries the neo-Malthusians worry about) are trying to limit their growth through policies opposite to those of the U.S.S.R. or France. Here the objective is the smaller family and fewer children, and large families are discouraged by insufficient tax relief for the parents, and even by the ready availability of information on birth control and contraceptive devices. Of course the economic structure of the state is always a more effective deterrent than the availability of birth control methods, but unfortunately these countries that show themselves willing to limit population growth and which make contraceptives available are not always capable of effecting the necessary socioeconomic changes. Thus India still has an average annual rate of increase of 2.3 percent (and with 500 million people, this means over 11 million people are added every year) despite this country's efforts to reduce this rate, compared to 0.7 percent for the United Kingdom and 0.6 percent for Italy. Such *restrictive* population policies, therefore, are hardly successful as yet.

Thus underpopulated countries are likely to have expansionist population policies, while overpopulated ones make attempts to restrict population growth. A third kind of policy exists, namely one that favors one particular segment of the population—usually one racial group over other racial sectors of the people. In South Africa, for example, the political and economic situation strongly favors the white minority, with the African majority in a comparatively poor position and the people of mixed blood (the "Coloreds") and the Asians somewhere in between. Government officials have publicly exhorted the white population to produce more babies "in the interest of the security of the state." Something similar, though much more extreme of course, was seen in Nazi Germany during the recognition of a Germanic "master race." Population policies favoring one racial, cultural, linguistic, religious, or other group within a state are *eugenic* policies, and in a very mild way the Soviet Union's favoring of Russian language and culture falls into this category.

Distribution. Official policy can also concern itself with the horizontal distribution of population. In the Soviet Union, such policy has been aimed at an eastward movement of the people, toward the Asian Heartland, although World War II did more to stimulate this eastward shift than Soviet policy has done. For as long as the country has existed, the great majority of the population has been concentrated to the west of the Volga River (or west of the Urals; herein lies one sound reason for the otherwise rather unfortunate choice of the Urals as the eastern "boundary" of Europe). The eastern conquests were made by relatively few pioneers whose total impact on the population distribution was minimal, and, as we know, the economic contributions of the Russian East long were slight. Indeed, the major direction of population in Russia before the twentieth century was southward rather than eastward: from the forests of central Russia people moved into the better soils of the Ukraine and the area north of the Caucasus. This movement, much more than the eastward push of the Cossacks and fur traders, resembles the American occupation of its Western "frontier;" the eastern conquests of the Russians were more in the nature of a colonization process.

About a century ago, the eastward shift of the Russian population finally commenced. Its vanguard was still the search for land by peasants: having occupied the Ukraine and the areas west of the Volga, they now crossed the river and began to move toward the southern Urals. By the time World War I broke out, this agricultural phase of the eastward shift had increased the population east of the Volga River by some 12 million people.

The most active phase of population movement, however, came with the first Five-Year Plan, after 1928. This period witnessed the U.S.S.R.'s rapid industrialization and truly spectacular urbanization, and a great increase in the economic contributions made by areas located east of the Volga. Magnitogorsk, Karaganda, and Komsomolsk were planned and laid out from scratch (among many other new urban centers); and although these developments affected the west as well as the east, the rate of growth in the east soon exceeded that in the west. World War II, which ravaged western Russia, Belorussia, and the Ukraine, left the east comparatively unscathed—and vastly more important than before in the economic picture, since western productive areas had been destroyed or dismantled. Thus the year 1941 marks an important turning point. Counting war losses, deficit births, and emigration, the population of the Soviet Union west of the Urals and north of the Caucasus for two decades remained almost stationary: the 1939 and 1959 census reports both show a population between 135 and 136 million people. Actually, there was a loss during the war years that was only made up by about 1959, but the effect remains the same: the western U.S.S.R.'s population growth had been temporarily arrested.

No such impact was felt in the east. On the contrary, unprecedented growth rates were recorded there: over the same period the Urals area acquired one-third more people, Kazakhstan's population grew by over half, and the Kutznetsk–Baykal–Far East zone gained in excess of 30 percent. No part of the country experienced such a high rate of city development either; in fact, more people moved into eastern cities such as Novosibirsk, Chelyabinsk, and Omsk than entered the cities of the U.S.S.R. west of the Volga. Thus the eastern "wedge" of the great Soviet population triangle (Plate 6) gained greatly in strength. Prior to World War II, the population of the Urals and eastward constituted just over one-quarter of the total Soviet population, and today it is approaching 40 percent.

During the 1960's the Soviets have had to encourage a continuation of this eastward shift by providing incentives for families interested in making such a move. The war impact is long behind, and the initial rush to industrialize and urbanize, which produced rates of city growth that exceeded even those of boomtime California, is now also over. A labor shortage exists in some eastern areas, and as the citizen's life in the Soviet core area improves, fewer people seem to be willing to join an eastward migration. However impressive the developments in Middle Asia and the Far East may be, one needs a certain pioneering spirit to leave the comparative comforts of the west for the isolation and often harsh climate of the east. In such cities as Khabarovsk and Vladivostok, investments are being made in high-rise apartment houses with good facilities, hopefully to attract those who might have a long wait for comparable housing in the west. Factories are also authorized to offer somewhat higher wages than are paid for similar work in the west. But in the late 1960's the eastward march of Soviet population was numerically not as great as Soviet planners would like it to be. It may be, then, that the eastern expansion of the Soviet Heartland is slowing down in the face of distance and environment, two old enemies of this giant country.

ADDITIONAL READING

Several regional geographies of the Soviet Union are now available. The volume by P. E. Lydolph, *Geography of the U.S.S.R.*, was published by Wiley in 1964. The book by R. E. H. Mellor is also called *Geography of the U.S.S.R.*, and it was published by St. Martin's Press, New York, in 1965. Another standard work, though older, is T. Shabad's *Geography of the U.S.S.R.; a Regional Survey,* published in New York by Columbia University Press in 1951. J. P. Cole's book, not surprisingly entitled *A Geography of the U.S.S.R.,* appeared in 1967 as a Penguin Books edition, in Baltimore. The volume by D. J. M. Hooson is entitled *The Soviet Union: People and Regions,* and it was published in 1966 by Wadsworth in Belmont. In the Van Nostrand Searchlight Series, No. 15 is *The Soviet Union,* by W. G. East. Still another work is that by J. C. Dewdney, *A Geography of the Soviet Union,* published in 1965 by Pergamon Press, New York. Very recently J. S. Gregory's book appeared, entitled *Russian Land; Soviet People: A Geographical Approach to the U.S.S.R.;* it was published in London by Harrap & Co., in 1968. In 1967, Wiley published a translation of G. Jorré, *The Soviet Union;* the translation was done by E. D. Laborde and revision work by C. A. Halstead.

Numerous works in English are also available on more specialized topics. A good way of reviewing Russian history is by looking at one of several historical atlases of the country, for example A. F. Chew's *Atlas of Russian History: Eleven Centuries of Changing Borders,* published by Yale University Press in 1967. On the planned economy, see the translation by I. Nove of the book by P. J. Bernard, *Planning in the Soviet Union,* published in 1966 by Pergamon Press in New York. On Soviet agriculture and the impact of Soviet practices in the East European satellites, J. F. Karcz edited a volume entitled *Soviet and East European Agriculture,* published in 1967 by the University of California Press in Los Angeles. Also see, on this subject, the book edited by R. D. Laird, *Soviet Agriculture: The Permanent Crisis,* a Praeger publication, New York, 1965. With reference to the physical environment, it is probably best to consult the relevant chapters in the regional volumes mentioned; one book not yet listed is N. T. Mirov's *Geography of Russia,* a 1951 Wiley publication which is strongly oriented toward the physical features of the country. Many details about permafrost, soils, vegetation and so forth appear in a book by S. P. Suslov, *Physical Geography of Asiatic Russia,* a translation from the Russian published by W. H. Freeman in San Francisco in 1961. On the population question, see a publication by the Milbank Memorial Fund, *Population Trends in Eastern Europe, the U.S.S.R. and Mainland China,* published in New York in 1960, and an older work by F. Lorimer, *The Population of the Soviet Union; History and Prospects,* published by Cambridge University Press in 1946. On population policies and problems see the work edited by E. Goldhagen, *Ethnic Minorities in the Soviet Union,* published by Praeger, New York, in 1968. Another revealing book is the Oxford University Press paperback by G. Wheeler, *Racial Problems in Soviet Muslim Asia,* published in London in 1962.

Descriptions of individual Soviet regions are also available. Columbia University Press in 1967 published *Central Asia: A Century of Russian Rule,* edited by E. Allworth. A Volume by T. Armstrong, *Russian Settlement in the North,* was published by Cambridge University Press in 1965. The eastern realm of the Soviet Union is covered by E. Thiel in *The Soviet Far East: A Survey of Its Physical and Economic Geography,* published by Praeger, New York, in 1957. Also in this connection, see W. Kolarcz, *The Peoples of the Soviet Far East,* a Praeger publication of 1954.

Transportation and general economics in the vast Soviet realm are discussed in a University of Chicago publication, 1960, by R. N. Taaffe, *Rail Transportation and the Economic Development of Soviet Central Asia.* Another valuable source is J. A. Hodgkins, *Soviet Power; Energy Resources, Production and Potentials,* published by Prentice-Hall in Englewood Cliffs in 1961.

On the Heartland question and related matters, much reading is available. Mackinder's own statements have been collected in a volume entitled *Democratic Ideals and Reality,* published in 1962 by Norton, New York. Another good source is D. J. M. Hooson, *A New Soviet Heartland?,* which is No. 21 in the Van Nostrand Searchlight Series. An earlier book in this series is W. A. D. Jackson, *The Russo-Chinese Borderlands: Zone of Peaceful Contact or Potential Conflict?,* published by Van Nostrand in 1962. These three books are all paperbacks, and all are available. In the library, it is worth looking at R. J. Kerner's controversial but interesting *The Urge to the Sea; the Course of Russian History,* published in 1942 by the University of California Press, Berkeley.

6

NORTH AMERICA

WITH THE exception of one small country, the average citizen in the United States currently possesses a higher standard of living than the comparable citizen of any other country in the world. More significant than this average of the citizens' well-being is the fact that this level of welfare is enjoyed by a relatively large portion of the population. Each man, woman, and child in the United States averages an annual income of $2893, second only to oil-rich Kuwait's $3184 per capita income (see Table 6-1). Canada is also relatively well placed within a listing of annual per capita incomes by country. It is easy to see why most of the population of North America often find it difficult to relate to the much lower levels of welfare so common in the rest of the world, and even in portions of the U.S. and Canada.

Perhaps more important for a substantial proportion of the population is a consideration of how well distributed the personal wealth in North America is. Obviously, wealth is not shared equally by regions any more than it is shared equally by individuals. There is a discernible spatial pattern to the differences in personal well-being. Every reader will be familiar with the term "pockets of poverty." This term has spatial significance in that it is an attempt to describe in a dramatic manner the fact that in the United States, populations with low per capita income often tend to be found in clusters. Within the pattern of income for the United States, for example, such regions as Appalachia, the Southeast, the Ozarks, and others, are clearly in visible contrast to the higher average incomes elsewhere in the country. Some of the spatially relevant sources and consequences of this pattern will be examined later.

Examination of the map of North America's population distribution, without regard to cultural or economic differences, will also illustrate

171

Table 6-1

Country	Population (Mid-1968 Estimate) (Millions)	Life Expectancy at Birth (Years)	Per Capita Incomes (Dollars)
Canada	20.7	72	1825
United States	201.3	71	2893
Sweden	7.9	74	2204
United Kingdom	55.8	71	1451
France	50.4	71	1436
Germany, West	60.3	71	1447
Czechoslovakia	14.4	71	804
U.S.S.R.	239.0	70	928
Portugal	9.5	64	351
Mexico	47.3	58–64	412
Guatemala	4.9	50–60	281
Peru	12.8	55–60	218
Venezuela	9.7	65–70	745
Brazil	88.3		217
United Arab Republic	31.8	50–55	130
Ghana	8.4	40–45	245
Nigeria	62.0		63
Kenya	10.2	40–45	77
Zambia	4.1	40–45	174
Republic of South Africa	19.2	50–60	509
Israel	2.8	72	1067
Jordan	2.2		179
Kuwait	0.5		3184
India	523.0	45	86
Pakistan	126.0	45	89
Indonesia	112.8		85
Japan	101.0	71	696

Source: "1968 Population Fact Sheet," Population Reference Bureau, New York.

a definite lack of spatial homogeneity (see Plate 6). It can be clearly seen that virtually the entire Canadian population resides in a narrow strip of land immediately north of the United States. It is not difficult to explain this preponderance of population along the southern margin of the country, but there are more interesting questions which should follow: the effect the population pattern has had on economic and political development; the added effects which follow from the generally low density found even within the population "corridor"; the effect on the United States immediately to the south, that is, the effectiveness of the political boundary separating the two countries.

It is of major significance that both the United States and Canada are what sociologists call "plural societies." A plural society is one in which there has been extended cultural contact between two or more ethnic groups, presumably within a single political unit, without any real cultural mixing of these groups having taken place. At first, this term might seem inapplicable to both the U.S. and Canada. Each country has received immigrants from many cultures in large numbers for over a century (although immigration to the U.S. declined sharply after restrictive legislation was passed following World War I). These immigrants have, to a varying degree, been absorbed into an American culture which was different from any of the parts, earning the United States the label of "melting pot." Three or four generations were usually sufficient for American (or Canadian) cultural characteristics to prevail over those of the grandparents' home country. Absorption was occasionally retarded by choice, as in the case of concentrations of farmers descending from German settlers in Pennsylvania, Michigan, and Wisconsin, or the population enclave of those with Polish background in Hamtramack, Michigan, a city within the city of Detroit. It may be severely retarded by the non-European background of the immigrants, leading to a Chinatown in San Francisco and in Chicago, Puerto Rican-American ghettos in New York, and Mexican-American ghettos in the Southwest. In spite of the fact that each of these culture groups is still recognizable after several generations, members of these populations and others like them also gradually develop cultural characteristics which identify them as distinctly American or Canadian to an observer from Europe, Africa, Asia, or Latin America.

Both countries are pluralistic, however, in that both states contain sizable minorities which differ significantly from the dominant culture, even after centuries of residence on the continent. In Canada, the great majority of the inhabitants of the Province of Quebec are of French ancestry and still maintain many cultural similarities with the French. In the United States, although possessing many similarities with the rest of the population, that portion with dominantly African ancestry have found it necessary to develop a culture of their own within the larger culture of the dominantly white population. This black culture must be considered here independently of the white for two important reasons. First, it differs from the other immigrant cultures in that it is no more like the white culture than it is like its parent set of cultures, the African. Second, the fact that for many decades, even centuries, the black population was denied access to the white culture has made the presence of a dual culture in the U.S. a divisive factor in the country as a whole. An awareness of the spatial character of this pluralism is helpful in developing an understanding of the present geographic patterns and future development in both Canada and the United States.

Cultural and racial differences, income differences both within and between the racial and cultural groups, the maintenance of cultural pluralism within two states unusual in their populations' ability and willingness to absorb other cultures without totally destroying them: all of these factors have had significance in the gradual spatial development of Canada and the United States politically, economically, and culturally. The complex geographic character of these countries can be illustrated by tracing the development of these population patterns and their distinct areal organization.

Pre-European Patterns

The pre-European inhabitants of North America possessed varying degrees of sophistication in economic and political organization, but technologically they were basically of the Stone Age. Although not known for some time following the earliest European arrival, there was apparently nothing technologically similar to the involved civilizations located in Middle and South America anywhere on the North American continent.

Archaeological evidence indicates the presence of man in North America at least 20,000 years ago. It has been difficult to grant Stone Age man sufficient technological skill for him to have traversed either the Pacific or the Atlantic Ocean, although there are proponents of theories arguing that the Americas were peopled by vessels from Eurasia via each ocean. At the present time, however, the most accepted theory holds that man arrived in the Western Hemisphere by moving overland from northeastern Siberia to Alaska during one or more periods of the recent

MODIFIED AFTER: CARL SAUER, LAND AND LIFE

Map 6-1 EARLY MIGRATION ROUTES INTO THE AMERICAS.

In spite of the modifications which have taken place, these Navaho Indians are clearly representatives of the Mongoloid racial group. The quality of the land from which they may make their living can be seen in the background. (Burk Uzzle/Magnum)

Ice Age. During maximum glacial coverage, a tremendous quantity of the earth's water was removed from the ocean basins and was deposited on North America and northern Eurasia in the form of ice. Of course, this had the effect of lowering sea level as much as several hundred feet, thus extending continental coastlines by many miles. Those locations along the coasts where the oceans are now less than several hundred feet were at that time dry land, above the depressed sea level. The Bering Strait, now separating Asia and North America, was one such area of exposed land. Undoubtedly following

game or game trails, hunting and gathering people would have been able to move southeastward through the largely ice-free channels between the western mountain ranges (Map 6-1).

The movement of the early American Indians from Northeast Asia is also supported by the similarity in physiological characteristics with Mongoloid peoples vis-a-vis either Caucasoid or Negroid peoples. There is, in fact, a wide range in physical appearance of the American Indian, but this is not unexpected. Low population densities were prevalent over much of the continent, and obviously, it is intermarriage that leads to a merging of physical appearance among members of distinct groups.

Although some unity of physical characteristics was present among the North American Indians, the complex tribal organizations and methods of livelihood were considerably different. This variety of societal organization existing among the Indians during the fifteenth and sixteenth centuries provides interesting illustrations of the argument over the validity of *environmental determinism*. (This concept is discussed more fully in Chapters 1 and 7.)

Remember that environmental determinism, or environmentalism, in its simplest form argues that man's behavior, his success as an individual, and the level of his civilized development are determined by the type of environment in which the man or group lives. The inhabitants of eastern North America lived in an environment similar in many respects to that found in Eastern Asia at comparable latitudes. (Note the patterns on Plate 2.) That is, the technologically unparalleled Chinese Empires were developed and advanced in environments similar to that found by European explorers to contain men with much less control of their environment. Granted, there are differences between the conditions, but it remains that civil organization and levels of technology in North America were far from those sought by Columbus.

More significant for the present population of North America was the type and degree of development that *had* taken place prior to the European invasion. Although the earliest white settlers found maize to be grown along the East Coast, tribes of the Northeast possessed materials originating in the Southeast, and vice versa (Map 6-2). The Indian tribes, then, were essentially engaged in subsistence activities, while some long-distance trade was conducted. The

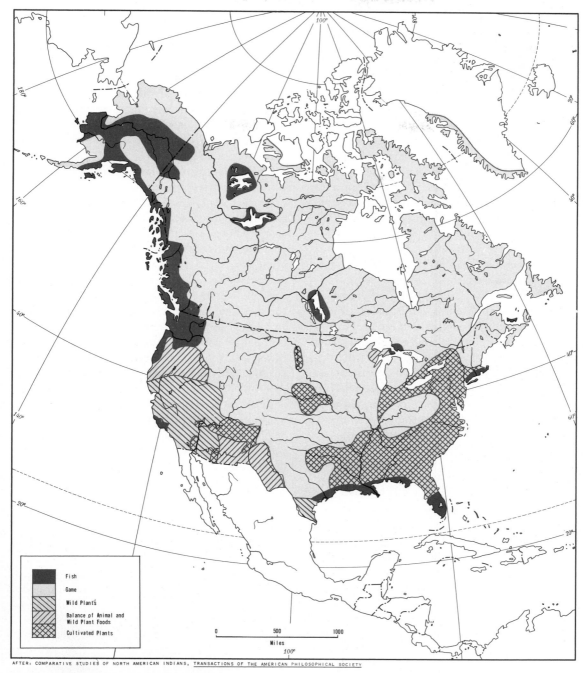

Map 6–2 DOMINANT PRE-EUROPEAN SUBSISTENCE.

AFTER: COMPARATIVE STUDIES OF NORTH AMERICAN INDIANS, <u>TRANSACTIONS OF THE AMERICAN PHILOSOPHICAL SOCIETY</u>

Legend:
- Fish
- Game
- Wild Plants
- Balance of Animal and Wild Plant Foods
- Cultivated Plants

0   500   1000

Miles

Great Lakes were used by Indians of the Northeast for trade of products of the West (copper and desirable stones) with those of the East (tobacco and coastal products). In the Southeast, spatial complementarity was marked by trade between the coastal peoples of the lowlands and the interior tribes of the highlands. In many instances, it was along the trade routes established by the Indian that European goods and migrants later moved.

Complementarity of economic activities was reinforced by similar differences between cultures. These distinctive differences may be seen as variable adjustments to differing environments. The greater the disparity between environments, often, the greater the differences between types of housing, dependence on hunting, modes of transportation, clothing styles, and so forth. This does not necessarily support the environmentalist position but only indicates the relatively low level of technology held by all tribes. If technology is viewed as a method of "compensating" for environmental deficiencies, that is, of adjusting the environment to suit the knowledge at hand rather than the reverse, the response of man to the character of his surroundings is more related to his developed capacities (and values) than it is to some immutable "force" limiting him to only a few alternatives. Man is limited by himself.

Because the variations were many, and because they were primarily adjustments to the local environment, Indian contributions to the incoming Europeans have become increasingly obscured. Linguistic and cultural contributions are numerous, although much is distorted to fit the mythology associated with westward expansion of the white population. To the early settler, with technology less adapted to North American conditions than that of the Indian, the assistance provided was often of immeasurable help. The manner in which this assistance was rewarded can be seen in the decline in Indian population in North America (about one million in 1492 but only 200,000 by 1900).[1]

Patterns in Colonial America

For fully one hundred years following Columbus's initial voyage, there was almost no penetration of North America. Spain's American empire was the most extensive by the end of the sixteenth century but even this was still limited to the continent's southern fringe, immediately north of Mexico and the Gulf of Mexico. Spain was the only European power with the Pope's approval in North America, as will be discussed later. Throughout the sixteenth century, Spain was also one of the two most powerful maritime powers in Europe.

This official approval from the Pope and Spanish naval strength, however, were not sufficient to eliminate explorations by the other nations of Europe. In the same year that Vasco da Gama rounded the southern tip of Africa to be the first European to reach India directly by ship within the modern era, thus effectively stealing much of European attention from Columbus, John Cabot was pursuing the first of two exploratory trips from England to North America. It was on the basis of these early voyages that later English claims were made. France sent out explorers in the early sixteenth century as did England again later in that century and the Netherlands early in the seventeenth century.

The period of time elapsing between first discovery and first settlement is significant. It indicates the colonial goals of the Spaniards (exploitive and extractive), which in turn reflect the economic and often the political and cultural patterns which developed in Spanish (and Portuguese) America over the following centuries, patterns very different from those which developed in North America. It also meant that by the time the continental margins were settled by significant numbers of Europeans, the economic and political foundations for the industrial revolution and the related Representative, or "Democratic," Revolution were much closer to completion. The increasing tempo of commercial activity in Europe was an important stimulus to the early settlers' successes.

The French were able to claim, and initially control, the two major water routes into and out of the continent. French interest in "New France," however, was largely as pecuniary as Spanish interest in New Spain. Furs were the primary economic reason for control of the northerly portion of New France which meant that the population of this segment of the empire remained sparse. As the Mississippi River and its tributaries were also seen as an approach to the fur-trapping regions and it was only the exclusion of other, competing European interests that were of primary concern, major efforts were needed only at the mouth of the rivers and selected, strategic locations along their routes. Montreal and Quebec on the St. Lawrence and New Orleans on the Mississippi became the French strong points, both militarily and culturally. Thus, French influence in North America, as Spanish influence earlier, was localized and basically peripheral.

In contrast to the French penetration, the general thrust of English settlement in North America was coastal throughout the seventeenth century. Granted, initial movement was up the major rivers emptying into the Atlantic Ocean;

[1] For a more complete discussion of the many contributions of the pre-European cultures in North America to the present inhabitants, see Harold E. Driver, *Indians in North America*, University of Chicago Press, Chicago: 1961, Chapter 26, "Achievements and Contributions," pp. 583–612.

nevertheless, there were enough separate towns and associated farming settlements established to provide an extremely broad front for inland penetration by the first half of the eighteenth century. The French holdings were greatly over-extended for the manpower available, while the English developed a broad coastal base which could not be stifled by the loss of only one or two sites.

Sharp differences can also be seen between the several English colonies. The first formally established, permanent colony was begun on the James River in Virginia in 1607. Initial settlement was along the short navigable stretches of the many rivers crossing the flat coastal plain both north and south of Jamestown. By learning the techniques of growing tobacco from the Indians (once it became apparent that gold and gems were not locally available), the plantation economy became quickly established in Virginia and the newly formed colonies of Maryland and the two Carolinas. As in Virginia, tobacco was the basis of the economy in early Maryland while to the south, commercial rice, indigo, and sea-island cotton provided the base for the plantations of North and South Carolina and northern Georgia.

There were several significant results which followed the continuation of this form of economic development in the southern colonies. One was the sharp distinction which developed between northern and southern colonial economies. This distinction remained after the American Revolution and in many ways laid the bases for noneconomic differences which have carried to the present. Another is related to the character of a plantation economy. The plantation system is basically a feudal agricultural economy with a few owners of large properties each employing a considerable amount of labor. Each of the major crops grown in these southern colonies required great amounts of hand labor. Although apparently introduced accidentally in 1619, slaves from Africa came to be a major source of this labor. The Africans were supplemented by bonded, or "indentured" servants from Europe. These latter were virtually short-term contract slaves in spite of the careful distinction (carried down even to the present) between white "servants" and black "slaves." And a third result, the settlement pattern in the southern colonies was relatively dispersed with only a few large coastal ports. The dominating plantation econ-

omy, again, did not lead to dense settlement or require numerous urban centers. Any reasonably prosperous plantation was nearly self-sufficient with respect to the goods and services ordinarily obtained in such urban centers.

The middle colonies were different in many respects from those to the south. Taken from the Netherlands in mid-seventeenth century, Delaware and the much larger Pennsylvania were granted to William Penn, and New Jersey and New York, where the primary Dutch influence had been felt, were granted to the Duke of York. In New York, the Dutch "patroons," or large landholders were allowed to remain in control of their holdings and similar sized grants were given to English friends of the colony's governors. Thus, there were similarities to land ownership in the southern colonies along much of the Hudson River valley but a sharp contrast to the landlord pattern developed immediately to the east in New England. Pennsylvania, on the other hand, was opened by the Quaker governor to small farmers from many countries regardless of their religion. Therefore, in economic organization and cultural makeup, it tended to have more in common with the New England colonies than any of its major colonial neighbors.

As to the south, entry to the interior was gained primarily along the major rivers. North of Maryland, however, the coastal plain narrows sharply so that penetration was not as easy as southward. Even an outline map of the east coast of North America illustrates this narrowing of the flat coastal plain; there are no major river *estuaries* north of the Delaware River comparable to those to the south, such as Chesapeake Bay (Susquehanna and Potomac Rivers) and the James River estuary. Penetration inland by large, deep-draft vessels is therefore possible in the southern coastal region (where sand bars do not block the river mouths as in North Carolina) whereas movement inland by such ships is sharply limited in New England by rapids and the like.

The New England colonies were still different in their settlement patterns, political foundations, and economic organization. Three of the four New England colonies (New Hampshire the exception) were founded on dissension and separatism. Political fragmentation was reflected in New England by the economic organization which developed. From the beginning, the set-

This town in northeastern Vermont is an excellent example of a small community nestled in the irregular New England topography. Note the small farm field size, the forested hillslopes, the typical New England church, and the fact that this "town" is largely a cluster of farm houses. (Grant Heilman)

tlers who migrated to New England came for the express purpose of living according to one's own, or one's group's, standards. The primary goal, therefore, was not commercial gain but subsistence. One result was that during the early years, many more small farmers entered New England than did initially in the Southern colonies. This approach to farming also was sustained to a great extent by the environment, relatively inhospitable to large-scale commercial farming. Cold, wet winters and cool, often wet summers in a region with virtually no coastal plain, hilly topography and thin, rocky soils, meant that the individual surviving in agriculture was one accustomed to self-sustaining efforts vis-a-vis operations with hired or slave labor. These conditions resulted in a much higher average population density in the northern colonies. And equally important, the population lived in many small, nucleated settlements which increasingly became interconnected and interdependent.

During the decades immediately preceding the American Revolution, the character of the colonies became rather more similar in detail. While the coastal southern colonies remained a region of plantation agriculture, the *Piedmont Region* became increasingly populated with small-scale farmers, in many respects similar to the New Englanders. The Piedmont lies between the flat coastal plain and the Appalachian Mountains. It is a hilly region with rocky soil, much of which is better left in forest than cultivated. The farmers settling the Virginia and Carolina piedmont were immigrants avoiding wage labor on the coastal plantations or were indentured servants who had served their bond period and were free to leave the coast. Needless to say, except for the low percent freed, the black slaves remained with the plantation economy until they died and were replaced. With this major regional exception of south coastal plantations, the dominant form of agricultural pursuit became increasingly small-scale, one-family farms throughout most of British America.

The colonists' success in the American Revolutionary War had several significant results. Of course, this success heralded the beginning of political and economic expansion across the breadth of the continent, but the expansion required several generations to complete. Within the context of the immediate aftermath of the war, a number of implications may be drawn from the geographic character of the population and settlement patterns as well as the 1790 economic and political organization of the continent.

First, there was little likelihood of restraining the westward movement of farming population. Before the revolution, in 1763, the Indian peoples had been guaranteed by proclamation that European settlement would not expand beyond a line approximately following the front range of the Appalachians through Maryland and the North and Northeast irregularly, allowing for the European farmers, at most, several hundred miles between the coast and the "Indian Reserve." Although this mild restraint was not accepted by all colonists as binding even before the war, it was legally removed after independence from Britain was achieved. Furthermore, the Indians were generally too loosely organized to resist the westward expansion. The only peoples having sufficient organization to possibly have managed such resistance for some time, the Iroquois, had fought for and with the British. (This is not to imply that allegiance with the colonial revolutionaries had significant benefit for those Indians who fought on the winning side; "human rights," it was clear were relevant only for "civilized" men, and the Indians, like their black counterparts within the system, were hardly civilized in the eyes of most European-Americans.)

Second, the economic activities of the successful revolutionaries were likely to continue the immediate prewar patterns of development. The small-farm operator continued to move onto the piedmont, between coastal plain and mountain barrier. The plantation economic organization, however, was no longer restricted by political considerations to the plain along the eastern margin of the continent. It could now proceed westward onto the lower piedmont, and south of the mountains onto the Gulf of Mexico coastal plain, restricted to the south for only a few decades by the Spanish claims to Florida. The Appalachians had served the purpose of allowing considerable consolidation of economic strength along the eastern margin during the early years of settlement. Had such consolidation not been forced, rapid westward expansion could have resulted in a weak *urban* and *agricultural* foundation, much in the manner that the French exploration penetrated too rapidly into the interior leaving too weak a base on which to build. After fulfilling this function, the Appalachian Mountains were breached in several places and skirted on the south to open what must have seemed to be limitless land opportunities for agricultural development.

Third, the very diversity of the separate colonies in economy, in outlook, in population composition, in religion, and in other ways, required a great deal of compromise and conciliation in order to form a single country. The form used, indeed the only form liable to succeed, was a *federation*. A federation is a form of political organization wherein a number of political units are joined by mutual consent with certain powers residing in a central government, binding on all member states, but with each state maintaining political control (power) over the other governmental and administrative activities not agreed upon as delegated to the central government. The federal form of organization is contrasted with the *unitary* type, wherein all powers focus upon a single governmental hierarchy. This distinction will be discussed further in Chapter 11 when dealing with India. That the United States was from the beginning a federation, however, is of great significance in illustrating the diversity of the separate states which agreed to form a union and also in leading to the events of the mid-nineteenth century. These events, together with the heritage of the federal form of government, have had considerable bearing on the social spatial structure of present-day America.

And fourth, the results of the American Revolution carried much significance for the development of the remainder of the continent's lands. The war was not popular with all settlers in the British colonies. Three colonies, Quebec, Nova Scotia, and Newfoundland, did not revolt, and many citizens in the other new lands remained loyal to the established order. Many such "loyalists" were expelled from the United States during and after the war. Those that stayed in North America migrated to the British portions of the continent, that is, Canada. This, of course, provided a sudden influx of English population into an area which for some time had been predominantly French, in spite of the recent control established by Britain. This influx, in turn, affected the political and economic organization of Canada, as will be seen later.

Early Black America

The American success in throwing off British hegemony over the land between the Great Lakes and Spanish America east of the Mississippi River also carried implications for a significant minority of the new country's population, both within the short run future as well as the nearly two centuries of development which have followed the outbreak of hostilities. The descendants of the black slaves (and the black freedmen as well) would certainly have fared differently, although not necessarily better, under the British than they have in American society. It would be fatuous to hypothesize in detail about the possible developments if the crown had maintained control of its colonies, but the developing pattern of Negro population, both before and since the American Revolution does provide a clear illustration of the many associated patterns, economic, political, cultural, urban, which make up the changing geography of North America.

It has already been mentioned that black men were among the first immigrants to early British America. That they were involuntary immigrants does not alter the fact of their early arrival. Although Africans had been brought to Spanish colonial holdings in the Caribbean Sea and in Florida for nearly a century, successful English settlement in North America was not begun until 1607 along the James River in Virginia. Within five years, it was clear that gold and gems were not readily available while Indian-domesticated tobacco could be grown and sold to England. For profitable cultivation, tobacco demands many man-hours of labor, a long growing season, fertile sandy soil, and moderate amounts of rainfall well distributed over the growing season. The environment in coastal Virginia met these requirements but the sparsely populated colony was in severe need of labor. This labor had to be cheap and not necessarily skilled, if the production of tobacco was to meet demands from rapidly expanding markets in Europe. By the middle of the seventeenth century, it was clear that Negroes, held in hereditary

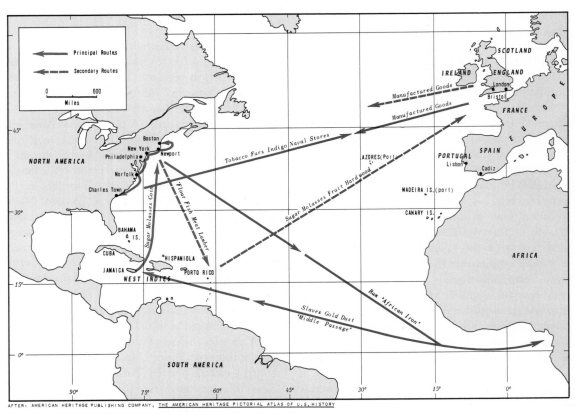

AFTER: AMERICAN HERITAGE PUBLISHING COMPANY, THE AMERICAN HERITAGE PICTORIAL ATLAS OF U.S. HISTORY

Map 6–3 THE TRIANGULAR TRADE.

lifetime slavery, were looked upon as the optimum source of the needed labor.[2]

The number of black slaves in North America expanded slowly from the first twenty in 1619 to about three hundred in Virginia by 1649. During the next 140 years, however, as European immigration accelerated, as agricultural production of plantation crops such as indigo, rice, and some cotton in addition to tobacco expanded rapidly southward along the coastal plain, and as it became even clearer that the temporary bondage of indentured whites was not as satisfactory as the permanent bondage of blacks, the Negro population in America grew rapidly.

Trade routes became established as the demand for slaves increased. The pattern of these routes illustrated the growing complementarity between sections of colonial America, and also between the new colonies and other parts of the world. The notorious "triangular trade" became established between North American colonies, Caribbean colonies, and Africa (Map 6-3). Rum and "African iron" (i.e., small iron bars) were shipped from New England to the Guinea Coast of Africa where they were exchanged for human beings. The Africans, who may have originated anywhere between the Congo and the Senegal Rivers, were carried to Jamaica or another Caribbean island, or to Charles Town (now Charleston), South Carolina. The trips' survivors were sold for Caribbean sugar or molasses, or for one of the plantation crops or furs in North America. If the slaves were taken to Jamaica, the ships proceeded to New England to complete the triangle. If their enslavement was to continue on the continent, the goods for which they were traded were shipped to England and exchanged there for manufactured goods desired in the frontier colonies.

The diversity of the North American colonies was clear even at this early period, a regional diversity which remains today. In terms of the pattern of black population, differences in colonial economies resulted in an overwhelming preponderance of slaves in the southern four colonies (Maryland, Virginia, and North and South Carolina). The middle and New England colonies had little need for the kind of labor supplied by slaves (Map 6-4). Small farms, indi-

Map 6-4 COLONIAL ECONOMIC DIVERSITY.

vidually owned and worked, dominated the northern rural economy. While extra hands could nearly always be used, crops grown in the North, chiefly grains, supplemented by cattle and fur sales, were not sufficiently remunerative to recover the payment necessary to purchase a man or woman. In addition, religious scruples, beginning with the Quakers, raised doubts in many minds concerning the moral issue of slavery. An increasing divergence of social and political attitudes as well as economies followed. The changing attitudes in the North culminated

[2] For an excellent history of the early developments of black-white relationships in America, see Winthrop D. Jordan, *White over Black: American Attitudes Toward the Negro 1550–1812,* The University of North Carolina Press, Chapel Hill, 1968.

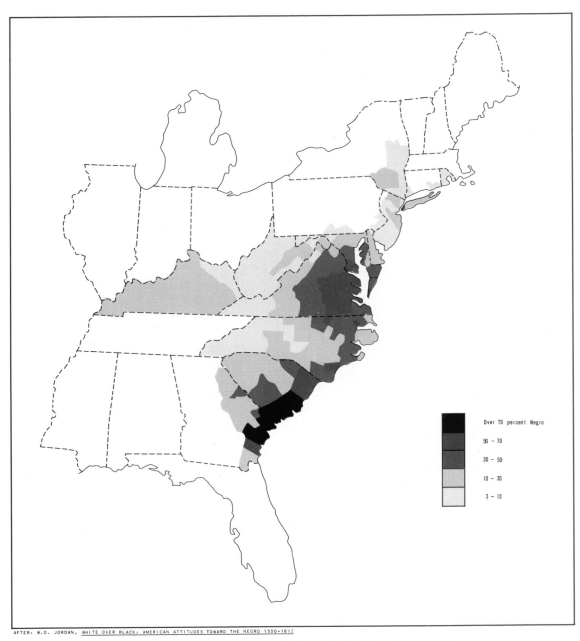

AFTER: W.D. JORDAN, <u>WHITE OVER BLACK; AMERICAN ATTITUDES TOWARD THE NEGRO 1550-1812</u>

Map 6–5 DISTRIBUTION OF NEGROES, 1790, AS A PERCENTAGE OF TOTAL NONABORIGI-
NAL POPULATION.

in the state-by-state abolition of slavery in the middle and northern states, beginning with Vermont in 1777 and ending with New York and New Jersey in 1799 and 1805, respectively. With this, the seeds were planted for the American Civil War and the regional separatism which has smoldered for a century after the war's formal conclusion (Map 6-5).

Westward Expansion and the Emergence of National Patterns

The end of the Revolutionary War heralded the beginning of a century of settlement expansion in the United States. Although the Appalachian Highlands had been breached even before the war, settlement in large numbers did not begin in earnest until after 1790 (see Map 6-6). Primary expansion during the first twenty years of nationhood was in settlement of the Ohio River valley and the valleys of its major tributaries east of the Tennessee River. There was also a major two-pronged thrust toward the Great Lakes, one northward from Fort Pitt (Pittsburgh) to Lake Erie and the other westward through the great Mohawk Valley in central New York State to Lake Ontario.

It is easy to note the importance of the river system to the settlers migrating into this region. As earlier with the first colonists, rivers provided the only easy transportation; they were the only means of ready access to and from the developed portion of the country. A farm and homesite located on a navigable river during the early nineteenth century is closely analogous to a commercial farm or merchant located a short distance from a railroad depot in the middle of the century.

The use of, and dependence on, rivers as means of extending the settlement frontier during the period 1810–1830 was not lessened. In addition to filling in the earlier areas of sparse settlement, there was considerable population movement north from the Ohio River, up the Scioto and Miami Rivers into Ohio, up the Wabash into Indiana, and north from the Ohio along the Mississippi River. Even the beginnings of settlement north and west from St. Louis are visible by 1830, west along the Missouri River and north along the Illinois and Mississippi. In the South, too, settlement was densest along the major rivers, primary expansion occurring along the southern Mississippi, up the Arkansas, and along the Alabama River and its tributaries. By 1850, little in the east could be called "frontier" settlement. And by 1870, the forest margins had finally been left behind for the frighteningly open grass plains, with settlement expanding toward central Texas, Kansas, and Nebraska. The striking explosion of settlement, 1870–1890, along the west coast, along the front of the Rocky Mountains, and throughout the dry West was the final rapid movement of the frontier, with settlement expansion in the twentieth century primarily in urban clusters, especially in the major developing core areas of the nation.

Although many families took upon themselves the rigors of long distance migration during the middle and late nineteenth century, a sizable proportion of the population moved only short distances, or did not move at all. What was it that led some to move and some to remain? Why was Pennsylvania satisfactory to one family, while others felt the need to move to Michigan, Illinois, Iowa, Nebraska, or even the forbidding distance to Oregon or California?

Hardly unique to this period, the migrants' motivations were undoubtedly a complex of "push" factors from the local environment and "pull" factors from the general direction toward which the migrants moved. What comprises these "push" and "pull" elements in the environment offers a series of extremely interesting, although difficult, questions which should be considered by those interested in why people move (and why they do not). Another way of posing the problem is to consider the individual's perception of the environment, both at the original homesite and at the final residence. How does such perception lead to a decision to move or stay? How does an individual "see" or perceive his environment? It is, after all, the variations in perception of the environment which lead to variations in the evaluation of the local environment. And it is these individual differences which cause one man to be "pushed" into migrating while his neighbor remains satisfied with local conditions. Although geographers are increasingly interested in the question of *environmental perception* and its relation to behavior, these questions are much too difficult to answer here. Nonetheless, it is possible to examine the environmental patterns which were presented to the nineteenth century American, those landscape and climatic features encountered by the migrating population.

Population density of 2 or more persons per square mile

to 1790	1830 - 1850
1790 - 1810	1850 - 1870
1810 - 1830	1870 - 1890

Miles

0   100   200   300   400   500

MODIFIED AFTER: KARN AND SALE, AMERICAN EXPANSION · A BOOK OF MAPS

Map 6-6 THE EXPANSION OF AMERICAN SETTLEMENT, 1790 TO 1890.

Physical Setting

In considering topographic regions in North America, it is important to keep the scale of the topic in mind. Considerably larger than their equivalent in Europe, and yet somewhat smaller than the vast distances dealt with in the preceding discussion of the Soviet Union, the major physiographic divisions of North America are of considerable size in terms of the time taken to traverse them, even with present modes of surface transport. To the nineteenth century traveler, moving by foot, wagon, or, later, railway, the mountain chains could only have seemed more imposing barriers and the flatness of the Great Plains more interminable than they do today.

In spite of the scale of relief, the topography of North America is not as confusing as that of either Europe or the U.S.S.R. Major topographic features over the continent may be rather cleanly divided into *physiographic provinces,* or regions with considerable internal uniformity of relief character and morphology (see Map 6-7). In broadest terms, the continent is composed of a vast interior plain or lowland nearly triangular in shape with one edge of the triangle facing north and the other edges bounded by mountains, with the older and lower chain in the east. The triangle is not "closed" since the eastern boundary does not extend to the Gulf of Mexico.

The earliest migrants from Europe in the seventeenth and eighteenth centuries settled, as we have seen, in two distinctly different environments. Those remaining north of New York were virtually without coastal plain while those to the south found a plain which gradually increased in width with increased distance from New York. The extremely irregular topography of New England is, in fact, the northern end of the *Appalachian Highlands.* Penetration of these highlands was very difficult, except along the few major rivers (the Hudson, the Merrimac, and the Connecticut). With notably few exceptions, much of this region remains isolated from the vigorous activity of the coastal cities (Portsmouth, Boston, Providence, and of course on the edge of New England, New York City). To the northeast, the same picture holds true in the Maritime Provinces of Canada with only Halifax (Nova Scotia), St. John's (Newfoundland), and St. John (New Brunswick) having even moderate population size. On the hilly, rocky, tree-covered land in New England and the Maritimes, farm-

ing has been difficult. Because of this, shipbuilding, small home industries, and fishing were long the traditional activities of this region.

The *Coastal Plain* from Maryland, Delaware, and Virginia southward was nearly ideal, as we have seen, for large-scale commercial farming with tobacco, indigo, peanuts, and cotton being the most important early crops. Between the flat coastal region and the Appalachian barrier lies the *Piedmont,* underlain by the roots of a once large mountain range now reduced by millions of years of weathering to rolling, irregular hills. The land is generally not too steep to forbid cultivation of hilltops, valleys, and scattered areas of low relief, but it is liable to erosion if care is not exercised. Much of the Southern Piedmont, therefore, remains forested.

The contact line between the hilly Piedmont and the flat Coastal Plain is striking. In such cities as Washington, D.C., Richmond, Va., Columbia, S.C., or Augusta or Columbus, Ga., what appears to be a ridge lies immediately to the north or west of the city, or is found within the city boundaries. Ships and barges traveling upriver from the Atlantic Ocean were stopped by this ridge (or series of hills) and the rapids and falls which are formed by the rivers' flow over it; for this reason, the contact line was termed the "Fall Line." The change of transport necessitated by the rapids led to the development of urban centers which provided the services required when freight was "broken" from large, bulk units to other units more easily moved by land transport. More important for city growth than freight handling, however, are the manufacturing and processing activities which are often most efficiently performed where freight is handled. Not only does processing at these points minimize total handling costs, Fall Line rivers were harnessed to obtain water power, an important factor during the nation's early years.

Although the Appalachian Highlands may be seen as beneficial to the young nation's development (for without this barrier it is possible that the initial energies of exploration and exploitation might not have led to rapid settlement as, for example, in the French experience), the Appalachians *were* a barrier. Once penetrated, once reports of large basins, plains, and river valleys reached the increasing flow of immigrants, and once the troublesome agreement between the British and the Indians had been annulled by Cornwallis' surrender, settlers streamed across,

MODIFIED AFTER: FENNEMAN, LOBECK, AND OTHERS

Map 6-7 PHYSIOGRAPHIC PROVINCES OF NORTH AMERICA.

through, and around this barrier and onto the *Interior Plains* of the Middle West.

For those moving from the Coastal Plain to the Interior Plains, the change of relief variation need not have been great. There is sufficient variation within both provinces, of course, to provide for detailed differences but in general, both are composed of flat to rolling countryside with a few areas of moderately hilly land. For those arriving from New England or other portions of the Appalachian Highlands, the low relief must have appeared welcome to those interested in agriculture. Climate in the Interior Plains was not greatly different, either, from what had been experienced closer to the Atlantic at comparable latitudes. Winters are more severe and summers are perhaps a bit more irregular than near the coast but not much so. In this, the climate is *continental* in character rather than *humid subtropical* as found in the southeast.

It would be incorrect to assume that settlers pouring out of the eastern regions of the newly established United States migrated and intermingled throughout the Interior Plains without regard for their region of origin. Settlers of the Ohio River Valley and southern Ohio, Indiana, Illinois, and areas to the south moved overwhelmingly from the Virginia, Carolina, and Georgia Coastal Plain and Piedmont. Migration routes through the Appalachians were most traveled but others skirted the southern end of

the mountains. From New York, Pennsylvania, and the New England states came the settlers of northern Ohio, Indiana, and Illinois, and southern Michigan (see Map 6-8). Thus, the different patterns of economic organization and general outlook prevalent along the coastal belt maintained themselves in the interior by the association of the origins and settling points of the population of the eastern Interior Plains. This point will be pursued later.

Strictly in terms of physiography, there is little to break the flatness of the Interior Plains between the Appalachian Highlands in the east and the Rocky Mountain front range in the west. With one exception, these plains also extend

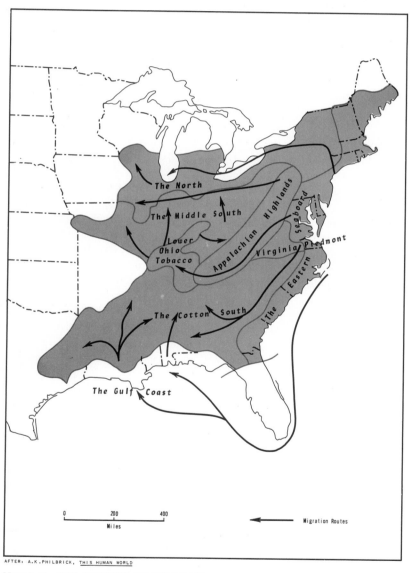

AFTER: A.K.PHILBRICK, THIS HUMAN WORLD

Map 6–8 THE PATTERN OF MIGRATION TO THE INTERIOR.

The low relief of the Interior Plains allows large farm fields almost without exception. The rectangular pattern of these fields may be seen in this photograph, though this pattern should not be confused with the three- or four-crop rotation being practised on the farm in the foreground. Note the scattered farmsteads and the presence of trees only around buildings and along field boundaries and stream courses. (Lathrop)

from the Gulf of Mexico in a narrowing arc to the Arctic Ocean. The exception is the smallish *Interior Highlands* in southern Missouri and northern Arkansas (consisting of the Ouachita, Ozark, and Boston Mountains).

More important for the westward expanding white population than the extent of this physiographic region was the climatic gradation and its related vegetation and soil changes. Once west of the Mississippi River, there is a gradual decrease in the average annual precipitation received. Except for the grass plains of northern Illinois, the replacement of trees with grass as the dominant vegetative form began about the 92nd meridian. By the time the 95th meridian was reached, trees were found almost exclusively along the narrow river courses with grasses of medium height on the plains between. By the 100th meridian west, average annual precipitation dropped to 20 inches and a short grass dominated the landscape. For nearly half a century, the barren-appearing grasslands held little promise for settlement to families raised in the shelter of the eastern forests. If trees could not grow in this region, the argument must have gone, then the environment was obviously too hostile for successful agricultural development. The label "The Great American Desert" was placed on these grasslands and on the fertile

black *chernozem* soil underlying the grasses. During the mid-nineteenth century the thousands of farming families, on their more than 1500 mile treks to Oregon and California, *perceived* this rich-soil region as undesirable. The "Great American Desert" was not settled until such technological innovations as barbed wire, deep-well drilling, the self-polishing steel plow, and the windmill were introduced to the region and railroads were built connecting the west coast as well as this midland with the eastern markets. Settlement also expanded as people began to realize that the grasses made ideal pasture for cattle, just as they had for the buffalo, and that the environment was suitable for cultivation of moderately drought-resistant grains, such as wheat.

To the north of the Interior Plains, the *Canadian* (or Laurentian) *Shield* extends northward in all its monotony to the Arctic. Topographically, the shield may be no more rolling than the Interior Plains, but because this was the region which spawned the continental glaciers of North America during the recent Ice Age, the soil is very thin and unproductive. As in Scandinavia, the bedrock was scoured clean during the glaciers' advances, and temperatures were too low and effective precipitation too slight to have allowed much replacement of soil in the last 6000 to 8000 years. For Canada, the result has been that population has remained concentrated within the St. Lawrence River Valley and the lowlands around the eastern Great Lakes. North of Lake Superior, settlement is very sparse and even moderate population densities are not resumed until the shield gives way to the Interior Plains in southwestern Manitoba and beyond.

The paucity of population north of Lake Superior reflects the poor agricultural resource base even today. This poor environmental base retarded the westward expansion of Canadian population relative to that in the United States, although the expansion south of the border was an indirect stimulus to expansion to the north. For over 100 years after U.S. independence, relations between these neighbors were marked by suspicion and occasional "jingoism" on the part of the United States, which in turn appeared to justify Canadian suspicions. Many English-speaking inhabitants of Canada during the nineteenth century were descendants of the loyalists who fled the revolting colonies at the end of the eighteenth century. New Brunswick, for example, was formed from part of Nova

Scotia by about 35,000 immigrant loyalists after the American Revolution, and Ontario's population was similarly increased.

Even by the 1880's, however, settlement had not spread in significant quantities onto the grass plains of Canada, that is, the "prairie provinces." Although there were still fewer than a thousand settlers on the prairies, Canada's transcontinental railroad was constructed during this decade, following by more than ten years the completion of the United States' first east-west rail connection. In addition to the important function of connecting for the first time both coastal extremities of this very broad country, the railroad also stimulated settlement by making possible the development of a wheat economy. In addition, in what may have been one of the primary factors of the day, it was believed that the railroad and the accompanying population increase prevented a much-feared infiltration of the western territories by the energetically expansionist American population.

Beyond the Interior Plains, the physiographic provinces are associated with the great western *cordillera*, the continental "spine" which extends from Alaska into Mexico and beyond. The *Rocky Mountains*, the coastal *Pacific Mountains and Valleys*, and the spatially intermediate *Intermontaine Basins and Plateaus* have less topographic homogeneity than most of the previous provinces discussed. Their general north-south orientation meant severe difficulties for the railway engineers, but the lack of relief consistency meant that a number of routes through the mountains, or around the mountains, were present and only needed to be found. Of the major mountain ranges, the Sierra Nevada-Cascade Mountains proved the most continuous and difficult to penetrate. Once explored, however, the settlement in the Sacramento-San Joaquin Valley which had begun with the Spanish, and the settlement in the Willamette Valley (begun substantially in the 1840's), both increased rapidly in population.

In these westernmost provinces, however, population remained highly clustered in river valleys, coastal aggregations, and adjacent to exceptional sources of water, for example, along the rivers flowing into the Great Salt Lake. As severe a barrier as the mountains were to the nineteenth-century migrants, equally severe was the barrier of aridity, for between the Pacific and Rocky Mountains lies a zone receiving less than an average 20 inches of precipitation annually,

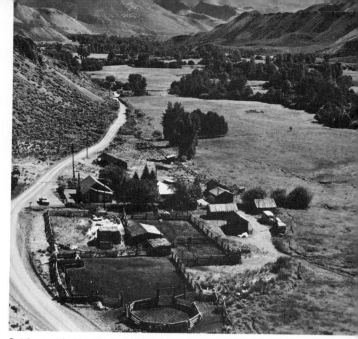

Settlement in the Rocky Mountain Province is sparse, but much of what is there can be found in small, isolated valleys such as this. The ranch shown is in fact only one of several in this narrow valley. (Charles E. Rotkin)

much of the area receiving only half of this inadequate amount (Plate 1). The severity of this barrier can be seen in the present low average population density of the region. In spite of rather sophisticated "environment compensating" technology, population densities remain well below those of the humid east. Lack of water, therefore, remains the region's most basic problem. Irrigation is not likely to affect the settlement pattern to any significant degree in the near future. Water is obtained from mountain-fed streams and rivers and from occasional ground water sources, both severely limited sources of supply. Indeed, as population grows rapidly in the western urban clusters, competition will intensify further between rural and urban interests, and between the many urban interests, for the scarce water supplies.

This, then, was the general physiographic and climatic pattern faced by the waves of settlers that moved across North America east to west. The entire eastern half of the United States and southern fringe of eastern Canada are humid with frost-free periods of 120 days or more, a situation adequate for dependable agriculture. This vast region of agricultural potential, the second largest in the world, is penetrated by the Appalachian Highlands, but in general is a flat to rolling lowland. The western half of North

America is too dry or too mountainous for all nonirrigated agriculture except drought-resistant grains and controlled animal grazing. In Canada, the thin-soil Laurentian Shield extends south to the U.S. border (and beyond), thus restricting agriculture as surely as does the northerly country's growing season.

The geographic patterns of North America in the twentieth century, however, cannot be generated solely from the continent's environmental patterns with an overlay of population density developments to the end of the nineteenth century. Also important are technological changes and the patterns of economic, political, and social organization which were carried with the migrants. The patterns of human activity and the differing attitudes which developed within regions of environmental similarity have resulted in a complex set of multifactor regions. The continent's present geography, its present potentialities and problems cannot be understood without some consideration of these regions.

Economic Patterns and Regional Separatism

As was discussed earlier, differences in the coastal states' regional economies were carried into the interior with the migrating settlers. Tobacco was reported as early as 1751 under cultivation by Indians in the Ohio Valley portion of Kentucky. White settlers from Virginia and the Carolinas were quick to exploit the environmental possibilities for commercial tobacco production shown them by the Indians. Settlers at Boonesboro (near present-day Lexington), for example, produced commercial tobacco as early as 1783 in spite of the formidable transportation difficulties. Familiar farming possibilities were undoubtedly attractive to potential migrants in the east, thus draining white labor from the plantation and specialty cash crop economies found from Maryland and Virginia south. The primary change in attitude, perhaps, was that most frontiersmen who left the east coast for the interior were small-scale farmers, and therefore more similar in economic organization to the transplanted New England farmers settling the Middle West north of the Ohio River than either were to the southern plantation owner. The initial attitudes of the small farmers, however, were different, for example, toward slavery, the southerner holding that it was tolerable and occasionally necessary, while the northerner was more likely to judge slavery unnecessary and, on occasion, intolerable.

While farmers left the eastern Coastal Plain and Piedmont for the less-crowded, often richer farmlands of the Interior Plains, and residents of the long-established New England towns and farming regions similarly migrated westward, the economic character of both the northeast and the southeast began to change, becoming increasingly polarized in their differences. In New England, a commercial-industrial economy had been established during the colonial period based on the timber (shipbuilding) and fishing resources of the region. Rum distilleries, small-scale iron works, and similar activities focused on the ports of the Northeast. This in turn developed the region into the import-export and financial core for the colonies. During the early years of independence, the textile industry became increasingly important as it was this region that contained the needed capital for industrial establishment as well as the technical ability and abundant labor in the rapidly growing cities. Urbanization and industrialization grew apace as immigrants from crowded European cities and towns brought with them the skills and knowledge (or at least the attitudes) which were driving the industrial revolution in the Old World.

In the Southeast, the plantation economy was well established as a productive form of commercial enterprise. Cotton production expanded, both in area under cultivation and in revenue earned, moving south along the Piedmont and Coastal Plain of the Carolinas into Georgia. Cotton, however, is a crop that absorbs soil nutrients without much return to the soil after the crop has been harvested. This required cotton growers who did not have the technology for soil refertilization to expand the area of production rapidly into equally suitable areas elsewhere in the South. As of 1810, however, most of the land comprising the present states of Georgia, Alabama, and Mississippi was "owned" and occupied by numerous Indian peoples. As in the Northwest Territory, land was ceded piecemeal by the Indians until 1830, when the Indian Removal Bill was passed by Congress. This led to a series of forced mass migrations during the 1830's from the Indians' traditional homes to the relatively barren Indian Territory (Oklahoma) on the margin of the "Great American Desert." The removal of the Indians, accompanied by great hardship (4000 Cherokee, for example, died on the trip west from northern Georgia), was completed by the middle of the decade (Map 6-9). Millions of acres of new land were thus made available to the southern cotton planters.

In an irregular, patchwork pattern of expansion, plantation cotton production gradually shifted westward across the Gulf Coastal Plain. By 1850, large-scale cotton cultivation had spread onto the *chernozem* plains of Texas, and during the decade to 1860, cotton consolidated its position of importance in the national economy as the country's primary export product. So important was cotton to the United States as a source of foreign exchange ($192 million out of a total $316 million worth of goods exported in 1860) that the southern states believed the North would have difficulty surviving without them. In any case, it was thought, European aid would be forthcoming if the South was forced to assert its separate identity from the North. The South was not the first region to overestimate the economic security provided by a single agricultural commodity.

A major misconception which has continued to the present concerns the regional homogeneity of the southern cotton economy. The misleadingly termed "Cotton Belt" was not (and is not)

a distinct area in the country where cotton cultivation was nearly ubiquitous. First, corn, wheat, cattle and other livestock were produced throughout the Deep South even at the height of cotton's dominance. Second, in addition to large expanses of woodland, the average farm or plantation produced foodstuffs for local consumption on about half of the land cultivated. And third, other cash crops, such as tobacco, hemp, sugar cane, and rice were grown over significant areas within the slaveholding South. In geographic terms, the spatially discontinuous features comprising the "Belt" constitute a *formal* region, homogeneous primarily in the pervasiveness of cotton's dominance in the lives of the region's inhabitants.

Accompanying this southerly and westward movement of the plantation system was a marked change in the population pattern in the country. This change has been influential in forming the composition of regional character even into the mid-twentieth century. Reference is being made, of course, to the large-scale transfer of slaves from the east coast to the deep cotton South.

The primary agent for moving the slave population from the east coast into the lower South was not merely the westward migration of plan-

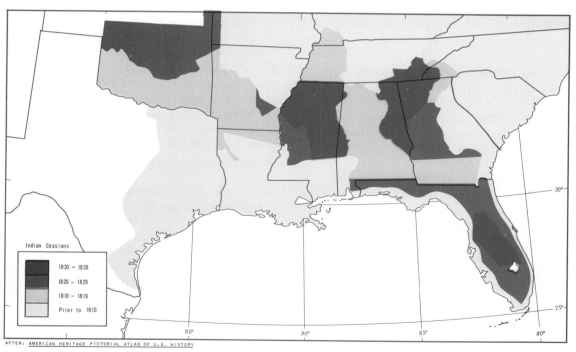

Map 6–9 INDIAN LAND CESSIONS IN THE SOUTHEAST, TO 1840.

tation owners carrying with them their large slave holdings. Nor was it a matter of a change of focus for slave importation from Charleston, South Carolina to, say, Mobile, Alabama. Generally speaking, during the years immediately following the close of the Revolutionary War, there was almost no importation of newly acquired slaves from Africa. Indeed, several poor cropping years plus a burgeoning natural growth rate among the black population raised considerable fear among whites in Virginia and the Carolinas of a general slave revolt. On occasion, this fear was justified, although none of the revolts had lasting success.

The problem in the 1790's, then, was how to "dispose" of a serious financial investment which also contained increasing potential for disaster. An initial answer, revived repeatedly, lay in the exportation of "excess" slaves to an African colony, a Caribbean island, or perhaps even a portion of the vast Louisiana Purchase. The actual answer followed from Eli Whitney's 1793 invention of the cotton gin. Although cotton had been grown inland for years, the variety which did best there contained many seeds which were very difficult to extract from the lint. Whitney's invention seeded the cotton quickly and easily, thereby opening vast areas to cotton production. (Note that this is only one more example of an innovation altering man's relation to the environment rather than the environment controlling human behavior as the environmental determinist would hold.) As cotton production moved westward across the lower South, the threat of black dominance in the older colonies was reduced (or so it was believed) by the transfer of slaves into these developing areas. The slave trade, legally, became an internal affair in 1808 when a federal ban on the importation of new slaves went into effect.

Regional differences that had existed from the beginnings of European settlement in North America, that could be discerned in the lower, middle, and New England colonial characteristics, and that were maintained in a modified form in the Interior Plains between the North and the upper South, gradually polarized the two regions over several issues with the moral issue of slavery being central. First faced by the nation during the Constitutional Convention, slavery was permitted to continue for the sake of unity among the colonies. The issue (and the divisiveness) was raised again in 1803 when

South Carolina began to import slaves, but the official ending of this importation in 1808 ended the debates and seemingly ended the division. By 1819–1820, however, westward expansion of the population resumed the heated debates between North and South as territories not clearly positioned in either the South or the North were placed first in the free and then in the slave category. The Missouri Compromise "settled" the question for several decades and slaves were shipped into Missouri, Arkansas, and Texas while emotions in both the North and South regained strength and heat over the issue. Several more "compromises" during the 1850's added considerable area to the slave category (a ridiculous feature of the last compromise was that almost all of the land was much too dry for the activities for which slave labor was profitable), but the verbal and military skirmishes during this decade finally led to secession and war by 1861.

By 1860, the major transformation in the pattern of Negro population in the South was virtually complete. Although the actual numbers moved are not known, it was undoubtedly many hundreds of thousands. Most trade was overland with chained slaves marched in gangs the hundreds of miles from Kentucky and Virginia into Alabama, Mississippi, and beyond. The change in the pattern over two generations, 70 years, is startling (compare Maps 6-10 and 6-5). While in 1790 slaves were concentrated in coastal Virginia through northern Georgia, by 1860 a linear region through Georgia and Alabama, the Mississippi "Delta" from the Tennessee border to the Gulf, and portions of Texas and Florida contained counties with over 50 percent of the total county population slave. Also striking is the great similarity between this 1860 distribution of Negro population and its counterpart for 1960.

The changing pattern of nonwhite population provides a ready key to understanding the geography of the U.S. up to the mid-nineteenth century, just as it illuminates the more important features of Man's geographic organization at the beginning of the final third of the twentieth century in the United States.

The presence of high densities of Negroes before 1860 meant, almost without exception, a predominance of economic activities for which slave labor was suitable, that is, activities demanding large amounts of cheap, largely un-

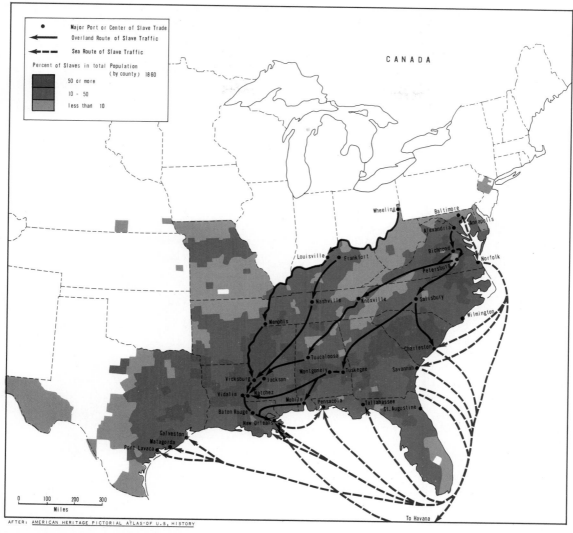

Map 6–10 DOMESTIC SLAVE TRAFFIC AND THE RESULTANT DISTRIBUTION OF SLAVES, 1860.

skilled hand labor. Considering that the slaves themselves were totally resentful of their servitude, the activities for which they were used also had to be such that sabotage and work slowdowns could be guarded against by an overseer. In sum, these requirements limited most slavery to plantation agriculture of labor intensive, high value-per-acre crops. (Output per man-hour was obviously less important although still relevant.) The continuation of slavery in the South reflected the continued importance of specialized agricultural activities in this region rather than manufacturing, as developed in the Northeast, and commercial food crop production, as increased in the Middle West.

In mid-century, the dominance of specialized agriculture utilizing slave labor was also reflected in the changing pattern of total population densities and in the construction of new transportation lines. Immigrants from Europe during the first half of the nineteenth century tended to avoid settlement in the South. Employment opportunities in the southern urban centers could not compare with those in the Northeast. The opportunities in agriculture were limited in the Middle West almost solely by land availability while to become part of the "ruling class" in the South, capital had to be available for the purchase of at least half a dozen black men in addition to other framing "necessities." The

Map 6-11 RAILROAD EXPANSION PRIOR TO THE CIVIL WAR.

AFTER: AMERICAN HERITAGE ATLAS OF THE U.S. HISTORY

Railroad in operation 1850
Lines Added 1850 – 61

ATLANTIC OCEAN

194

population north of Virginia and Kentucky grew rapidly during the first six decades of the century.

Transportation lines are reflections of the pattern of a society's organization. As such, they are also a reflection of some minimum organization that the economy requires for operation. Links between areas engaged in subsistence activities are weak or nonexistent; they are unnecessary because subsistence activities produce no surplus. Exchange economies, on the other hand, do require means for the movement of goods and people. The more intense the exchange, the more dense will be the transport network allowing the exchange.

The production of specialized crops requires the crops' sale, but more importantly, the sale will be to a region with a significant demand for the crops, usually one that does not produce these crops itself. In other words, the regions must enjoy *spatial complementarity*. In terms of the transportation pattern, large specialized production regions are concerned with export more than internal exchange. Transport lines will be primarily from the interior of the region to the coast with few interconnections. This was the railroad pattern in the South immediately before the Civil War. Although the South had greater east-west than north-south extent, and although there was no major east-west river to provide an alternative means of movement, there was only one east-west railroad line; the Memphis-Chattanooga line separated at Chattanooga, the northern fork connecting with Norfolk and Richmond and the southern with Savannah and Charleston via Atlanta. North-south connections in the South were equally tenuous (Map 6-11).

The density of the rail network in the North by 1860 reflects the region's different economic bases. The pattern east of Ohio was built largely before 1850 for both the manufacturing activities and the urban centers had developed early. Most of the railroads built in Ohio, Indiana, Illinois, Michigan and Wisconsin, however, were constructed during the decade immediately preceding the Civil War. Because of the railroad construction, the growing population in this old Northwest Territory, decreasingly subsistence farmers, were able to exchange more easily with each other the commodities they produced. They were also better able to engage in commerce with the manufacturing centers of the Northeast. The farmlands of the Northwest Territory had sup-plied the growing eastern market with foodstuffs for decades, but before the 1850's, it had been necessary to ship much of it via the Mississippi River and the Atlantic. By 1860, however, Interior Plains agriculture no longer depended on the South-controlled Mississippi while the South remained dependent on an open coastline for export of its cash crops and import of the manufactured items not produced locally.

In each of these ways, the distinctive pattern of Negro population accurately reflected the changing geography of the United States. Present virtually from the beginning of British settlement on the continent, these Africans and their descendants participated fully in the economic development of the southern half of the United States and in less rigid, more imaginative, individual ways in the U.S. as a whole. Their presence in the South in increasing numbers reflected regional political as well as economic differences. This led to intense sectionalism and finally national fragmentation. The depth of feeling over this separation and its causes are such that their effects are still being felt, and have a bearing on the geographic patterns of the country, more than one hundred years after formal reunification of the briefly separated territory.

The initial cultural heritage of the United States was overwhelmingly English. Although millions of immigrants from other European cultures were accepted into the country and the national character changed slightly with each wave, the process was primarily one of assimilation into the modified English culture rather than a mutual change. Even the African cultures carried to North America were altered by the dominant English culture just as they generated alterations in it. Generally, these contributions were neither desired nor recognized by the Europeans. Canada, on the other hand, was distinctly bicultural virtually from its beginning.

The pattern of Canadian biculturalism began to develop with the British military victory over the French, finally settled in 1763. The French had been firmly in control of the St. Lawrence River valley since Champlain's explorations in the early seventeenth century. French settlement along the St. Lawrence, mostly by farmers from northern and central France, had become sufficiently well established for these farmers to stay with their land even after Canada became a British possession.

In spite of the fact that New France had never been settled as densely as the British colonies, the Francophone community in Canada managed to remain distinct in the face of increasing political and numerical dominance by the Anglophone community. It is important to remember that the French in Canada had several distinguishing characteristics, only one of these being linguistic. They were also Catholics, exclusively. The Protestant French Huguenots had been prohibited from settling in French Canada. Their migration to British America had contributed to the low population of New France. Religious and linguistic differences and the complete range of cultural associations which distinguish one such group from another permitted and encouraged the French Canadians to maintain their group solidarity in the face of increasing formal and informal pressures for assimilation by the British Canadians.

By the end of the eighteenth century (1791), there were three distinct regions in Canada: Upper Canada (Ontario—English), Lower Canada (Quebec—French), and the Maritimes (Nova Scotia, New Brunswick, and Newfoundland—mixed, but mostly English). Although the French settlers had numerical superiority during the years immediately following the Treaty of Paris (1763), this changed rapidly during and after the American Revolution. Loyal British subjects fled the revolting colonies in large numbers, many migrating to Nova Scotia, there to be given their own, separate colony (New Brunswick) while many others moved to settle upriver from Quebec in Ontario.

As early as two hundred years ago, then, Canada indicated that it had potential for internal fragmentation. The maintenance of a bicultural country, with all its concomitant *centrifugal* (i.e., pulling apart) forces, was made even more difficult by the agricultural resource pattern. Population expansion northward was severely limited by climate and soil (see Plates 2 and 3). Although westward expansion was not hindered by a mountain barrier as in the United States, the long distances separating the productive east from the potentially productive plains, across the agriculturally barren land north of Lakes Huron and Superior, were perhaps an even more formidable barrier to migration.

Remaining under British control, this country was much less inviting than the United States to European immigrants seeking life away from European hegemony. Canada appeared colder and the "best" (i.e., eastern) agricultural lands were settled early by British and French. The great plains in the U.S. were better connected to the larger eastern markets than were the plains in Canada. As discussed earlier, settlement of the Canadian plains followed American settlement of its plains and was encouraged primarily to prevent American expansion north of the agreed upon international boundary along the 49th parallel. All of these elements contributed to the maintenance of a small Canadian population and slow expansion of settlement westward across the continent. The slow growth of total population, in turn, permitted French Canadians to remain a large, coherent national minority. Implications of this pattern of population growth and the difficulties of internal spatial integration will be discussed again later in this chapter.

Development of Present Patterns

The half century following the end of formal Civil War hostilities was one of economic transformation and growth for much of the United States. In the North, centers of mercantilist activity, basing their *raison d'etre* on commerce, increasingly became manufacturing instead of trade centers. In the South, however, this was a period of bitter and stubbornly slow economic, political, and social reorganization. This was also the period during which the West was populated and the basic patterns of continent-wide human organization laid. The northern development was powered on mineral resource use, and absorbed great numbers of immigrants from Europe. It was further stimulated by increasingly efficient means of interaction and the growing markets, East and West, which could be reached by the expanding network of railroads. After a relatively brief period of attention from the rest of the country, during which both exploitative and egalitarian forces were present, the South was allowed by a preoccupied North to revert to many of the antebellum forms of economic, social, and political exploitation.

The fifty years following World War I has also been a period of national economic transformation. At the root of changing spatial patterns during this period are (1) basic alterations in agricultural organization, (2) the increased importance of service activities complementing the previously developed manufacturing-industrial sector, (3) a series of migration flows—

South to North, East to West, and rural to urban, (4) the changed methods and rates of movement for both people and goods by air and highway carriers, and more recently (5) the revolution in social attitudes required of the white population majority by a large, hitherto excluded, black minority. Each of these changes has spatial effects, all multiple and many interacting with one another. The existing geographic patterns must be viewed in light of the involved social and economic trends which generated them, keeping in mind the nonhuman resource framework within which Man must work.

North America contains two of the most urbanized and industrialized countries in the world. In 1960, the United States contained 24 metropolitan areas (Standard Metropolitan Statistical Areas, or SMSA's) exceeding one million inhabitants each and another 30 with over half a million each. These 54 urban regions alone (of the total 224 SMSA's in the country) contained nearly half of the population of the entire country (46 percent), with most of the remaining half residing in cities over 10,000 population. In 1963, manufacturing firms in the U.S. employed over 12 million production workers, most of these workers living with their families in the SMSA's across the country. The manufacturing in which these workers were engaged added over $192 *billion* in value to the raw materials consumed by these processes. Canada, with a population approximately one-tenth the size of the United States, contains only two metropolitan areas exceeding one million population (Montreal and Toronto) and also possesses a much smaller manufacturing base. Nevertheless, it is increasingly urban and has received substantial industrial investment from American and Canadian investors in recent years.

It seems obvious that urbanization and industrialization are interdependent and mutually supportive. Granted, not all major cities in North America are based on industry even today (e.g., Washington, D.C.), but most important urban centers on the continent enjoyed their greatest periods of population growth during the late nineteenth and early twentieth centuries when industrial activities were dominant. Urbanization and industrialization in North America must be treated together, for the impetus behind one was very often a direct or indirect cause in the growth of the other.

Forms of industrial activity generally are di-vided into three categories: *primary,* or extractive industries (e.g., mining, lumbering, agriculture), *secondary,* or manufacturing industries (e.g., steel, chemicals, automobiles, furniture), and *tertiary,* or service industries (e.g., insurance, entertainment, wholesaling and retailing, transportation). It can be shown that the primary sector typically employs a larger proportion of the active labor force than do the secondary and tertiary sectors in countries which are in their earlier stages of organizational development, what is currently called "underdeveloped." As the economy becomes more complex and the per capita standard of living higher, the proportions of the economy devoted to secondary and tertiary activities increase with the latter gradually coming to dominate the former. In 1960, for example, the percent of U.S. labor force engaged in primary, secondary, and tertiary industries was 7, 39, and 54, respectively, while these proportions in 1880 were 50, 24, and 26 percent, respectively. The change of economic emphasis in Canada, although several decades later than that in the United States, is equally striking.

The secondary sectors of the North American economies are engaged in the conversion of raw materials into products consumable either by the general population or by other industries. Their development has been assisted by what must be called an exceptional, if not unique, combination of raw material resource quantity, quality, and location. Quantity and quality of resources may be measured in economic-geologic terms, but location is most meaningfully described in this context as *accessibility.* Both Canada and the United States are "resource blessed" in this manner, but the latter's advantages are much the greater of the two.

North American Mineral Resources

The patterns of both metallic minerals and the mineral fuels reflect the subsurface rock structure. Sedimentary rock, remember, is laid down by the settling of solid, water-borne particles in areas where the bodies of water are relatively undisturbed for long periods of time. It was the sedimentation that occurred during the Carboniferous geologic time period which contains the world's supply of coal. It was primarily during this time that conditions were correct for the formation of major coal deposits in much of the world.

North America contains vast regions of sedimentary lowlands of the Carboniferous Period. Indeed, nearly the entire Interior Plains physiographic province is underlain by such sediments with a folded portion of the sediments also on the western side of the middle Appalachian Highlands. It is in this latter region, especially in the eastern half, that tremendous quantities of bituminous ("soft") coal have been mapped and mined (Map 6-12).

Including a small field of hard, anthracite coal in eastern Pennsylvania, the extensive Appalachian fields are the most eastern. They are easy to work, lying in seams between one and nine feet thick, in many areas close to the surface, and

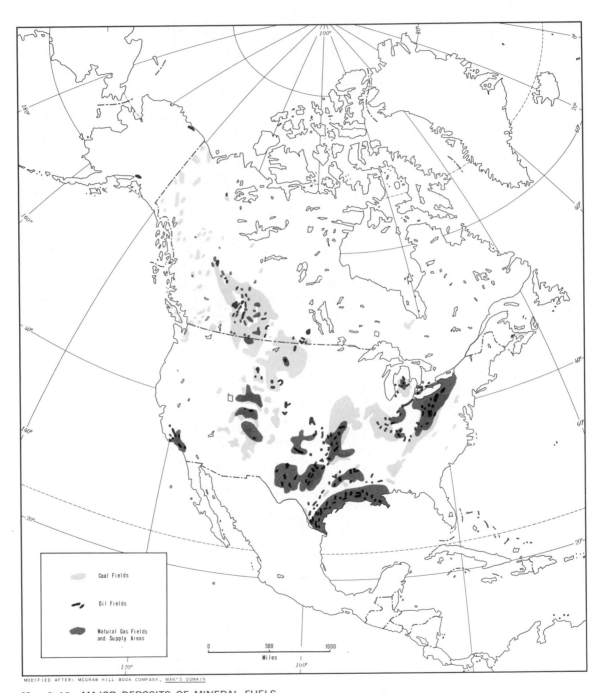

Map 6–12 MAJOR DEPOSITS OF MINERAL FUELS.

MODIFIED AFTER: MCGRAW HILL BOOK COMPANY, MAN'S DOMAIN

Wide areas of the Appalachian landscape have been scarred by strip-mining. The tailings (or refuse) from the mining activities can be seen extending from the horizon almost into the back yards of many of the homes in the foreground. Strip-mining is very efficient in terms of its low labor requirements per ton of coal mined, but this often means low average income levels because of extensive unemployment. And once the coal is gone and the mining activities have moved elsewhere, the land is rarely suited to any productive use. (Grant Heilman)

nearly continuous from Kentucky northward. The mines in this field, primarily eastern Kentucky, West Virginia, and western Pennsylvania, account for over 75 percent of the nation's coal production. Although mined for decades, the reserves remaining are still large. Even now they comprise about 20 percent of the nation's total supply of bituminous coal. Most of the remaining bituminous coal presently mined in the United States is obtained from the Eastern Interior field underlying most of Illinois and extending into western Kentucky. Of slightly poorer quality than the excellent coal found in the two eastern fields, the Western Interior field is also extensive and as yet, little mined. Many small, somewhat lower quality coal deposits are found throughout the Rocky Mountains, but because of the abundance of good, easily obtainable coal in the east, the tilted and fragmented reserves

in the Rockies are not likely to be much used in the foreseeable future.

Although not restricted to Carboniferous sediments, petroleum is generally found in sedimentary basins and often in association with natural gas deposits. Major petroleum deposits have been developed along the Gulf of Mexico coastline, both on land and on the continental shelf, in southern California, and in several areas of the interior in Texas, Oklahoma, and Kansas. Recent drilling has also indicated extensive deposits, expected to exceed those of the Gulf, along the coastal plain of northern Alaska, well above the Arctic Circle. It is a reflection of the desirability of petroleum that men will tolerate extremely hostile environments, such as that in northern Alaska, in order to obtain this fuel, lubricant, and chemical base. Also important are the very large deposits of oil shale in the Rocky

Mountains, mostly within the United States. Although still somewhat expensive to recover commercially, oil shale may become a major source of petroleum products in North America because of the present high rate of consumption.

In contrast to the formation of sedimentary rocks, *metamorphic* rocks are those which have been structurally transformed by tremendous heat and pressure applied over great lengths of time. It is primarily because of this transformation that metallic minerals, found in complex compounds in other rocks, may often be concentrated into economically extractable quantities in metamorphic rocks. As might be expected, then, North America's deposits of metallic minerals are located in the three zones

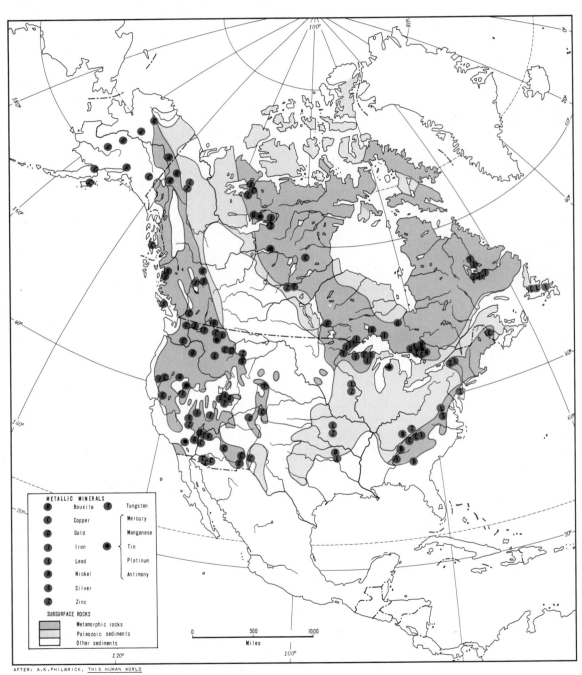

Map 6–13 MAJOR METALLIC MINERAL DEPOSITS AND SUBSURFACE ROCK ASSOCIATIONS.

containing significant concentrations of metamorphic rock: the Laurentian Shield, sections of the Appalachian Highlands, and scattered zones throughout the western mountain ranges (Map 6-13).

The shield's minerals, for example, extend in a broad arc from the Quebec and Labrador deposits of iron ore at Schefferville, Wabush, and Gagnon, through the Sudbury (Ontario) nickel deposits, the copper and iron deposits in northern Michigan and Minnesota, iron at Steep Rock in western Ontario, and copper and zinc at Flin Flon (Manitoba) to the major uranium and copper deposits from extreme northern Saskatchewan to the Arctic waters. This long arc of developed mineral deposits approximates the contact zone between the shield and the surrounding nonmetamorphic lowlands.

The second major zone of metallic minerals is located throughout the Appalachian Highlands, from the recently closed iron ore mines at Wabana, Newfoundland, to the Alabama iron ore used in the industrial development at Birmingham. Although this mineralized district is often forgotten by American students, it was very important during the early rise of United States industry, and remains a significant contributor of selected minerals in the East.

The third region containing extensive concentrations of metamorphic rock is the western cordillera, or mountain region. No student familiar with North American frontier lore can be unaware of the gold rushes to California and later to Alaska, the similar silver "boom" in Nevada, and the many frontier tales associated with mineral prospecting in the western United States. Also of great economic importance are the less romantic but substantial deposits of copper, zinc, lead, silver, molybdenum, uranium, and other minerals scattered throughout this region from Alaska into Mexico.

Much of the mineral wealth of North America can be attributed to the intensity with which it has been sought during the last century and more. It cannot be disputed, however, that the United States and Canada contain fuel and metallic minerals in an abundance and quality found in few other countries. Few minerals required by modern industry are not found there; only high-grade bauxite (for aluminum), tin, manganese, and several less important minerals are absent or in small quantity. This does not mean that the United States is self-sufficient in virtually all its minerals. Much is imported to supply the immense manufacturing-industrial complex which has developed since the Civil War. But the abundance of minerals was a great help to the development of this complex.

Location as a Resource

As important as the quantity and quality of resources have proved for economic growth in the United States and Canada, the spatial association of agricultural and industrial resource patterns is equally fortuitous. The great Interior Lowlands comprise, with the southeastern Coastal Plain, the second most extensive agricultural region in the world, and they are practically surrounded by metallic and nonmetallic mineral deposits. Furthermore, the Lowlands are underlain by tremendous mineral fuel deposits which are cheaply mined and are the primary power source for the numerous urban and industrial clusters in the northeastern U.S. and southeastern Canada.

These factors—resource quantity, resource quality, general locational associations—however, cannot be well utilized without a developed accessibility network. The two nations of North America are extremely fortunate to possess great natural transportation resources within the very zone containing the spatial association of rich agricultural and industrial resources. Long used by American Indians for interregional trade before the arrival of Europeans, the Great Lakes are an internal waterway broken only at Niagara Falls, between Lakes Ontario and Erie, and at Sault St. Marie, between Lakes Superior and Huron. With these exceptions, the Great Lakes offer over 1000 miles of East-West transportation routes and cover over 600 miles North-South at their greatest extent. They connect the southern margin of the Canadian Shield with the coal-rich and agriculture-rich Interior Lowlands.

Transportation routes are the physical expression of a society's social and economic organization. When these routes are built, as railways or highways, then they express either the existing needs for interaction or the patterns which are desired. A railroad may be constructed to connect a localized resource supply with the industrial consumer of that resource, or it may be built because the region through which it is to pass is believed to possess sufficient economic potential to justify the construction costs. In either

Much heavy industry in North America is located along the shores of the Great Lakes. The iron and steel works at Gary, Indiana, shown here, are one example. Note the long ore carriers in the artificial channel, the intensity of land use, and the quality of the atmosphere. (Charles E. Rotkin)

case, capital is required and the economic system must have developed to the point where potential can be recognized as such. The great advantage of internal waterways, whether connected lakes or an extensive system of navigable rivers, lies in their existence exclusive of previous human organization. They are not resources until man uses them, to be sure, for "resource" is a man-related concept rather than an absolute. The sequential growth of settlement and economic interaction in North America, however, definitely follows the pattern of natural accessibility provided by these waterways, especially during the early stages of economic and technological development.

The presence of industrial and agricultural resources adjacent to the Great Lakes lent strong support to the rapid economic and urban development of the northeastern Interior Lowlands. The Great Lakes, of course, are not the only transportation resource. They are connected to the early core region of colonial America, the Philadelphia–New York–Boston region, by a marked gap through the Appalachian bar-rier—the valleys of the Hudson and Mohawk Rivers. It was through this gap that the Erie Canal was constructed to permit cheap, pre-railroad accessibility for New York with the developing interior. It was undoubtedly New York's presence at the head of this natural route to the interior which gave it the impetus and the advantage over other important colonial ports to become the country's primate city.

Of more direct importance to the long-term growth of the Interior Lowlands was the accessibility provided by the Mississippi River and its many tributaries. The Great Lakes are peripheral to the interior plains. The fact that the interior is largely of low relief means that the river systems contained within the region can be extensive; they are not restricted by irregular topography to narrow channels and few tributaries. Similarly, because the region traversed is of relatively even relief, the river system is navigable over great distances, at least by low draft boats.

As one of several major tributaries of the Mississippi, for example, the Ohio River and *its* tributaries drain an area of over 150,000 square miles. More crucial than the extent of the area drained by these rivers is the quality of the land comprising this area. And with such tributaries entering the Ohio River from the north as the Wabash, the Scioto, and the Allegheny, and those entering the Ohio from the south, such as the Tennessee, the Cumberland, the Kentucky, the Kanawha, and the Monongahela, the Great Lakes are well complemented in their contribution to intraregional accessibility. The agricultural portions of the regional resource potential are therefore well connected by cheap means of movement to the mineral fuel and metallic mineral concentrations which support the industrial base.

The primary Canadian core immediately north of the U.S. core, in fact, the Canadian portion of the North American continental core region, has also benefited from the accessibility provided by the internal water transport system. The primary development in Ontario is associated with transport routes on the Great Lakes and the upper St. Lawrence River, while Quebec's development remains closely tied to the lower reaches of the St. Lawrence. So important are the Great Lakes and the St. Lawrence River to Canada, in spite of navigation problems on the St. Lawrence, that it proposed and argued

for the development of this waterway into a system capable of handling ocean-going vessels, that is, the St. Lawrence Seaway, decades before the United States acceded and cooperated in its construction. The St. Lawrence, and by implication Canada itself, was simply too peripheral to U.S. national interests for seaway construction expenses to be considered until after World War II.

There are numerous major tributaries of the Mississippi entering the main channel from the west and it might be asked why these rivers did not prove as important in extending the core region west of the Mississippi as the Ohio River and the Great Lakes did in integrating the resources to the east. The answer, of course, lies in the quality and quantity of resources found to the west. Although agricultural potential remains high 300 miles west of the Mississippi, the gradual reduction in average annual precipitation, declining to 20 inches at about the 100th meridian, seriously restricts the possibilities for productive agriculture. Only drought resistant crops, such as wheat, or high-value crops suitable for irrigation, such as cotton, or farming which does not require cultivation of the soil, such as cattle ranching, can survive economically in the arid West. Agriculture no longer contributes

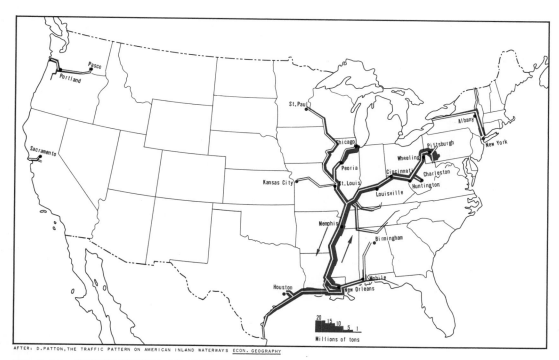

AFTER: D. PATTON, THE TRAFFIC PATTERN ON AMERICAN INLAND WATERWAYS ECON. GEOGRAPHY

Map 6–14 FREIGHT TRAFFIC ON INTERNAL WATERWAYS.

greatly to freight shipments on internal waterways. Add to this the relative lack of developed mineral deposits, the low level of industrial and urban development relative to the East, and most importantly for the movement of goods, the extremely seasonable flow of the rivers (late summer will often find dry the beds of such major rivers as the Platte, the upper Arkansas, and the upper Canadian), and it is not difficult to understand the low traffic volume which is carried on these western rivers (Map 6-14).

The Urban-Industrial Core

The North American core region (see Map 6-15) contains an unusually beneficial combination of economic and geographic resources. The post-Civil War development of industrialization and urbanization in this core may be illustrated by reference to the advantages and effects of *agglomeration* on the cities, on manufacturing, and on the pattern of agricultural production.

Agglomeration is the clustering or grouping of phenomena from a relatively dispersed spatial pattern to one which has little areal extent. Geographers view agglomeration forces as those which lead to this clustering, whether the phenomenon is people (urban places), economic

activities (e.g., retail, wholesale, or production concentrations), political control (nodal structure of political power), or whatever. Because the spatial result is the same (i.e., clustering) regardless of what is being agglomerated, some have argued that it would be of value to discover what it is that these human activities share, leading to agglomeration, regardless of their obvious differences. It has been further urged that those forces leading to the dispersion of activities and objects also be studied insofar as they contribute to this dispersion.

During the last third of the nineteenth century and the beginning of the twentieth, American cities grew not only as population concentrations, but also, and in general proportion to, the growth of manufacturing. There are many factors which contribute to the growth or stagnation of an urban area, but during this period, one of the more important supports of growth was the increased concentration of industrial activities in metropolitan regions.[3]

[3] For a careful introduction to the complexities of the interdependent support of urban and industrial growth, see Allan Pred, "Industrialization, Initial Advantage, and American Metropolitan Growth," *Geographical Review*, Vol. 55, No. 2, April, 1965, pp. 158–85.

AFTER: E.L. ULLMAN

Map 6–15 RAILWAY FREIGHT TRAFFIC AND THE NORTH AMERICAN CORE REGION.

It is clear that individuals have long lived together in small areas because of the advantages which accompany this clustering. Economic efforts are concentrated, settlement is more defensible, and unless the cultures of the population are too diverse, the socialization afforded by proximity is generally more pleasant than individual isolation. The advantages related to the clustering of economic activities are of three types: economies of scale, complementarity, and mutual use.

Strictly speaking, economies of scale are a result of large volume production rather than clustered production. Once the initial investment in basic manufacturing apparatus has been made by a firm, a more complete use of this apparatus permits the initial investment costs to be absorbed more gradually by the many units of output. It is advantageous, therefore, to use machinery to its productive capacity where handcrafting is less desirable and the market remains open. In addition, volume production has meant large demands for labor accessible to the production plant and also a large available market for the goods. Large-scale industrial operations, therefore, became established in already developed population clusters, or they drew population to them.

Economies of scale also have meant specialized production. Separate companies manufactured product components in volume and sold them to other firms for use in the assembly of finished items. It proved advantageous to these manufacturers, therefore, to locate close to one another. The complementarity of such activities is often made more efficient by minimizing the cost of interindustry commerce. This clustering of interdependent and interacting industry multiplies the need for labor and markets and reflects the shared growth patterns of urbanization and industrialization.

The agglomeration of industrial activities also requires and leads to an increase in many political, social, and economic facilities. At minimum, there must be a *means* for interaction, that is, transportation and communication lines. When one views urban places as well-established *nodes* within a transport network, the advantages of such locations for manufacturing (and most other human activities) are obvious. There are many additional requirements for effective production. Large firms have special needs in water supply, sewage, electrical power, and so forth. Their employees require schools, recreation, food, and clothing, among many other items. Urban places have been best able to supply these functions. Those places already possessing these service functions attracted industry, and the arrival of industry required the expansion of these functions.

Industrial location is much more complicated than presented here, of course, but these "agglomeration economies" have contributed to the clustering of much manufacturing activity in a relatively few metropolitan regions. During the transformation of the American economy from a mercantilist to a capitalist-industrial economy following the Civil War, cities in the Northeast and eastern Interior Plains grew rapidly as the manufacturing core of the country developed.

As Professor Pred has pointed out, however, several other concepts must be considered for a more complete understanding of this period of urbanization and its interdependence with industrialization. Insights gained from study of Christaller's *central place theory*, discussed in Chapter 2, would lead one to expect the largest manufacturing concentrations to contain the greatest variety of industrial and other supporting activities, just as the largest urban centers possess the widest functional support for their population. Urban and industrial concentrations, then, clearly provide mutual growth support. Furthermore, once a city reached a certain size, manufacturing activities requiring a local market of that size could locate within the metropolitan region. This propelled not only further urban growth and the entry of other manufacturing dependent on the recently introduced industry, it also brought still other industry which required the even larger urban market generated by the new manufacturing firms. It should also be easily seen that those metropolitan regions having the *initial advantage* of growth from developing industry are the most likely to continue to grow through such stimulation. It was this process of mutual support which led to the development of the North American manufacturing core region.

The half century prior to World War I was also the era of greatest railroad expansion, for not only was it necessary for the core region to become fully integrated economically and spatially, but there were smaller secondary cores

Table 6-2 Population Growth, Selected North American Cities, 1880-1960

City		1880	1890	1900	1910	1920	1930	1940	1950	1960
Montreal[a]	Population	140,747	216,650	267,730	470,480	618,506	818,577	903,007	1,021,520	1,191,062
	% Increase		53.9	23.6	75.7	31.5	32.5	10.3	13.1	16.6
Toronto[a]	Population	86,415	181,220	208,040	376,538	521,893	631,207	667,207	675,457	672,407
	% Increase		109.7	14.8	81.0	38.6	20.9	5.7	1.2	−0.5
Boston	Population	362,839	448,477	560,892	670,585	748,060	781,188	770,816	801,444	697,197
	% Increase		23.6	25.1	19.6	11.6	4.4	−1.3	4.0	−13.0
New York	Population	1,911,698	2,507,414	3,437,202	4,776,883	5,620,048	6,930,446	7,454,995	7,891,957	7,781,984
	% Increase		31.2	37.1	39.0	17.7	23.3	7.6	5.9	−1.4
Philadelphia	Population	847,170	1,046,964	1,293,697	1,549,008	1,823,779	1,950,961	1,931,334	2,071,605	2,002,512
	% Increase		23.6	23.6	19.7	17.7	7.0	−1.0	7.3	−3.3
Baltimore	Population	332,313	434,439	508,957	558,485	733,826	804,874	859,100	949,708	939,024
	% Increase		30.7	17.2	9.7	31.4	9.7	6.7	10.5	−1.1
Cleveland	Population	160,146	261,353	381,768	560,663	796,841	900,429	878,336	914,808	876,050
	% Increase		63.2	46.1	46.9	42.1	13.0	−2.5	4.2	−4.2
Detroit	Population	116,340	205,876	285,704	465,776	993,678	1,568,662	1,623,452	1,849,568	1,670,144
	% Increase		77.0	38.8	63.0	112.9	57.9	3.5	13.9	−9.7
Chicago	Population	503,185	1,099,850	1,698,575	2,185,853	2,701,705	3,376,438	3,396,808	3,620,962	3,550,404
	% Increase		118.6	54.4	28.7	23.6	25.0	0.6	6.6	−1.9
St. Louis	Population	350,518	451,770	575,238	687,029	772,897	821,960	816,048	856,796	750,026
	% Increase		28.9	27.3	19.4	12.5	6.3	−0.7	5.0	−12.5
Cincinnati	Population	255,139	296,908	325,902	363,591	401,247	451,160	455,610	503,998	502,550
	% Increase		16.4	9.8	11.6	10.4	12.4	1.0	10.6	−0.3

City		1881	1891	1901	1911	1921	1931	1941	1951	1961
Atlanta	Population	37,409	65,533	89,872	154,839	200,616	270,366	302,288	331,314	487,455
	% Increase		75.2	37.1	72.3	29.6	34.8	11.8	9.6	47.1
New Orleans	Population	216,090	242,039	287,104	339,035	387,219	458,762	494,537	570,445	627,525
	% Increase		12.0	18.6	18.1	14.2	18.5	7.8	15.3	10.0
Houston	Population	16,513	27,557	44,633	78,800	138,276	292,352	384,514	596,163	938,219
	% Increase		66.9	62.0	76.6	75.5	111.4	31.5	55.0	57.4
Dallas	Population	10,358	38,067	42,638	92,104	158,976	260,475	294,734	434,462	679,684
	% Increase		267.5	12.0	116.0	72.6	63.8	13.2	47.4	56.4
Seattle	Population	3,533	42,837	80,671	237,194	315,312	365,583	368,302	467,591	557,087
	% Increase		1112.5	88.3	194.0	32.9	15.9	0.7	27.0	19.1
Portland	Population	17,577	46,385	90,426	207,214	258,288	301,815	305,394	373,628	372,676
	% Increase		163.9	94.9	129.2	24.6	16.9	1.2	22.3	−0.3
San Francisco	Population	233,959	298,997	342,782	416,912	506,676	634,394	634,536	775,357	740,316
	% Increase		27.8	14.6	21.6	21.6	25.2	0.02	22.2	−4.5
Los Angeles	Population	11,183	50,395	102,479	319,198	576,673	1,238,048	1,504,277	1,970,358	2,479,015
	% Increase		350.6	103.4	211.5	80.7	114.7	21.5	31.0	25.8

[a]Canadian data are for 1881, 1891, 1901, . . . , 1951, and 1961.

Sources: United States—*Statistical Abstract of the United States*, 1928, 1952, 1961.

Canada—*The Statesman's Yearbook*, 1883, 1894, 1903, 1913, 1924, 1933, 1948, 1953, 1968.

developing in the West, as well. As Pred has written,

". . . the extension of markets and the alterations in average scale of production, which were the outward expressions of rapid industrial growth, can be seen in their proper perspective only in relation to certain railroad developments: intensification of trunk lines, spread of feeder lines, and integration of fragmented operating units; technical improvement of freight-car carrying capacity, motive power, rails, roadbeds, and terminal facilities; financing innovations adopted by industrial enterprises; precipitation of an enormous demand for steel rails, steel engines, and rolling stock; and elaboration of consumer demand through wages paid to line employees and construction workers."[4]

Transportation, then, not only was necessary to meet increasing demands for spatial interaction between manufacturing firms and between growing urban nodes, but it also was a generator of certain kinds of industrial development and, therefore, urban development.

City growth has not been evenly distributed across the continent since the late nineteenth century. As Table 6-2 indicates, each section of the United States contained regionally distinct city growth patterns. The northeastern coastal cities grew relatively slowly and at a diminishing rate; New York is an exception. The primacy of New York City is illustrated by its growth relative to the growth of Philadelphia, with New York increasingly overshadowing its nearby competitor as the present century proceeded. In the interior manufacturing core, Cleveland, Detroit, and Chicago continued to grow in population until the decade of the Great Depression. St. Louis and Cincinnati, marginal to the manufacturing core region and dependent primarily on the advantages of riverside location, grew rather more slowly until after World War II. Southern cities also grew rapidly during this period and differ from the midwestern cities by gaining population through the Depression years and returning to a rapid growth rate after World War II; the port city of New Orleans was an exception and is more similar to St. Louis and Cincinnati in its population trends than to the other southern cities mentioned. The western cities experienced the greatest population growth rates since 1880, with San Francisco the excep-

[4] Pred, *op. cit.*, pp. 163–64.

tion, but to some extent this reflects the low initial population of these cities for the period tabulated. Montreal and Toronto have grown irregularly although with general strength. However, Toronto's central city has not been able to maintain its rate of population increase, while Montreal's has. The *metropolitan* populations are more nearly equal, with Montreal's at 2.1 million and Toronto's at 1.8 million in 1961.

Table 6-2 also illustrates other phenomena of urbanization in North America. After the generally low population increase during the 1930's (with notable exceptions in the South and West), the 1940–1950 rates of increase rose slightly only to return to conditions of net *decrease* in population during the 1950–1960 decade. This decrease in population reflects the flight of the central cities' populations to the surrounding suburbs, a trend continuing through the 1960's. The generally white populations moving out of the cities are leaving at a faster rate than the blacks and others are moving in. Some of the geographic implications of this population change will be discussed later. It should be noted, however, that in contrast to the others listed, southern cities, and Seattle, Los Angeles, and Montreal have not declined in central city population. Much of the population growth in the southern cities, especially those in Texas, reflects the sharp rise in economic activity across the South in the last two decades. City population growth has also been stimulated by the extreme rural poverty of much of the South. Some of the growth, however, is illusory. In contrast to the North and West, some southern cities do not have suburbs; the cities are legally permitted to annex adjacent unincorporated population clusters without the consent of the residents of these clusters. Until recently, therefore, southern city population figures often represented metropolitan areas, not merely the central cities. Seattle contained space for expansion within the city boundaries and has been strongly influenced by the rapid growth of Boeing Aircraft production facilities. And although the city of Los Angeles is surrounded by distinct suburbs as are the eastern cities, its population has had room to increase within the city boundaries by gradually increasing the degree of vertical development. Even with this recent trend, Los Angeles has one of the continent's lowest urban profiles. With these exceptions, however, central city populations have not increased significantly in spite of continued total metropolitan population growth. Some of the spatial implications of current metropolitan growth and central city stagnation will be discussed later.

Agriculture's Changing Role

At the same time that many North American cities experienced extremely rapid growth and many others grew substantially from industrial capital and labor inputs, agricultural output declined in relative importance. The key here is the term "relative," for the total output from the agricultural sector has increased tremendously both in volume and in value. It has not grown as rapidly as the secondary and tertiary industries, however, and agriculture now contributes a much smaller proportion to the Gross National Products of the United States and Canada than it did a century ago.

More important than the relative decline of agriculture is the absolute decline in the number of people directly engaged in agricultural production. In 1880, fully 72 percent of the total U.S. population could be classified as rural, most of these either cultivating land for themselves, acting as tenant farmers for others, or practicing animal husbandry. By 1900, this percentage had declined to 53, and in 1920 it was clear that less than half of the population was engaged in rural activities. Except for the period of near economic stagnation during the 1930's, the proportion of population directly involved in agricultural production diminished further each decade. By 1960, only some 30 percent of the United States' labor force could be classified as residing in rural areas, with approximately half of these actually engaged in farming or farm industries. The flight from agriculture has continued through the 1960's also, so that by 1970, the estimated 205 million inhabitants of the country will be fed by the efforts of only three million farm families.

An important element of the relative decline of agriculture's contribution to the national economy should not be overlooked. With fewer people actually engaged in agriculture, the absolute increase in agricultural output reflects a tremendous rise in productivity per man-hour. Today, it takes fewer men less time to produce a given quantity of food and plant fiber than even a decade ago, and such increasing efficiency shows few signs of moderation. This efficiency has, in turn, forced marginally productive regions of the country to change their agricultural

emphasis. The areas most suited to corn production, for example, are the primary sources of corn.

The great increase in agricultural productivity is a reflection of the many technological innovations which have been accepted and used by North American farmers. Farm machinery, hybrid seeds, and a wide variety of fertilizers, insecticides, and herbicides, are some of the more obviously important innovations that have revolutionized agriculture. Also of great impact have been modifications in crop handling and storage, in the care and feeding of farm animals, and in the marketing of farm produce. However, each refinement costs money. With the increased efficiency of production, small farm operations are increasingly difficult to maintain. The farm population has decreased rapidly as average farm size has increased to levels at which the new technology can be used.

This rapid decline in the farm population relative to the urban population has raised an important political-geographic problem in the United States. Political representation based on areal units of approximately equivalent population size is generally accepted as equitable. Most United States Representatives are elected from districts that have been adjusted several times during this century to account for changing population patterns.[5] The members of the Senate, of course, are elected two from each state with no pretense to equalizing state populations, areal size, or economic contribution to the nation.

Many states also established bicameral legislatures, one house based on districts of approximately equal population size and the other house using the county as the basic unit for representation. In those states where the latter house was more important than the former, perhaps holding veto power, changing population patterns have resulted in a few dominantly urban counties with constituents having urban needs indirectly controlled by the representatives of much smaller and declining populations in largely rural counties. The urban regions have felt that their needs were not being met by rural-controlled state legislatures, and have turned increasingly to the federal government for

[5] Sometimes, of course, the areal units are designed by the party in power to minimize the electoral strength of the party out of power, a practice called "gerrymandering," but this is another, although no less interesting, problem in geography and politics.

direct assistance. (Interestingly, even the federal government tended to overemphasize agriculture relative to urbanization, for Agriculture has long had a place on the President's Cabinet while a representative for Housing and Urban Development was not placed until 1965.) Rural interests at the state level, on the other hand, are fearful that control of the state legislature by the majority urban interests will delegate farming and related activities an insignificant voice in state affairs. After the criteria are settled on, the rearrangement of state political districts, or redistricting, is a political geography problem, for it deals with regionalizing political phenomena.

Major Crop Regions

By far the most important crops grown in North America, whether measured in value of production or acreage devoted to their growing, are corn and wheat. Corn production is heavily concentrated in a region extending from western Ohio, through northern Indiana and Illinois, Iowa, and into southern Minnesota and extreme eastern South Dakota and Nebraska (Map 6-16a). This region, with most production in Illinois and Iowa, is often called the "Corn Belt." It is often incorrectly concluded (1) that

this region is the only source of corn in North America, (2) that it produces nothing but corn, (3) that most of the corn produced is for human consumption, and (4) that corn is grown throughout the region with equal intensity. Each of these conclusions is incorrect. As indicated on Map 6-16a, corn is also grown throughout the eastern half of the United States, with substantial acreages devoted to its production along the interior Coastal Plain from Maryland and Delaware through the Carolinas into Georgia and southern Alabama. Numerous other crops are grown throughout the Corn Belt, with soybeans, wheat, oats, and alfalfa being among the more important. Most of the corn grown in the United States is grown as feed for cattle and hogs, these animals comprising the primary source of meat for the large markets in the northeastern core region. Although Americans do consume corn more readily and in greater volume than many other peoples of the world, its primary use is as feed. That corn is used for feed rather than one of the other feed crops, for example, hay, timothy, clover, or alfalfa, illustrates several advantages that it holds over most of its crop competitors. The entire plant may be used for silage or the high yielding and nutritious grain can be used

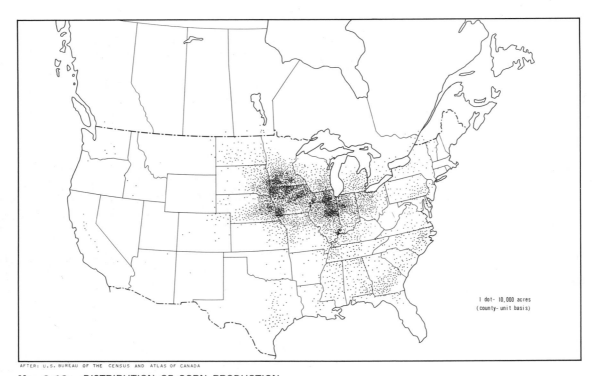

AFTER: U.S. BUREAU OF THE CENSUS AND ATLAS OF CANADA

Map 6–16a DISTRIBUTION OF CORN PRODUCTION.

for feed. The result is greater flexibility and higher value per acre to the farmer.

In contrast to the variety of uses for corn, wheat is primarily consumed by humans. Its areas of production are even more widespread than those of corn, although similarly, there are several regions containing the overwhelming majority of production (see Map 6-16b). A winter wheat region, growing wheat sown in the late fall and allowed to remain dormant over a mild winter before early spring growth, is found in extreme northern Texas and through Oklahoma and Kansas into southern Nebraska. Spring wheat, sown after the harsher winters of the North have moderated into spring, is grown from the Dakotas into the Canadian Plains of western Manitoba, southern Saskatchewan, and into Alberta. Although smaller than these two regions of production, wheat is also grown in the Columbia–Snake River–Polouse region in Washington and Oregon. A substantial amount of wheat is also produced from the mixed farming region in the East. In addition to the differences of human consumption and greater dispersion of production for wheat vis-a-vis corn, the winter wheat belt is in fact a virtual one-crop region. Few other crops can approximate wheat's value-

per-acre returns in this region's inhospitable environment.

It should be noted, however, that the Canadian portion of the spring wheat belt is one of only two major food-producing regions in that country. Food crops are grown on the Ontario peninsula and along the St. Lawrence River to Quebec. Wheat is annually grown on the Canadian Plains in quantities far beyond the needs of the 20 million inhabitants of the country, and so it is an obvious and reasonably dependable source of export funds. Purely in locational terms, the best buyer for this surplus wheat might be the United States, a wealthy country of over 200 million people. Of course, the environmental conditions permitting a wheat surplus in Canada are also present south of the border, and so Canada must locate buyers for this surplus overseas. A difficult (and possibly irritating) factor for Canada is that it is often in competition with the United States on the world wheat market, and while wheat export sales comprise only a small portion of the total U.S. export income (5 percent in 1964), wheat sales are somewhat more important to the less diversified Canada (12 percent of the total in 1964). When Canada, therefore, engages in trade with a

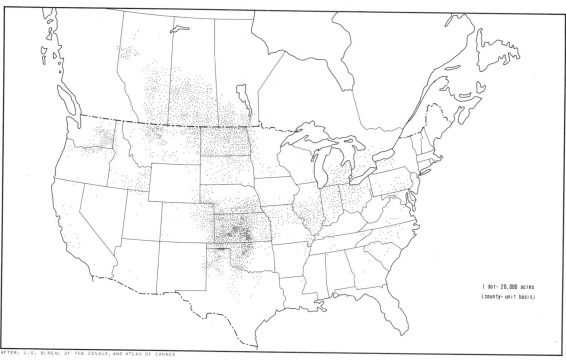

Map 6–16b DISTRIBUTION OF WHEAT PRODUCTION.

country considered by the U.S. to be an undesirable trade partner, for example, China, Canada is not merely exerting its independence or attempting to irritate its southern neighbor; it is conducting commerce more important to its total economy than an equivalent amount of trade is to the economy of the United States.

During most of the first century of the United States' existence, the populated region east of the Appalachians contributed substantially to the food needs of the country. As agriculture became firmly established in the much more suitable environment of the Interior Plains, the margins of this new agricultural core developed complementary, rather than competitive, products. The land immediately north and east of the mixed farming "Corn Belt" produces vegetable and fruit specialty crops, or "truck" crops—so called because they are sufficiently perishable to require production within trucking distance of the consumption centers. This area also contains a sizable quantity of dairy production (Map 6-17a). It is no geographic accident that the Wisconsin dairy "belt" is located adjacent to the Chicago metropolitan region and the agriculturally specialized Corn Belt. Because of storage and shipment requirements, milk production must be,

and is, proximate to nearly all major urban centers regardless of the relative environmental suitability.

The patterns of production for several of the specialty crops grown in North America also warrant attention. Cotton and tobacco, especially the former, were long the major agricultural income earners in the South. However, the primary areas of cotton production have shifted westward, out of much of the region which depended so heavily on its production during the nineteenth century. Examination of the distribution of cotton production (see Map 6-17b) shows the eastern-most region of concentrated output in what is called the upper Mississippi "delta"; acreage of cotton grown through the eastward states of the former Confederacy is sparse, dispersed across the Piedmont. More important than the Piedmont regions are those in eastern and northwestern Texas, central California, and southern Arizona along the Gila River. With the exception of eastern Texas, nearly all cotton grown in these western regions is grown under irrigation, and is therefore of high quality. Governmental restrictions on production acreages, because of surpluses similar to those of wheat, and the overuse of soils before the need

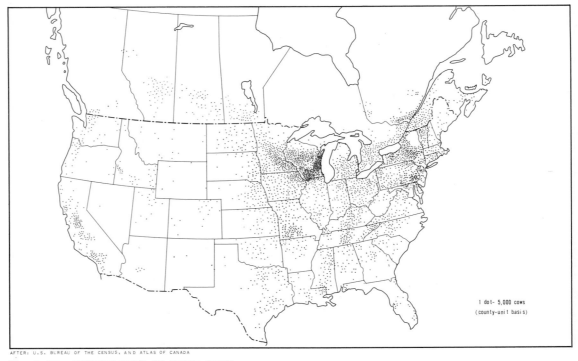

AFTER: U.S. BUREAU OF THE CENSUS, AND ATLAS OF CANADA

Map 6–17a DISTRIBUTION OF DAIRY COWS.

for soil conservation was recognized have both led to a sharp reduction in cotton acreage in the last several decades. Production remains high, however, as farmers use the best of their land to grow this lucrative crop. Thus, in spite of a drawing inward of production regions, cotton is still second in value of output only to corn. Tobacco, too, has become more concentrated through stringent government controls in the United States. North Carolina and Kentucky contain most of the tobacco production, as they have since the turn of the century (Map 6-17*b*). Unless there is a marked rise in tobacco exports, the production of this crop is very likely to decline in the future. Significant amounts of tobacco are also grown on the Ontario peninsula in Canada.

The general economic-geographic patterns for North America, then, are clear. Situated astride major deposits of coal, adjacent (via the Great Lakes) to large deposits of a wide variety of metallic minerals, connected to large petroleum and natural gas deposits by pipeline, and also containing much of the high quality agricultural land on the continent, North America's core region extends from St. Louis to Montreal to Boston to Baltimore. This region contains nearly

The photograph clearly illustrates the high productivity per man typical of much modern North American agriculture. Several decades ago, the man driving the cotton-picker would have been joined in the field by his family and perhaps several other families in picking the cotton by hand. (Grant Heilman)

60 percent of the United States' manufacturing activity and about 80 percent of Canada's, just under 50 percent of the U.S. population and some 60 percent of the total Canadian population, and yet occupies only seven percent of the area of

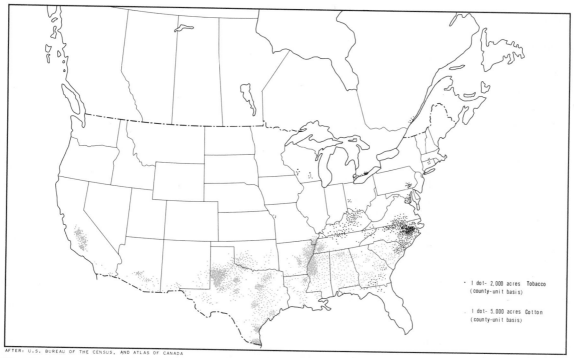

AFTER: U.S. BUREAU OF THE CENSUS, AND ATLAS OF CANADA

· 1 dot= 2,000 acres Tobacco
(county-unit basis)

. 1 dot= 5,000 acres Cotton
(county-unit basis)

Map 6–17b DISTRIBUTION OF COTTON AND TOBACCO PRODUCTION.

the United States and less than one percent of Canada's. In many respects, because of commercial ties with most of Latin America, this region is the economic core of the entire hemisphere. A west coast core is in the process of development, extending from San Francisco to San Diego, as are numerous other secondary nodal regions across the South and around such cities as Seattle, Washington, and Vancouver, British Columbia. The locational inertia, however, for both North American manufacturing and the population which produces and consumes its products maintains the primary continental core in the northeastern United States and adjacent Canada.

Patterns of Black America Since 1860

Just as the changing spatial pattern of the Afro-American population in the United States provided insights into much of the basic human geography of the country prior to the Civil War, changes in the black population distribution also illustrate many geographic developments after the war. Negroes in America have participated in, and helped cause, many of the more significant features of North America's current geography.

During the twenty-five years following the end of the Civil War, while the North was undergoing a major industrial and urban transformation and the West became settled in at least the outline form which has carried to the present, economic progress was largely absent from the South. Production increases lagged behind those of the rest of the country as a result of the war. This is not to say that there were not concrete changes, for members of the black population began to participate in virtually all economic, political, and social aspects of southern life. There was some migration from rural to urban areas within the South, and a small stream from the South to northern cities, but most Afro-Americans remained close to their residences of 1860. Many became freeholders with small farms, but an even larger number drifted into sharecropping and tenant farming as southern landowners attempted to meet tax demands levied after the Civil War. During these brief two decades or so, then, the plural society which had separated black and white showed signs of melting; although much ill-feeling remained between members of these groups, many of the formal restrictions on full participation by blacks

in white society were diminishing.

During the 1890's, however, restrictions and repression were rapidly reintroduced throughout the South in the form of Jim Crow laws. In the space of a few years, the racial populations in the South were separated physically and socially. Whatever positive cultural mixing had been taking place ended. In addition to the personal suffering caused by the indignities of *de jure* segregation, direct and indirect geographic changes were also caused in this way. Housing zones became fixed. Black educational and health facilities, such as they were, had to be duplicated and spatially distinct from the white communities'. Because of socioeconomic restrictions, the geographic patterns of income, health, employment, government expenditures, and of course, population, and many others became more sharply associated with racial differences. Violence against Negroes because they were black was never minor in the South. (Anti-black violence was present in the North as well, but the smaller number of Negroes meant that it was less noticed.) This violence increased, however, as the political, social and economic forces restricting such activities during Reconstruction and after were refocused on Northern industrialization, urbanization, and the continuing flood of migrants from Europe.

Negroes had migrated north since well before the Civil War, via the "underground railroad," and the numbers participating in this movement increased somewhat in the decades following the end of the war. The increase was not substantial, however, for the trip was long, the end uncertain, and black resources relatively few. As long as possibilities for personal improvement were formally limited by the individual's capacity and motivation, the South remained "home" for the vast majority of Afro-Americans. As Jim Crow laws were introduced in state after state across the old Confederacy, the northward migration of Negroes began to increase in volume.[6] By 1914, black Americans were almost completely disenfranchised in the South, thus losing legal political recourse for the correction of racially motivated injustices. Schools for Negroes were poor or nonexistent, wages were low (fifty to sixty cents per *day* in Alabama as opposed to thirty

[6] For a good, short article on the motivations for this migration, see Dewey H. Palmer, "Moving North: Migration of Negroes During World War I," *Phylon*, Vol. 28, No. 1, Spring, 1967, pp. 52–62.

Table 6–3 Negro Population of the Conterminous United States by Major Regions, 1870 to 1960, Showing Totals and Percent Distribution

Year	Total	Northeast	North Central	South	West
Negro Population in Thousands					
1960	18,860	3,028	3,446	11,312	1,074
1950	15,042	2,018	2,228	10,225	571
1940	12,866	1,370	1,420	9,950	171
1930	11,891	1,147	1,262	9,362	120
1920	10,463	679	793	8,912	79
1910	9,828	484	543	8,749	51
1900	8,834	385	496	7,923	30
1890	7,489	270	431	6,761	27
1880	6,581	299	386	5,954	12
1870	5,392	180	273	4,933	6
Percent Distribution of Negro Population					
1960	100.0	16.1	18.3	60.0	5.7
1950	100.0	13.4	14.8	60.8	3.8
1940	100.0	10.6	11.0	77.0	1.3
1930	100.0	9.6	10.6	78.7	1.0
1920	100.0	6.5	7.6	85.2	0.8
1910	100.0	4.9	5.5	89.0	0.5
1900	100.0	4.4	5.6	89.7	0.3
1890	100.0	3.6	5.8	90.0	0.4
1880	100.0	3.5	5.9	90.5	0.2
1870	100.0	3.3	5.1	91.5	0.1

Source: C. Horace Hamilton, "The Negro Leaves the South," *Demography*, Vol. 1, 1964, Table 1.

to forty cents per *hour* in the North), and lynching common (54 Negroes in 1915).

In addition to these undeniably strong "push" factors, the North vigorously attempted to "pull" black labor from the South during the years of World War I. National industrialization before World War I depended heavily on millions of European immigrants to meet its growing labor requirements. When the war interrupted this flow of labor, an alternative supply was needed. The increasingly discriminated-against southern Negro met this need nicely. Press campaigns were waged to encourage northward migration and labor recruiters were sent to southern states to provide the means for this move. The initial southern feeling seemed to be relief at the departure of large numbers of "troublemakers." The relief quickly changed to anger and apprehension as emigration continued and it was realized that the major source of cheap southern labor was departing by the hundreds of thousands. In spite of increased white opposition, however, black Americans continued to leave the South in increasing numbers during the 1920's, primarily for the manufacturing core region. There was a decline in Negro migration during the depression years of the 1930's, but a return to still larger numbers during the 1940's and 1950's. By 1960, 40 percent of the Afro-American population of the United States lived outside the South in contrast to only ten percent in 1900 (Table 6-3).

In addition to reflecting the sharp drop in European immigration and comprising a major population shift within the United States, this migration also reflects other important changes that were taking place in the economy. Urban employment opportunities, in manufacturing, in service, and in government, continued to draw large numbers of workers from rural areas. The increasing use of machinery in agriculture permitted farmers to produce more per man and per man-hour. As previously discussed, agriculture has become increasingly efficient, producing more food and industrial crops with less labor and on less land in order to supply the increasingly urban population. Since most rural Negroes did not own the land they worked, many moved to urban areas when mechanized farming became more important in the South. Many others remained on the land to live without a regular source of income.

Table 6-4 Urban Residence of the Negro Population of the Conterminous United States, 1920-1960, by Major Region

Year	Total	Northeast	North Central	South	West
Urban Negro Population in Thousands					
1960	13,801	2,896	3,297	6,608	1,000
1950	9,371	1,897	2,082	4,880	513
1940	6,254	1,234	1,261	3,616	142
1930	5,194	1,021	1,108	2,966	99
1920	3,559	589	661	2,251	58
Percent Distribution, Urban Negro Population					
1960	100.0	21.0	23.9	47.9	7.2
1950	100.0	20.2	22.2	52.1	5.5
1940	100.0	19.7	20.2	57.8	2.3
1930	100.0	19.6	21.3	57.1	1.9
1920	100.0	16.5	18.6	63.2	1.6

Source: C. Horace Hamilton, *op. cit.,* Table 2(a).

The population of black America has in fact become urbanized more rapidly than has the white population. By 1960, over 73 percent of the Negro population were city dwellers while the comparable proportion for the white population was only 69.5 percent.[7] Furthermore, the South continued to contain 93 percent of the total Negro population which could be classified rural, and 95 percent of the non-South black population was urban in 1960 (Table 6-4). Clearly, the Negro American is highly urbanized even for an urban-dominant country such as the United States.

That Negro migrants from the South have moved predominantly into the cities of the North is clear. Of equal spatial significance is the pattern of settlement *within* the major cities receiving these migrants. Incoming Afro-Americans have invariably settled in one or two extremely concentrated residential zones, usually in the older and more central portions of the cities where the resident black population already lived. The result since World War I has been both an intensification of habitation and a gradual areal increase in the sections containing the cities' Negro populations. The higher densities are caused by much greater inflow of migrants than can be accommodated by the slowly expanding residential zones in which it has been possible for Negroes to buy homes. The *de facto* residential segregation in northern and western cities has been as effective as the *de jure*

separation of racial groups long maintained in the South.

Two important results have followed from the spatial separation of blacks (and other identifiable ethnic groups) and whites in major cities. The high population densities which follow from restricted housing have led to severe structural overcrowding and accompanying environmental deterioration. "If the population density in some of Harlem's worst blocks obtained in the rest of New York City, the entire population of the United States could fit into three of New York's boroughs."[8] When a low-income population lives in old structures under such overcrowding, extremely unfavorable living conditions invariably follow. And as the Negro population has been forced to limit residential expansion almost exclusively to the margins of existing ghettos rather than into an open, widely dispersed residential pattern, the separate cultures are maintained (Map 6-18). The black urban culture remains essentially closed to whites just as the white neighborhood remains virtually closed to blacks. This lack of cultural mixing is a concrete spatial example of the maintenance of a plural society in the United States.[9]

[7] Data for this paragraph and some others have been taken from C. Horace Hamilton, "The Negro Leaves the South," *Demography,* Vol. I, No. 1, 1964, pp. 273-95.

[8] Civil Rights Commission, 1959, quoted in M. Harrington, *The Other America: Poverty in the United States,* Penguin Books, Baltimore, 1963, p. 64.

[9] Two articles by geographers that have dealt with this process are Harold M. Rose, "The Development of an Urban Subsystem: The Case of the Negro Ghetto," *Annals of the Association of American Geographers,* Vol. 60, No. 1, March, 1970, pp. 1-17, and Richard L. Morrill, "The Negro Ghetto: Problems and Alternatives," *The Geographical Review,* Vol. 55, No. 3, July, 1965, pp. 339-61.

These Detroit expressways exemplify the spatial separation of black and white cultures in much of the United States. The John Lodge Expressway, in the center of the photograph, is used each day by tens of thousands of white commuters as they enter the business and retail core of the city and return to their suburban homes. Little contact is necessary with the black residents of houses and apartments along and above the expressway. Windsor, Ontario, can be seen in the distance across the Detroit River. (Charles E. Rotkin)

Map 6–18   GENERALIZED CHANGES IN URBAN POPULATION DISTRIBUTIONS.

MODIFIED AFTER: U.S. BUREAU OF THE CENSUS, MAP GE-50,NO.16 AND V.PUGLISI,(TIME (1969)

Map 6-19 DISTRIBUTION OF NEGRO POPULATION AND OF POPULATION WITH SPANISH SURNAME, 1960.

Table 6–5 Negro Urban Population, 1910–1960, for Selected Northern United States Cities

Year	New York	Philadelphia	Chicago	Detroit	Cleveland
	Total Negro Population				
1960	1,141,322	535,033	837,656	487,174	253,108
1950	723,725	362,455	489,956	272,233	141,044
1940	477,494	252,757	282,244	150,790	84,919
1930	341,069	222,155	239,345	122,006	72,596
1920	160,585	135,599	112,536	41,613	34,815
1910	97,721	85,637	46,226	5,840	8,738
	Percent Increase of Negro Population, by Decade				
1950–60	57.7	47.6	70.9	78.9	79.5
1940–50	51.6	43.4	73.6	80.5	66.1
1930–40	40.0	13.8	17.9	23.6	17.0
1920–30	112.4	63.8	112.7	193.2	108.5
1910–20	64.3	58.3	143.4	612.6	298.4
	Negro Population as Percent of Total City Population				
1960	14.7	26.7	23.6	29.2	28.9
1950	9.2	17.5	13.5	14.7	15.4
1940	6.4	13.1	8.3	9.3	9.7
1930	4.9	11.4	7.1	7.8	8.1
1920	2.9	7.4	4.2	4.2	4.4
1910	2.0	5.5	2.1	1.3	1.6

Sources: *Statistical Abstract of the United States,* 1967, 1950, 1930, 1928.

Although on the average there was a sharp relative improvement in opportunities for Afro-Americans when they left the rural South for the urban North, it should be remembered that blacks were not unreservedly welcomed by white city-dwellers in the North. Riots clearly based on racial animosities occurred in East St. Louis, Omaha, Washington, and Chicago before 1920 and in other cities, for example, Detroit's destructive 1943 race riot, before the end of World War II. Migration to the cities increased, however, reflecting the unfavorable conditions being left behind (Table 6-5). It would seem, then, that once the movement was initiated, it responded to relative economic opportunities rather than solely to the repressive conditions of the Jim Crow system.

Many Negroes did not leave the South (compare Maps 6-10 and 6-19), and the average well-being of this population remained low. In this poverty, many Afro-Americans share a condition with several other major population subgroups in North America, some white and some not white. Because these groups together contain many millions of people, they must be included in any treatment of the geographic patterns of the continent. When it is realized that there are more Afro-Americans in the United States than

there are Canadians in Canada, the importance of this portion of the continental population, and others receiving unequal treatment, may be clearly appreciated.

Distribution of Poverty

As discussed at the beginning of this chapter, the United States and Canada are clearly countries of great wealth. Per capita incomes are among the highest in the world. The results of public services, such as health and education, indicate average levels of personal welfare found in only a few other select regions where population densities are moderate to high. It is equally clear that this wealth is not distributed evenly either per capita or spatially. It can also be shown that public services are not evenly spread among the populations of both countries. The causes of low population welfare are as varied as the populations themselves, but the results have great similarity.

The distribution of per capita money income in the United States illustrates several strong associations between patterns of income and either race or economic organization (Map 6-20). The most extensive region of very low per capita income is the Southeast. With the exceptions of Texas and most of Florida, each state of the old

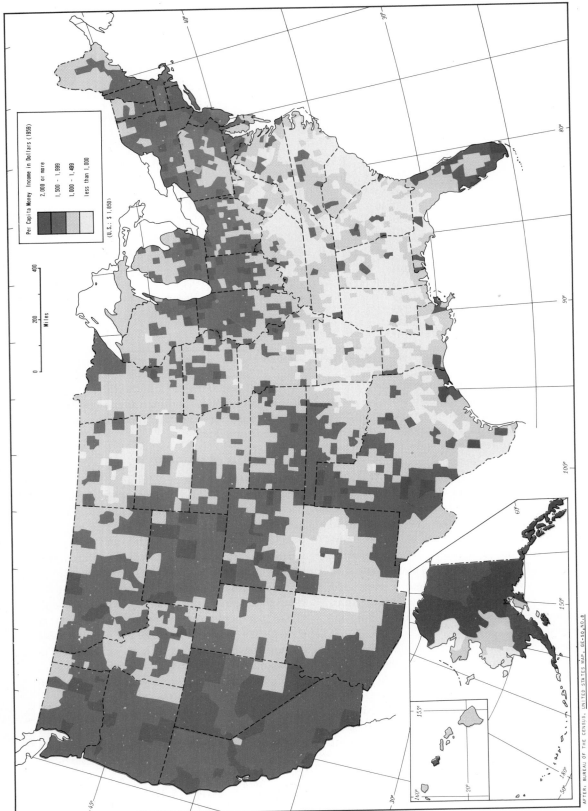

Per Capita Money Income in Dollars (1959)

2,000 or more
1,500 - 1,999
1,000 - 1,499
less than 1,000

(U.S.: $ 1,850)

400

200

0

Miles

AFTER: BUREAU OF THE CENSUS, UNITED STATES MAP, GE-50, NO.8

Map 6-20 DISTRIBUTION OF INCOME PER CAPITA, 1959.

220

Confederacy, plus Kentucky and West Virginia, contained more counties in 1960 with average per capita incomes below $1250 than above this amount. Many of these states were dominated by counties with less than $1000 average per capita annual incomes.

In the states of the Confederacy, this general low income reflects the continued regional importance of agriculture during a period of national economic transformation away from such an emphasis. Almost without exception throughout the South, counties with *high* average per capita incomes are those few with major urban centers. It is not merely concentration upon agriculture which continues low incomes in the South, however, for many counties in Ohio, Indiana, and Illinois contain no large cities and yet have much higher per capita incomes than do southern counties. And the same may be said for much of the High Plains and the far West. It could be argued that the destruction of the social and economic organization of the South during and after the Civil War lowered the base from which the region had to "rise again," and economically at least, it was not able to do so. Perhaps an equally convincing argument, considering the century that has passed since the end of formal hostilities, is related to the size and racial composition of the South's population.

The South has never had as large a population as the North. Many of the reasons for this, from the colonial period to the Civil War, have already been discussed. The importance of large-scale, cash crop farming in the South vis-a-vis the North's emphasis on small, single family farms contributed greatly to this pattern. In addition to the large no-income (i.e., slave) population in the South, there were many small-scale, essentially subsistence white farmers among the real and aspiring agricultural aristocracy. After 1865, many of the wealthiest aristocrats were taxed into poverty while the freed slaves and low income whites did not gain much in the way of income. And although after Reconstruction many whites were able to recover and reaccumulate wealth and power, the poor white farmer could not. The Negroes who remained in the South were faced with increasingly repressive Jim Crow conditions under which only the truly exceptional could achieve even moderate success. Since black Americans were in the majority in many counties of the South, even in 1960, the

low average per capita incomes there are maintained by the extreme poverty in which the blacks (and many whites) live. In addition, as long as much of the region's population possessed little income beyond that needed for the barest of necessities, there was an equivalently reduced market for local manufacturers. In sum, the result of these and similar factors has led, in Leake County, Mississippi, to a 1960 median male annual income of only $1215 for the total working population and only $772 for the black male population. In the poorer Lowndes County, Alabama, $865 was the median annual income for the total male working population and $709 for the median black man. And these figures are not indicative of special conditions for there are many such counties elsewhere in the South.

It is clear from a comparison of the patterns of income variability and proportion of the population Negro, however, that extreme poverty is not limited to regions with large numbers of black Americans. Wolfe County, Kentucky, for example, in the eastern part of the state, contained no Negroes at the time of the 1960 census and yet the median annual income for the county's male labor force was only $989. The "poverty pocket" in eastern Kentucky and adjacent areas of Tennessee and West Virginia, com-

This small house in Mississippi is an excellent example of the structures inhabited by the rural poor in the Southeast. Giving the impression that it may collapse at any moment, the house obviously provides only a minimum of shelter for a family without the financial resources to repair it or to move. (Grant Heilman)

The row of houses in need of repair facing a poorly maintained, narrow dirt road is a scene typical in eastern Kentucky. Note the surrounding hills (indicating poor agricultural resources), the discarded automobile, and what appear to be outdoor toilets on the extreme left and the middle right of the photograph. (United States Dept. of Agriculture)

prising the largest single cluster of low income counties in Appalachia, has a very different complex of conditions behind its poverty than does the South.

Early farmers migrating from the eastern Coastal Plain and Piedmont usually did not settle until they reached the Interior Plains. Some farmers, however, must have been sufficiently attracted to the many small valleys and hollows between the low mountains of the Appalachian Highlands to remain within the Highlands. Ten to twenty acres of bottomland were enough for subsistence plus sale of whatever hand-manufactured items could support the high transport costs necessary to bring the products to market. Corn mash whiskey, of course, is one high-value per unit weight example.

As national industrial development proceeded, the coal reserves in this region were mined bringing in much nonfarm population. Communication links with the rest of the country, never great, were improved only in the sense of connecting mine heads and the small mining communities with the points or regions of coal consumption. Following the depression in the 1930's, during which the demand for coal was low, the need was multiplied by American involvement in World War II, a period of low male

labor availability. As in many other industries, this combination of high demand and low labor supply resulted in a sharp rise in production efficiencies by the use of labor-substituting machinery. After the war, highly productive underground mining and strip mining continued to increase regional overpopulation. Farming was also often pushed below the competitive margin as mechanization of agriculture in the Interior Plains, where large farms are possible, proceeded rapidly. To make matters worse, alternative economic pursuits were often not feasible because of the extremely poor transportation network. Roads are rarely built without some economic justification, but development in a modern sense also cannot proceed without these roads. In eastern Kentucky, the road net is sparse, the local terrain rarely permits low-cost road construction, the expected economic support for roads is often indefinite, and the region remains poor. In many respects, except for the role played by coal in Appalachia, the situation is similar in the Ozark and Ouachita Mountains of Arkansas and southern Missouri.

There are other small clusters of low per capita income counties in the United States, several of which are associated with the presence of a distinct ethnic group. In the extreme south of Texas, for example, in the group of counties west and south of San Antonio to the Rio Grande, average annual per capita income is under $1000 in spite of substantial irrigated fruit and vegetable farming in some of these counties. In these counties, at least, the countywide average incomes are lowered by the minimal, often submarginal incomes earned by the numerous Mexican-American laborers living there. Several other regions with a combination of unsuitable agricultural environments and concentrations of Mexican-American or American Indian populations also may be seen on the income map (compare Maps 6-19 and 6-20).

It should be made clear that poverty in the United States, extensive as it is shown to be by Map 6-20, is not restricted to the Southeast, to eastern Kentucky and northern Arkansas, and so forth. The economic core of the country the northeastern manufacturing region, is not a formal, that is, homogeneous, region with respect to the single criterion of income. This portion of the country also contains individuals, families, and entire communities with average incomes comparable to those of the more clearly distin-

This collection of poles, scrap lumber, and sheeting of various kinds could be described as typical of many Indian homes on Arizona reservations. The surrounding land is obviously not suited to cultivation. (Paul Conklin/PIX)

guishable poverty regions. Very few cities above 10,000 to 20,000 population are without small clusters of low income residents; certainly none above 100,000 population is without its "ghetto" or some equivalent. Generalized explanations of the presence of the poor, why they are poor as well as why they are where they are, have proved unsatisfactory. The explanations are unsatisfactory because the root causes are of great complexity and each cause has variable relevance for different groups of people and for different environments.[10] What can be discussed here, however, is the fluid, ever-changing pattern of population associated with perceived and real income differences.

Migration Within North America

Large-scale movements of people within the North American continent have usually resulted from economic causes. Occasionally, other factors will generate migration, for example, political disagreements during the American Revolution or the intense racial hostility experienced by black Americans, but most often the motivation for migration is dominantly economic. Settlement of the Interior Plains farmland was carried out by families seeking improved agricultural opportunities in the early nineteenth century. The west coast was reached by ship and

[10]For a good introduction, see Harrington, *op. cit.*

by wagon again because of the promise of good, cheap land in the Willamette and Sacramento Valleys or quick riches in the California gold fields. The rich Canadian prairies were not settled until they were connected with the eastern Canadian population centers; the transport connection was not needed to carry the migrants but it was necessary to make the land economically attractive.

Of the three distinct population movements that have dominated the spatial patterns in North America during the twentieth century, two have been interregional. It is the third, however, that has led to the most significant changes in the geography of the United States and Canada. One of the interregional migrations began several decades before 1900. Southern blacks and southern whites had moved north in moderate numbers before World War I. This migration was stimulated when the war and subsequent restrictive legislation virtually ended the influx of European migrants. Rapid expansion of the United States' manufacturing base continued, with periodic interruptions, until the Great Depression of the 1930's. Following World War II, industry in the national manufacturing core as elsewhere in the country, resumed its growth. Much of the rural Southeast, on the other hand, has remained poor in spite of substantial economic activity in many of the region's metropolitan areas. The presence of this regional

Table 6-6 Income and Net Migration in Canada, by Region

Province	Earned Income per Capita, 1960 (Canadian Dollars)	Net Inter-Provincial Migration 1951–1961
Maritime Provinces	818.89	−80,702
Quebec	1043.56	−34,122
Ontario	1427.75	110,387
Prairie Provinces	1210.15	−127,860
British Columbia	1413.23	133,400

Source: George W. Wilson, Scott Gordon, and Stanislaw Judek, *Canada: An Appraisal of Its Needs and Resources*, Twentieth Century Fund, New York, 1965, from Table 3.27 and Table 3.28.

differential in economic activity and income levels has led a considerable number of people, black and white, to leave the South for the cities of the national core. The additional stimulus of Jim Crow segregation and all of the indignities associated with this "institution" have already been discussed, but it should also be remembered that many of the northbound migrants were white. The states of the Southeast, then, have long been typified by extremely low average population growth; some states have actually declined in population between the dicennial censuses (e.g., during the 1950–1960 decade, the population of Arkansas decreased six percent). In Canada, although the recent migration has not been northward, large-scale population movement is also characterized by migration from low income to higher income regions (see Table 6-6).

In spite of this inflow of residents from the southern United States, the Northeast has not been the most rapidly growing section of the country. During the same decades that thousands of Americans were moving South to North, many other thousands were migrating westward. Some of these moved onto the Great Plains or to California, Oregon, or Washington because of the apparent attractiveness of these areas. They were "pulled" by the promise of increased personal welfare. Others drifted westward out of desperation generated by overcrowding or environmental deterioration such as that described by John Steinbeck in his *Grapes of Wrath*. Of course, this was a continuation of the traditional trend in the population movement, and many went from the Southeast as well as from the Northeast. One result has been a tremendously complex mixture of attitudes, values, and approaches to the solution of problems in the West, and another is an extremely dynamic society. It is not a coincidence that the west coast, usually

California, is increasingly the source of national trends. Spatial concentration of representatives from various regions has heightened the potential for change via the greatly stimulated interaction made possible by just such a diversified population.

The third major shift in population that dominates the human geography of North America is the continuing flow of individuals and families from farms and small towns into large urbanized areas. The fact of this migration has already been discussed in several contexts. The long-term causes of farm population loss and urban population gain have been the great labor demands made on the two countries' populations by expanding manufacturing and service industries *and* the increased use of mechanization in agricultural production. As farm machinery was used increasingly by American and Canadian farmers, less man-hours of labor have been required to produce a given quantity of agricultural produce. The "surplus" population generated in this way has been absorbed by the growing industrial complexes. Although clearly the industries benefited from this large source of domestic labor, it should also be remembered that without the expansion of the economy in nonagricultural sectors, much of the surplus farm labor would have remained in agriculture, reducing the average farm income beyond its present moderately low level. Mechanization has had the further effect of making large-scale operations more efficient and thus driving small farm operators out of agriculture.

In addition to the obvious spatial changes generated by this rural to urban shift in population, for example, increasing farm size, increasing importance of urban foci in national activities, etc., there has been significant development in the cities' spatial arrangement because of the great numbers of people who have come to live and work there. The cities have absorbed incoming population either by substantial vertical development of structures or by horizontal expansion beyond their initial political limits. Vertical development has occurred, and is occurring to some extent in most cities, but by far the greatest spatial change resulting from rural to urban migration has been the outward explosion of urbanization since the turn of the century.

For several decades the cities contained much of their population growth within the central city limits, as indicated in Table 6-2. Since World War II, however, city population growth has not been able to regain prewar levels. Much of the additional urban population has settled outside the formal boundaries of the city, often in politically independent suburban communities. The complex, overlapping patchwork of governmental jurisdictions within metropolitan areas which exists today (literally hundreds of such units in the larger SMSA's) has resulted from an undirected outward growth of populated areas from the central city cores. Serious political and economic, and therefore social, problems have developed within the central cities because of this multitude of independent municipal units. These local and regional units control various aspects of a suburban population that works, recreates, or passes through the central city without contributing to its financial administration. When a city's white population flees from the interior of that city for one or another of the all-white suburbs, it takes with it not only most possibilities for effective cross-cultural contact but also a good deal of the city's taxable personal income and the businesses depending on the expenditure of this income. With an unsatisfactory tax base, the city's municipal services suffer in quality or completeness. And as services deteriorate, the suburbs attract further movement by those city-dwellers able to shift their residences.

Megalopolitan Growth

A much more frequently discussed result of the population growth of North American cities has been the great increase in the spatial extent of the urbanized areas, that is, "horizontal" expansion. This aspect of urban growth, popularly referred to as "urban sprawl," has proceeded to the point in some sections of the continent that the extensive metropolitan areas of several nearby cities have merged. The result has been the development of a number of "supercities" far larger even than the largest single city on the continent. The largest of these consolidated metropolitan areas, extending from Boston to Washington, D.C. and beyond, has been studied in detail by the French geographer Jean Gottmann.[11] Gottmann chose the name *"Megalopolis"* for this vast urban area and the term has been applied as a label to other smaller or less well-connected coalesced cities within North America and around the world.

The Boston–Washington Megalopolis, being the first and largest such urbanized area formed by the merging of distinct metropolitan units, bears great significance for the geography of the continent. This massive urban region is significant both in terms of its importance relative to the remaining cores of urban activity and its indication of future problems which will require solution in other developing megalopolitan areas. As discussed earlier, the northeastern manufacturing core contains a majority of the continent's industrial and financial activity and a sizable proportion of the continent's population. Megalopolis is the eastern "hinge" of this continental core region, and therefore, of the United States and Canada as a whole. It contains a greater density of high income population and high income-producing activities, for example, research and laboratory industries, financial and managerial operations centers, than any other equivalent area in the continental core. Because of this tremendous concentration of population, economic potential, and cultural and intellectual leadership, Megalopolis is an area of intense human activity. It is also an urban "laboratory" possessing problems in a variety and seriousness present only on a much smaller scale in other megalopolitan and metropolitan areas. The importance of these problems is magnified by the presence of the United States' political and economic foci in Megalopolis (Washington, D.C. and New York City, respectively).

The problems facing Megalopolis and other huge urbanized regions are many, but the most crucial may be treated under one of three major headings: political viability, livability, and accessibility. Some of the increasing difficulties of central city political viability have been men-

[11] Jean Gottmann, *Megalopolis: The Urbanized Northeastern Seaboard of the United States,* The Twentieth Century Fund, New York, 1961.

The increasingly dispersed population of Megalopolis has led to increasingly congested routes of access between the urban cores and their peripheries. This complex road system is part of the attempt to relieve this congestion. It is New Jersey's entry and exit system for the George Washington Bridge. Well over 150,000 vehicles cross the Hudson River each day via this bridge. (Grant Heilman)

tioned. The governmental operations and organizations of these nodes within megalopolitan regions are restricted by fantastically complex patterns of jurisdiction. Regionwide planning and policy coordination are hampered. As sources of central city revenue diminish and the black population increasingly indicates its dissatisfaction with existing occupational and residential discrimination, the urban political stability becomes ever more difficult to maintain.

Megalopolitan regions have developed from the coalescence of many cities of varying size. The resulting land-use pattern is extremely complex with innumerable tertiary and secondary activities interspersed among large residential areas. Such is the spatial extent of these various functions that large urban areas possess their own climates, slightly different from the general regional climatic pattern. The efficiency of atmospheric circulation is reduced in such urban areas. Used air is not adequately replaced by fresh air. Air pollution is supplemented by noise pollution to further reduce local environmental quality, that is, the region's "livability."

The atmosphere is not the only polluted portion of the megalopolitan environment. Most internal waterways are fouled by industrial, municipal, and even recreational wastes. Pollution makes more serious the problem of water supply, for great concentrations of people and industry demand large and regular quantities of fresh water for survival. Furthermore, the greater the amount of water needed, the larger the catchment area required. The greater the catchment area, the less likely cheap and noncompetitive sources of water will be available. This may not *appear* as serious in Megalopolis, for example, as in dry southern California. On the average, the Northeast receives more dependable and greater quantities of precipitation than does the Southwest. The much larger water consumption in the former, however, makes the water supply problem crucial in both regions.

In addition to environmental pollution and the demands placed on the deteriorating environment by the population and businesses of megalopolitan regions, livability is also related to the quality and availability of recreational facilities. With the land between urban cores continually being converted to urban uses, less and less land is available for recreation by the increasing population. Competition for land and water resources between these functions has become sharp, with the less-remunerative use of recreation often losing to either residential, industrial, or retail activities, or to pollution. In order to obtain the benefits of recreation, great numbers of people from Megalopolis and other large urbanized areas must travel increasing distances each year. This, in turn, encourages further the geographic growth of Megalopolis.

Many of these factors are related to the problem of constrained accessibility. Cities, or urban agglomerations, exist because there are advantages which accrue to urban functions by a geographic clustering of the functions. Some share inputs, some use the outputs of other activities as input, some compete for the labor or the concentration of personal income found in urban regions, and some desire the advantages of location near a major transportation focus. In each of these cases, accessibility is crucial, for the interdependency described is not possible without the movement of either goods or people from a multitude of origins to a similarly large number of destinations. When the number of people and activity sites is as numerous as in Megalopolis, when the distances involved are as great

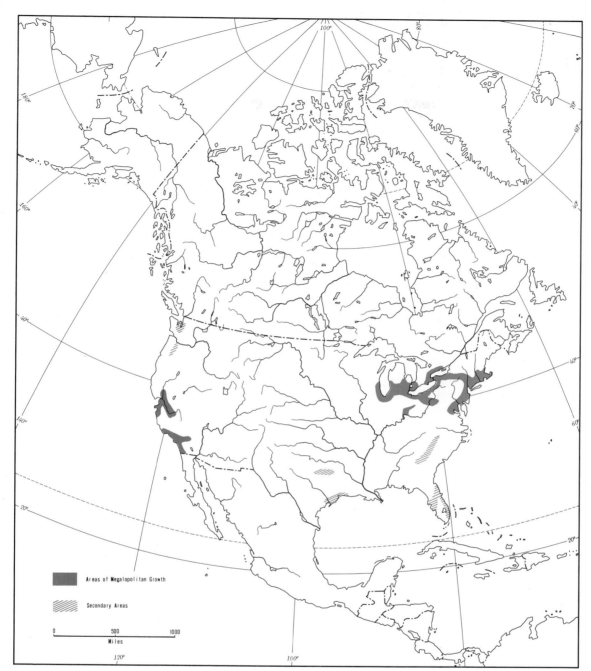

Map 6–21 PATTERNS OF MEGALOPOLITAN GROWTH.

Legend on map:
- Areas of Megalopolitan Growth
- Secondary Areas

0   500   1000
Miles

as they have become, and when the primary centers of daily activity continue to increase the volume of interaction within fairly constant areas (e.g., Manhattan, downtown Philadelphia, the governmental complex in Washington, D.C.), then accessibility eventually decreases and movement is constrained. When this happens, the essential purposes for agglomeration are undercut. Tales about one final, permanent, immovable traffic jam in the major urban centers are not total fantasy.

The characteristics and problems in Megalopolis are of interest because it is clear from urban expansion elsewhere on the continent (and in the world) that other megalopolitan clusters are in the process of formation (Map 6-21).

Although it is possible to foresee in the distant future numerous coalesced metropolises in North America approaching megalopolitan dimensions, only two such regions presently warrant consideration. The more extensive of the two lies between Chicago and its nearby cities to Pittsburgh via Cleveland and to Toronto via Cleveland and Buffalo. Some observers also predict a not-too-distant connection of urban settlement between the Chicago–Pittsburgh–Toronto megalopolis and the Boston–Washington megalopolis through the cities of the Mohawk Valley in New York (Rochester, Syracuse, Utica, Schenectady–Troy–Albany). The other major megalopolis has its beginnings in the urbanized connections developing between Los Angeles and San Diego. This rapidly growing region is not likely to reach the urban development extending southward from San Francisco Bay for some time, but such a merging is not impossible (pending solution of problems of water supply and land deterioration) if recent rates of metropolitan expansion continue. It has been estimated that by the year 2000, over half of the approximately 300 million people in the United States will live and work in these three megalopolitan regions. This aspect of the continent's changing geography, therefore, will have great impact upon human spatial or-ganization in North America.

One might be motivated to ask what changes led to the vast sprawl that has become characteristic of North American urban centers. As seen previously, in Table 6-2, the populations of the continent's major cities grew at rapid rates until the depression of the 1930's and yet the great areal expansion of urban areas is more closely identified with the post-World War II period. One of the major contributors to this urban development had its beginnings before the war but reached unprecedented proportions only after 1945.

The more universal use of the automobile in the United States and Canada after World War II produced a series of changes in human activity and activity patterns that had not been clearly foreseen. Along with increased use of the personal automobile came a reduction in dependence on alternative methods of travel (public transit within the city and railroad between cities). The urban population began to live greater distances from work and shopping areas downtown. Employment and retail functions gradually moved outward toward the more dispersed population, clustering in small shopping centers and industrial "parks." This deprived the traditional areas within the city of significant

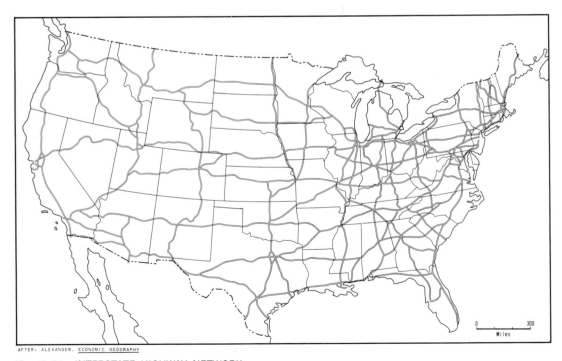

AFTER: ALEXANDER. ECONOMIC GEOGRAPHY

Map 6–22 INTERSTATE HIGHWAY NETWORK.

proportions of their support activity. Conversely, as it became clear that retail outlets could appeal to much larger markets because of the shoppers' increased mobility, the advantages of large-scale operation (or economies of scale) were applied to this aspect of American culture and small, local community stores gradually closed. The increasing volume of long-distance intercity movement of people by automobile led to the planning in 1956 (and eventual construction) of more than 41,000 miles of limited access highway in the Interstate Highway System (Map 6-22). With increased ability to move, urban populations have become even more dispersed, placing higher demands upon existing transport facilities.

It should not be concluded from this that the automobile was the primary generator of urban sprawl. It was the technological means for dispersion, not the motivation. More important, in that they were more basic in urban geographic expansion, were higher incomes more generally dispersed among the population, the desire to escape from the unpleasant aspects of city life, and continuing national economic prosperity. Although none of these is specifically geographic, they had direct geographic effects on spatial dispersion of the urban population. "Space-shortening" techniques (i.e., the automobile and airplane) allowed the dispersed population to live, in many respects, as though the greater distances did not exist. Of course, there are still substantial numbers of people within cities who do not share the benefits (and liabilities) of increased mobility. They do not have equivalently higher incomes; they are barred from moving into most suburban communities by their skin color and the fear it arouses; or they may not have benefitted directly from the continued national prosperity.

The Interactance Hypothesis

It should be clear from the preceding discussion that cities are not static features of the landscape. The physical structure of a city is a constantly changing pattern of land use and human activity sites. Changes in the patterns reflect varying degrees of interaction both within the city and between the city and nearby metropolitan areas. Since this interaction involves people and distance, sociologists, demographers, and geographers have long been interested in determining its precise nature. A general model

has been formulated attempting to include the major determinants of human interaction between spatially distinct population groups. This model is called the *interactance hypothesis*.

The interactance hypothesis is a statement of the relation between populations, the distance which separates them, and the relevant activities contributing to the interaction of the populations. The general form of the hypothesis may be stated in the following equation: $I = KTP_a{}^m P_b{}^m A_a A_b / D^n$, where I is the amount of interaction between population P_a and population P_b over the length of time, T, for the relative activity intensities, A_a and A_b, and with a distance separating the populations of D. K is a constant, and m and n are the appropriate exponents for the situation under consideration. Basically, the hypothesis, or model, states that interaction between two groups increases as the sizes of the groups increase but is less the farther apart the populations are located from one another, other elements of the population being equal. In this abbreviated form, the model is reduced to the more familiar $I = K(P_a P_b / D^n)$.

As a model of reality, the interactance hypothesis makes up in succinctness and clarity what it lacks in precision. It is clear from the equation that twice as much interaction may be expected between cities A and B, say, twenty miles apart, as between cities C and A, also twenty miles apart but with C only half the population of B. Or generally, the fewer the people, the less interaction expected. Similarly, the greater the distance between the city pairs, the fewer the interactions. By testing this relationship in numerous real situations, it is possible to compare the differences that exist in the results of the tests. The "other things" assumed equal by the model are, of course, not equal in the real world. The variable impact, or importance, of each set of other things contributes to our understanding of the variability of the real world. The interactance hypothesis is of interest to geographers in this manner because it is possible to discover just how important the "friction of distance" can be under varying conditions and for different types of spatial interaction. The spatial interaction, for example, may be economic (trade), political (dispersion of decisions), educational (university attendance), social (migration), and so forth.

The interactance relationship has been used by a Canadian geographer to illustrate the effectiveness of political and cultural boundaries in North America.[12] By arranging the number of interactions between known pairs of cities in Canada and the United States for given values of the distance exponent, n, and the constant, K, it was possible to observe variations for different types of city-pair situations. Interactions between pairs of cities *within* the dominantly French Canadian Quebec Province were compared with interactions between Quebec cities and non-Quebec (i.e., English-speaking) cities and also with interactions between Canadian and U.S. cities. The interactions used in this study were telephone calls between city pairs, and marriages between city-pair residents. The results indicate that Ontario cities and cities of other English-speaking Canadian regions are five times as far from Quebec cities as they would be if there were no language-cultural differences. United States cities are 50 times as far from Quebec cities as if there were no cultural and no political differences. In other words, real distance, as measured on a map or on the ground, can be shown to be misleading when the separation of cultures and political units is being discussed. Implications concerning the effective distances separating black and white populations in the United States should be considered in light of the results of this study. Distance and geographic patterns arranged on the basis of a single distance scale may be deceptive if there is no interaction or communication.

Circulation and Continental Integration

French geographers have long used a concept that is applicable here, the concept of *circulation.* Although generally considered nearly synonymous with spatial interaction because it deals with movement aspects of human society, "circulation" carries implications of organization within a complex spatial *system.* Without the factors of movement to define the extent and degree of spatial organization, there is no system. Just as "resources" cannot be defined without human utilization of the items to be so considered, organization cannot be defined without the circulation which binds together the elements of a specific system. This holds true regardless of what is circulating, (goods, people, troops, ideas,

messages, etc.) or of the type of system. Spatial interaction does, in fact, include such movement within its conceptual framework. To nongeographers, however, implications of spatial organization might be more easily seen in their broadest sense under the term "circulation."

When the geographic patterns of North America are viewed in light of the circulation concept, development of these patterns and many of the problems associated with them may be seen more clearly. The French Canadian community, for example, remains distinct to a great extent because it became a "closed" society after the British gained hegemony over Canada. Members of this community concentrated on those aspects of French culture which provided a firm base for the continuation of the group in spite of increasing dominance by English-speaking Canadians. That is, in those matters

This scene illustrates the typical form of French Canadian agriculture in the St. Lawrence lowland. When this land was distributed by the French crown, it was parceled out in narrow, parallel strips extending inland from the river. Each "long lot," as the fields are called, has access to the river, or to a road built parallel to the river at some distance to the interior. This small, linear village (about 65 miles from the city of Quebec) lies along the first road inland from the river, visible at the bottom of the photograph. The rather irregular topography and the presence of numerous rocks in virtually every field indicate that farming is not easy here, nor is it especially remunerative. (Laurence R. Lowry)

[12]J. Ross MacKay, "The Interactance Hypothesis and Boundaries in Canada: A Preliminary Study," *Canadian Geographer,* Vol. 2, 1958, pp. 1–8.

which would erode the community's distinctness, circulation was internalized. Interaction with the population outside the community was minimized. In this case, the community was large enough and strong enough to resist the dispersion of French Canadian culture and its assimilation into the more numerous remainder of the Canadian population. Although this culture remains concentrated within the Quebec Province, it has long been officially recognized as an "equal" partner in the development of this country. Considering its present vigor, there is little chance that the French Canadian role will be diminished in the foreseeable future. The existence of a plural society in Canada may not lead to political fragmentation, in spite of strong, sometimes violent, pressures for such separation from a very small number of French Canadians. This is because most circulation within modern Canada has developed without regard for the cultural differences present.

The plural society of the United States presents more serious problems. Movement of people, goods, and ideas has at times been restricted between the black and the white cultural communities as, for example, in the Jim Crow South and the post-World War I North. There has, in fact, been little interaction, little communication. Black American culture developed separately from white American culture largely because there was little social interaction between the two groups. Each culture does contain major elements of the other, but this is a result of the overlapping activity space. White and black Americans live near one another in terms of earth distance, whether in rural communities in the South or urban regions in the North and West. The cultural distance might be expected to be several times as great, considering Professor MacKay's results discussed earlier.

In terms of national integration vis-a-vis national fragmentation, demands for the interaction and communication necessary to maintain national unity have increased sharply since World War II. There are some indications that such interaction has increased. That the two largest cultures in the United States are spatially interdigitated reflects their interdependence in spite of a tradition of spatial segregation based on racial differences. This makes actual spatial fragmentation both less likely and more disruptive of the national fabric than if the two cultures were located in distinct regions. (Compare this

Urban environments can be as harsh as rural conditions for the poor. This neighborhood in Philadelphia is in desperate need of structural rehabilitation. The rubbish-strewn yards, alleys and streets indicate insufficient public services. The volume of unused automobiles and the poor condition of many buildings indicates a generally low per capita income. The environment is one eminently suited to rats, disease bearing insects, and intestinal parasites. (Grant Heilman)

possibility with the actual separation of India and Pakistan—Chapter 11.) The communication and spatial interaction required to overcome cultural distance, of course, will not be possible without the full assumption of equality between the cultures.

The changing pattern of circulation in North America is also related to urbanization, industrial aggregation, national economic integration, and the spatial arrangement of personal well-being. The patterns of resources and people which provide the geographic *raison d'etre* to the North American manufacturing core also generated a circulation pattern to organize the resources and population. The circulation pattern has spawned the largest contiguous urban region in the world and is likely to make a second such region in the near future. Interaction between clusters of population and industrial activity within the manufacturing core lends mutual support through interdependence, an interdepend-

Map 6–23 HIGHWAY TRAFFIC FLOW IN THE NORTH AMERICAN CORE REGION.

Thousands of Vehicles per Day

Lake Ontario

Lake Erie

Lake Huron

Lake Superior

Lake Michigan

ence and support reflected in patterns of railroad and highway traffic (Map 6-15 and 6-23). The increase in high-speed land transport has encouraged some disaggregation of industrial activity with an equivalent rise in the importance of production advantages other than those leading to agglomeration. Thus, shifting market patterns have drawn some industry westward while an increased importance placed on labor costs has led to a southward migration of other industries. Such changes, in turn, alter the pattern of national industrial circulation.

Both the United States and Canada are very large countries, each extending well over 2500 miles east and west. The dominant physiographic trend of the continent is north-south. The national integration of each country was achieved in spite of this trend by the early construction of transcontinental railroads. Strengthening circulation patterns have provided the key to the past century of integration. Conversely, to a great extent, it has been the lack of interaction, the exclusion from the dominant circulation patterns, which has maintained poverty in east Kentucky, the rural Southeast, the Ozarks, the Maritime Provinces, and so forth. The poverty within urban ghettos is similarly accentuated and perpetuated, not by lack of facilities for interaction, but by the absence of residents' participation in the flow of national activity. The reasons for such nonparticipation are many, but they are supported by spatial separation and isolation, whether physical, economic, or social.

ADDITIONAL READING

Standard works on North America include P. F. Griffin, D. L. Chatham, and R. N. Young, *Anglo-America: A Systematic and Regional Geography,* published in a second edition by Fearon Publishers, Palo Alto, in 1968, and C. L. White, E. J. Foscue, and T. L. McKnight, *Regional Geography of Anglo-America,* published in a third edition by Prentice-Hall in Englewood Cliffs, in 1964. An older work is by E. B. Shaw, *Anglo-America: a Regional Geography,* a Wiley publication of 1959. Very readable is J. W. Watson's *North America, Its Countries and Regions,* published in New York by Praeger in 1967. Recently, O. P. Starkey and J. L. Robinson produced *The Anglo-American Realm: A Geographical Analysis of the Economies of the United States and Canada,* published in New York by McGraw-Hill, in 1969. Specifically on Canada, see an edited volume by R. L. Gentilcore, *Canada's Changing Geography,* published by Prentice-Hall in Scarborough, Ontario, in 1967, and another edited book, by W. J. Megill, called *Patterns of Canada,* published by Ryerson Press in 1967. A standard geography on Canada is the work by D. F. Putnam and D. P. Kerr, *A Regional Geography of Canada,* published by Dent and Sons, Toronto, in 1965. Some 23 geographers participated in a volume edited by J. Warkentin, *Canada: A Geographical Interpretation,* published by Methuen in Toronto, in 1968. On the United States, see Max Lerner's *America as a Civilization: Life and Thought in the United States Today,* published by Simon and Schuster, New York, in 1957. Another standard book is R. H. Brown, *Historical Geography of the United States,* published by Harcourt, Brace, and World in 1948. A paperback still available is F. J. Turner's *Frontier in American History,* published in New York by Holt, in 1949.

On the physiography and physiographic provinces of North America, see W. W. Atwood, *The Physiographic Provinces of North America,* published in 1940 by Ginn, in Boston, and two monumental volumes by N. Fenneman, *Physiography of the Western United States* and *Physiography of the Eastern United States,* published by McGraw-Hill in 1931 and 1938, respectively. C. B. Hunt's *Physiography of the United States,* published in San Francisco by Freeman, appeared in 1967. A more difficult book is *Regional Geomorphology of the United States,* published in New York by Wiley, in 1965.

On the problems of cities and mega-cities, see O. D. Duncan, et al., *Metropolis and Region,* brought out by Resources for the Future and Johns Hopkins Press in 1960. Another worthwhile publication is E. C. Higbee's *The Squeeze: Cities Without Space,* published by Morrow in New York in 1960. Also consult the works of E. A. Gutkind and L. Mumford, and, of course, the volume entitled *Megalopolis: the Urbanized Northeastern Seaboard of the United States* by J. Gottmann, published by the Twentieth Century Fund in 1961.

MIDDLE AMERICA AND THE LEGACY
OF MESOAMERICA

NORTH and South America are connected by what physical geographers call a *land bridge*—a relatively narrow stretch of territory connecting major landmasses of the world. Europe and Africa were at certain times in Earth history connected by land bridges across what is today the Mediterranean Sea. Geologists have postulated such land connections in many places on Earth: across the Bering Strait between Alaska and Asia, across the Moçambique Channel between East Africa and Madagascar, and even across the whole Atlantic Ocean between Africa and South America. They presume that these "isthmian links" have now disappeared because of geologic processes, but that they greatly affected the spread of animals and vegetation during the period of their existence.

Certainly a land bridge exists today between North and South America; whether it was very effective in promoting the spread of flora and fauna through the Americas still remains a question. At least it is likely that man used it in his first occupation of these continents, for indications are that he came across the Bering Straits and proceeded southward through North America, perhaps to enter South America via the corridor provided by what we refer to here as Middle America (Map 6-1). As the map shows, this land bridge is rather funnel-shaped —widest in Mexico, then narrowing irregularly to its smallest width in Panama, in the vicinity of the Canal Zone, where it is a matter of 40 miles or so. The link persists for some 2500 miles—and the only place where it is broken by water is by the work of Man, in the Panama Canal.

Middle America as a geographic term means different things to different people. Some geographers argue that Mexico cannot be included in any region of which the smaller republics of Central America (Guatemala, El Salvador, Honduras, Belize, Costa Rica, Nicaragua, and

Panama) are a part, and that by its size and strength, Mexico stands alone. Others feel that the recognition of a Middle American region entails the inclusion not only of all these republics, including Mexico, but also the islands of the Caribbean and even five South American countries that face the Caribbean Sea and Atlantic Ocean: Colombia, Venezuela, Guyana, Surinam, and French Guiana. Still others prefer to use the term *Central* America. Our discussion of Middle America will confine itself to that part of the mainland and those islands that lie between the United States of America and the mainland of South America.

The Legacy of Mesoamerica

Middle America has an immensely varied population. This aspect, as much as anything else, gives this area its distinctiveness: it is a patchwork of peoples, cultures, and economies. Thousands of years ago its Indian peoples arrived and began to organize themselves into larger communities; during the two millenia preceding the arrival of the Europeans, they developed great empires with large cities, highly organized economies, and very advanced cultures. Then the Spanish conquest took place, and soon a *mestizo* population (of mixed Spanish and Indian descent) began to emerge. But the Spanish were not the only Europeans interested in Middle America: the French, British, Dutch, and Danes came to the Caribbean islands as well. And from Africa these colonizers brought Africans in bondage, to work on the plantations that thrived on the productive soils. Eventually this black slave population was freed, but not before slaves outnumbered Europeans in most of the islands of the Caribbean. In Jamaica, for example, more than three-quarters of the total population of two million is black, and most of the remainder is of mixed European-African descent. And when the emancipation of the slaves produced critical labor shortages, indentured labor came from still another part of the world, namely India: nearly 150,000 to Trinidad and over 30,000 to Jamaica. Most of these Asians arrived during the second half of the nineteenth century. The ethnic diversity of Middle America is probably unparalled in the world.

Thus the American Indian, the original master of Middle America, was crowded out by the European and his serfs. Today Indians are in a majority in only one Middle American country, Guatemala, where the Maya are the principal element in 55 percent of the population of under 4 million. Elsewhere they were either annihilated, as they were for all practical purposes throughout the Caribbean, or lost their identity through mixture with the European invaders. In Mexico, less than 30 percent of the population of over 40 million is classified as Indian; in Honduras and Nicaragua people of pure Indian descent make up only about seven and five percent of the respective totals. And yet both the mainland and the islands were quite densely populated when the Europeans first made contact. Estimates of the island populations vary widely, but probably there were between one and two million Indian inhabitants in the Caribbean around 1500 A.D., the majority of them located in Cuba, Hispaniola, and Puerto Rico. On the mainland there may have been as many as 15 million people, most of them concentrated in what is today central and southern Mexico; some estimates are considerably higher. At that time, then, this was one of the great population concentrations, not only of the Americas, but of the world.

Today, virtually nothing remains of the original peoples and cultures of the islands. When the Europeans arrived, the Caribs, latest arrivals on the scene, were in competition with the Arawaks for the available land. Though less numerous than the Arawaks, the Caribs were great boatsmen and sailed the Caribbean in huge canoes capable of carrying several dozen men. The Arawaks, a more sedentary and agricultural people (they grew root crops such as cassava and sweet potatoes, as well as tobacco and cotton), were losing ground. In terms of social organization neither people had achieved anything like the great civilizations of the mainland, but both were well advanced over their predecessors, the Ciboneys, who were a people that relied exclusively on hunting and gathering.

When the Europeans arrived the Arawaks and Caribs were destroyed almost overnight. After half a century only a few hundred survived, some of them only because they had fled to smaller islands where the arrival of the white man was somewhat delayed. They had neither the weapons nor the numbers to hold out long; on the sugar plantations, where they had been ruthlessly pressed into service, they failed to adapt to the extreme rigors of labor to which they were subjected and they perished by the thousands. They possessed no immunity to the diseases the

Europeans brought to the islands, and smallpox, for one, took a heavy toll. The white man took their land as well; there was no alternative to serfdom, save a flight into the dense forests of the island interiors. There a few survived, later to be joined by runaway slaves. With their mixture, the last pure Caribbean Indian strain disappeared.

Mesoamerica. In mainland Middle America, the Indian populations were far more numerous and much more effectively organized than their Caribbean contemporaries. Of course, this was not so throughout all of Middle America: even here there were areas where subsistence farming supplemented by hunting and fishing formed the way of life for village-based communities. But in the midsection of Middle America, from central Mexico to southwest Costa Rica, lay a region of highly developed culture with agricultural specialization, urbanization, permanent transport networks, writing, various art forms, religion, and many other features reflecting great advancement. This was no recent or sudden emergence: this region for many centuries sustained cultural progress in countless spheres. Anthropologists refer to it as *Mesoamerica,* and it is recognized as

one of the world's true culture hearths (Map 7-1). Here lay the great Mayan and Aztec civilizations.

Mesoamerica was one of the two areas of great cultural advancement in all of the Americas (the other was the Inca domain in the South American Andes Mountains), and one significant aspect of these areas is that both were located within the tropics. There was a time when some geographers argued that real political and cultural progress and achievement could only be expected under certain environmental conditions. Pointing to the present concentration of wealth and power in the middle latitudes, especially in the Northern Hemisphere, they theorized that the tropics were simply not conducive to human productiveness. Mostly they related their views to one particular element of the environment, namely climate. Tropical climates, with their enervating monotony, were supposed to be a retarding influence; the varied climates of the middle latitudes with their high and low pressure systems, fronts, and variable weather, stimulated human achievement. One of the most famous— and prolific—"environmentalists" or "environmental determinists" was E. Huntington, whose

Map 7-1 MESOAMERICA.

1940 edition of a book called *Principles of Human Geography* could still include statements such as the following:[1]

"The people of the cyclonic regions rank so far above those of other parts of the world that they are the natural leaders." (p. 471)

And again:

"The well-known contrast between the energetic people of the most progressive parts of the temperate zone and the inert inhabitants of the tropics and even of intermediate regions, such as Persia, is largely due to climate." (p. 339)

Apart from the odious character of such generalizations regarding human quality, it is unlikely that they hold any scientific value. While climatic changes have certainly occurred in various parts of the world, and sand-covered Middle Eastern cities that now lie in ruins probably thrived under rather different conditions, there is also much evidence that Man made a good deal of progress in the tropics. One body of

[1] E. Huntington, *Principles of Human Geography*, Wiley, New York, 1940.

evidence for this comes from this very region—Mesoamerica.

With the exception of modern-day northern Mexico, Middle America lies within the tropics. Its physiography is dominated by a huge plateau which, flanked by two major mountain ranges, occupies almost all of northern and central Mexico (Map 7-2). This mountain-rimmed tableland lies at an average elevation of some 8000 feet in the south, near Mexico City, and although it is somewhat lower to the north, it still reaches nearly 4000 feet at the Mexico–U.S. boundary. To the east of this great *Meseta Central* lies the Sierra Madre Oriental, and to the west, predictably, the Sierra Madre Occidental. Great volcanoes, some reaching over 17,000 feet in height, form part of the southern rim of the plateau. The surface of the Meseta Central is divided into a number of basins of present or recent internal drainage, some still partly filled with lakes, some possessing the fertile soils of former lake bottoms. One of these is the Valley of Mexico about which more will be said later. The remainder of Middle America is marked by a persistent mountainous backbone of varying height, studded with volcanoes and dotted by

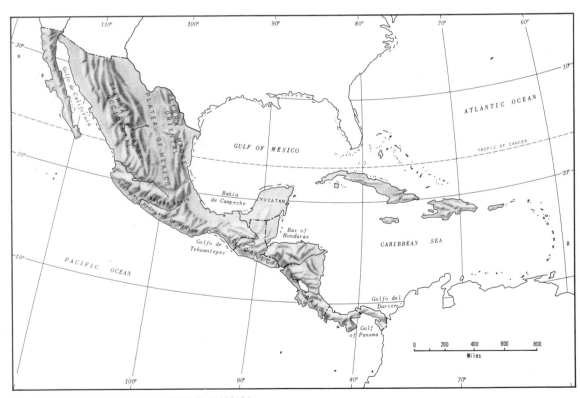

Map 7-2 PHYSIOGRAPHY OF MIDDLE AMERICA.

many lakes (of which the largest is that of Nicaragua). Among these volcanoes, many are active and emit volcanic ash which recurrently covers the countryside and quickly weathers, replenishing the fertility of the soil. In southern Mexico, Guatemala, Honduras–El Salvador, and Nicaragua, this "backbone" is in fact a jumbled mass of large and small ranges, but Costa Rica and Panama have one dominant spine, although there are some smaller outlying *cordilleras* even here. Low-lying coastal areas are larger and wider on the Caribbean than the Pacific side of Middle America. The Yucatan Peninsula, northern Guatemala, and northern Belize constitute the largest contiguous coastal lowland; next comes the "Mosquito Coast" of Nicaragua and Honduras. The Caribbean Islands constitute the highest parts of submarine mountain arcs that connect with mainland Middle America and off-shoots of the Andes in northern Venezuela.

Mesoamerica thus was a largely tropical realm, and consisted not only of the plateau of Mexico, but also of the hot and humid lowlands of eastern Mexico, Yucatan, Belize, and Guatemala. In this area, which falls in part in the tropical rain forest category (see Plate 2), flourished the civilization of the Maya. Its beginnings lie in the second century A.D., and from the fourth to the seventh century it reached and sustained its zenith in Guatemala. Then began a period of decline, followed by a second rise in northern Yucatan. However, internal disruption and civil war had, by the end of the fifteenth century, resulted in disintegration of the empire into more than a dozen individual tribal states.

The Maya civilization unified an area larger than any of the modern Middle American states except Mexico. Its population probably was somewhere between two and three million; the Mayan language, the local *lingua franca,* and some of its related languages remain in use in the area to this day. This was a theocratic state with a complex religious hierarchy, and the great cities which today lie in ruins, overgrown by the tropical vegetation, served in the first instance as ceremonial centers. Their structures testify to the architectural capabilities of the Maya, and include huge pyramids and magnificent palaces. Intricate stone carvings, impressive murals, and detailed ideograms hewn into palace walls tell us that the Maya culture had its artists and writers. We know also that they were excellent mathematicians, had accumulated much knowledge of astronomy and calendrics, and were well ahead of other advanced peoples in the world in these fields. No doubt the Maya civilization had its poets and philosophers, but it had its practical side too: in agriculture and trade they made major achievements. They grew cotton, had a rudimentary textile industry, and even exported finished cotton cloth by seagoing canoes to other parts of Middle America, in return for, among other things, cacao, which was prized by the Maya. This involvement in foreign trade is another reflection of the degree of organization that was achieved by this empire.

Of course the mainstay of the Maya Empire was the great variety of food crops produced by the farmers in this tropical land. While impressive, the Mayan cities never contained as many people as did the cities of later Mesoamerican civilizations. They functioned as headquarters for the priest-rulers and their numerous vassals, and were surrounded by rural service areas. Only on ceremonial occasions would thousands of people gather in these places; the great majority of the Maya lived on the land. Here they cultivated a truly impressive array of root crops, vegetables, and fruits. Like their Mesoamerican contemporaries and successors, the Maya were very good farmers. They practiced methods of soil conservation and fertilization, experimented with new crops, and sought to improve existing strains. When the Spanish arrived on mainland Middle America's shores, some of the Mayas' cities lay in ruins, but their farms and trading activities still thrived.

Thus tropical lowland Middle America was the scene of impressive cultural achievements— but no more so than the highlands. At about the same time that the Maya civilization emerged in the forests, the Mexican highland also witnessed the rise of a great Indian culture, similarly focused on ceremonial centers, and likewise marked by major developments in agriculture as well as architecture, religion and ritual, and the arts. Certainly these achievements were comparable to those of the Maya, but they were to be overshadowed by what followed.

One of the important successors to the early highland civilization were the Toltecs, who moved into this area from the north, conquered and absorbed the local Indian peoples, and formed a powerful state centered on one of the first true cities of Middle America, Tula. The

Many impressive structures have been left behind by the ancient Maya. The top photograph shows the Maya Temple at Chichén Itźa, in Yucatán, Mexico. Below left is the Castle at the same site, and to the right the *Caracol* or Astronomical Observatory. (*Above:* Richard Davis/DPI; *below:* Anita M. Beer/DPI).

Toltecs' period of hegemony in the region was relatively brief, extending for less than three centuries after their rise about 900 A.D. But during this period they conquered parts of the Mayan domain, absorbed many of the Mayas' innovations and customs, and introduced them on the highland. When their state was in turn penetrated by new elements from the north, it was already in decay. But Toltec technology was not lost; it was readily adopted and developed by their successors, the Aztecs.

The Aztec state, the pinnacle of organization and power in Middle America, is thought to have originated in the early fourteenth century when a community of Nahuatl (or Mexicano)-speaking Indians founded a settlement on an island in one of the many lakes that lie in the Valley of Mexico. This village and ceremonial center, named Tenochtitlan, was soon to become the greatest city in the Americas, and the capital of a large and powerful state. Through a series of alliances with neighboring peoples, the early Aztecs gained control over the whole Valley of Mexico, the pivotal geographic feature of Middle America and today still the heart of the modern state of Mexico. This region is in fact a mountain-encircled basin positioned about a mile and a half above sea level. Elevation and interior location both affect its climate: for a tropical area it is quite dry and very cool. A number of lakes lie scattered through the region, and these water bodies formed valuable means of internal communication to the Aztec state. The Indians of Middle America never developed the wheel, and so they relied heavily on porterage and, where possible, on the canoe for the transportation of goods and people. The Aztecs connected several of the Mexican lakes by canals, and maintained a busy canoe traffic on their waterways. Tens of thousands of canoes carried

agricultural produce to the cities and tribute paid by its many subjects to the headquarters of the rulers and nobility.

Throughout the fourteenth century the Aztec state strengthened its position, developing a strong military force, and organizing its territory into provinces with their respective governors and district commissioners. Each province was allotted a certain agricultural production requirement. In the capital the bureaucracy grew ever more complex, and society soon became stratified into several classes. Systems of tax collection, law enforcement, rapid messenger communication, and other trappings of the state came into being. By the early fifteenth century the Aztecs were ready to begin with the conquest of neighboring peoples.

The Aztec drive to expand the empire was directed primarily eastward and southward. To the north, the land quickly became drier, inhabited by a sparse nomadic population, and very unproductive. To the west lay a powerful competing state, that of the Tarascans (Map 7-3), with whom the Aztecs sought no quarrel in view of the open way to the east and south. Here, then, they penetrated and conquered almost at

MODIFIED AFTER: WEST & AUGELLI, MIDDLE AMERICA, ITS LANDS AND PEOPLES

Map 7-3 THE AZTEC AND TARASCAN STATES.

will. But the Aztec objective was not the acquisition of territory, but the subjugation of peoples and towns for the purpose of exacting taxes and tribute. They did not introduce their concepts of religion, nor their language: they were not "colonists" in the European sense. On the other hand, they carried off thousands of people for purposes of human sacrifice in the ceremonial centers of the Valley of Mexico, a practice which would have made colonization rather difficult and "pacification" a self-defeating aim. The Aztecs needed a constant state of enmity with weaker people in order to take their human prizes.

Just how far the tentacles of Aztec power might have spread through Middle America will never be known, for the Spanish invaders interrupted the Aztec drive early in the sixteenth century. But even then Aztec traders were already dealing with peoples as far away as Panama, and as a rule, legitimate trade was soon followed by conquest and demands for tribute. Certainly the Aztecs seem to have had the numbers and capacities to bring most of Middle America under their control. As their influence spread, so the volume of goods streaming back to the Valley of Mexico increased. Gold, drawn largely from the Balsas Basin, cacao beans, mostly from coastal areas, cotton and cotton cloth, the feathers of tropical birds, and the skins of wild animals were among the items that were carried back to Tenochtitlan and its surroundings. The state grew ever richer, its population mushroomed, the cities expanded. Tenochtitlan probably had over 100,000 inhabitants; some put the estimate as high as a quarter of a million. These were not just ceremonial centers, but true cities, with a variety of economic and political functions and large populations among which there were labor forces with particular skills and specializations. Unlike the Mayan cities, which really stood alone amidst service areas (so that the Maya Empire has been described as a set of city states), the Aztec cities were true central places, providing services and goods to their hinterlands and competing with each other for primacy. They were more effectively interconnected and thus formed part of a state framework that represented a considerable advance over that of the Maya.

Aztec civilization produced an enormous range of impressive accomplishments, though the Aztecs seem to have been better borrowers and refiners than they were innovators. They practiced irrigation by diverting water from streams via canals to farmlands, and built elaborate walls to terrace hillslopes where soil erosion threatened. Along the shallow margins of their lakes they developed an amazingly sophisticated method of irrigation. This involved the alternate layering of floating aquatic vegetation and lake mud until the bottom layer rested on the lake floor and the top layer was just above the water surface. Then the highly fertile alluvial material of the lake floor itself was scooped up and spread thickly across this mass of aquatic vegetation or *chinampa*. Thus a soil layer was provided that was so fertile that several crops could annually be grown there; the fertility could be maintained by renewing the application of lake mud. Small fields, separated by narrow canals (for purposes of drainage) and anchored by willow trees, produced corn, beans, and other crops. The method is still in use today.

Indeed, when it comes to measuring the legacy of the Mesoamerican Indian to his successors and, indeed, to the world, then the greatest contributions surely come from the field of agriculture. If his methods of cultivation were not unique, many of his domesticated crops certainly were, and they spread around the world to alter the diets of hundreds of millions of people. Corn (maize), various kinds of beans, the sweet potato, a variety of manioc, the tomato, squash, the cacao tree, tobacco—these are just a few of the crops that grew in Mesoamerica when the Europeans first made contact. It is hardly necessary to suggest what these mean in other parts of the world today: the staple food for many millions of Africans is corn; many countries around the world produce cacao and some of them depend on it as their chief source of export revenue; hundreds of millions of people smoke tobacco in various forms.

Three distinct types of land tenure existed in the Aztec state, each of which has some relevance to the fortunes of the state and the post-Aztec period in Middle America. Despite the great cities and (for the time) high degree of urbanization of the population, by far the majority of the Indians of Mesoamerica were villagers who lived a subsistence life of agriculture. Each village owned its surrounding cultivable area, not through the property of individuals, but communally. Only the village council could assign those who might farm parcels of land. In Aztec

The old Aztec builders left large structures in Mesoamerica, many with stonework of meticulous detail. The top photograph shows the roofed palace at Palangúe, while the picture at the lower left shows massive stone warriors guarding the citadel of Tula, capital of the ancient Aztec predecessors, the Toltecs. Among the Teotihuacan ruins (lower right) is the Pyramid of the Sun. (Richard Davis/DPI—Federico Pateliani of Camera Press/PIX—Arnold Weichert/DPI)

villages, these assignments were made on the basis of kinship groups, each family within the group then being given a part of the allotted land. This principle survived centuries after the demise of the Aztec state: where the Indians were left to run their own affairs they maintained the practice of the landholding village, notably in Mexico and Guatemala. Today, when Middle American governments are trying to implement land reform programs, a modified version of this principle is applied in the redistribution of land to landless peasants.

A second kind of agriculture existed near the major cities, where small garden plots were used to cultivate a variety of crops that were produced for trade in the urban markets. The lake-shore *chinampa* system was a part of this. But more often these plots were heavily fertilized, meticulously cared for, and based in the dry-land soils around the cities. What is remarkable about these plots is the variety of crops grown on them; together and at the same time corn, vegetables, fruits, tobacco, and even flowers were raised. No doubt the intensity of this cultivation, reminiscent of some Mediterranean gardens, and the proximity of the small farms to the cities, had much to do with the great size to which these cities could grow. Fresh produce was constantly available.

The third kind of land ownership and tenure that existed in the Aztec state had in it the seeds of disaster. In this stratified society, where traders' guilds, the military, the priests, and the nobility all held special and privileged status, aggrandizement was easy. Wealthy priests and nobles through various means acquired great pieces of land from village holdings, and then, rather like the nobility of Europe, had this land tilled by slaves, or by serfs and their families who were granted the use of a small parcel for their own subsistence. This was done in the Valley of Mexico, and also in some of the conquered territories, where conditions promised quick rewards. When the Spanish invaded Middle America and overthrew Aztec authority, some of these became the first *haciendas*. This is not to suggest, of course, that Spanish practices of land alienation would not have occurred but for the Aztec estates—but they provided an immediate objective for the land-hungry invaders.

In the Aztec cities the civilization reached its zenith. They were monuments to the abilities of Aztec architects and planners, and they were veritable warehouses of wealth and splendor. Aqueducts brought drinking water from miles away to the city centers. Open-air ball courts provided the privileged with recreational facilities. Magnificent palaces and great pyramids dominated the urban scene, which was further embellished by great, wide avenues and impressive public squares. Residential suburbs crowded around the city centers where the ceremonial structures were concentrated. A constant flow of luxury goods came in from distant parts of the domain, while volumes of farm produce was brought in, by human carrier and canoe, from the surrounding agricultural areas. In the cities, thousands of people specialized in particular crafts: there were stonecutters, goldsmiths and silversmiths, tailors, cloth weavers, potters, basket makers, canoe builders—in addition to the artists and musicians, philosophers and mathematicians, teachers and priests who made up this complex society. Certainly the Aztecs forged a true and distinct cultural landscape, elements of which still exist in present-day Middle America, five centuries later.

Confrontation of Cultures

We in the Western world all too often are under the impression that history began when the Europeans arrived in some area of the world, and that the Europeans brought such superior power to the other continents that whatever existed there previously had little significance. Middle America seems to bear this view out: the great, feared Aztec state fell before a relatively small band of Spanish invaders in an incredibly short time. But let us not lose sight of the facts. At first, the Spanish were considered to be "White Gods," whose arrival was predicted by Aztec prophecy; having entered Aztec territory, the earliest Spanish visitors were able to determine that great wealth had been amassed in the Aztec cities. And Hernán Cortés, for all his 508 soldiers, did not singlehandedly overthrow the Aztec authority. What Cortés brought on was a revolt, a rebellion by peoples who had fallen under Aztec domination and who had seen their relatives carried off for human sacrifice to Aztec gods. Led by Cortés with his horses and artillery, these Indian peoples rose against their Aztec oppressors and followed the Spanish band of men toward Tenochtitlan; thousands of them died in combat against the Aztec warriors. They fed and guarded the Spanish soldiers, maintained connections for them with the coast, carried supplies from the shores of the Gulf of Mexico to the point Cortés had reached, secured and held captured territory while the white men moved on. Cortés started a civil war; he got all the credit for the results. But it is reasonable to say that Tenochtitlan would not have fallen so easily to the Spaniards without the sacrifice of many thousands of Indian lives.

Actually, many sixteenth and seventeenth century Europeans, were they to come back to life today, no doubt would be amazed at our ethnocentric and superior view of what happened when their culture met those of the Middle American Indian, and the Asian. They themselves sometimes reported back to their capitals with awe and admiration for much of what they had seen. In America as well as Africa, the Spanish, Portuguese, Hollanders, and other European visitors considered many of the peoples they confronted to be equals—equals to be invaded, attacked, and, if possible, defeated— but equals nevertheless. The cities and farms of Middle America, the urban centers of West Africa, the great Inca roads of South America all reminded the Europeans that technologically they were, if anything, a mere step ahead of their new contacts. It is a common misconception relating to the history of African-European contact that the Europeans more or less neglected Africa because there was so much to be gained elsewhere. In fact African organization often was too strong for the Europeans to destroy. If a gap developed between the European powers and the indigenous peoples of many other parts of the world it only emerged clearly when the industrial revolution came to Europe—centuries after Columbus, Cortés, and Vasco da Gama.

In Middle America, the confrontation between Spanish and Indian cultures spelled disaster for the Indian in every conceivable way. The quick defeat of the Aztec state (as well as that of the Tarascans) was followed by a catastrophic decline in the population. Whether there were 15 or 25 million Indians in Middle America when the Spaniards arrived, after only one century just 2.5 million survived. It is not that the Spanish were particularly ruthless colonizers—they were ruthless to be sure, but not much more so than other European powers that subjugated other cultures—even a systematic campaign of extermination would not have succeeded so dramatically. True, the Spanish at first enslaved the

Indian and Spanish determination to destroy the strength of Indian society was clear. Writing about early Guatemala, Kalijarvi states:

"All Indian culture was wiped out as far as possible, even to the point of burning the books that could be seized (in July 1562 at an *auto da fe*, the Spaniards burned 5,000 idols and 27 hieroglyphic rolls as the works of the devil). Indians were slaughtered and burned alive in groups. Nobles, priests, and warriors were exterminated. Only the common people and the workers were allowed to survive. The strong were enslaved, the weak killed, families split, some to work in the mines, some to serve as beasts of burden, others to toil away their lives in different ways for the Spaniards. Young women became concubines . . . and bore the Spaniards children of mixed blood (mestizo), although it should be noted that the Indians had no objection to miscegenation."[2]

But biology did what even such ruthlessness could not have achieved in so short a time. As in the Caribbean islands, the Indians were not immune to the diseases the Spaniards introduced by their presence: smallpox, typhoid fever, measles, influenza, mumps; neither did they have any protection against disease introduced by the white man through his African slaves, such as malaria and yellow fever, which took huge tolls in the hotter and more humid lowland areas of Middle America. Great epidemics struck the Indian population at regular intervals of 10 to 15 years throughout the sixteenth and the first half of the seventeenth centuries; the first, a vicious outbreak of smallpox, hit Middle America in 1520, even before Tenochtitlan fell. The Indians died by the millions, and in the 150 years after 1500 nearly 90 percent of the Indian population of Middle America disappeared.

Thus the mainland American Indian fared little better than his Caribbean counterpart. As in the Caribbean, the Spanish soon (1542) outlawed the traffic in Indian slaves, but this had little effect in stemming the tide of depopulation. Small incidents of culture contact had major effects: the Indians used a sweat-bath for purposes of religious purification, but the Spaniards opposed this practice on the grounds that it had associations of paganism and because the Islamic

[2]T.V. Kalijarvi, *Central America: Land of Lords and Lizards,* Van Nostrand, Princeton, 1962, pp. 55–56.

Moors, who had so long held southern Iberia, also used it. By discouraging the sweat-bath, the Spaniards thus removed yet another obstacle to the rapid spread of germs and disease.

What happened to Middle America's cultural landscape, to the great cities, to the *chinampas,* the terraced fields, the dispersed villages of the Indians? It was drastically modified. The Indian cities ceased to function as they had before, and the Spanish brought to Middle America new traditions and innovations in urbanization, agriculture, religion, and other areas. Having destroyed Tenochtitlan, the Spaniards nevertheless recognized the attributes of its site and situation, and chose to rebuild it as their mainland headquarters. But whereas the Indians had used stone almost exclusively as their building material the Spaniards, for their churches, public buildings, and residences, used great quantities of wood. Soon the forests of oak and pine that stood on Middle America's highland slopes began to disappear—just as the forests of the Mediterranean region had been destroyed. The Spaniards, like other Europeans at that particular stage of development of energy use in Europe, utilized charcoal for purposes of metal smelting, heating, and cooking, and thus the onslaught on the forests was such that great rings of deforestation quickly formed around the major Spanish towns. And not only around the Spanish towns: the Indians soon adopted Spanish methods of house construction and charcoal use and they, too, contributed to forest depletion. Soon the scars of erosion began to replace the stands of tall trees.

The Indians had been planters, but had no domestic livestock that would make demands of the original vegetative cover. Only the turkey, the dog, and the bee (for honey and wax) had been domesticated in Mesoamerica. The Spaniards, on the other hand, brought with them cattle and sheep in numbers—numbers that multiplied rapidly and made increasing demands not only on the existing grasslands, but on the cultivated crops as well. And again the Indians soon adopted the keeping of livestock, especially the remaining nobles, putting further pressure on the land. The net effect on food availability was not favorable. Cattle and sheep were avenues to wealth, and the owners of the herds benefited. But the livestock competed with the people for the available food, and they contributed to a gross disruption of the food balance that existed in the region. Hunger became a major problem in Middle America during the sixteenth century, and no doubt contributed to the susceptibility of the Indian population to the diseases that threatened them.

The Spaniards also introduced their own crops, notably wheat, and their own farming equipment, of which the plow was the most important. Soon large fields of wheat began to make their appearance alongside the small plots of the Indian cultivators. And, inevitably, the wheat fields encroached on the Indians' lands, reducing them further. The wheat was grown by and for the Spaniards, so that what the Indians lost in farmlands was not made up in available food. And neither were their irrigation systems spared. The Spaniards needed water for their fields and power for their mills, and they had the technological know-how to take over and modify the regional drainage and irrigation systems. This they did, leaving the Indian fields either waterless or insufficiently watered, and thus the Indian's chances for an adequate supply of food were diminished further.

But the most far-reaching changes in the cultural landscape brought by the Spaniards had to do with their traditions as town dwellers, as people accustomed to living in concentrated towns and villages. As we noted previously, the Indians, their large cities notwithstanding, were mainly rural folk who lived in widely dispersed hamlets and villages amid their farmlands. The Spanish found this situation undesirable; it made control difficult. Hence the decision was made to bring the Indians from their land into nucleated villages and towns established and laid out by the Spaniards. In these settlements the kind of government and administration to which the Spanish were accustomed could be exercised. The focus of each town was the Catholic church; indeed, until 1565 the resettlement of Indians was mainly the responsibility of missionary orders. The location of each of these towns was chosen to lie near what was thought to be good agricultural land, so that the Indians could go out each day and work in the fields. Unfortunately, the selection was not always a good one, and a number of villages lay amid land that was not suitable for Indian farm practices. Here, food shortages and even famine resulted—but once created, the village must try to survive. Only rarely was a whole village abandoned in favor of a better situation.

In the towns and villages, the Indian came face to face with Spanish culture. Here he learned the white man's religion, paid his taxes and tribute to a new master, found himself in prison or in a labor gang according to European laws and rules. Packed tightly in a concentrated settlement, he was rendered even more susceptible than he was in his own village to the diseases that regularly ravaged his people. But despite all this, the nucleated Indian village survived. Its administration was taken over by the civil government, and later by the independent governments of the Middle American states; today it is a feature of Indian areas of the Mexican and Guatemalan landscape. Anyone who wants to see remnants of the dispersed Indian dwellings and hamlets must travel into the most isolated parts of Middle America's remaining Indian areas.

Thus the Indian was relocated in a network of rural villages and towns. What of the Spanish towns and cities, the urban places of the conquerors? These in the first instance were administrative centers, located in the interior to function as centers of control over trade, tax collection, labor recruitment, and so on. At times the Spaniards, recognizing an especially salutory location chosen by their Indian predecessors, would build on the same site: Tenochtitlan became the mainland headquarters of Spanish control and later the capital of the state of Mexico. Other sites were chosen because they served Spaniards better than pre-existing Indian sites would; within a half century of the Aztec defeat the Spanish conquerors founded more than 50 new towns, some of which have become major modern cities. Along the Caribbean and Pacific coasts, cities emerged that were part of the globe-girdling chain of cummunications within the rising Spanish empire: Veracruz on the Gulf of Mexico rose to prominence in this way, as did Acapulco on the Pacific coast, the leading station on the way to Manila, the Philippines, and the westward link to the mother country.

The Spanish in Middle America had not come simply to ransack the riches in gold, silver, precious stones, works of art, and other valuables that had been accumulated by Indian kings, priests, and nobles. That aftermath of conquest was soon over. Next the Spaniards wanted to organize the sources of this wealth for their own benefit. Mining, commercial agriculture, live-

stock ranching, and profitable trade were the avenues to affluence, and among these, mining held the greatest promise. The first mining phase, during the first half of the sixteenth century, involved the "washing" of gold from small streams and carrying gold dust and nuggets—called *placering*. This is how the Indians had found their gold, but it was (and still remains) a small-scale operation. The Spaniards simply used the Indian placer-workers for their own profit, something that was especially easy during the period of Indian slavery. But the placers produced less and less gold, Indian slavery was abolished, and Indian labor soon became more scarce as the depopulation cycle began to take effect. Thus Spanish prospectors looked for other, larger mineral deposits—not just of gold, but of other valuable minerals as well. They were quickly successful, especially in finding lucrative silver and copper deposits. Between 1530 and 1570 a number of mining towns were founded that have developed into modern urban centers, and a new phase of mining began. The initial focus lay in the southwest and west of the Valley of Mexico, where erosion exposed veins of silver and other ores (such as tin) at places like Taxco, Sultepec, and Temascaltepec (Map 7-4), and farther away from Mexico City around Guadalajara. But the most significant finds were made somewhat later and farther to the north, along the eastern foothills of the Sierra Madre Occidental, from Guanajuato and Zacatecas northward to the Chihuahua area. This became known as the Silver Belt, but it was not the only mineralized area in the rich north: the Sonora district in the northwest and the fields north of Monterrey in the northeast also yielded silver.

The mining industry set in motion a host of changes in this part of Middle America. The mining towns drew laborers by the thousands. The mines required equipment, timber, mules; the people in the towns needed food. Since many of the mining towns, especially those of the north, were located in dry country, irrigated fields were laid out wherever possible. Mule trains and two-wheeled carts connected farm supply areas to the mining towns, and the mining towns to the coasts. More than a network of administrative towns could have done, the mining towns integrated and organized the Spanish domain in Middle America. More than government and church could have done, the mining towns brought effective and permanent

Map 7-4 MINING AREAS OF MEXICO.

Spanish control to some very far-flung parts of the New Spain. Mining, indeed, was the mainstay of colonial Middle America.

Mainland and Rimland

Little has been said, in the preceding discussion, of areas in Middle America lying outside of Mesoamerica. Mesoamerica was the center stage for both Indian civilization and Spanish colonization, although the Spaniards quickly recognized the strategic nature of Panama's narrow "waist" as an easy connection between the Caribbean and the Pacific Ocean. It is easy to see why the Spanish colonists should have centered their activities in Mesoamerica rather than elsewhere. To the north and northwest lay the drier country of nomadic and seminomadic Indian peoples whose material property had little value and whose land held little promise, at least for the sort of farming the Spaniards practiced. And to the south and east of Mesoamerica lived peoples whose culture and technology was far less developed than that of the Mayas and Aztecs. Many of them were forest dwellers whose farming methods involved the clearing of a patch of vegetation, its burning, and the planting of crops—mostly root crops—in the cleared soil. This shifting agriculture was accompanied by hunting and fishing. There was very little to compare to the great buildings and cities of Mesoamerica; dwellings were made of branches and thatch, villages were the largest agglomerations of people. All this is more reminiscent of the forests of South America than the culture hearth of Mesoamerica, and the only places where things were any different were in the volcanic mountainlands of Costa Rica and the Azuero Peninsula and adjacent coastal areas of Panama. There somewhat more complex societies had developed, whose technological achievements included metallurgy (especially gold-working), pottery, and weaving.

Outside of Mesoamerica, only Panama, with its twin attractions of transit function and gold supply, became an early focus of Spanish activity. The Spaniards founded the city of Panama in 1519, and apart from their use of the area of the modern Canal Zone as an interocean link, their primary interest lay on the Pacific side of the isthmus. From here, Spanish influence began to extend northwestward into Middle America; Indian slaves were taken in large numbers from the densely peopled Pacific lowlands of Nicaragua, to be shipped to South America via Panama. The highlands, too, fell into the Spanish sphere, and before the middle of the sixteenth century Spanish exploration parties based in Panama met those moving southeastward from Mesoamerica.

But the primary center of Spanish activity was in what is today central and southern Mexico, and the major arena of international competition in Middle America lay not on the Pacific side, but on the islands and coasts of the Caribbean. Here the Spaniards faced the English, French, and Dutch, all interested in the lucrative sugar trade, all searching for quick wealth, all hoping to expand their empires. Only the English gained a real foothold on the mainland, which otherwise remained an exclusively Spanish colonial domain. British Honduras (now called Belize) is all that remains of a coastal sphere of British influence that extended from the Yucatan Peninsula to the Nicaragua–Costa Rica boundary (Map 7-5). For understandable reasons the Spaniards were more interested in securing their highland holdings than in the rainy, disease-ridden, forested, swampy coasts, and they formally recognized lowland British interests in 1670. Later, after centuries of European colonial rivalry in the Caribbean, the United States entered the picture and made its influence felt in the coastal areas of the mainland—not through colonial conquest, but by the introduction of large-scale and widespread plantation agriculture. The effects were as far-reaching as was the colonial impact on the Caribbean islands. The economic geography of the Caribbean coastal areas was transformed as hitherto unused alluvial soils in the many river lowlands were planted with thousands of acres of banana trees. Since the diseases the Europeans had brought to the New World had been most rampant in these hot, humid areas, the Indian population that survived was small and provided an insufficient labor force. Tens of thousands of black laborers came to the coast from Jamaica and other islands, completely altering the demographic situation. In many physical ways the coastal belt already resembled the islands more than the Middle American plateau, and now the economic and cultural geography of the islands was extended to it.

These contrasts between the Middle American highlands on the one hand, and the coastal areas and Caribbean islands on the other hand, were

MODIFIED AFTER: WEST & AUGELLI, MIDDLE AMERICA, ITS LANDS AND PEOPLES

Map 7-5 BRITISH INFLUENCE IN MIDDLE AMERICA.

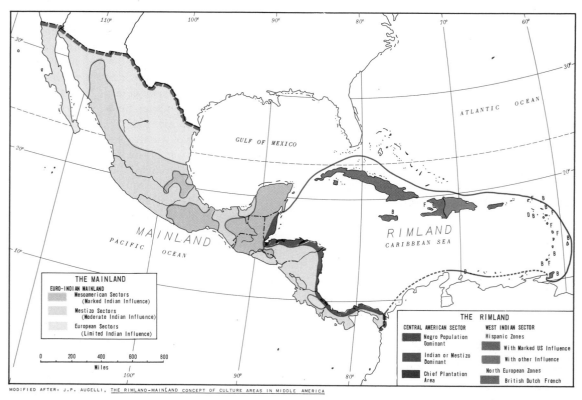

MODIFIED AFTER: J.P. AUGELLI, THE RIMLAND-MAINLAND CONCEPT OF CULTURE AREAS IN MIDDLE AMERICA

Map 7-6 MAINLAND AND RIMLAND IN MIDDLE AMERICA.

conceptualized by Professor Augelli into a mainland—rimland framework.[3] Professor Augelli recognizes a Euro-Indian *Mainland,* consisting of mainland Middle America from Mexico to Panama but with the exception of the Caribbean coast from Yucatan southeastward, and a Euro-African *Rimland,* comprising this coastal zone and the islands of the Caribbean (Map 7-6). The terms Euro-Indian and Euro-African suggest the cultural heritage of each region: in the Mainland, European (Spanish) and Indian influences are paramount, and in the Rimland, the heritage is European and African. As Map 7-6 shows, the Mainland is subdivided into several areas on the basis of the strength of the Indian legacy: in southern Mexico and Guatemala Indian influences are marked; in northern Mexico and parts of Costa Rica, Indian influences are limited. Between these areas lie sectors with moderate Indian influence. The Rimland, too, is subdivided: the most obvious division is into the mainland-coast plantation zone on the one hand and the islands on the other, but the islands can themselves be grouped according to their cultural heritage. Thus there is a group of islands with Spanish influence (Cuba, Puerto Rico, and the Dominican Republic on old Hispaniola), and another group with North European influences, including the former British West Indies, the French islands, and the Netherlands Antilles.

Obviously this is a generalized framework, but it is a very useful one for the purpose of identifying the major lines of human development in Middle America. A good deal of evidence can be marshalled to substantiate its essential validity. Thus the contrasts of human habitat described above are supplemented by regional differences in outlook and orientation. The Rimland was an area of sugar and banana plantations, of high accessibility, of seaward exposure, and of maximum contact and mixture. The Mainland, on the other hand, was farther removed from these contacts; it has been an area of greater isolation, of greater distance from all these forces. The Rimland was the region of the great plantation, and thus its commercial economy was susceptible to fluctuating world markets and tied to overseas capitals; the Mainland was

the region of the *hacienda,* more self-sufficient, less dependent on outside markets.

In fact, this contrast between the *plantation* and the *hacienda* in itself constitutes strong evidence for a Rimland-Mainland division. There are important differences between these forms of land-labor organization, despite the fact that the terms are sometimes used interchangeably. True, there are similarities too: both need large amounts of land and much cheap labor; both have occupied the best farmlands and have inhibited the emergence of any sizable class of well-to-do small farmers; both were introduced by European forces. But the *hacienda* was a Spanish institution, and the plantation, Professor Augelli argues, was the concept of Europeans of more northerly origins; later it also became an Anglo-American one. In the *hacienda* the Spanish landowner possessed a domain whose productivity he might never push to its limits: the very possession of such a vast estate brought with it the prestige and the comfort and style he sought. The workers would live on the land—it may once have been *their* land—and had plots where they could grow their own subsistence needs. They might help themselves by selling handcrafts in the same nearby towns where the *hacienda*'s owner sold the produce of his lands. Traditions survived: in the Indian villages incorporated into the early *haciendas,* in the methods of farming that were used, in the means of transportation of the produce to the markets. All this is written as though it is mostly in the past, but the legacy of the *hacienda,* with its inefficient use of land and labor, is still there in Middle America. When peasants are crowded on poor or mediocre land, there may be unused land on a nearby *hacienda,* whose successive owners never sought to extract maximum profit from their land, never achieved the agricultural specialization that was possible on their soils. They were satisfied with self-sufficiency and security.

The plantation, on the other hand, was conceived as something very different. In their volume entitled *Middle America, Its Lands and Peoples,* Professors Robert C. West and J. P. Augelli list five characteristics of Middle American plantations which always apply.[4] These quickly illustrate the differences between *hacienda* and plantation: (1) plantations are located in the humid coastal tropical lowlands of the region; (2) they

---

[3] J. P. Augelli, "The Rimland—Mainland Concept of Culture Areas in Middle America," *Annals of the Association of American Geographers,* Vol. 52, No. 2, June 1962, pp. 119–129.

[4] R. C. West and J. P. Augelli, *Middle America, Its Lands and Peoples,* Prentice Hall, Englewood Cliffs, 1966, p. 15.

produce for export almost exclusively, usually a single crop; (3) capital and skills are often imported, so that foreign ownership and an outflow of profits occur; (4) labor is seasonal, that is, it is needed in large numbers during harvest time and may be idle at other times, and such labor has been imported because of the scarceness of Indian labor; and (5) with its "factory in the field" characteristic the plantation is more efficient in its use of land and labor than the *hacienda.* The objective was not self-sufficiency but profit, and wealth rather than prestige was a leading motive in the plantation's establishment. During the past century, both *hacienda* and plantation have changed a great deal. The vast American investment in the Caribbean coastal zones of Guatemala, Honduras, Nicaragua, Costa Rica, and Panama transformed that area and brought a whole new concept of plantation agriculture to the region; in the Mainland the *hacienda* has been under pressure from governments that view them as political and social (not to mention economic) liabilities. Indeed, some *haciendas* have been parcelled out to smallholders, while others have been pressed into greater specialization and productivity. Still other land has been placed in *ejidos,* and is communally owned by groups of families (see *Mexico*). But both institutions for centuries contributed to the different social and economic directions that have given the Mainland and Rimland their respective individuality.

Political Differentiation

Mainland Middle America today is fragmented into eight different countries, all but one of which (Belize, the former British Honduras) have Hispanic origins. Largest of them all—in fact, the giant of Middle America—is the United States of Mexico, whose 762,000 square miles constitute over 70 percent of the whole land area of Middle America (the islands included), and whose more than 40 million people outnumber those of all the other countries and islands of Middle America combined. The cultural variety in Caribbean Middle America is much greater. Here Cuba dominates: its area is larger than that of all the other islands together, and its population of over 7 million is nearly twice that of the next ranking country (Haiti with about 4 million). But the Caribbean is hardly an area of exclusive Spanish influence: whereas Cuba has an Iberian heritage, its southern neighbor, Ja-

maica (population 2 million, mostly black) has a legacy of British involvement, and Haiti's strongest imprint has been French. The crowded island of Hispaniola is shared between Haiti and the Dominican Republic, where Spanish influence survives; it is strong also in Puerto Rico where, however, it is modified by the impact of the United States.

These four islands with their five countries (Cuba, Jamaica, Hispaniola with Haiti and the Dominican Republic, and Puerto Rico) are referred to as the *Greater Antilles;* the arc of smaller islands extending in a semicircle from the Virgin Islands to the Dutch ABC Islands (Aruba, Bonaire, Curacao) are called the *Lesser Antilles* (Maps 7-7 and 7-8). Here, too, the cultural diversity is great. There are the American Virgin Islands, French Guadeloupe and Martinique, a group of British-influenced islands including Dominica, Barbados, St. Lucia, and St. Vincent, and Dutch Saba, St. Eustatius, and the ABC Islands off the Venezuelan coast. Standing apart from the Antillean arc of islands is Trinidad, another formerly British dependency which, with its smaller neighbor of Tobago, became a sovereign state in 1962. While the mainland countries (except Belize) ended their colonial status at a relatively early stage, the Caribbean islands remained colonized for a much longer time. No doubt this was due in large measure to their smaller size and insularity; Spain, which had to yield to demands and struggles for independence on the mainland, could hold Cuba and Puerto Rico, where there was much agitation for reform as well, until the Spanish-American War at the end of the nineteenth century. Spain ceded western Hispaniola to France in 1697, Frenchmen having developed sugar plantations there, and in 1795 gave up the rest of the island to the French. One of the earliest independent republics to emerge in the Caribbean was Haiti, whose black (95 percent) and mulatto (5 percent) population fought a successful slave revolt. In 1804 the Republic of Haiti was established, and the Haitians, having defeated the French—and destroyed a good part of the economic structure of their country—now turned on their Dominican neighbors. From 1822 to 1844 Haiti ruled all of Hispaniola; then, after the Dominicans had fought themselves free, Spain briefly reestablished a colony there (1861–1865). The most recent alien intervention in Hispaniola came during the twentieth century, when the United

Map 7-7 THE GREATER ANTILLES.

Map 7-8 THE LESSER ANTILLES.

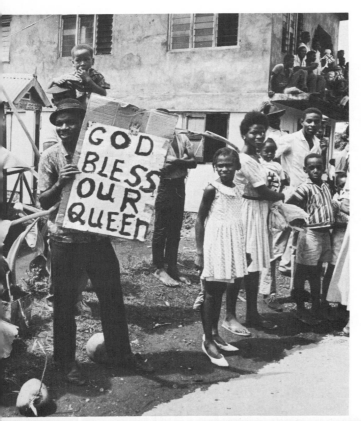

States occupied both Haiti (1915–1934) and the Dominican Republic (1916–1924).

In the Greater Antilles, then, Spanish influence was strong. By the late nineteenth century Cuba and Puerto Rico resembled the mainland republics in the composition of their population and in their cultural imprint. But Spain's colonial archrival, Britain, also had a share of the Greater Antilles in Jamaica, and it gained a large number of footholds in the Lesser Antilles. The sugar boom and the strategic character of the Caribbean area brought the European competitors to the West Indies; in the twentieth century the United States of America made its presence felt as well—though the plantation crop was now the banana and the strategic interest even more immediate than that of the European powers. In parts of Caribbean Middle America the period of colonial control is only just ending. Jamaica attained full independence from Britain in 1962, as did Trinidad and Tobago. An at-

Acculturation in the Caribbean: the legacy of Britain. In the top picture, crowds line the streets of St. George, capital of Grenada, to cheer the British queen. In the lower picture, Jamaicans play cricket—Jamaica's ''national'' game. (Camera Press/PIX—Jamaica Tourist Board)

tempt by the British to organize a Caribbean-wide West Indies Federation failed, but other long-time British dependencies—Barbados, Grenada, St. Lucia, and Dominica among others—were nevertheless steered toward a precarious independence as well. Elsewhere the colonial situation continued; as late as 1969 there were no signs of significant changes in the political status of French Guadeloupe and Martinique or the Netherlands Antilles.

Caribbean Patterns

Caribbean America today is a land crowded with so many people that, as an area, it is the most densely populated part of the Americas. It is a land of poverty and, in all too many places, much misery and little chance for escape. In many respects Cuba and Puerto Rico constitute exceptions to any such generalizations made about Caribbean America. But in most of the other islands, life for the average person is difficult, often hopeless, and tragically short.

All this is in almost incredible contrast to the early period of riches based on the sugar trade. But that initial wealth was gained while one whole ethnic group (the Indians) was being wiped off the Caribbean map, while another (the Africans) was being imported in bondage; the sugar revenues always went to the planters, not the workers. Then the economy faced the rising competition of other tropical sugar-producing areas, and it lost its monopoly of the European market. The cultivation of sugar beets in Europe and America also cut into the sales of tropical sugar, and difficult times prevailed. Meanwhile the Europeans helped stimulate the rapid growth of the population, just as they did in other parts of the world. Death rates were lowered, but birth rates remained high, and explosive population increases occurred. With the declining sugar trade, millions of people were pushed into a life of subsistence, malnutrition, and hunger. Many sought work elsewhere: tens of thousands of Jamaican laborers went to the mainland coast when the Panama Canal was constructed and when the United Fruit Company's banana plantations began to arise there. Large numbers of West Indians went to Britain in search of a better life, while Puerto Ricans have for decades been coming to the United States. But this outflow has failed to stem the tide of population growth: today there are well over 22 million people on the Caribbean islands.

The trouble is that the Caribbean islanders just have not had many alternatives in their search for betterment. Their habitat is fragmented by water and mountains; even on the smaller islands the amount of good, flat, cultivable land may be only a small fraction of the whole area. And although there has been some economic diversification, agriculture remains the area's mainstay and sugar still must be the leading product: it heads the exports of Cuba, the Dominican Republic, Jamaica, and Puerto Rico. In Haiti, coffee has taken the place of sugar as the chief export (sugar now ranks second), while Trinidad, just off the Venezuelan coast, is one of the fortunates of the Caribbean in possessing sizable oil fields, and petroleum accounts now for nearly 80 percent of that country's exports. In the Lesser Antilles, sugar has retained a somewhat less prominent position, having been replaced by such crops as bananas, sea island cotton, limes, and nutmegs. But even here, in Barbados, St. Kitts, St. Lucia, and Antigua, sugar remains the leading revenue producer.

Those islands that continue to depend on the sugar industry face intense competition for the world's great markets. The largest producer, Cuba, recently lost its largest nearby market, the United States, through the political events that were set in motion by the ascent to power of the Castro regime; although little reliable information has been available concerning this country, it is clear that the industry has been adversely affected. One set of by-products that have proved lucrative for sugar-producing countries are molasses and rum, which are exported wherever sugar is grown. Another crop that has had some success is tobacco, and Cuban cigars have world fame. But those crops that are grown in the Caribbean—Haiti's coffee, Jamaica's bananas, the Dominican Republic's cacao, the Lesser Antilles' citrus fruits—face severe competition from other parts of the world and have not become established on a scale that could begin to have a real effect on standards of living. And those minerals that do exist in this area—Jamaica's bauxite, Cuba's iron and chromium, Trinidad's oil—do not support any significant industrialization within Caribbean America itself. As in other parts of the underdeveloped world, these resources are exported for use elsewhere.

And so the vast majority of the people in this area continue to eke a precarious living from a

The Netherlands Antilles have a great locational advantage off the coast of Venezuela, one of the world's major oil-producing areas. Shown here are the refineries that have sprung up on Curacao. (Kurt Scholz/PIX)

small plot of ground, mired in poverty and threatened by disease. Farm tools are still primitive, and methods have undergone little change over the generations. Inheritance customs have divided and redivided peasant families' plots until they have become so small that the owner must sharecrop some other land or seek work on a plantation or estate in order to supplement his harvest. Bad years of drought, cold weather, or hurricanes can spell disaster for the peasant family. Soil erosion threatens: much of the countryside of Haiti is scarred by gulleys and ravines. Where soils are not eroded their nutrients are depleted, and only the barest of yields are extracted.

With such problems it would be unlikely that Caribbean America has many large cities; after all, there is little basis for major industry, little capital, little local purchasing power. And indeed the figures reflect this situation, for barely over one quarter of the population is classed as urban. The two largest cities, Cuba's Havana (1.5 million) and Puerto Rico's San Juan (750,000) owe

a great deal of their development to American influences. Jamaica's capital, Kingston, has a population of under a half million; Santo Domingo has about 400,000 inhabitants and Port-Au-Prince less than a quarter of a million. In the mid-1960's it was estimated that Haiti had less than 10 percent of its population of about 4 million living in towns larger than 5000.

Still, the cities are one hope for Caribbean America, for they can become, with the area's magnificent beaches, an attraction for the increasingly prosperous North American tourist. Both Havana and San Juan have benefited greatly from the tourist trade, and tourism is growing in importance as a source of revenue. The restrictions on American travel to Cuba proved a boon to the Jamaican tourist industry, and in that country tourism now ranks third as a source of income.

Origins. Writing of the Haitian people in a recent issue of *Focus,* Professor R. Y. Gildea Jr. comments that the Haitian is "outwardly a simple and destitute person but inwardly he is rich

A Haitian scene that could have been taken almost anywhere in West Africa: an informal market in a village. (Fujihira/Monkmeyer)

in folklore, music, and tradition."[5] No country in Middle America has a larger percentage of black people in its total population, and Professor Gildea, then, is referring to folklore, music, and tradition of largely African origins. These traditions were carried on slave ships from Africa, mainly West Africa, to every inhabited island of the Caribbean. More than three-quarters of Jamaica's population is black; over half of Trinidad's, nearly half of Cuba's, about one-fifth of Puerto Rico's, and about one-eighth of Santo Domingo's. Exactly where in Africa these people's ancestors came from is known in only very few instances. But there is no doubt that many of them were uprooted from stable, urbanized, culturally rich societies. In the process, the African slave lost a great deal—apart from his freedom and his connections with his family, he lost his language and his ability to practice those capacities and talents he could display in his own social environment. What

[5]R. Y. Gildea, Jr., "Haiti," *Focus,* American Geographical Society, Vol. XVII, No. 9, May 1967, p. 4.

happened was that some of these customs and traditions were nevertheless carried over to the Caribbean, where they reappeared in modified but often still recognizably African forms. In the construction of village dwellings, the operation of rural markets, the role of women in rural life, the preparation of certain kinds of food, methods of cultivation, the nature of the family, artistic expression, and a host of other areas the African heritage of Caribbean America can be read.

The African heritage of the Caribbean, like that of black America, has not received nearly the attention it merits. In part this is due to the initial master-slave relationship and later to the generally underprivileged position of the black population in Caribbean America compared to lighter-skinned people. In general terms, it is still possible to argue that the European or white man is in the best position in this area politically and economically, the mixed-blood or mulatto ranks next, and the black man ranks lowest. In Haiti, for example, where 95 percent of the population is "pure" black and 5 percent mulatto,

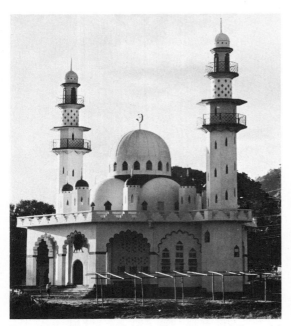

East meets West in Trinidad. This is one of the most cosmopolitan islands in the whole varied Caribbean scene: one third of Trinidad's population is originally from the Orient. Many mosques and temples reflect this circumstance. (Trinidad and Tobago Tourist Board)

it is this mulatto minority that holds the reigns of power. In Jamaica the British have only recently given up control—and the 17 percent "mixed" sector of the population plays a role of prominence in island politics far out of proportion to its numbers. In Santo Domingo, the pyramid of power puts the 25 percent white sector at the top, the 60 percent mixed group next, and the 15 percent black population at the bottom; here is a country that holds tenaciously to its Spanish-European legacy in the face of a century and a half of hostility from neighboring, Afro-Caribbean Haiti. In Puerto Rico, likewise, Spanish values are adhered to in the face of American cultural involvement and a nonwhite sector counting just under one-fifth of the island's 2.5 million people. In Cuba, too, the 15 percent of the population that is black has found itself less favored than the 15 percent mestizo sector and the 70 percent white Cuban population. In very general terms it is still possible to see on the map what were the original functions of Caribbean America's black peoples; there is yet a correlation between the distribution of sugar-producing areas and black people on the islands.

The composition of the population of the islands is further complicated by the presence of Asians from both China and India. During the nineteenth century, the emancipation of slaves and the ensuing localized labor shortages brought some far-reaching solutions. To Cuba, during the third quarter of the nineteenth century, came over 100,000 Chinese as indentured laborers (their number has dwindled considerably since then), and Jamaica, Trinidad, and Guadeloupe-Martinique saw nearly a quarter of a million Indians arrive for similar purposes. To the Afro-modified forms of English and French heard in the Caribbean, therefore, can be added several Asian languages; Hindi is especially strong in Trinidad. The ethnic and cultural variety of Caribbean America is endless.

After 1821 a number of efforts were made on the mainland to achieve some form of political unification. First among these was the Mexican Empire, ruled by Iturbide and intended to control all of mainland Middle America except Belize and Panama. But this venture was short-lived; several countries entered the Empire in 1822 and were out again by 1823. Then followed a Central American Federation, with a constitution resembling that of the United States, but it, too, fell apart in the face of hatreds, rivalries, and special interests. Finally, the Central American Republics declared themselves independent one by one, Costa Rica and Nicaragua in 1838, Honduras also in 1838, after having separated from Guatemala, El Salvador in 1841, and Guatemala in 1847. In several instances the independence period has been a turbulent one. Nicaragua, an early crossing point for Americans seeking a way to California's gold, faced intense foreign interference and in 1856 found itself with one President Walker, an American adventurer; in the twentieth century United States marines for some years occupied the country. Panama, endowed with its canal and with independence in 1903, has seen continued occupation by the United States of the Canal Zone—which cuts the country into two halves—become an explosive political issue. Enormous American investments in the banana plantation ventures along the Caribbean coast produced charges of neocolonialism. Boundary disputes bedeviled the relationships between the Central America Republics. And, as in the Caribbean area, poverty prevails.

Mexico. Mexico is a land of distinction. Not only is this the only "latin" American country to have a common boundary with the Giant of the North (and an exceptionally long one: 1600 miles along the Rio Grande and across the west), but in terms of population this is Spanish-influenced America's largest state. In terms of territory Argentina is larger, but Argentina's population is little more than half of Mexico's. Mexico, on the other hand, is set apart from the other large Spanish-American states by its isthmian location, the cultural contrasts across its northern boundary, and the ineffectiveness of the narrowing Middle American landbridge as a functioning avenue of contact with South America.

Mexico's population of over 40 million counts 24 million mestizos, perhaps 4 million whites, and less than a half-million blacks; the remainder, some 12 million people, are Indians. Thus Guatemala has the largest *percentage* of Indians in its population, but Mexico has the greatest *numbers,* so that the Indian imprint upon Mexican culture is very strong. This is regionally expressed by the language map (Map 7-9). The majority of this population is concentrated in south-central parts of the country, where the heartland of Mexico lies in the basins and valleys

One of the strategic foci of Middle America is the Panama Canal, built during the decade 1904-1914. It takes ships about 8 hours to make the 50-mile trip through the locks. (United Nations)

of the Mexican Plateau. The focus, of course, is Mexico City, a conurbation of over 5 million people; about 300 miles to the northwest, Guadalajara is the country's second urban center at

Areas where over half the population speaks Indian languages

MODIFIED AFTER: WEST & AUGELLI, MIDDLE AMERICA, ITS LANDS AND PEOPLES

Map 7-9 INDIAN LANGUAGES IN USE.

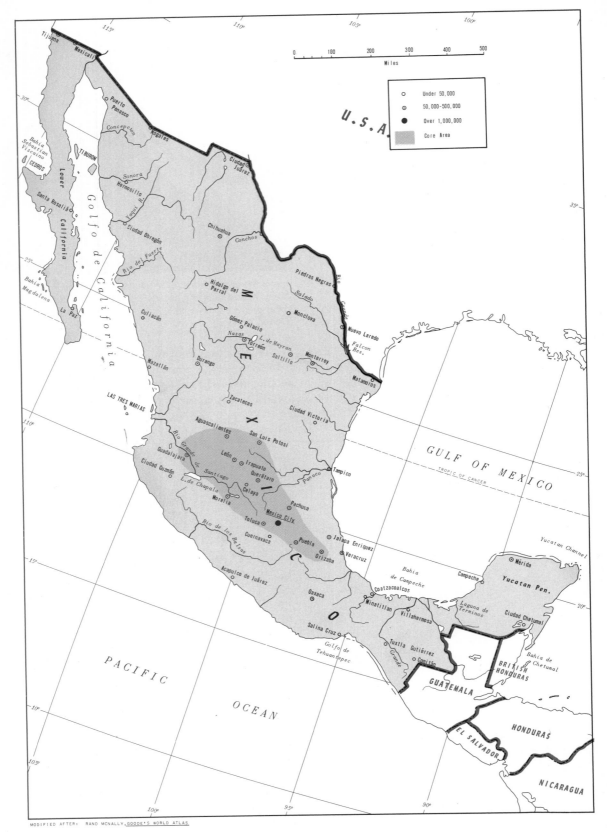

MODIFIED AFTER: RAND MCNALLY, GOODE'S WORLD ATLAS

Map 7-10 MEXICO: LOCATION MAP.

under 1 million. In and around these cities clusters nearly half the Mexican population within a broad belt of relatively high density that extends from Veracruz in the east through Mexico City and Guadalajara to the vicinity of Campostela (Map 7-10). This distribution reflects to a considerable extent the country's climate and relief, for subsistence agriculture is still the livelihood for a majority of the people, and northern Mexico is dry or semiarid, the tropical areas (especially Yucatan) support few people, and excessive relief is a problem elsewhere.

In a country so obviously underdeveloped, the size of Mexico City is remarkable; it alone accounts for some 15 percent of the whole Mexican population and it contributes in a major way to the rather high degree of urbanization that exists in Mexico (estimated at 40 percent of the population in towns of over 5000 in 1970). This is in contrast to other Middle American coun-

tries, whose leading cities, with few exceptions, are of modest size and whose next ranking towns are often quite small. While Mexico City was the headquarters of the colonial viceroyalty of New Spain, its growth was by no means spectacular, and during the first century of independence its development was still not abnormally rapid. But the past half century has witnessed unprecedented expansion, indeed, a transformation of the whole urban scene. In about 1930 Mexico City had just 1 million inhabitants; shortly after 1950 it reached 2 million; by 1965 the figure had climbed to 5 million, and by 1970 it will approach 6 million. True, Mexico City, like Havana until the late 1950's and San Juan, benefits from the tourist trade, and increasingly so; but the major bases for the city's growth appear to lie in its centrality to the country, its function as the location for the administrative headquarters of virtually every mining

An air view of Mexico City, Middle America's largest urban center and the capital of the region's largest national state. (Air France)

and manufacturing concern in Mexico, its position as the capital city and the cultural focus of the state, and its recent emergence as an important industrial center. Mexico City seems to have that same self-perpetuating quality which was referred to in Chapter 3 as marking some of Europe's great cities.

But the majority of Mexicans live in the rural areas, not in the cities, and although some impressive developments have taken place in agriculture, there still are enormous difficulties to be overcome. After achieving independence, Mexico failed for nearly a century to come to grips with the problems of land distribution that were a legacy of the colonial period. In fact, the situation got worse rather than better, and during the 34 years of rule by strongman Porfirio Díaz (1877–1910) matters came to a head. In the first years of the twentieth century the situation was such that 8245 *haciendas* covered about 40 percent of Mexico's whole area; approximately 96 percent of all rural families owned no land whatsoever and these landless people worked as *peones* on the *haciendas*. There was deprivation and hunger, and the few remaining Indian lands and the scattered *ranchos* (smaller private holdings, owned by mestizo or white families) could not produce enough to satisfy the country's needs. Meanwhile, thousands of acres of cultivable land lay idle on the *haciendas,* which occupied just about all the good farmland of the country.

The revolution began in 1910, and set in motion a sequence of events that is still going on today. One of its major objectives was the redistribution of Mexico's land, and a program of expropriation and parcelling out of the *hacienda* was made law by the Constitution of 1917. Since then, about half the cultivated land of Mexico has been redistributed, mostly to peasant communities consisting of 20 families or more. Such communally owned lands are called *ejidos,* and they are an adaption of the old Indian village farm community; the land is owned by the group, and parcels are assigned to each member of the group for cultivation. Not surprisingly, most of the *ejidos* carved out of *haciendas* lie in central and southern Mexico, where Indian traditions of land ownership and cultivation survived and where the adjustments were most successfully made. A few *ejidos* are true collectives in that the members of the group do not farm assigned parcels but work the whole area in return

for a share of the profits. These are located mainly in the oases of the north, where cotton and wheat are grown, and in some of the former plantations of Yucatan.

With such a far-reaching program it is not surprising that productivity temporarily declined; the miracle is that it has been carried off without a major death toll, and that the power of the wealthy land-owning aristocracy could be broken without ruin to the state. Mexico alone among the countries with large Indian populations has made major strides toward solving the land question, although there is still much malnutrition and poverty in the country. But the revolution that began in 1910 did more than that. It resurrected the Indian contribution to Mexican life, and blended Spanish and Indian heritage in the country's social and cultural sphere. It brought to Mexico the distinctiveness that it alone possesses in "Latin" America.

The revolution could change the distribution of land, but it could not change the land itself, nor the methods with which it was cultivated. Corn, beans, and squash continue to form the main subsistence food of most Mexicans, with corn the chief staple and still occupying over half the cultivated land. It used to be more than that; even this proportion suggests that some crop diversification is taking place. But corn is grown all too often where the conditions are not right for it, so that yields are low; wheat might do better, but the people's preference determines the crop, not soil suitability. And if the people's preference has not changed a great deal, neither have farming methods over much of the country. Apart from sections of the densely populated heartland, much of the cultivable land is either too dry or too wet (see Plates 1 and 2 in rear of this volume), and steep slopes make farming difficult and erosion a frequent problem. The Spanish recognized three altitudinal zones in tropical America, namely the hot, usually coastal *Tierra Caliente,* the temperate *Tierra Templada,* from about 2500 to 6000 feet, and the cool *Tierra Fría.* Most of Mexico's subsistence farming occurs on the plateau at *templada* and lower *fría* levels, which, in addition to corn, are good for wheat, rice, barley, and beans. But the pockets of good land in the basins and valleys are surrounded by steep, uncultivable slopes, soils are depleted, and the wooden plow is still in use as it was in colonial times. And along the escarpment bordering the central plateau and in the *caliente* lowlands, slash and burn farming continues as in pre-Spanish times, but with the important difference that the greater number of mouths to be fed is leading to a shortening of the fallow period. It will be a long time before the lot of the Middle American subsistence farmer can materially improve.

Commercial agriculture in Mexico has made major strides in recent decades both with respect to the home market and for export. The greatest productivity is still in the hands of private cultivators, although much of the land has been subdivided into *ejidos*. The central plateau is geared mainly to the domestic market and produces food crops, but in the north, large irrigation projects have been built on the streams coming off the interior highlands. Cotton is cultivated for the home market as well as export (Mexico leads Middle and South America in this product), and for the home market, wheat and winter vegetables are grown. The boom in large-scale farming in the unlikely area of Mexico's arid north is due in large measure to the adoption of American technology in the fields of irrigation and mechanized agriculture. By contrast, the *caliente* and lower *templada* areas of southern Mexico—on both the Gulf and Pacific sides of the country—still produce the range and type of crop they did during colonial times: sugar, coffee, rice, bananas, and other fruits.

The importance of manufacturing in the Mexican economy also is rising. Mexico has a wide range of raw materials, many of which are located in the north and northeast. Quite early, an iron and steel industry was built at Monterrey, using iron ore located near Durango and coking coal from the Sabinas Basin in northern Coahuila. A second plant was later built near the coke, at Monclova. But the majority of Mexico's manufacturing takes place in the cities, and that means that there is a heavy concentration of manufacturing in and around Mexico City. Indeed, by some measures nearly two-thirds of all Mexican industry is located here. This is Mexico's main market and its chief labor pool. The full range of manufacturing is represented here, from textiles (Puebla-Orizaba, for example) to chemicals and from food processing to beer brewing.

Mexico's mining industries are less important today than they once were. The country still exports about one-quarter of the world's silver, but other products now rank higher in terms of the revenues they produce, including lead and zinc; some gold is still produced, and some copper as well. The mines lie scattered throughout northern and north central Mexico, but many of the mines that were important in the colonial period—and near which urban centers of some size developed—have been worked out. On the other hand, a recent success has been the petroleum industry, centered on the Gulf coastal city of Tampico. Production began just after the turn of the century, and it grew, sustained by American and British capital; in 1938 the industry was nationalized. After Venezuela, Mexico is now Middle and South America's largest oil producer, and this source of power has been invaluable to both industry and transportation. Associated with the petroleum deposits are natural gas, another important domestic power source, and sulphur, most of which is sold to the United States. Along the Gulf coast and on the plateau, a number of industries related to petroleum and its by-products have arisen, and these developments are reflected by the coastward expansion of the country's heartland.

Thus Mexico in some ways resembles its Middle American partners, albeit on a larger scale; in other ways it is distinctly different. Foreign investment in the mining industry and the export of half-treated raw materials, as well as a few commodities dominating the export list—these are some of the reminders of underdevelopment. But Mexico is industrializing; it has addressed itself with determination to agrarian reform; it is seeking to integrate all sectors of the population into a true Mexican nation. After a century of struggle and oppression the country has taken long strides toward representative government, progress which will hopefully be secure. In these respects, Mexico leads all of Middle America and much of South America as well.

The Republics. Crowded on the narrow eastern section of the Middle American isthmus are seven countries, sometimes referred to in combination as the Central American Republics. Territorially they are all quite small—only one, Nicaragua, is larger than the Caribbean island of Cuba. Populations range from Guatemala's 4.5 million down to Panama's 1.5 million in the six Hispanic republics, while the sole British territory, Belize (formerly British Honduras) has barely over 100,000 inhabitants. As elsewhere in

Middle America the ethnic composition of the population is varied, with Indian and white minorities and a mestizo majority. The exceptions are in Guatemala, where something like 55 percent of the population remains relatively "pure" Indian (Maya and Quiché) and another 35 percent is mestizo with strong Indian character, and in Belize where two-thirds of the population is black or mulatto, a situation resembling that prevailing in the Caribbean islands. The least racial complexity exists in Costa Rica, where there is a large white majority of Spanish and relatively recent European immigrants; in a population of over 1.5 million less than 5000 Indians are counted and the black population constitutes less than 2 percent of the total.

The narrow land bridge on which the republics are situated consists of a highland belt flanked by coastal lowlands on both the Caribbean and Pacific side; from the earliest times the people have been concentrated in the *templada* zone. Here, tropical temperatures are moderated by elevation, while rainfall is adequate for the cultivation of a variety of crops. As noted earlier, the Middle American highlands are studded with volcanoes, and areas of volcanic soils are scattered throughout the region (individual areas are too small to be outlined on Plate 3). The old Indian agglomerations were located in the fertile parts of the highlands, and the Spanish period confirmed this distribution. Today, the capitals of Guatemala (Guatemala City, 500,000), Honduras (Tegucigalpa, 200,000), El Salvador (San Salvador, 300,000), Nicaragua (Managua, 250,000), and Costa Rica (San Jose, 300,000) all lie in the interior, most of them at several thousand feet elevation; only Panama City (300,000) and Belize (40,000) are coastal capitals in mainland Middle America. (Map 7-11) The size of these cities, in countries whose population averages about 2 to 2.5 million, is a reflection of their primacy, just as was Mexico City in Mexico. On an average the next ranking town is only one-fifth as large (again, just as in Mexico).

The distribution of population in the repub-

MODIFIED AFTER: RAND MCNALLY, GOODE'S WORLD ATLAS

Map 7-11 CENTRAL AMERICA: LOCATION MAP.

A farm family stops for the noon meal near the village of San Andreas Semetaba, Guatemala. Their fields are nearby. He is a member of a buying and marketing cooperative which has sold him seed, fertilizers, and insecticides, and which will buy his crops after the harvest. (Paul Conklin/PIX)

lics, in addition to its concentration in the area's higher sections, also shows greater densities toward the Pacific than toward the Caribbean coastlands (Plate 5). The most densely populated country, and one where the average has some meaning because the people are rather evenly spread, is El Salvador, mainland Middle America's smallest country (8260 square miles) and the only republic without a window on the Caribbean. El Salvador, with Belize and to a lesser degree Panama, is an exception to the rule that people in mainland Middle America are concentrated in the *templada* zone; most of El Salvador is *tierra caliente,* and the majority of its 3 million people are crowded, hundreds to the square mile, in the intermontane plains below 2500 feet above sea level. In Nicaragua, too, the Pacific areas are the most densely populated; the early Indian centers lay near Lakes Managua and Nicaragua and in the adjacent highlands. The frequent volcanic activity in this area is accompanied by the emission of volcanic ash, which settles over the countryside and quickly weathers into fertile soils. This has been happening for uncounted centuries, and the soil here thus has had a natural fertilizing system. The Spaniards centered their colonization in the same area where the Indians were farming these soils, and recently population growth has begun to fill the Pacific coastal zone. By contrast, the

Caribbean coastal lowlands, hot, wet, and leached, support comparatively few people. In the most populous of the republics, Guatemala, the heartland also has long been in the southern highlands, and although the large majority of Costa Rica's population is centered in the central highlands around San Jose, the Pacific lowlands have been the scene of major immigration since banana plantations were established there. Even in Panama there is a strong Pacific concentration: more than half of all Panamanians (and that means over 70 percent of the rural folk) live in the southwestern lowlands and adjoining mountain slopes. Another 25 percent live and work in the Canal Zone, and of the remainder, a majority, many of them descendants of black immigrants from the Caribbean, live on the Caribbean side.

With the single exception of Costa Rica, Middle America's mainland republics face the same problems as Mexico, only more so; they also share many of the difficulties confronting the Caribbean islands. Although efforts are being made to improve the land situation, the colonial legacy of *hacienda* and peon hangs heavily over the area. Earlier we referred to the brevity of life for the peoples of the Caribbean: the most recent United Nations statistics indicate that life expectancy in Guatemala, where nearly one-third of the mainland's people outside Mexico live, is 37 years. It is only 41 years in comparatively well-off Costa Rica, and 51 in El Salvador. Generations of people have seen little change in their way of life and have had even less opportunity to effect any real improvement: dwellings are still made of mud and straw, sanitary facilities have not reached them, schools are overcrowded, hospital facilities are inadequate. As for their livelihood, each year brings a renewed struggle to extract a subsistence, and no hope that the next year will bring something better. One of the staggering contrasts in Middle America is that between the attractive capital city and its immediate surroundings and the desolation of the distant rural areas; another is that between the splendor, style, and refined culture of the families who own the coffee *fincas* and the destitute, rag-wearing peasant.

Coffee—and not bananas—is the most important export crop for all but three of the republics (in Honduras and Panama bananas lead, and in Belize, lumber, naval stores, and chicle). Though introduced as early as the 1780's, major coffee plantings were first made between 1855 and 1860, mostly in Guatemala. This is a *templada* crop that does best on the mountain slopes, and it spread quickly to other countries, especially El Salvador and Costa Rica. But while

coffee gave the republics a valuable source of income, it did little to transform the social situation—indeed, in some ways it made it worse, for it increased the value of landholdings suitable for cash agriculture and entrenched deeper the *finca* system. True, Indian labor that otherwise might never have come into contact with the commercial economy now did so, but to the laborers the returns were something less than the profits that went to the large landowners. Then, during the second half of the nineteenth century, the banana plantation began to make its appearance along the Caribbean coast, and with it came an involvement of American big business in the affairs of Middle America. Black laborers came from Jamaica and elsewhere to take the newly available jobs (as they did when the Panama Canal was dug), and a Caribbean-ward flow of people was stimulated on the mainland. This was especially pronounced in Honduras, where the influx continued even after the banana industry began to decline as a result mainly of disease. In Caribbean Honduras, the banana revolution set in motion a whole train of associated events: not only were health and sanitation conditions improved, adequate housing built, agricultural research pursued, and so on, but the whole region was transformed: ports were laid

The fertile farmland of comparatively prosperous Costa Rica is shown here. (Paul Conklin/PIX)

Large-scale banana cultivation in Middle America. Modern methods of disease-prevention and watering are applied now by an old industry that was a vehicle for the penetration of U. S. influence in Middle America. (Standard Fruit & Steamship Co.)

out, roads built, land cleared for cultivation, and service industries stimulated (such as livestock raising, sugar cultivation, food processing). When the banana industry began to fail, the land that fell vacant and the facilities that became available were in some cases turned over to local farmers, and assistance was given in the introduction of potentially cultivable crops.

The companies that introduced banana plantation agriculture in Middle America—among them United Fruit, Standard Fruit and Steamship, and Boston Fruit—were not, as the preceding might suggest, merely benevolent contributors to better days for Middle America. There was a good deal of interference in domestic affairs, and considerable resentment in what the

Americans sometimes referred to as the "banana republics." But as time went on, positive contributions were indeed made—although the industry has sustained major setbacks. As we noted, disease caused a large-scale decline in the Caribbean area, and the focus shifted to the Pacific coastal areas, which, in contrast to the always-wet Caribbean shores, are seasonally dry. In recent years, however, new, disease-resistant varieties have been developed and reintroduced in the Caribbean zone (which had not lost its locational advantages with reference to the United States market), and there are signs that the old pattern of Caribbean dominance in the banana industry of Middle America is on the way back.

ADDITIONAL READING

A book that appeared quite recently has already taken its place as the standard work on Middle America. This is *Middle America, Its Lands and Peoples* by R. C. West and J. P. Augelli, published by Prentice-Hall, Englewood Cliffs, in 1966. On the precolonial period, see E. Wolf, *Sons of the Shaking Earth,* a Phoenix paper-back of the University of Chicago Press, published in 1959. Also see a volume edited by H. E. Driver, *The Americas on the Eve of Discovery,* a Prentice-Hall publication of 1964, and R. Wauchope (Ed.), *Handbook of Middle American Indians,* University of Texas Press, Austin, 1964. On the problems and potentials of development there are many sources; see E. Staley, *The Future of Underdeveloped Countries,* a Praeger paperback published in 1961, and J. H. Kautsky's *Political Change in Underdeveloped Countries,* published by Wiley in 1962, also in paperback. On plantations, see a publication by the Pan American Union, *Plantation Systems of the New World,* Social Science Monograph No. 7, published in Washington. Another useful work is H. Mitchell's *Caribbean Patterns: A Political and Economic Study of the Contemporary Caribbean,* published in London by Chambers in 1967.

Individual countries and groups of countries have also been discussed by geographers. On Mexico, see D. D. Brand, *Mexico: Land of Sunshine and Shadow,* No. 31 in the Van Nostrand Searchlight Series, and, on the Indian peoples, the chapter entitled "The Indians in Mexico" in *Minorities in the New World: Six Case Studies,* by C. Wagley and M. Harris, published by Columbia University Press in 1958. On Central America see Thorsten V. Kalijarvi's *Central America,* No. 6 in the Van Nostrand Searchlight Series, and on the Caribbean islands, consult D. Lowenthal (Ed.), *The West Indies Federation: Perspectives on a New Nation,* a Columbia University Press publication of 1961. The perspective has since changed, but the book remains of interest. Also see *The West Indian Scene* by G. Etzel Pearcy, No. 26 in the Van Nostrand Searchlight Series. Puerto Rico is discussed by E. P. Hanson in another Searchlight Book, No. 7, entitled *Puerto Rico: Ally for Progress.* On El Salvador, there is a book by D. R. Raynolds, *Rapid Development in Small Economies: The Example of El Salvador,* a Praeger publication of 1967. For information on Guatemala, consult the volume by N. L. Whetten, *Guatemala: the Land and the People,* published by Yale University Press in 1961. On other Middle American countries, consult the appropriate editions of *Focus,* published by the American Geographical Society in New York.

8

"LATIN" SOUTH AMERICA

NO CONTINENT has as familiar a shape as South America, that giant triangle that is connected by Middle America's tenuous land bridge to its sister continent in the north. But South America might be more appropriately called *Southeast* America, for most of it lies not only south, but also east of its northern counterpart. Lima, the capital of Peru, lies farther east than Miami, Florida. Thus South America juts out much farther into the Atlantic Ocean than does North America, and South American coasts lie much closer to Africa and even to southern Europe than coasts of Middle and North America. It is worth looking at this situation with the help of a globe: the distance from London to New York is about the same as that from Lisbon to northeast Brazil. No part of Middle and Caribbean America lies nearer the Iberian Peninsula than does northern Brazil, and the distance from Madrid to Mexico City is about 800 miles greater than that from Lisbon to Rio de Janeiro. And when it comes to Africa, South America is virtually next door; from West Africa to Brazil is about as far as it is from Boston to Denver—well under 2000 miles. As the makeup of the Brazilian population of today still proves, many Africans made the journey.

Lying so far eastward, South America on its western flank faces a much wider Pacific Ocean than does North America. From its west coast to Australia is nearly twice as far as from San Francisco to Japan, and South America has virtually no interaction with the Pacific world of Australasia—not just because of the vast distances, but also because it lies opposite the insular and less populous south, whereas North America faces Japan and the crowded East Asian mainland. The additional factor of culture comes into play, for Australia and New Zealand, South America's nearest Pacific neighbors, share with

269

Map 8-1 PHYSIOGRAPHY OF SOUTH AMERICA.

North America the imprint of English and northwest European culture.

As if to confirm South America's northward and eastward orientation, the western margins of the continent are rimmed by one of the world's great mountain ranges, the Andes, a giant wall that extends from the very southern tip of the triangle to Colombia in the north (Map 8-1). Every plate in the rear of this volume reflects the existence of this mountain chain: in the alignment of the isohyets (lines of equal precipitation, Plate 1), in the elongated area of "H" highland climates (the gray area, Plate 2), in the mountain soils (18–22, Plate 3), and in the zonation of vegetation (Plate 4). And as Plate 6 indicates, South America's largest population agglomerations are positioned along the east and north coasts, overshadowing those of the west. The northern third of South America is dominated by the vast Amazon Basin, which is bounded by plateau areas in Venezuela and Guiana to the north, and by the Mato Grosso ("great woods"—great *bush* might be a better connotation) and the Brazilian Highlands to the south. These Brazilian Highlands are the heart of South America's largest shield (see Chapter 2), and they form a divide between the continent's two great river systems; the Amazon and its tributaries in the north and the Paraguay-Parana river system in the south, which empties into the Plata estuary. Southern South America east of the Andes is dominated by the *Gran Chaco,* an extensive savannaland covering much of Paraguay, southern Bolivia, and northern Argentina, and the famous Pampa, Argentina's rich grain-producing, cattle-ranching area of good soils and excellent pastures.

The Human Sequence

Although the major population clusters and largest countries of South America today lie on the continent's east coast, there was a time when the Andes Mountains contained the most densely peopled and best organized state on the continent. The origins of this, the Inca Empire, are shrouded in mystery and legend, but its Indian inhabitants were descendants of people who came to South America via the Middle American land bridge. For thousands of years before the first Europeans came on the scene, Indian communities and societies had been developing. Probably the earliest modes of life, hunting, fishing, and gathering (these still prevail in parts

of the Amazon Basin today) soon began to be supplemented, in certain areas, by the cultivation of crops. Archaeological excavations have proved that such plants as squash, beans, and a type of cotton were being grown at least 4000 years ago along the north coast of Peru. Gradually technological innovations made their appearance: ceramics, weaving, metallurgy, irrigation. About 1000 years ago a number of regional cultures thrived in the valleys between the Andean mountains, and along the Pacific coast. The llama had been domesticated and was a beast of burden, a source of meat, and a producer of wool. Religions flourished and stimulated architecture and the construction of temples and shrines. Sculpture, painting, and other art forms were practiced. It was over these cultures that the Incas extended their authority from their headquarters of Cuzco in the Cuzco Basin (Peru), beginning late in the twelfth century, to forge the greatest empire in the Americas prior to the arrival of the Europeans.

Nothing to compare with the cultural achievements of the Andean area existed anywhere else in South America. In addition to the Andean civilizations, anthropologists recognize three other groups of peoples: those of the Caribbean

Map 8-2   INDIAN AREAS OF SOUTH AMERICA.

fringe, those of the tropical forest of the Amazon Basin and other lowlands, and those called "marginal," whose habitat lay in the Brazilian Highlands, the headwaters of the Amazon River, and Patagonia (Map 8-2). The people of the Caribbean fringe show signs of influence from the Inca realm, from Middle America, and from the tropical forest as well. They were closest in cultural development to the peoples of the Central Andes, in that they had knowledge of metallurgy (especially in gold and copper), of road and bridge building, of irrigated agriculture, and hillslope terracing; they practiced human sacrifice and built stone temples. But in their settlements they resembled their neighbors in the tropical forest: villages were palisaded, dwellings were made of thatch and branches, and their household furniture, too, was like that of the forest peoples. Quite possibly these cultures of the Caribbean fringe could have been integrated in the Inca realm, but they were rather remote from the main center of Inca organization, and shortly after the Incas made a major northward thrust and established Quito as a second headquarters, the Spaniards interrupted the course of events in the Andes.

The tropical forest cultures, with their river-edge villages, slash-and-burn agriculture (mainly manioc, but some corn and beans as well), hunting, and fishing, were far removed from the Andean civilizations. The numbers of people in this lowland group (which includes mainland Carib and Arawak people, whom we first encountered, ill-fated, in the Caribbean islands), were much smaller than those in the intermontane valleys. Neither was the "marginal" group, consisting mainly of nomadic bands of people who depended largely on hunting, fishing, and gathering, very numerous. It has been estimated that the Caribbean, forest, and "marginal" peoples together constituted only about one-quarter of the continent's total population. Thus, during the Indian era, the center of human organization and development in South America lay very much in the mountainous west. Here the Incas forged an empire and their Andean cohabitants, the Chibchas (of what is today Colombia) also created an advanced civilization.

The Inca Empire. What the Incas achieved in the Andes is a miracle. It is very easy to consider superficially the major environments of South America, and then to dismiss the Amazon lowlands as too hot and humid for the kind of agriculture than can sustain millions, to do the same with the dry lands of the Brazilian northeast, the Brazilian Highlands, and interior Argentina, and to view as incompatible the habitat of the Pampa and the agricultural methods known to the Indian peoples just identified as the "marginal" group. Obviously, then, that leaves the Andes. But was there anything inevitable about the evolution of great civilizations in Andean South America? Was there anything comparable in North America?

When the Inca civilization is compared to that of ancient Mesopotamia, Egypt, the old Asian civilizations, and the Aztecs' Mexica Empire, it quickly becomes clear that this was an unusual achievement. Everywhere else, rivers and waterways provided avenues for interaction and the circulation of goods and ideas; apart from the ready availability of the water itself as an essential ingredient in urbanization and agriculture. One of the great advantages enjoyed by the Aztecs in their early expansion through the Valley of Mexico was the ease of communication and contact in this huge plateau-basin. Later, when the Mexica Empire spread across much more of Middle America, distance and physiographic impediments became problems in the maintenance of the state. But compare these problems to those faced by the Incas, and they will seem mild indeed. Here, the empire was forged out of a series of elongated basins in the high Andes, basins called *altiplanos,* which are created when mountain valleys between parallel and converging ranges are filled with erosional materials, volcanic debris, etc. These *altiplanos* are often separated from each other by some of the world's most rugged terrain, with high, sometimes snowcapped mountains alternating with precipitous canyons. Individual *altiplanos* accommodated regional cultures; the Incas themselves were first established in the intermontane basin of Cuzco. From these headquarters they proceeded, by military conquest, to extend their authority over the peoples of coastal Peru and other *altiplanos*. Their first thrust was southward, and it seems to have occurred toward the close of the fourteenth century. But more impressive than the Incas' military victories was their subsequent capacity to integrate the peoples and regions of the Andean realm into a stable and well-functioning state. All the odds would seem to have been against them; as they progressed farther southward, into northern and central

A view over an *altiplano* in Bolivia. This is the Indian village of Peñas. (Paul Conklin/PIX)

Chile and northwestern Argentina, their domain became ever more elongated, making effective control more and more difficult. The Incas, however, were expert road and bridge builders, colonizers, and administrators. In an incredibly short time they consolidated their southern territory; shortly before the Spanish arrival (1531) they conquered Ecuador and a part of southern Colombia.

The early sixteenth century was a critical period in the empire, for the conquest of Ecuador and nearby areas for the first time placed stress on the existing administrative framework. Until that time, the undisputed center of the state had been Cuzco, but now it was decided that the empire should be divided into two units, a southern one to be ruled from Cuzco, and a northern sector to be centered on Quito. This decision was related to the continuing problem of control over the dissident and rebellious north, and the possibilities of further expansion into central Colombia. Now the empire was beset by a number of internal difficulties: an uncertain frontier in the north, tensions between Cuzco and Quito. And just as the Aztec Empire had been ready for internal revolt when Cortés and his party entered the country, so the Spanish arrival

in western South America happened to coincide with a period of stress within the Inca Empire.

In many respects the Aztec and Inca Empires differed greatly from one another. The Aztecs conquered distant peoples not to make them and their lands a part of the Aztec state, but to establish a system for the payment of tribute and taxes. The Incas, on the other hand, were true colonizers. To every part of the empire they introduced their religion, worship of the sun— not to the extent that local religious practices were forbidden, but to bring the essentials of the state religion to all the provinces. And the Incas engaged in large-scale transfers of population, to settle empty areas, to stabilize rebellious areas, to develop underdeveloped areas. Farming families were moved to lands where soils of good quality lay unused. Inca engineers terraced the hillslopes and built connecting roads. Elaborate arrangements were made for the storage of surpluses and their transportation to parts of the empire where crops might have failed and hunger threatened. Whereas the Aztecs did little to bring their culture to the entire sphere of Aztec influence, the Incas introduced their culture and technology, with benevolent objectives, wherever they ruled. In this respect the Inca state was

The ancient heritage of Andean America. The left photograph shows the impressive buildings constructed by the Incas (see the terraces high on top of the mountain, above the clouds).

The right photograph shows the modern descendants of the Andean Incas, threshing their grain in the same old way against a backdrop of terraces built by their ancestors. (Air France—Paul Conklin/PIX)

much more a real empire than that of the Aztecs. In the outlying Aztec areas, Tenochtitlan was the distant capital of ruthless invaders who demanded tribute and carried people away for purposes of human sacrifice; distant peoples under Inca rule were made subservient to Cuzco's rigorous discipline, but they received much in return.

When it was at its zenith the Inca Empire may have counted as many as 25 million subjects; estimates range from 10 to 30 million. Of course the Incas themselves were in a minority in this huge state, and their position eventually became one of a ruling class in a rigidly class-structured society. The Incas, representatives of the emperor in Cuzco, formed a caste of administrative officials who implemented the decisions of their ruler by organizing all aspects of life in the conquered territories. They saw to it that all harvests were divided between the church, the community, and the individual family; they maintained the public granaries; they recognized and reported the need for investments in improved roads, road maintenance, the terracing of hillslopes, and the layout of irrigation works.

The life of the subjects of the Inca empire was strictly controlled by this huge bureaucracy of Inca administrators, and there was little personal freedom. Farm yields were predetermined, and there was no market economy to speak of: the produce, like the soil on which it was grown, belonged to the state. Marriages were officially arranged, and families could live only where the Inca supervisors would permit. Indeed, the family (as a productive entity within the community) was considered to be the basic unit of administration, and not the individual. Inca rule was amazingly effective, and obedience was the only course for its subjects. Everything was geared to the well-being of the state and its emperor, the earthly representative of the Inca god, the sun. So highly centralized was the state and so complete the subservience of its effectively controlled population that a takeover at the top was enough to gain power over it—as the Spaniards proved in the 1530's.

Of course the peoples over whom the Incas extended their authority already possessed their own cultures, but the effect of Inca administration was to bring new development and much

Two photographs illustrating the detailed workmanship that went into the Inca's buildings. At left is a part of the mountaintop structure shown previously, and at right is a detail of the famous Inca fort, Sacsayhuaman, near the Inca capital of Cuzco, Peru. (Air France— D. Grabitzky/Monkmeyer)

uniformity to these societies. Specifically, they introduced their sun-god religion and built temples of stone in the image of Cuzco, expanded agricultural production by terracing and irrigation techniques and by introducing new crops, domesticated the llama and alpaca and used them not only as beasts of burden but also as producers of wool and hence textiles, and brought a uniform pattern of social life to all parts of the empire. Earlier we commented on the difficulties they faced in the creation of a road network, in view of the mountains and canyons that separated the many *altiplanos;* in agriculture they also faced major problems, for much of their domain lay in the *tierra templada* and *tierra fría,* with water shortages and excessive slopes. They overcame the transport problem through fantastic feats of engineering, building paved roads that were sometimes hewn out of the very bedrock they traversed and crossing canyons with great suspension bridges; similarly they vastly increased the agricultural productivity of their realm by their irrigation and terracing programs. Only in this way could the empire have sustained so many people. As in Middle

America, corn was the staple food, and along with beans and pumpkins it was grown in the lower *templada* zone. In *caliente* areas many tropical fruits were cultivated, as well as sweet potatoes; and in the higher templada zone potatoes (a staple at these levels) and quinoa, a grain crop, were grown. By increasing the arable land, maintaining a tight control over planting and harvesting, and developing efficient gathering and distribution systems the Incas produced something of a population explosion.

The Inca Empire, which had risen to greatness so rapidly (some argue that it may have taken less than a century and not, as others believe, several centuries) disintegrated abruptly with the impact of the Spanish invaders. Perhaps it was the swiftness of its development that contributed to its weakness. But it left behind many social values that have remained a part of Indian life in the Andes to this day—and which continue to contribute to fundamental divisions between the Iberian and Indian populations in this part of South America. The most basic of these relates to the ownership of land and other property. Even before the advent of the Inca Empire, the

Indians of the various regional cultures practiced communal land ownership—if not by the state, then by villages or groups of villages. The Inca period confirmed and even intensified this outlook by rigid state control of all land and resources. Personal wealth simply could not be achieved through the acquisition of land or the control of resources such as minerals or water supplies; not only did the system not permit it, the concept itself was more or less unknown. Even less was prestige associated with land ownership. The Spaniards, we know, held almost precisely opposite views and values. Today, more than four centuries after the fall of the Inca Empire, these conflicting outlooks still divide Iberian and Indian South America.

The Iberian Invaders. In South America as in Middle America, the location of Indian peoples determined to a considerable extent the direction of the thrusts of European invasion. The Inca, like the Maya and the Aztec, had accumulated gold and silver in their headquarters, they possessed productive farmlands, and they constituted a ready labor force. Not long after the defeat of the Mexica Empire of the Aztecs, then, the Spanish conquerors crossed the Panamanian isthmus and sailed down the west coast. Francisco Pizarro on his first journey heard of the existence of the Inca Empire, and after a landfall in 1527 at Tumbes, located on the northern coast of Peru very near the Ecuador boundary, he

returned to Spain to organize a penetration of the Inca realm. He arrived at Tumbes with 183 men and a couple of dozen horses in January, 1531, at a time when the Incas were occupied with problems of royal succession and strife in the northern provinces. The events that followed are well known: less than three years later the party rode, victorious, into Cuzco. Initially the Spaniards kept the structure of the empire intact by permitting the crowning of an emperor who was, in fact, in their power, but soon the land- and gold-hungry invaders were fighting among themselves and the breakdown of the old order began.

The new order that eventually emerged in western and southern South America placed the Indian peoples in serfdom to the Spaniards. Land was alienated into great *haciendas,* taxes were instituted, and a forced labor system was introduced to maximize the profits of exploitation. As in Middle America, the Spanish invaders mostly were people who had little status in Spain's feudal society, but they brought with them the values that prevailed in Iberia: land meant power and prestige, gold and silver meant wealth. Lima, the coastal headquarters of the Spanish conquerors, was founded by Pizarro in 1535 (it is positioned approximately northwest of the Andean settlement of Cuzco, and before long it was one of the richest cities in the world, reflecting the amount of riches being yielded by the ravaged Inca Empire. Soon it was the capital of the viceroyalty of Peru, as the Spanish authorities in Spain began to integrate the new possession in the colonial empire. Later, when Colombia and Venezuela became Spanish-controlled and Spanish settlement progressed in the coastlands of the Plata Estuary (in present-day Argentina and Uruguay), two viceroyalties were added: those of New Granada in the north, and Río de la Plata in the south (Map 8-3).

Thus the major thrust from Spain into South America initially was via the Panama isthmus to Peru. Expeditions set out from Lima to Chile and into Bolivia; Spanish settlers entered the Argentina-Paraguay lowlands; and the Colombia-Venezuela area fell into the Caribbean sphere of Spanish influence; but the early focus of Spanish South America unquestionably was Lima.

Meanwhile, another vanguard of Iberian invasion was penetrating the eastern part of the

A **view** of modern-day Cuzco among the high Peruvian Andes. The cathedral represents the superimposition of an alien culture upon an Indian empire which was soon virtually wiped out. (A. Gregory of Camera Press/PIX)

Map 8-3 COLONIAL UNITS IN SOUTH AMERICA.

continent—the coastlands of present-day Brazil. This area had become a Portuguese sphere of influence almost by default. It was visited by Spanish vessels early in 1500, and later that year the Portuguese navigator Cabral saw the coast on his way to the Far East. The Spaniards did not follow up their contact, and the Portuguese in 1501 sent Amerigo Vespucci with a small fleet to investigate further. After this journey the area remained virtually neglected by the Portuguese for nearly 30 years; they were absorbed in the riches of the East Indies. And the Spaniards, we know, had other interests in the Americas. Additionally, the two countries had agreed, in 1494 by the Treaty of Tordesillas, to recognize a line drawn 370 leagues west of the Cape Verde Islands as the boundary between their New World spheres of influence. This north-south border ran approximately along the line of 50 degrees west longitude (see Map 8-3), thus cutting off a sizable triangle of eastern South America for Portugal's exploitation.

The Africans. As the map shows, the Spaniards got very much the better of the territorial division of the New World—not just quantitatively but, initially at least, qualitatively as well. No rich Indian states could be conquered and looted; no productive agricultural land was under cultivation. As for the Indians, they were comparatively few in number and constituted no usable labor force. It has been estimated that the whole area of present-day Brazil may have been inhabited by about one million people. When finally the Portuguese began to look with renewed interest on their American sphere of influence they turned to the same lucrative business that was sustaining their Spanish rivals in the Caribbean: the plantation cultivation of sugar for the European market. And they found their labor force in the same region, as millions of Africans were brought in slavery to the Brazilian northeast and north coast. Again, estimates of the total number of African forced immigrants vary, but over the centuries the figure probably exceeded 6 million. Today, with the population of Brazil approaching 100 million, more than 11 percent of the people are black and another 30 percent are of mixed African,

Many Africans were forcibly brought to South American shores, principally Brazil, to work as slaves. This population of African descent is still concentrated in the continent's northeast. Here, at Salvador, Brazil, Afro-Brazilians load coffee on a freighter. (Jerry Frank/DPI)

white, and Indian ancestry.

Here, then, is the third major immigration of foreign peoples into South America; with the Portuguese came the Africans. Along the north coast, where the British, Dutch, and French gained footholds between the Portuguese Brazilian and the Spanish Granadan domains in the "three Guianas," smaller numbers of African slaves were also introduced on the sugar fields, especially by the Dutch and British. These coastal areas thus resemble the "Rimland" culture sphere of Middle America to some extent (see Chapter 7). Since the Portuguese had footholds in Africa, many Africans were brought from these areas to Brazil, because no middlemen were needed—no *European* middlemen, at least—to import them. But the Portuguese also acquired slaves taken across the Atlantic by the Dutch, the British, and Spaniards, and others, and it is no easier to reconstruct the African heritage of Brazil than it is to determine that of Caribbean America.

In any comparison between Spanish and Portuguese developments in South America it is important to consider the contrasts between the two societies that prevailed in the Iberian Peninsula itself. Spain still consisted of an amalgam of individual feudal units, some stronger, some weaker, with a rigid authoritarianism and a wide gap between the lords, the landowners, the wealthy on the one hand, and the landless and poor on the other. The Spaniards who came to South and Middle America, especially those who came early, as adventurers and conquerors, were out for personal gain, prestige, and wealth; often, when laws and decisions made in Spain were extended to their conquered lands, there was conflict. In Peru, for example, laws relating to the *economiendas* were so resented by the conquerors that there was nothing less than civil war over them. But Portugal was a somewhat more cohesive and unified state, and in Brazil the role of the individual in these terms was much less pronounced. In fact, in 1532, while King John III was on the Portuguese throne, efforts were made to organize an administration of Brazil which gave 15 coastal stretches of 200 miles each (called captaincies) to individuals favored by the king, in the hope that these people, who had virtually unlimited powers in their territories over men and resources, would set about developing them. Actually, only two of them showed any real progress, and the system failed com-

pletely. Some of the appointed rulers did not even show up in their captaincies. Finally the king abandoned the whole plan and instituted a centralized administration for all of Brazil under a governor general who directly represented the crown. This is not to suggest, of course, that there were no land-hungry adventurers in Portugal, but the situation both at home and in Brazil differed essentially from that in Spain.

A brief look at the map of South America shows that the Treaty of Tordesillas did not succeed in limiting Portuguese colonial territory to the agreed line of 50 degrees west longitude. True, the boundaries between Brazil and its northern and southern neighbors (French Guiana and Uruguay) both reach the ocean very near the 50 degree line, but then the boundaries of the country bend far inland, to include almost the whole Amazon Basin and a good part of the Parana-Paraguay Basin as well. Portugal's push into the interior was reflected by the Treaty of Madrid (1750) whose terms included a westward shift of the 1494 Papal Line. In fact, Brazil alone, with nearly 3.3 million square miles, is only slightly smaller than all the other South American countries combined (3.6 million square miles). In population, too, Brazil has about half the total of all South America; history dealt the Portuguese sphere a fairer hand than any treaties did. This enormous westward thrust was the work of many Brazilian elements—missionaries in search of pastures, explorers in search of quick wealth—but no group did more to achieve it than the so-called *Paulistas,* the settlers of São Paulo. From early in its colonial history São Paulo had been a successful settlement, with thriving plantations and an ever-growing need for labor. The *Paulistas* organized huge expeditions into the interior, seeking Indian slaves, gold, and precious stones, and, incidentally, intent on reducing the influence of Jesuit missionaries over the Indian population there. The *Paulistas'* expeditions ranged far and wide; they reached the Peruvian and Colombian Andes and thus traversed almost the whole continent, in the process establishing a Brazilian sphere of influence extending thousands of miles from their home base. Eventually this sphere became the legitimate territory of Brazil, and it is remarkable how little conflict there was with the Spaniards, who left the trans-Andean area pretty much alone. Only when the *Paulistas* moved southward, in the direction of Uruguay and the Plata

Estuary, was there a Portuguese-Spanish confrontation—a confrontation out of which the buffer state of Uruguay was born.

Isolation

The activities of the *Paulistas* point up some characteristics of South America in the seventeenth and eighteenth centuries that still mark the continent today. They ranged far and wide through a virtually empty interior; today the Amazon Basin and its margins still constitute one of the world's largest, most sparsely populated areas (Plate 5). Indeed, the Indian population of Brazil, which in the sixteenth century was over 1 million, perhaps 1.5 million, is now less than 100,000 and it continues to decline. The *Paulistas* found their Spanish co-colonizers concentrated in areas of maximum productivity (in either gold and other mineral: or arable land and Indian labor), while between the clusters of European conquerors and their Indian serfs lay major physical barriers or large unpopulated tracts. Again the present-day situation is still more or less the same: from Venezuela to Chile and from Argentina to the Guianas core areas of relatively dense population lie separated from each other by desert, mountain range, or tropical forest. The *Paulistas* began their journeys in a coastal area of plantation agriculture whose chief urban center was the town of São Paulo; to this day the major population agglomerations of South America lie along the coast, mainly the east and north coast. Of all the continents only Australia has a population distribution that is more markedly peripheral, but there are only some 13 million people in Australia against nearly 200 million in South America.

But even with 200 million people South America may be described as an underpopulated area, not just in terms of its low total for a continental area of its size, but also in view of the resources available or awaiting development. The continent never drew as large an immigrant European population as did North America. The Iberian Peninsula could not provide the numbers of people that Western and Northwestern Europe did, and colonial policy, especially Spanish policy, had a restrictive effect on the European inflow. The American viceroyalties existed primarily for the purpose of the extraction of riches and the filling of Spanish coffers; in Iberia there was little interest in developing the American lands for their own sake. Only after those who had made America their permanent home and who had a stake there rebelled against Iberian authority did things begin to change, and then very slowly. South America was saddled with the values, the economic outlook, the social attitudes of seventeenth and eighteenth century Iberia—not the best equipment with which to begin the task of forging modern states.

Despite their common cultural heritage (at least so far as their European-Mestizo population is concerned), their adjacent location on the same continent, their common language, and their shared national problems, the states that arose out of South America's Spanish viceroyalties have existed in a considerable degree of isolation from each other. Distance, physiographic barriers, and other factors have played their role in this. Some of these factors had their effect even during the wars for independence, which came after long and often desperate struggles for self-determination by the American-born Spaniards. Spanish military strength was always concentrated at Lima, and those territories that lay farthest from the center of power—Argentina and Chile—were the first to establish their independence from Spain, in 1816 and 1818 respectively. While the famous Argentine general, José de San Martín, led the combined Argentinian and Chilean armed forces to the coast of Peru, another famous South American, Simón Bolivar, was leading the north, New Granada, in its fight for independence. Eventually Bolivar organized an assault on the remaining Spanish forces, fortified still in the Andean mountains; in 1824 two decisive battles ended Spanish power in South America. Thus in little more than a decade the Spanish countries fought themselves free; the significance of their cooperation in this effort can hardly be overstated. But the joint struggle did not produce unity. Nine countries emerged out of the three viceroyalties, including Bolivia, formerly known as Upper Peru and named after Bolivar when it was declared independent in 1825. Bolivar's Colombian Confederacy, which had achieved independence in 1819, broke up in 1831 into Venezuela, Ecuador, and New Granada, which in 1861 was renamed Colombia. Uruguay was temporarily welded to Brazil but it, too, attained separate identity. Paraguay, once a part of Argentina, also appeared on the map as a sovereign state.

It is not difficult to understand why this fragmentation should have taken place: with the

Andes intervening between Argentina and Chile, and the Atacama Desert between Chile and Peru, distances seem even greater than they are and the obstacles to contact are very effective. Thus the countries of South America began to grow apart, separated by sometimes uneasy frontiers. Friction and even wars over Middle and South American boundaries have been frequent occurrences, and a number of boundary disputes remain unsettled to this day. Bolivia, for example, at one time had a direct outlet to the sea, but lost this access in a series of conflicts involving Chile, Peru, and, indirectly, Argentina. Chile and Argentina themselves were long locked in a dispute over their Andean boundary, while Peru and Ecuador both laid claim to the upper Amazon Basin and the town of Iquitos (Peru). But these conflicts are only one manifestation of the divisions that grew in Spanish America. As time went on, the embryo core areas–each of them focused on one of South America's few large urban centers–developed in a greater or lesser degree of isolation. In any case, the contact and cooperation that marked the period of emancipation from Spanish overlordship soon gave way to go-it-alone policies; each of South America's core settlements faced its own peculiar set of problems. As the hostility toward Spain dimmed with time, the Iberian heritage of each country brought about renewed associations with the "motherland," and before long the South American republics had better contacts with Spain than they had with each other.

Brazil attained independence from Portugal at about the same time the Spanish settlements in South America were struggling to end overseas domination, though the sequence of events was quite different. In Brazil, too, there had been revolts against Portuguese control—the first as early as 1789—but the early 1800's, instead of witnessing a steady decline in Portuguese authority, actually brought the Portuguese government (Prince Regent Dom João and a huge entourage) from Lisbon to Rio de Janeiro. Thus Brazil in 1808 was suddenly elevated from colonial status to the seat of empire, and it owed its new position to Napoleon's threat to overrun Portugal, which was allied with the British. At first it seemed that the new era would bring progress and development to Brazil. The Portuguese (like the Spanish) had maintained a trade monopoly in their colonies; now the ports and businesses of Brazil were opened to foreign merchants and European investment was welcomed. British trading companies soon opened offices in the colony, and the whole commercial situation changed. At the same time local industrial development was encouraged, educational institutions were set up, governmental organization in various spheres was streamlined, and for the first time European immigrants from areas other than Iberia began to arrive. In 1815 Brazil's new status was formalized in a decree which made it an equal partner in the United Kingdom of Portugal, Brazil, and the Algarve (European Portugal's south, long under Moslem domination).

Although there was some agitation against the regime based in Rio de Janeiro, notably in the Brazilian northeast, where Pernambuco (now named Recife) was the center of a revolt in 1817, the real causes of Brazilian independence lay in Portugal, not Brazil. Dom João did not return to Lisbon immediately after the departure of the French, and by the time he did, it was in response to a revolution there; the Napoleonic period had left behind it a great deal of dissatisfaction with the *status quo*. Worse, the regime in Lisbon wanted to end the status of equality for Brazil and once again make it a colony. Thus Dom João appointed his son, Dom Pedro, as regent and in 1821 set sail for Portugal. It was to no avail. The national assembly of Portugal was determined to undo Dom João's administrative innovations, and Dom Pedro was ordered to return to Lisbon as well. This he refused to do, and in 1822 he proclaimed Brazil's independence and was crowned emperor. He had overwhelming support from the Brazilian people in this decision; the loyalist Portuguese forces still in the country were forced to return to Lisbon.

Thus Brazil joined its Spanish sister states in achieving sovereignty during the same momentous decade of independence. And although the road to self-determination for Brazil had been a somewhat different one from that of the Spanish republics, the post-independence situation was quite similar, if not in terms of its form of government (Brazil until 1889 remained a monarchy), then at least where its relations with its neighbors were concerned. Brazil, with its enormous area, was in periodic conflict with a number of its neighbors over the boundaries in the upper Amazon Basin, but the most serious confrontation also was the oldest one, involving the lands between the country's southernmost core

(São Paulo) and the Plata Estuary. As early as 1726 Spanish interests on the Plata had founded Montevideo as an outpost against southward Portuguese encroachment, since there was a Spanish settlement of some consequence in Paraguay and the Paraguay River itself served as a secondary access route to the Peruvian Andes; the Spaniards could not afford to see the Plata entry threatened. But neither was the area north of the Plata Estuary effectively settled; there was no gold, there were few Indian villages, and it was not good country for sugar cane. Thus when Brazil became independent it renewed its claim to territory as far as the northern banks of the Plata. Settlers from Portugal and elsewhere in Europe entered the country; many were directed to the empty lands of the south. In turn, the Argentinians supported an armed repulsion of Brazilian encroachment, and only after British intervention was the matter settled by a peace treaty which in 1828 established Uruguay as a republic independent of both Brazil and Argentina. It was, in effect, a Spanish republic, and Brazil continued to seek influence there; in 1865 such intervention brought war with another neighbor, Paraguay.

Thus the post-independence relationships of Brazil to its Spanish-influenced neighbors have been rather similar to the relationships between the individual Spanish republics themselves. Distance, physical barriers, and culture contrasts served to inhibit contact and interaction of a positive kind. Brazil's orientation toward Europe, like that of the republics, remained stronger than its involvement with the states on its own continent.

Plural Societies and Culture Spheres

When we speak of the "orientation" or "interaction" of South American states, it is important to keep in mind just who does the orienting and interacting, for there is a tendency to generalize the complexities of these countries away. The fragmentation of colonial South America into 10 individual states and the nature of their relationships was the work of a small minority of the people in each country. The black people in Brazil at the time of independence had little or no voice in the course of events; the Indians in Peru, numerically vastly in the majority, could only watch as their European conquerors struggled with each other for supremacy. It would not even be true to say that the European mi-

norities *in toto* governed and made policy: it was the wealthy, landholding, upper class element that determined the posture of the state. These were—and in some cases still are—the people who made the quarrels with their neighbors, who turned their backs on wider American unity, and who kept strong the ties with Madrid and Paris (Paris long had been a cultural focus for "Latin" America's well-to-do; many children were sent to French rather than Spanish schools).

So complex and heterogeneous are the societies and cultures of Middle and South America that practically every generalization has to be qualified. Take the one just used: the term *Latin* America. Apart from the obvious exceptions that can be read from the map, such as Jamaica, Guyana, and Surinam, which are clearly not "Latin" countries, it may be improper even to identify some of the Spanish-influenced republics as "Latin" in their cultural milieu. Certainly the white, wealthy upper classes are of Latin European stock, and it is they who have most influence at home and are most visible abroad; they are the politicians and the businessmen, the writers and artists. Their cultural environment is made up of the Spanish language, the Catholic church, the Mediterranean architecture of Middle and South America's cities and towns. These things provide them with a common bond and with strong ties to Iberian Europe. But in the villages of Guatemala and in the mountains of Ecuador and Peru live millions of people to whom the Spanish language is still alien, to whom the white man's religion is another element of deculturation, and to whom decorous Spanish styles of architecture are pretty meaningless when a decent roof and a solid floor are still unattainable luxuries. There is little that is "Latin" about those parts of America. And these are not small areas or insignificant minorities; in Peru, of 13 million people, some 45 percent are Indians, nearly half of Ecuador's population of 5 million is Indian, as is half of Bolivia's 4 million inhabitants. Indeed, if there is a minority involved here it is the "Latin," pure European sector, which in Ecuador, for example, accounts for a mere 10 percent of the total population.

South America, then, is a continent of plural societies, where Indians of different cultures, Europeans from Spain, Portugal, and elsewhere, Africans from the west coast and other parts of tropical Africa, and Asians from India, Java, and

Japan have produced a cultural and economic jigsaw of almost endless variety. Certainly to call this "Latin" America is not very meaningful, but is there a more meaningful way to arrive at a regional differentiation that would represent the continent's cultural and economic spheres even approximately?

One such attempt was made in 1963 by Professor Augelli, who also authored the Rimland—Mainland concept for Middle America. His map (Map 8-4) shows five culture spheres in South America.[1] The first of these, the *Tropical Plantation* region, in many respects resembles the Middle American *Rimland*. It consists of several areas, of which the largest lies along coastal northeast Brazil while four others lie along the Atlantic and Caribbean north coasts of South America. Location, soils, and tropical climates favored plantation crops, especially sugar; the small indigenous population led to the introduction of millions of African slave laborers, whose descendants today continue to dominate in the racial makeup and strongly influence the cultural expression of these areas. Later the plantation economy failed, soils were exhausted, the slavery system was terminated, and the people were largely reduced to poverty and subsistence, conditions which now mark much of the region mapped as tropical plantation.

The second region on Professor Augelli's map, identified as *European–Commercial,* is perhaps the most truly "Latin" part of South America. Along the east coast this region extends from Rio de Janeiro southward, to include southern Brazil, Uruguay, and most of Argentina. It connects across the Andes to the most populous, middle section of Chile on the west coast. Initially this region did not attract many European colonizers (except along the Brazilian and Plata coasts), since there were no precious metals to be found, few Indian people to be exploited, and few opportunities for plantation agriculture. Large-scale European immigration here came relatively late, and the developments that have put this region ahead of all others in South America in economic terms have taken place mostly during the past century. They were based on the favorable conditions prevailing here for commercial agriculture and grazing. Much of the region lies in South America's middle latitudes, outside the

[1] J. P. Augelli, "The Controversial Image of Latin America: a Geographer's View," *Journal of Geography,* Vol. LXII, March 1963, pp. 103–112.

tropics. Rainfall in the Argentine Pampa and in southern Uruguay ranges from 20 to 40 inches (see Plate 1 in the rear of this volume), and the soils are prairie and chernozem (see Plate 3), while the monthly temperature range is from about 45 to 75 degrees and the climate classifies as *C* or warm temperate (see Plate 2). Conditions become cooler and drier to the south and warmer and more moist toward the north, while the ideally flat land on both sides of the Plata estuary gives way, several hundred miles to the west, to the foothills of the Andes. But within a radius of about 350 miles from Buenos Aires the combination of conditions is ideal for grain cultivation (wheat, corn, and alfalfa) and pastoralism (cattle and sheep), and this includes the core areas of Argentina and Uruguay. When the Pampa was opened up and the innovation of meat refrigeration occurred, Europeans (mostly Spaniards, but Italians in large numbers and Poles, French, and Germans as well) flooded to southeastern South America. After Brazil, Argentina with 23 million people is South America's second most populous state, and two-thirds of this population is concentrated in the Pampa core area, where also the cities of Buenos Aires (over 7 million) and Rosario (1 million) are located.

Although Argentina and Uruguay, each with a population that is between 80 and 90 percent "pure" European and with a strong Spanish cultural imprint, constitute the bulk of the European–Commercial region as defined by Professor Augelli, two other areas form part of it: the southern section of Brazil's core area and central Chile, also the core of that country. Southern Brazil shares the temperate grasslands of the Pampa and Uruguay (see Plate 4), and this area has importance as a zone of livestock raising as well as corn growing; Brazil fostered European settlement here at an early stage for strategic reasons. Middle Chile is an old Spanish settlement, and Chile is much more a mestizo country than either Argentina or Uruguay. The one-quarter of the Chilean population that claims pure Spanish ancestry, like 90 percent of all Chilean people, is concentrated in the valleys between the Andes and the coastal ranges, and between the Atacama Desert of the north and the mountainous, forested, sea-indented south. Here, in an area of Mediterranean climate (Plate 2), pastoralism (sheep and cattle) and mixed farming are practiced. In general, then, the

Map 8-4 SOUTH AMERICA: CULTURE SPHERES.

Tropical Plantation

European- Commercial

Indo-Subsistence

Mestizo-Transitional

Undifferentiated

EQUATOR

TROPIC OF CAPRICORN

Longitude West of Greenwich

Miles

0   200   400   600   800

European–Commercial region is economically more advanced than most of the rest of the continent. A commercial economy prevails rather than subsistence modes of life, living standards are higher, literacy percentages are better, transportation networks are superior, and, as Professor Augelli points out in his article, the overall development of this region is well ahead of that in several parts of Europe.

The third region is identified as *Indo–Subsistence,* and it forms an elongated area along the central Andes from southern Colombia to northern Chile and Argentina, an area that coincides approximately with the old Indian empires. There is a small outlier in southern Paraguay. The feudal socioeconomic structure that was established by the Spanish conquerors still survives. The Indian population forms a large, landless peonage living by subsistence or by working on the *haciendas,* far removed from the Spanish culture that forms the primary force in the national life of their country. This region includes some of South America's poorest areas, and what commercial activity there is tends to be in the hands of the white or mestizo.

The fourth region, *Mestizo–Transitional,* surrounds the Indo–Subsistence region, covering coastal and interior Peru and Ecuador, much of Colombia and Venezuela, most of Paraguay, and large parts of Brazil, Argentina, and Chile. This is the zone of mixture between European and Indian (or African, in Brazil, Venezuela, and Colombia). The map thus reminds us that countries like Bolivia, Peru, and Ecuador are dominantly Indian and Mestizo; in Ecuador, for example, these two groups make up nearly 90 percent of the total population, of which a mere 10 percent can be classed as white. The term "transitional" has an economic connotation, because, as Professor Augelli puts it, this region "tends to be less commercial than the European sphere but less subsistent in orientation than dominantly Indian areas."

The fifth region on the map is marked as *Undifferentiated,* because it is difficult to identify its characteristics. Some of the Indian peoples in the interior of the Amazon Basin have remained almost completely isolated from the momentous changes in South America since the days of Columbus, and isolation and lack of change are two of the dominant aspects of this region. The Amazon Basin and the Chilean and Argentinian southwest, also, are sparsely populated and have only very limited economic development; poor transportation and difficult location have contributed to the unchanging nature of this region.

Like the Mainland–Rimland concept as applied to Middle America, the framework of culture spheres described in the preceding pages is obviously a generalization of a very complex situation, and it is easy to think of exceptions to it. But even so, it has much merit, for even in its simplicity it underscores the diversity of South American peoples, cultures, and economies. And Map 8-4, showing the culture spheres superimposed on the boundary framework of the continent, permits us at a glance to recognize regional contrasts within and between South American countries.

Urbanization

As in other parts of the world, the population of South America to an ever greater degree is concentrating itself in the cities and towns. This process has speeded up in recent decades; the growth figures for cities such as São Paulo and Caracas after World War II are much higher than those of previous years. In South America, of course, large numbers of people still live a life of subsistence on farmland that is owned by someone else, and industrial development cannot compare to that of Europe or North America; hence the percentage of the people living in cities and towns is still far below that of those two regions. But the gap is narrowing: in 1925, just one third of South America's population was classed as urbanized, and in 1965 the figure was 50 percent. Contributing to this continuing urban growth are (1) a very high rate of population expansion, (2) the accelerated development of industry and commerce, and (3) the ever-present attractiveness of the cities to people who leave the land to seek a new and better life there.

By any standard of measurement South America has some of the world's greatest cities (Map 8-5). Three of them—São Paulo and Rio de Janeiro (Brazil), and Buenos Aires (Argentina)—rank among the 25 largest cities, and by some calculations the first two of these rank number 6 and 10, respectively. Here once again we run into problems associated with definitions and criteria. If we are to state the population of a city, we need to know where the boundaries of that city are—where the *urban* center ends and the *rural* area begins. But is it realistic to create

Map 8-5 SOUTH AMERICA: LOCATION MAP.

such a division? Often, as one travels from the center of a city toward the surrounding area, the great density of buildings and people in the heart of the urban center give way gradually to more dispersed arrangements, to spacious suburbs, parks, and, almost imperceptibly, to "rural" areas. Where, then, did the city end? Quite probably it is still defined according to an old boundary, which included most of the built-up area many years ago, but across which the city now sprawls. Thus the population of the city itself may be very much less than that of the whole urban agglomeration of which it now forms the core. The "official" city population of Buenos Aires in 1967 was 3,350,000 but the whole urbanized area—the *metropolitan* area—counted between 7 and 8 million people. The most recent census for São Paulo listed 3,164,804 people, but the metropolitan area included nearly 5,400,000 inhabitants. To give an example in North America: the 1960 census population of Boston was 697,197, when in fact the Boston urban area incorporates a total of between 2.5 and 3 million. In the United States a distinction is made between the corporate city and the Standard Metropolitan Area, the latter including the whole conurbation for which the central city forms the focus. When reading population figures for cities or their ranking according to size, it is important to know whether the totals refer to the metropolitan areas or to the cities alone.

South America's three largest cities—each a true metropolis—all lie on or near the east coast. Rio de Janeiro, Brazil's chief port and transportation center and for two centuries its capital, and São Paulo, the rapidly growing manufacturing city, are the twin foci of Brazil's core area. Buenos Aires, a diversified city with manufacturing plants and extensive commercial areas, is the focus not only of the Argentinian Pampa but of the whole country. The next ranking cities are Santiago, Chile (2.5 million) and Lima, Peru (2 million). Both these urban centers lie inland from the west coast and are served by ports that also constitute the next largest cities in their respective countries: Valparaíso (Chile) and Callao (Peru); it is a measure of the primacy of Santiago and Lima that these capitals are about ten times as large as their coastal neighbors. As the headquarters of thriving colonies they got their start; later they became the indus-

trial and commercial as well as the political and cultural centers of their respective states.

Caracas and Bogotá, the capitals of Venezuela and Colombia, also lie away from the coast—in this case, the Caribbean, north coast. Caracas and Venezuela are served by La Guaira, and Bogotá and the Magdalena Valley use chiefly the Caribbean outlet of Barranquilla-Puerto Colombia. These cities, whose populations are somewhat over 1.5 million (Caracas is the larger), illustrate the self-perpetuating character of the primate city. Caracas, at over 3000 feet in elevation the residential choice of the sugar planters, has difficult, mountain-obstructed connections with the interior; but when it was made the capital of Venezuela its superiority over Valencia, a better-positioned competitor, was assured. Caracas reflects the enormous growth rates of urban South America: from 1950 to 1955 its population increased by over 40 percent. Bogotá, also, is in a difficult location from the point of view of intra-Colombian communications, but in this country the several ranges of the Andes form such formidable barriers that a really favorable location does not exist. Again, Bogotá (from the Chibcha, *Bacata*) has retained its primacy since being made the capital of the Viceroyalty of New Granada, although it has competition from Cali, inland from the port of Buenaventura on the Pacific coast, and Medellín, the manufacturing center, each with over 750,000 inhabitants.

The only other South American cities with a population well in excess of one million are Uruguay's capital, Montevideo, and Brazil's industrial center, Belo Horizonte. Montevideo is another east coast city, whose chief functions, in addition to those of politics, are in commerce and transportation; its port handles about 90 percent of all Uruguayan external trade. Thriving, prosperous, and cosmopolitan, Montevideo presents a striking contrast to the urban foci of the three remaining republics: La Paz (Bolivia), Asunción (Paraguay), and Quito (Ecuador), each with well under a half million population. In the case of Ecuador, the coastal city of Guaquil is larger than the mountain capital of Quito; it is the commercial and transport center and the focus of a rich agricultural region of rice and sugar production.

Hierarchy and Function

In the preceding paragraphs, several of the major urban centers of South America were discussed with reference to two of their properties: size and function. The question of city size came up once before, in our discussion of central place theory (Chapter 3), in which a network of cities and towns and their associated functions implied the existence of several discrete size categories. But as we know, there are cities and towns of almost all sizes, from the very largest, such as Tokyo and New York, to small villages that do not even merit the description of "town." If we made a list of all the world's urban places, ranking them by size, we would produce a giant tabulation in which practically every population figure represents one or more cities or towns. There would be very few gaps, and all those urban places would seem to lie along a *continuum;* only near the top of the table would there be sizable breaks. On this basis, we might argue, to speak of any *hierarchy* of cities and towns would be unreasonable, for a hierarchy suggests that urban places tend to cluster strongly in various discrete size categories, leaving gaps or lower incidences between the clusters.

But is it unreasonable? In our everyday language, do we not create a hierarchy when we speak of a metropolis (as we did in referring to Rio de Janeiro and São Paulo), a city, a town, a village, a hamlet? One might even refine this a bit and speak of a megalopolis (a term brought into general use by Professor J. Gottmann to describe the coalescing cluster of metropolises from Boston to Washington, D.C., in the eastern United States of America), a *national* metropolis and a *regional* metropolis, a regional city and a small city, and then, lower on the scale, a town, village, and hamlet. São Paulo–Rio de Janeiro is a megalopolis in the making; Buenos Aires is clearly a national metropolis, as are Santiago, Lima, Caracas, Bogotá, and Montevideo. Lower on the scale, however, identifications become more uncertain. On the basis of size alone, one might group these eight cities together; similarly there is a cluster of cities with populations, in the late 1960's, of between three-quarters of a million and one million, including Rosario (Argentina), Recife (Brazil), Cali and Medellín (Colombia). Again, Guayaquil (Ecuador), Maracaibo (Venezuela), Quito (Ecuador), La Paz (Bolivia), and Asunción (Paraguay) cluster around the half million mark. Are these cities just as "important" as their larger counterparts or have we reached a lower ranking on the scale?

Here may lie a solution of some value. Not infrequently one sees a city described as "important" ("Medellín is an 'important' manufacturing city in Colombia; Rosario is an 'important' center for the corn district of Argentina"). What does this mean other than that these cities possess *functions* and perform services of significance to their surrounding regions and possibly to the world? Now, perhaps, we have something measurable How many, and what kind of, establishments (businesses, stores, banks etc.) do the various urban centers have to offer? On our common-sense scale stated above, the hamlet would have least, the national metropolis most. One would not go to a hamlet to find a restaurant or a grocery store—a hamlet is so small that it may provide only one establishment, and it may be neither of these—but to a village. For a furniture store or a dentist one might look in a town rather than in a village, and for a florist or some specialization (such as women's clothing, men's shoes) one might have to travel to a small city. Thus, on the basis of the urban center's *functions,* it is possible to rank it in a hierarchy with other urban centers, not arbitrarily according to the size of its population, but rather as a reflection of its strength as a place of trade and commerce. Hence the concept of *threshold* emerged: for an urban place to be, say, a town, it must have the functions of a village *plus* certain functions that are not provided by villages.

A number of geographers have devised means to identify the various levels in the urban hierarchy, and systems that work well in Europe or North America may not be applicable at all in Africa or Asia—or in parts of South America. But at least this kind of effort to recognize a hierarchy takes into account something that mere population numbers do not, namely the interaction between the urban center and its trade area (or hinterland, as it is also called). In some cases the procedure has gone the other way: rather than measuring the available goods and services in the urban centers themselves, the surrounding trade areas are identified, and thus the *economic reach* of the cities and towns is determined, and the ranking established accordingly. The greater this economic reach, the higher is the degree of centrality a city possesses.

If it is not easy to discover hierarchies of urban

centers, neither is it simple to classify cities and towns, regardless of size, not according to the *number* of their functions, but according to their *nature*. What, for example, do Valdivia (Chile), Callao (Peru), Santos (Brazil), and La Guaira (Venezuela) have in common? They all have retail establishments, a commercial district, a certain economic reach—but *primarily* they are port cities, places where the transportation and transfer of goods form the leading activity. Another fairly clear-cut class of urban centers is that consisting of mining towns, places that developed and continue to be sustained primarily by mining activities. Not many towns in South America belong in this category, unless the Andean mining "camps" are included with such cities as Cerro de Pasco (Peru), Potosí and Oruro (Bolivia).

These, however, are functions that are easily recognized, but most cities are multifunctional; that is, they possess more than one sphere of economic activity, including retailing, wholesaling, and manufacturing. In these cases it is not so easy to identify the dominant activity, and so their classification becomes a difficult problem.

Is it worth the effort? Urban geographers feel it is; when cities are classified and then represented on maps, the resultant patterns of distribution bring out regional variations that lead to fruitful lines of inquiry and analysis. In 1943, Professor C. D. Harris addressed himself to the classification problem with specific reference to the cities of the United States. He used census data on the employment of labor in the cities in each of the major economic activities. Thus in an urban center he would classify as Manufacturing City, the total employment in manufacturing would have to be at least 74 percent of the total employment in manufacturing, retailing, and wholesaling combined.[2] In addition to Manufacturing Cities (M), Harris recognized Retail Centers (R), Wholesale Centers (W), and Diversified Cities (D), as well as Transportation Centers (T), Mining Towns (S), University (E) and Retirement Towns (X). Later, geographers who sought to modify and improve the Harris classification thought of some other types (such

as Public Administration—Brazil's new capital of Brasilia would fall into this category), and also suggested some alternatives to the percentages Harris had proposed, since these percentages were quite arbitrary. But the map based on the Harris classification was but little changed by these modifications. It showed a great concentration of Manufacturing Cities (M) in the core area of the United States, the Northeast, and a striking north-south belt of (R) and (W) cities through the middle of the country from North Dakota to Texas.

For any such classification as has been proposed, detailed census data on employment are obviously needed, and while such statistics may be available in the United States, they may not be so easily obtained in some other parts of the world. Thus some of our work on urban centers unfortunately is still of limited applicability, and may require further qualification in a worldwide context. In South America, as in the United States, some cities clearly fall into certain categories, as do Brasilia, Callao, and Cerro de Pasco. São Paulo no doubt is a manufacturing city—as we identified it earlier—and Buenos Aires, with its huge conurbation but rather limited industrial development, is Diversified (D). Up the Paraná River, Rosario not only is the prime service center for Argentina's corn district, but also the chief transshipment point for this export commodity, and it may classify as (T). On the other hand, although Montevideo is for all practical purposes Uruguay's only port, handling almost all the exports of wool, meat, and hides and most of the imports (such as motor cars and machinery of various kinds) as well, other activities, in terms of employment, are much more important, and the city classifies as (D).

Structure and Region

If we keep in mind the definition of the regional concept as outlined in Chapter 1, it is obvious that cities have regions. In fact, we realize this without any need for complicated theories: when we go "downtown" we have in mind the central city, where the tall buildings, the shops, offices, and theaters are concentrated. Without perhaps being able to draw any precise distinctions between what is "downtown" and what is not, we still know what we mean. Sometimes there is an assist: in Chicago one goes to "The Loop," in New York, to Manhattan, in São Paulo, to the *Triangulo*, the high-rise heart of that

[2] In fact Harris recognized two types of Manufacturing Cities (M' Subtype and M Subtype). For a complete listing of the criteria, see Chauncy D. Harris, "A Functional Classification of Cities in the United States," *Geographical Review*, Vol. 33, 1943, pp. 86–99.

The *Triangulo* of Sao Paulo, Brazil. (Jerry Frank/DPI)

city, and in Santiago, to the *Centro*. Other urban regions that are easily brought to mind are the suburbs, the industrial areas, the slum areas that so often seem to ring the downtown region, the parks, and so forth.

All these urban regions or zones, of course, lie adjacent to one another and together make up the total city. But how are they arranged? Is there any regularity, any recurrent pattern to the alignment of the various zones of the city, perhaps reflecting certain prevailing processes of urban growth? In other words, do the city's regions constitute the elements of an urban *structure* that can be recognized to exist in every city, perhaps with modifications related to such features as the city site, functional class, and so forth? Again, common sense suggests that this may indeed be the case: one would look for the downtown area in the heart of the city rather than on its periphery, and for sprawling suburbs on the edges of town rather than near the center, where the price of land is highest. The crowded ghettos and slums that often adjoin the downtown shopping areas of United States cities also exist in South America: in São Paulo the Triangulo is encircled by areas of decay and poverty. In Rio de Janeiro, the slums have crept up the slopes of the hills that form part of this city's magnificent site, producing a stark contrast to the high-rise apartments and seafront hotels of the Copacabana District.

In their search for recurrent patterns of city structure, urban geographers have come up with three different theoretical explanations (and, as usual, many modifications of each). The first of these is the *Concentric Zone Theory,* proposed by a sociologist, Professor E. W. Burgess, after a study of the structure of Chicago.[3] According to this idea, the city, in America at least, has five concentric zones. In the center lies the central business district (1) (Map 8-6), with its shops, offices, banks, theaters, and hotels; the larger the city, the more likely is the central business district (CBD) to divide into several districts of specialization, such as a theater district, a financial district, and perhaps an outer zone of wholesaling and warehousing. Surrounding the CBD is the so-called transition zone of residential deterioration, marked by the encroachment of

[3] Among various statements and restatements of the theory, the most accessible is Chapter 2, "Growth of the City," in Robert E. Park, Ernest W. Burgess, and Roderick D. McKenzie (Eds.), *The City,* University of Chicago Press, Chicago, 1925, pp. 47–62.

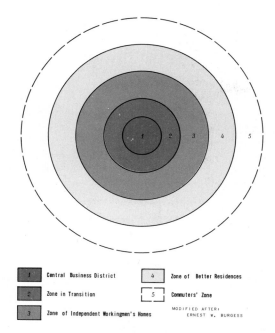

1	Central Business District
2	Zone in Transition
3	Zone of Independent Workingmen's Homes
4	Zone of Better Residences
5	Commuters' Zone

MODIFIED AFTER:
ERNEST W. BURGESS

Map 8-6   THE CONCENTRIC THEORY.

business and light manufacturing (2). Here lie the main slum areas, the areas of urban blight. Farther out lies the "zone of working men's homes" (3), a ring of closely built but adequate residences of the urban labor force. The fourth zone consists of middle-class residences (4), suburban areas that are characterized by greater affluence and spaciousness; at this distance from the CBD, local business districts (shopping centers) make their appearance. The outermost zone is the commuters' zone (5), consisting of communities that are in effect "dormitories" for the CBD, where most of its inhabitants work. Here lie some of the city's high-quality residential areas.

An important feature of the Burgess theory is the mobility of these urban zones. As a city grows, it is postulated that each zone encroaches on and actually invades the next outward zone, so that the whole system is in a constant state of flux. Thus the CBD expands into zone (2), and the suburban areas of zone (5) are located even farther away from the central city. This, certainly, is something we have seen occurring in many cities, but the Burgess idea also has its shortcomings when it is put to the test of real situations. For example, it fails to account for areas of heavy industry in the city; the industry mentioned in zone (2) is light manufacturing. Certainly heavy industrial areas do not form a ring surrounding the whole city; anyone who

knows an American city with big industrial plants knows that these tend to be concentrated in one or two places, related mostly to transportation routes. Also, the wholesaling district, part of zone (1), does not surround the entire retail center of the city, but also focuses on one part of the city's core, again with respect to the major transport arteries that penetrate the urban area. Again, while Burgess' concept of an outward mobility may apply to residential areas, other urban elements, such as railroad yards and stations, port facilities, and manufacturing plants become entrenched in the urban landscape and resist this locational change. In fact, as Chicago itself has shown, this is true even of residential areas, especially where they have a strong national-cultural flavor. Furthermore, the Chicago study dealt with a city located on almost flat land, while many other cities lie in hilly or even mountainous areas where relief has a great deal to do with the positioning of urban zones.

This last point is of considerable relevance to South America, where many large cities—including coastal ones—lie on hilly, and even mountainous, sites. Both Rio de Janeiro and São Paulo have been affected in their development by steep slopes: in the former capital, some of these are steep, rocky, remote from the main roads, and unserved by city service; slums have developed there. In Rio de Janeiro, elevation is less desirable than proximity to the cooling bay and ocean. In São Paulo, on the other hand, the higher elevations soon attracted high-class residential development; this city lies more than 30 straight-line miles inland, and here it is height, not seaward location, that affords coolness. Two cities that resemble Chicago somewhat in terms of their site are Buenos Aires, flat, monotonous, and sprawling endlessly, and Montevideo, smaller, and more compact. Both these cities show elements of concentricity in their city plans, and in both, heavy industry is rather less developed than it is in most large United States cities, so that the pattern is less interrupted. In the case of Buenos Aires, the port stretches for several miles along the Plata Estuary, and the outer suburbs—zone (5)—lie some 20 miles from the city center. In Montevideo, the more concentrated port opens into the Old City—the *Ciudad Vieja*—which still remains the core of the urban area; a newer commercial district has developed adjacent to it, and a transition zone to outer residential areas is clearly visible.

As in every city, the central business districts of South America's cities are served by major transport arteries, which radiate outward and eventually turn into intercity highways. As everyone who has entered a big city by road knows, these major arteries tend to draw certain economic activities to them. One is unlikely to find expensive homes along truck-congested, four-lane roadways; rather, one expects factories. Thus any such artery interrupts Burgess' concentric scheme—so much so that it is possible to argue that the major structural features of a city are pie-shaped rather than concentric. This is the second theoretical explanation of urban structure: the *Sector Theory,* whose author was economist Homer Hoyt (Map 8-7). Though based on data related only to residential land use, the theory has wider application to the entire urban area. Hoyt showed that much could be learned from an analysis of one aspect of residential life: the rent people paid. Following Burgess' theory, we would conclude that rent consistently increases with distance from the CBD, since the quality of housing is lowest in zone (2) and highest in zones (4) and (5). Hoyt, however, concluded that a low rent area could extend all the way from the center of the city to the outer edge. Similarly, high-rent "sectors" could be recognized in the cities he studied, flanked by intermediate-rent areas. Thus, a high-rent sector might extend along a very fast, convenient route into the city center (train or superhighway), but where communications were somewhat less efficient this would be reflected by intermediate-quality residential areas.[4]

Looking at all the elements of the urban structure, and not just residential areas, a pie-shaped arrangement is often clearly present. Industrial plants may be concentrated along one or several arteries that connect the city to raw material supply areas and markets. A low-rent residential sector may adjoin this industrial complex; a high-quality residential sector may fan outward from the central city in another part of town. Major transport lines may be bunched so closely together along a certain belt that a transport sector can be recognized. In South America, a sectorial arrangement is favored in such cities as Bogotá (Colombia) and Santiago (Chile), as well as Rio de Janeiro, by the topog-

[4]The theory is stated in H. Hoyt, *The Structure and Growth of Residential Neighborhoods in American Cities,* United States Federal Housing Administration, Washington, D.C., 1939.

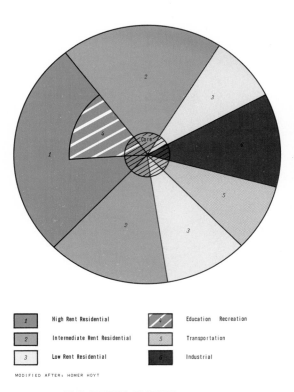

High Rent Residential	Education Recreation
Intermediate Rent Residential	Transportation
Low Rent Residential	Industrial

MODIFIED AFTER: HOMER HOYT

Map 8-7 THE SECTOR THEORY.

raphy of the site. Rio de Janeiro is limited by the Atlantic Ocean and the bay on two sides (the south and east); to the west, the twin, 3000 foot peaks of Corcovado and Tijuco inhibit both urban growth and land communications. Thus Rio's three major routes to the interior all leave the congested city in a northwesterly direction, and the urban structure has a sectorial rather than a concentric character.

The third general explanation for the structure of cities is the *Multiple Nuclei Theory,* an idea first alluded to by McKenzie but formulated most effectively by Professors C. D. Harris and E. L. Ullman.[5] Once again, the concept arises from something we can observe every day as we live in or visit cities: the CBD is losing its dominant position as the undisputed nucleus of the city, and it is getting competition from other, vigorously growing business centers that constitute secondary nuclei. Professors Harris and Ullman point out that a city, especially a large city, may grow *not* around just one focus, but around several nuclei. These multiple nuclei include not only the CBD and subsidiary busi-

[5]Chauncy D. Harris and Edward L. Ullman, "The Nature of Cities", *Annals of the American Academy of Political and Social Science,* Vol. 242, p. 14.

1	Central Business District
2	Wholesale, Light Manufacturing
3	Low-class Residential
4	Medium-class Residential
5	High-class Residential
6	Heavy Manufacturing
7	Outlying Business District
8	Residential Suburb
9	Industrial Suburb

AFTER: HARRIS AND ULLMAN

Map 8-8 THE MULTIPLE NUCLEI CONCEPT.

ness areas such as large shopping centers, but any other element—a university, a port, a government complex—that can stimulate urban growth. In their diagram (Map 8-8) they suggest schematically what a city might look like, with none of the regularity of the Burgess and Hoyt models. One thinks immediately of Los Angeles, but nuclear development of this sort is occurring in most of the world's larger cities. In South and Middle America, the Spaniards initially laid out their cities and towns around a central square, dominated by a church or cathedral and flanked by imposing government buildings. Lima's *Plaza de Armas,* Montevideo's *Plaza de la Constitución,* and Buenos Aires' *Plaza de Mayo* are famed examples of this; streets led outward from these central squares at right angles, enclosing square or rectangular city blocks. In earlier days the *plaza* did indeed form the hub of the city, surrounded as it was by shopping streets and arcades. But modern times have seen the emergence of newer commercial districts, both in the city center and in outlying areas, and the original *plaza* often remains but a link with the past.

South America's largest cities, of course, are more than the commercial and industrial centers of their respective countries. Cities such as Lima, Santiago, and Bogotá are also the headquarters of government, the seats of important universities, and the cultural foci of national states. As such, they generally fit very well the Harris-Ullman model of urban structure. Santiago, Chile's capital, has a well-defined commercial core (the *Centro*), a distinct cluster of government buildings, an eastward, upslope development of higher-quality residential areas, and a western zone of industrial activity that is flanked by lower-quality residential areas and slum development. Although one can recognize a sector arrangement in this, especially in the manufacturing zone that is expanding along the artery leading southward to San Bernardo and to southern Chile, each of these functional zones also has its individual nucleus. As the city continues to grow and diversify, these nuclei will emerge ever more strongly as cohesive units within the whole urban framework.

The Republics: Regional Groupings

In the preceding discussion, the republics of South America have been grouped in the broadest possible way, that is, with reference to their Iberian heritage. Thus we referred to the Spanish-influenced countries on the one hand, and to Portuguese Brazil on the other, even though the republics and Brazil have a great deal in common. Certainly Brazil by itself constitutes a region in South America on several grounds: in terms of sheer size and population it ranks in a class by itself, and it is the eminent non-Andean country of the continent. It incorporates almost all of the drainage basin of South America's greatest river, as well as the whole eastern upland of the continent. Closer than any other South American country to Europe and Africa, it has received more immigrants from both. Never densely populated by Indian peoples, it now has fewer Indian inhabitants than most of its neighbors. No South American country exemplifies to a greater degree the peripheral location of population that is a characteristic of the continent as a whole.

Can the other countries be grouped regionally? Some geographers have suggested that South America's "Caribbean" countries, Venezuela and Colombia, together constitute a regional unit (others have added Guyana, Surinam, and

Map 8-9 CARIBBEAN SOUTH AMERICA.

MODIFIED AFTER: RAND MCNALLY, GOODE'S WORLD ATLAS

Andes and Spurs

Highlands

"THE GUIANAS"

FRENCH GUIANA

Cayenne

SURINAM

Paramaribo

GUYANA

Georgetown

GUIANA HIGHLANDS

BRAZIL

Amazon

Orinoco R.

Cerro Bolívar

V E N E Z U E L A

Caracas

La Guaira

Valencia

Barquisimeto

VENEZUELAN HIGHLANDS

TRINIDAD

BONAIRE

CURACAO

ARUBA

SA. NEVADA DE MERIDA

San Cristóbal

L. Maracaibo

Maracaibo

Barranquilla

Puerto Colombia

Cartagena

PANAMA

COASTAL RANGE

CORDILLERA OCCIDENTAL

Buenaventura

Cali

Cauca R.

Medellín

Manizales

Magdalena R.

CORDILLERA CENTRAL

CORDILLERA ORIENTAL

Bogotá

C O L O M B I A

ECUADOR

P E R U

A N D E S

Longitude West of Greenwich

Miles

500 400 300 200 100 0

EQUATOR

294

French Guiana); in fact, these countries are sometimes included in discussions of the economic geography of *Middle* America! On the basis of their Indian heritage and present population the republics of Ecuador, Peru, Bolivia, and Paraguay can be viewed as constituting a regional grouping. And on the grounds of mid-latitude location and European-commercial development (see p. 283), Argentina and its neighbors Chile and Uruguay also have some joint regional identity.

"Caribbean" South America. As another look at the map of South American culture spheres (Map 8-4) confirms, the countries of the north coast have something in common other than their coastal location: each has an area of "tropical plantation," signifying early European plantation development, the arrival of black labor, and the absorption of this African element in the population matrix. Not only did black workers arrive: many thousands of Asians (Indians) came to South America's northern shores as contract laborers. In the Republic of Guyana, which has a population of well under one million, more than half the inhabitants are of East Indian descent, and another one-third are of African ancestry. In Surinam, also, about one-third of the population is black, and in addition to some 70,000 East Indians, the Dutch brought Javanese to this dependency, who now number about 60,000.

The pattern is familiar: in the absence of large local labor sources, the colonialists turned to slavery and indentured workers to serve their lucrative plantations. But between Venezuela and Colombia on the one hand and the "three Guianas" on the other there is this difference: in the former, the population center of gravity soon moved into the interior, and the plantation phase was followed by a totally different economy, while in the Guianas, the plantation economy still dominates. Again, the map gives us a clue, for in Guyana, Surinam, and French Guiana the "plantation" region gives way immediately to the "undifferentiated" zone that marks the Amazon interior of South America, while both Venezuela and Colombia have large regions characterized as "mestizo-transitional." Apart from large-scale bauxite deposits that provide a mainstay in the export economies of Guyana and Surinam, there is little economic alternative to subsistence for the people of these small countries. A measure of the contrast with Colombia is the fact that some 75 percent of Colombia's over 20 million people are mestizo (and only 5 percent are black), while a mere 12 percent of Guyana's population is categorized as mestizo and mulatto. And while most of Colombia's inhabitants live in the interior highlands, the concentration of population in Guyana, Surinam, and French Guiana is still heavily coastal. In Venezuela and Colombia, new products have long since replaced the old plantation crops, but in the Guianas, they still rank high on a short list of exports.

Venezuela and Colombia have what the Guianas lack (Map 8-9). Their territories and populations are much larger, their physiographies more varied, their economic opportunities greater. Each has a share of the Andes Mountains (Colombia's is the larger), and each produces oil from an adjacent joint reserve that ranks among the world's leading fields (here Venezuela is the major beneficiary). Much of what is important in Venezuela is concentrated in the northern and western parts of the country, where the Venezuelan Highlands form a spur of the Andes system. Most of Venezuela's 10 million people are concentrated in the highlands, which include the capital, Caracas, its early rival, Valencia, the commercial and industrial center of Barquisimeto, and San Cristobal, near the Colombian border. The Venezuelan Highlands are flanked by the Maracaibo Lowlands and Lake Maracaibo to the northwest, and by a region of savanna country (see Plate 2) called the *Llanos* to the southeast. The Maracaibo Lowland, once a disease-infested, sparsely peopled coastland, is today one of the world's leading oil-producing areas; much of the oil is drawn from reserves that lie beneath the shallow waters of the lake itself. Actually, "Lake" Maracaibo is a misnomer, for the "lake" is open to the ocean and is in fact a gulf with a very narrow entry. Venezuela's second city, Maracaibo with over a half million people, is the boom center of the oil industry, which has transformed the Venezuelan economy; in the 1960's over 90 percent of the country's annual exports by value were crude oil and petroleum products. Large refineries on the Dutch islands of Curacao and Aruba for many years have refined the Venezuelan crude oil prior to transportation to United States and European markets, but the capacity of local Venezuelan refineries is steadily increasing.

The *Llanos,* on the south side of the Venezuelan Highlands, along with the Guiana Highlands in the country's southeast, are two of those areas that contribute to South America's image as "underpopulated" and "awaiting development." Although the Llanos does share in Venezuela's oil boom (reserves have been discovered here), the agricultural potential of these savannas—and of the *templada* areas of the Guiana Highlands— has hardly begun to be realized. There are good opportunities for a major pastoral industry in the Llanos, and for commercial agriculture in the Guiana Highlands. Of course, the Llanos and Guiana Highlands are Venezuela's interior regions, and transportation remains a problem. The discovery of rich iron ores on the northern flanks of the Guiana Highlands (chiefly near Cerro Bolivar) has begun to integrate one part of this region with the rest of Venezuela. A railroad was constructed to the Orinoco River, and from there the ores are shipped directly to the steel plants of the coastal eastern United States. Ciudad Guayana, less than a decade old, has over 100,000 people. The Guri Dam on the Orinoco has been put into service and will soon supply all Venezuela with electricity. Two new, paved roads link this eastern frontier with Caracas.

Colombia, too, has a vast area of Llanos, covering about 60 percent of the country, and in Colombia as in Venezuela this is comparatively empty land, outside the national sphere, far less productive than it could be. Eastern Colombia consists of the headwaters of the Orinoco and the Amazon Rivers, and it lies partly under savanna and partly under rainforest conditions (Plates 2 and 4); in recent years the Colombian government has begun in earnest to promote settlement east of the Andes. But it will be a long time before any part of eastern Colombia matches the Andean part of the country, or even the Caribbean Lowlands of the north. In these regions live the vast majority of all Colombians, and here lie the major cities and productive areas.

Western Colombia is dominated by mountains, but there is some regularity to this mountainous topography. In very broad terms there are four parallel mountain ranges, generally aligned north-south, separated by wide valleys. The westernmost of these four ranges is a coastal belt, less continuous and lower than the other three. These latter are the real Andean mountain chains of Colombia, for in this country the Andes separates into three ranges: the Eastern, Central, and Western Cordillera. The valleys between the Andean Cordillera open into the Caribbean Lowland, where two of Colombia's important ports (Barranquilla and Cartagena) are located.

Colombia's population of over 20 million consists of a set of clusters (there are different ways of identifying them, but in any case there are more than a dozen), some of them in the Caribbean Lowlands, others in the valleys between the great Cordillera, and still others in the intermontane basins within the Cordillera themselves. This was so even before the Spanish colonizers arrived: the Chibcha civilization existed in the intermontane basins of the Eastern Cordillera. The capital city, Bogotá, was founded in one of the major basins in this same range, at an elevation of 8700 feet. For centuries, the Magdalena Valley between the Cordillera Oriental and the Cordillera Central was one link in a cross-continental communication route that began in Argentina and ended at the port of Cartagena, and Bogotá benefited by its position adjacent to this route. Today the Magdalena Valley is still Colombia's major transport route, but Bogotá's connections with much of Colombia still remain quite tenuous, and to a considerable degree the population clusters of this country, like those of Venezuela, exist in isolation from each other.

Colombia's physiographic variety is matched by its demographic diversity, as Map 8-4 indicates. In the south it has a major cluster of Indian inhabitants, and the country begins to resemble its southern neighbors. In the north vestiges remain of the plantation period and the African population it brought. Bogotá is a great "Latin" cultural headquarters whose influence goes beyond the country's borders. In the Cauca Valley (between the Cordillera Central and Occidental), Cali is the urban commercial focus for a *hacienda* district where sugar, cacao, and some tobacco are grown. Farther to the north, the Cauca River flows through a region comprising the Departments (Provinces) of Antioquia and Caldas, whose urban focus is the manufacturing city of Medellín, but whose greater importance to the Colombian economy lies in a large production of coffee. With its extensive *templada* areas along the Andean slopes, Colombia is one of the world's largest producers of coffee, and in Antioquia-Caldas it is grown on

Programs are now in effect to assist the Indian sector of South America's population in a variety of ways. Here, members of an Indian farmers' cooperative in Colombia get their land ready for a potato crop, after having harvested grain. (United Nations)

small farms by a remarkably unmixed European population cluster; elsewhere it is produced on the large estates that are so common in Iberian America.

Coffee and oil are Colombia's two leading exports, with coffee accounting for about 60 percent by value and oil about 15 percent. Colombia's oil fields are extensions of Venezuela's reserves, and the oil is piped to its Caribbean ports. The importance of Colombia's window on the Caribbean can be read from the map; two major and several minor ports handle goods brought by land and water from near the southern boundaries of the country, which in total far overshadow the volume of goods transferred at the lone Pacific coast port of Buenaventura. Buenaventura, however, does hold the position of Colombia's leading single seaport.

Venezuela and Colombia, then, share a number of characteristics, though in varying degrees. Both have a marked clustering of population, and the *templada* areas are the more densely peopled sections. They share a relatively empty interior, where development is possible. Both countries depend on a single product for the bulk of their export revenues, though the coffee industry of Colombia involves many more Colom-

bian producers than the Venezuelan petroleum industry involves local Venezuelan workers; the petroleum industry is to a larger extent a foreign-dominated enterprise. At the same time, it must be remembered that in both Colombia and Venezuela the majority of the people live an agricultural subsistence, and that the social and economic inequities common to most of Iberian America prevail here as elsewhere.

"Indian" South America. The second regional grouping of South American states includes Peru, Ecuador, Bolivia, and Paraguay. This is a contiguous group of countries, including South America's only two landlocked entities. Map 8-4 indicates the common Indo-Subsistence sphere extending along the Andes Mountains and a similar area in southern Paraguay. These are the countries of South America that have large Indian sectors in their populations. Nearly half the people of Peru, numbering 13 million, are of Indian stock, and in Ecuador and Bolivia, too, the figure is near 50 percent. In Paraguay it may be over 60 percent, but all these percentages are only approximate, since it is often impossible to distinguish between Indian and "mixed" people of strong Indian character. But there are other similarities among the four coun-

tries considered here. Their incomes are low, they are comparatively unproductive, and, unhappily, they exemplify the grinding poverty of the landless peonage, a problem that looms large in the future of Ibero-America. These, too, are Iberian South America's least urbanized coun-tries; the capitals of three of the four states have under a half million inhabitants and only Lima, the capital of Peru, ranks with Bogotá and Santiago as a major-scale urban center.

In terms of territory as well as population Peru is the largest of the four. Its half million

MODIFIED AFTER: KEYSER LAUTENSACH, ATLAS ZUR ERDKUNDE

Map 8–10 INDIAN SOUTH AMERICA.

square miles divide both physiographically and culturally into three regions: (1) the desert coast, the European-Mestizo region, (2) the Andes Mountains or Sierra, the Indian region, and (3) the eastern slopes and the Montaña, the sparsely populated Indian-Mestizo interior (Map 8-10). It is symptomatic of the cultural division still prevailing in Peru that the capital, Lima, is located not in a populous basin of the Andes, but in the coastal zone. In Mexico, Colombia, and even Ecuador the Spaniards placed their headquarters in the populous Indian interior, bringing their culture to the Indian sphere and promoting intercultural contact. Yet here they avoided the greatest of the Indian empires and chose a site some 8 miles inland from a suitable anchorage (now the port of Callao). The city, we know, soon came to reflect in its splendor the riches pillaged from the Inca interior, but eventually the Andean sources of wealth were exhausted, the empire destroyed, and the Indian population reduced to serfdom on alienated lands. From an economic point of view the Spanish choice of a coastal headquarters proved to be a sound one, for the coastal region has become commercially the most productive part of the country. A thriving fishing industry based on the cool, productive waters of the Humboldt (Peru) current offshore contributes a quarter of all exports by value. Irrigated agriculture in some 40 oases distributed all along the coast produces cotton, sugar, rice, vegetables, fruits and some wheat; the cotton and sugar are important export products, and the other crops are grown mostly for the domestic market. Some of the oases also form a link in a transhumance process involving cattle, which are driven up the Andean slopes when pasture is available there and brought to the oases at other times. It is not impossible that here, too, the Spaniards were heirs to earlier Indian systems of irrigated agriculture. Ideally, the oases benefit from alluvial soils swept down from the Andes by rivers which annually supply sufficient water for their cultivation, but the fact is that many of the rivers that flow off the Andean escarpment never manage to reach the sea. Hence water must be drawn from beneath the surface, and this is done by an elaborate system of underground channels similar to those used in the Middle East. It may be that the Spaniards themselves contributed this innovation to the Peruvian desert (having learned it from the Moors), but possibly the coastal Indian peoples had developed this technique themselves. In any event, Indian irrigated agriculture of the oases was very intensive before the Spanish colonizers arrived, and the ruins of ancient cities suggest that a major civilization may have thrived here.

The Andean region occupies about one-third of the country, and here are concentrated the majority of the country's Indian peoples, most of them Quechua speaking. But despite the fact that this is one-third of Peru and nearly one-half of its people, the political influence of this region is slight and its economic contribution, the mines excepted, only minor. In the valleys and intermontane basins the Indian people are either concentrated in isolated villages around which they practice a precarious subsistence agriculture; in the more favorably located and more fertile areas they are tenants, peons, on white- or mestizo-owned haciendas. Such is the situation, for example, around the town of Huancayo, where the haciendas produce much of Peru's wheat crop, and around the old Inca center of Cuzco. Thus the vast majority of Peru's Indians remain mired in a poverty from which there is no escape, though it must be said that where the government has attempted to initiate reform, it has not always met with a favorable response: the impact of the injustices of centuries cannot be swept away overnight with a gesture of redress, however magnanimous. Most of the Indian people never in their lifetime get an adequate daily calorie intake or a balanced diet of any sort; the wheat produced around Huancayo, for instance, is sent to Lima's European market and would be too expensive for the Indians themselves to buy. Potatoes (which can be grown to 14,000 feet), barley, and corn are among the subsistence crops, and in the high *altiplanos* the Indians graze their llamas, alpacas, and cattle and sheep. But from the point of view of any direct contribution to the national product, the Indians of the Andes are virtually excluded; such products as the wheat and tea for the local market and *templada*-grown coffee for the export trade almost invariably come from alienated lands. So, of course, do the major products that are derived from the Sierra: copper, silver, lead, and several other associated minerals that are taken from mines in a number of districts, of which the one centered on Cerro de Pasco is the chief.

In all Andean countries of South America,

The Indian population of South America has been on the sidelines as far as the region's increasing prosperity is concerned. This photograph shows the house of an Indian family in the Barriada district of Comas, near Lima, Peru. It consists of straw mats on a bamboo frame, has no conveniences or modern facilities. (Paul Conklin/PIX)

transportation is a major problem, and Peru is no exception. Of the country's three regions, the east—the eastern slopes of the Andes and the Amazon-drained, rainforest-covered Montaña—feels isolation the most. A look at the map of permanent (as opposed to seasonal) routes, including railroads, shows how disconnected Peru's regions still are: however marvellous an engineering feat, the railroad that connects Lima and the coast to Cerro de Pasco and Huancayo in the Andes does not even begin to join the country's east and west. Peru has no longitudinal railroad; in fact, it has no railroad network in the true sense of that term. Road connections to the Montaña, which, though sparsely populated, constitutes fully half of the territory of Peru, are also poor. The focus of the region, in fact, is a town that looks east rather than west: Iquitos, which can be reached by ocean-going vessels sailing up the Amazon River. Iquitos grew rapidly during the Amazon wild rubber boom, and then declined; but now it is growing again, reflecting Peruvian plans to begin development of the east. Meanwhile, traders plying the navigable rivers above Iquitos collect such products as chicle, nuts, rubber, herbs, special cabinet

woods, and small quantities of coffee and cotton.

Ecuador, smallest of the four republics, on the map looks to be just a corner of Peru. But that would be a misrepresentation. Ecuador has the full range of regional contrasts: it has a coastal belt, an Andean zone which may be narrow (under 150 miles) but by no means of lower elevation than elsewhere, and an *Oriente*—an eastern region that is just as empty and just as undeveloped as that of Peru.

As in Peru, the majority of the people of Ecuador are concentrated in the Andean intermontane basins and valleys, and the most productive region is the coastal belt. But here the similarities end: Ecuador's coastal region consists of a belt of hills interrupted by lowland areas, of which the most important one lies in the south between the hills and the Andes, drained by the Guayas River and its tributaries. The largest city and commercial center of the country (but not the capital), Guayaquil, forms the focus for this area. Ecuador's lowland west is not desert country—it is a fertile tropical lowland not bedeviled by excessive rainfall (see Plates 2 and 4). Neither is Ecuador's west really a "European" region as is Peru's, for the white element in the total pop-

On the shores of Lake Titicaca, Peru. While tending her sheep, this woman spins wool by hand. At the water's edge is a canoe woven from totora weeds. (United Nations)

ulation of 5.5 million is a mere 10 percent, in part engaged in administration and hacienda ownership in the interior, where most of the 50 percent who are Indian also live. Of the remainder, over 10 percent are black and mulatto, and the rest are mixed Indian-white, many with strong Indian ancestry. The products of this region, too, differ from those of Peru: Ecuador is the world's top banana exporter, and bananas account for 50 percent of its exports by value; small farms owned by black and mulatto Ecuadorians and located in the north, in the hinterland of Esmeraldas, contribute to this total, as do farms on the eastern and northern margins of the Guayas Lowland. Cacao (12 percent) is another lowland crop, and coffee (22 percent) is grown on the coastal hill-slopes as well as in the Andean *templada* areas. Cotton and rice are also cultivated, cattle can be raised, and in recent years the production of petroleum from Santa Elena has helped reduce the quantity of imports of this commodity required.

Thus Ecuador is not a poor country, and the coastal region, especially in recent years, has seen vigorous development. But the Andean interior, where the white and mestizo administrators and hacienda owners are outnumbered by their Indian countrymen to the tune of three or four to one, is a different story—or rather, a story similar to that of other Andean regions. Quito, the capital city, lies in one of the several highland basins in which the Andean population is clustered. Its functions remain primarily administrative; there is not enough productivity in the Andean region to stimulate commercial and industrial development. The Ecuadorian Andes do differ from those of Peru in that they are without known major mineral deposits. Despite the completion of a railroad linking Quito to Guayaquil on the coast, the interior of Ecuador remains isolated and economically comparatively inert.

From Ecuador southward through Peru, the Andes Mountains broaden until, in Bolivia, they reach a width of some 450 miles. In both the Cordillera Oriental and the Cordillera Occidental, elevations in excess of 20,000 feet are recorded; between these two great ranges lies the Altiplano proper. On the boundary between Peru and Bolivia, fresh-water Lake Titicaca lies at over 12,500 feet. Here, in the west, lies the heart of modern Bolivia; here, too, lay one of the centers of Inca civilization—and, indeed, of pre-Inca cultures. Bolivia's capital, La Paz, is an Altiplano city; at 12,000 feet this urban center, still of well under a half million people, is the world's highest city. It is Lake Titicaca that helps make the Altiplano livable, for its large body of water ameliorates the coldness on the plateau in its vicinity, and the surrounding lands are cultivable. Grains can be grown in the Titicaca Basin to the amazing elevation of 12,800 feet, and have been for centuries; to this day the Titicaca area, in Peru as well as Bolivia, is a major cluster of subsistence-farming Indians.

The population figures will prove that Bolivia is an Indian country, for not only can 50 percent—perhaps more—of its four million inhabitants be classed as of pure Indian descent, but the 35 percent mestizo sector is of strong Indian inheritance. But modern Bolivia is the product of the European impact, an impact that has passed some of the Indian population clusters by. Of course the Bolivian Indian no more escaped the loss of his land than did his Peruvian or Ecuadorian counterpart, especially east of the Altiplano. What made the richest Europeans in Bolivia wealthy, however, was not land, but minerals. The town of Potosí in the Eastern Cordillera became a legend for the immense deposits of silver nearby; copper, zinc, and several alloys also were discovered. Most recently Bolivia's tin deposits, among the richest in the world, have yielded some two thirds of the country's annual export income. Prior to nationalization of the tin industry,

"... the fabulous profits from the mines at Potosí, Uncia, and Oruro went to three families, Patiño, Aramayo, and Hochschild, who controlled more than 80 percent of the country's tin output, and invested almost nothing in the form of wages, dividends, railroads, highways, factories, farms, or other *local* wealth-producing ventures. The income of Patiño alone was greater than the entire federal budget of Bolivia."[6]

Bolivia has had a turbulent history. Apart from internal struggles for power, the country lost its seacoast in a disastrous conflict with Chile, lost its territory of Acre to Brazil in a dispute involving the rubber boom in the Amazon Basin, and then lost 55,000 square miles of Gran Chaco territory to Paraguay in the war of 1932–1935. By far the most critical was the loss of its outlet to the sea; although Bolivia has rail connections to the Chilean ports of Arica and Antofagasta, it is permanently disadvantaged in its landlocked situation. Since the Cordillera Occidental and the Altiplano form the country's inhospitable western margins, one might suppose that Bolivia would look eastward, and that its *Oriente* might be somewhat better developed than that of Peru or Ecuador, but such is not the case. The densest settlement clusters occur in the valleys and basins of the Eastern Cordillera, where also the mestizo sector is stronger than elsewhere in the country. Cochabamba, Bolivia's second city, lies in a basin that forms the country's largest concentration of settlement; Sucre, the legal capital, lies

A **view** over La Paz, the capital of Bolivia, and its mountainous surroundings. (United Nations)

[6] Raymond E. Crist, "Bolivia," *Focus,* American Geographical Society, Vol. XVI, No. 7, March 1966, p. 4.

in another. Here, of course, lie the chief agricultural districts of the country, between the barren Altiplano to the west and the savannas to the east.

Paraguay is the only non-Andean country in this group, but it is no less Indian. Of some 2.5 million people, perhaps 60 percent are Indian or, by other definitions, mestizo with so strong an Indian element that any white ancestry is almost totally submerged. Although Spanish is the country's official language, Guaraní is more commonly spoken. By any measure, Paraguay is the poorest of the four countries of "Indian" South America, although it does have opportunities for pastoral and agricultural industries that have thus far gone unrealized: Plates 1 and 2 show a rainfall of about 40 inches in the east and a *Cfa*, humid subtropical climate in the same area (the west, rather drier, is BSh—steppe); Plate 3 indicates that the soils are similar to those that yield well in Argentina. Yet it is in the south and to a lesser extent in the east where most of the people are living. What has held Paraguay back?

Isolation must be one of the answers—the country's landlocked position. Paraguay's exports, in their small quantities, must be exported via Buenos Aires, a long river haul from the Paraguayan capital of Asunción. Meat (dried and canned), timber (sold to Uruguay and Argentina), oilseeds, quebracho extract (for tanning leather), cotton, and some tobacco reach foreign markets. Grazing is the most important commercial activity, but the cattle generally do not compare well to those of Argentina. With respect to its dominantly subsistence economy, Paraguay resembles the Andean Indian clusters of Peru, Ecuador, and Bolivia—with this difference, that the possibilities of attainable alternatives exist here.

Mid-Latitude South America. Several justifications exist for the grouping of South America's three southern countries, Argentina with its neighbors Chile and Uruguay, in one region. One of these relates, once again, to Map 8-4, on which all of Uruguay, most of Argentina, and the most populous part of Chile are all identified as "European-Commercial." And indeed, more than 90 percent of the total population of these three countries is either European or mestizo, and their economic development is quite unlike that in "Indian" South America, just discussed. These are not "underdeveloped" countries; they

do include areas that are awaiting development, but economically these three states are quite advanced. And, on a more direct basis, the three countries lie almost completely outside the tropics; with their temperate climates and mid-latitude agricultural products they constitute a rather distinct unit in South America.

By far the largest in terms of both population and territory is Argentina, whose 1.1 million square miles and 25 million people rank second only to Brazil in South America. Argentina has four major physiographic regions and a great deal of physical variety within its boundaries, but the vast majority of the population— three-quarters of it—is concentrated in one of these, the Pampa. Plate 6 indicates the degree of clustering of Argentina's inhabitants on the land and in the cities of the Pampa and the relative emptiness of the other three regions, the scrub-forest Chaco in the north, the mountainous Andes in the west (along whose crestlines the boundary with Chile runs), and the arid plateaus of Patagonia south of the Río Colorado (Map 8-11).

Some of the circumstances that led to the rapid development of the Pampa were discussed earlier (pp. 283 to 285); Argentina is the product of the last hundred years. It was during the second half of the nineteenth century, when the great grasslands of the world were being opened up (including those of the United States, Russia, Australia, and South Africa), that the economy of the long-dormant Pampa began to emerge. The food needs of industrializing Europe grew by leaps and bounds, and contributions of the industrial revolution—railroads, more efficient ocean transport, refrigerator ships, and agricultural machinery—helped make large-scale commercial meat and grain production in the Pampa not only feasible but very profitable. Large haciendas were laid out, farmed by tenant workers, who would clear the virgin soil and plant it with wheat and alfalfa, harvesting the wheat and leaving the alfalfa as pasture for livestock. Thus the Pampa was soon brought into production, and railroads radiated ever farther outward from Buenos Aires; today Argentina has South America's densest railroad network, and the once-stagnating city is now one of the world's largest. Yet even today the Pampa has hardly begun to fulfill its productive potential. More efficient and intensive agricultural practices could probably double the Pampa's yields; here

BOLIVIA

BRAZIL

PARAGUAY

La Paz

Arica

DESERT

NORTH

TROPIC OF CAPRICORN

Antofagasta

CHILE

Chuquicamata

Asuncion

ATACAMA

ANDES

Tucuman

CHACO

Resistencia

Corrientes

El Tofo

SLOPES

NORTH

Cordoba

Santa Fe

MIDDLE

San Juan

URUGUAY

CHILE

Valparaiso

Mendoza

Rosario

Santiago

San Rafael

Buenos Aires

RANGE

Montevideo

ARGENTINA

La Plata

Plata Estuary

Concepcion

LEBU PEN

COAST

SOUTH

P A M P A

MIDDLE

CHILE

VALLEY OF CHILE

Rio Colorado

Bahia Blanca

CHILE

CHILOE

ATLANTIC OCEAN

PATAGONIA

Comodoro
Rivadavia

Gulf of
San Jorge

SOUTH

CHILE

FALKLAND ISL. (Br.)

Main Railroads

0   100   200   300   400   500

Miles

TIERRA DEL FUEGO

MODIFIED AFTER: KEYSER LAUTENSACH, ATLAS ZUR ERDKUNDE

Map 8–11 MID-LATITUDE SOUTH AMERICA (CHILE–ARGENTINA–URUGUAY).

again South America's low man-resource ratio is exemplified.

Over the decades, the Pampa has developed several areas of specialization, of which one, the corn district around Rosario, was mentioned earlier. As we would expect, a zone of vegetable and fruit production has become established near the huge conurbation of Buenos Aires. In the southeast is the most exclusively pastoral district, where beef cattle and sheep (for both mutton and wool) are raised. To the west, northwest, and southwest, wheat becomes the important commercial crop, but half the land still remains devoted to grazing. Among the exports, meat usually leads by value, followed by cereals, wool, and some relatively minor exports such as vegetable oils and oilseeds, hides and skins, and quebracho extract. In 1967, as a result of a rare event, a rather low wheat harvest, Argentina was forced to import for home use some 150,000 tons of Spanish and Bulgarian wheat.

Europe, which buys more than half of Argentina's exports, has contributed in several ways to the development of the country's commercial and industrial base. British investment and know-how helped build the extensive railroad network as well as the port facilities and some industrial establishments; from Germany and Italy came hundreds of thousands of immigrants. Today, the United States is Argentina's leading individual trade partner, but in combination the European countries are still more important. Argentina's wealth and vigor are reflected in its fast-growing cities; depending on the criteria used, nearly two-thirds of the population may be classified as urbanized, which is an exceptionally high figure for South America. Well over a quarter of all Argentinians live in the conurbation of Greater Buenos Aires alone, and here, too, are most of the industries, many of them managed by Italians, Spaniards, and other recent immigrants. The Cordoba area, too, is a focus of industrial growth. The majority of the manufacturing in the major cities is the processing of Pampa products—the canning of meat, milling of flour, treatment of hides, processing of wool—and, of course, the production of consumer goods for the domestic market. Of all the wage earners in the country, between one in six and one in seven is engaged in manufacturing, which is another indication of Argentina's advanced economic position.

Argentina, nevertheless, is yet another country

The Argentine Pampa has been Argentine's breadbasket and economic mainstay. This picture shows *gauchos* herding cattle into paddocks at the Balcarce Station for weighing and inoculation. (United Nations)

in South America whose population shows a high degree of clustering and a strong peripheral location. By the widest of definitions, the Pampa region covers only a little over 20 percent of Argentina's area, and with three-quarters of the people here, the rest of the country cannot be densely populated. Outside the Pampa, pastoralism is an almost universal pursuit (except, of course, in the high Andes), but the quality of the cattle is much lower than in the Pampa. In dry Patagonia, sheep are raised. Some of the distant areas are of actual and potential significance to the country: an oilfield is in production near Comodoro Rivadavia, on the San Jorge Gulf, and in the far north, in the Chaco, Argentina may share with Paraguay a larger oil reserve that is yet to be brought into production. Yerba maté, a local tea, is produced in the northeast (Paraguay has a small output of this product as well), and the quebracho extract that figures in both Argentina's and Paraguay's exports comes from the valley of the Paraguay—Paraná River. In addition, the streams that flow eastward off the Andes provide opportunities for irrigation. At Tucumán (300,000), Argentina's major sugar-producing district developed in response to a unique set of physical circumstances and a rapidly growing market in the Pampa cities, which the railroad made less remote; at Mendoza (315,000), and San Juan to the north and San Rafael to the south, vineyards and fruit orchards have been established. But despite these sizable Andean outposts, effective Argentina still remains the area within a radius of 350 miles from Buenos Aires.

Uruguay, unlike Argentina or Chile, is compact, small, and quite densely populated. The buffer state of old has become a fairly prosperous agricultural country, in effect a smaller-scale Pampa; Plates 1 to 4 show the similarity of physical conditions on the two sides of the Plata Estuary. Montevideo, the coastal capital with nearly 1.5 million residents, contains well over 40 percent of the country's population, and from here railroads and roads radiate out into the productive agricultural interior.

In the immediate vicinity of Montevideo lies Uruguay's major farming area, and here vegetables and fruits are produced for the city, as well as wheat and fodder crops. Just about all of the rest of the country is used for grazing sheep and cattle; wool constitutes half of the annual exports by value, and meat about a quarter. Of course Uruguay is a small country, and its 72,000 square miles, less even than Guyana, do not leave much room for population clustering. But it is nevertheless a special quality of the land area of Uruguay that it is rather evenly peopled right up to the boundaries with Brazil and Argentina. No interior city has even 10 percent of the population of Montevideo. Of all the countries of South America, also, Uruguay is the most truly "European," without the racial minorities that mark even Argentina and Chile but with a sizable non-Spanish European component in its population. Undoubtedly these conditions have been helpful in the creation here of something unusual in Middle and South America: a real and fairly stable parliamentary democracy. Political change, such as the 1967 governmental reorganization from a nine-man National Council to a single-executive presidency, has long gone on without violent conflict, although Uruguay has not escaped the problems of economic mismanagement so common in Middle and South America, and which in 1966 led to an inflation rate of 70 percent.

For 2500 miles between the crestline of the Andes and the coastline of the Pacific lies a narrow strip of land that is the Republic of Chile. On the average just 100 miles wide (and only rarely over 200), Chile is the textbook example of what political geographers call an "elongated" state, whose shape is one that tends to contribute to external political, internal administrative, and general economic problems. In the case of Chile, the Andes Mountains do form a barrier to encroachment from the east, and the sea constitutes an avenue of north-south communication; history has shown the country to be well able to cope with its northern rivals, Bolivia and Peru.

As Plates 2 and 6 as well as Map 8-4 indicate, Chile is a three-region country. About 90 percent of Chile's 10 million people are concentrated in what is called Middle Chile, where Santiago, the capital and largest city, and Valparaíso, the second city and chief port, are located. Middle Chile coincides largely with the zone of Mediterranean (Csb) climate on Plate 2, although it also includes a small northern part of the Marine West Coast (Cfb) area that prevails in the south. North of Middle Chile lies the Atacama Desert, wider and colder (BWk instead of BWh) than the coastal desert of Peru. South of Middle Chile the coast is broken by a number of fjords and islands, the topography is mountainous, and the climate, wet and cool near the shore, soon turns drier and colder against the Andean interior. South of the latitude of the island of Chiloé no permanent land transport routes of any kind exist, and hardly any settlement.

Chile is a mestizo country with a larger Indian minority than has Argentina (8 percent) and a much smaller "pure" European sector (about 25 percent). Map 8-4 suggests a culture-sphere breakdown for the country that involves three major regions: a mestizo north, a "European–Commercial" zone in southern Middle Chile, and an "undifferentiated" south. In addition, a small Indo–Subsistence region in the northern Andes is shared with Argentina and Bolivia. The Indian element in the two-thirds of the Chilean population that is mestizo largely came from the million or so Indians who lived in Middle Chile, on one of the frontiers of the Inca Empire, when the Spaniards arrived. The early Spaniards naturally landed between the deserts of the north and the fjords and forests of the south; moving inland they found in Middle Chile a valley located longitudinally between the Andes to the east and a coastal range of lesser proportions to the west. This, the Central Valley of Chile, was destined to become the core area of the modern state. In the customary fashion, land was granted in haciendas and other units to a relatively small number of the Europeans, but for a long time little agricultural development took place, the environmental similarities between the Central

Valley and parts of Iberia notwithstanding. Chile's economy really began to emerge in the nineteenth century, when agricultural productivity in the valley increased, German immigration opened up the southern part of Middle Chile, and nitrates and copper, along with several other minerals, began to be mined in northern Chile.

But despite the absence of a large, landless Indian class, Chile nevertheless had—and continues to have, though in decreasing numbers—its tenant farmers on the haciendas and estates. Though they are not *peons* in the strict sense of the word, in that there is no debt bondage in Chile, these people, the *rotos,* are little if any better off than their counterparts in other South American countries, and there always were few means of escape from the system. The army provided one of these means, and after service in the Atacama campaign in the late nineteenth century several thousand of these people were settled on land in the open areas of southern Middle Chile. Another means was to leave the land and head for the cities, there to seek work; this migration still goes on. In recent decades the land situation has become a major political issue in Chile, and gradually the large landholdings are being converted into smaller farms. In 1967 the long-debated Agrarian Reform Law, permitting government expropriation of large haciendas, was passed.

Some regional differences exist between northern and southern middle Chile. Northern middle Chile, the land of the hacienda and of Mediterranean climate with its dry summer season, is an area of irrigated crops that include wheat, corn, vegetables, grapes, and other Mediterranean products; some are grown without irrigation. Livestock raising and fodder crops still take up much of the productive land, and agricultural methods are not particularly efficient; the region could yield a far greater volume of food crops than it does. Southern middle Chile, into which immigrants from both the north and from Europe (Germany especially) have pushed, is a moister area where cattle raising predominates but where wheat and other food crops, including potatoes, are also cultivated.

Few of Chile's agricultural products reach external markets, however. Some specialized items such as wine and raisins are exported, but Chile remains a net importer of foods. Thus, despite the fact that 9 out of 10 Chileans live in what has been defined here as middle Chile, it is the north, the Atacama region, that provides most of the country's revenues. First, the mining of nitrates, of which the Atacama includes the world's largest exploitable deposits, provided the country's economic mainstay, but the industry declined after the discovery of synthetic production methods (about the time of World War I). Subsequently, copper became the chief export; it occurs in several places, but the main concentration lies on the eastern margin of the Atacama near the town of Chuquicamata, not far from the port of Antofagasta. At present copper in various forms—some of it as pure bars, some as a concentrate refined at Chuquicamata, and some raw ore—constitutes about 70 percent by value of Chile's exports. Of course the country is especially vulnerable to fluctuations of prices on the world market for this product.

Copper has been mined for over a century; iron ore was found in the southern Atacama Desert at El Tofo. Most of the ores are sent to the United States for conversion, as is most of the copper, and American investment in the mining industries of Chile is heavy, which is another sensitive political issue. The Chileans are seeking to reduce this foreign influence, and have bought a controlling share of the El Teniente copper mine and refining complex. Near Concepción, Chile is developing an iron and steel plant with a projected million-ton capacity, using El Tofo iron ore and low-grade Lebu Peninsula coal. The iron ore travels some 500 miles by coastwise transportation from the Atacama to the south coast.

Thus Chile, though a mid-latitude South American country, differs substantially from its neighbors east of the Andes. As Map 8-4 suggests, southern middle Chile most resembles commercial-agricultural Argentina and Uruguay, but as a whole the country's ethnic, social, and economic condition sets it considerably apart.

Brazil. By itself, Brazil constitutes a region in Ibero-America. It is almost as large as all the rest of the South American republics combined, and its population, approaching 100 million, is larger than that of the rest of the entire continent. Brazil has common boundaries with all but two of the republics of South America (Ecuador and Chile), yet it nowhere touches the Andes. Its Indian population is a mere 1 to 1.5 million; on the other hand, there are more people of

African descent in Brazil than in all other South American countries together. And of course Brazil is Portuguese-influenced and not Spanish America.

The most striking spatial feature of Brazil is the high degree of concentration of human activity along and near its east coast, the rapid decline of population densities toward the interior, and the virtual emptiness of those distant areas that the *Paulistas* struggled so hard to secure

(see Plate 6). And while the low population of the vast Amazon Basin (which constitutes the northern half of Brazil) can perhaps be explained in terms of the environmental problems and the lack of alternatives it presents, the same is not true for the highlands that form the south. Here Brazil possesses a huge area for future settlement and development; the country could sustain more than double its present population. And the Brazilians know it and have shown their

Map 8-12 BRAZILIAN REGIONS.

desire to get away from their coastal orientation and peripheral position in their country. They built a new capital city, from scratch, 700 miles inland from the old capital of Rio de Janeiro, in a newly laid-out Federal District. They called it Brasilia, and the winning architectural design in a national competition showed the city, wings spread like a giant airplane, pointed toward the Brazilian interior. In 1960, the capital functions were transferred amid expressions of determination that Brazil's 'empty heart' would remain empty no longer.

But of course the process will take much time, although extensions of the railroad networks focused upon São Paulo and Rio de Janeiro can be seen reaching out into the Mato Grosso interior; whether the Amazon area will ever become integrated is doubtful (Map 8-12). The Amazon Basin is the world's largest region of tropical rainforest (Plates 2 and 4), and in the dense *selva* (jungle) communication is reasonably easy only along the waterways, the Amazon itself and its numerous tributaries. The Amazon is navigable for most of its great length—well beyond Brazil's western borders in any case, but it had only a brief period of commercial significance during the second half of the nineteenth and the first years of the twentieth centuries, when a demand for wild rubber arose and the Amazon Basin was the world's leading source. In the heart of the Basin, the city of Manaos experienced a boom as the chief point of transfer of this commodity from river boats to ocean-going ships. Ornate public buildings and private homes in the city are a reminder of this brief period; Caruso sang on the opening night of the magnificent opera house. Even during its time of prosperity, however, the rubber industry was beset by the problems of the Amazon Basin: the wild rubber trees were scattered through the forest, local labor was scarce and inefficient, and rubber-gathering became impossible not far inland from the river courses—it was simply not possible to maintain the organization for it. With the advent of the rubber plantation, the wild-rubber industry of Brazil declined precipitously, and attempts to establish rubber plantations in the region failed; conditions in the Amazon Basin were not suitable. Thus ended the moment of promise and prominence for the region, which on Map 8-4 is very appropriately designated as "undifferentiated." It remains the domain of the widely

South America's forgotten ancestors: the Indians of the Amazon. Decimated and isolated, these people number only in the hundreds of thousands now. Here a group of forest-dwellers make use of a pond near the Amazon River. (Paris-Match/Pictorial Parade)

scattered Brazilian Indian tribes, who have little or no contact with the country's government; only along the Amazon, from its mouth to Manaos (which actually lies a few miles up the Rio Negro) do settlements of a permanent nature justify the ribbon of mestizo influence shown on the map.

When it comes to the maltreatment of Indians and the alienation of their land, it is usually the Spanish-American countries that receive criticism; little is heard about Brazil. Unfortunately the record shows that despite their small (and declining) numbers, Brazil's Indian peoples have suffered perhaps even more severely than their Andean counterparts. In March, 1968, Brazil's Ministry of the Interior shocked the nation by reporting that, since World War II, thousands of Indians had been systematically exterminated by white men for the acquisition of farm lands and mineral properties. Brazil's Indians are not confined to the Amazon Basin; Indian groups live also on the grassland plains of the southern interior, where the land is more desirable to the prospective white settler. In June, 1968, newspapers around the world told the story of the fate of Indian peoples in Brazil; in the United States the report was carried by the *Los Angeles Times-Washington Post* News Service under the authorship of Dom Bonafede. Figueirado Correia, who headed the Ministry's investigating commission, was quoted as follows:[7]

". . . the Cinta-Larga (Wide Belt) tribe of Mato Grosso was massacred by dynamite thrown from low-flying aircraft; those who survived were hunted down and machine-gunned. Another tribe, the Tapaiuna, was wiped out by arsenic planted in 'gifts' of sugar . . . members of the Pataxo tribe in the state of Bahia were eliminated by inoculating them with smallpox. Others were slain by infecting the rivers in which they bathed with deadly bacteria.

". . . the overriding purpose of this inhuman treatment was to keep the Indians in subjugation, to control their use as slave laborers, and to get rid of them so as to take over their lands.

". . . the decrease in the population of Indian tribes was reflected in recently-released statistics comparing their numbers 20 years ago to today: Munducurus, 19,000 to 1,200; Nhambiquaras, 10,000 to 1,000; Carajas, 4,000 to 600; Xocrens, 800 to 200; Cinta-Largas, 10,000 to 400."

To say, then, that "the Indians of Brazil live an isolated existence in the rainforests of the Amazon Basin," as we sometimes do in our textbooks, is rather misleading; true, the forests of the Amazon and the unattractiveness of the Basin in general have to some extent shielded those Indian tribes and clans who live there, but there are tens of thousands of Indians elsewhere who have been less fortunate.

The Amazon Basin covers most of the northern half of Brazil; the southern half of the country is occupied by the Brazilian Highlands, generally a plateau surface between 1000 and 2500 feet above sealevel. This rather uneven plateau is tilted from east to west, that is, it reaches its greatest elevations very near the Atlantic coast, and from there declines steadily toward the Amazon Basin, under whose sediments it eventually disappears. This means, of course, that there is a considerable escarpment leading down from the eastern margins of the plateau to the coast. This escarpment—locally called the Great Escarpment, just as a similar terraced scarp is called in Africa—is especially well defined south of the coastal town of Vitória; in the immediate hinterland of Rio de Janeiro

[7]Quoted from the *Lansing State Journal*, Sunday, June 2, 1968, p. E-4.

lie some mountains over 9000 feet high within 100 miles of the coast. North of Vitória the coastal plain widens and the escarpment is less pronounced. In the far north, between the mouth of the Amazon River and the bend in the coast, the coastal plain has its greatest width, and here the first sugar plantations, which gave Brazil its initial economic stimulus, were located. Today, the northeastern area still contributes about one-third of the country's sugar and cotton crops, but the center of economic activity in Brazil has moved far to the south.

Brazil, like the United States, has a federal organization, and the country is divided into 21 states and seven federal territories. The three largest states lie in the sparsely peopled interior: Amazonas, Mato Grosso, and Pará, and the smallest lie in the northeast, the scene of early European settlement. As Plate 6 shows, there is still a fairly heavy concentration of population in the northeast, where the African element is quite strong. But the two most populous sets of states are those of the east (Rio de Janeiro and its District, Minas Gerais, Espirito Santo, Bahia, and Sergipe) and south (São Paulo, Paraná, Santa Catarina, and Rio Grande do Sul). The southern states correspond approximately to that part of Brazil mapped as "European–Commercial" on Map 8-4, and in this part of the country the European element in the population is strongest (Map 8-13). Here, on the narrow coastal plain, in the eastern highlands, and beyond the highlands to the south, lie Brazil's most productive agricultural areas—and Brazil, for all its size and its great cities, is still primarily an agricultural country. In the southern part of the highlands, in the state of São Paulo and adjacent areas, lies an area that is particularly well suited for the cultivation of the crop that is today Brazil's leading source of foreign revenues: coffee. Here is grown about half of the world's coffee, and for well over a century the area has been developing and expanding, attracting hundreds of thousands of immigrants, not only from Portugal, but from Italy, Germany, and several other European countries as well. Settlers came even from Japan: today one percent of Brazil's population is of Japanese ancestry. São Paulo, which lies at the focus of a dense network of transport lines that extends into the coffee-producing hinterland, has emerged as the country's largest and fastest-growing urban center; it is the gathering point for the coffee harvest,

MODIFIED AFTER: RAND MCNALLY, GOODE'S WORLD ATLAS.

Map 8-13 EASTERN BRAZIL: THE CORE AREA.

311

A **view** over Rio de Janeiro, Brazil. This is the central section, with municipal buildings and churches. (D. Grabitzky/Monkmeyer)

which is then transported to Santos for overseas shipment.

The contribution of recent-immigrant Europeans to the agricultural development of southern Brazil, as in Uruguay and Argentina, has been considerable. Many came not to the coffee areas of São Paulo, but to the available lands farther south. Here they occupied fairly discrete areas; for example, Portuguese rice farmers clustered in the valleys of the major rivers of Rio Grande do Sul, and the state today produces between one-quarter and one-fifth of Brazil's annual rice crop. The Germans, on the other hand, occupied the somewhat higher areas to the north and in Santa Catarina, where they were able to carry on the type of mixed farming with which they were familiar: corn, rye, potatoes, and hogs, as well as dairying. The Italians selected the highest slopes and established thriving vineyards. The markets for this produce, of course, are the growing cities to the north.

Thus Brazil can produce an enormous range of crops. Corn, rice, beans, and wheat are the staples; cotton has become an additional important commercial product of the São Paulo area, citrus fruits are grown in the low areas near Rio de Janeiro and along the whole east and south coast. Cacao figures in the exports, as do tobacco and wine. But we should not lose sight of two characteristics of agriculture in Brazil. In the first

place, despite this impressive list of farm products, there is a terrific amount of waste—of land, of labor—in Brazil's agriculture. Literally millions of families still survive on subsistence agriculture, practicing shifting cultivation sometimes where the environmental conditions do not really require it. In 1964, less than one-third of Brazil's area was in farm holdings, and of this total, only 10 percent was in cultivation—or just over 3 percent of all of Brazil. It was this small percentage which in the mid-1960's was producing most of Brazil's export revenues and farm produce for the internal markets as well. Second, Brazil, after all, is a "Latin" American country, and it has not escaped the hacienda system; here it is the *fazenda* that takes the place of the hacienda. The system is more or less the same, with tenant workers or less frequently hired labor doing the farm work. The São Paulo coffee area is expanding much in the way the Argentine Pampa did, in that tenants are assigned land to clear, on which they then grow food crops; subsequently they plant the coffee trees. While the trees grow, the tenants tend them while still planting their own food crops in the same soil. Eventually, after about 5 years, the maturing trees begin to yield a harvest, and the tenants move on to a new area for clearing. The tenants of the coffee lands generally are considerably better off than those still tied to the much older *fazendas* in the north.

There was a time when Brazil seemed destined to be a country whose wealth would come from mines, not *fazendas*. Just as the sugar plantations of the northeast were in decline, gold was discovered in the highlands, and diamonds as well, and the rush to Minas Gerais was on. During the eighteenth century Brazil may, as some geographers believe, have been the world's leading gold producer, but the boom did not last, and coffee soon took over as the next source of wealth. But mining in Brazil today is by no means an unimportant industry. Brazil has large iron ore deposits, principally near Itabira, not far from Belo Horizonte in the state of Minas Gerais, and in Santa Catarina and Rio Grande do Sul it has coal, though of rather low quality. A major iron and steel complex has developed at Volta Redonda, located not only between the main market centers of Rio de Janeiro and São Paulo, but also between the iron ore, which is railed south, and the coal, which is shipped north. Still, Brazil has to import both metal products (machinery) and sources of power (coal and petroleum); along with most other South American countries it does not have either adequate coal or oil reserves, although there is some development of the latter in the state of Bahia. Brazil does have a considerable hydroelectric power potential, of which major use is already made: in the mid-1960's three-quarters of its electricity came from this source.

Brazil has a deserved reputation for its great cities, and especially the magnificent former capital, Rio de Janeiro, still the cultural focus of the state. But it also has a reputation for its industrial strength, and here São Paulo leads. In no South American country is a greater percentage of the labor force (18) employed in manufacturing, and no country produces a wider range or a greater volume of manufactured goods. Of course Brazil has some advantages: its domestic market is nearly four times as large as that of any other country in South America, there is no shortage of labor, urbanization is increasing (and the bulk of the urbanized population is concentrated in a small part of the country, facilitating distribution), the purchasing power of the market is growing, and foreign investment in Brazilian enterprises has been huge. As in many other parts of the world, World War II did its share to stimulate Brazilian industries, since the country was cut off from supply areas, and the Brazilian government has consistently assisted local industries through tight import controls.

As might be expected, Brazil's manufacturing is still dominated by the production of consumer goods, but the balance between the output of consumer goods and capital goods is constantly improving. Brazil is on the way toward becoming South America's first industrial power.

ADDITIONAL READING

Several geographies of South America and Middle and South America together are available. One of the oldest is still full of interest: that of C. F. Jones, *South America*, published by Holt in New York in 1930. Another, less durable volume has been F. A. Carlson's *Geography of Latin America*, last printed in 1952 by Prentice-Hall, Englewood Cliffs. The book by R. S. Platt, *Latin America: Countrysides and United Regions*, published by McGraw-Hill, New York, in 1942, consists of field studies and still retains much interest as well. More modern geographies abound. A simple one is G. J. Butland's *Latin America: A Regional Geography*, published in a second edition by Wiley in 1966. J. P. Cole's book, *Latin America: An Economic and Social Geography*, was published by Butterworth in Washington in 1965. P. James' *Latin America* has gone through several Odyssey Press editions. I. Pohl and J. Zepp have edited *Latin America: A Geographical Commentary*, published in London by Murray in 1966. H. Robinson's *Latin America: A Geographical Survey* was published by Praeger, New York, in 1967. Also see C. Wagley, *The Latin American Tradition: Essays on the Unity and Diversity of Latin American Culture*, published by Columbia University Press in 1968.

Under the editorship of J. H. Steward, the U.S. Bureau of American Ethnology has published a series of volumes under the general title *Handbook of South American Indians*. There are 6 volumes and an index, but the best way to study the central issues involved is to consult Steward's summary of these works, entitled *Native Peoples of South America*, edited in cooperation with L. C. Faron and published in 1959 by McGraw-Hill. On individual countries and regions in South America, see T. R. Ford, *Man and Land in Peru*, University of Florida Press, 1962; L. Linke, *Ecuador; Country of Contrasts*, Oxford University Press in New York, 1960; H. Osborne, *Bolivia, a Land Divided*, Oxford University Press in London, 1964; C. Wagley, *An Introduction to Brazil*, Columbia University Press, 1963; G. J. Butland, *The Human Geography of Southern Chile*, published by Philip in London in 1958; and T. F. McGann, *Argentina: The Divided Land*, No. 28 in the Van Nostrand Searchlight Series.

On urbanization and urban problems in South America, see P. M. Hauser (ed.), *Urbanization in Latin America*, Columbia University Press, 1961, and G. H. Beyer's edited volume, *The Urban Explosion in Latin America: A Continent in Process of Modernization*, Cornell University Press, 1967. For a very comprehensive list of journal references on this subject see L. E. Guzman, *An Annotated Bibliography of Publications on Urban Latin America*, University of Chicago, 1952.

"AFRICA FORMS THE KEY"

CONSIDER a map of the physiography of Africa (Map 9-1). Is there anything unusual about the continent's physical geography? That is, does Africa have physical features that do not occur on other continents, and are certain elements that are present on other continents absent in Africa? Africa, after all, is the second largest continental landmass, and it constitutes close to one-fifth of all the land on earth. We might reasonably expect the full range and variety of landscapes to be present here.

If there was any physical feature that dominated the discussion of South America in the preceding chapter, it surely was the Andes, that great backbone of crushed and folded rocks that extends from the southern tip of that continent to Venezuela. It is hard to imagine a North America without the Rocky Mountains, or a Europe without the Alps—and the fabled Himalayas are the spine of Asia. Less famous, perhaps, but nonetheless prominent enough, is the Great Dividing Range of eastern Australia. And in Africa—where is Africa's comparable mountain chain? It is not the Atlas of the northwest, which constitutes only a corner of its vast landmass. Neither is it the Cape Ranges of the far south, which again have local rather than continental dimensions. What about the mountains of Ethiopia, the great volcanoes of East Africa (including Kilimanjaro), and the sometimes snow-clad Drakensberg in Southern Africa? Could this be the African equivalent of the world's great mountain chains? Certainly Africa seems to have an axis in the east, with elevations in excess of 10,000 feet in Ethiopia, East Africa (Kilimanjaro reaches 19,340 feet), and southeastern Africa. But closer inspection reveals that these mountainous areas are quite different. It can be deduced even from an atlas map: there are not the parallel ranges, there is not the regularity and elongated, "chain" character of an Andes, an Alps. Africa's elevated eastern axis is made up of high plateaus, eroded by rivers into mountainlike terrain; the highest elevations are those of volcanic mountains that generally stand alone, towering over a surrounding table-landscape. Geologists refer to Ethiopia's highlands as the Ethiopian "massif"—a very appropriate term for a massive block of the earth's crust, and one that could not be applied to the weaker, folded, and broken rocks of the Andes or the Himalayas.

This discovery—that Africa lacks a great mountain chain—ought to stimulate us to scrutinize other aspects of the African physiography. In East Africa lies a set of great lakes. With the single exception of Lake Victoria, these lakes are markedly elongated, from Lake Malawi (formerly Nyasa) in the south to Lake Rudolf in the north. What causes this elongation, and the persistent north-south trend that can be observed in these lakes? Again, the map gives a good part of the answer. The lakes lie in deep trenches cut

Map 9-1 PHYSIOGRAPHY OF AFRICA.

The "wall" of a rift valley in the eastern part of Africa. This particular rift valley lies in Southeastern Africa. The wall is about 2000 feet high, and the surface on which the car stands has dropped down from the elevation along the top of the scarp. If the photographer looked behind him, he would see an identical sight. (Harm J. de Blij)

through the East African plateau, trenches that can be seen to extend *beyond* the lakes themselves. Northeast of Lake Rudolf such a trench cuts the Ethiopian massif into two sections, and the Red Sea itself looks very much like a continuation of it. On both sides of Lake Victoria smaller lakes lie in similar, huge ditches, of which the western one runs into Lake Tanganyika and the eastern one cuts completely across Kenya and then peters out in Tanzania. There is a technical term for these trenches: these are *rift valleys,* and, as the name implies, they are formed when huge parallel cracks or *faults* appear in the earth's crust, and the in-between strips of land sink or are pushed down to form great valleys (Map 9-2). Rift valleys are not unique to Africa, of course. But in no other continental landmass do they extend, as they do here, for 6000 miles from beyond one end to the other. And in terms of scale they are impressive—these are not features that appear on the map but have to be looked for in the field. The two "walls" may be 40, 50, or more miles apart, and the dropped plateau "floor" often lies thousands of feet below the original surface.

Several great rivers drain Africa's landmass, some of them, such as the Nile and the Congo,

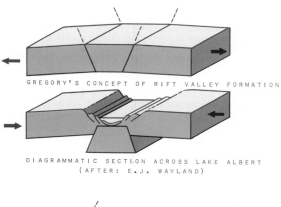

GREGORY'S CONCEPT OF RIFT VALLEY FORMATION

DIAGRAMMATIC SECTION ACROSS LAKE ALBERT
(AFTER: E.J. WAYLAND)

STAGES IN DEVELOPMENT
OF RIFTS UNDER
COMPRESSION
(AFTER: E.C. BULLARD)

Map 9-2 FORMATION OF RIFT VALLEYS.

among the major rivers of the world. But consider their courses. The Niger starts in the far west of Africa, on the slopes of the Futa Jallon Highlands, and then flows *inland* toward the Sahara Desert. Then, after forming an interior delta, it suddenly elbows southward, leaves the desert area, plunges over falls as it cuts through the plateau area of Nigeria, and creates another large delta at its mouth. The Congo begins as the Lualaba River on the Congo (Katanga)–Zambia boundary, and for some distance it actually flows *northeast* before turning north, then west, then southwest, finally to cut through the Crystal Mountains to reach the ocean. It seems as though the *upper* courses of these two rivers are quite unrelated to the continent's coasts. In the case of the Zambezi River, whose headwaters lie in Angola and northwestern Zambia, the situation is the same: the river first flows south, toward the inland delta known as the Okovango Swamp; then it turns northeast, eventually to reach its delta immediately south of Lake Malawi. We may learn something by looking at the course of the Kafue River, the Zambezi's chief tributary. It flows southwest, also toward the Okovango Swamp, but then abruptly vacates its course to turn due east, as though the Zambezi "captured" and diverted it. Finally there is the famed erratic course of the Nile River, which braids into numerous channels in the Sudd area of the southern Sudan, and in its middle course actually reverses direction and flows southward before resuming its flow toward the Mediterranean delta. With so many course peculiarities, could it be that all of Africa's rivers have been affected by the same event at some time in their history? Perhaps—but let us first look further at our map.

All continents have low-lying areas; witness the coastal plain of North America and the coastal and river lowlands of Eurasia and Australia. But as the map shows, coastal lowlands are few and of small extent in Africa. In fact, it is reasonable to call Africa a *plateau* continent; except for some low-lying areas in coastal Moçambique and Somali, and along the north and west coasts, nearly all of the continent lies above 1000 feet in elevation and fully half of it is over 2500 feet high. Even the Congo Basin, equatorial Africa's tropical lowland, lies well over 1000 feet above sea level, in contrast to the much lower Amazon Basin across the Atlantic. Africa is sustained by a vast *shield* (see Chapter

2); a great part of the continent—proportionately more than any other landmass—has a crystalline foundation.

But although Africa is mostly plateau, this does not mean that the surface is completely flat and unbroken. In the first place, the rivers have been attacking the surface for millions of years and they have made some pretty good dents in it, as the Zambezi's Victoria Falls, a mile wide and over 300 feet high, can attest. Volcanoes and other types of mountains, some of them erosional "leftovers," stand above the landscape in many areas—even in the Sahara Desert, where the Ahaggar and Tibesti Mountains both reach about 10,000 feet. And in several places the plateau has sagged down under the weight of accumulating sediments—in the Congo Basin, for example, rivers for tens of millions of years brought sand and silt downstream and for some reason dropped their erosional loads, apparently in a giant lake, almost an interior sea that seems to have existed here. Today the lake is no longer there, but the thick sediments that press the African surface into a giant basin are proof that it was there. And it was not the only one (Map 9-3). To the south the Kalahari basin was filling with sediments that today provide the desert's sand, and to the north, three basins lay in the Sudan, in Chad, and farther west in what is today Mali.

The margins of Africa's plateau, too, are of significance. In Africa—especially in Southern Africa—the term "Great Escarpment" is as commonplace as, say, "The Rockies" is in the United States. And it is not surprising, for the plateau (the *Highveld*, as it is sometimes called) over many hundreds of miles drops precipitously down from more than a mile in elevation to a narrow, hilly coastal belt. From the Congo to Swaziland, and intermittently on or near most of the African coastline, a scarp leads to the interior upland. Other parts of the world also have such escarpments; Brazil at the eastern margins of the Brazilian Highlands, India at the western edge of its Deccan Plateau. But Africa, even for its size, has a disproportionately high share of this phenomenon.

All of this amounts to quite a list of what might be termed as unusual physical characteristics, and perhaps it would be instructive to see how Africa relates to the other continents in global terms. Most world maps, notably those that do not cut Asia in half in order to put

America in the middle, show Africa squarely in the center, with the Americas to the west, Eurasia to the north, Australia to the east, and Antarctica to the south. There is a good reason for this: the earth divides into a Sea Hemisphere, which comprises, mostly, the Pacific Ocean, and a Land Hemisphere, at the heart of which Africa quite clearly lies. A globe will not only verify this—it will quickly suggest why projections of the land areas of the world tend to reflect Africa's central location.

Now this central position of Africa may be meaningless, but on the other hand it may have something to do with the physiographic details just described. Consider the following conclusions, which we can again draw from any world physical map: (1) there is a really amazing gap-and-bulge similarity between the western coast

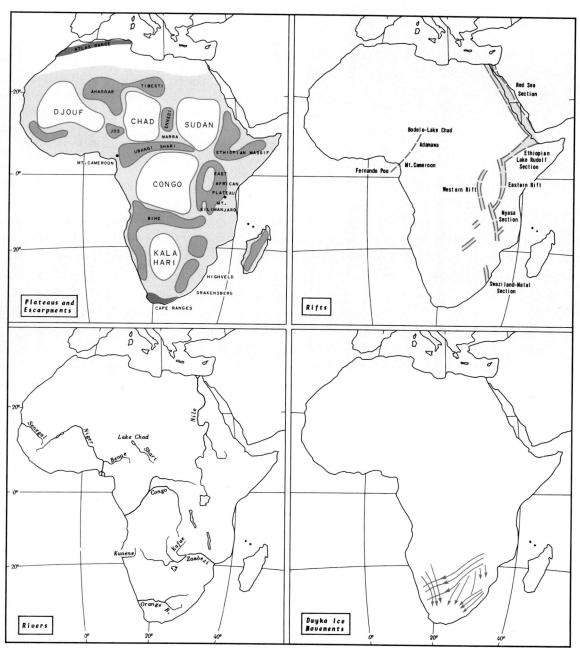

Map 9–3 PHYSICAL FEATURES OF AFRICA.

The Great Escarpment in Southern Africa. To the right lies the African plateau, to the left the lower country that slopes toward the Indian Ocean coastline. The vertical drop in this area exceeds 5000 feet. As the isolated portions of the basalt-supported plateau show, erosion is forcing the escarpment to retreat. (Satour)

of Africa and the east and north coasts of South America; (2) not only do the great mountain chains of the earth miss Africa, but the mountains seem most often to lie on the *away* side from Africa: on the *west* side of South and North America, and on the *east* side in Asia and Australia; and (3) Africa's rift valleys are antipodal to a line that cuts the Pacific Ocean midway.[1] Could it be that what those south Atlantic coasts suggest—that Africa and South America once lay close together or possibly formed part of one enormous continent—did in fact happen? Could

[1] The American, Asian, and Australian mountain chains form the so-called Pacific "Ring of Fire," a belt of crustal instability complete with a high incidence of earthquakes and volcanic activity. The east Asian archipelagos, which are structurally a part of this "Ring," may be evolving mountain chains. The Alpine–Himalayan systems do not conform to this spatial pattern.

it be that not only Africa and South America, but Antarctica, Australia, and perhaps even other land-masses formed part of this one-time supercontinent?

Continental Drift

Here, now, we have a hypothesis, a hypothesis that emerged out of an analysis of the physical properties of the African landsurface and a consideration of the relative location of the African continent. True, Continental Drift as an idea is not new, but it is worth recording the extent to which *geographical, spatial* considerations—direction, distance, shape, relative location—can contribute to the raising of very relevant questions in this connection. And make no mistake about it: physical geographers have contributed very significantly to the concept's development. This is not entirely the realm of geology and geophysics. The behavior of Africa's rivers, of which we can only make bare mention here, has been intensively pursued by geographers, not only from the point of view of course direction, but also in terms of their genetics. Professor J. H. Wellington, pointing out that the Zambezi, of all of Africa's great rivers, is the only one to flow eastward, into the Indian Ocean, wondered whether this might be related to four other peculiarities of southeast Africa: the island of Madagascar, the lowering of the Great Escarpment opposite this island, the existence of an escarpment on the eastern coast of the island whose dimensions are virtually identical to those of the Drakensberg, and the development of one of Africa's few coastal plains in Moçambique. He then proceeded to make a strong case for a former reassembly of Madagascar and Southeast Africa, arguing that the breakaway of the island allowed the Zambezi, which had until then been flowing into a huge inland lake through the Okovango Delta, to crash through to the sea, draining the lake, cutting deep gorges in the process, and completely reorienting the drainage pattern in this area.[2] Thus, for example, did the Kafue River acquire its unusual course. And Professor Wellington implies that there may be something significant also about the close juxtaposition of the Zambezi mouth and delta, the Lake Malawi rift, and the westward "bulge" of Madagascar. Was Madagascar an early rift block to separate from Africa, and will another part

[2] J. H. Wellington, *Southern Africa*, Cambridge University Press, Vol. 1, 1955, pp. 460–473.

of East Africa break off in the future, perhaps along the Malawi–Tanganyika fracture?

This, then, is a specific suggestion. But what can continental drift as a hypothesis do to help explain the general physical characteristics of Africa, and can the hypothesis be tested in any way? First, let us consider the basic idea, which is that Africa, along with South America, Antarctica, Australia, Madagascar, and even southern India at one time formed a supercontinent (called Gondwana or sometimes Gondwanaland), as illustrated by Map 9-4. The significant point of this map is that Africa occupied the *central* position in Gondwana and was thus surrounded by the other landmasses. It may be that these various landmasses in an even earlier geologic era came together, or perhaps the great continent emerged in this way as the planet evolved. In any event, Gondwana after a long period of unity began to break up about 80 to 100 million years ago, and the various fragments have moved radially away from Africa, and are

continuing to do so. Africa itself has moved least of all; according to the hypothesis it was situated near the South Pole, as shown on the map.

In the hypothetical reassembly of Gondwana, then, Africa was the core-shield, it occupied the heart of the supercontinent, and as such it did not have the coasts it has today. The rivers that arose in the interior failed to reach the sea: the upper Niger flowed into Lake Djouf, the Shari River into Lake Chad, the upper Nile into Lake Sudan, the Lualaba into Lake Congo, and the upper Zambezi into the Okovango Delta on the shores of Lake Kalahari. Other rivers also drained toward the great basins, which filled with sediments. In Southern Africa and adjacent areas (southern South America, Antarctica, and India) a great phase of deposition was in progress which ended when the loaded crust cracked and lavas flowed out (mostly basalts); these basalts today still sustain the Drakensberg, the Deccan Plateau of India, parts of the southern Brazilian Highlands, and sizable sections of Antarctica.

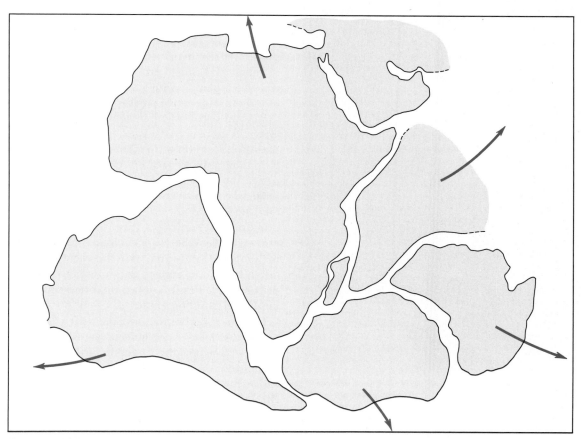

Map 9-4 GONDWANA.

Eventually the great Gondwana landmass, for reasons still in doubt, began to break up. South America began to drift westward, and the West African coast formed. Antarctica moved away to the south, Australia to the east, India to the northeast, and Africa itself drifted slightly northward. Of course, this was not just a simple break: it was, rather, an oscillatory process with outward as well as inward forces at work, the outward forces overcoming the inward ones. Now the coasts were created all around the continent, where huge escarpments marked the great continental fractures (probably looking a good deal like today's rifts), and rivers began to attack the Great Escarpment. Before long these rivers, by headward erosion, reached the long-isolated lake-basins in Africa's interior. One by one they were drained: the lower Niger, joined by the Benue, emptied Lakes Djouf and Chad; the Nile drained the Lake Sudan, the lower Congo cut through the Crystal Mountains, the Zambezi cascaded to the Indian Ocean. With the huge volumes of lake water these rivers cut deep gorges and formed fast-retreating waterfalls; it is no accident that Africa, not exactly a moist continent today, still has nearly half of the hydro-electric power potential of the world. According to the drift explanation, then, Africa's major rivers have a double history: their upper courses are predrift and filled the basins, and their lower courses are younger, and resulted from the release of the pent-up lakes.

It is tempting to carry the explanation further, to argue that since landmasses moved away from Africa both east and westward, Africa was under great tensional stress and began to yield along the rift valleys we can observe in the landscape today. In fact it is probably not so simple, although the rifts are certainly related in some way to the drift process, and those who say that these rifts are the new fractures along which Africa will someday split may yet be proved correct. And the absence of mountains—again a deceptively simple answer appears: all the other landmasses moved many hundreds, even thousands, of miles, and in the process their leading edges were crumpled up into giant folded and crushed mountain arcs, while Africa, which moved only somewhat to the north, shows such evidence only along its northwestern margin. This would explain the location of the American and Australian mountain ranges on the away-side from Africa, but it would also mean that mountains form when continental landmasses move laterally, and there are other theories of mountain building that contradict this. In North America, for example, the Appalachians predate the breakup of Gondwana, and elsewhere there are other mountain ranges that are probably not drift-related. Possibly more than one process of mountain-building exists.

In any case, consider how the drift hypothesis ties the African—and, indeed, the whole Gondwana—physiography together. It relates the rivers and their peculiarities to the Great Escarpment, the Escarpment to the rift valleys, the rift valleys to the absence of fold-mountains, the absence of fold-mountains to Africa's limited lateral movement, and the factor of lateral movement to the distribution of major world

The immense Victoria Falls and the "smoke that thunders," on the Zambezi River. The railroad bridge in the right background connects Rhodesia and Zambia. The Victoria Falls may be related to the diversion of the Zambezi drainage when Madagascar was severed from the African mainland. (J. Allan Cash/Rapho Guillumette)

mountain chains. It provides an explanation for the similarities between Africa's west and South America's east, for the stratigraphic (rock-layering) similarities between Africa, southern South America, Madagascar, India, and Australia. It has even made some predictions possible: on the grounds of rock successions in South Africa, geologists predicted that certain beds would be found beneath the surface in Antarctica—and they were. It suggests that the imbalance inherent in the concentration of the continents in the Land Hemisphere and the virtual emptiness of the Pacific Sea Hemisphere is being corrected by a slow redistribution of the landmasses into the Pacific Basin—and the Ring of Fire marks the forward push of the drift process.

Can the drift hypothesis be put to any sort of test? We know that the South Pole was positioned not far from Durban, South Africa. If this Gondwana continent existed for as long as is postulated, then it must have been glaciated, just as we saw evidence of a much more recent glaciation (the Pleistocene) in the Northern Hemisphere continents quite recently (Chapter 2). Thus, beneath those lavas that spread across India, Africa, South America, and Antarctica just before the breakup, there must lie evidence of this glaciation: glacial drift, till, scoured rocks, and so forth. And this is exactly what has been discovered: hundreds of feet of glacially deposited sediments, now buried deep, but incontrovertible evidence that an ice age marked Gondwana, from the Cape to the Congo in Africa, from Tierra del Fuego to Brazil in South America. When the glacial deposits are mapped and the sections reassembled on a Gondwana base, the glaciated area is almost identical in extent to that of the Northern Hemisphere Pleistocene. And there is more: fossil remains of some species—one a small reptile—have been found in a region which is contiguous on the Gondwana map, but is interrupted by thousands of miles of ocean today.

Does all this constitute proof? Not as long as there are valid objections to the drift hypothesis. Other answers could be found for some of the points raised here; to take the last one, land bridges could be postulated to account for the fossil distributions for which Gondwana also accounts. Actually, the evidence for continental drift has, in the view of some geographers and geologists, long been overwhelming, except that

no one has produced a really reasonable explanation for the mechanism involved—what forces can move continents? But in recent years research in geophysics has begun to indicate that continental drift *may* be possible. Meanwhile, it is difficult to look at the physical map of Africa without trying to find another piece to fit into the puzzle. In Gondwana as in the hypothesis, Africa's position is unique; as the famed geologist A. L. du Toit wrote on the frontispiece of his 1937 book *Our Wandering Continents,* "Africa forms the Key."

The African Past

We turn now to Africa south of the Sahara, and it is paradoxical that more is known, relatively speaking, of prehistoric Africa than of Africa in more recent history, but such is the current state of knowledge. In East and South Africa such a complete fossil record of early Man has been built by the archaeologists that some consider it likely that Man originated here. The most productive area has doubtlessly been east of Lake Victoria, where Dr. Louis S. B. Leakey has made very significant discoveries of early Man and his tools, especially at the famous Olduvai Gorge in northern Tanzania. All the evidence suggests that the early Africans achieved various specializations in toolmaking no later than their contemporaries elsewhere in the world, and in some parts of the continent they seem to have been well ahead. Evidence exists in Nigeria and in Sudan that iron tools were in use there more than 2000 years ago, which means that in this respect the Africans in these areas were on a par with the Europeans of their time.

However, knowledge of the precolonial history of Africa is still very slight and, in general, it becomes less toward the south and away from the coasts, into the interior. This is only partly due to the colonial period itself, during which African history was neglected and many misconceptions about African cultures and institutions arose and became entrenched. It is also a result of the absence of a written history over most of Africa south of the Sahara until the sixteenth century, and over a large part of it until much later than that. The best records are those of the savanna belt immediately to the south of the Sahara Desert, where contact with North African peoples was greatest and where Islam made a major impact. But the absence of a

Zimbabwe consists of two major and numerous minor structures. In the distance lies the so-called temple, an oval-shaped building with walls that are several feet thick and 30 feet high in places. In the left foreground is a portion of the acropolis, a building that has the appearance of a fortress, positioned on a hill and possessing narrow entries and steep, sheer walls. The origins and purpose of Zimbabwe and its related Khami ruins are still not known with certainty. (United Nations)

Bushmen and their house, in the dry country of Southern Africa. The oldest surviving inhabitants of the southern part of the continent, their numbers are dwindling. (Courtesy of the American Museum of Natural History)

written record does not mean, as some historians have suggested, that Africa therefore does not have a history as such prior to the coming of Islam and Christianity. Nor does it mean that there were no rules of social behavior, no codes of law. Modern historians, encouraged by the intense interest shown by the Africans themselves, are trying now to reconstruct the African past, not only where this can be done from the meager written record, but also from folklore, poetry, art objects, buildings, and other such sources. But much has been lost forever. Almost nothing is known of the farming peoples who built well-laid terraces against the hillslopes of north-eastern Nigeria and East Africa, or of the communities that laid irrigation canals and constructed stone-walled wells in Kenya; and very little is known with any certainty about the people who, perhaps a thousand years ago, built the great walls of Zimbabwe in Rhodesia. Porcelain and coins from China, beads from India, and other goods from distant sources have been found at Zimbabwe and other points in East and Southern Africa; but the trade routes within Africa itself, and even the products that moved on them and the people who handled them, still

remain the subjects of guesswork.

Language. Inevitably the question arises: where was the African culture hearth? Did African peoples migrate from somewhere toward the south, into the South African *cul de sac,* or did they spread in all directions from the early evolutionary site in East Africa? This is a question that comes up time and again, and the response always seems to be the same: the authorities are reluctant to talk about places of origin, but they are quite willing to postulate routes of inmigration. So it was, as will be recalled, with the South American Indians, who, we reported, probably came "from Asia" across the Alaskan land bridge and then across the Middle American land bridge. So, too, it goes with Africa's oldest surviving peoples, the Bushmen; they also are thought to have moved southward into the Cape, but from where is a matter for debate. The Bushmen are famous for their rock paintings, and there have been suggestions that their art can be traced all the way from the Sahara to South West Africa, revealing a migration route.

Just when the Bushmen entered Southern Africa, or when the other very ancient peoples of Africa south of the equator appeared on the scene, is not certain. It is thought that the Bushmen, the Hottentots (who also moved into Southern Africa) and the Kushites (who appear to have extended from the Nile valley into Kenya) were all peoples of the later Stone Age. The Kushites are especially important, for it is they who probably brought cattle-keeping to East Africa, from where the practice diffused outward. Little is yet known of agricultural practices and techniques during the Stone Age, and there is not much archaeological evidence for any other peoples that may have lived in the southern half of the continent.

One line of evidence that may give a hint of what happened in Africa is provided by something Black Africa has plenty of—languages. A language map of Africa looks like a giant jigsaw puzzle; literally hundreds of languages are still spoken by the peoples of the continent, and these are not merely dialects of one dominating tongue, but discrete and individual languages. For a long time linguists have been trying to classify and group these many languages into larger units, analyzing them for common roots, and seeking evidence for shared origins. The language of the Bushmen and the Hottentots, still in use by the survivors of these peoples, share a peculiar "click" sound and are called the *Click* or *Khoisan* languages of Africa (Map 9-5). But, as Professor Greenberg's map shows, these languages, if they did once prevail over most of Eastern and Southern Africa, are no longer of great importance: the area of distribution corresponds largely to the sparsely peopled Kalahari Desert. The language map tells us that a new wave of peoples spread over much of Africa south of the Sahara, peoples whose languages are related, some distantly, some still quite closely. These are the *Niger-Congo* languages, and they extend from western West Africa to southeastern South Africa, so that they are spoken by the vast majority of the 250 million or so inhabitants of Africa south of the Sahara. In very general terms this Niger-Congo "family" of languages can be subdivided into seven subfamilies, of which one consists of the numerous Bantu languages which are spoken in almost all of Africa south of the equator.

How can these languages tell us anything about the movement of peoples in Africa? Consider what has happened in Europe where, as we saw in Chapter 1, the Romanic language

Map 9-5 AFRICAN LANGUAGES.

family has differentiated into, among others, Italian, French, Spanish, and Portuguese; within these languages we can even recognize differentiation, such as between Castilian and Catalan Spanish, and Walloon, northern and southern French. Such differentiation occurs over time, and it is perhaps reasonable to assume that the more time that elapses, the greater will be the differences between languages. If, therefore, the peoples of a large region speak languages that are somewhat different, but still closely related, it is reasonable to argue that they have migrated into that region or emerged there relatively recently. On the other hand, languages that are of recognizably common roots yet strongly different must have undergone modification over a long period of time. Thus it was deduced that the Bantu peoples of Central and Southern Africa are of much more recent origins than West Africans, since the languages of the Bantu subfamily are far more closely related to each other than the languages of West Africa.

Of course there is some other evidence to support the conclusion that the peoples of older West and more recent Bantu Africa have affinities to each other, such as prevailing matrilineal inheritance, and certain kinds of societies, cults, and art forms. In fact there is little doubt any longer that the Bantu peoples of Central Africa are most directly related to the peoples of the West African forest, rather than with the Nilotic peoples (who speak Macro- or Central-Sudanic languages, Map 9-5). But this does not solve the problem of timing, nor does it exactly define any migration routes. For a long time anthropologists and anthropolinguists believed that the Bantu "culture hearth" might have been positioned in the highland area that forms the boundary between Nigeria and Cameroun, and that the West African peoples moved into the forests and savannas of West Africa long before the Bantu migration into Central and Southern Africa began. A plausible and spatially reasonable hypothesis, but one that has recently been discarded. In the first place, linguists have come to realize that their assumption that the Bantu emerged in Central and Southern Africa within the past several centuries has been a gross underestimate. Even two millenia is too brief a period, some of them feel, for the degree of linguistic differentiation that exists today to be achieved. Hence the conclusion is that even *while* the Kushites, the Bushmen, and the Hottentots occupied much of Eastern and Southern Africa, there already was a Bantu hearth in existence.

But where? If not in the Nigeria-Cameroun area, where did the center of Bantu development and dispersal lie? The British linguist, Professor M. Guthrie, believes that there may have been two "nuclear areas," the savanna area of what is today the central Katanga and the forest-savanna area north of the Ubangi River in what is now the Central African Republic. This might help explain some of the apparently anomalous evidence that has come to light, for example, that the Bantu entry into parts of the southern half of Africa had a *northward* rather than a consistently southward direction. But the riddle is far from solved, and undoubtedly those who follow with any interest the exciting developments in African history are in for a few surprises.

Livelihoods. A great many Africans today live as their predecessors did, by subsistence farming, herding, or both. The oldest means of survival, hunting and gathering, still sustain the Bushmen of the Kalahari and the Pygmies of the Congo, but for many centuries the majority of Africans have been farmers. Not that the African environment is particularly easy for the farmer who tries to grow food crops or raises cattle: tropical soils are notoriously unproductive and much of Africa suffers from excessive drought. It seems strange, a continent located astride the equator, most of it in the tropics, flanked by two oceans; how can water supply possibly be a problem here? Consider Plate 1: only small parts of West Africa and the interior Congo have really high annual precipitation totals, and the rainfall drops off very quickly both northward and southward. In West Africa only a comparatively narrow coastal zone is well-watered; East Africa is actually quite dry, with steppe conditions (compare with Plate 2) penetrating far into Kenya; and in Southern Africa the rainfall is a modest 20 to 40 inches only in the east and south, while westward the Kalahari steppe and desert take over. And what the map does not tell us is that the hot tropical sun, by evaporating a good part of that rainwater which does reach the ground, reduces the moisture available for plant growth even further. Moreover, much of the rainfall of Africa outside the wettest areas comes concentrated in one or perhaps two seasons, and the intervening periods may be bone-dry; before long the herds of livestock have used up the reserve from the last wet season and their owners must drive them in search of water somewhere else. This is not just a problem faced by the Masai pastoralists of Kenya and Tanzania; it confronts also the modern white rancher in South Africa. The Masai drives his cattle in a never-ending search for pasture and water; the South African rancher may rail his cattle out of the drought area before it is too late.

Africa's soils are mostly tropical soils, with all the problems they present to the farmer. On Plate 3, it is quickly clear how much of the continent lies under latosolic soils (15 and 16) and desert "lithosols" and sand (12). Tropical soils are not completely unproductive, of course, but they are heavily leached of nutrients and will not normally sustain crops year after year; the cultivator who grows corn or sorghum must frequently clear a new patch of land to get a harvest. It is a precarious subsistence, made more so by the high variability of the annual rainfall, that is, its unreliability. In the savanna areas, especially toward the steppe margins, the rains may come in great quantity in one year and stay away almost completely the next; both conditions are bad for the farmers, who will have trouble harvesting a grain crop in the excessively moist year and may not be able to grow one at all in the following year; their kinsmen will have to come to the rescue. Still, grains have for centuries been the major staple crops in these areas: rice in the western part of West Africa, in Senegal, Sierra Leone, Liberia, and Ivory Coast; sorghum and millet in the interior savanna to the east; teff and several other cereals in Ethiopia and Sudan; and corn and sorghum in East and Southern Africa.

Just how some of the crops that are now grown in Africa came into domestic cultivation is still uncertain. It is easy to assume that they were gradually introduced from other parts of the world, but in many cases this was clearly not so. Professor G. P. Murdock has no doubts about it:[3]

"It has hitherto escaped attention . . . that agriculture was independently developed . . . by the Negroes of West Africa. This was, moreover, a genuine invention, not a borrowing from another people. Furthermore, the assemblage of cultivated plants ennobled from wild forms in Negro Africa ranks as one of the four major agricultural complexes evolved in the entire course of human history. Interestingly enough, the innovators have belonged to four distinct races. Along with the Caucasoids who developed the Southwest Asian complex, the Mongoloids who achieved the Southeast Asian complex, and the American Indians who elaborated the Middle American complex, we must now align the West African Negroes as one of mankind's leading creative benefactors."

Indeed, a number of African crops spread to *other* parts of the world: millet to India, varieties of sorghum to India and China, the watermelon to virtually all parts of the world where it can be grown. Other crops, such as a variety of rice, several root crops, and oil plants (the oil palm, sesame plant, and shea tree) have remained primarily of local importance. And, of course, some very important crops did come to Africa from other parts of the world: the Portuguese brought corn from the Americas, along with cassava and sweet potatoes; from southeast Asia came the banana and a variety of yam. Via the Middle East and North Africa, cattle raising was introduced.

[3] George P. Murdock, *Africa, Its Peoples and Their Culture History*, McGraw-Hill, New York, 1959, pp. 64–65.

The vast majority of African families, then, still depend on subsistence farming of one kind or another for their living. Only very few still survive on hunting and gathering alone; even the Pygmies trade for vegetables with their neighbors. Along the coast, especially the west coast, and the major rivers, some communities depend primarily on fishing. But otherwise the principal mode of life is farming. And in the methods of farming, the sharing of the work between men and women, the value and prestige attached to herd animals, and other cultural aspects, subsistence farming gives a great deal of insight into the Africa of the past. Moreover, the subsistence form of livelihood was little changed by the colonial impact; tens of thousands of villages all across Africa never were brought into the economic orbit of the European invaders, and life in these settlements went on more or less unchanged. In the wetter and warmer parts of the continent, corresponding fairly closely with a combination of the *Af* climatic region (Plate 2) and the latosolic soils (16 on Plate 3), root crops form the staples: yams, cassava, sweet potatoes. In some areas of the Congo a variety of the banana is a staple food. It is too humid here to cultivate and harvest cereals. In the savanna and steppe regions, and on the *Highveld* of South Africa, grains predominate, and the methods by which they are grown are time-honored. In some areas, especially the east and southeast, the ox-drawn plow has made its appearance, but elsewhere the basic farm implement remains the simple hoe, with which the soil is turned after the land has been cleared, or the digging-stick, with the pointed end of which small holes are made in the ground, into which the seed is then dropped. The amount of labor that goes into this kind of farming is huge, and yields are low, but the African farmer does not have much alternative. There have been several occasions when the Europeans, with their advanced equipment and agricultural know-how, went into a subsistence area, ripped out the vegetation, bulldozed and tractored the soil, planted tons of seed, and then, when the crops failed to produce, left the area a dustbowl, in economic ruin. Quite apart from the African farmer's problems relating to distance from markets and transport facilities, he faces a very direct obstacle in the environmental complex with which he has to contend. It is unwise to criticize patch agriculture as "primitive."

Africa's herders more often than not mix farming with their pastoral pursuits, and very few of them are actually "pure" herdsmen. In Africa south of the Sahara there are two belts of herding, one extending along the West African savanna-steppe and connecting with the East African area (where the famous Masai drive their herds), and the other centering on the plateau of South Africa. Especially in East and South Africa, cattle are less important as a source of food than they are as a measure of the wealth and prestige of their owners in the community; hence African cattle-owners in these areas have always been more interested in the size of their herds than in the quality of the animals. And as far as the staple food in these areas is concerned, these are generally the grain crops with which the herding areas overlap. Probably a majority of Africa's cattle owners are sedentary farmers, although some peoples—such as the Masai—engage in a more or less systematic cycle of movement, following the rains and seeking pastures for their livestock. In West Africa the pastoralists who will sell their cattle for the meat face the problem of the considerable distances they are located from the major, coastal markets. If the animals are railed or taken by truck the several hundred miles from the savanna-steppe margins to the coast, the cost is high but they do not lose much weight between farm and market; on the other hand, if the animals are driven to the markets on the hoof, the cost of moving them is low but their weight loss is considerable. And the cattle herder, too, faces environmental problems. Not the least of these is the dreaded tsetse fly, which carries sleeping sickness to man, and which ravages Africa's animals—wild as well as domestic. Large parts of Africa that might otherwise be usable cattle country are infested by the tsetse fly, and are rendered practically useless for pastoralism.

Cattle, of course, are not the only livestock in Africa. Quite literally there are millions of goats; they are everywhere, in the forest, in the savanna, in the steppe, even in the desert; they always seem to survive and no African village would be complete without a few of these animals. Where conditions are favorable they multiply very fast, and then they denude the countryside and promote soil erosion; in Swaziland, for example, they constitute both an asset to their individual owners and a serious liability to the state. But elsewhere, goats are an important and valuable property.

States and Tribes. Africa on the eve of the colonial period was in many ways a continent in transition. For several centuries the habitat in and near one of the continent's culturally and economically most productive areas—West Africa—had been changing. For 2000 years and probably more, Africa had been creating as well as adopting ideas. In West Africa, cities were developing on an impressive scale; in Central and Southern Africa, peoples were moving, readjusting, sometimes struggling with each other for territorial supremacy. The Romans had penetrated to the Southern Sudan, North African peoples were trading with West Africans, Arab *dhows* were plying the eastern coasts, bringing Asian goods in exchange for gold, copper, and a comparatively small number of slaves.

West Africa's environmental changes were of profound importance, not only for the peoples and states of the time but also for Africa as a whole, for centuries to come. Archaeologists, physical geographers, paleobotanists, and other authorities believe that the Sahara Desert over the past several thousand years has been expanding; it is still encroaching on the steppe areas of West Africa and the steppe, in turn, on the savanna. It may be that the increasing isolation of West Africa led to the agricultural achievements described earlier by Professor Murdock; certainly trade and commerce across the Sahara became ever more difficult. Rock paintings in the Tibesti depict cattle; today the whole region is desert-dry. Old lake beds have dried up. Oases have disappeared. But the strands of commercial contact between the peoples of the Mediterranean and those of savanna West Africa remained intact—indeed, they seem to have become stronger.

What were the bases for this trade across the desert? Consider the environmental situation in West Africa. As Plates 1–4 indicate, the environmental regions in this part of the continent have a strong east-west orientation. The isohyets run parallel to the coast (Plate 1); the climatic regions, now positioned somewhat differently from where they were two millenia ago, still trend strongly east-west (Plate 2). Soil regions are similarly aligned (Plate 3), and the vegetation map, though very generalized, also reflects this situation (Plate 4), with a coastal forest belt yielding

Arab *dhows* in port at Mombasa, Kenya. For many centuries these small ships have carried on trade between Arabia, India, and East Africa. They once carried thousands of slaves away from their East African homes, and many tons of ivory. Now they carry spices and coffee from East Africa, and from the Persian Gulf they return with carpets and carved chests. (Marc and Evelyne Bernheim/Rapho Guillumette)

to savanna (tall grass in the south, short grass in the north), which in turn gives way to steppe and desert. About 2000 years ago the forest belt was probably wider than it is today, the savanna extended into the *BSh* area of Plate 2, and the steppe and desert were thus somewhat more confined. But even so, the desert and steppe covered most of the western "bulge" of Africa, and the area to the east as well, and the strip of moister, coastal Mediterranean conditions in the far north was—as it is today—quite narrow.

Essentially, then, the situation in West Africa was such that over a north-south span of not too many hundreds of miles there was a terrific contrast in environments, economic opportunities, modes of life, and products. Obviously the people of the tropical forest produced and needed goods that were quite different from the products and requirements of the peoples of the dry, distant north. To give an example: salt is a prized commodity in the forest, where the humidity precludes its formation, but salt is in plentiful supply in the desert and steppe. Hence

the desert peoples could sell salt to the forest peoples. What could the forest peoples offer in exchange? Ivory and spices could be sent north: there were elephants in the forest, and certain plants that yield valuable condiments. Thus there was a degree of *complementarity* between the peoples of the forests and the peoples of the dry lands. And the peoples of the savanna—the in-betweens—were beneficiaries of this situation, for they found themselves in a position to channel and handle the trade, and that activity is always economically profitable.

This, then, was another stimulus for indigenous development in West African savanna-lands, but there were others. In North Africa there was a growing demand for the products from the south as the Moors rose to power, and it is always good to have strong and wealthy trading partners, although, as the West Africans found out, you may eventually have to protect your lands and resources against them. And the savanna areas themselves were not devoid of their own economic assets: sedentary agriculture could

As they have done for uncounted centuries, camels carry their cargo across the Sahara Desert. This scene could have been pictured in the days of the old Ghana empire, nearly two thousand years ago; still today camel trains converge on markets located on the Sahara's margins. (Anthony Howarth/Camera Press—PIX)

be carried on, along with herding—but most important of all, the rivers of the southern savanna and the forests beyond carried considerable quantities of alluvial gold. Gold, ivory, spices, hides and skins, certain kinds of wood, and slaves passed through the markets of the savanna on their way north; from the desert and beyond came salt, copper goods, cloth, dried Mediterranean fruits, and even some European products. The markets on which these goods were exchanged prospered and grew, and so, in the savanna belt of West Africa, there arose a number of true cities. There was a time when one of these old cities, now an epitome of isolation, was a thriving center of commerce and learning, and one of the leading urban places in the world: Timbuktu. Others, predecessors as well as successors of Timbuktu, have declined, some of them into oblivion. Still other savanna cities continue to have considerable importance, like Kano in the northern part of Nigeria.

West Africa's economic mainstays were gold and salt; its political foundations were the divine kingship, a concept introduced from Egypt, and a knowledge of iron working, which came from the famous Kushitic city of Meroë on the Nile—whose iron workings were so extensive that it has been called the Birmingham of ancient Africa. Iron working contributed to the political development of West Africa because the technique remained exclusively in the hands of the rulers and their representatives; it provided them with weapons that had no equal and hence a means of control to which the tribal peoples of

the region had no answer. As for the concept that the king was himself of divine powers or had direct access to such powers, this was at once an asset as well as a liability in political West Africa. It had the advantage of strengthening the authority of the king amid the ruling group in the state, but invariably it created problems of succession. Elaborate rituals were developed to transfer divine powers to a new king from a dead one, but more often than not the new king first had to assert himself through a down-to-earth power struggle. Needless to say, this did not have a very positive effect on the continuity of West Africa's savanna states.

Nevertheless, states of impressive strength and truly amazing durability arose in the West African culture hearth. The oldest state about which anything at all concrete is known is Ghana. Ancient Ghana was located to the northwest of the coastal country that has taken its name in modern times, covering parts of present-day Mali and Mauritania, along with some adjacent territory (Map 9-6). It lay astride the Upper Niger River and included gold-rich streams coming off the Futa Jallon, where the Niger has its origins. For a thousand years, perhaps longer, old Ghana managed to weld various groups of people into a stable state. The country had a large capital city, complete with markets, suburbs for foreign merchants, religious shrines, and, some miles from the city center, a fortified royal retreat. There were systems of tax collection for the citizens and extraction of tribute from subjugated peoples on the periphery of the territory, and tolls were levied on goods entering the Ghanaian domain. An army kept control, and even after the Moslems from the northern dry lands invaded Ghana in about 1062, the state continued to show its strength: the capital was protected for no less than 14 years, and a decade after its fall, a successful rebellion brought the old royal dynasty back. However, the invaders had ruined the farmlands of the country, and the trade links with the north were, of course, destroyed. Ghana could not survive, and it finally broke up into a number of smaller units.

Ghana, then, was located in the western part of West Africa, and in the centuries that followed, the center of politico-territorial organization in the West African culture hearth shifted almost continuously eastward—first to the successor state of Mali, which was centered on Timbuktu and the Middle Niger River, and which became consolidated in the 13th century, and then to Songhai, whose focus was Gao on the Niger—and still on the map today. Eventually certain states in northern Nigeria rose to prominence. One possible explanation for this eastward movement may lie in the increasing influence of Islam; Ghana had been a pagan state, but Mali and its successors were Moslem and sent huge and rich pilgrimages to Mecca along the savanna corridor south of the desert. Indeed, hundreds of thousands of citizens of the modern state of Sudan trace their ancestry to Nigeria, their forefathers having settled down there while journeying to or from Mecca. Whether Islam is indeed a major cause for West Africa's eastward shift is not certain; quite possibly the answer may lie in some other area, for example, that of intervening opportunity—the development to the east of shorter and better trade routes and larger trade volumes to the northern markets.

In any event, the West African savanna was

Map 9–6 ANCIENT STATES OF AFRICA.

the scene of momentous cultural, political, and economic developments for many centuries—but in its progress it was not alone in Africa. In what is today southwestern Nigeria, a number of urban farming communities became established, the farmers being concentrated in these walled and fortified places for reasons of protection and defense; surrounding each "city of farmers" were intensively cultivated lands that could sustain thousands of people clustered in towns. In the arts, too, southern Nigeria produced some great achievements, and the bronzes of Ife and Benin are true masterworks. In the region of the Congo mouth, a large state named Kongo existed for centuries. In East Africa, trade on a large scale with China, India, Indonesia, and the Arab world brought crops, customs, and merchandise from these distant parts of the world to the coast, to be incorporated in African cultures. In Ethiopia and Uganda, populous kingdoms emerged. And much of what Africa was in those earlier centuries has yet to be reconstructed. But with all this external contact, it was clearly not isolated, as the historical hiatus might suggest.

On the other hand, early Africa, no less than early Europe, Asia, or America, was a continent of many communities or tribes, some of them expanding and encroaching on their neighbors, others dwindling and decaying, still others living in isolation, with little contact even with their vaguely known neighbors. When it comes to judging who "knew" more of the outside world, say, by 1400 A.D., it is relevant that Chinese, Arab, and Roman sea captains had seen and reported to their homelands on African coasts, whereas Africans, if they did travel to these lands, normally did so on a one-way ticket—as slaves. However, we may seriously doubt whether Europeans in 1400 A.D. knew much more of Africa south of the Sahara than Africans knew of Europe, for Roman times were far behind and anyway the Romans saw relatively little of Black Africa. As for their knowledge of the Middle East, literally hundreds of thousands of Africans visited parts of the Arab world after the rise of Islam. The "known world" of the Africans may have been somewhat smaller than that of some Eurasian peoples, but not nearly as much smaller as sometimes is suggested. And for the vast majority of black Africans the "known world" in 1400 was the same as it was for millions of Europeans and millions of Asians: the local community or tribe.

Neither was Africa one vast mass of warring tribes. True, peoples were migrating, and Europe's *völkerwanderung* certainly did not pass without a few conflicts—nor did Africa's. For people in the Western world to make too strong a point of conflicts in other cultures is perhaps injudicious, considering what Western wars have cost in human lives over the past three generations alone; but even with this consideration aside, the tribal strife in Black Africa has been exaggerated. Certainly stronger peoples were ousting or absorbing weaker ones, and wars were no less prevalent than they were elsewhere in the world. But neither were they more so.

The Colonial Transformation

During the fifteenth century, thus, there were states, kingdoms, and tribal groups in Africa. In West Africa lay a culture hearth; from the Congo margins the Bantu had radiated outward into the east and south. In Southern Africa the Bushmen and Hottentots still held off the Bantu invasion, though they had lost the more favorable southeast to their competitors. In some parts of the continent stable states had existed for centuries, and elsewhere culture groups were still adjusting to their territories.

Now began the period of European involvement in Black Africa, a period that was to alter irreversibly the entire cultural, economic, political, and social makeup of the continent. It started quietly enough, with Portuguese ships groping their way along the west coast, rounding the Cape of Good Hope not long before the turn of the sixteenth century, and finding a route across the Indian Ocean to the spices and riches of the Orient. Soon other European countries sent their vessels to African waters, and a string of coastal stations and forts sprang up. In West Africa, the nearest part of the continent to European spheres in Middle and South America, the initial impact was strongest. At their coastal points the Europeans traded with African middlemen for the slaves that were wanted on American plantations, and for the gold that had been flowing northward across the desert, and for ivory and spices. All of a sudden the centers of economic activity lay not in the cities of the savanna, but in the foreign stations on the coast. As the interior declined, the coastal peoples thrived. Some of the small forest states rose to power and gained unprecedented wealth, transferring and selling slaves captured in the interior to the white men on the coast. Dahomey and Benin were slave-trade-built states, and when eventually the practice of slavery came under attack in Europe, abolition was vigorously opposed by those who had inherited the power and riches it had brought.

Although it is true that slavery was not new to West Africa, the kind of slave trading introduced by the Europeans certainly was. In the savanna states, African families who had slaves usually treated them comparatively well, permitting marriage, affording adequate quarters, and absorbing them into the family. The number of slaves held in this way was quite small, and probably the largest number of persons in slavery in precolonial Africa were in the service of kings and chiefs. In East Africa, however, the Arabs had introduced—long before the Europeans—the sort of slave trading that the white man brought first to the west: African middlemen from the coast raided the interior for slaves, marched them in chains to the Arab *dhows* that plied the Indian Ocean, and there, packed by the hundreds in specially built vessels, they were carried off to Arabia, Persia, and India. It is sad but true that European, Arab, and African combined to ravage the black continent, forcing perhaps as many as 30 million persons away from their homeland in bondage, destroying families, whole villages, and cultures, and bringing to those affected an amount of human misery for which there is no measure.

The European presence on the West African coast brought about a complete reorientation of trade routes, it initiated the decline of the interior savanna states and strengthened coastal forest states, and it ravaged the population of the interior through its insatiable demand for slaves. But it did not lead to any major European thrust toward the interior, nor did it produce colonies overnight. The African middlemen were well organized and strong, and they managed to maintain a standoff with their European competitors, not just for a few decades, but for centuries. Although the European interests made their appearance in the fifteenth century, West Africa was not carved out among them until nearly four centuries later; the British could not establish control over the Ashanti group of states (in present-day central Ghana) until the first years of the twentieth century. Once again this points up the lack of superiority of the Europeans in their contacts with non-European peoples, to

which reference was made in Chapter 7. For centuries Europe held a series of uneasy beachheads along Africa's coasts, and the African interior felt its impact only through the terrors of the slave trade. Even the Portuguese, earliest of Africa's European colonizers, just a century ago had effective control over only about one-quarter of their present African provinces.

As fate would have it, European interest was to grow strongest—and ultimately most successful—where African organization was weakest. In the middle of the seventeenth century the Hollanders chose the shores of Table Bay, where Cape Town lies today, as the site for a permanent settlement, though not for purposes of colonization. The objective, rather, was to establish a revictualling station on the months-long voyage to and from Southeast Asia and the Indies; the tip of Africa, where the Atlantic and Indian Oceans meet, was the obvious halfway house. African considerations hardly entered into the choice, for there was no intent to colonize here. Southern Africa was not known as a productive area, and the more worthwhile east lay in the spheres of the Portuguese and the Arabs. Probably the Hollanders would have elected to build their station at the foot of Table Mountain whatever the indigenous population of the interior, but as it happened they picked a location about as far away from the major centers of Bantu settlement as they could have found. Only the Bushmen and their rivals, the Hottentots, occupied Cape Town's hinterland. When conflicts developed between Amsterdam and Cape Town and some of the settlement's residents decided to move into this hinterland, they initially faced only the harassment of groups of these people, not the massive resistance that might have been offered by the Bantu states. To be sure, a confrontation eventually did develop between the advancing Europeans and the similarly mobile Bantu Africans, but it began several decades after Cape Town was founded and hundreds of miles from it. Unlike some of the West African waystations, Cape Town was never threatened by African power, and it became a European gateway into Southern Africa.

White people came to the southern part of Africa in numbers far greater than elsewhere, and in the two centuries that followed the founding of Cape Town, the Cape of Good Hope was the chief focus of European immigration. In Africa between the tropics the white man's

Cape Town and Table Mountain. This was the site of the first permanent South African white settlement. On the slopes of the mountain, and with its drainage runoff, the Hollanders planned to farm; old Cape Town developed on the waterfront. Today the urban sprawl is wiping out the last remnants of that early period. (Courtesy of Satour)

impact was felt through the coastal trading stations whose economic influence was of course very great, but the hinterlands of these settlements saw only small-scale European penetration; Europe's ignorance of Africa was slowly reduced through the often unreliable reports of individual travelers, missionaries, and traders. For centuries the African interior, though thrown into turmoil by the slave trade, remained the "dark" continent—"dark" not in any real sense except in the minds of Europeans.

South of the Tropic of Capricorn, however, events from the very beginning followed a different course. The Cape had not been conceived as a European colony, but soon white settlers began to move into the valleys between the Cape Ranges. Dissident Dutchmen, displeased

with the administration of the Dutch East India Company, soon were joined by religious refugees from France (the *Huguenots*); later the British arrived. In Cape Town and its environs, considerable intermarriage took place between white and Hottentot, and thus began to emerge a "mixed" population not altogether unlike the Iberian-American mestizo; today these people are officially referred to as "colored" in South Africa and their number approaches 2 million. From the Indies, the European vessels brought to the Cape slaves from Malaya and other parts of Southeast Asia, and a Moslem Malay community has survived to this day in Cape Town. The Dutch pushed ever farther into the interior: first to the eastern Cape, where they met a Bantu people, the Xhosa, and later they climbed the

Great Escarpment and drove their cattle into the plateau Highveld. By the late 1830's the white invaders had fought and won their inevitable battle with the center of Bantu power, the Zulu Empire. They took the land and farmed and ranched it, they searched for and found minerals, and they built villages and towns. The British, who during the nineteenth century took a major interest in Southern Africa as well, founded the city of Durban and penetrated Natal. When they recognized the agricultural possibilities of this area and were confronted by a shortage of labor, they brought tens of thousands of indentured workers from India to the sugar plantations they had established. Today this Asian population sector, centered on Durban, approaches 1 million. Comparisons between the course of events in Southern Africa and Caribbean America immediately present themselves, and by the middle of the nineteenth century, before the period of exploration in tropical Africa had ended, before a Belgian king had even thought of appropriating the Congo, before there was a Rhodesia or a Uganda, before Nairobi was founded—indeed, before the whole colonial invasion of Africa had really started— the white man's conquest and domination of Southern Africa was an established fact and the confrontation of cultures, still delayed in most of tropical Africa, had long been in full swing.

South Africa, thus, presented—and continues to present—a striking contrast to tropical Africa. South and east of the Kalahari Desert the area was effectively controlled by white men; by 1850 the British were in charge along the coast, from Cape Town eastward and north to Durban and beyond, and in the interior the Dutch (or *Boers,* farmers) were in the process of organizing their own republics. The Xhosa, Zulu, and other Bantu peoples were confined in the territories they occupied when they were subjugated, and although they offered some resistance, the Zulu defeat of 1838 really marked the end of any African power in this part of the continent. And during the past century the early individuality of South Africa was confirmed by the discovery of the continent's richest mineral resources and the realization of the country's considerable agricultural potential. The diamonds of Kimberley and the gold of Johannesburg, in the late 1860's and mid 1880's, respectively, led to great waves of white immigration, not only from Europe, but

from America, Australia, and elsewhere as well. Africans left their tribal homelands and sought jobs and incomes in the mines, wealthy Asian merchants arrived in Durban to try their luck in the booming economy. Railroads connected the ports to the interior towns, farmers saw markets for their produce grow by leaps and bounds. Industries were needed to serve the mines and the needs of the mushrooming population. And much of this was happening while tropical Africa still awaited the real onslaught of European colonialism.

The Scramble. After more than four centuries of contact, Europe finally laid claim to all of Africa during the second half of the nineteenth century. Adventurous men like Livingstone, Park, Speke, Burton, and Grant had "explored" parts of the continent. Now individual representatives of European governments sought to expand or create African spheres of influence for their homelands: Rhodes for Britain, Peters for Germany, De Brazza for France, and Stanley for the king of Belgium were among those who helped shape the destiny of the continent. The British talked of an all-British axis from Cairo to the Cape; the French desired to create a vast colonial empire across West and North Africa from Dakar to Cape Guardafui in the east, or at least to Djibouti. The Portuguese suddenly sought to connect their Angolan and Moçambique possessions across South-Central Africa, and the Germans, coming late to the colonial scene, showed up in part to acquire colonies, as they did in West, South, and East Africa, but also to obstruct the colonial designs of their European rivals. In some areas, as along the lower Congo River and in the vicinity of Lake Victoria, the competition between the European powers was very intense. Spheres of influence began to crowd each other; sometimes they even overlapped. And so, late in 1884, a conference was convened in Berlin to sort things out. At this conference the groundwork was laid for the now-familiar boundary lines of Africa.

As the 20th century opened, Europe's colonial powers were busy organizing and exploiting their African dependencies. The British, having defeated the Boers, came very close to achieving their Cape-to-Cairo axis: only German East Africa interrupted a vast empire that stretched from Egypt and the Sudan through Uganda, Kenya, Nyasaland and the Rhodesias to South

Africa (Map 9-7). The French took charge of a vast realm that reached from Algiers in the north and Dakar in the west to the Congo River in equatorial Africa. King Leopold II of the Belgians held personal control over the Congo. Germany had colonies in West Africa (Togo), equatorial Africa (Kamerun), South Africa (South West Africa) and East Africa (German East Africa, later Tanganyika). The Portuguese controlled two huge territories in Angola and Moçambique, both much larger than the original spheres of influence along the Atlantic and Indian coasts; in West Africa, Portugal also held a small entity known as Portuguese Guinea. Italy's possessions in tropical Africa were confined to the "Horn," and even Spain was in the act with a small dependency consisting of the island of Fernando Poo and a mainland area called Rio Muni. The only places where the Europeans showed some respect for African independence were Ethiopia, which fought some heroic battles against Italian forces, and Liberia, where Afro-Americans retained control.

The two World Wars had some effect on this politico-geographical map of Africa. In World War I, defeated Germany lost its African possessions altogether, and they were placed under the administration of other colonial powers by the League of Nations' Mandate System. In World War II, fascist Italy launched a briefly successful campaign against Ethiopia, but the ancient empire was restored to independence when the

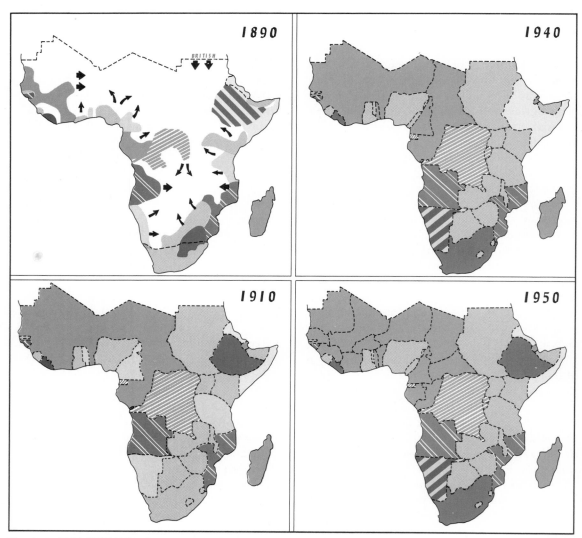

Map 9-7 THE SEQUENCE OF COLONIZATION AND DECOLONIZATION.

Allied forces won the war. Otherwise the situation in colonial Africa in the late 1940's—after a half century of colonial control and on the eve of the Wind of Change—was still quite similar to that which arose out of the Berlin Conference.

Colonialism as a Concept. Is Czechoslovakia a colony of the Soviet Union? Is Tibet a colony of the People's Republic of China? Can a dependency that lies adjacent to and is contiguous with the controlling state be referred to as a colony, or is a colony always physically separated from the ruling power? These are some of the questions that have led to attempts to give colonialism as a term more exact meaning. Of course, colonialism is most often associated with Europe's great expansion in recent history, but history shows that non-European states also were—and are—capable of colonial subjugation. The Aztec State established spheres of influence in Central America; the Inca Empire, as we saw, developed an elaborate scheme for the colonization and integration of distant peoples and areas. It is difficult to view Indonesia's acquisition of West Irian (former Dutch New Guinea) as anything but an exercise in colonial expansion. It may be that Tibet was within a Chinese sphere of influence anyway, but there are many parallels between the Chinese takeover in Tibet and the Italian conquest of Ethiopia. And perhaps Eastern Europe is an established sphere of Soviet influence, but Hungary in the 1950's and Czechoslovakia in the 1960's had experiences

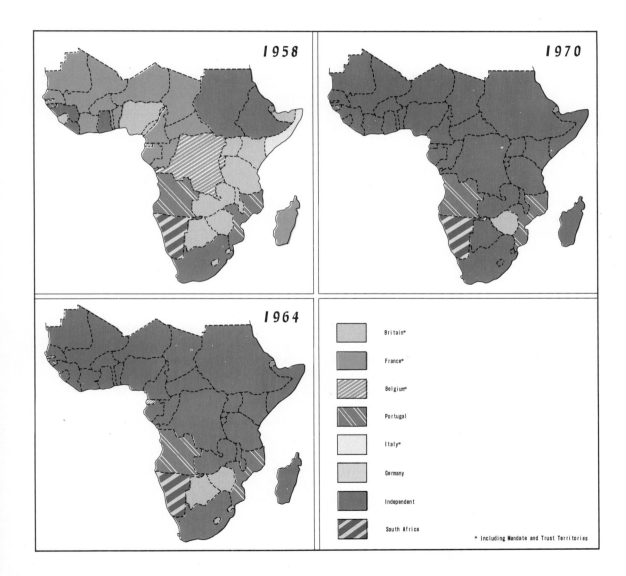

with Moscow that are quite similar to experiences had by African countries with London and Paris.

What, then, is colonialism? One suggestion is that we recognize *primary* and *secondary* types of colonization.[4] *Primary* colonization is defined as the occupation of the lands of indigenous peoples and their domination by the colonizing power, while *secondary* colonization involves the acquisition of virtually empty territory by a colonial power. Another suggestion is that we recognize colonialism as something different from imperialism: "colonialism, as a process, involves the settlement from a mother country, generally into empty lands and bringing into these lands the previous culture and organization of the parent society. Imperialism, as distinct from colonialism, refers to rule over indigenous people, transforming their ideas, institutions, and goods."[5] By this latter definition, then, imperialism is more or less the same as primary colonialism as Frankel defines it, and the problem is far from solved. A look at some present or recent colonies suggests how difficult the application of such definitions is: in Kenya, for example, many thousands of square miles of virtually empty land were "colonized," but the real thrust of British rule lay in the southwestern quarter of the country, where Kenya's productive capacity and most of its indigenous peoples are concentrated. Perhaps it is unwise to try to identify types of colonization or to differentiate colonialism from imperialism. In the broadest sense, colonialism is the application in practice of the international designs of imperialist powers, whereby the peoples of weaker states or territories are subjected to rule and control by the governments of these more powerful states, for purposes of exploitation of resources, imposition of alien values, and the safeguarding of the colonizing power's national interest. Whether this national interest demands involvement in distant lands or adjacent ones is really immaterial; the objectives and the methods are much the same. African countries long saw their raw materials hauled to European factories; East European countries now see their economies geared in a service function to the economic machine of the Soviet Union.

Colonial Policies. From a geographer's point of

[4] S. H. Frankel, *The Concept of Colonization.* Clarendon Press, Oxford, 1949.

[5] S. B. Cohen, *Geography and Politics in a World Divided,* Random House, New York, 1963, p. 204.

view, one of the most interesting aspects of colonialism is the way in which the colonizing power has set about organizing—in a spatial sense—the dependencies under its control, and how various colonial powers differed in their objectives and methods. What the colonial power contributed in each African colony, in both a positive and a negative way, has had much to do with the degree of success with which those colonies later faced independence. When the (former Belgian) Congo achieved sovereignty, the fact that but a few dozen Africans had been trained to university level was bandied about as a major contributing cause to the troubles the young state experienced. In those days, Nigeria was often used as an example of a country where the colonial power (Britain) had produced not only a sizable trained African civil service, but a large number of African univeristy graduates, and where the course of independence would doubtlessly be smoother. And yet, by the late 1960's, the Congo seemed to have overcome its initial crises, while Nigeria was plunging deeply into a full-scale civil war. One important factor was that in the Congo, the Belgian administrative structure had been abandoned in favor of a wholly new layout, while Nigeria continued until 1967 to struggle along with a federal framework that was Britain's legacy but which had proved to be unworkable.

Geographers—and especially political geographers—are interested in the way in which the philosophies and policies of the colonial powers were reflected in the spatial organization of the African dependencies. These colonial policies were often expressed in just a few words. For example, Britain's administration in many parts of its vast empire was referred to as "indirect rule," since indigenous power structures were sometimes left intact and local rulers were made representatives of the Crown. Belgian colonial policy was called "paternalism" in that it tended to treat Africans as children, to be tutored in Western ways, although slowly. While the Belgians made no real efforts to make Belgians of their African subjects, the French very much wanted to create an "Overseas France" in their African dependencies. French colonialism has been identified as a process of "assimilation," the acculturation of Africans to French ways of life. Always France made a strong cultural imprint in the various parts of its huge colonial empire. Portuguese colonial policy has had similar objectives, and even today the African dependencies of Portugal are officially "Overseas Provinces" of the state. If you sought a one-word definition of Portuguese colonial policy, however, the term "exploitation" would probably emerge most strongly. Few colonies have made a greater contribution (in proportion to their productive capacity) to the economy of their colonizing masters than have Moçambique and Angola.

But these terms—"paternalism," "indirect rule," "assimilation"—refer to *institutional* features of colonial rule. Can they also have *spatial* expression? Indeed they can, and do. Consider the former British empire in Africa. It consisted of numerous parts: four separated territories in West Africa (Nigeria, Gold Coast, Sierra Leone, and Gambia), a huge area in East Africa (Kenya, Uganda, and the mandated territory of Tangan-

yika), the heart of South-Central Africa (Northern and Southern Rhodesia and Nyasaland), the so-called High Commission territories of South Africa (Bechuanaland, Basutoland, and Swaziland), and a number of other entities.[6] Thus Britain ruled over a wide variety of indigenous cultures and peoples, and it responded by trying to adjust to each situation in a practical, individual way, instead of by imposing one, uniform type of colonial rule everywhere. One characteristic of Britain's African empire was the existence of different categories of dependencies, including colonies, protectorates, mandate (trust) territories, and, in the Sudan, a condominium. In British colonies, the white settler population enjoyed a considerable degree of self-determina-

[6]The names used here are those in effect during the colonial period.

Map 9–8 TYPES OF COLONIAL ADMINISTRATION.

tion, as was the case in Kenya and Southern Rhodesia. In *protectorates,* as the name implies, indigenous peoples and their rulers gained guarantees of inviolability from the Crown, as did the people of Bechuanaland (now Botswana) and Uganda. In *mandate* (later *trust*) territories, including Togo and Tanganyika, the British undertook to abide by the stipulations of the League of Nations and the United Nations, to guard the interests of the indigenous peoples. And in the *condominium* of the Sudan, the British shared the responsibilities of administration with Egypt (Map 9-8).

The principle of indirect rule was involved in the administration of all four of these types of dependencies; to a greater or lesser degree the indigenous peoples in British Africa retained some autonomy. But what might seem to be a laudable objective had its negative aspects too. In Nigeria, for example, while the south received the full impact of British colonization and experience much political, economic, and social change, the north remained more distant, isolated, and far less affected by these processes. The north's *emirs* continued to rule their people, Islam was little affected by any competition from Christian churches, the role and status of women stayed much the same as it was before the British came, few inroads were made into the prevailing mass illiteracy. All this is not necessarily bad, but the fact is that in the south, everything was changing. Eventually, when independence was achieved, these differences between north and south, long dormant under the umbrella of colonial rule, became powerful, disruptive forces that threatened the very existence of the state. Indeed, in northern Nigeria the British had even used southerners—the Ibo from the southeast—to do many of the jobs that needed to be done, such as running the trains and building the roads. After independence northern Nigerians vented their resentment against these Ibo, many were killed, and many thousands were forced to return to their homeland in the southeast. In Nigeria, indirect rule helped intensify the contrasts between the colonized peoples. As Chief H. O. Davies has put it, "the North and the South . . . remained almost complete strangers to each other. The . . . South was looking towards England and Western Europe. The Islamic North fixed its gaze on distant Mecca." . .[7]

[7] H. O. Davies, Q. C., *Nigeria: The Prospects for Democracy,* Weidenfeld and Nicholson, London, 1961, p. 92.

Elsewhere in British-ruled Africa, too, such regional contrasts emerged and remained for the state to cope with after independence. In Uganda, for example, the old Buganda Kingdom had a special status under British administration, a status it lost, amid a great deal of conflict, after the British had departed. In other territories the problems were less and indirect rule worked better, as for instance in Basutoland (now Lesotho) and Bechuanaland (now Botswana). But where there were large white settler populations, as in Rhodesia and Kenya, the Africans found themselves under more direct European control. Here Britain faced its most critical problems: whatever might be the desire in London to look after African interests, it was overridden by the pressure of the white settlers. These white colonizers took huge amounts of land, confined the Africans to reserves, set up governments, and in effect became the masters of these territories. Through various acts of legislation they entrenched themselves, and made it difficult for London to interfere on the Africans' behalf even if it wanted to. In Kenya the Mau Mau revolt brought the beginning of the end of white rule, but not before years of strife had passed; in Rhodesia the white population in 1965 forestalled any British intentions to grant sovereignty based on democratic principles by a unilateral declaration of independence on its own part. In the late 1960's the Rhodesian question seemed as far from any reasonable solution as ever.

In general, thus, British colonial policy was adjusted to each individual situation. In some places it worked extremely well and clearly to the benefit (in Africa's total colonial context) of the territory and its people. Elsewhere there was a curious mix of success and failure, and at times the most incredible errors were made. In Nyasaland (now Malawi), a protectorate, British administration was quite successful; the white population at its maximum was about 10,000, and although there naturally were problems, these were capable of solution. In Northern Rhodesia (now Zambia), there was a white settler population about as large as that of Kenya (70,000), but Northern Rhodesia was a protectorate, and the whites were largely concentrated in the northwestern mining region known as the Copperbelt. Again, major conflicts were avoided. In Southern Rhodesia, on the other hand, there was a long history of strife and subjugation, land

A traditional spectacle in northern Nigeria: an *emir* travels his domain. British colonial rule left Niger's north relatively unchanged. Under indirect rule, there was no threat by Christian missionaries to the hegemony of Islam, nor was there any major impetus toward modernization. Upon independence, the contrasts between north and south in Nigeria came into sharp focus. (Dr. Georg Gerster/Rapho Guillumette)

alienation and oppression. In view of the relative success of British administration in Nyasaland and at least partial success in Northern Rhodesia, it is almost unbelievable that London in the early 1950's decided to go ahead with the federation of these territories with "white" Southern Rhodesia, against the almost universal African opposition in the two protectorates, but with the enthusiastic support of the Rhodesian whites. It severely set back the relations between Britain and Black Africa, hurt the reputation and credibility of British intentions in Africa at a crucial time, and was doomed to failure—the "Central African Federation" survived just 10 years.

It is not surprising that Britain, with its far-flung and widely varying types of dependencies, in places turned to the federal idea to solve the problems that arose with the onrush of inde-pendence. While the Central African Federation was clearly designed to help maintain white supremacy in that part of Africa and thus was a special case, there were other countries in which the regional contrasts were so great that only a federal constitution would seem to have any chance of survival. In Nigeria, Britain organized a three-region federation, with a Western and Eastern Region constituting the south, and a Northern Region to represent the Moslem north. In Uganda, the British granted the old Buganda kingdom a special status within the federal state. Unfortunately, the federal framework as conceived by the British was not strong enough to withstand the devisive forces that emerged after independence. Even before Nigeria became free, Nigerian statesmen were warning Britain that the planned three-region federation would not last; in Uganda, the Ganda were displeased with what they considered the insufficient privileges of their Buganda kingdom, while nationalist leaders voiced disapproval of the special status to be accorded to the traditionalist kingdom. Before long, a bloody civil war was tearing Nigeria apart, and the *Kabaka* or king of Buganda barely escaped with his life to an exile of poverty in London. It would almost seem as though federalism is just not a workable political philosophy in present-day Africa.

In contrast to the British, the French put a cloak of uniformity over their colonial realm in Black Africa. Contiguous and vast, though not very populous, France's colonial empire extended from Senegal eastward to Chad and southward to the former French Congo. This huge area was divided into two units, French West Africa (centered upon Dakar) and French Equatorial Africa (whose headquarters was Brazzaville). After World War I, France was granted a mandate over the former German colony of Kamerun in equatorial Africa and over a part of Togo in West Africa; its only other dependencies were French Somaliland, the gateway to Ethiopia, and the island of Madagascar. France itself, we know, is the textbook example of the centralized, unitary state, whose capital is the cultural, political, and economic focus of the nation, overshadowing all else. The French brought their concept of centralization to Africa as well. In France all roads led to Paris: in Africa all roads were to lead to France, to French culture, to French institutions. For the purposes of assimilation and acculturation, French West Africa, half the size of the entire United States of America though with a population of only some 30 million in 1960, was divided into administrative units, each centered on the largest town, all oriented toward the governor's headquarters at Dakar. As the map shows, great lengths of these boundaries were delimited by straight lines across the West African landscape; history tells us to what extent they were drawn not on the basis of African realities but on grounds of France's administrative convenience. The present-day state of Upper Volta, for example, existed as an entity prior to 1932, when, because of administrative problems, it was divided up among Ivory Coast, Soudan (now Mali), and Dahomey. Then, in 1947, the territory was suddenly re-created. Little did the boundary makers expect that what they were doing would one day affect the national life of an independent country.

Of all the former French dependencies in Africa, only one has become a federal state, namely Cameroun, the former German colony and mandate. And Cameroun opted for federal organization not because of any French political legacy, but because after independence it acquired an additional area from the former British Cameroons, an area where the European language was English. The people in this former British area voted to join Cameroun, and Cameroun, in response, created a political framework in which the regional identity of the English-speaking area could be maintained. But elsewhere the pattern has been more according to the French plan: during their period of tenure the French created a French-speaking, acculturated African elite, trained at French universities and often experienced in French politics through direct representation in Paris. Out of these elites, headquartered in the colonies' capital cities, the governments of the now-independent states were forged. These have retained their ties with France, and virtually every modern institution, including the political machinery, is based on French models. Heavy investments are made, where possible, to make the capital a true primate city—Ivory Coast's Abidjan is a good example—and to improve the connections between the various regions of the country and the capital and core area. France has even stepped in with an armed force to support a government in trouble (Gabon), and it continues through aid programs, loans, educational assistance, and various other programs to maintain its presence in what is now called Francophone Africa.

Belgian administration in the Congo provides another insight into the results, in terms of spatial organization, of a particular set of philosophies of colonial government. Unlike the French, the Belgians made no effort to make Belgians of their African subjects. The policies identified under the catchword "paternalism" actually consisted of rule in the Congo by three sometimes competing interest groups: the Belgian government, the managements of huge mining corporations, and the Catholic Church. And, as it happened, each of these groups had major regional spheres of activity in this vast country. As the map shows, the Congo has a corridor to the ocean along the Congo River, between Angola to the south and the former French Congo to the north. The capital, long known as Leopoldville (now Kinshasa), lies at the eastern end of this corridor; not far from the ocean lies the country's major port, Matadi. Here has developed the Congo's administrative and transport core area; this was the place from where the decisions made in Brussels were promulgated by a governor-general; here now rules the Congolese government. But the great mining concerns operated elsewhere, mainly in the southeast in a province known as the Katanga. There, centered on the economic capital of the Congo, Elizabethville (now Lubumbashi), emerged the economic core area of the country. From hundreds of miles around, and even from outside the Congo, African labor came to the mines of the Katanga, attracted by the wages and the new way of life of the cities and towns. Finally, there was the enormous Congo Basin itself, the heart of the country, once the profitable source of wild rubber and ivory. Here the church took over when the companies that had ruthlessly exploited people and resources alike closed down, and often the Africans' only real contact with their European rulers was through the mission station, the medical dispensary, or perhaps the mission school. Of course, the church had a vested interest in Belgian colonial policy; here was an immense field for the spread of the Catholic faith, and in the nature of Belgium's paternalist policies lay a guarantee that the field would be open for many decades to come.

Belgium's administrative program was reflected in another way in the Congo. The country is shaped rather like a giant, wide-rimmed saucer, and it is along the rim that most of the resources and productive capacity are located. Here lay the revenues, the assets; in the deep basin lay the liabilities, the costs of administration. Hence the country was divided in such a way that each of the six provinces had a share of the rim as well as a section of the interior basin. The province of Leopoldville was situated in the corridor to the ocean and included the capital, but also the lowland around Lake Leopold. In Kasai the south (near the Angola border) produced diamonds, but the north was the very heart of the basin. In Katanga were the rich copper mines, and rather less of the low interior. In Kivu were fertile slopes and healthful recreation areas, but also a forested, hot, humid area west of the Congo River. Northern Oriental produced gold, cocoa, and cotton, but the province extended well into the low heart of the country. Least well endowed was Equator, which

formed the northeastern corner of the Congo and lay largely within the basin itself.

This administrative framework was developed not merely to facilitate the governing of the Congo, but also to maximize the exploitation of Katanga. It was in Katanga that Belgium reaped the fruits of King Leopold's colonial designs, and Belgian administration was designed to develop this mineral complex (including the world's richest copper ores) as fully as possible. Investments were made to overcome Katanga's problems of inaccessibility and to improve connections with the Atlantic ports of Lobito and Matadi; roads, schools, hospitals, housing, and other facilities were built. For all this, the rest of the Congo paid—not just by losing in the disproportionate appropriations that favored the Province of Katanga all the time, but also directly, in that moneys earned by the colonial government in the other provinces were spent in the development of the Katanga economic region. Small wonder that, when independence came, the Congolese government was determined not to let the Tshombe-led and corporation-supported secession effort of Katanga succeed.

Whereas British, French, and Belgian colonialism in Africa are matters of the past, and we can now witness the efforts of African states to capitalize on whatever assets were created (and struggle with the many liabilities that are part of the colonial legacy), Portugal in the late 1960's still remained a colonial power on the black continent. The Portuguese themselves do not agree: they claim that their African dependencies are in fact full-fledged "provinces" of the state of Portugal, integral parts of a country that just happens to be scattered all over the world. Thus Angola is just as much a part of Portugal as Alaska is a part of the U.S.A. Greater Portugal, then, is a community of political entities in which none stands below any other; the cultural focus and political center lie in the European entity, but all territories and all citizens are, theoretically, equals.

After many decades of inflexible, iron-handed rule over its African dependencies, Portugal has recently shown that it, too, feels the Wind of Change. Its response has been to change colonial policies somewhat rather than to begin preparations for a transfer of sovereignty; in the Portuguese program for its African "provinces" there is no room for any talk of independence. Angola, the richest of the three territories, and Moçambique have long contributed heavily to the Portuguese economy, and the loss of its African empire would be a blow that the state probably could not sustain. Thus Portugal is fighting a large-scale rebellion in Angola and a smaller one in northern Moçambique. The cost of these wars is an enormous burden, but Lisbon feels that there is no choice: the empire must be maintained.

Until the early 1960's, Portugal's colonial policies were designed almost solely to extract from the African domain the greatest possible profits. These profits were based on (1) the favorable location of coastal Angola and Moçambique relative to the developing Katanga, Copperbelt—Great Dyke, and Witwatersrand, (2) the African labor force, especially that of southern Moçambique, (3) the agricultural potential of certain areas, and (4) the presence of some mineral deposits, primarily in Angola. For many years most of Katanga's copper was hauled along the Benguela Railroad across Angola, a joint Belgian-Portuguese venture. In Moçambique, Beira lies opposite Rhodesia, and Lourenço Marques is the nearest port to the Witwatersrand, Johannesburg, and South Africa's core area. As the interior developed, with its rich minerals, good farm land, growing urban centers, and ever-increasing volume of trade, the ports of Portuguese Africa prospered. And the mines, especially those of South Africa, needed labor—something that was plentiful in Moçambique. The Portuguese and the South Africans made a deal whereby the Portuguese induced its African subjects to seek work at the recruiting stations of the gold mines, in return for which the Portuguese government received a guaranteed percentage of the external trade of South Africa as well as a payment for every laborer counted at the border. The way in which Africans were induced to volunteer for such labor had much to do with the way in which Portuguese territorial organization was designed, as is described below. The agricultural potential, too, was realized in part through a system of enforcement—a system that became infamous as the "forced crop" program. By this scheme, African peasants were compelled to place part of their land under cash crops, especially cotton. Food supplies were severely affected and famines occurred, but the Portuguese still punished those who refused to grow cash crops. Later, the Lisbon government began to encourage the voluntary emigration of Portuguese peasants to the better lands of Angola and Moçambique, where, along with African farmers, they joined in the creation of cooperative-type settlements. This program is still going on. Of course, Portuguese Africa has its share of huge European owned estates: in the hinterland of the Angolan capital, Luanda, these produce a large coffee crop. Finally, although Portuguese Guinea is quite poor and Moçambique has only minor known mineral deposits, Angola does have very productive diamond fields on the Kasai River boundary with the Congo, copper ores in the hinterlands of both Luanda and Lobito, iron ore at Cassinga, petroleum, and a number of other paying mineral resources.

The spatial features of Portugal's colonial administration have been much less obvious than those of Britain or Belgium. In Moçambique, for example, there was always an official map, showing the territory simply divided into a number of districts, each with a Portuguese official in charge. But the drive to round up labor in the south, and the "forced crop" system in the north led to the superimposition of a wholly different administrative structure. An anthropologist, Professor M. Harris, was the first to describe this system in detail.[8] While engaged in field research, he realized that while the official maps of Moçambique might show the *de jure* colonial administrative framework, the *de facto* arrangement was something else. What was needed to serve the objectives of Portuguese colonial exploitation was control at the local level—not through indirect rule or through missionary stations and schools, but through the direct imposition of governmental authority. And for this purpose the Portuguese colonial administrative framework was designed.

As Professor Harris describes it:

". . . Moçambique is divided into approximately eighty *conçelhos* (urbanized districts) and *circunscriçoes* (rural districts). Each of these administrative units is governed by an administrator assisted by a few 'chiefs-of-post.' Within the administrative corps, the highest rank held by Africans is that of interpreter. The median pop-

[8] M. Harris, *Portugal's African Wards,* American Committee on Africa, New York, 1958.

ulation under the control of a single administrator is 63,000, while fifteen of the districts contain more than 100,000 people. Over their African wards, the administrators and to a lesser extents, the chiefs-of-post, exercize broad discretionary powers.

"Within his district, the administrator executes virtually all police and judicial functions. He has the power to accuse, apprehend, try, and sentence delinquents. No *indigena* can enter or leave his district without his or his chief-of-post's permission. He has the power to regulate all commercial transactions involving *indigenas* and 'civilized' persons, including the sale and purchase of land, machinery and consumer goods. Only with his permission may his wards exercise a particular occupation or profession. He controls all bank deposits and is the executor of all legacies. He can draft laborers for public service and directs the labor of those in penal servitude. He assigns areas and types of crops to be planted and orders or prohibits the sale of cattle. All large-scale feasts and ceremonies must receive his authorization. And finally, he is in practice, the sole judge of whether or not an *indigena* possesses the qualifications for becoming a legal citizen.

"In Moçambique, the administrative districts are understaffed and the responsibilities are overwhelming. Most administrators have no choice but to exact stern, almost military discipline from their wards. In the rural areas, native men and women must salute or remove their hats when any white person passes. Everywhere that the administrator goes within his district, he is greeted with profuse displays of respect, mixed occasionally with genuine admiration but more usually with fear. Yet, despite their key position in the architecture of the *indigenato*, administrators and chiefs-of-post are notoriously underpaid. The fact that they command the lives of thousands of Africans does little to raise their status among the European colonists. The latter tend to regard the administrators as opportunists who have not exiled themselves into the bush without expecting to be paid for it. Obviously men upon whom such extraordinary powers are conferred, do not lack opportunities for personal gain. Although there are many administrators whose personal standards are above reproach, the system invites many others who readily succumb to its built-in temptations."[9]

This system, which drove young men away from their homes and villages to seek work in South Africa elsewhere beyond Portuguese Africa's boundaries, which disrupted family and tribal life, and destroyed the whole fabric of African society, finally was legally abolished in the early 1960's. Northern Moçambique's "forced crop" system also was modified, but the consequences of many decades of this kind of colonial rule cannot be erased overnight, and it has left indelible marks on the colonies' political and cultural geography.

[9] *Ibid.,* pp. 8–10.

AFRICAN REGIONS

On the face of it, Africa would seem to be so massive, so compact, and so unbroken that any attempt to justify a regional breakdown appears doomed to failure. No deeply penetrating bays or seas create peninsular fragments, as in Europe; no major islands (other than Madagascar) provide the sort of fundamental regional contrasts we see in Middle America; neither does Africa taper southward to the peninsular proportions of South America. Nor is Africa cut by an Andes or a Himalaya Mountains. Does any clear regional division nevertheless exist?

Indeed it does. To begin with, the Sahara Desert, extending as it does from west to east across the entire northern part of the continent, constitutes a broad transition zone between Arab Africa and Black Africa. This is not to suggest that the desert separates totally contrasting cultures and is therefore an absolute dividing line; we know that busy trade routes crossed it for centuries, and that millions of people who live to the south of the Sahara share the religious outlook of the Arab world. Political ideas and agricultural innovations, goods and people came from north of the Sahara to the south, and vice versa—and yet Black Africa retains a cultural identity quite distinct from that of the Arab realm.

Within Africa south of the Sahara regional identities exist as well, although the regional boundaries involved are not easily defined and have at times been the subjects of debate (Map

9-9). Three such regions are in common use: *West* Africa, which includes the countries of the west coast from Senegal to Nigeria, *East* Africa, by which is normally understood the three states of Kenya, Uganda, and Tanzania, and *South* Africa (or *Southern,* to get away from the political connotation), of which South Africa and Rhodesia are parts. Less clear has been the use of such regions as *Equatorial* Africa, by which is generally meant the Congo (Kinshasa) and the countries that lie between it and Nigeria. Although the appellation "Equatorial" presumably has to do with the location of these countries on or near the equator, the similarly positioned countries of East Africa are practically never referred to in this way. In fact, "equatorial" has come to mean hot tropical and low-lying rather than equatorially located: thus Gabon is a part of Equatorial

Africa, while Uganda is not. A similar problem has arisen with the use of *Central* Africa. Presumably the Congo, again, would lie in the heart of "Central" Africa, since, not unreasonably, we might associate "central" with "middle." But anyone who looks into this will find that the regional term Central Africa has been applied, at one time or another, to practically every noncoastal part of the continent. Within the past two decades two states have even adopted it in their names: the Central African Republic (formerly the French dependency of Ubangi-Shari) and the now-extinct Central African Federation (also known as the Federation of the Rhodesias and Nyasaland). And, as will be clear from the following discussion, even the dimensions of the more accepted regional divisions of Africa are still being debated.

Map 9–9 AFRICAN REGIONS.

West Africa

West Africa extends from the margins of the Sahara Desert south to the coast, and from Lake Chad to Senegal. Politically, the broadest definition of the region includes all those states that lie to the south of Morocco, Algeria, and Libya, and west of Chad (itself sometimes included) and Cameroun. Apart from Portuguese Guinea and long-independent Liberia, West Africa comprises only former British and French dependencies: four British and nine French. The British-influenced countries (Nigeria, Ghana, Sierra Leone, and Gambia) lie separated from each other; Francophone West Africa is contiguous. As Map 9-10 shows, political boundaries extend from the coast into the interior, so that from Mauritania to Nigeria the West African habitat is parcelled out among parallel-oriented states. Across these boundaries, especially across those between former British and former French territories, moves very little trade. For example, in terms of value, Nigeria's trade with Britain is about one hundred times as great as its trade with nearby Ghana. The countries of West Africa are not interdependent economically, and their income is to a large extent derived from the sale of their products on world markets. But the African countries do not control the prices their goods can demand on the world markets, and when these prices fall, they face serious problems. So it was with Ghana. Ambitious development programs begun under the leadership of President Nkrumah collapsed when the world price of cocoa—the country's chief export product—declined sharply.

If the economic, and in some cases the political, contacts between the former colonial powers and the West African countries remain stronger than the ties between the African states themselves, what are the justifications, if any, for the concept of a West African region? First, there is the remarkable cultural and historical momentum of this part of Africa. The colonial interlude failed to extinguish West African vitality, expressed not only by the old states and empires of the savanna and the cities of the forest, but also by the vigor and entrepreneurship, the achievements in sculpture, music, and dance of peoples from Senegal to Iboland. Second, West Africa from Dakar to Lake Chad carries a set of parallel ecological belts, so clearly reflected by Plates 1-4, whose role in the development of the region is pervasive. As a map of transport routes in West Africa indicates, connections within each of these belts, from country to country, are quite poor; no coastal or interior railroad was ever built to connect the coastal tier of countries. Communication and contact across these belts, on the other hand, is better, and some north-south economic exchange does take place, importantly in the coastal consumption of meat from cattle raised in the northern savannas. Third, West Africa received an early and crucial impact from European colonialism, which with its maritime commerce and the slave trade transformed the region from one end to the other. It was an impact that was felt all the way to the heart of the Sahara, and it set the stage for the reorientation of the whole area—out of which emerged the present patchwork of states.

The effects of the slave trade notwithstanding, West Africa today is Black Africa's most populous region (Plate 6). In these terms, Nigeria, with perhaps 55 million people, is Africa's largest state, and Ghana, with 9 million, also has a high ranking. As Plate 6 shows, West Africa also has regional identity in that it constitutes one of Africa's five major population clusters (the others: northern Morocco and Algeria, the Nile valley and delta, the Lake Victoria environs, and eastern South Africa). The southern half of the region understandably carries the majority of the people: Mauritania, Mali, and Niger include too much of the unproductive steppe and desert of the Sahara to sustain populations comparable to Nigeria, Ghana, or Ivory Coast.

This is not to say that only the coastal areas of West Africa are densely populated. True, from Senegal to Nigeria there are large concentrations of people in the coastal belt, but the interior savannalands also contain sizable clusters. The peoples along the coast reflect the new era brought by the colonial powers: they prospered in their newfound roles as middlemen in the coastward trade. Later they were in position to undergo the changes the colonial period brought: in education, religion, urbanization, agriculture, politics, health, and many other fields they adopted new ways. The peoples of the interior, on the other hand, retained their ties with a very different era in African history; distant and often aloof from the main scene of European colonial activity, they experienced very much less change. But the map reminds us that Africa's boundaries were not drawn to ac-

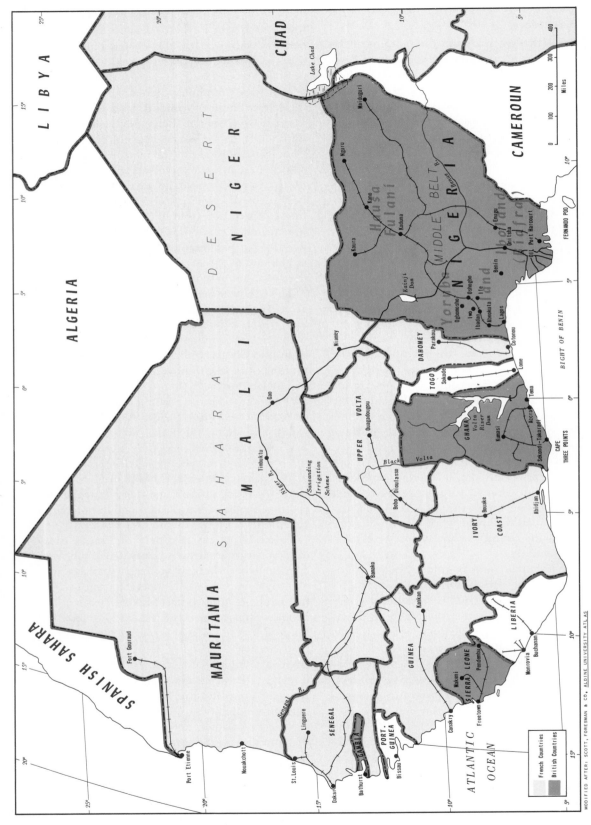

MODIFIED AFTER: SCOTT, FORESMAN & CO., ALDINE UNIVERSITY ATLAS

Map 9-10 WEST AFRICA.

French Countries

British Countries

Map 9-11 NIGERIA.

commodate such differences. Both Nigeria and Ghana have population clusters representing the coastal as well as the interior peoples, and in both countries, the wide gap between north and south has produced political problems.

When, in 1960, Nigeria (Map 9-11) achieved full independence, it was endowed with a federal political organization based on the three major population clusters within its borders—two in the south and one in the north. Around the Yoruba core in the southwest lay the Western Region. The Yoruba are a people with a long history of urbanization, but they are also good farmers; in the old days they protected themselves in walled cities around which they practiced intensive agriculture. The colonial period brought coastal trade, increased urbanization, cash crops (the mainstay, cocoa, was introduced from Fernando Poo in the 1870's), and, eventually, a measure of security against encroachment from the north. Lagos, the country's capital and leading port, developed on the region's south coast; Ibadan, one of Black Africa's largest cities (1 million), evolved from a Yoruba settlement founded in the late eighteenth century. At independence the Western Region counted some 8 million inhabitants, and more than any other part of Nigeria it had been transformed by the colonial experience. East of the Niger River and south of the Benue, the Ibo population formed the core of the Eastern Region. Iboland, though coastal, lay less directly in the path of colonial change, and its history and traditions, too, differed sharply from those of the West. Little urbanization had taken place here, and even today, while over one-third of the Western Region's people live in cities and towns over 20,000, a mere 10 percent of the Eastern population is urbanized. With some 12 million people, the rural areas of Eastern Nigeria are densely peopled, and the local economy has long been based on the oil palm. This tree yields oil and kernels, both export products, and its leaves and fibers are used as building materials and for matmaking; for home use, soap, lamp fuel, cooking oil, and other commodities are extracted from the oil. The oil palm thus is a basis for subsistence in the East. In recent years, the discovery of significant oil fields in and near the Niger delta have given the economy of this part of Nigeria— and, indeed, of the country as a whole—a boost. Still, many Ibo over the years have left their crowded habitat to seek work elsewhere, in the west, in the far north, in Cameroun, and even in Fernando Poo. The third federal region at independence was at once the largest and the most populous: the Northern Region, with some 30 million people. It extended across the full width of the country from east to west, and from the northern border southward beyond the Niger and Benue Rivers. This is Nigeria's Moslem North, centered on the Hausa-Fulani population cluster, where the legacy of a feudal social system, conservative traditionalism, and resistance to change hangs heavily over the country. The core area lies focused on Kano, the shipping center of the region's chief agricultural product, peanuts. Kano is one of those old savanna cities that bear witness to the long urban traditions of the savanna: it was a town even in the days of the old Ghana empire, a dozen centuries ago.

Nigeria's three regions, then, lay separated—not only by sheer distance, but also by tradition and history, and by the nature of

A street scene in Lagos, Nigeria. Lagos, on the coast west of the Niger River, became the federal capital upon independence. It is a cosmopolitan city with impressive government buildings and reminders of a varied past: some of the three-story houses along this street were built by freed slaves who returned from Brazil and became part of Nigerian high society. (Marc & Evelyne Bernheim/Rapho Guillumette)

colonial rule. In Nigeria, even physical geography and biogeography conspire to divide south from north: across the heart of the country (and across much of West Africa in about the same relative location) stretches the so-called Middle Belt, poor, unproductive, disease- and tsetse-ridden country that forms in effect an empty barrier between the regions. Plate 6 reflects the "empty heart" of Nigeria—and Ghana as well—and suggests the extreme northerly location of the Northern Region's population cluster. For example, pastoralism, an important livelihood in the north, is made impossible by the conditions that prevail in the Middle Belt. But even more difficult than the economic situation have been the political conflicts that emerged after independence. With its numerical superiority, the Northern Region could dominate federal politics; parties that had had "national" allegiances prior to independence soon polarized into regional bases. Latent hostilities between the Hausa and the numerous Ibo whose families had gone north to seek work led to murders and a mass evacuation of Ibos to an Eastern Region many of them had never seen. Even before independence, Nigerians had warned the British that too small a number of regions could lead to successful secession attempts: Azikiwe wanted 8 regions for Nigeria, and another prominent Nigerian, Awolowo, thought in terms of 30 to 40. But this seemed to be a reversal to the tribalism the federal framework was designed to counteract, and the three-region federation became reality, though one modification—the creation of a Midwest Region out of the Western Region—was made.

It was not enough. Not long after the attacks on the Ibo in the Northern Region, and a series of political conflicts, the Eastern Region declared its secession from the Nigerian state, and pronounced itself the sovereign state of Biafra. Actually, each of the Nigerian regions incorporates not only the core clusters of people (such as the Yoruba and Hausa-Fulani), but numerous minorities as well; in the case of Biafra, the dominant Ibo made the decision to secede, but they took several million non-Ibo with them in a civil war that was still in progress in late 1968. In early 1968, while the war raged, the decision was made to redivide Nigeria into 12 regions, in the hope that such a framework would help prevent any recurrence of Biafra-type conflicts.

Of all the states of coastal West Africa, Nigeria

A view over Ibadan, Nigeria. This Yoruba city is one of the largest in Africa. This photograph reflects the changes brought in recent times: the corrugated iron roofs are now the nearly universal roofing material in the larger towns and cities; the Christian church that dominates the landscape represents the new religious order and the alien power that introduced it. (Marc & Evelyne Bernheim/ Rapho Guillumette)

extends farthest into the savanna-steppe interior, and so it has the greatest range of internal contrasts. Ghana, too, has a middle belt and a Moslem cluster in the north, but unlike Nigeria, Ghana is a single-core country—and the core area lies in the forest belt, in the south. Ghana and its French-influenced western neighbor, Ivory Coast, share an area of productive capacity that lies within a semicircle with a radius of about 300 miles from Cape Three Points. In Ghana, this includes the cocoa farming areas of the southwestern forest, the mineral zone between Takoradi and Kumasi, where manganese ore, bauxite, and gold are mined, the timber areas in the hinterland of Takoradi, and the new Volta River Dam; it includes also the whole railroad system of the country, its capital (Accra), the old port of Takoradi and the new one at Tema, and the historic capital of the Ashanti empire, Kumasi. In Ivory Coast, where the southeast forms the core area, lie coffee and cocoa farms; banana and cotton cultivation are on the increase here. The focus of Ivory Coast is its impressive capital, Abidjan (300,000).

Where the West African coast turns northward lie two countries whose history is bound up with the slave trade and its termination. Sierra Leone was founded in 1787 by the British as a place of settlement for freed African slaves, and Liberia had similar origins: it was selected by the American Colonization Society to serve as a home for slaves who had been in bondage in America. Liberia became independent in 1847, but it was always a troubled country. It was hard pressed to maintain itself in the face of the nineteenth century European scramble for Africa and lost territory in the process. There were conflicts between the "Americo-Liberians" and the indigenous peoples of the interior, and in 1928 Liberia itself was charged with slavery by the League of Nations in an issue that involved the allegedly forced transportation of several thousand young men from the interior to the Spanish island of Fernando Poo. Economically the country was poor, and not until the twentieth century rubber boom did it enter a phase of growth, a phase that was later strengthened by the exploitation of rich iron ores in the Bomi Hills and Bong Mountains. Sierra Leone remained a British dependency; the colonial presence was deemed necessary in order to protect its territory against European encroachment and to maintain stable relationships between the new black settlers and the indigenous peoples. In fact, after Liberia, Ghana was the first black African state to attain sovereignty (1957); Sierra Leone's independence (1961) came even after that of Nigeria.

The nine states of former French West Africa have a combined population that is less than that of Nigeria. Reflecting France's decision to divide its vast West African empire into convenient administrative units, there is no great range in the population figures of the nine states: Upper Volta and Mali have between 4 and 5 million people, Ivory Coast, Guinea, Senegal,

and Niger have between 3 and 4 million, Togo and Dahomey, between 1.5 and 3 million, and Mauritania, about 1 million. These are poor countries; by comparison, Nigeria and Ghana are prosperous states. The vast majority of the people eke out a life of difficult subsistence which, in Mauritania, Mali, and Niger, involves nomadic or seminomadic pastoralism. In the countries of the dry zone, the population concentrates near the water courses (the Senegal River in Mauritania, and the Niger River in Mali and Niger) and in the southernmost areas, where there is perhaps adequate moisture for farming. The cash crop that provides the greatest opportunities for income is peanuts, grown from Gambia and Senegal, through Mali and Upper Volta to Niger and Nigeria. In Francophone Africa, Senegal is the leading producer. In coastal countries, coffee, cocoa, bananas, and oil palm products are exported, but again the cash crop farmers are far outnumbered by people who live an agricultural subsistence. Almost everywhere in former French West Africa, the per capita annual income is under $100, and the young states depend on French loans and assistance for their very survival. Best off is Ivory Coast, where the agriculturally based economy is doing fairly well by the prevailing standards; Guinea may have the ingredients necessary for a sound economy, with its banana and coffee farms and its bauxite and iron mines.

Even an ordinary atlas map suggests the state of things in former French West Africa: the only two cities of any size are Abidjan and Dakar (350,000), the erstwhile headquarters of the French colonial administration and now the capital of Senegal. The railroad lines are obviously designed to extract raw materials, not to interconnect the individual states. From Mali's capital, Bamako, a single line leads to Dakar; from Ouagadougou, Upper Volta's capital, a similar line leads to Abidjan. No external railroads at all exist in Niger (or, for that matter, in the French-influenced countries of Chad and Central African Republic). Towns are mainly ports and/or capital cities, and there is little urbanization. But all this does not mean that Francophone Africa has little in the way of culture and tradition. Above all it has produced a number of great writers and poets, who have articulated what has become the pride of this part of West Africa, the concept of *negritude*. By this is meant all that is good in being black, with an emphasis on the long history, the arts and music of Africa. Negritude is a state of mind, a self-confidence that comes with self-knowledge, and it recognizes the kinship, the spoken and unspoken ties among black people everywhere. In fact, the concept seems to have originated in the French possessions in the Caribbean, from where it was carried to France, there to be adopted by students from West Africa working and writing at French universities. In Francophone West Africa it found its fullest expression, fired by the achievement of independence and by the desire to rediscover a history obscured by the colonial period. Thus the richness of African writing and art is specifically employed to sustain people faced with the bleak realities of life.

Periodic Markets. It is easy, in a discussion of continental regions, to give undue emphasis to the export products of the several countries involved, and to negate the local, domestic exchanges of goods. Yet these local transactions are of great importance, for more than anything else they bring people together and promote interaction of many kinds. In West Africa, as in many other parts of the "developing" world (including China, Korea, and Middle and South America as well), there are *periodic markets* at which these local exchanges take place. As the term implies, these are market places—sometimes located *in* villages, but also *between* villages—that operate at intervals. These intervals vary. In China there are markets which are open on every 3rd, 5th, 6th, 10th, or 12th day. In West Africa, the 4-day market is a basic temporal unit, especially in southern Nigeria, where there are 2-, 4-, and 8-day cycles, although in northern Ghana the 3-day cycle is basic.

Periodic markets, then, form a sort of interlocking network of exchange places that serves even areas where there are no roads; as each market in the network gets its turn, it will be near enough to one section of the area so that the people who live in the vicinity can walk to it, carrying what they want to sell or trade. In this way, small amounts of produce filter through the market chain to a larger regional market, where shipments are collected for interregional or perhaps even international trade.[10] What is

traded, of course, depends on where the market is located. A visit to a market in West Africa's forest zone will produce impressions that are very different from a similar visit in the savanna zone. In the savanna, sorghum, millet, and shea-butter (an edible oil drawn from the shea-nut) predominate, and one will see some Islamic influences here; in the south, such products as yams, cassava, corn, and palm oil change hands. In the southern forest zone, too, one is more likely to see some imported manufactured goods passing through the market chain, especially near the relatively prosperous cocoa, coffee, rubber, and palm oil areas. But in general the quantities of trade are very small—a bowl of sorghum, a

A periodic market in Mali, West Africa. This is a weekly market; people are arriving with their goods and are finding a spot to sell their wares. Some have already taken up position and the day's business is just about to begin. As the photograph shows, things are carried to the market mostly on foot; headloading is common. (Pierre Pittet/ World Health Organization)

[10] For details of this process see J. M. Hunter, "Seasonal Hunger in a Part of the West African Savanna: a Survey of Bodyweights in Nangodi, North East Ghana," *Institute of British Geographers, Transactions and Papers,* 1967, No. 41, pp. 167–185.

bundle of firewood—and their value is low; these markets serve people who mostly live near the subsistence level.

This does not mean that West Africa's markets fail to generate excitement, color, and jostling crowds. In fact, the markets are the main arenas not only for trade, but also for social and cultural exchange. In preliterate society they take the place of newspapers, radio, and television: here the latest news can be heard. Elders meet together on market day to discuss matters of local traditional government. Young men and women customarily use market days for courtship. Men attend market, especially in the savanna zone, to sit around and drink millet beer; the women do the bulk of the trading. All day, on market day, people barter, discuss, gossip, argue, eat and drink—and then everyone goes home and the market place is deserted until, three or four days later, the market has another turn.

East Africa

The narrowest definition of the East African region is that it includes only Kenya, Uganda, and Tanzania—the three states that emerged out of British East Africa. Some geographers, however, view East Africa as larger than that, to incorporate also Ethiopia and the Somali Republic, Rwanda and Burundi, Moçambique, and even Zambia and Malawi. Southern Moçambique and western Zambia, however, can hardly be considered to be parts of the same region as Ethiopia; rather, these areas are involved in the Southern African sphere, while Ethiopia, the Somali Republic, and the Sudan form a transitional region between Black and Arab Africa. More justification exists for the inclusion of Rwanda and Burundi. These countries lie east of Lake Tanganyika and Kivu, and they formed part of German East Africa. They were severed and placed under Belgian mandate after the First World War, and thus came to be oriented westward, to the Belgian Congo, rather than eastward, to British East Africa. But in effect these are East African countries, and they will be considered as such.

Despite the fact that they are neighbors, the three formerly British territories of East Africa display numerous differences. These differences have arisen less out of any contrasts in colonial rule than out of the nature of the East African habitat, the course of African as well as European settlement, and the location of the areas of productive capacity. This, of course, is highland, plateau Africa, mainly savanna country that turns into steppe toward the northeast. Great volcanic mountains rise above the plateau, which is cut by the giant rift valleys; the pivotal physical feature is Lake Victoria, in which the three major countries' boundaries come together (Map 9-12), and on whose shores lie the primary core area of Uganda and secondary cores of both Kenya and Tanzania. With limited mineral resources (the chief ones: diamonds in Tanzania, not far south of Lake Victoria, and copper in Uganda, to the west of the lake), most people depend on the land—and on the water that allows crops to grow and livestock to live. And in much of East Africa, rainfall is marginal or insufficient. The heart of Tanzania is dry, tsetse- and malaria-ridden, and occasionally still the scene of local food shortages. Eastern and northern Kenya largely consist of dry steppe. As Plate 1 shows, the wettest areas lie spread about Lake Victoria; in terms of rainfall received, Uganda is best endowed.

Thus Tanzania has been described as a "country without a primary core," for its areas of productive capacity (and its population, Plate 6) lie around its margins. Dar es Salaam (150,000), the capital, largest city, and chief port, lies on the coast, at the end of a cross-country railroad that reaches to Lake Tanganyika and was built by the Germans. Along the coast lie sisal plantations; in the northeastern corner of the country, in the hinterland of the port of Tanga, conditions are good for sisal and tea cultivation. Then the slopes of Mount Kilimanjaro have good soils which are farmed by the Chagga coffee farmers; the mountain lies on the Kenya border. On the shores of Lake Victoria, cotton is grown, and the lake port of Mwanza is connected by rail to the main line. On the northern shore of Lake Tanganyika, coffee is being tried; at the northern end of Lake Nyasa (Malawi), Mbeya is the center of a thriving agricultural district where tea, coffee, and pyrethrum are grown. The southern districts of the country, along the Rovuma River boundary with Moçambique, are mainly subsistence areas, but the southeast corner was the scene of an ill-fated colonial project, the Groundnut Scheme—an effort to establish large-scale peanut production. Thus Tanzania's cash cropping areas, and the subsistence farming as well, lie in an intermittent belt around the rim of this

SUDAN

ETHIOPIA

Nile River

Gulu

UGANDA

Pakwach

Lake Albert

Bunyoro

Lake Kyoga

Kitale

K E N Y A

Toro

Kampala

Buganda Ganda

Entebbe

Kisumu

Solai

Thomson's Falls

KENYA CORE AREA

Kasese

Luo

Kíkuyu

Nairobi

C O N G O

Lake Edward

Ankole

LAKE

VICTORIA

Lake Magadi

Tana River

Lake Kivu

RWANDA

Kigali

TSAVO NATIONAL PARK

Masai

BURUNDI

Mwanza

KILIMANJARO 19,565'

Bujumbura (Usumbura)

Shinyanga (Diamonds)

Arusha

Moshi

Voi

Malindi

Mombasa

Kigoma

Tabora

Tanga

PEMBA ISL.

Lake Tanganyika

T A N Z A N I A

ZANZIBAR

INDIAN

OCEAN

Mpanda

Morogoro

Dar-es-Salaam

Lake Rukwa

Mikumi

Rufiji River

KILOMBERO VALLEY

Mbeya

ZAMBIA

Lindi

Core Development

0   50   100   150   200
Miles

MALAWI

L. Malawi

Rovuma River

MOCAMBIQUE

Map 9-12 EAST AFRICA.

The Kenya Highlands. This region was known as the "White" Highlands during the colonial period, for here lay the farms of the white colonizers, who had reserved large areas of the Highlands for themselves. This photograph, taken in the vicinity of Eldoret, shows the modern aspect of the region, with its large parcels of land, contour plowing, and tree-lined windbreaks. (Marc & Evelyne Bernheim/Rapho Guillumette)

country. But Tanzania is not a country of one tribe or people; it is a country of many. Hence the disadvantages of physical and economic geography are counterbalanced in an important way by a human geography in which no single people has the overwhelming numbers and power, and neither do two or three potentially warring groups. Depending to some extent upon the kind of measure used, Tanzania may be said to have about 10 major and several dozen minor population groups, and in general the political relationships among these many peoples have been such that progress toward a stable state has been made.

Tanzania, southernmost of the three major East African countries, is also to the greatest degree a Bantu country; the main non-Bantu are the Masai of the north and, of course, the 100,000 Asians, 27,000 Arabs (mostly on the island of Zanzibar, which in 1964 was joined with Tanganyika), and perhaps 15,000 whites. With over 12 million people, Tanzania is also East Africa's most populous state, but, as the map quickly shows, this population lives in a country that is larger than all four remaining East African countries combined.

In some ways, Kenya, with a population of 10 million, is the direct opposite of Tanzania. It has a well-defined primary core area, centered

upon East Africa's largest city, Nairobi (350,000); much of what matters in the country is concentrated in the southwestern quarter of the country, including the Kenya Highlands (formerly known as the White Highlands and one issue of contention in the Mau Mau uprising of the 1950's). Unlike Tanzania, Kenya is dominated by one African people, the Kikuyu, but the Kikuyu have competitors, especially the Luo, a Nilotic people who live near Lake Victoria. Also unlike Tanzania, Kenya attracted a large (70,000) European settler population, which alienated a great deal of good farm land, especially in the Highlands, and established a thriving agricultural economy producing corn and wheat, tea, coffee, sisal, tobacco, pyrethrum, and other crops, in addition to meat and dairy products. Kenya also experienced a large influx of Asians (200,000) and, not counting Zanzibar, a larger Arab immigration as well.

Without major mineral resources, Kenya depends heavily on its agricultural economy—but not entirely. The drier bush areas are places where Africa's wildlife still survives, and Kenya is reaping an ever greater harvest of revenues from its game reserves. One of these, Amboseli, is located in Masai country on the Tanzania border, and since independence it has undergone some deterioration. This, fortunately, is not the general picture, and in Tsavo National Park and numerous other locations tourists come in growing numbers to view Africa's wildlife heritage. With the decreasing prices of intercontinental airline trips, this may become East Africa's greatest fortune; with reference to South African competition for the European market, this region is in an excellent position of intervening opportunity. In recognition of this, facilities are being expanded and access routes improved.

Coastal Kenya, Tanganyika, and Zanzibar from very early times lay in the Arab maritime sphere just as West Africa lay in that of Europe after the fifteenth century. Arab *dhows* carried trade and slaves to and from African coasts, and Arabs founded numerous settlements, including Kenya's only effective port, Mombasa. This city from the late fifteenth century became the scene of intense rivalry and conflict with the Portuguese, who arrived to challenge Arab mastery over the east coast. Eventually it became a bridgehead for British colonial activity, and with British pressure, the slave trade was finally ended

Kampala, a traditional capital in Uganda, East Africa. This was for many generations the seat of government of the king or *kabaka* of the BaGanda, whose Buganda Kingdom lay on the shores of Lake Victoria. Buganda was absorbed in the British Protectorate of Uganda; after independence it failed to maintain its identity and the king was forced to leave the country. Kampala, however, remains the capital of the state. (Marc & Evelyne Bernheim/Rapho Guillumette)

late in the nineteenth century. With Asian labor, the railroad was begun to connect Mombasa to the interior (the ultimate objective was Lake Victoria). Nairobi developed out of a repair and storage station on the railroad, which eventually reached Kisumu on the lake in 1901. This, of course, connected Uganda to the outside world, for goods could be shipped by water across Lake Victoria and then onward by rail. Ever since, Mombasa has been not only the sole outlet for Kenya, but the major ocean port for Uganda as well.

Uganda was a primary objective of British colonialism, for it was known that a large indigenous state existed there, namely Buganda, centered on the royal capital, Kampala. Uganda was an objective of German colonial expansion too, but it fell into the British sphere, and a protectorate was established there.

In many ways Uganda is East Africa's most interesting state. With 7.5 million people and only 91,000 square miles (20,000 of it in lakes, swamps, and marshes), Uganda is the most densely populated of the three major states of the region. By order of the British government, land alienation was prohibited, and only a fraction of Uganda's farmland is not owned and worked by African smallholders themselves. Cotton and coffee were introduced to the Baganda people whose kingdom, Buganda, occupied the southern part of the country; soon comparative economic prosperity was added to the prevailing political stability. And Uganda is truly a transitional country: its natural vegetation includes the savannalands of the east and the forests of the Congo margins in the west. The swampy lake regions of the south give way to the dryness and rain-deficient conditions that characterize the Sudan Basin to the north. Uganda is also situated astride ethnic transition zones: south of Lake Kyoga it is generally Bantu country, in the north it is mainly Nilotic, and in the northeast, Nilo-Hamitic. The south is dominantly agricultural; much of the north, pastoral. From the east have come the Arabs, Europeans, and Asians, and from the north came many of the political ideas out of which Uganda's strong traditionalism emerged.

The Kingdom of Buganda was the most important African political entity of the lakes region when the British entered the scene. It was not the only kingdom in the area, nor had it always been dominant; but the British found it in a position of primacy, on the shores of Lake Victoria, centered on an impressive capital at Kampala, stable, large, and eminently suitable for indirect rule. The British established their headquarters at Entebbe on the lake (thus adding to the status of the kingdom), and proceeded to solidify their hold over the Uganda Protectorate.

Uganda affords an excellent illustration of an idea published in 1954 by Professor S. B. Jones entitled *A Unified Field Theory of Political Geography*.[11] Professor Jones argues that whenever we see a politically organized area, we are looking at the end result of a number of "hubs" of activity that have succeeded one another over time. At the beginning, he argued, there had to be an idea—a political idea—from which the political framework eventually was to emerge. Thus he arrived at a 5-step model:

This became known as the "idea—area chain," but of course it is not a simple sequence in which one step regularly gives rise to the next. Rather, while the progression in general is from left to right (in the diagram), the several elements in the model interact with each other all the time, backward as well as forward. For example, when a certain decision occurs, such as the abolition of slavery, this may give rise to a form of movement—or pressure for movement—which in this case was a desire on the part of both black slaves

[11] S. B. Jones, "A Unified Field Theory of Political Geography," *Annals of the Association of American Geographers,* Vol. XLIV, 1954, p. 111.

and white societies to make possible an African homeland for repatriation. This led to a journey to West Africa by members of the American Colonization Society to seek such a home, and this, in turn, produced a new decision, namely to create the settlement in Liberia. A new phase of movement began with the journey of thousands of former slaves to West Africa (the field), and the political area, the state of Liberia, eventually emerged.

In Uganda, the fundamental political idea had both general and specific aspects. The general political idea, of course, was the prevailing imperial philosophy of the time, which brought not only Uganda but practically all parts of Africa under colonial control. Specifically with reference to Uganda, however, it involved the welding together of the Kingdom of Buganda, a number of smaller kingdoms, and numerous tribal peoples to the north. The conspicuous decisions were to establish a protectorate in Uganda, to introduce a form of indirect rule, and—most importantly—to accord Buganda a special status within the country. It was through this last idea that the significant movement came about. The British decided not only to use Buganda forms of government in the other parts of the territory, but they actually employed Buganda personnel to represent their authority elsewhere, outside Buganda. Hence people from Buganda went outside their kingdom, in fact, outside the Bantu part of Uganda, to help consolidate the new government. This was not just a movement of people; it also involved the introduction of new ideas and practices of many kinds. In the context of the field theory model, movement does not mean only the migration of people, but also the spread of new ideas, the introduction and diffusion of new ways of doing things. Obviously the imposition of British control and the use of BaGanda administrators led to many kinds of movement. From their capital at Entebbe and from the Buganda headquarters of Kampala, roads began to radiate outward to other parts of the country, and traffic on the many waterways was regularized. Police stations, post offices, and many other facilities associated with the new form of government made their appearance. All of this was related to the movement phase.

The field in which this sequence of events occurred, in the broadest terms, was East Africa.

But, anomalously, the outlines of the political area had already been established—unlike the Liberian example, where the "field," West Africa, was long the scene of an expanding coastal settlement which grew by purchase and conquest, and eventually attained the characteristics of a state system, complete with stable boundaries. The politically organized area of Uganda came about long after a boundary girdle had been thrown around it, evidence that the model's chainlike sequence should not be viewed as a straightjacket for politico-geographic development.

The privileged position of Buganda during the British administration, and the wealth and productivity concentrated there, produced political problems at independence. The king and his people demanded a federal constitution whereby Buganda would have a certain autonomy within the state of Uganda, and when the British plans were seen as an insufficient guarantee, Buganda threatened to secede. Eventually independence did come (1962), but the latent conflict between traditionalist Buganda and the non-Buganda sections of the country remained. Eventually, the political struggle turned into a crisis, which led to some fighting and to the forced exile of the king; in the new Uganda, the days of a politically dominant Buganda are at an end.

Rwanda and Burundi. The two states that lie in the northwestern corner of Tanzania, are the products of World War I. Belgian Congolese forces attacked German East Africa from the west and contributed to the Germans' defeat; after the war, the League of Nations at first failed to grant Belgium a mandate over any part of the German colony. The Belgians resented this action, and appealed it, indicating a desire for control over the two territories where they had carried the battle most strongly. These were known as Ruanda and Urundi, and they lay adjacent to the Congo though on the east side of Lakes Kivu and Tanganyika. The League of Nations, which had granted Britain a mandate over all German East Africa, gave Belgium the right to negotiate with the British for these territories, and so "Ruanda-Urundi" fell into Belgian hands.

Territorially these countries of mountain slopes, highlands, and valleys are small, but their population, as Plate 6 shows, is great—between 6 and 7 million on their combined area of about 20,000 square miles. Even in the 1920's the population already was some 3 million, and this was what attracted the Belgians: here lay a source for Katanga labor, comparatively near the mines yet outside the Congo proper. Just a generation after Belgium had been subjected to world criticism over the infamous Congo atrocities, the country once again faced censure, this time over the way it was transferring large numbers of residents from Ruanda-Urundi to the Katanga under conditions which seemed to contravene the terms of the League of Nations' mandate. But the Belgian administration had a twofold objective: not just to supply labor to the copper mines, but also to alleviate somewhat the enormous population pressure in their new dependency. Even today the average population density per square mile of arable land (including pasture, which is of good quality) approaches 400, and in central Rwanda and northern Burundi it is two to three times that high. Few alternatives to subsistence agriculture exist; coffee and cotton seed are exported but in very limited quantities. The transport system, rudimentary as it is, reflects the low productivity as well as the low demand for external goods in Rwanda and Burundi; the very low degree of urbanization describes the internal economic situation. Here, in two countries numbering 6 to 7 million inhabitants, Bujumbura (Burundi's capital) is the largest town at 70,000 people, and Kigali (Rwanda's) is *second* at 7000. Only eight other towns have more than 1000 residents.

The social situation was never promising, and after independence it produced ugly conflicts. Three distinct peoples live here: the famous WaTusi (Tutsi), the tall pastoralists, whose number at independence was perhaps 15 percent of the total, but who have long ruled over the "middle" group, the BaHutu, who form the bulk (over 80 percent) of the population and who are related to other Bantu farmers in the Congo. At the bottom of the scale are some thousands of BaTwa (pygmies), who are generally in the service of the BaHutu. After independence, the BaHutu rose against the WaTusi, especially in Rwanda; Belgian administration had done little to reduce the intensity of the latent conflict that lay beneath the surface throughout its tenure. The result was a disastrous decimation of the WaTusi people. Many died; still more fled their country into Uganda and Tanzania.

"Equatorial" Africa

As the term is often used, "Equatorial" Africa lies to the west of highland East Africa, and consists of the Congo (Kinshasa), the "other" Congo, whose capital, Brazzaville, lies opposite Kinshasa across the Congo River, Gabon, the Central African Republic, Cameroun, and Equatorial Guinea (Map 9-13). Chad, to the north of the Central African Republic, was a part of French Equatorial Africa and is often still included in this region, but it lies between Niger and Sudan, in the West Africa-related savanna belt, and only by the most tenuous argument can it be considered a part of Equatorial Africa.

The giant of Equatorial Africa, of course, is the former Belgian Congo which, with not much

Map 9-13 EQUATORIAL AFRICA.

less than a million square miles and 15 million people, has the bulk of the region's human and natural resources, its largest cities, its best communications, and undoubtedly its greatest opportunities and potential. The Francophone countries of Equatorial Africa have less—much less—and from the beginning of the colonial era this has been obvious in their development. Brazzaville, the old French headquarters, was not connected by rail to the port of Pointe Noire until more than 30 years after Leopoldville had been linked with Matadi. Brazzaville always lagged behind Leopoldville in terms of size and growth rate. In fact, France managed to develop only the most minimal communications in its equatorial domain, and in no way was French Equatorial Africa an interconnected whole.

Independence has not made the economic picture much brighter, although a few resources have been brought into production in recent years. Gabon, a land of forests, has long depended on wood products such as mahogany, ebony, and okoumé (a plywood material) for its meager export revenues, but now a mineral phase seems to be in the offing with oil production near Port Gentil and manganese and iron deposits in the interior. The Congo (Brazzaville) is a resource-poor country, whose major asset can be identified in one word: location. It occupies the lower section of a major Equatorial African exit and entry route, the Ubangi River, possesses the chief port of Francophone Equatorial Africa, Pointe Noire, the largest city, Brazzaville, and the railroad that crosses the intervening mountains. The Pointe Noire–Brazzaville–Bangui route used to be known as the Federal Artery, since it was the best entry into French Equatorial Africa; hence the Congo was endowed with the colonial headquarters. One hope for the Congo's future lies in the perpetuation of its break-in-bulk in Equatorial Africa. The landlocked Central African Republic has better agricultural possibilities; it can produce cotton and coffee, and cattle can be raised on the highland areas—but its colonial history was one of conflict and its location is a great obstacle. *Together,* the three countries just mentioned have a mere 3 to 4 million people, another measure of their limitations.

Cameroun, with a population of 6 million, is the wedge-shaped country that adjoins Nigeria. Both its location and its shape have some relevance: its shape reflects the northward drive of German colonialism at a time when it hoped to obstruct the designs of Britain and France, and its location adjacent to Nigeria bequeathed it with a section of English-speaking Africa. Upon independence, the so-called Cameroons—a stretch of territory between Cameroun and Nigeria that had been under British mandate—was given an opportunity to choose its alignment. The northern portion went with Nigeria, but the south, despite its British legacy and Nigerian proximity, voted to join the French-influenced Cameroun; hence the sharp bend in the boundary between the two countries. In response, Cameroun adopted a federal political framework, to accommodate the English-influenced province. Compared to its equatorial neighbors, Cameroun is relatively well off. It received better treatment as a German colony than did any other German territory in Africa; the Germans laid out an economic blueprint that included transport development, research on crop suitability, the involvement of African farmers in a cash crop economy, and limitations on land alienation by Europeans. The French, during their mandate, did not capitalize to any great degree on these early advantages, but Cameroun stayed ahead of the other countries in the region. Physically the country is quite varied, so that peanuts can be grown in the far north, cattle can be raised on northern and western slopes, and such crops as coffee, bananas, cocoa, and cotton can also be cultivated. The most impressive development has taken place in the west, where the major port, Douala, the capital, Yaounde, the Edea hydroelectric project, and industrial complex and the Sanaga River Valley agricultural development are all located.

Southern Africa

Southern Africa is the richest region of the continent. From the Katanga southward, through Zambia's Copperbelt, Rhodesia's Great Dyke, and South Africa's Witwatersrand to the gold fields of the Orange Free State lies one of the world's great mineralized belts, from which over the past century untold wealth has been derived. And here, too, lie the vineyards of the Cape, the orange groves of Natal, the maize triangle—the corn belt—of the Highveld, the rich pastures of Swaziland, the good soils of Rhodesia's plateau. Here lies much of Africa's wealth, and here the white man's control still prevails. The gravity center of the region, of course, is powerful South Africa; Rhodesia, too, lies in it (Map 9-14). For reasons of European control and, especially in the case of Moçambique, economic involvement with South Africa and Rhodesia, the Portuguese "Provinces" are also considered part of the Southern African region. The location of Botswana, Lesotho, and Swaziland leaves little doubt about their inclusion, and neither can South West Africa be viewed as anything other than a Southern African territory. This leaves only two problem countries: Zambia and Malawi, not long ago caught up in the Central African Federation, of which (Southern) Rhodesia also was a part. Can Zambia and Malawi, independent black African states, be viewed as part of a Southern African region?

In several ways they may be so considered. Zambia, formerly Northern Rhodesia, for three-quarters of a century was bound up with the development of Southern Africa, an involvement which eventually led to the territory's inclusion in the Central African Federation. Though a protectorate during the colonial period, the

Map 9-14 SOUTHERN AFRICA.

country nevertheless attracted a considerable white settler population (70,000), and its Copperbelt, the northern mining area that borders on the Katanga, is a small-scale replica of the mining complexes for which the rest of Southern Africa is famed. Copper brings in between 80 and 90 percent of Zambia's external revenues, but although the Copperbelt has attracted a sizable labor force and carries cities such as Ndola, Kitwe, and Chingola (all somewhat under 100,000), the majority of the country's 4 million inhabitants subsist off the land. For its electric power supply and its fuel, as well as its outlets, Northern Rhodesia also has had to look south; but a railway from the Copperbelt across the east and through Tanzania to the coast is now in the survey stages. But this railway is the product of politics: Zambia's nearest exits lie in Moçambique south of the Zambezi River, and in Angola.

Malawi (formerly Nyasaland) also has to look south for its external trade. Like Zambia, Malawi (population 4 million) is a landlocked country, though its elongated shape creates a penetration into Moçambique to within 200 miles of Beira (Map 9-14). For many years the most active, accessible, and populous part of the country has been the south, where the largest urban center (Blantyre-Limbe, 100,000) and the old capital of Zomba are located. In an effort to reorient the country, and to give the central and northern sections more visibility, the capital after independence was moved to Lilongwe, but most of the productive tobacco farms lie south, in the Shire Highlands, and the tea plantations on the slopes of Mount Mlanje. It will be difficult to overcome Malawi's connections with and orientation to Southern Africa, and one indicator of its outlook in 1968 was its decision to establish formal diplomatic relations with the Republic of South Africa—the first black African state to do so.

Southern Rhodesia (now, in the absence of a *Northern* Rhodesia, simply called Rhodesia) was the dominant partner in the ill-fated federation that briefly welded colonial Zambia and Malawi together. And while we just enumerated some reasons why Zambia and Malawi might be included in the Southern African region, Rhodesia quickly reveals to what extent things change at the Zambezi River. Rhodesia is white man's country, despite the fact that its nearly 4.5 million Africans far outnumber the quarter million

whites. Malawi was never white man's country (of 4 million inhabitants there never were more than 10,000 whites), and Zambia's black-white relationships, though not always smooth, were never bedeviled by wholesale land alienation and determined repression. Zambia was a protectorate: it was adopted. Rhodesia was a colony; it was conquered.

Of the belt of countries that stretches across Southern Africa south of the Congo and Tanzania and north of South Africa, Rhodesia is the richest. Its Great Dyke, a mineralized belt over 200 miles long that stretches north-northeast through the center of the territory, has yielded gold (which at one time brought the country one-third of its export revenues) and produces asbestos, chrome, copper, and associated minerals. Not far from the two ends of the Great Dyke, Rhodesia's impressive cities, Salisbury (300,000) and Bulawayo (200,000) have arisen, and between these two lie good plateau farmlands on which tobacco is the chief cash crop; there is good grazing and the staple crop, corn, is harvested in such quantities that exports can be made in some years. As the map shows, a quite well-developed transport net has evolved in this central part of the country, which constitutes a clearly defined core area. Between Salisbury and Bulawayo lie the cotton mills of Gatooma and the iron and steel plant of Que Que; the coal supplies of Wankie lie to the west, and the electricity supply comes from the famous dam in the Zambezi River at Kariba. With its compact shape and its physiographically reinforced boundaries (the Zambezi to the north, the scarp into Moçambique in the east, the Limpopo in the south, the Kalahari to the west), Rhodesia gives the impression of a fortress—not altogether inappropriate in view of its modern problems.

These problems stem from the history of conquest of Rhodesia by the white man, and the subsequent period of deteriorating relationships between the country's quarter million whites and over 4 million black Africans. By a Land Apportionment Act, the whites reserved for themselves much of the territory's most productive farmland; restrictions were placed on the cash crops Africans could grow. The division of land had the effect of crowding the majority of Africans on one-third of the land, while the comparatively few Europeans got half of it; Africans who went to the mines and cities in search of work found South African-style segregation and job restric-

Salisbury, capital of Rhodesia. The shining capital saw great development during the 1950's, but then the political storm clouds gathered. The Central African Federation broke up, and Rhodesia forestalled any British plans for a Kenya-type independence through a unilateral declaration of independence, under white control. (Harm J. de Blij)

tion. And yet Rhodesia's race relations never were as intensely hostile as those of South Africa; after World War II, there was talk of "partnership" and provisions existed for "qualified" Africans to vote for members of the Rhodesian parliament. Unlike South Africa, whites sat with black Africans in the government, although there was no question, of course, of proportional representation. White Rhodesia was the driving force behind the creation of the Central African Federation, but north of the Zambezi African opposition to this political union was almost universal, and soon the federal structure foundered. Then, with British principles of African self-determination about to affect Rhodesia, the white minority government, led by Ian Smith, declared unilateral independence for Rhodesia in 1965 rather than submit to British demands for movement toward African majority rule. As of 1969, Rhodesia was isolated by world criticism and trade boycotts, but with its friendly neighbor, South Africa, at its side, it continued to resist British pressure for change.

South Africa. The southern tip of Africa is a great natural storehouse of wealth. Its possibilities are almost endless; the range and variety of its products are unequaled. South Africa is also rich in its diversity of cultures: its 19 million people include Africans (13 million) of several great nations and many smaller groups, Europeans (3.5 million) of Dutch, British, and other ancestries, Asians (under 1 million) from India, Pakistan, and other parts of that continent, and people of mixed ancestry who are called Coloreds (under 2 million), the products of Eurafrican intermarriages. In a country of nearly a half million square miles, these peoples have

forged Africa's most modern state. Their combined efforts have brought into production the famous gold mines of the Witwatersrand and the diamond mines of Kimberley, the vineyards of the Cape and the sugar plantations of Natal, the enormous ranches of the Highveld plateau, the farms of the Maize Triangle—South Africa's Corn Belt—and the factories of Johannesburg, Durban, and Cape Town. From the Mediterranean Cape and the subtropical east coast to the temperate Highveld and the tundralike crests of the Drakensberg, South Africa presents a wide range of environments. In winter, while the Natal coast enjoys its near-tropical warmth, the Highveld shivers as the chilled air from southern latitudes penetrates to the very Tropic of Capricorn. While the pastures of the plateau turn gray with drought, snows mantle the nine and ten thousand foot levels of the Drakensberg. While the winter rains come to the Cape, the Kalahari Desert lies parched under a cloudless sky.

South Africa is a land of opportunity. Its modern history has set it apart from the rest of Black Africa. The old Boer republics were founded before the colonial scramble elsewhere really gathered momentum; the diamonds of Kimberley attracted thousands of white men long before individual European travelers first saw much of tropical Africa. While African laborers from Lesotho made their way in the 1860's and 1870's to the mines of Kimberley to seek work and wages, Asians from India were arriving by the thousands on the shores of Natal, under contract to work on the sugar plantations there. Together, black, white, Asian, and Eurafrican contributed to the emergence of Africa's most developed country. White men located the

A white-owned, "European" farm in the productive Paarl Valley, South Africa. The architectural style of the farm houses is distinctly "Cape Dutch," and relates to the seventeenth and eighteenth century, when the Cape frontier was penetrated. French settlers contributed their skills of viticulture, and the Cape is one of the world's productive wine districts. (Satour)

mineral resources of South Africa and provided the capital for their exploitation, but black men did the work in the mines at wages low enough so that their extraction was an economic proposition. White men laid out the farms and plantations that produce so wide a range of crops, but Africans and Asians constituted the essential labor force to make them successful. White men built the factories of Johannesburg, but black men form the majority of the wage earners that work in them. White men built or bought the fishing boats which annually bring in a catch large enough to rank South Africa among the world's top ten fishing nations, but the nets are manned mostly by Coloreds.

With such interracial cooperation in common economic pursuits, one might expect that South Africa's plural society over the years would have become more and more integrated. But in fact this has not been the case. South Africa stands today as a world symbol of racial discrimination, an outpost of white minority rule on a black continent. It acquired this reputation largely after World War II. While much of the rest of the world reflected on the racial injustices involved in that war, and while the decolonization

period of the 1950's transformed the international political scene, South Africa became known for its official policy of racial separation as summarized by one word: *apartheid*. With this policy, South Africa moved in a direction that was more or less opposite to those ideals that were championed—if not necessarily practiced—in other parts of the world.

There was a time when South Africa, with its wealth and its power, played a leading role in Africa, albeit colonial Africa. The Boers had lost their war (1899–1902) with the British and had agreed to join the victors in a Union, which was established in 1910. At that time, the timid moves toward multiracialism that had been made at the Cape and in Natal were aborted; the British, concerned that they might lose the cooperation of the Boers, sacrificed nonwhite aspirations in the Constitution of the Union of South Africa. Nevertheless, as time went on, the nonwhite peoples of South Africa could see small instances of interracial cooperation. The country's universities began to produce some nonwhite graduates, many of whom took positions at colleges and universities elsewhere in Africa. While residential suburbs in the urban areas were largely segregated, there were nevertheless some areas of contact and even interdigitation. While public transportation was segregated, non-whites could usually find a place on a "European" bus. This trend, if it can be so called, always disturbed the more conservative Afrikaners (Boers) in the white population, and after World War II the growing strength of this element was proved by a stunning election victory (1948), in which the British-oriented United Party Government of General J. C. Smuts was defeated by the man who became known as the architect of *apartheid*, D. F. Malan.

The platform on which Malan's National Party ran included an enunciation of the *apartheid* doctrine, but within South Africa and in the world at large, the concept has often been incorrectly defined. *Apartheid* involved not only the separation of the black peoples of South Africa from the white, but also the Colored from the white, and the Asian from the African. Behind it lay a vision of South Africa as a community of racial states, in which each racial sector would have its own "homeland," although residents of one such "homeland" could travel to another to work. Anyone who may have thought that such a plan would be unworkable after three centuries of interracial contact and mixture soon found that the new government was serious: it immediately appointed a commission to draw up a blueprint for a territorially segregated South Africa. In the mid-1950's this commission produced its report, which proposed that a number of Bantu homelands or Bantustans should be created, positioned in a horseshoe-shaped belt around the South African core area (Map 9-15).

What gave South Africa its international reputation was less this scheme than the *ad hoc*, "petty" *apartheid* that was designed to set the wheels of segregation in motion. The number of "Europeans Only" signs on public places, park benches, bus stops, and so forth multiplied. Discriminatory pass laws appeared. An "Immorality Act" was passed, designed to stop interracial sexual contact. Violators were severely punished and suspects harrassed by decoys. Nonwhites began to disappear from the white universities. Police powers increased; persons of liberal polit-

Map 9-15 SEPARATE DEVELOPMENT IN SOUTH AFRICA.

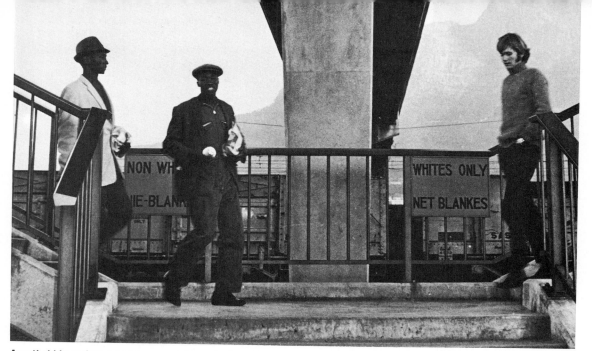

Apartheid in action: a scene on a stairway in the railroad station of Cape Town, South Africa. Trains, as well as buses, public conveniences, and countless other facilities are rigidly segregated. Out of the idea of *apartheid* came the concept of Separate Development, which is intended to provide each racial group in South Africa its own homeland, carved out of the present state. (Camera Press—PIX)

ical views lost their freedom. African political organizations were banned, their leaders exiled. Eventually, a protest at Sharpeville led to a mass killing of Africans by police—an event that evoked world condemnation.

Nevertheless, South Africa, with the view that the end justifies the means, is proceeding in its program of *separate development*—now the official name for the *apartheid* scheme. The first of the African "homeland states," the Transkei, has been functioning for some years; others, such as Zulustan, was being prepared for "independence" in the late 1960's. Initially there were plans for a massive relocation of industry, from the core areas of the country to the "borderlands," the boundaries between the Bantustans and the white areas. If industries could be located there, it was argued, then skills and capital could come from white South Africa and labor from the black states, without any need for racial mixing. This part of the program has proved unworkable on any large scale, but other aspects of it are being vigorously pursued. And, taking into consideration the spatial characteristics of the country, it is not surprising that a program that might seem totally unworkable in most other countries is viewed in South Africa as the ultimate and inevitable solution to racial problems

in plural societies. South Africa has a history of regional racial separation. Quite apart from the obvious associations between Bantu peoples and certain areas (such as the Zulu in Natal, the Xhosa in the eastern Cape, and so on), there are other, pervasive regional contrasts. The Colored people, for example, are largely concentrated in Cape Town and its environs; they are, indeed, called the Cape Colored in South Africa. In Cape Town itself, a city of some 800,000 people, the majority of the people are Colored (just over 50 percent); there are some 300,000 whites and only about 80,000 Africans. In Cape Town, then, the black population forms but a small minority; in the western Cape Province the Colored have a clear regional focus. South Africa's Asians, on the other hand, came mainly to Natal and to Durban. Many of them went to other parts of the country, of course, but Natal remains the South African province with the largest Asian population. Again, the urban area of Durban, with some three-quarters of a million people, has over one-third Asians, about one-third whites, and slightly under one-third Africans. Coastal Natal is the center of Asian South Africa, and once again a racial-regional association exists. Not unexpectedly, South African planners have considered a Colored "homeland" in the western

The skyline of Johannesburg, Southern Africa's greatest city. The city lies a mile high on the plateau *highveld* of South Africa, and had its origins in the gold of the Witwatersrand. Although the mine dumps (background) still form part of the urban landscape, Johannesburg today is a diversified city, with major industries as well as mines. (Camera Press—PIX)

Cape and an Asian one in Natal.

But the majority of the people in South Africa are black, and the separate development program's first objective has been the allocation of "homelands" for these people. When the Boers penetrated the Highveld, they left behind them the Xhosa and other peoples of southern Natal and the eastern Cape, who were hemmed in by the Great Escarpment on one side and the ocean on the other. On the plateau itself they found relatively little effective opposition, and before long virtually the whole of the Highveld was white man's country, with the Africans confined in reserves around the drier Kalahari margins, the rugged northern and eastern Transvaal, and below the escarpment in Natal and the eastern Cape. Thus the horseshoe arrangement of reserves existed before the separate development program came into being; the white man's invasion had jelled the Bantu migrant peoples in their domains.

But what made the whole program viable was the economic development of the Highveld. Johannesburg grew from a mining town into an important industrial center and one of the financial capitals of the world. East and west, the productive Witwatersrand sustained satellite towns. To the north, the old Boer capital of Pretoria, now the country's administrative capital, found itself located between coal deposits (east, at Witbank) and iron ores (northwest, at Thabazimbi), and the beginnings of a steel industry sprang up there. To the south, the Vaal

River was dammed near Vereeniging, to supply water to the City of Gold. Thus South Africa's core area emerged, with its twin axes along the Witwatersrand and from Pretoria to Vereeniging; by the late 1960's this economic region was the home of one-quarter of the country's people. And this core itself is surrounded by other productive areas: to the north, the Bushveld Basin with its chromium and platinum ores; to the east, the coal of Natal and the asbestos of Swaziland; to the south, the new gold fields and coal of the Orange Free State, the growing industrial complex of Vanderbijlpark, and the Vaal-Orange project of hydroelectric power and irrigation; to the west, the asbestos and manganese of the northern Cape and the uranium of the West Rand. The Highveld also is a productive farming region, and the staple, corn, is harvested in quantities that permit exports.

From Swaziland, Zululand, Lesotho, the Transkei, and every other African territory, black laborers have come to the farms, the factories, and the mines of the Highveld. They came by the hundreds of thousands: the population of Johannesburg proper today is about 1.2 million, nearly two-thirds of it African. All too often African families would find themselves in the miserable shantytowns, without adequate shelter, food, medical care, schools, or other necessary facilities. Building programs could not keep pace with the influx; an ugly juvenile delinquency problem arose. Yet the government could not prohibit the immigration of workers, for they were always needed. This situation was high on the list of problems the separate development program was designed to solve. If an African from an autonomous "homeland"—say the Transkei—wanted to work in a white-owned mine or factory, he would apply for a permit to enter white South Africa. He would then go to that country on a temporary visa, would leave his family in the Transkei (and would send his wife and children part of his income for their subsistence) and would live in a dormitory-type residence in the city, near the mine or factory. Thus the municipal government would no longer be faced with the task of making available the facilities necessary for the man's family, while the economy does not lose its labor supply.

Naturally, such a program could not be implemented overnight. The homelands must be established and given the trappings of semi-autonomous states, governments installed, arrangements for the accommodation of repatriated families must be made—and, of course, people who may have been living in the big cities for generations must be classified according to their "race." In this process, once again, the South African government paid little heed to the sensitivities of many of its subjects. People were classified and—regardless of their race—were made to carry identity cards to prove their racial heritage. In many cases, of course, there was little doubt; a man's language quickly tells whether he is a Zulu or a Xhosa. But race in South Africa has long been correlated with status, and if a man can prove he is white rather than Colored, his chances for a decent life are immeasurably improved; if a man can prove that he is Colored rather than black, he will be substantially better off than otherwise. Some of the saddest tales ever to come from Africa tell of people desparately trying to disavow their own ancestries, hoping to climb the social ladder one rung, to give themselves and their children a better chance in life.

All this, separate development is designed ultimately to solve. Rather than for a black man or a Colored man to struggle for advancement in the white society, he will do so in his *own* society, without racial barriers, job restrictions, or other racially imposed obstacles. Thus separate development is *parallel* development, and eventually all the states of South Africa are destined to have every facility, including universities, hospitals, transport services, and so

Slums in the big cities: every large city in Africa has attracted more people than it can accommodate, and ugly slums have developed. This one, known as Cato Manor in Durban, South Africa, is representative of the problem. Various solutions, including limitation of the urban-ward migration and rehousing of the slumdwellers, have been applied. (Camera Press—PIX)

forth. But the critical flaw in the design appears to lie in the allocation of the country's areas of productive capacity to the homelands. Of South Africa's nearly 500,000 square miles of territory, only about one-sixth is presently being considered for the African homelands; yet the black peoples constitutes two-thirds of the total population. Four-fifths of the territory of the country will become "white" South Africa—to be controlled by the one-fifth of the population that is of European ancestry. Obviously, these figures by themselves mean very little; in many of the countries of the world, most of the productive lands lie on less than one-sixth of the total area. But a look at maps of the distribution of resources, cities, transport lines, and the other assets of South Africa quickly shows that the African homelands, far from having a real share of South Africa's prosperity, are among the poorest and most isolated parts of the country. No major mineralized areas lie in the homelands: from 1955 to 1965 only about 3 percent of South Africa's income from its mineral resources came from the Bantu areas. As far as agricultural areas are concerned, there is good farmland in the eastern homelands—in Natal and the eastern Cape—but these areas, especially the Transkei and Zululand, also face a serious overpopulation problem. Those reserves that lie in the northern Cape and on the western and northern Transvaal fringe suffer from drought and terrain problems. In terms of industrialization, the homelands have hardly begun to change; secondary industries are practically absent there. They have few resources, have never had much capital, and lie far from the markets and sources of power supply. South Africa's railway network and the country's major roads skirt the homelands or cross them only to connect cities of white South Africa; internal communications remain poor. And none of South Africa's major cities lies in a present or potential Bantu homeland. Indeed, when the separate development program was instituted, plans were laid to actually create—from the ground up—several dozen towns within the African homelands, to relieve the pressure on the land and to act as dormitory settlements for the border industries scheme.

But towns as well as states need *raisons d'être,* and in the foreseeable future the African homelands of South Africa appear to be little other than a political gamble. Despite the optimistic implications of the term, it is hard to conceive of anything parallel between the development of white South Africa and that of the impoverished Transkei or Zulustan—and these may be the best endowed of the projected black states. With such limited opportunities, the labor forces of the homelands will have to seek work in the mines and factories of white South Africa, and this of course strengthens the white sector's hand in its dealings with the "autonomous" black states. And while it is possible that the African homelands will be nearly exclusively Bantu in their population, it is inconceivable that white South Africa will ever find itself without a resident (as opposed to temporary migrant) black population sector. Today, 8 out of 10 farm laborers in "white" South Africa are black Africans, as are 9 out of 10 miners and 9 out of 20 factory workers. Geography and history may have conspired to divide South Africa's peoples, but economics has thrown them irretrievably together.

ADDITIONAL READING

Many journal articles have been written about the hypothesis of continental drift, but the best source books remain A. L. du Toit's *Our Wandering Continents,* published in 1937 by Oliver and Boyd in Edinburgh, and L. C. King's *Morphology of the Earth,* published in New York by Dutton, 1963. Also see L. C. King's *South African Scenery,* a book much broader in scope than its title suggests, published in Edinburgh by Oliver and Boyd, in 1951. Volume I of J. H. Wellington's 2-volume set entitled *Southern Africa,* published by Cambridge University Press in 1955, is mainly a physical geography, and it, too, contains materials relevant to the drift argument.

Many regional geographies of Africa have appeared in recent years. Without specifying the similar titles in each case, there are books by Hance, Harrison-Church, Carlson, Stamp, Fitzgerald, Embleton and Mountjoy, Jarrett, Grove, Kimble, de Blij, and others. A very useful introductory book is P. Bohannon's *Africa and Africans,* published by Natural History Press, Garden City, in 1964 and available in paperback. A great deal of information also can be found in M. Herskovits, *The Human Factor in Changing Africa,* published by Knopf, New York, in 1962. On the peoples of Africa, G. P. Murdock's 1959 volume entitled *Africa: Its Peoples and Their Culture History,* published by McGraw-Hill, is still one of the best places to start reading up on the continent. Also see R. Oliver and J. D. Fage, *A Short History of Africa,* available as a Penguin edition published in Baltimore, in 1962. On trade, see P. Bohannon and G. Dalton (eds.), *Markets in Africa,* published in paperback by Northwestern University Press and Doubleday, 1962.

Topical and regional works also are available. R. J. Harrison-Church wrote *West Africa: a Study of the Environment and Man's Use of It* for publication by Wiley in 1961 in a second edition; Pedler's *Economic Geography of West Africa,* though published in 1955, still is useful. Another book by R. J. Harrison-Church, *Environment and Policies in West Africa,* is No. 9 in the Van Nostrand Searchlight Series. On East Africa, there are International Bank for Reconstruction and

Development studies on each of the three largest countries. Also see A. M. O'Connor's *Economic Geography of East Africa,* a Praeger volume published in 1966. On Equatorial Africa, see A. Merriam's volume entitled *Congo: Background to Conflict,* published by Northwestern University Press in 1961, and the work by V. Thompson and R. Adloff, *The Emerging States of French Equatorial Africa,* published by Stanford University Press in 1960. And on Southern Africa, see M. M. Cole's encyclopedic work called *South Africa,* published in New York by Dutton in 1961, the second of Wellington's two-volume set entitled *Southern Africa,* and his more recent book on *South West Africa and Its Human Issues,* published by Clarendon Press, Oxford, in 1967. For both substantive and bibliographic materials on colonialism, see Chapter 17 in H. J. de Blij's *Systematic Political Geography,* a 1967 Wiley publication. And on Separate Development see G. M. Carter, T. Karis, and N. M. Stultz, *South Africa's Transkei,* published by Northwestern University Press in 1967.

10

NORTH AFRICA AND SOUTHWEST ASIA

FROM Morocco on the shores of the Atlantic Ocean into Afghanistan in Asia, and from Turkey between the Black and Mediterranean Seas to the Somali Republic in the "Horn" of Africa lies a vast realm of enormous historical and cultural complexity. It lies juxtaposed between Europe, Asia, and Africa, and it is a part of all three: throughout history its influences have radiated to these continents and to practically every other part of the world as well. This is one of Man's source areas: on the banks of the Euphrates and the Nile arose civilizations that must have been among the earliest; in its soils plants were domesticated that are now grown from the Americas to Australia; its paths were walked by prophets whose religious teachings are followed by hundreds of millions of people to this day.

It is tempting to characterize this realm in a few words, to stress one or more of its dominant features. It is, for example, often called the "dry world," containing as it does the Sahara and Arabian Deserts. As Plate 2 shows, the majority of the region lies under *BWh* and *BSh* climatic conditions, with a fringe of *Csa* (Mediterranean) climate along the sea. But a comparison with Plate 6 suggests that most of the people in the region live where there is water: in the Nile Delta, in the coastal strip (or *tell*) of Tunisia, Algeria, and Morocco, along the eastern and northeastern shores of the Mediterranean Sea, in the Tigris-Euphrates basin, in the desert oases, and along the mountain slopes south of the Caspian. Although the Nile Delta lies in a region typified as desert (Plate 2), the 30 million people who are concentrated there live in a patchwork of cropland and waterways, and the adage "dry world" would hardly apply here. True, we know this region as one where water is almost always at a premium, where peasants often struggle to make soil and moisture yield a small harvest, where nomadic peoples follow their animals

MODIFIED AFTER: MCGRAW HILL BOOK COMPANY, MAN'S DOMAIN

Map 10-1 ISLAM AMONG WORLD RELIGIONS.

across dust-blown plains, where oases are islands of sedentary farming and trade in a sea of aridity. But it also is the land of the Nile, the lifeline of Egypt, and the great Gezira irrigation scheme, the mainstay of the Sudan, and the crop-covered tell of Algeria. This is a dry world—but not everywhere.

North Africa and Southwest Asia also are often referred to as the "Arab world." Again, this implies a uniformity that does not actually exist. In the first place, the name "Arab" is given loosely to the peoples of this area who speak the Arabic language (and some related languages as well), but ethnologists normally restrict it to certain occupants of the Arabian Peninsula, the Arab "source." Anyway, the Turks are not Arabs, and neither are the Iranians, nor the Israelis. Second, while it is true that Arabic is spoken over a wide region that extends from Mauritania in the west across North Africa to the Arabian Peninsula, Syria, Iraq, and southwestern Iran in the east, there are many areas in North Africa and Southwest Asia where it is not used by most of the people. In Turkey, for example, Turkic is the major language, and it has Altaic rather than Semitic or Hamitic roots. In Iran, the Iranian language belongs to the Indo-European family. In Ethiopia, Amharic is spoken by the ruling plateau people, and while it is more closely related to Arabic than is Iranian, it nonetheless remains a distinct language as well. The same is true of Hebrew, spoken in Israel; other "Arab world" languages that have their own identity are spoken by the Tuareg of the Sahara Desert, the Berbers of Northwest Africa, and the peoples who live in the transition zone between North Africa and Black Africa to the south. In parts of the "Arab world" there are elements that are distinctly non-Arab.

Another name often given to this region is the "world of Islam." The prophet Mohammed was born in Arabia in A.D. 570, and in the centuries that followed his death in 632, Islam spread into Africa, Asia, and Europe. This was the age of Arab conquest and expansion, during which their armies penetrated Southern Europe, caravans crossed the deserts, and ships plied the coasts of Asia and Africa. Along these routes they carried the Moslem faith, converting the ruling class of the states of savanna West Africa, threatening the Christian stronghold in the highlands of Ethiopia, penetrating the deserts of inner Asia, pushing into India and even Indonesia. Islam

was the religion of the marketplace, the bazaar, the caravan. Where necessary it was imposed by the sword, and its protagonists aimed directly at the political leadership of the communities they entered. Today, the Islamic religion extends well beyond the limits of the region discussed here: it is the major religion in northern Nigeria, in Pakistan, and in Indonesia; it has strength in parts of the Soviet Union and survives even in Eastern Europe, notably in Albania and Yugoslavia (Map 10-1). On the other hand, the "world of Islam" is not entirely Islamic either. In Israel, Judaism is the prevailing faith. In Lebanon, perhaps as much as half the population adheres to an old form of Christianity. In Ethiopia, the Amharic-speaking ruling class, having managed centuries ago to stave off the Islamic onslaught, also practices an ancient Christian religion. Coptic Christian churches still exist in Egypt. Thus the connotation "world of Islam" for North Africa and Southwest Asia is far from satisfactory: in the first place, the religion prevails far beyond these areas, and in addition, there are two or three countries within the "Islamic world" where Islam is not the faith of the majority.

Finally, this region is sometimes called the "Middle East." This must sound quite odd to someone, say, in India, who might think of a Middle West rather than a Middle East! The name, of course, reflects its source: the "Western" world, which saw a "Middle" East in Egypt, Arabia, and Iran, and a "Far" East in China and Japan and adjacent areas. Still, the term has taken hold, and it can be heard in general use by members of the United Nations. In view of the complexity of the region, its transitional margins, and its far-flung areal components, at least the name "Middle East" has the merit of being imprecise. It does not make a single-factor region of North Africa and Southwest Asia, as do the terms "dry world," Arab world, and world of Islam.

A Greatness Past

Today, when we think of the "Middle East," the things that come to mind are its ever-present political conflicts, its difficult environments, its poverty, and, perhaps, its single rich resource—oil. What we sometimes forget, as we watch a region swept up in the international political struggles of our day, is its great heritage of civilization and culture. For many centuries—

MODIFIED AFTER: OLIVER & BOYD, COMPARATIVE ATLAS

Map 10–2 EARLY CIVILIZATIONS.

indeed, for thousands of years—the Middle East was one of the foci of the civilized world. Then as now, the key to human life was water. The great centers of early progress lay in the valley and delta of the Nile, in the basin of the Tigris and Euphrates Rivers, and in the Indus valley in what is today West Pakistan (Map 10-2). It is tempting, of course, to postulate that man's early achievements in sedentary agriculture here were "taught" by nature through the periodic flooding of the riverine lowlands—a sort of natural irrigation process soon augmented and then imitated by the peoples who lived in these areas. But it may not be quite so simple. Archaeologists and anthropologists suggest that the farmers who brought the lands of the Nile Delta, Mesopotamia, and the Indus lowland into cultivation got their food-producing techniques from elsewhere, probably from the slopes of the hills in Anatolia and the area known as the Fertile Crescent (Map 10-2).

It is not known with certainty when the innovations in agriculture that were later to be diffused into the areas of the Nile and Mesopotamia began to emerge. It was a gradual process which, as we have already seen, occurred independently in several parts of the world, including Middle America. It probably was in progress ten thousand years ago, and it is likely that the first steps involved the protection of certain wild plants by the removal and control of weeds. In the Fertile Crescent, some varieties of wild wheat and barley still grow at elevations between two and four thousand feet, and perhaps these were among the earliest "domesticated" crops. They were, in effect, forest weeds, not plants of a dry environment. Of course, the protection of wild plants is one thing; keeping the seed for controlled cultivation and harvesting is another. This requires planning, organization, soil preparation, storage and distribution practices, and much more. It took several thousand years for these techniques to become established in the small communities of the Fertile Crescent, and it is from there that they were probably diffused to the Tigris-Euphrates basin, the Nile Delta, and the Indus lowland.

Only the most approximate dates can be reported for the evolution of the Fertile Crescent's farming villages, but even the widest margin of error fails to obscure the primacy of this area in the world picture. By about 5000 B.C. some of the settlements that had been founded in the basin of Mesopotamia, adjacent to the hills of the Fertile Crescent, were showing signs of growth into larger towns. Farming techniques slowly improved as the farmers developed systems of irrigation; the rich alluvial soils produced yields that helped make larger population agglomerations possible. In the settlements, a division of labor began to become established, and soon, metals came into use. Again, metals come from highlands, not alluvium: the basis of the innovation lay far from the alluvial lowlands. First it was copper, which can be smelted fairly easily and occurs in quite pure form in the bedrock of the region. But copper—in use as early as 4000 B.C.—is rather soft and not very useful, and it seems mainly to have had ornamental functions. Then the discovery was made that copper could be smelted in combination with tin, producing something much harder and more practical—bronze. The Bronze Age thus began in Mesopotamia between 1000 and 1500 years earlier than it did in Europe, and the metal played a crucial role in the history of the region. Swords and other war equipment were made of it; tools of all kinds were enormously improved by it. Meanwhile, still another innovation appeared: writing. It is not surprising that the increasing trade and transportation between the towns, the specialization of their inhabitants, and the problems of government and administration should have led to improved means of record-keeping and long-distance communication. By 3000 B.C. writing was part of urban life in the Middle East. Nowhere in the world could anything approach Middle Eastern developments in agriculture, metallurgy, urbanization, and political consolidation. For the first time, man was living in cities, and was facing the problems city life brings—problems he still faces today. Mesopotamia, the Nile area, and the Indus lowland were more than a millenium ahead of their potential rivals: to the east, the Yellow River basin of China, where by 2000 B.C. agriculture and urbanization revealed technological progress, and to the west, the pre-Grecian Bronze Age civilizations of the Aegean shores. The Middle East was even farther ahead of West Africa and Middle and South America: the early Mesoamerican "cities" were ceremonial centers amidst a dispersed rural population, and true cities did not appear until the first millenium A.D.

Diffusion of Innovations. The centers of civilization of the Middle East for a long time were vanguards of human progress, and just as techniques of farming had spread to the river lowlands from the hillslopes of the Fertile Crescent, so the many innovations of Egypt, Mesopotamia, and the Indus were carried to many parts of the world. There are those anthropologists who argue that West Africa was the scene of independent plant and animal domestication, but others view the developments in savanna West Africa in the context of stimuli received from the Nile region. Quite possibly the growth of urban agglomerations in the bend of the Yellow River in northern China also was aided by knowledge that came from the Middle East. Europe, of course, received a multitude of innovations from the Middle East. Today we and most other peoples in the world continue to benefit from the endless list of contributions made by the ancient Mesopotamians and Egyptians: for example, the cereals we eat, such as wheat, rye and barley; the vegetables we are used to, including peas and beans; the fruits (grapes, figs, olives, apples, peaches and many more); and the domesticated farm animals —oxen, pigs, sheep, and goats. The horse, too, was first put to human service here. And of course the range of indirect contributions is indeed beyond measure. The ancient Mesopotamians made progress not only in irrigation and agriculture, but also in calendrics, mathematics, astronomy, government and administration, engineering, metallurgy, and a host of other fields. As time went on many of their innovations were adopted and then modified by other peoples and cultures in the Old and eventually even in the New World. Europe was the greatest beneficiary of the legacies of Mesopotamia and Egypt, whose achievements constitute the very foundations of "Western" civilization.

The story of the Middle East's role in the dawn of human history quickly leads to a question of immense geographic interest: given the innovations made by the ancient Mesopotamians and their contemporaries, just how were these innovations transmitted to other peoples and other cultures—in short, how were they *diffused?* This is a question that can also be asked in a modern context: given a technical innovation or some other new idea, how does it gain acceptance by the people to whom it is introduced, and how are the decisions to adopt or not to adopt arrived at? Take, for example, the production of a new, advanced, and more effi-

cient type of plow. Some farmers will no doubt acquire and use such a new piece of equipment almost as soon as it becomes available; others will wait to see how it performs; still others will stick to what they know best and will resist the new invention. What determines the patterns of adoption—what are the processes involved? The answers to these questions are obviously of great interest, for they will show the way of progress in many spheres. They may indicate how best to introduce improved farm tools in rural Africa, how to persuade the people of Peruvian Indian villages to boil their water, how to encourage the use of contraceptive devices in India.

Geographers, of course, are not the only social scientists who have been interested in the processes of diffusion. Anthropologists have worked in this field; as long ago as 1923, Professor Clark Wissler in his book *Man and Culture* discussed the diffusion of the use of the horse from the Spanish invaders to the Indian peoples of America.[1] Sociologists, political scientists, economists, and others have worked on diffusion as well, but in recent years a group of geographers have developed a whole subfield of geography that focuses on diffusion—always a spatial process—and they have published a large body of research. Of course, most of their studies focus on contemporary problems in diffusion; it would be very difficult indeed to gather the data required to unravel the diffusion processes that brought the knowledge of the hillslope farmers of the Fertile Crescent to the lowlands of Mesopotamia! But what they discover by studying the present may help us assess what happened when the Middle East witnessed the laying of the foundations of a world civilization.

It is clear that Mesopotamia, Egypt, and the Indus lowland were the scenes of contemporaneous civilizations—that is, they overlapped in time, but their beginnings may not have been simultaneous. There is always the problem of the enormous telescoping of centuries of history, but it is not surprising that archaeologists and others have still sought to find out which area was in the lead. Looking at the situation on Map 10-2, it is relevant that Mesopotamia was not only the "land between (*meso*—middle of) the rivers," but also the land positioned between Egypt to the west and the Indus lowland to the east. The Tigris-Euphrates region, moreover, lay immedi-

[1] Clark Wissler, *Man and Culture,* Crowell, New York, 1923, pp. 111–121.

ately adjacent to what appears to have been the largest cluster of hillslope farming settlements in the Fertile Crescent. And in addition, it lay at the head of the Persian Gulf, and thus astride the main trade route between the Orient and the Levant (the lands on the eastern coast of the Mediterranean). Mesopotamia, thus, seems to have been positioned closest to the sources of the initial stimuli of progress; it was in contact eastward with the Indus and westward with the Nile; it was by far the most exposed to trade and outside influences. It is likely that Mesopotamia was indeed the earliest of the Middle Eastern civilizations to emerge, but the developments in Egypt and the Indus lowland did not simply result from a one-way diffusion of Mesopotamian knowledge. Probably Mesopotamia contributed more to the two other regions than they did to the land between the rivers, but Egypt and the Indus had their own identity and achievements as well. And not all of Mesopotamia's innovations found their way quickly to the other areas: the wheel and the chariot, in use in Mesopotamia not long after 4000 B.C., found adoption in Egypt fully two thousand years later—and then only as a result of the only outside invasion and conquest to interrupt Egyptian continuity—that of the Hyksos (1730–1570).

The map tells us more. The middle and lower Nile lie enclosed by inhospitable country, open to the Mediterranean, but otherwise rather inaccessible to overland contact. To the south, the Nile is interrupted by a series of cataracts that begins near the present boundary of Egypt and the Sudan. To the east, the southwestern arm of the horseshoe-shaped Fertile Crescent lay some distance away. To the west lies the endless desert. The ancient Egyptians had themselves a natural fortress, and in their isolation they converted security into progress. Internally, their Nile waterway was the country's highway of trade and association—its lifeline. Externally, the Egyptians needed trade, for they had little wood and required metals, but they left most of this trade to the Phoenicians and the Cretans. True, Mesopotamia was more open, more active in interregional trade, and it lay closer to the original stimuli of the Fertile Crescent, and so it forged ahead of Egypt. But there were disadvantages in all this. Mesopotamia's openness to trade was also an invitation to invasion and conquest. Its proximity to the hills of the Fertile

The splendor of Pharaoic Egypt is reflected by these enormous sculptures, hewn out of rock and often in meticulous detail. They remain as witnesses to the wealth and security of ancient Egypt; for scale, see the man walking just to the right of the pillar in the lower part of the photo. (Air France)

Crescent also meant that its covetous enemies were never far away.

In physical terms the lands of the middle and lower Nile in many ways were similar to those of Mesopotamia, but again there were very important differences. The Nile is the product of headwaters that rise in two different parts of Africa: Ethiopia (the Blue Nile) and the lakes region of East Africa (the White Nile). This assured a fairly regular flow—man has built dams in the Nile since then—and it made the annual floods predictable both in terms of timing and extent. The Nile is at its lowest level in May and June, but rises during July, August, and September to its flood stage in October, when it is some 25 feet above its June level. After a rapid fall during November and December, the river declines gradually until its low stage in May. The Tigris and Euphrates, on the other hand, rise in Turkey within a few hundred miles of each other, and their headwaters share the same rainfall regime. Hence the ancient Mesopotamian farmers always had to cope with strong spring floods of widely varying volume,

Persian Empire

Ottoman Empire

Assyrian Empire

Alexander's Empire

MODIFIED AFTER: PALMER, HISTORICAL ATLAS OF THE WORLD

Map 10-3 EMPIRES OF THE MIDDLE EAST.

and in some years these floods were very destructive. Certainly Mesopotamia did not have all the advantages in the Middle East.

Today, as we look back at the ancient heritage of this culture realm, we are struck by the continuity of culture and civilization in Egypt, and by the succession of states and the discontinuity of political consolidation in Mesopotamia. We also know of the magnificent stone monuments created during the three millenia of ancient Egyptian culture (from the First Dynasty to the Roman conquest) and the great pyramids that have stood now for nearly five thousand years. We know of nothing comparable in Mesopotamia, and again we wonder—why? The question of political succession has already been answered in part, and it has to do with Mesopotamia's vulnerability to outside interference. From the earliest times the land of the two rivers was the marching ground of peoples and armies. Eventually two empires arose, Babylon in the south and Assyria in the north. Babylon arose first to prominence; it received a major stimulus through the arrival of the Sumerians in 3500 to 3000 B.C. Subsequently it was afflicted by successive invasions by Semitic-speaking peoples, but although the center of political power repeatedly changed hands, the line of cultural progress was not destroyed. To the north, the Assyrians had their own history of trade and agricultural development, but Assyria also shared in the culture of Babylon, especially after 2000 B.C. Eventually Assyria briefly gained primacy in the Middle East, as the empire during the seventh century B.C. expanded to include Mesopotamia, the Levant, and an internally weakened and divided Egypt as well (Map 10-3). In short, Mesopotamia was something of a marchland; Egypt, on the other hand, protected by distance and terrain, faced foreign invaders only rarely and asserted its strength in foreign areas only infrequently. But what of the stone structures and monuments? Certainly the Babylonians and Assyrians were good builders and artists. But the Egyptians again had the natural advantage; rock beds suitable for quarrying lay near the Nile—upon which the cubic pieces could be transported to the building sites. The alluvial bed of the Tigris and Euphrates Rivers is much wider than that of the Nile, and hard building stone in Mesopotamia was to be found much farther from where it was needed. Hence the ancient Mesopotamians built much more of brick, often made of sun-dried mud and not very resistant to weathering and erosion—not, at least, over five thousand years.

A Greatness Lost

Everyone is interested in the question of why it is that states and empires may rise with great flourish, and then decline, sometimes precipitously. In the case of Mesopotamia and Egypt, the rise was not sudden, though it was attended by unprecedented progress in a wide range of fields of human endeavor. And the final decline of the states of Mesopotamia and the great dynasties of Egypt came after many centuries—indeed, several millenia—of sustained political and cultural identity. Even if we consider Egypt's period of true empire to have been between 2650 and 1100 B.C., that still is longer than the Roman Empire survived as a unit, and, in other terms, it is six times as long as there has been a United States of America. What gave the ancient empires their strength? What weaknesses finally led to their demise? Are the rules still the same?

The rise of states and empires—and the attendant forward push in human achievement and innovation—has been the subject of study by many social scientists. In his book *The Culture of Ancient Egypt,* Professor J. A. Wilson points to three concepts that have been contributed.[2] Probably the most famous of these three concepts is that of the historian, Arnold J. Toynbee, who has introduced the principle of "challenge and response." This involves a situation in which people are confronted by a challenge (in this case, the clearing of river jungle, bringing into production the alluvial soils, and the control of flood waters) which demands a response that takes the form of improved or new techniques in various spheres of activity. Obviously, the challenge must be sufficiently serious that man will go about finding solutions; on the other hand, it must neither be too great nor should it be too minor to require real changes in existing modes of life. It is an interesting idea, and one that can be —and has been—applied to many cases of marked human progress. But Professor Wilson feels that it does not help much in any assessment of the rise of Egypt. Why, he asks, did the prehistoric peoples of Egypt respond to the challenge, whereas their southern neighbors in the Sudan did not respond? How was it that the

[2] John A. Wilson, *The Culture of Ancient Egypt,* The University of Chicago Press, Chicago, 1951, pp. 32 ff.

ancient Egyptians ignored that fertile soil so long, and then undertook to cultivate it? These questions remain unanswered by the Toynbee principle. Obviously something had to be changing in the communities that existed at the time—perhaps population growth and food demands, some external influence, or something else. Anyway, the challenge and response principle leaves many questions unanswered.

A second idea discussed by Professor Wilson is one to which we have already referred in previous chapters, though we have not identified it as such: the "urban revolution" concept as defined by Professor V. Gordon Childe.[3] This relates to the way people, at a certain stage in history, tend to begin to cluster around villages, thus setting in motion a process of agglomeration that eventually makes urban centers out of some of the more favored agricultural villages. Two products of the process are a surplus of wealth and a surplus of population. The surplus of wealth produces a ruling class with free time for leisure, the arts, and other avocations; the surplus of population stimulates the professional specialization that is a hallmark of the urbanizing society. But Wilson warns against the indiscriminate use of both terms: "urban" and "revolution." Despite our eagerness to recognize urban centers in ancient Mesopotamia and Egypt, he argues that the first *true* cities in Egypt appear as late as the Eighteenth Dynasty—about 1500 B.C. Like many other writers, Wilson does not propose just where the threshold lies beyond which a growing village or town becomes a city or urban center. But the point is unmistakable and relevant: the concentrated settlements of the Middle East grew gradually, along a continuum; deep into the period of empire they differed from their earliest predecessors not in kind, but only in degree. Childe's concept, he says, is a good one—so long as "it is understood that it was not "urban" and not a "revolution" in the case of Egypt, at least.[4] The "urban revolution" concept, then, may have less than general applicability.

Third, Professor Wilson reminds us of Redfield's "folk society" concept. This is another attempt to evaluate the process of urbanization in the evolution of human society, in this case

by comparative methods. Folk society is homogeneous, its numbers are small, its sense of community very strong. It has a subsistence economy, is preliterate, deeply religious, strongly traditional. Now this construct is placed opposite urban society, with its complexity, large numbers of people, commercialism, secularism, literacy, and constant change. Is this what happened in the Middle East? Wilson reports that predynastic Egypt was not really a folk society, in that the people experimented with new techniques and adopted innovations that reached them, that there already was a market orientation despite the prevailing self-sufficiency of most communities. Thus Egypt may not have commenced its journey toward empire as a folk society, and it never really achieved an urban society. Once again the concept is instructive, but it does not provide answers to those questions we raised about the rise of states.

What about the decline of states? In the case of the Middle East a number of answers have been given. Newspaper reports constantly tell us details of urban sites, being excavated by archaeologists, located in presently drought-stricken areas. These stories lead us to assume that when those towns were thriving centers of government and trade, they must have enjoyed a moister climate. And indeed, climatic change has been one of the major theories to account for the decline of Middle Eastern power, especially in Mesopotamia and North Africa. But climatic change is not the only answer. True, there was a general drying-up in the Middle East during the last ten thousand years B.C., a process that was related to the retreat of the Würm glaciation in northern Eurasia. In general, this was felt most strongly in the west, in North Africa (the Sahara dried and widened markedly during this period), but undoubtedly it had its effect in the Levant and Mesopotamia also. In fact, here may lie the very stimulus that led to the agricultural innovations in riverine Mesopotamia and Egypt: perhaps the people were forced to congregate there because other parts of the realm lost their rainfall and groundwater. But it would be a mistake to tie the whole history of civilization in the Middle East to a tyranny of rainfall. It is clear, for example, that a great deal of physical change took place in the drainage systems of Mesopotamia during recorded history. The Euphrates appears to have built a

[3] See V. Gordon Childe, *What Happened in History,* Penguin Books, Hammondsworth, 1950, pp. 89–112.

[4] Wilson, *op. cit.,* p. 34.

Ancient ways still remain in Egypt and its neighbors. Here an ox walks the endless circles around a *saqiya* or water-wheel, which raises water across the levee and sends it into the irrigated fields nearby. The wooden equipment is centuries old; the scene could have been taken a thousand years ago. (Paul Almasy/World Health Organization)

sizable alluvial deposit at its mouth in recent times, and in the days of Babylon the Persian Gulf may have extended many miles into what is today Iraq. Farther inland, the Tigris and Euphrates repeatedly changed their channels, and it is evident that the area between the rivers was much narrower several thousand years ago than it is today—making irrigation an easier proposition, since it was based on connecting canals between the Tigris and Euphrates. Lateral movement of the rivers must have caused the Mesopotamian farmers major problems—and it may have left some settlements without a basis for existence.

Man, himself, also contributed to the reduction of available water supplies. He cut the vegetation and promoted erosion, permitted overgrazing and allowed the sand of the desert to encroach on his farmland. All too often the nomadic invaders who entered the scene could not maintain the farm and irrigation systems to which they were heirs, and allowed them to go to waste. Salinization and siltation slowly spoiled the soils and clogged the waterways of Mesopotamia—not only because of natural conditions, but also because of Man's failure to adhere to principles of conservation. This, in turn, was made worse by the recurrent political instability of the area, which had bad effects on agricultural practice.

All this may apply to Mesopotamia, but Egypt is clearly a different case. The Nile continued to serve as it always had, the soils were no less productive, invaders were few. And yet, when the Assyrians made their conquest in the seventh century B.C., Egypt was already in decline, as it had been for several centuries. How it happened is fairly clear. The empire's time-honored ways of life were breaking down; it was losing its spiritual foundations, its confidence, its cohesion. Inflation was rampant and the economy was in trouble, despite some belated expansionist efforts in what is today Israel and Jordan (where the conflict with the Assyrians first arose). But *why* it happened is something else. Part of the story may be read from the activities surrounding the copper mines in Sinai, and the fate of the grain-for-silver exchange the Egyptians had long carried on, through their Phoenician and Cretan intermediaries, with Turkey. By about 1000 B.C., the copper mines were more or less abandoned: a new age had started—the Iron Age. But while Egypt had copper mines nearby (in the Sudan as well as Sinai) and could also trade for this metal in other parts of the Middle East, it depended completely on faraway sources of iron. And these sources were cut off just before the last millenium B.C. opened. Egypt's primacy had coincided with the Bronze Age, and it ended when the Iron Age dawned.

But, as we said, this was only part of the story. Egypt was already in decay before the coming of iron struck what may have been the crucial blow. And for the early phases of decline, no easy explanations exist. As with the rise of states, we are given to wondering whether there is something inevitable about the end of empire—whether, for example, it would all have happened even if Egypt had possessed its own workable sources of iron ore. States often seem to rise on the basis of a flourish of innovations and technical and organizational progress at their beginning, then to "level off," maintaining their advantages without sustaining that forward drive. Is there a universal principle involved here? True, the decline of practically every political state—ancient Rome included—can be accounted for on economic grounds. But just as we objected to the Toynbee "challenge and response" principle on the basis of the argument that the challenge existed so long before the response occurred, so we can point to cases where the economic problems that foundered the state also existed before the collapse began, were successfully coped with for some time, and only later gained the upper hand. In the case of Egypt, the economic crisis was real, and it was worsened by the competing, would-be rulers who spent the resources of the state in their struggles for control over land and people. But again—in a state so long unified, so tradition-bound, so conscious of its history, what could start a breakdown such as this?

There is, of course, another way of looking at this whole question of decline. It is not difficult to make a case for the position that it is not the Middle East and its empires that declined, so much as other parts of the world, taking Middle Eastern innovations, rose to greater heights. In our descriptions of the glories of ancient Egypt and Mesopotamia we could lose sight of the fact that these were glories in a backward and primitive world, created by a small minority · elite whose ruthless use of manpower is silently evinced by the monuments they left behind. The peasants, the workers, the slaves—they would have a different judgment of Egypt's grandeur. Today, while it is probably true that Mesopotamia supports far fewer people now than it did in the days of empire, Egypt is more productive, its population is much larger, its cities greater in size and number than ever in the days of the dynasties. But standards have changed, and Egypt has yielded world leadership to a long series of successors in the Old and the New World. The "decline" of states and empires is a relative term.

The Successors

The end of the primacy of Mesopotamia and the Nile valley did not signal the end of empire building in the Middle East. The Assyrians, who managed to incorporate a large part of the Egyptian sphere of influence in their own, were soon replaced by the Persians. The Persians, whose capital was located at Persepolis, little more than 100 miles inland from the Persian Gulf (not far northeast of Shiraz), were received as liberators by the Egyptians in 525 B.C.; Assyrian rule had been ruthlessly suppressive. The Persian empire was never as highly centralized as Dynastic Egypt or Mesopotamia's states had been, and it did not last very long: Alexander the Great found it less than two centuries later cracking into a number of semi-independent fragments. Alexander's empire, the beginning of the Hellenistic Age and the culmination of Greek power, was established during the great ruler's brief lifetime (356–323 B. C.). The consolidation of the Greek city states came about in the face of the Persian threat from the east, and before Alexander's death Persian power had been broken and his empire extended from Turkey to Egypt and from the Levant to India. Egypt, which was taken in 332 B.C., again welcomed the new foreign invasion, as Persian overlordship had lost its popularity.

Again, the empire of Alexander fell apart soon after his death, and now it was the turn of the Romans to bring their empire to the Middle East. As we noted in Chapter 1, Rome made the Mediterranean an internal sea, and Roman rule was instituted from Morocco to Egypt in North Africa, and through Greece and Turkey into Mesopotamia. Roman rule in Mesopotamia, however, was never very effective nor was it long-lived. The eastern territories were destined, in the Roman scheme, to provide food and raw materials for the core of the empire, and Mesopotamia—especially the area of Babylon—was far away from the Mediterranean coast and not very productive. Later, we know, this eastern part of the Roman realm itself became the surviving remnant of the state, and Constantinople took over from Rome itself the functions of its capital. The traditions of the fallen headquarters were preserved by the Byzantine kings who ruled there.

After the "indigenous" empires of Mesopotamia, Egypt, and Persia, the Middle East thus had experienced two alien powers: Greece and Rome. Rome's power lingered on the shores of the eastern Mediterranean. Now came the turn of a Middle Eastern empire to export its force to alien lands. This occurred with the rise of the Arabs and Moors. Before the birth of the prophet Mohammed, there was little to unify the peoples of Arabia or those of the Middle East not subject to the Egyptian or Mesopotamian empires. In the seventh century A.D., Mohammed brought the Middle East not only a religion, but a whole new set of values, a new way of life. He prescribed on family relations, moral behavior, drinking, gambling, the treatment of servants and slaves, and, of course, worship. Moslem mosques made their appearance not only as places of prayer, but also as gathering places to bring the community closer together; Mecca became the spiritual center for a widely dispersed people for whom a joint focus was something new. The stimulus given by the prophet Mohammed was such that the Arab world was mobilized almost overnight. He died in 632 A.D., but his faith and fame spread like wildfire. Arab armies invaded and conquered, and Islam was carried throughout North Africa, where the Islamic Moors joined the Arabs and organized a successful invasion of Iberia and even southern France (Map 10-3). Eastward, the faith was carried to present-day West Pakistan and to the shores of the Aral Sea in what is now Soviet Asia. By the middle of the eighth century A.D. the Arab empire extended from western India to Morocco and from Turkey to eastern Ethiopia. The original capital was in Arabia, at Medina, but it was moved, first to Damascus in what is now Syria and then to Bagdad, which is not far from the ruins of Babylon in Mesopotamia (Iraq). This was a time of great achievements and innovations. The distinctive architecture of the time is still imprinted on the urban landscapes of Spain and Portugal. In mathematics and science, the Arabs made major strides, and they established universities in many cities, including Bagdad, Cairo, and Toledo. It was a time of splendor; the Arabs now were supreme in the trade routes of the Middle East. But it was also a time of emerging internal conflict in the empire, which soon broke up into numerous individual Moslem states. Nevertheless, the vigor of the Islamic drive was hardly diminished: the Moors hung on in Iberia until just before 1500, in West Africa the ruling elite of the savanna

states was converted to the faith, in Ethiopia the Christians who sought refuge on the high plateaus were under constant challenge. By trading ship, Islam reached and took root in Indonesia.

Despite the vigor and the unifying properties of the Moslem faith, the Arab empire never again achieved its eighth century unity. Outsiders appeared once more to test the Middle East's weaknesses. From Europe in the eleventh century came the first of a series of Christian crusades, to challenge Islamic superiority on the shores of the eastern Mediterranean. From Asia came the fabled horsemen, the Mongols, who in the thirteenth century imposed their empire (which stretched from China to eastern Europe) upon Mesopotamia. The heyday of Arab power in the "Arab world" was indeed brief.

The last of the great empires to emerge in the Middle East was centered in Turkey. This, the Ottoman Empire, began its rise as early as the thirteenth century with the appearance of Osman I as the first sultan of the empire's embryo, a Turkish sultanate. The name, Ottoman, is a corruption of "Osmanli." Over the centuries, the sultanate evolved gradually into a sizable state centered on the city of Constantinople (Istanbul); by the sixteenth century it was in full development. An Islamic state, the Ottoman Empire incorporated parts of North Africa, Southwest Asia, and Southeast Europe. This included northern Algeria, Tunisia, northern Libya, Egypt, a part of the Sudan, and the western sector of the Arabian Peninsula, Mesopotamia and the Levant, and, in Europe, the southern countries of Eastern Europe as well as Black Sea portions of Russia. In the middle of it all lay the Turkish homeland, surrounded by the Aegean Sea to the west, the Black Sea to the north, the eastern Mediterranean to the south, and the Caucasus and the mountains of Iran to the east. Thus for a second time in one thousand years of European history, Europeans lived under the domination of Islamic rulers. Again, the empire in its three centuries of full power made a permanent impact on the European scene. Not only did it affect the urban landscape in the cities it controlled, but it also converted many East Europeans to the Moslem faith, especially in Albania and Yugoslavia. But the Ottoman Empire was not a progressive force, certainly not during its later period. Its rule was despotic, prevailing standards of living were low, change was resisted, education lagged, religious intolerance prevailed, corruption was rife. Even before World War I, the empire was breaking apart as Greece, Bulgaria, and Serbia fought the Ottoman armies, Italy invaded Libya, and France incited the Algerians to resist Turkish rule. Despite a rejuvenation in the Turkish leadership in the first years of the twentieth century, the remnants of the Ottoman Empire came out on the German-Austrian side as World War I erupted. The Turks hoped that a victory would resurrect and revitalize the state while repelling its most threatening and nearest enemy: Russia. Instead, the Ottomans had picked the losing side, and the end of the war signaled the end of the empire in Europe as well as in the Middle East.

From the fragments of the Ottoman Empire, the modern states of the Middle East emerged. Turkey, the old heartland, was constituted as a separate entity. Under the Mandate system of the League of Nations, the administration of various territories was turned over to the victorious powers; other parts of the Empire became outright colonies. The French obtained mandates over Syria and Lebanon in addition to their dependencies in Algeria, Tunisia, and Morocco. Britain administered Palestine, had

responsibilities in Transjordan and Iraq (Israel and Jordan are the two states presently in this area); the British also had a hand in the government of Egypt and shared with Egypt the administration of what used to be known as the "Anglo-Egyptian Sudan." Italy held sway in Libya, and, in Africa's Horn, over Eritrea and Italian Somaliland as well. Thus the pieces of the Ottoman Empire acquired new individual identity, and in several cases independence was not far away. Today, just a half century later, all Middle Eastern countries—save some Spanish territory in Northwest Africa, several oil-rich territories on the Arabian Peninsula, and France's small dependency in the Horn—are sovereign states. And once again there is a dream in the Middle East of Arab unity, strength, and influence in the affairs of the world.

Boundary Morphology

A look at the political map of the modern Middle East (Map 10-4) points up the numerous and lengthy straight-line boundaries by which the realm's states are delimited. Most Middle Eastern states share such boundaries with their neighbors, and some of them, notably Egypt, are to

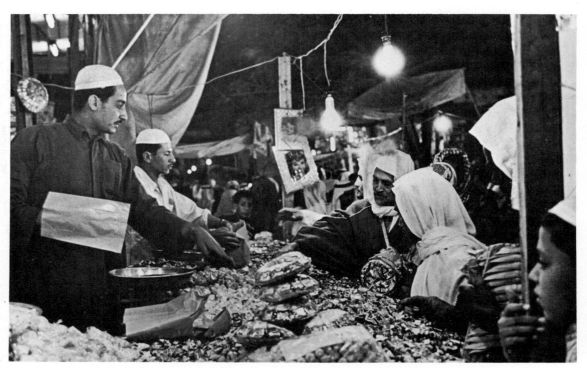

A *bazaar* in an Arab city. This particular market stand is busy, for it is the end of the fasting period of Ramadhan, and people are buying candies and sweets to celebrate. (Arabian American Oil Co.)

a large extent enclosed by straight lines. From the viewpoint of treaty-makers, these are convenient boundaries, for they are drawn from place to place without the need for detailed terrain descriptions. Political geographers refer to them as *geometric* boundaries, to emphasize that they were defined by mathematic methods, not by any consideration for land or people between the end-points.

The term *geometric,* thus, is simply derived from the appearance of the boundary on the map, and it puts these boundaries in a morphological class. In the Sahara Desert, there was ample justification for their use, since the barren terrain through which the geometric boundaries lie is virtually unpopulated. But there are parts of Africa where geometric boundaries served the convenience of the colonial powers, at the dire expense of the local people. Examples include the boundary between Kenya and Tanzania, between South West Africa and Angola to the north and northern Botswana to the east, and the border between the Somali Republic and Ethiopia. Several of these boundaries have since become serious political liabilities, and there is little doubt that Africa will see numerous boundary disputes in the coming decades.

In the morphological classification of boundaries, then, geographers simply consider whether the boundary lies along a prominent physical feature, whether it conforms to any distinct cultural breaks in the landscape, whether it does neither, or is geometric. Boundaries that lie along physical features are called *physiographic* boundaries, but this is a term that has to be used with care, since physical geographers use it too—for something quite different! To give an example, the present boundary along the Jordan River between Israel and Jordan is a physiographic political boundary. The boundary between the Sudan and Ethiopia, which skirts the lowest slopes of the giant Ethiopian mountainland (and in places follows a river) is in the same class.

A boundary that coincides with a cultural discontinuity is identified as an *anthropogeographic* boundary. This is what the Somali peoples of the Horn of Africa want: to have their political boundary moved to incorporate all those people now separated from the Somali Republic. Thus all Somali will be unified in a Somali state, with non-Somali country across the border. In the Middle East, with its enormous ethnic complexity and scattered core areas, few anthropogeographic boundaries exist; those between Israel and its Arab neighbors are increasingly anthropogeographic, but there is a sizable Arab population even within Israel itself. In Europe, on the other hand, there are many anthropogeographic boundaries, for example that between Spain and France, whose culture contrasts are many.

For want of a better term, we must refer to boundaries that coincide neither with physiographic prominences nor cultural breaks in the landscape as *indeterminate,* so long as they are not geometric. Large stretches of the boundary between Tunisia and Algeria are in this class, as

MODIFIED AFTER: RAND MCNALLY, GOODE'S WORLD ATLAS

Map 10–4 BOUNDARIES OF NORTH AFRICA AND SOUTHWEST ASIA.

are portions of the eastern boundary of Turkey.

The boundary framework of the Middle East, then, is a relatively recent creation, and modifications of it are still going on. There are boundary questions still unsettled, and almost invariably these disputes hark back to the colonial period or the Mandata administrations. Morocco has laid claim to a huge portion of Mauritania, insisting on historic rights there. The ephemeral Somali–Ethiopia conflict, dormant in 1968, contains elements of serious trouble. Syria has never been satisfied with its boundary with Turkey, and blames the Mandate powers for its misfortune. And it is hardly necessary to emphasize the conflicts between Israel and its Arab neighbors, which go far beyond boundary struggles—but it is exactly along the boundaries that the pressures are most actively exerted.

On the Arabian Peninsula, a very unusual situation exists. There, a number of semi-autonomous sheikdoms and sultanates have only tentative boundaries with their giant neighbor, Saudi Arabia—and along that country's northern border lie two "neutral" territories which are, in effect, properties of Western oil interests. Thus the political processes which were deemed indispensable by the European powers who gathered to determine the Middle East's future after World War I were quite readily shelved by the same powers in the interest of facilitating the exploitation of the petroleum of the Persian Gulf area. To many Arabs, especially those who do not share in the rich petroleum harvest's revenues, this is merely on a par with the promises that were made in regard to Palestine during the last years of the Ottoman Empire—promises that were, in their eyes, shamelessly broken.

Arabian Bonanza

Under different circumstances, the oil fields of the Middle East could have been a basis for general progress and development in the region. In ancient times, the availability of tin and copper within the Middle East itself sustained the old empires. Europe's industrial revolution was based on resources that were locally mined. But when petroleum became a vital resource in the world, the countries of the Middle East were not at a technological level that would have enabled them to exploit their own fields; neither did they have an internal market for it. Other countries,

especially those of Western Europe, needed oil in increasing quantities. These were the same countries that had defeated the Ottoman Empire; as Mandate powers they had considerable influence in the Middle East. British, French, and later American companies had the capital, the technical know-how, and the necessary political influence to do what the people of the Middle East could not themselves do. And so, along the fringes of the Persian Gulf, began the Western exploitation of the Middle East's single rich resource—the greatest oil fields in the world.

The first major oil discoveries, actually, occurred not on the very shores of the Persian Gulf, but about a hundred miles inland from the head of the Gulf, at the foot of the Zagros Mountains in what was at the time Persia (Iran). From there, oil was being exported as early as 1912. It now travels by pipeline to refineries at coastal or near-coastal locations that can be reached by ocean tankers, such as Abadan (Map 10-4). But this is an expensive and, as the political events of the 1950's and 1960's have shown, a precarious route, and the European powers soon developed more direct means of transporting the oil to its destination. Iraq's principal field, which began production in 1927, is one of the leading oil fields in the Middle East, and, indeed, in the world. It lies not far from the Tigris River, near Kirkuk. But rather than shipping the oil down the Tigris to Persian Gulf refineries, it is sent by several pipelines to shipping points on the Mediterranean coast in Syria and Lebanon. In fact, oil that is brought to the surface on the very shores of the Persian Gulf itself (in fields located in the Hasa sheikdom of Saudi Arabia) is now piped all the way across the Arabian Peninsula to the coastal Lebanese station of Saida.

But even as the pipelines went into operation, tankers continued to file through the Suez Canal to supplement what the pipes could carry. The European presence in the Middle East—and, increasingly, the American presence as well—has been a by-product of the growing importance of petroleum to the Western world. Britain maintained military forces on Cyprus and until 1956 along the Suez Canal itself; when, in 1956, Egypt announced its intention to nationalize the Canal, the then British Prime Minister called it a threat to "Europe's lifeline." Along the Persian Gulf coast, Western involvement has produced a peculiar political situation. Kuwait, a small country of vast oil wealth, remained a British protectorate until as late as 1961. On the northern boundaries of Saudi Arabia lie territories that are marked as "neutral" on the atlas map; in effect they are oil dependencies of the West, the properties of huge oil companies. Saudi Arabia has traded its oil not only for American money, but also for some of its sovereignty as the United States maintains military installations at Dhahran.

Until quite recently, the Persian Gulf states and territories have had the undisputed lead in oil production in the Middle East. Saudi Arabia and Kuwait have been almost legendary names for the riches in "black gold" they have brought forth. The Persian Gulf reserves, moreover, are reputed to be the greatest in the world, far exceeding those of the United States and Venezuela–Colombia. Iran is a major oil exporter. But in the past few years a North African state, Libya, has seen its production rise rapidly, and late in 1968 its Ministry of Petroleum Affairs announced that monthly outputs now exceeded those of Kuwait and were near those of Saudi Arabia.[5] This has placed Libya among the world's first three exporters, after Venezuela and Saudi Arabia. The Saharan fields may have a great future: Algeria's production, too, is on the rise.

But what does it mean for a Middle Eastern state to possess riches in oil? Does it mean that workers find jobs, that annual incomes improve? Does it mean that at last the country is developing? Does the petroleum industry stimulate other industries? Does it make the country stronger, more independent? Unhappily the answer to most of these questions must be negative. Countries that must export their raw materials in crude or even partially refined form do not fare well in the economics of the deal. They really do not have much choice: they do not have the capital or technology to extract the resources themselves, and so they must either attract these or leave the resources untouched. In attracting foreign companies they have to make concessions, knowing that all too often the company can turn elsewhere if conditions are not favorable to it. Standing behind the companies are the interested governments, and in the politics of the deal the countries may fare poorly too. Some countries, like Chile, are attempting to mobilize

[5] See *Africa Report*, Vol. 13, No. 8, November, 1968, p. 28.

Wealth from the desert. The impact of underground "black gold" is well illustrated by this photograph, which shows the refinery and storage facilities and pipelines through which the petroleum goes on its way to the coastal terminal and ocean tankers. This scene is from the Arabian Peninsula. (Arabian American Oil Co.)

local capital to get a larger share of the income and are insisting that the raw material (in Chile's case, copper) shall be treated in Chilean plants prior to export. Iran in 1951 nationalized its oil facilities but soon found that the united opposition of Western oil interests (which refused to buy Iranian petroleum) could bring even the most determined nationalist government to its knees.

Of course, the picture is not all one-sided. The oil installations have stimulated impressive local development; Dhahran, for example, is by all standards a boom town. And the countries whose resources are being tapped do receive large royalties—which do not always find their way to places where they might benefit the economy most. But in common with other countries in the "developing" world which have little alternative but to sell their raw materials to foreign interests which set the terms and control the price, the overwhelming majority of the people of the oil-rich states of the Middle East still find themselves among the very poor.

Modern Egypt (United Arab Republic)

The Nile dominates Egypt today as it did in ancient history. Its waters still irrigate the farmlands, though the great river is controlled now by a series of dams and reservoirs, and motor-driven pumps are used to raise the water to soils the river cannot reach by itself. But otherwise Egypt is still the Nile, its valley and delta, and the *fellah* of today still struggles to make a living for his family, as did the peasant in the Egypt of five thousand years ago. In a land which at one time was the source of countless innovations and the testing ground for many others, the tools of the farmer often are as old as any still in use: the hand hoe, the wooden, buffalo-drawn plow, the sickle. Water is still drawn from wells in buckets attached to a wheel that is kept in motion by the endless, circular walk of a farm animal; the Archimedean screw can still be seen in service. The peasant still finds himself in a small, mud-made dwelling, which would look very familiar to the farmer of many centuries ago; neither would the distant ancestor be surprised at the poverty, the diseases, the high rate of death among children, the lack of change, or hope of alternative ways of life.

And yet, of all the countries in the Middle East, Egypt is the largest—in terms of population—and, by most measures, the most influential. Its two nearest rivals in size, Turkey and Ethiopia, also are significant forces in the realm, but each has directed its energies toward non-Arab as well as Middle Eastern areas. Turkey's twentieth century position has to a considerable extent been one of orientation toward Europe, and Ethiopia has sought to attain a position of strength in the affairs of Black Africa. Egypt alone is spatially, culturally, and ideologically at the heart of the "Arab world." What factors have combined to place Egypt in this position, despite the unchangeable aspect presented by its rural areas, the backwardness and poverty of its mass peasantry?

Location obviously is one of these factors. Egypt's position at the southeastern corner of the Mediterranean Sea provided an early impetus: in ancient times the Phoenicians and the Cretans connected the Nile valley with the Levant and other parts of the Mediterranean world. Egypt lies immediately opposite the Arabian Peninsula, and in the centuries of Arab power and empire, it received the full thrust of the new wave. The Arab victors founded Cairo in A.D. 969; the city has benefited from its location near the place where the Nile valley spreads into the delta (Map 10-5). Egypt, too, lies astride the land bridge between Africa and southwest Asia (between the Mediterranean and the Red Sea), a route which at another time of history had great importance. More recently this land bridge has become an even greater asset to Egypt: cut in 1869 by the Suez Canal, it became the vital link in the shortest route between Europe, the Mediterranean, and South and East Asia. The canal brought Egypt revenues, but it also brought the European presence to many spheres of life. Beginning in the 1880's British interest stimulated modernization of the country's transport and communication systems, of agriculture and irrigation practices, of trade and commerce, and many other spheres of activity. Now Cairo's fortuitous location really came to play its role, and the city began its development into what is today the largest urban center in a huge area of the world, including all of Africa, the whole of Southwest Asia, and much of Southern Europe.

Eventually Britain's role in Egypt ceased with the miscalculated 1956 invasion of the Sinai Peninsula (which was intended to guarantee European security of the Suez "lifeline"). But during that century, modern Egypt had emerged, and Egypt's prominence in the Middle East had been established. Cairo (4 million) at the head of the delta, and Alexandria (2 million), on the Mediterranean coast and the leading seaport, reflected the new, commercial era in Egypt. Port Said and Suez, at the northern and southern end of the Suez Canal, respectively, grew with the ever-rising tonnage that passed through the waterway. The biggest boost came with the discovery of the Persian Gulf oilfields about a half century ago, and soon large tankers were moving petroleum to Europe. Recently, when the Canal functioned normally and without wartime interruptions, oil still remained the commodity to pass through it in greatest quantities, despite the use of pipelines across Arabia to the eastern Mediterranean coast.

British and Anglo-French involvement in Egypt may not, in Arab eyes, have been a salutory experience, but it nevertheless left an enormous legacy in material assets. Apart from the achievement of the Suez Canal, which of course was built by European (especially British) capital and French technicians, other contributions were

Map 10-5 EGYPT AND THE NILE.

AFTER: MURPHEY, INTRODUCTION TO GEOGRAPHY

391

Scene near an Egyptian village. The men in the foreground are turning the ancient Archimedan screw, raising water from the little stream at the left to the land at right. Washing is being done by the women. Mud houses stand in the upper left of the photo, with a field of grain nearby. Typically, the sun beats down from a cloudless sky. (World Health Organization)

made. In agriculture, additional land was brought under cultivation by the mechanization of pumps and by the construction of reservoirs. This had the effect of converting hundreds of thousands of acres of land under flood irrigation (or *basin irrigation,* as it is called) to *perennial irrigation,* which means that the stored-up water is made available to the plants whenever it is needed. Fertilizing methods were improved, and the total yields of corn—the staple for most of the country—as well as average yields per acre increased considerably. In addition, the 1860's saw a boom in a commercial agricultural product, grown with British stimulation, namely cotton, which still today accounts for over three-quarters of Egypt's income from exports. And, Egypt's poverty in natural resources notwithstanding, it has emerged as the leading industrial country in the "Arab world." A steel plant has been built about 20 miles south of Cairo, where iron from a deposit in southern Egypt is smelted with imported coke shipped up the Nile. Al-

though the local market is comparatively small and quite poor, a textile industry has developed. The large agricultural output also has given rise to a sizable food processing industry. But, like many other Middle Eastern countries, Egypt on a world scale has barely begun to industrialize, and its position of leadership in this sphere in the Middle East only serves to emphasize the problems of development facing the whole region.

Population/Resource Regions. Egypt may not have much in the way of natural resources, but it does have very good and intensively cultivated soils, and somehow the country now sustains 35 million people. Why, with the assets of the Suez Canal, salable cotton, high farm yields, an important tourist trade, and more—why are so many people still mired in the grinding poverty, even misery, of the rural areas? Time and again improvements have occurred that have raised harvests, expanded irrigable acreages, increased farm efficiency. What has happened to the benefits derived from all this? When a project such as that of the High Dam at Aswan is completed at vast expense and with enormous benefits, how can so many people remain untouched by the march of progress?

The answer is just that—so many people. The century of modernization in Egypt has also been a century of population explosion. It is estimated that the country's population increased six-fold during this period; hence, every time productivity was raised, the people were there in greater numbers to absorb the increase, keeping the available amount of food per person at the existing low level, or even lowering it a bit. Today, Egypt occasionally has to import rice and other foodstuffs even to maintain its low per capita calorie intake, and the Aswan High Dam, once the great hope for a substantial step forward, is now expected to do only what other improvements have done—help maintain the unsatisfactory status quo.

This situation, of course, is not unique to Egypt. As we shall see when discussing parts of Asia that are in certain ways comparable to Egypt, similar conditions exist elsewhere. This being so, some insights might be gained from the study of what has been called the population/resource ratio of various regions of the world. In 1959, Professor E. A. Ackerman proposed a set of five such world regions (Map 10-6).[6] The types involved are self-explanatory (see the map legend), but in a later commentary, Professor W. Zelinsky modified one of them, the *China type,* and identified it instead as the *Egyptian type,* since the situation in the relevant areas of China appeared to be changing.[7] In discussing the Egyptian type, Professor Zelinsky tells us something about the state of things in this region, and what he says is directly and specifically applicable to Egypt itself:

"The Egyptian Type region . . . is most discouraging in its imbalance of excessive numbers of consumers and shortages of immediately available resources, both physical and social. These Egyptian lands are densely populated, not only in terms of sheer numbers, but also in ratio

[6] E. A. Ackerman, "Population and Natural Resources," in Philip M. Hauser and Otis D. Duncan (Eds.), *The Study of Population: An Inventory and Appraisal,* University of Chicago Press, Chicago, 1959, pp. 108–109 (see footnote 7).

[7] W. Zelinsky, *A Prologue to Population Geography,* Prentice-Hall, Englewood Cliffs, 1966. See pp. 106 to 116 for a discussion of the concept and its applications.

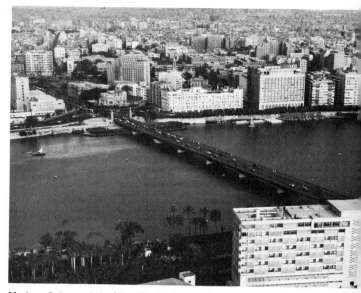

Modern Cairo and the Nile. The high-rise character of this part of Cairo stands in sharp contrast to the clutter of low-level dwellings that marks most of the remainder of this, the region's largest city. (Almasy/World Health Organization)

of inhabitants to accessible means of subsistence or employment . . . (there is) scant hope for any qualitative improvement in the welfare of these groups—unless there is a revolutionary, and thus quite painful, transformation of society and economy by either external or internal forces. In some instances, the physical endowment is such that even the most radically thoroughgoing approach to a community's development would seem to offer little chance of general betterment."

Egypt experienced a revolution in 1952, when the government of King Farouk was overthrown. Soon afterward, the revolutionary government announced a sweeping policy of land reform, whereby the many large estates (often owned by city-based absentee landlords and worked by sharecroppers or wage labor) were expropriated and parcelled out among landless peasants on long-term loans. It was a step in the right direction and there were some good results, but so severe are Egypt's problems that the ultimate impact on the state of affairs was only slight—just as Professor Zelinsky suggests. He continues:

"Egypt illustrates the general problems of . . . technologically backward countries. The ecumene of this ancient land has always been confined to the alluvial valley of the Nile. Nearly

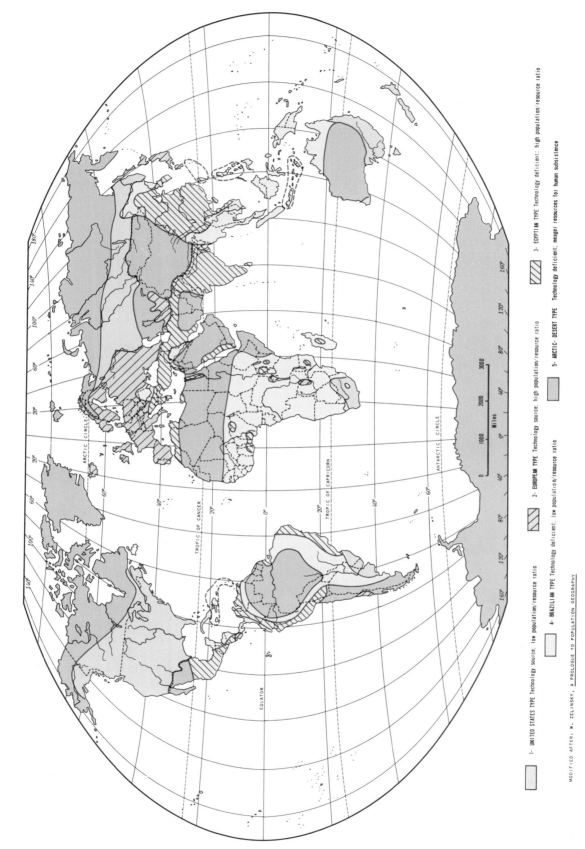

1- **UNITED STATES TYPE** Technology source; low population/resource ratio

2- **EUROPEAN TYPE** Technology source; high population/resource ratio

3- **EGYPTIAN TYPE** Technology deficient; high population/resource ratio

4- **BRAZILIAN TYPE** Technology deficient; low population/resource ratio

5- **ARCTIC- DESERT TYPE** Technology deficient; meager resources for human subsistence

MODIFIED AFTER: W. ZELINSKY, A PROLOGUE TO POPULATION GEOGRAPHY

Map 10-6 POPULATION-RESOURCE REGIONS.

all this precious soil is given over to raising food crops for a population that is now doubling in size roughly every 25 years; only a minor fraction of it is allotted to exportable crops such as cotton. . . . Virtually no outlets exist in other parts of the world for Egyptians who might want to emigrate, and military conquest of neighboring lands, with the dubious exception of Sudan, would open no new horizons for colonization. Integrating the Egyptian economy with that of neighboring states, an expedient now being explored, offers only slim hope for deliverance from the population/resource crisis, since these neighbors are almost as inadequately outfitted for economic advancement as is Egypt. Equally grave—and particularly ironic in a land that was once the unquestioned leader of civilized nations—are the human and institutional barriers to general progress: the low incidence of educated and literate persons; the general absence of advanced skills; conservative, even negative, social attitudes; the present incapacity of a large peasant population for anything beyond its current hand-to-mouth existence . . . ."

Nevertheless, Egypt has become the cultural and political focus for the modern Middle East; the capital, Cairo, also is the ideological headquarters of the "Arab world," the site of frequent Arab League conferences, the embodiment of Arab aspirations to supranational unity. It is, in a sense, a hollow primacy, for the weaknesses of the Arab countries of the Middle East have been repeatedly revealed in the continuing conflict with Israel. Indeed, Israel's existence is one of the major contributors to such Arab unity as the modern Middle East has witnessed, and certainly Egypt, Israel's southwestern neighbor, has taken a lead among Arab states in this contest. But there are other grounds for Egypt's leadership. The country occupies a pivotal position at the center of four wings of the varied Middle Eastern realm (see Map 10-4). To the west, the countries of North Africa share a number of characteristics; they form a "bloc" within the Arab world. To the northeast lie the countries of the Levant, with which Egypt has had millenia of association. To the southeast lies the vast Arabian Peninsula, and Egypt has involved itself in the political development of such countries as Yemen. And to the south lie the Sudan, the Somali Republic, and Ethiopia; in this direction, too, Egyptian influence has been felt—as in the Condominium it shared with the British in the former Anglo-Egyptian Sudan. With none of its former glory, Egypt still maintains a modern prominence.

The "Middle East" in North Africa

West of Egypt lie the states of the *Maghreb*—the western part of the Arab realm (Map 10-7). Morocco, Algeria, Tunisia, and Libya share a number of characteristics with Egypt, including the high population/resource ratio that marks region 3 on Map 10-6. Plates 1 to 4 in the rear of this volume reflect the environmental circumstances that are common to the Mediterranean states of North Africa, and Plate 6 demonstrates the littoral concentration of a combined population which now numbers about 37 million, barely more than Egypt's alone. As Plate 6 shows, the areas of most dense population lie in northern Algeria and Tunisia, but the population per country declines from west to east. Morocco is in the lead with 15 million; Algeria, the largest state territorially, is next with 14 million; Tunisia, the smallest state territorially, has about 5.5 million people, and Libya has only some 2 million inhabitants.

Topographically, the pivotal feature is the great Atlas, whose mountain slopes wrest vital moisture from the air, whose streams irrigate the lands of the *tell,* and whose porous rock layers feed the underground channels or *qanats* that supply wells and reservoirs. The Atlas Mountains extend from western Morocco through northern Algeria into Tunisia, but they do not reach Libya. In Libya as in Egypt, then, this barrier between sea and desert is absent, and the two countries are largely desert; in the countries of the Atlas the coastward slopes and the Mediterranean climatic regime combine to produce between 15 and 30 inches of rain annually (in a very few places, more), which is what has made agriculture possible.

The people of the Maghreb are of many ancestries. The Berber people, who still make up almost half of the area's population, were in this region before other arrivals. Phoenician and Greek colonies existed along the North African coast. In the days of the Roman Empire, the city of Carthage (a few miles north of modern Tunis) was a southern European outpost. Later Germanic elements arrived in the confusion that followed the fall of Rome, and then came the Arab conquest and the spread of Islam. During this period, black African strains also were fused

Map 10-7 THE MAGHREB.

MODIFIED AFTER: RAND MCNALLY, GOODE'S WORLD ATLAS

in the mixture. The last major invasion came quite recently, during the nineteenth and twentieth centuries, when France established and maintained control over Algeria, Tunisia, and Morocco, while Italy took control of Libya. During the colonial era between 1 and 1.5 million Europeans came to North Africa (most of them to Algeria), and of course these immigrants soon dominated commercial life. They stimulated the renewed growth of the region's towns, and Casablanca (over 1 million), Algiers (1 million), and Tunis (approaching 1 million) rose to become the foci of the colonized territories. But while the Europeans dominated trade and commerce and integrated the North African countries with France and the European Mediterranean world, they did not confine themselves to the cities and towns. They recognized the agricultural possibilities of the favored parts of the tell, and established thriving farms. Agriculture here, naturally, is of the Mediterranean variety, and Algeria became known for its vineyards and its wines, its citrus groves, and its dates; Tunisia has long been the world's leading exporter of olive oil; Moroccan oranges found their way to many European markets. The staples—wheat and barley—long were among the exports.

During the colonial era, then, the Maghreb presented some aspects that were unusual for the Middle East: a large European settler population, considerable land alienation, and thriving commercial agriculture. Algiers, Casablanca, and Tunis were "European" towns—more so, in many ways, than Alexandria and Cairo. The contrasts between wealth and poverty were not intra-Arab, but rather interracial. In Egypt, the rich landowner living in the city and drawing income from his estate was likely to be an Egyptian; in Algeria he was likely to be a Frenchman. What the colonized Maghreb did have in common with Egypt was the abject poverty of its peasantry—and it still does. In the drier parts of North Africa, the peasant can hope to grow some cereals in poor soils and under the threat of a very variable rainfall regime; he will have a few sheep or perhaps goats. And, as many still do, he may live a nomadic life in a constant search for the means of survival.

Despite the proximity of the Maghreb to France and Southern Europe, and the tight integration of the region's territories within the French political framework, nationalism emerged as a powerful force in Morocco and Tunisia as well as in Algeria. Morocco and Tunisia secured independence mainly through negotiation, but in Algeria a revolution began in 1954. This costly war did not end until a settlement was reached that took effect in 1962. It was not difficult for the nationalists to recruit followers for their campaign; the justification for it was etched in the very landscape of the country, in the splendid, shining residences of the landlords and the miserable huts of the peasants. But the revolution's success brought new troubles. Hundreds of thousands of Frenchmen left Algeria, and the country's agricultural economy fell to pieces; an orderly transition was not possible. Productive farms went to ruin, exports declined, badly needed income was lost. As Professor Zelinsky pointed out in his discussion of the Egyptian-type population/resource region, such a transformation is painful, and there is little chance for the people's general betterment after it has occurred.

But there are some bright spots in the Maghreb. Algeria has one resource to compensate for its losses in agriculture—oil, the same oil that led the French in the 1950's to resist Algerian independence. Petroleum has now risen to the top of Algeria's export commodities; it is won from the Sahara Desert south of Touggourt and piped to Bougie in Algeria and La Skhira in Tunisia for export to Europe. Libya's oil production has risen spectacularly to place the country among the world's leading exporters; again, the fields are located in the Sahara. And, quite unlike most of the remainder of the Middle East, the countries of the Maghreb have a rather varied set of mineral resources. Chief among these is phosphate of lime, used in the manufacture of fertilizer. Morocco is the world's leading exporter of this commodity, which occurs in a layer of limestone in the immediate hinterland of Casablanca and Safi. It is found also in Algeria, and in substantial quantities in Tunisia, where it ranks as the most valuable export as well. Both Morocco and Algeria have Atlas-related iron ores, which are exported to the United Kingdom from their favorably located sources. Manganese, lead, and zinc also are mined in all three countries, but none has sufficient coal locally available. Despite the new power source in the oil fields (and the associated natural gas), this not inconsiderable range of minerals has failed to stimulate much domestic industrial development.

The Interior. There was a time when the Sahara, its oases, its caravan routes, and its faraway fringes were of intense importance to the Maghreb. Morocco's old cities, Fez and Marrakech, are reminders of this bygone era. Fez for a long time was the country's capital; it was founded in 801 A.D. and about five centuries ago may have had double its current population (250,000). It lay at one of North Africa's most important crossroads: one route stretched from Tangier (near the tip of the Iberian Peninsula) via Fez, through passes in the Atlas, and along a line of Sahara oases to Timbouctou. The other stretched across the Maghreb from Tunisia and points east to Marrakech and the Atlantic portion of Morocco. Marrakech itself, founded somewhat later than Fez, is today the foremost interior and really indigenous city of the country. It, too, grew at the intersection of trade routes of old, and as an oasis in a dry region it has maintained its functions of market place, industrial center (like Fez, it is known for its carpet weavers), and focus of an intensively cultivated area of Mediterranean agriculture.

But the old Saharan connections (see Chapter 9) have faded, and modern times have brought a new orientation to the Maghreb. Once the cities of the north traded across the unorganized interior with the states of the southern savannas, but today there is little contact even with those transitional states that now occupy the territories adjoining the Maghreb—Mauritania, Mali, Niger, and Chad. Nevertheless, Morocco for one has not forgotten its ancient sphere of influence: it is one of the bases for its claim that most of Mauritania should become Moroccan national territory. And, of course, with the rising production of oil in the Sahara, the peoples of the Maghreb countries now have a renewed interest in the dry lands across the Atlas mountains. To the state of Libya, Saharan petroleum has been a veritable lifesaver. Deprived of a tell region, Libya has scant agricultural possibilities even on its Mediterranean coast, and the country long seemed doomed to almost total dependence on foreign aid and the income derived from foreign military bases on its soil. In any event, Libya appeared to have little reason for political existence, since it was created in 1951 under United Nations auspices—a matter of great-power convenience. It was made out of three distinct units: coastal Tripolitania, which had a long history of connections with and orientation toward the western Maghreb; Cyrenaica on the eastern coast, whose outlook traditionally was toward Egypt and the Levant; and the interior Fezzan. The union survived, albeit with three capitals (Tripoli, Benghazi, and Beida); its economic future could hardly have been guessed in 1951. In Libya, the interior has truly unprecedented significance.

The Sudan and Africa's "Horn"

To the south of Egypt lies another part of the Middle East that has its own regional qualities. There are three states: the Sudan, a giant country that extends from the Egyptian border into Black Africa; Ethiopia, another large unit whose inclusion as a Middle Eastern country is open to debate; and the Somali Republic, the desert fringe of the "Horn" of Africa. Here we are in the transition zone between Arab Africa and Black Africa. Of the three countries, only Somali is an overwhelmingly Moslem state; in the Sudan, the black population of the three southern provinces constitutes at least a quarter of the country's total, and peoples of dominant Black African ancestry live in large numbers in other parts of the country as well. The Sudan's northern two-thirds came under the influence of Islam, but the southern black population was exposed to Christianity during the colonial period. And in Ethiopia a centuries-long struggle was waged between the Christian kings of the high, protective plateaus and the sultans of the coastal areas and the lower east—a conflict in which the Christians managed to hold their own. A recent estimate suggests that Ethiopia is only 31 percent Moslem (the figure is probably low), the Sudan, 80 percent (this may be high), and Somali, 99 percent.[8]

Territorially, the Sudan is the largest country in the Middle East. As Map 10-8 shows, its entire length is traversed by the White Nile, which is fed by the waters of Lake Victoria and maintains a rather even flow. The Blue Nile, which rises in Ethiopia at Lake Tana, joins the White Nile at Khartum and there contributes its floodwaters to the great river. Above Egypt, therefore, almost the whole length of the two Nile Rivers combined lies in the Sudan: small wonder that Egypt over the years has been a concerned participant in the affairs of its southern neighbor. Early in the nineteenth century, Egypt sent its

[8] J. Baulin, *The Arab Role in Africa,* Penguin Books, 1962, pp. 12–13 (map).

MODIFIED AFTER: RAND MCNALLY, GOODE'S WORLD ATLAS

Map 10–8 THE SUDAN AND AFRICA'S HORN.

forces southward against the Fung Empire, an indigenous Sudanese political entity that had arisen at the confluence of the two Niles; by 1830 the Empire was defeated and an Egyptian government was established at Khartum. Then Britain entered the scene, in part to counter the slave trade in the area; the British sphere of influence eventually extended over Egypt as well as the Sudan. But there was only nominal British supervision over Egyptian rule in Khartum, and the Egyptian overlords were hated and despised by the Sudanese. In 1881 came the predictable revolution, led by Muhammad Ahmad, a man who proclaimed himself the Mahdi, a messenger of the Prophet's Islam. As is true of so many indigenous rulers of Africa and Asia, history has not treated the Mahdi kindly, but he was undoubtedly a man of great capabilities. He rallied countrywide support, and soon his Dervishes controlled large parts of the country. Despite the

role of the Bristish in all this, made famous by Gordon's stand at Khartum, the campaign of the Mahdi and his followers was fundamentally an anti-Egyptian rather than an anti-British one. The Mahdi died soon after his victory at Khartum, but under his successor, Abdullahi, the Sudan fell apart. This set the stage for the colonial intrusion: in 1898 a British-Egyptian force, assisted by Sudanese elements, finally defeated the Dervish forces. Now began the period of Condominium, which lasted until independence came to the Sudan in the mid-1950's.

The Sudan, with 14 million people, has been called a bridge between Arab and African, but this appears to be an inappropriate designation. It was based on some obvious features of the country: not only does it have an "Arab" north and a black African south, but even in the north there are many people of pronounced black African ancestry. The "north" is generally taken

In the Gezira, the giant irrigation project between the Blue and the White Nile in the Sudan, a vast area has been brought into cotton production. These men are inspecting one of the canals (foreground) that have been constructed to divert the Blue Nile's waters into this area; in the background is cotton land coming into production. (Eric Schwab/World Health Organization)

to mean the area north of 12° north latitude; of the Sudan's 9 provinces, 6 are in the north and 3 comprise the south. In the old days of power and wealth in West Africa, millions made the pilgrimage to Mecca; perhaps as many as a half million Africans—most of them of Yoruba heritage—stayed in the Sudan to make a living. But the Sudan has come to terms neither with its regional ethnic contrasts nor with its local racial problems. Ever since independence, more or less, the south has been in revolt against the Khartum government. Charges have been made that the Sudanese are engaged in a campaign of suppression and extermination in the south, and that the "blacks" of the north are eventually to be ousted.[9] Whatever the degree of exaggeration in these claims, it is unlikely that official descriptions of successful integration and economic development for the south are any nearer the truth.

The Sudan, then, is another of those countries in the excolonial world that found, upon independence, a legacy of trouble and conflict. The colonists did not *make* all the difficulties, but neither did they manage to lessen them very much. True, the northern Sudanese had been slavers, and the black southerners, the slaves. True again, the British destroyed the slave trade. But to

[9]For example, in J. Oduho and W. Deng, *The Problem of the Southern Sudan,* Oxford University Press, 1963, p. 8.

encourage the Christianization of the black pagan peoples of the south in a country destined to be Moslem-oriented was to add yet another divider between peoples and regions.

In terms of economic development, too, the contrasts between north and south are strong, and the south shows the effects of long-term neglect as well as limited potential. The heart of the Sudan—and here lies one major reason for considering this a Middle Eastern country—lies in the area of the old Fung Empire, including Khartum and its sister cities, Omdurman and Khartum North (total: 500,000), and the great irrigation project between the Blue and White Nile. This is the famous Gezira ("island"), a triangle of irrigation canals and farmland, watered by the Blue Nile, whose higher gradient permits a gravity flow to the White Nile. In the 1920's, the British introduced cotton here, and this has become the country's mainstay; it contributes about three-quarters of the annual export revenue. But more important was the way it was done: not by huge estates and at the expense of landless peasants, but through a land-renting, profit-sharing small farm scheme in which three parties—the Sudan government, the two capital-investing concession companies, and the farmers—shared in the income and paid off the debt. The success of the whole idea can be read off the maps: planted acreages increased from about one-quarter of a million in the mid-1920's to over 1.5 million today. Of course, Egypt has been very interested in this project, since the waters of the Nile are diverted here: there is a treaty which stipulates that expansion of the irrigated areas can be carried out only with the consent of Cairo.

The Khartum–Gezira area thus is the historic as well as the modern core of the Sudan; today, there are rail connections from here to the main outlet at Port Sudan on the Red Sea as well as to Nyala in the west, Wadi Halfa near the Egyptian border in the north, and most recently to beyond Wau in the south. The first stimulus for the building of modern communications came with the rapid growth of the Sudan's other industry, the extraction of gum arabic (used in medicine, confectionery, and textile manufacture) from the ever-present acacia tree on the Kordofan Plateau. At this early state, El Obeid became the chief auction center, and the Kosti–Kassala–Port Sudan line was laid. But the gum arabic industry has been overshadowed by the Gezira's cotton. More recently, railroad building has taken on another aspect: the southern line to Wau is hardly economical, but it was built and extended to improve the effectiveness of the Sudanese presence in the sometimes secessionist-minded south. Communications to this part of the country have always been difficult; the Nile has been the main link, and Juba the leading regional center. The many-channeled Sudd through which the Nile flows is a vast morass, subject to flooding over hundreds of miles in both directions from the river. Here is yet another divisive element in the spatial makeup of this huge country.

Earlier in this chapter we made reference to physiographic-political boundaries as elements in the Middle Eastern cultural landscape. Over a considerable part of its course, the Sudan–Ethiopia boundary can be identified as such. For hundreds of miles, the Sudan ends where the land begins to rise to the enormous, abyss-fractured Ethiopian Plateau. Ethiopia seems to be a natural fortress, its walls overlooking the plains of Sudan to the west, and those of Somali to the east.

Such an impression is to some degree consistent with the facts. Modern Ethiopia's beginnings lie in the Aksum Kingdom, an entity which centered on the still-extant town of that name not far from the Red Sea coast (Map 10-8). From the first to the seventh century, A.D., this was the leading political entity in the region, and it was in touch with the outside world through the visits of Greco-Egyptian and Roman ships. Christian missionaries settled in Aksum and began the conversion of the ruling dynasty. But then Islam arose, and Aksum declined; as the forces of Mohammed crossed the Red Sea, the remnants of Aksumite power retreated into the Ethiopian mountains. There, an almost ceaseless struggle with the Moslems began. The fortunes varied; at times it seemed that the Christians were surely doomed, only to survive again. A powerful Moslem sultanate arose centered on Harar, still a focus of Islam today, and in the sixteenth century its power was too much for the Christians. Just as the final blow was about to be struck, an outside power with interests in the Red Sea appeared: Portugal. In 1541, Portuguese forces, combined with those of the Ethiopians, defeated the Moslems not far from Lake Tana.

But the struggle was far from over. An African people, the Galla, now migrated into Ethiopia, and their infusion caused mass political disorganization. Once again the Ethiopians, using the

The unending quest for water. When, in dry periods, the wells of an area cease to produce, the villagers start their search for alternate sources. Here, in Sudan's Kordofan province, two men are digging for water by the oldest of methods—by hand. The earth is handed up on a platter; eventually there may be several men stacked in this narrow hole, each handling such a platter. The hole may reach 30 feet in depth. (Eric Schwab/World Health Organization)

protection of the highlands, withdrew. And they fought among themselves as they did jointly against their enemies; by the early nineteenth century, there were no less than six "kingdoms" in the country, each ruled by a man who considered himself the emperor of all Ethiopia. All this time, of course, the Ethiopians had been separated from Europe, and from religious change (except for an attempt by Jesuit missionaries to bring Roman Catholicism there). Thus the rulers of the old feudal states, as well as the emperor of today, laid claim to direct descent from King Solomon himself. In their isolation, Ethiopia's Christians developed their own religious dogma and ritual, and their own distinctive interconnections between church and state.

Out of the chaos of the nineteenth century modern Ethiopia emerged. Truly incredible were the political entanglements: Egypt had entered the scene along the Red Sea coast and had established a sphere of influence as far even as Harar in the east. The Mahdist revolution in the Sudan brought incursions into western Ethiopia. The opening of the Suez Canal brought European influence that much closer, and a quarrel with the British led to an invasion. The French were interested in the southern bottleneck of the Red Sea and based themselves at Djibouti. The Russians were interested in Ethiopia as "protectors" of the Coptic Orthodox Church. The Italians came to establish coastal stations in Eritrea. Feudal divisions continued on the plateau. Ethiopia seemed doomed to suffer the same fate as the rest of Africa: it would surely fall, to be divided among the colonial interests.

This was not to happen, mainly because of the leadership of the emperor Menelik, who came to power in 1889 while he was involved in a power struggle with another claimant to the Ethiopian throne, John IV. Menelik, seeing his country in danger of partition and colonial occupation, immediately began his own campaign of expansion, playing off the colonial powers against each other with great skill. He had founded the capital, Addis Ababa, in 1883, and negotiated with the French for the building of a railroad to their port at Djibouti. He settled a dispute with the Italians (over the language of a treaty he had signed) by defeating the Italian forces. He pushed Ethiopia's borders far out from the plateau, and, as the map of today still shows, he managed to make many of his claims stick. He also began the first modernization program for Ethiopia: schools, postal services, public utilities. He was in some ways the father of Ethiopia.

But neither Menelik nor his successors could quickly overcome the problems with which the country has always been faced. The huge plateau, with its high surface (much of it over 9000 feet) was an excellent protector, but with its deep ravines and precipitous escarpments it also has been an unyielding divider. Its former name, *Abyss*inia, was a very appropriate one. What had made the survival of several feudal entities possible also kept the peoples of Ethiopia in isolation. Indeed, the greatest step forward in the integration of the country's regions was made not by the Ethiopians, but by the Italians during their brief occupation of it (1936–1941). They initiated a vigorous program of road building and economic development which still has favorable effects today. But even this period of investment (and more recent programs as well) barely constituted a beginning. In Ethiopia, the communications problem is enormous, and the map shows how quickly the road system that radiates out of Addis Ababa decreases in density. Huge parts of the country can still be reached only by track or path, and goods still travel by pack mule.

And yet Ethiopia may have a bright future, for its physiographic diversity is matched by an environmental variety unequalled anywhere in the Middle East—or, for that matter, in most of Africa. As plate 1 shows, there is a good deal of rainfall, and from the low-lying, hot areas around the plateau margins and in the deep valleys to the high pasturelands on the upper plateaus there are opportunities for a wide range of agricultural pursuits. Bananas and dates will grow in the low regions; coffee, the leading export product, grows in the intermediate zone, where Mediterranean fruits do well also; wheat and barley grow up high, where sheep and cattle use the extensive pastures. Ethiopia is sometimes called a potential breadbasket for the Middle East, and the country awaits the necessary reforms—in education, in land ownership, in land tenure, in conservation practices, in irrigation methods, in farm technology, in seed improvement, and many other spheres—to make its potential come true.

Present-day Ethiopia is one of the most populous countries of Africa and the Middle East. Its exact population is not known, and estimates vary widely, but there are probably around 25 million Ethiopians, of whom 90 percent still remain illiterate. It is still a country of isolated villages and rural stagnation. Neither have the political problems—internal or external—been solved. Internally, there is no escaping the fact that the Christian, Amharic-speaking ruling group is a minority, and that the Moslem and pagan-African majority are not much involved in the administration of their country. Externally, there has been trouble with the Somali Republic, which has not forgotten Menelik's expansionist campaign and wants the Moslem–Somali "Horn" united in a Greater Somali Union. But Ethiopia has cast its lot with Africa rather than the Moslem world, and has sought to attain a position of leadership by making the capital, Addis Ababa (450,000) the headquarters of the Organization of African Unity. Again, it is the strong leadership of a single man that has been instrumental in much that has been achieved: the emperor Haile Selassie, who was crowned in 1930 at the age of 39, has been a second Menelik to the country. He led Ethiopia to membership of the League of Nations and fought there for its survival when Italian fascism threatened it. He negotiated successfully after World War II for the inclusion of the Italian dependency of Eritrea (on the Red Sea) with Ethiopia, ending its landlocked condition and its total dependence on the French at Djibouti for external surface connections. And he led his country in its push toward influence in African affairs, a move that may yet prove of the greatest importance to the well-being of the state.

The coastal part of Africa's Horn is occupied by a country that comprises two former colonies, British and Italian Somaliland. The combination of these two units was a recognition of the essential ethnic unity of the colonized peoples, the Somali; the Somali would like to see that recognition extended into Ethiopia, and into the French territory around Djibouti as well. It was Menelik who described the Somali as the cattle-keepers of the Ethiopians, and who insisted that the land of these people should be under the rule of Addis Ababa. After World War II, Haile Selassie said the same thing—that the whole Horn of Africa should be unified under Ethi-

opian rule. But the Somali had other ideas about unity, and ever since Somali independence (1960) there has been sporadic trouble between the two countries.

Actually, what is involved is more than a drive for Somali unification. Most Somali are indeed pastoralists, and their country is very dry, as Plates 1 and 2 indicate. Thus the Somali herders are tempted to follow the rains into Ethiopia's higher and moister foothills, and in so doing, they cross the international boundary. This has of course given rise to conflict, but the Somali have a point when they talk about the territorial interdependence involved in the system of transhumance just described: like the Kenya–Tanzania boundary, the Somali–Ethiopia border interferes with an ecologic whole.

The Somali Republic is much smaller than the Sudan and Ethiopia, but it is Texas-sized and not a small country by any means. Its population, too, is much less than that of either Ethiopia or the Sudan, but the three million inhabitants have many kinsmen on the other sides of political boundaries—even in Kenya. More strongly Moslem even than the Sudan, there is almost total adherence to Islam. And undoubtedly this is the poorest country of the three. Most Somali live a life of pastoral subsistance, often nomadic or seminomadic. What urban development there is can be summed up by the capital, Mogadishu, with about 100,000 people, and Kismayu, at the mouth of the Juba River, which has a better port. The Juba River valley provides some opportunities for agriculture, and what little export there is consists of bananas, some sugar and cotton, and, of course, hides and skins. Neither does Somali have any apparent advantages of location. It occupies the coastal strip adjacent to much of Kenya and Ethiopia, but Kenya's traffic goes through its own port at Mombasa, and Ethiopia's external orientation has been to Djibouti, the French port, and increasingly through the outlets of Massawa and Assab, obtained through the acquisition of Eritrea. Here lies one reason for the Somalis' interest in the territory formerly known as French Somaliland (now the Territory of the Afar and the Issa); Djibouti, with its rail link to Addis Ababa, would greatly improve Somali's economic as well as political position in the Horn.

Map 10–9 ISRAEL AND THE PIVOT.

MODIFIED AFTER: SCOTT, FORESMAN &CO., ALDINE UNIVERSITY ATLAS

Israel and the Pivot

The Fertile Crescent and Egypt certainly have been the pivotal areas of the Middle East, temporally as well as spatially. As we have loosely defined the Middle Eastern realm, this is the heart of it: to the west lies the Mahgreb, to the south, the Sudan and Arabian Peninsula, to the north, Turkey, to the east, Iran and Afghanistan. And here, in the middle of the very core of the Middle East, lies Israel, whose neighbors are Lebanon and Syria to the north, Jordan to the east, and Egypt to the southwest—all of them in some measure resentful of the creation of the Jewish state in their midst. Since 1948, when Israel was created as a homeland for Jewish people upon the recommendations of a United Nations commission, the Arab–Israeli conflict has overshadowed all else in the Middle East.

Indirectly, Israel was the product of the collapse of the Ottoman Empire. Britain gained control over the mandate of Palestine, and it was British policy to support the aspirations of European Jews for a homeland in the Middle East. These aspirations were embodied in the concept of Zionism. In 1946, the British granted independence to the territory lying east of the Jordan River, and "Transjordan" (now the state of Jordan) came into being. Shortly afterward, the territory west of the Jordan River was partitioned by the United Nations, and the Zionist Jews got the bulk of it—including, of course, land that had been occupied since time immemorial by Arab and other Semitic peoples. The original U.N. plan proposed to allot 55 percent of all Palestine to the Jewish sector, although only 7 percent of the land was actually owned by Jews (about one-third of the total population of Palestine was Jewish), but this Partition Plan was never implemented as it was intended. As soon as the Jewish people declared the independent state of Israel, the new state was attacked by its Arab neighbors, who rejected the scheme. In the ensuing battle, Israel not only held its own, but gained some territory in the southern Negev Desert (Map 10-9). At the end of the first war (1949), the Jewish population controlled 80 percent of what had been Palestine. Of course, this success was not won by overnight organization: at the time of Israeli independence, there already were three-quarters of a million Jews in Palestine. Indeed, the world Zionist movement had been assisting Jews in their "re-turn" to Palestine since the late nineteenth century.

For more than twenty years now, a state of latent—and at times actual—war has existed between Israel and the Arab world. In 1967, a week-long conflict led to the Israeli occupation of Syrian, Jordanian, and Egyptian territory, including the Sinai Peninsula to the edge of the Suez Canal—a facility which the Egyptians did not allow Israeli vessels to use. The brief war flared up after the Egyptians allegedly made moves to close the Gulf of Aqaba to Israeli shipping as well. As the map shows, Israel has a coast on the Mediterranean and was provided with a corridor via Aqaba to the Red Sea, thus obviating the need for use of the Suez Canal. Threatened closure of the Gulf was enough to escalate the continuous exchange of terrorist attacks into a brief but full-scale war. Since then, with Arab pride newly wounded, a new mass of Arab refugees, and the captured areas a further issue, the level of hostility has remained dangerously intense. In addition, the area is a power vacuum, subject to Soviet and United States moves and pressure.

So many issues divide the Jews and the Arabs that a permanent solution (other than the destruction of the state of Israel itself) or even a peace imposed by the Great Powers does not seem to be possible. One prominent issue has been the refugee problem: when Israel was created, over three-quarters of a million Palestinian Arabs were forced to leave the area to seek new lives in Jordan, Egypt, Lebanon, or Syria. Jordan could not absorb such an influx, and neither could Egypt; many of the refugees found themselves in squalid camps, surviving with United Nations aid. The Gaza strip, a narrow belt of Egyptian-administered territory along the Mediterranean coast off Sinai, became a focal point for refugees and for the political arguments that have accompanied the refugee problem. Arab propaganda agencies have complained about the treatment of those Arabs who remained in Israel (there still are about 250,000 Christian and Moslem Arabs among Israel's 3 million inhabitants, and 1 million counting the conquered territories of 1967); Israel has brought up the issue of the treatment of Jews living in Arab countries, especially Iraq. Another problem has been the city of Jerusalem, a holy place for both Jew and Moslem, as well as Christians. In the original blueprint for Palestine, the city was to have become internationalized; then Jerusalem was divided between the Jewish state and Jordan. This was a point of friction, with neither side satisfied. But the 1967 June War saw Jordanian lands west of the Jordan River fall to Israel, which as a result came to control all of the holy city. Still a further problem has been the use of the waters of the Jordan River. In 1964, Israel unilaterally diverted over half of the Jordan's flow for its own use. The river's headwaters lie in Lebanon and Syria, and these two countries have joined with Jordan in plans to divert the water completely away from its present course. But the river is crucial to Israel's economy, and the Israelis have threatened to destroy any diversion works or dams built to cut the lower Jordan's water level.

One major irritant in the whole matter has been Israel's rapid rise to strength and prosperity amid the poverty so common to the Middle East. There is nothing particularly produtive about most of Israel's land, and neither is it very large (7815 square miles, smaller than Massachusetts). But Palestine has been transformed by the energies of the Jewish community and, importantly, by heavy investments and contributions made by Jews and Jewish organizations elsewhere in the world, especially the United States. It is often said that in Israel, desert has been converted into farmland, and it is partly so: the irrigated acreage has been enlarged to many times its 1948 proportions, and water is carried even into the Negev Desert itself. Israel's best land lies along its north coast and in Galilee, between the Mediterranean and the Sea of Galilee. Here, as Plates 1 and 2 indicate, rainfall is about 20 inches, and the Mediterranean climatic regime prevails: citrus fruits form the country's major agricultural export, and wheat and barley are grown for the local market. Water is piped from these moister regions southward to Beersheeba and the Negev; rainfall decreases quite sharply south of Tel Aviv. Israel would like to develop a hydroelectric and irrigation project on the Jordan River, but of course cooperation with Jordan cannot be achieved.

With 3 million people, however, Israel's future obviously does not lie in agriculture—for export or self-sufficiency. Already, the country must import a large volume of wheat. Despite the general intensity of agriculture, and the dairying and vegetable gardens that have grown up around the large cities and towns, the gap be-

tween local supply and demand is widening. No effort is spared to maintain as great a degree of self-sufficiency as is possible, but despite the *kibbutzim* (collectivized farm settlements) and other innovations designed to combine communal living, productivity, and defensibility, Israel depends increasingly upon external trade and support.

Without an appreciable resource base, industrialization also presents quite a challenge to Israel. Evaporation of the Dead Sea waters has left deposits of potash, magnesium, and salt, and there is rock phosphate in the Negev. But there is very little fuel available within Israel: no coal deposits are known, though some oil has been found in the Negev. To circumvent the Suez Canal, an oil pipeline leads from the port of Eilat on the Gulf of Aqaba to the refinery at Haifa; a second, larger pipeline is under construction to link Eilat and Ashkelon. The oil comes from Iranian fields. The small steel plant at Tel Aviv uses imported coal as well as iron from foreign sources, and was built for strategic reasons; it is not an economic proposition. Thus the only industry for which Israel has any domestic raw materials at all is the chemical industry, and it, in response, has seen considerable growth. But for the rest, Israel must depend mostly on the skills of its labor force. Many technicians and skilled draftsmen have been among the hundreds of thousands of Jewish immigrants who have come to Israel since its creation, and the best course, naturally, is to make maximum use of these people. This is being done: the diamond cutting and polishing industry has risen to first position by value of its production.

In effect, then, Israel is a Western-type, developed country in the Middle East. It is highly urbanized, with two-thirds of its population living in towns and cities of 10,000 or more residents. Israel's core area includes the two major cities, Tel Aviv-Jaffa (750,000) and Haifa (350,000) and the coastal area between them; in total this core region incorporates over three-quarters of the country's population, though the proportion is less if the territory conquered in the 1967 War are added (including Jerusalem's Jordanian sector of 80,000 population).[10] With the cities surrounded by dairy and poultry farms, vegetable gardens, and dense road networks, the impression given by the core area is indeed

[10] Jerusalem's Israeli sector at the time of the 1967 June War had about 200,000 people.

Western, which is confirmed by the factories of Tel Aviv and the busy seaports of Haifa and Jaffa. In more ways than one, Israel is an island in the culture realm called the Middle East.

The Adversaries. Not only does Israel have the misfortune of being located in the heart of the Middle East, rather than in one of the region's peripheral or transition zones, but has an inordinate number of neighbors for so small a country with so much coastline—five (counting Saudi Arabia). Of these five, three lie along Israel's northern and eastern boundary: Lebanon, Syria, and Jordan. Each of these countries was created out of the wreckage of the Ottoman Empire, went through a period of mandate, and achieved independence relatively recently, between 1943 and 1946. But there the similarity stops. Lebanon, Israel's coastal neighbor on the Mediterranean Sea, is one of the exceptions to the rule that the Middle East is the world of Islam: nearly half the population of over 2.5 million adheres to the Christian rather than Moslem faith. Smaller still than Israel (in fact, only a little over half of Israel in terms of territory), Lebanon has a long history of trade and commerce, beginning with the Phoenicians of old—this was their base. Like Israel, Lebanon must import much of its staple food, wheat, for the coastal belt below the mountains, though intensively cultivated, cannot produce sufficient food for the population. So Lebanon continues to be a country of trade and commerce, and its success is reflected in the impressive capital, Beirut (350,000). The country's largest port and leading tourist attraction, Beirut is also a financial headquarters for the Middle East and one of its busiest break-of-bulk, entrepôt, and transit points. Damascus, the capital of Syria, lies in its hinterland, and Israel cuts Jordan off from the Mediterranean: hence Beirut serves as the major outlet for both countries. And as in the case of Israel, money comes to Lebanon from the many Lebanese people who have over the centuries left their homeland to seek work in other parts of the world; this, too, helps Lebanon in its efforts to counter the effects of its limited productive capacity.

Syria in a sense is a remnant of a greater sphere of influence which included Jordan and Israel and, at certain times in history, Lebanon as well. It was the land of the upper Euphrates, the western arm of the Fertile Crescent, the legacy of the greatness of Assyria. But today Syria is cut off by geometric boundaries from its former sphere, it has lost territory to Turkey, and it does not incorporate Lebanon. Its association with Egypt and Yemen in the United Arab Republic has failed. It is now a rather poor country, wracked by political instability.

Syria, like Lebanon and Israel, has a Mediterranean coast, where unirrigated agriculture is possible, but this coast does not provide a harbor comparable to Beirut in Lebanon or Iskenderun in Turkey; the latter was a part of the French Syrian mandate but was allotted to Turkey in 1939. Behind this densely populated coastal belt, Syria has a much larger interior than its neighbors, but the areas of productive capacity are quite dispersed. Damascus, the capital (600,000), was built on an oasis and is one of the Levant's ancient cities; though on the dry, lee side of the coastal mountains it is surrounded by a district of irrigated agriculture. It lies in the southwestern corner of the country (Map 10-9), in close proximity to Israel; in a sense it fills the role here of a "forward" capital. In the far north, near the Turkish boundary, lies another old caravan-route center, Aleppo (500,000), now on the little-used railroad from Damascus to Turkey and at the northern end of Syria's important cotton-growing area; here the Orontes River is the chief source of irrigation water. Syria's wheat belt, east along the northern border, also centers on Aleppo. In the eastern part of the country, the Euphrates valley and the far northeast, known as the Jezira, are being developed for large-scale, mechanized wheat and cotton farming with the aid of pump-irrigation systems. Production of these crops is rapidly rising; more than half the Syrian harvest now comes from this region. Recent discoveries of oil deposits add to the potential importance of this long-neglected part of Syria.

Southward and southeastward, Syria turns into desert, and the familiar sheep herders and goat herders move endlessly across the countryside. There may be as many as a half million of them, a sizable proportion of Syria's 6 million people. In contrast to Israel, and, to a lesser degree, to Lebanon as well, Syria is very much a country of farmers and peasants; only about one-third of the people live in cities and towns of any size. But again, Syria produces adequate harvests of wheat and barley and normally does not need such staple imports: in fact, it exports wheat from its northern wheat zone. It exports

Urban landscape in Syria. This photograph could have been taken almost anywhere in the Middle East: the mosque dominates a congested urban agglomeration with one- or two story houses and narrow alleys, with little apparent planning. In the upper part of the photograph lies the more modern section of this, the city of Homs, about a hundred miles north of Damascus. (Arab Information Center)

barley as well, but its biggest source of external revenue remains cotton. It is, in addition, a country where opportunities for the expansion of agriculture still exist; their realization will improve the cohesion of the state and bring its separate regions into a tighter framework.

None of this can be said for Jordan, the desert kingdom that lies east of Israel and south of Syria. It, too, was the product of the Ottoman collapse, but it suffered heavily when Israel was created—more so, perhaps, than any other Arab state. In the first place, Jordan's trade used to go through Haifa, now an Israeli port, and so Jordan has to depend on Beirut or the tedious route via Aqaba. Second, Jordan's final independence in 1946 was achieved with a total population of perhaps 400,000, including nomads, peasants, villagers, and a few urban dwellers: it was a poor country. Then, with the partition of Palestine and the creation of Israel, Jordan received more than a half million Arab refugees and found itself responsible also for another half million Palestinians who, though living on the western side of the Jordan River, were incorporated in the state. Thus refugees outnumbered residents by more than 2 to 1, and internal political problems were added to external ones—not to men-

tion the economic difficulties. Jordan has survived through United States, British, and other aid, but its problems have not lessened. Many Jordanian residents have little commitment to the country and do not consider themselves Jordanian citizens; they give little support to the embattled king. Extremist groups threaten constantly to drag the country into another war with Israel; the June War of 1967 was disastrous for Jordan. Where hope for progress might lie—as for example in the development of the Jordan Valley—political conflicts get in the way. In 1967, Jordan lost its second city, that is, its sector of Jerusalem; its capital, Amman (140,000), reflects the limitations and poverty of the country. Without oil, without much farmland (except along the Jordan River), without unity or strength, and overwhelmed with refugees, Jordan presents one of the bleakest faces in the Middle East.

By comparison, Iraq is well endowed. Iraq contains the lower Euphrates valley and also the Tigris River, and its agricultural potential is far greater than what is now used. Iraq is a rarity in the Middle East: it can be described as underpopulated, that is, it could feed a far larger number of people than it now does. This is a

Nomadic peoples of the hills and mountains: the Kurds in northern Iraq. This view shows the Kurds' temporary, tented settlement, and their horses as they drink in the stream nearby. Everything here—people, tents, utensils, bedding—is loaded on horseback when the group moves on. (E. Boubat/World Health Organization)

leftover from the decline that Mesopotamia went through during the Middle Ages, but the first steps are being taken to improve conditions and to raise standards of living. Iraq, as we saw earlier, is one of the major beneficiaries of the oil reserves of the Middle East, and it needs the income from the petroleum to make the necessary investments in industry and agriculture. Oil accounts for about 90 percent of the country's export revenues; the second export by value, dates, brings in a mere 3 percent, about the same as barley, which usually ranks third.

How can a country which exports food crops and which has a huge income from oil have so low a standard of living? There are many answers to this question. For nearly four decades since its independence (in 1932, after a decade as a British mandate) it has suffered from administrative inefficiency, corruption in government, misuse of the national income, inequities in landholding and tenure systems—a set of problems that add up to the concept of underdevelopment. When, in 1958, the Iraqi monarchy was deposed, there was perhaps a promise of better times, but since that event there has been political instability in addition to the other problems. And yet Iraq has much of what made

Mesopotamia great: there are enormous possibilities for more irrigated agriculture in the Tigris-Euphrates lowland, and improved flood control would do much to enlarge the productivity of the region. It is amazing to realize that much of Iraq's agriculture takes place on fields that are irrigated as they were thousands of years ago—by basin irrigation.

Apart from western Iraq, where nomads herding camels, sheep, and goats traverse the Jordan-Syrian-Iraqi desert, most of the people live in small villages strung along the riverine lowland, from the banks of the Shatt-al-Arab (the joint lower course of the Tigris and Euphrates) to the land of the Kurds near the Turkish border. The Kurds, who number over 1 million out of Iraq's 8 million people, have at times been opposed to the Bagdad government, and there still is a serious minority problem of strong regional character. But the general impression of rural Iraq reminds one of rural Egypt, though Egypt is ahead in terms of its irrigation works; the peasant finds himself facing similar problems of poverty, malnutrition, and disease. In the midst of all this stands the ancient capital, Bagdad (2 million), no more a link with the past than stagnant rural areas.

The Desert

South of Iraq and Jordan lies the huge kingdom of Saudi Arabia which, with its ill-defined southern and eastern margins, is something of an anomaly even in the Middle East (Map 10-10). Well into the twentieth century Saudi Arabia had seen very little change, as its Bedouin nomads roamed the vast desert searching for water and pasture for their herds. In the 1920's the area was politically consolidated by the energies of King Ibn Saud, but apart from the meager sedentary living provided by the coastal strip and scattered oases, there was little that could be done to stabilize the country. From the Persian Gulf to the Red Sea it is desert; the surface rises generally from east to west, so that the Red Sea is fringed by mountains that are 2000 feet near the Gulf of Aqaba and reach 7000 feet near the Gulf of Aden, with several peaks as high as 10,000 feet. There, as Plate 1 shows, precipitation is very slightly higher, but most of Arabia gets less than 4 inches annually.

In better times, Arabia was the source of Islam, the center of Arab power. Today, its 7 million people still are almost all of Arab stock, though there are many sects, and Arabic, in its several dialects, is the common language. In terms of wealth and income, there are immense contrasts between the rulers, those who have benefited from the discovery, in the 1930's, of oil reserves that rank among the world's largest, and the ruled—the nomad and the peasant.

Most of the human activity in Saudi Arabia occurs in a wide belt across the "waist" of the peninsula, from the oilfields between Qatar and

MODIFIED AFTER: RAND MCNALLY, GOODE'S WORLD ATLAS

Map 10-10   THE ARABIAN PENINSULA.

Kuwait to the Red Sea approximately between Medina and Mecca. South of this zone lies the wasteland of the Empty Quarter, called Rub Al Khali. In this belt lie not only the oil areas, but also the centrally positioned capital, Riyadh, the major Red Sea port, Jiddah, the great religious center of Mecca (250,000 and the largest urban center in the country), and Medina. Along the Red Sea coast, too, lie the few areas of agriculture that are not oases, such as the coffee farms in the higher elevations of Asir, just north of Yemen. Also in this central region lie the country's gold and silver mines; despite their modest production, they were at one time the chief sources of income. The country's only railroad connects Riyadh to the Persian Gulf port of Damman.

Around the eastern and southern fringes of the Arabian Peninsula lie a number of states, sheikdoms, and sultanates. By no means are all of these units territorially defined with any preci-

sion. This has been a British sphere of influence, but independence has not been easy to achieve. Egypt involved itself in Yemen on the side of the faction it wished to see in control after independence. More recently, serious internal troubles accompanied the departure of the British from South Yemen, of which the long-important port of Aden is a part. Eastward, there are the Sultanates of Muscat and Oman, and the Sheikdoms of Qatar and the Trucial states. In none of these entities has the process of boundary definition, delimitation, and demarcation taken place in its entirety, and for a long time it has not mattered—not in the desert lands of Arabia. But now, oil is being produced in all of these areas, and the quantities are rising; oil concessions require the kind of precision that has not prevailed here. This is likely to give rise to some interesting politico-geographical problems.

Cycle Theory. In the Middle East, we are able to witness the emergence of states, their develop-

Nomadic peoples of the desert: Beduins in Saudi Arabia. These Beduins are in the process of serving coffee to their visitors, which is their traditional way of showing friendship and hospitality. The implements of the coffee ceremony are among the Beduins' proudest possessions. (Arabian American Oil Co.)

ing relationships, the evolution of an ideological focus, and a series of attempts—still going on—to create supranational unions out of sovereign states (for example, the U.A.R.).

In 1939, Professor S. van Valkenburg suggested that it might be possible to identify a cyclic process of state development, and he introduced the *cycle theory* to test the observations to be made. As he put it, the theory is:

" . . . based on a cycle in the political development of nations, recognizing four stages, namely *youth, adolescence, maturity,* and *old age.* After completion the cycle may renew itself, possibly with a change in political extension, while the cycle can also be interrupted at any time and brought back to a former stage. The time element (the length of a stage) differs greatly from nation to nation and depends on the character of the state; correspondingly no forecast can be made on the time shift to a next stage."[11]

What Professor van Valkenburg is saying is that states in these various stages will exhibit certain characteristics, which will in turn be reflected in their external relationships. Thus a state in the stage of *youth* has just achieved sovereignty, and its greatest need is for internal organization or reorganization. It will consolidate its internal structure, and there will be no tendencies toward territorial expansion. Possibly the states of southern Arabia will go through such a stage, but of course many of the recently emergent states of Africa and the Middle East already had experienced a lengthy period of internal organization in preparation for complete independence, so that the characteristics of a youthful stage may sometimes be missing.

With *adolescence,* the state has accomplished its internal consolidation, and now its dynamic qualities begin to come through. Its aggressiveness may lead to conflict and hostilities, and these in turn may strengthen the nation-building process. In the Middle East, Syria and Iraq have shown such tendencies especially clearly: Syria in its aggressiveness (albeit fruitless) toward Israel, and Iraq in its expansionist attitude toward Kuwait. It is during this stage, argues van Valkenburg, that nationalism and patriotism are important psychological nation-builders, and

[11] S. van Valkenburg, *Elements of Political Geography,* Prentice-Hall, Englewood Cliffs, N.J., 1939, p. 5.

that national prestige ranks high as a concern of the citizenry.

When a state reaches *maturity,* it takes a different attitude toward territory; it may indeed be willing to give up some of it (or certain distant possessions) in order to reduce administrative costs or in order to preserve good relations with a neighboring state. The state now favors international cooperation, not with expansionist goals, but with such objectives as joint economic development and collective security. A case might be made for Turkey and Iran, especially if a one-stage rejuvenation is recognized in Turkey, which at the time of the collapse of the Ottoman Empire had clearly reached the state of old age.

This stage of *old age,* according to Professor van Valkenburg, is marked by declining power, that is, a failing internal organization. Often, a new ideology arises in one part of the country and there may be secessionist attempts, or the new force slowly but surely erodes the central authority, which eventually disintegrates: this may involve a rejuvenation. The "Young Turks" movement in Turkey just before the state's collapse was a failing attempt at rejuvenation; communism in China, on the other hand, disposed of the preceding regime and started the country over in what some would argue was a total reorganization—a state of youth. In the Middle East, apart from Turkey and potentially the difficult case of Ethiopia, states in this stage are few.

As anyone who has read any physical geography will have noticed, the terms used by van Valkenburg—youth, maturity, etc.—come from that field, where they were applied by W. M. Davis to the evolution of erosional landscape. That was not unusual in the 1920's and the 1930's, when physical geography had a dominating influence on other fields of geography. But today, it might be appropriate to revise the cycle theory using more current terminology. We could do this by concentrating on *type* rather than stage, and by stressing the behavioral characteristics in the various categories. Thus the stage of youth would become the *organizing type* of state, and perhaps it would be more appropriate to classify the Sudan here than to identify it as a youthful state. The Sudan is seeking to consolidate its internal structure, and to integrate the dissident south in the national framework. The stage of adolescence would become the *ex-*

panding type, indicative of states expressing claims, irredentist or otherwise, to territory of other states. Morocco, for example, in laying claim to part of Mauritania, is thus aiming at expansion, as are Syria and Iraq, as previously indicated. The mature stage is perhaps better identified as the *stable type,* and Iran would fit this category rather better than "mature." Turkey, also may be viewed as a stable state, and perhaps Egypt and Lebanon. Naturally these determinations are best made through the study of a number of indicators of each of these conditions. Finally, the state in the stage of old age may be said to be *disintegrating,* for its internal organization is failing. In the Middle East, no state seems at present to be in this situation, although the collapse of the Turkish Ottoman Empire clearly was a case of disintegration of organization in all spheres. The cycle theory may not have any predictive properties, but it does provide a useful model for comparative purposes.

Ataturk's Turkey

Like Ethiopia in the south, Turkey lies in a transition zone between the Middle East and another culture realm. As in Ethiopia, there is a history of involvement with the Arab world, but like the African mountain country, Turkey has in recent years decided to cast its lot in large measure with the non-Arab world. In Ethiopia, as we have seen, this involved the protection of a Christian minority against the Moslem advance. In Turkey, on the other hand, it was necessary to break the hold of old, traditional Islam—which pervaded the country—and to liberalize social institutions. Ethiopia has turned toward Africa; Turkey has turned toward Europe and the West.

As it entered the twentieth century, Turkey was at the center of a decaying, corrupt, and reactionary empire whose sphere of influence had extended across much of the Middle East and Eastern Europe. At that time conditions were such that it would undoubtedly have had to be considered a part of the Middle East, the world of its religious-cultural imprint, Islam. But the Turks are not Arabs (their source area lies in the Sino-Soviet border area in Central Asia), and in the early 1920's a revolution occurred which thrust into national prominence a leader who has become known as the father of Turkey—Ataturk. This revolution was, in fact, preceded by a fullscale war by the Turks against the Greeks in the west and the Armenians in the east; Mustafa Kemal Pasha (Ataturk's real name) was a leader in the Turkish effort to stave off the country's partition among European powers. With the new Turkey's power established (the Greeks were crushingly defeated, the Armenians decimated) and its new territory consolidated more or less as the modern map shows it (Map 10-11), Ataturk set about the country's modernization and reorientation.

What Turkey was left with after the demise of the Ottoman Empire and the hostilities following World War I was still a considerable territory (about 300,000 square miles, nearly the size of Spain and Italy combined), and, by East European and Middle Eastern standards, a large population, which now numbers about 34 million. Its area, surrounded on three sides by water, appears compact at first glance but is nevertheless significantly fragmented: its northwestern corner, which contains the historic old city of Istanbul (the Byzantine capital, Constantinople), is separated from the main area by three interlocking waterways. These waters—the Bosporus, the Sea of Marmara, and the Dardanelles—connect the Black Sea with its Soviet ports to the Mediterranean. For obvious reasons this is a sensitive route, over which the Russians have long wanted control, and in this context Turkey's fragmentation is especially consequential. Istanbul, renamed from Constantinople by the Turks after its defeat in 1453, lies on the western side of the Bosporus, and it was the headquarters of the Ottoman Sultans. But Ataturk had other ideas about a Turkish capital: the struggle for Turkey's survival had been waged from the heart of the country, the Anatolian Plateau, and here he wanted to place the seat of government. Ankara had many advantages: it would remind the Turks that they were, as Ataturk always said, Anatolians; it lay nearer the center of the country than Istanbul and thus might act as a stronger unifier; and strategically its position was much better than that of Istanbul, so near the Greek enemy and so exposed to the outside world. Ankara, much more a Turkish town than cosmopolitan Istanbul, recently passed the half-million mark in population, but since its administrative functions remain its chief basis for existence, it is unlikely soon to overtake the bustling straits city with its 1.5 million residents.

But although Ataturk moved the capital east-

414

MODIFIED AFTER: RAND MCNALLY, GOODE'S WORLD ATLAS

Map 10-11 TURKEY.

ward and inward, his orientation was westward and outward. To implement his plans for Turkey's modernization, he initiated reforms in almost every sphere of life in the country. Islam, the state religion, lost its official status. The state took over most of the religious schools that had controlled education. The Roman alphabet replaced the Arabic. Moslem law was replaced by a modified Western code. Symbols of old—the wearing of beards, wearing the fez—were prohibited. The emancipation of women was begun (in Moslem society women have been denied access to education, freedom of movement, or social contact, and in a few countries they still must cover their faces when in public), and monogamy was made law. The new government took pains to stress Turkey's separateness from the Arab world, and ever since it has remained quite aloof from the affairs involving the other Islamic states.

What Ataturk wanted for Turkey could not, of course, be achieved in a short time. Turkey is largely an agricultural country, and when three-quarters of the inhabitants of a country are subsistence or near-subsistence farmers, most of whom live in small villages in sometimes quite isolated rural areas, some opposition to change, especially radical change, has to be expected. So it was in Turkey; in fact, the government had to yield to the devoutly Moslem peasants on some issues. Still, Ataturk's directives are still pursued nearly a half century after he came to power. After World War II Turkey allied itself with Western states in both the now-defunct Bagdad Pact and the North Atlantic Treaty Organization, although some internal opposition to this association has been manifest. Again, Turkey has chosen to oppose the spread of Soviet influence in the Middle East, when a Middle Eastern type of neutrality might have been the easier course.

Like Ethiopia, Turkey is a mountainous country, but the relief in Turkey is generally much more moderate. The great mountain country lies in the east, on the borders of Soviet Georgia and Armenia. Here lies famed Mount Ararat, 16,946 feet, on the Turkish side of the border near Yerevan. Rugged country, also, lies in the area known as Kurdistan. This is the land of the Kurds, whose domain extends from Turkey into Iraq and Iran. From this high eastern part of the country, where elevations are mostly between 5000 and 10,000 feet and where level land is only

rare, the surface drops westward. But although the Plateau of Anatolia lies lower (between 2000 and 5000 feet), much of it is too dry for farming. Thus the most extensive cultivable areas are the relatively narrow coastal plains, notably those bordering the Aegean Sea and the Sea of Marmara. Plates 1 through 4 all reflect these environmental conditions: interior Turkey is dry enough to be classed as steppe (Plate 2), with mountain lithosols (note how the plateau stands out on Plate 3), and desert and semidesert vegetation (Plate 4).

In Turkey's coastal areas, then, lie the best farmlands; they receive more rainfall than the interior areas and they escape the more severe winters that occur there. There is little coastal lowland along the Mediterranean, where the areas around Antalya and Adana (northeast of Iskenderun) are the only places where the mountains retreat somewhat from the sea. The Black Sea coast, too, is quite narrow, but it is comparatively moist: it gets winter as well as summer rainfall (though with a winter maximum) totalling 30 to 105 inches—uncommonly high for the "dry world." Thus, as Plate 6 reflects, these coastal areas, and especially the Aegean–Marmara lowlands, are the most densely peopled zones of Turkey and the most productive ones.

If we carry the comparison between Turkey and Ethiopia still farther, we shall find that the Turkish peasant is considerably better off than his Ethiopian counterpart. He also lives somewhat better than the rural population of Egypt or Iraq and is more likely to own his land, farm animals, and farm equipment. Most Turkish peasants still live at or near the subsistence level, but the per capita average annual income of the country is nearly three times as high as that of Egypt, reflecting the generally better situation. In the farming areas where the export crops are grown, the association is the familiar Mediterranean type, with wheat and barley the staples (they occupy over three quarters of all the cultivated land in the country), and tobacco, cotton, hazelnuts, grapes, olives, and figs for the external market. Cotton, famous Turkish tobacco, and hazelnuts (the Black Sea coast is the world's chief source of these) form the three top exports by value in a country 90 percent of whose external revenues are derived from the land. As elsewhere in this part of the world, production could be increased, but unlike Iraq, the problems involved in irrigation expansion are serious: there is little level land and the streams are deeply incised in this mountainous country. And, despite the more optimistic picture painted of the rural areas, and notwithstanding the slight increase in mechanization of agriculture, the fact remains that by far the majority of the peasants are too poor to obtain the fertilizer that might improve their yields and similarly unable to buy farm equipment that would allow them to get rid of their antiquated tools. Meanwhile, in the absence of major mineral deposits except chromium, Turkey's exports are dominated by agricultural products.

On the Anatolian Plateau, the crops are the subsistence cereals—wheat and barley, and also some corn—which are grown under precarious climatic conditions, since the rainfall here diminishes to an average of 12 to 20 inches (less in some interior areas). Livestock takes on increased importance under such circumstances, and thus the farmer gets some returns for wool from his small flock of sheep and from the hides and skins of slaughtered animals. Mohair from the Angora goat is used in the manufacture of valuable textiles and rugs. Again, the map reveals that Turkey is served by a fairly intensive network of railroads and roads (the network is denser, at least, than that of the other Middle Eastern countries). In other ways, too, it continues to move ahead. Indeed, Turkey leads all the countries of the present region in industrialization. The production of cotton has stimulated a textile industry, and a small steel industry has been established based on a coal field near Zonguldak, not far from the Black Sea, and an iron ore deposit several hundred miles away in east central Turkey. One plant has been built at Karabuk, about 50 miles along the railroad from Zonguldak, and another has been erected at Eregli, on the Black Sea coast west of Zonguldak. In the southeast, Turkey has proved to possess some oil, and it may share in the zone of oilfields of which the famed Kirkuk reserve in Iraq is a part. With its complex geology, Turkey has proved to have a variety of mineral deposits if not a large mineral base, and the variety that does exist gives much scope for future development. Spatially, this development seems destined, at present, to confirm the country's westward orientation. Already the Mediterranean coastal zone is the economic focus of the country, and of what remains the western half is again the more developed. Turkey is quite

effectively separated from the rest of the Middle East by the mountains of its eastern sector and by the Mediterranean's northeast corner. Ataturk may have wanted to move his capital into the interior, but two-thirds of Turkey still lies to the east of it, and only one-third to the west. As the map shows, Turkey is, indeed, half an island.

Cyprus. In the northeastern corner of the Mediterranean, much farther from Greece than from Turkey or Syria, lies Cyprus—the problem of Cyprus. Since ancient times this island has been dominantly Greek in its population, but in 1571 it was conquered by the Turks, under whose control it remained until 1878. Then with the decay of the Ottoman state, the British first established an administration on Cyprus under an agreement with the Turks, but when World War I broke out they took full control of it. By the time the British were prepared to offer independence to much of their empire, they had a problem on their hands in Cyprus, for the Greek majority among the 600,000 population preferred *enosis*—union with Greece. It is not difficult to understand that the 20 percent of the Cypriot population that is Turkish wanted no such union; in fact, the answer the Turks gave was partition—the division of the island into a Turkish and a Greek sector. By 1955 the dispute had reached the stage of violence: differences between Greeks and Turks are deep, bitter, and intense. It was really impossible to find a solution to a problem in which the residents of an island country think of themselves as Greeks or Turks first rather than as Cypriots, and yet in 1960 Cyprus was made an independent country with a complicated constitution designed to permit majority rule but also to guarantee minority rights.

Cyprus, unfortunately, is not a rich land. In bygone millenia—into Roman times—it was an important source of copper, as its name suggests, and also timber. But today the biggest single industry on the crowded area of just over 3500 square miles is agriculture. With little good land—even steep slopes are cultivated—and few opportunities for irrigation, the population does not feed itself, and staple foods must be imported. For energy supply, too, Cyprus must depend on external sources. In exchange, it can export some Mediterranean fruits and wines, but its few mineral products leave the island raw and untreated, and this does not help the labor situation. In the mid-1960's trouble flared intermittently between the Greeks and Turks, and United Nations forces remained on hand to keep the adversaries apart. Cyprus, a pawn of outside powers, has been a casualty of history.

The East: Iran

East of the Middle East's pivot lies Iran, another prominent exception to the rule that the Middle East is the Arab world. Persia—as Iran was formerly called—is a populous country by Middle Eastern standards, and a large one, too: it is larger even than Saudi Arabia. Iran is mountain country and desert, physically resembling Utah and Nevada in some ways (Plate 2), and its political boundaries frequently are physiographic as well. To the northeast the land drops off from the Kopet Mountains into the low-lying Turkmen S.S.R.; to the northwest lie the river lowlands of the Azerbaydzhan S.S.R.; and to the

southwest Iran has but a small share of the Mesopotamian lowland. Eastward, the Iranian mountain and plateau country continues into Afghanistan and the Baluchistan region of Pakistan. Almost everywhere, dryness prevails. The eastern half of the land on an average receives less than 12 inches of rainfall annually (parts of it get less than 4), and even the moister north and west can only expect between 12 and 20 inches (Plate 1). With so little usable land—such as the moist, narrow ribbon along the Caspian Sea coast and the small area along the Shattal-Arab—and with most of the people dependent directly upon agricultural or pastoral subsistence, it is remarkable that Iran's population is as much as 26 million.

There was a time when Persia was a powerful empire in the Middle East, and in those days the center of its organization lay in the south, where the ruins of Persepolis still attest to its

Map 10-12 IRAN.

ancient greatness. Then as now, people clustered in and around the oases or depended on *qanats* for their water supply (as did Persepolis); some settlements were located on surface streams flowing off the somewhat better-watered mountains. The advantage of Persepolis obviously was that it lay protected on the plateau and yet was close to the lowlands of Mesopotamia, but the focus of modern Iran has moved far to the north. Today's capital, Tehran (2 million), lies on the southern flank of the Elburz Mountains, a range that rims the southern Caspian Sea (Map 10-12). Tehran was a significant settlement in ancient times, when it lay at the intersection of caravan routes to Anatolia and Bagdad in the west, Mashad near the Kopet Mountains in the distant east, and Persepolis and the Persian Gulf in the south. With the Elburz Mountains to the north, it had advantages of protection from the most likely source of external invaders. In time, with the decline of Mesopotamia and the growing agricultural importance of the Caspian coast and adjacent areas, Tehran's situation had important advantages: it was near the largest cluster of sedentary population in the country and yet open to the vast plateau of the south. Still sustained in part by *qanat*-supplied water, Tehran has grown from a caravan station to one of the largest cities in the Middle East.

As we noted earlier, Iran's national economy is based on the sale of its oil to overseas markets. From this point of view the Gulf area—where the oil is located and where the refineries of Abadan and other facilities have been built—is of great importance. But the overwhelming majority of Iranians are not involved in this development, and neither have they benefited appreciably from its revenues. More than three quarters of Iran's population are peasant farmers or seminomadic pastoralists, and they share a host of problems relating not only to the dry, earthquake-racked, hostile land but also to the political, economic, and social systems in which they are trapped. This, of course, is but another example of something we have come to know as symptomatic of the Middle East: the enormous contrast between the wealthy, influential, landowning elite and the grinding, hopeless poverty of the peasantry. In Iran these contrasts still prevail, although the government, under the shah's direction, has initiated a program of land redistribution as part of the "White Revolution." A large dam in Khuzistan and a So-

viet-subsidized steel plant at Ishfahan reflect the coming of a new day in Iran. Certainly Iran is not without its opportunities: the southwest, for example, shares with Iraq a condition of underdevelopment that is not imposed by the environment at all, but results rather from the usual combination of inadequate capital, lack of technical know-how and planning, and disinterest on the part of those who might initiate the desired progress. Meanwhile, there is disease, illiteracy, hunger, and misery on the land in Iran. Backs are still bent over farm tools that have not changed for centuries. Streams still flow through villages, arriving fresh and clear from the mountains but leaving foul and polluted, to be used again by people farther downslope. Herders still drive their animals across the searingly hot plateau in search of meager forage and water.

An Ecological Trilogy. In the Middle East, as we have seen, people live in cities and towns of considerable size, they live in villages, or they live a tribal life of a seminomadic or nomadic nature. We have come to know the region as one of sharp contrasts and strong differences, with its ornate architecture, palatial urban residences and its unchanged villages, its modern oil facilities and its ancient farm tools, its Western-looking upper class and its vividly garbed, picturesque nomadic peoples. So perhaps it is not surprising that we have tended to look at these elements in the Middle East to some extent in isolation—as though cities were islands in the barren region, and villages existed without outside connections. In 1966, Professor P. W. English wrote:

" . . . Villages are not isolated from the city or from one another. There are strong patterns of interrelatedness between every village and the city, and weaker social and economic ties among villages. The peasantry produces a large enough food surplus to support the heterogeneous occupational structures found in large population centers. Transportation and communication systems link all settlements into the larger pattern. Essentially the same cohesive social and economic region commonly found in more humid areas is found here, descriptions of isolated peasant villages, subsistence economies, and scattered oases notwithstanding."[12]

[12] Paul Ward English, *City and Village in Iran,* University of Wisconsin Press, Madison, 1966, pp. xviii, xix.

An oasis: Qatif, in the Eastern Province of Saudi Arabia. Dense stands of date palms and a cluster of houses reveal the presence of water—and form a stark contrast to the sandy flatland beyond. This is a quite substantial oasis; many are much smaller in both area and population. (Arabian American Oil Co.)

Tied to this situation is the concept of *urban dominance.* Professor English describes the relationships between the merchants, landowners, and entrepreneurs in Kirman and the region surrounding that town of over 60,000. The urbanites exercise control over most spheres of life in the "rural" areas: through land tenure contracts, whereby landowners determine how their sharecroppers will cultivate the soil and use the water; through herding contracts, a similar system whereby the city-dwelling herd owner controls the use of animals and pastures in the outlying areas; and through weaving contracts, involving the signing to contracts of expert weavers who work in the smaller villages and hamlets of the Kirman region. In this way the tentacles of the city penetrate every aspect of rural life. As Professor English put it:

"[There is] a delicately balanced, complicated, cultural adaptation to the land that we can call the *ecological trilogy.* In this system, society is divided into three mutually dependent types of communities—the city, the village, and the tribe—each with a distinctive life-mode, each operating in a different setting, each contributing to the support of the other two sectors and thereby to the maintenance of total society.

"Most Middle Eastern cities act as administrative, commercial, and cultural centers for large rural and nomadic populations for it is in the cities that the elite of Middle Eastern society, the wealthy, powerful, and literate decision-makers reside. These city-dwellers are principally engaged in collecting the processing raw materials from the hinterland—wool for carpets and shawls, vegetables and grain to feed the urban population, and nuts, dried fruit, hides, and spices for export. In return, the urbanites supply peasants and nomads with basic economic necessities such as sugar, tea, cloth, and metal goods as well as cultural imperatives such as religious leadership, entertainment, and a variety of services. This concept of urban dominance is basic to the idea of an interdependent urban trilogy; it replaces earlier notions which tended to stress the isolation of peasant and nomad from the civilizing influences of urban life."[13]

[13] Paul W. English, "Urbanites, Peasants, and Nomads: the Middle Eastern Ecological Trilogy," *Journal of Geography,* Vol. LXVI, No. 2, February 1967, pp. 54–55.

Eastward, Iran yields to a country with which it has some conditions in common, especially in terms of terrain: *Afghanistan.* Mountainous Afghanistan is really marginal to the Middle East. Unlike Iran it has no common boundaries with any "Arab" states; and it is the only state in this realm to be landlocked—a circumstance which ties it more to its eastern neighbors, the territory of Kashmir and the states of Nepal, Sikkim, and Bhutan. Afghanistan's population of about 16 million is of very mixed origins, though it is non-Arab. The Caucasian Tajik group of about 4 million is centered on Herat, not far from the Iranian border, and its language is Persian. The largest group is constituted by the Caucasian Pathan (who are known by other names, such as Pushtun or Pukhtun), who constitute about half the country's population and are concentrated in the capital, Kabul, and its surrounding area. There are also Mongolian

elements in the Afghanistan population, including Turkic-speakers in the northwest and Persian-speakers in the central parts of the country. These varied people do have one common trait, one they share with the vast majority of the inhabitants of the Middle East: adherence to Islam.

Afghanistan is cut by a great mountain chain which in the west is called the Paropamisus and which eastward becomes the famed Hindu Kush. The country's core area is split by this range: Kabul (350,000) lies to the south amidst intensively farmed, irrigated lands along the Kabul River, and Kunduz, the karakul-pelt center, lies to the north. The Hindu Kush is also a divider of ethnic sectors: to the south the core is occupied by the "real" Afghans, the Pathans, while to the north there are affinities with the peoples of Soviet Central Asia. Elevations are high, and surface communications are difficult. One major

Map 10-13   AFGHANISTAN.

road leads from Kabul through the Salang Pass to the Kunduz area, and from there, another skirts the northern slopes of the mountains to cross the Paropamisus near Herat. Herat in turn is connected by a circuitous route to the country's second city, oasis-centered Kandahar (Map 10-13). Without railroads, with just two routes leading to Iran and two to Pakistan, one of them through the Khyber Pass, Afghanistan's map reflects the small volume of its foreign trade and the difficulties involved in the current effort at national consolidation and territorial integration.

If any justification were needed for the inclusion of Afghanistan in a discussion of the Middle East, certainly its dry-world characteristics and the attendant problems of agriculture would be relevant. Even the wettest parts of the country—the Hindu Kush—get only about 20 inches of rainfall in an average year, and this rain comes concentrated in winter, much of it as snow in the higher elevations. Thus irrigation is necessary, though in the moister areas the winter rain is enough to sustain wheat, the local staple. The irrigated areas, especially those not based on rivers flowing off the Hindu Kush, are faced each year with a too-rapid early runoff as a result of melting snow and ice, and a subsequent lean summer. As in Iran, *qanats* contribute importantly to the sustenance of permanent settlements, and funds derived from aid programs sponsored by both the U.S.S.R. and the United States are used to control the waters of the all-important rivers, especially the Helmand. To the problems just mentioned must

be added another set of familiar Middle Eastern conditions: outdated farm equipment remains in general use, capital is very scarce, landlordism confronts the peasants (though not in degrees as serious as elsewhere in the realm), and the avenues of escaping from a life of subsistence have so far eluded most of the people. On the other hand, the Afghans state, possibly with justification, that there has never been starvation or serious malnutrition in the country, and that things are worse in parts of both the Middle East and India.

Afghanistan during the nineteenth century was caught between the expanding empires of tsarist Russia and colonialist Britain, but in 1907 the two powers agreed to recognize it as a separate political entity; Russia acquiesced to British paramountcy in Afghanistan in return for a sphere of influence in Iran. Afghanistan's eastern boundary was then extended to China to form the Wakhan Corridor, confirming its separating function as a buffer between British and Russian interests in the region. After World War II, Afghan Irredentism toward the Pushtunistan border area in Pakistan led to serious friction between the two countries, which in 1961 led to a severing of diplomatic relations and the closing of Afghanistan's main and most efficient route to the sea, via Karachi. Landlocked, caught in power struggles in both the nineteenth and twentieth centuries (American–Soviet competition here is currently strong) and at odds with its most useful neighbor, Afghanistan still pays the price of an historically strategic location.

ADDITIONAL READING

Probably the best place to start reading about the Middle East is in C. S. Coon's famous book *Caravan: the Story of the Middle East,* published in a revised edition by Holt, in New York, in 1958. A standard geography is G. B. Cressey's *Crossroads: Land and Life in Southwest Asia,* published by Lippincott in Philadelphia, 1960. Other geographies include the volume by W. C. Brice, *South-West Asia, A Systematic Regional Geography,* University of London Press, 1966; W. B. Fisher, *The Middle East: A Physical, Social, and Regional Geography,* a Dutton, New York, publication of 1963; and S. H. Longrigg, *The Middle East: A Social Geography,*

published by Aldine in Chicago in 1963.

Other useful works currently available focus on important topics and discrete regions within the Middle East. R. H. Nolte edited *The Modern Middle East,* a collection of articles published by Atherton Press in New York in 1963. The book by H. Frankfort, *The Birth of Civilization in the Near East,* a Doubleday Anchor paperback published in New York in 1956, says much about the diffusion theme. On the importance of petroleum, see C. P. Issawi's book called *The Economics of Middle Eastern Oil,* published by Praeger in 1963. On North Africa, an excellent source is N. Barbour, editor of *A Survey of North-West Africa (the Maghrib)*, published by Oxford University Press in London in 1962. On the Sahara's scattered peoples, see L. C. Briggs, *Tribes of the Sahara,* published by Harvard University Press in 1960. Other works by geographers are M. W. Mikesell's *Northern Morocco: A Cultural Geography,* University of California Press, 1961; and B. E. Thomas, *Trade Routes of Algeria and the Sahara,* University of California Press, 1957. Another standard geography is K. M. Barbour's *The Republic of Sudan: A Regional Survey,* published by London University Press in 1961. For an account of a Mediterranian city see J. Gulick, *Tripoli: A Modern Arab City,* a Harvard University Press publication of 1967. On Ethiopia, a standard work is a book by R. Greenfield, *Ethiopia: A New Political History,* published by Praeger in 1965.

On Southwest Asia, there are surveys by the International Bank for Reconstruction and Development on Iraq, Syria, Jordan, and Turkey. Also see a volume by R. K. Ramazani, *The Northern Tier: Afghanistan, Iran, and Turkey,* No. 32 in the Van Nostrand Searchlight Series. On Israel, see B. Halpern, *The Idea of the Jewish State,* published by Harvard University Press in 1961. A book by E. Orni and E. Efrat, *The Geography of Israel,* was published by Davey, New York, in 1966. A volume full of detail is by Paul Ward English, *City and Village in Iran: Settlement and Economy in the Kirman Basin,* published by the University of Wisconsin Press in Madison in 1966.

Boundary morphology is discussed in detail by J. R. V. Prescott in *The Geography of Frontiers and Boundaries,* an Aldine paperback published in Chicago in 1967. Other politico-geographical issues raised in this chapter can be followed up in the basic literature of that field, including the previously quoted volumes by Alexander, Pounds, and Kasperson and Minghi. Once again, use should be made of several relevant issues of the American Geographical Society's *Focus.*

11

INDIA AND THE INDIAN PERIMETER

THE SOUTH of Asia consists of three protruding landmasses. In the southwest lies the huge rectangle of Arabia, and in the southeast lie the slim peninsulas and elongated islands of Malaysia and Indonesia. And between these there is the familiar triangle of India, a subcontinent in itself. Bounded by the immense Himalaya ranges to the north, by the mountains of eastern Assam to the east, and by the rugged, arid topography of Iran and Afghanistan to the west, the Indian subcontinent is a clearly discernible physical region. Its population of two-thirds of a billion constitutes one of the great human concentrations on earth. The scene of one of the oldest known civilizations, it became the cornerstone of the vast British colonial empire. Out of the colonial period emerged two of the world's most populous states, India (500 million) and Pakistan (125 million).

The political framework of South Asia is such that India lies surrounded by several much smaller political units (Map 11-1). Pakistan is fragmented into two sectors which flank India to the west and east. To the north lie the disputed territory of Kashmir and the landlocked states of Nepal, Sikkim, and Bhutan. And off the southern tip of the Indian triangle lies the island state of Ceylon. The landlocked countries of the north lie in the historic buffer zone between the former British sphere of influence and its Eurasian competitors, and India, Pakistan, and Ceylon all formed part of what was once known as British India. British colonialism threw a false cloak of unity over a region of enormous cultural variety, and political fragmentation took place immediately upon British withdrawal in 1947. Since then, all three countries have had to contend with strong centrifugal forces. In India, a federal union of 17 states, 12 union territories, and a number of special-status territories, issues involving language, religion, and

423

Map 11-1 INDIA AND SURROUNDINGS.

424

employment opportunities have led to serious internal conflicts. Tensions between West and East Pakistan have threatened the political future of that state, which in early 1969 faced violent civil disorders. Ceylon's Buddhist Sinhalese majority has been at odds with the minority Hindu Tamils.

The Indus

The Indian subcontinent is a land of great river basins. Between the mountains of the north and the uplands of the peninsula in the south lie the broad valleys of the Ganges, the Brahmaputra, and the Indus. In one of these—the Indus—lies evidence of the region's oldest urban civilization, contemporary to and interacting with ancient Mesopotamia.[1] Unfortunately much of the earliest record of this civilization lies buried beneath the present water table in the Indus valley, but those archaeological sites that have yielded evidence indicate that here was a quite sophisticated urban culture, with large and well-organized cities in which houses were built of fired brick and consisted of two and sometimes even more levels. There were drainage systems, public baths, brick-lined wells. Protective fortifications were laid out around the major cities, of which Mohenjo-daro and Harappa are best known. As in Mesopotamia and the Nile valley, considerable advances were made in the technology of irrigation, and the civilization was based on the productivity of the Indus lowland's irrigated soils. In the pottery and other artistic expressions of this literate society lies evidence of contact—and thus trade—with the lowland civilizations of Southwest Asia. In its heyday, the Indus sphere extended as far east as the upper Ganges valley and as far south as the Kathiawar Peninsula (Map 11-2).

But the Indus valley did not escape the invasion of the Aryans any more than did Europe or the Mediterranean area. After about 2000 B.C., peoples began to move into the Indus region from Western Asia, through what is today Iran and through passes in the mountains to the north. Culturally these peoples were not as advanced as the Indus valley inhabitants, and they brought destruction to the cities of the

[1]Our use of the name "India" for the South Asian realm comes from the Sanskrit word "sindhu," which became "sinthos" in Greek descriptions and "sindus" in Latin. Corrupted to "Indus," it means "river"; subsequently "India" came into general use.

Ruins of Mohenjo-Daro, one of the headquarters of an ancient civilization in the valley of the Indus River. (Patellain/PIX)

Map 11-2   THE INDUS CIVILIZATION.

Indus. But they also adopted many of the innovations of the Indus civilization, and pushed their frontier of settlement out beyond the Indus valley into the Ganges area and southward into the peninsula. There, they absorbed the tribes they found through conquest and enslavement, and so the invaders from the northwest began to take on some Indian characteristics. Their rather light skin color darkened. The language they had brought to India, Sanskrit, began to differentiate into the linguistic complex that is India's today. Their village culture spread over a much larger part of the subcontinent than the urban civilization of the Indus had done.

The invaders who thus peopled and invigorated India have been described in many ways. Based on their common linguistic heritage they are called Aryans—Aryan is a linguistic rather than a racial term—but racially, they were so varied that it is hardly possible to identify them as a group. Still, they have been referred to as Indo-Aryan, as Caucasian, and even as Indo-European peoples. Yet the only thing that is certain is that the roots of their Sanskrit language were the roots of European as well as Indian languages; consult any atlas map and there will be a family of languages identified as *Indo-European*, of which Germanic, Romance, Slavic, Greek, Iranian, and Indo-Aryan languages are all a part. In India this language complex ends across the middle of the peninsula, where Dravidian languages take its place: the line suggests the limits of the Aryan penetration of the subcontinent (Map 11-3).

In the centuries during and following the Aryan invasion, Indian culture went through a period of growth and development. From a formless collection of isolated tribes and their villages, regional organization began to emerge. Towns developed; local rulers became something more as surrounding areas fell under their control. Arts and crafts blossomed once again, and trade with Southwest Asia was renewed. Hinduism emerged from the religious beliefs and practices brought to India by the Aryans; tribal priests took on the roles of religious philosophers shaping a whole new way of life. Social stratification evolved, with the ruling Brahmans, administered by powerful priests, at the head of a complex bureaucracy—a caste system—in which soldiers, artists, merchants, and others all had their place. Aggressive and expansion-

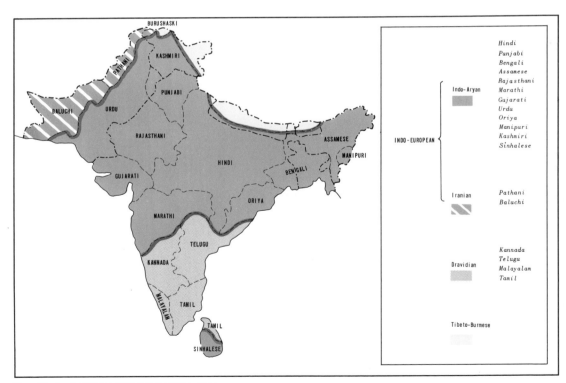

Map 11-3 LANGUAGES OF SOUTH ASIA.

minded kingdoms arose, always competing with each other for greater power and control. It was in one of these kingdoms, in northeastern India, that Prince Siddhartha, or Buddha, was born in the sixth century B.C. Buddha, we know, voluntarily gave up his princely position in his state to seek salvation and enlightenment through religious meditation. His teachings demanded a rejection of earthly desires and a reverence for all life. But although Buddha had his followers, the real impact of Buddhism was to come during the third century B.C., after an interval marked by the end of Persian domination in the northwest and a brief but intense intervention by Alexander the Great (326 B.C.). Alexander's Greeks pushed all the way to the Indian heartland in the Ganges valley.

Thus northern India was the theater of cultural infusion and innovation. The south lay removed, and protected by distance, from much of this change. Here a very different culture came into being. The darker skins of the people today still reflect their direct ties with ancient forebears such as the Negritos and Australoids, who lived here even before the Indus civilization arose far to the northwest. Their languages, too, are distinctive and not related to those of the Indo-Aryan region. Both the peoples and the languages of southern India are known collectively as *Dravidian*. The four major Dravidian languages—Telugu, Tamil, Kanarese (Kannada), and Malayalam—all have long literary histories, and they are not mere tribal tongues. Telugu alone is spoken by some 10 percent of all of India's population, and Tamil is the language of 8 percent. Together, these two languages are the medium of 18 percent or nearly one-fifth of India's 500 million inhabitants.

The Mauryan Era

The first Indian empire to incorporate most of the subcontinent emerged with the decline of Hellenic influence, shortly before 300 B.C. Its heartland lay in the middle Ganges valley, and quite rapidly it extended its power over India as far west as the Punjab and the Indus valley, as far east as Bengal, and as far south as modern Mysore (Map 11-4). The state was led by a series of capable rulers among whom the greatest no doubt was Asoka, who ruled for nearly 40 years during the middle period of the third century B.C. and who was a convert to Buddhism. In accordance with Buddha's teachings, Asoka diverted the state's activities from conquest and expansion to the attainment of internal stability and peace; had he not done so it is likely that all of the subcontinent would have fallen to the Maurya's rule. A vigorous proponent of Buddhism, Asoka sent missionaries to the outside world to carry Buddha's teachings to distant peoples; in so doing he contributed to the further spread of Indian culture. Thus Buddhism became permanently established as the dominant religion in Ceylon, and it achieved temporary footholds even in the eastern Mediterranean lands—although it eventually declined to minor status in India itself.

The Mauryan state represented the culmination of Indian cultural achievements up to that time. Centered on the city of Pataliputra (on the site of present-day Patna), it witnessed the growth of larger urban centers in a land of villages; once again Indian architects and builders used stone, as they had centuries earlier in the Indus valley. Agriculture was much improved, with a wide range of crops that included rice in the wetter areas and other grains, such as millet, sorghum, and wheat, in the drier zones. Cotton and sugar cane were grown as they were in the Indus. Internal communications were improving, with roads linking Pataliputra to ports both on the Bay of Bengal and the Arabian Sea. Slowly but surely the Indian was changing his environment: the forest was being cut back, land was opened for farming. The cultural environment also underwent change with the Buddhist challenge to Hinduism; political theories were evolved which were to play major roles in the India of centuries to come. In international commerce and foreign relations the Mauryan state made huge strides; it exchanged diplomatic and trade representatives with the Greek and Roman states. The easiest routes to the west—India's chief area of foreign contact—lay across the sea, and countless traders plied the waters between India, Southwest Asia, and Europe. Thus Christianity arrived on India's shores during the second century A.D., while the Indians themselves continued to carry their Hindu and Buddhist faiths to the outside world, including Southeast Asia.

The Mauryan achievement in India was not soon to be repeated. When the state collapsed late in the second century A.D., the old forces—regional-cultural disunity, recurrent and disruptive invasions, and failing central authority

GUPTA EMPIRE

States tributary to Empire

Samatata
Kamarupa
Nepal
Vaisali
Pataliputra
Nalanda
Champa
Bodh Gaya
Tamralipti
Sravasti
Ayodhya
Prayaga
Kausambi
Mahakosala
Bharhut
Kartripura
Indraprastha
Vakataka
Madraka
Mathura
Manyakubja
Eran
Devarashtra
Sakala
Padmavati
Sanchi
Amaravati
Vengi
Yaudheya
Malava
Ujjain
Kanchi
Pallava
Arjunayana
Bharukaccha
Nasik
Pratishthana
(Barygaza)
(Paithan)
Abhira
Konkan
Banavasi
Chola
Ganga
Surashtra
Supparaka
Kadamba
Chera
Valabhi
(Sopara)
Muziris
Pandya
Madura
Simhala
(Ceylon)
Anuradhapura

MAURYAN EMPIRE

Samatata
Pundra-
Vardhana
Vanga
Pataliputra
Pataan
Vaisali
Rajagriha
Champa
Tamralipti
Sravasti
Kosala
Ayodhya
Kasi
Magadha
Prayaga
Kausambi
Tosali
Kashmir
Gandhara
Indraprastha
Kalinga
Taxila
Mathura
Paropanisadai
Avanti
Sanchi
Ujjain
Andhra
Arachosia
Saurura
Surashtra
Bharukaccha
Kanchi
Pattala
Supparaka
(Sopara)
Maharashtra
Chola
Satiya
Kerala
Pandya
Anuradhapura
Tamraparni
(Ceylon)

Map 11-4 EARLY EMPIRES.

428

MOGUL EMPIRE
(about 1690)

Dacca

Patna

Madras (Br.)
Pondicherry (Fr.)

Agra

Shahjahanabad
(Delhi)

Marathas

Surat

Goa
(Port.)

Kabul

Rajputs

DELHI SULTANATE
(about 1320)

AFTER: SPENCER

Delhi

Map 11-5 LATER EMPIRES.

—again came to the fore. Of course the cultural disunity of India was a reality even during the Mauryan era; the Mauryans could not submerge it. And while India did not see anything like the Aryan invasions again, there were almost constant infusions of larger and smaller population groups from the west and northwest. Iranians, Afghans, and Turks entered the subcontinent, mostly along the obvious avenue—across the Indus, through the Punjab, and into the Ganges valley. Thus, in the tenth century, came the wave of Islam, which spread like a giant tide across the subcontinent from Persia and Afghanistan to the northwest. In the Indus region there was a major influx of Moslems and a nearly total conversion to Islam of the local population. In the Punjab perhaps two-thirds of the population was converted; Islam crossed the bottleneck in which Delhi is positioned and spread into the Indian heartland of the Ganges valley—Hindustan. While the Moslem impact in Hindustan was much less, about one-eighth of the population there became adherents to the new faith. In the delta region of the Ganges, where up to three-quarters of the people became Moslems, Islam seems to have spread at the expense especially of Buddhism. Southward into the peninsula, the force of Islam was spent quite quickly, and the extreme south was never under Moslem control.

Islam was an alien faith to India as it was to Southwestern Europe, and it brought great changes to the existing ways of life. It was superimposed by political control; early in the fourteenth century a sultanate centered at Delhi controlled all but the extreme southern, eastern, and northern margins of the subcontinents (Map 11-5). Constantly there were struggles for control, as challengers to existing authority came from the Afghan empire to the west and from Turkestan and inner Asia to the northwest. Out of one of these challenges arose the largest unified Indian state since Asoka's time, the so-called Mogul or Mughal Empire, which in about 1690 under the rule of Aurangzeb comprised almost the whole subcontinent from Baluchistan to the Ganges delta and from the northern foothills to Madras (Map 11-5). But Islam in India was neither the monopoly of the invaders from outside nor was it the religion only of the rulers and the new aristocracy. Islam provided a welcome alternative to Hindus who had the misfortune of being of low caste; it was an alternative also for Buddhists and others who faced absorption into the prevailing Hindu system. In the majority of cases the Moslems of the Indian subcontinent are indistinguishable racially from their non-Moslem neighbors; there is no correlation between race and religion. Most of the subcontinent's Moslems today are descendants of converts rather than descendants of any invading Moslem ruling elite.

European Involvement

Into this turbulent situation of religious, political, and linguistic disunity still another element was introduced, namely the power of Europe. Actually, the Portuguese had reached the shores of India even before 1500 (Vasco da Gama anchored off Calicut in 1498), and early in the sixteenth century they made their effort to intercept the ocean trade, then in the hands of the Arabs, by establishing fortified coastal stations at the source, in India. Thus the Portuguese enclave of Goa was founded; it became the headquarters of Portuguese interests in India through its central position on the Malabar Coast. But any really effective penetration of India by European influences was a long time coming. The Portuguese were succeeded by the Dutch, who, however, concentrated their efforts in the Indonesian archipelago; on the Indian coast the English East India Company now began to establish its holdings. During the second half of the seventeenth century the French made their try at an Indian foothold, and for a century they struggled to maintain it; after 1748 there was a veritable war in southern India between the English and French chartered India companies. By 1761 French power had been broken. In all these struggles and rivalries, the Europeans benefited enormously from the strife that was occupying the Indians themselves. Rulers of various territories battled for supremacy, and the Moslem–Hindu competition was in full swing at a time when India needed unity in the face of the new invaders.

Even the Mogul Empire could not withstand the European thrust, and for precisely the same reason. Mogul authority was never sufficiently strong to prevent local rulers from making private deals with the Europeans. The old quality of regional disunity in India once again played its role, and the white man reaped the harvest by exploiting local rivalries, jealousies, and animosities. Meanwhile, British merchants were taking over the trade, which was stimulated into ever-greater volumes: as the spice trade leveled off, cotton and silk goods became big business. This was the trade to Europe; there was also a busy trade to Southeast Asia, in which Arab, Indonesian, Chinese, as well as Indian merchants had long joined. This trade, too, the English took over. Company vessels carried textiles and metal wares manufactured in India to markets in Southeast Asia; thus the English benefited not only by direct exploitation of Asia for the benefit of European investors and stockholders, but also by involving themselves as middlemen in the profitable intra-Asian trade.

As time went on, the East India Company found itself faced with problems it could not solve: its commercial activities remained profitable, but it became entangled in an ever-growing effort to maintain political control over an expanding Indian domain. It proved an ineffective governing agent at a time when the increasing westernization of India brought to the fore new and intense kinds of friction. Christian missionaries were challenging Hindu beliefs, and many Hindus thought that the British were out to destroy the caste system. Changes came also in public education, and the role and status of women began to improve. Aristocracies saw their positions threatened; landowners had their estates expropriated. Finally, in 1857, a rebellion occurred that changed the entire situation. It took a major military effort to put down, and from that time the East India Company ceased to function as the government of India. The administration was turned over to the British government, the company was abolished, and India became formally a British colony. It retained this status until 1947, when the subcontinent, true to its tradition of division and disunity, fragmented into the states of India, Pakistan, and Ceylon.

The British Impact

Four centuries of European intervention in India greatly changed the subcontinent's cultural, economic, and political directions. Certainly the British made positive contributions to Indian life, but colonialism also had serious negative consequences. In this respect there are important differences between the Indian case and that of Black Africa, for when the Europeans came to India they found a considerable amount of industry, especially in metal goods and tex-

tiles, and an active trade to both Southwest and Southeast Asia in which Indian merchants took a leading part. The British intercepted this trade, and in the process the whole pattern of Indian commerce changed. India now ceased to be Southern Asia's manufacturing area: soon India was exporting raw materials and importing manufactured goods, from Europe of course. India's handcraft industries declined, and after the first stimulus the export trade in agricultural raw materials also suffered as other parts of the world were colonized and tied in trade to Europe. Thus the majority of India's people, who were farmers then as now, suffered a setback as a result of the colonial economy. Therefore, although in total *volume* of trade the colonial period brought considerable increases, the *kind* of trade India now supported was by no means a way to a better life for the people.

Neither did the British manage to accomplish what the Mauryans and the Moguls had also tried to do: to unify the subcontinent and to minimize its cultural and political divisions. When the Crown took over from the East India Company, there were about three-quarters of a million square miles of Indian territory still outside the British sphere of influence, and slowly the British extended their control over this huge, unconsolidated area, including several pockets of territory already surrounded but never integrated into the company's administration. Also, the British government found itself with a long list of treaties that had been made by the company's administrators with numerous Indian kings, princes, regional governors, and feudal rulers. These treaties guaranteed various degrees of autonomy for literally hundreds of political entities in India ranging in size from a few acres to Hyderabad's more than 80,000 square miles. The Crown saw no alternative but to honor these guarantees, and so India was carved up into an administrative framework in which, in the late nineteenth century, there were over 600 "sovereign" territories in the subcontinent. These "Native States" had British advisors; the large British provinces such as Punjab, Bengal, and Assam had British governors or commissioners who reported to the Viceroy of India who, in turn, reported to the Parliament and British Crown. It was a near-chaotic wedding of modern colonial control and traditional feudalism, reflecting and in some ways deepening the regional and local disunity of the Indian subcontinent. While certain parts of India quickly adopted and promoted the positive contributions of the colonial era, other units rejected and repelled them, thus adding still another element of division to the already complex picture.

And, indeed, colonialism did produce assets for India. As a glance at a world map of surface communications shows, India was bequeathed one of the better road and railroad transport networks of the colonial realm. British engineers laid out irrigation canals through which millions of acres of land were brought into cultivation. Settlements that had been founded by the British developed into major cities and bustling ports, as did Calcutta (5.5 million), Bombay (6 million), and Madras (2.5 million), three of India's largest urban centers. Modern industrialization, too, was brought to India by the British on a limited scale. In education an effort was made to combine English and Indian traditions; the westernization of India's elite was supported through the education of numerous Indians in Britain. Modern practices of medicine also were introduced. In addition, the British administration tried to eliminate features of Indian culture that were deemed undesirable by any standards, such as the burning alive of widows on the funeral pyres of their husbands, female infanticide, child marriage, and the caste system. Obviously the task was far too great to be successfully achieved in less than four generations of rule, but India itself has continued the efforts initiated by the British in these spheres.

But the Indian subcontinent is a region which alone contains one-fifth of the earth's people, and the majority of this huge population felt the colonial impact very little, if at all—except perhaps through the tax collectors and the new laws governing land ownership. India may be famous for its huge and crowded cities, but in 1969 there were still less than 2000 cities and towns of over 5000 people, and less than 20 percent of the population was urbanized. Thus the overwhelming majority of Indians continued to be what they have been for countless centuries: village dwellers. And in their villages, more often than not, they were remote from the progressive influences of the colonial period, unaffected by programs of social improvement, mired in a debt that had always been a way of life, confronted by the constant threat of starvation, and weakened by the ever-present condition of malnutrition.

Village Life

Take a train out of a major Indian city, and very soon the rural landscape takes on a character that remains the same for hour after hour, mile after mile. Scattered over the open country lie clusters of small dwellings, made of mud and covered by thatch or matting. These small homes generally do not even have a window; their whitewash is likely to be coming off in streaks and patches. The dwellings most likely consist of two rooms with a mud floor, and probably there are no sanitary facilities. Sometimes the central and largest building in the village is a Hindu shrine or a mosque, well-kept and starkly contrasting to the drabness of the surrounding scene.

This is a very different scene from what one sees during a train trip through the farmlands of the United States; individual farmhouses stand alone, and settlement is *dispersed*, but in India the village is occupied by groups of families—up to several dozen, in the larger ones—and the farmers go out each day into the surrounding lands, which they own or rent. Thus, in India, settlement is nucleated, and by latest estimates there are no less than three-quarters of a million of these small village nuclei. Not all of them, of course, are strictly agricultural villages: in many, there are small local handcraft industries such as weaving, pottery, metal work, and the like. Those villages which lie near larger urban centers tend to specialize in vegetable growing for sale in the city; most other villages have a few plots of vegetables for local use, to diversify a diet that is dominated by cereals. Rice is the staple grain in the Ganges–Brahmaputra and lower Indus valley (with wheat in the drier parts), wheat dominates the northwest, and millet is the main cereal over most of the peninsula proper, except along the moister coastal belt. Everything depends on the subcontinent's water supply, and although the region as a whole receives considerable rainfall (Plate 1), the northwest is quite dry (the lower Indus area receives 4 to 12 inches) and precipitation on the peninsular plateau is low (20 to 40 inches) and rather variable. As the map indicates, the best watered areas of India lie in the northeast and along the monsoon coast of the southwest. This distribution is reflected by that of the villages: Where there is ample moisture and good soil the villages are large and close together, the farmlands are tightly packed and

Hangala Pura, 60 miles south of Mysore, lies in southern India and is quite typical of India's half million villages. There are 500 inhabitants, and the village owns 1000 acres of good farm land for growing cereals and cotton and for raising cattle. It also has 300 cows, 80 pairs of oxen, 120 sheep, and 50 buffaloes. The scene shows the cattle being assembled in the center of the village before they are taken out to pasture. (Erich Schwab/World Health Organization)

small and very intensively cultivated. In the drier areas villages are fewer and farther between, and they are smaller as well.

In the Indian subcontinent as in other parts of the world Man has attacked the vegetative cover and, as in the Mediterranean area, destroyed it. As building material and fuel, the woodland helped make the peopling of India possible. But today, not much more than 10 percent of India still has a forest cover, most of it remote from settled areas and permanently uncultivable. In India, this destruction has serious consequences, for the absence of wood has led the villagers to turn to another material for burning: animal dung. Especially in the northern Deccan and in the northwest, animal manures—from water buffalo, horses, camels, cattle—are being burned rather than applied to the soil as fertilizers, so badly needed especially in these areas. In large parts of north-central and northwestern India as much as half the animal dung is used as fuel rather than fertilizer. And

of course the Indian peasant is far too poor to buy artificial fertilizers to replace what is thus lost. Where the animal manures are used, there is no doubt about their effect: almost every village in India is surrounded by a ring of darker, more productive land, where dung and other humus-producing waste is added to the soil. Limited use is made of human nightsoil; religious objections restrict the practice.

Another part of the Indian village scene is the huge number of cattle that wander about— perhaps as many as 250 million of them, mostly emaciated-looking, undernourished, weak animals. In the drier parts of the country, goats and sheep are most numerous, and there may be as many as 125 million of these. All this livestock is a liability rather than an asset, for there are religious obstacles to meat consumption, and by Hindu tradition the keeping of cattle for slaughter is forbidden. True, they produce milk and manure, but in their huge numbers they compete for the available food with a human population

A street scene in Agra, India. Cattle mix with people as the old competes with the new. (Paul J. C. Friedlander)

that cannot afford such competition. And anyway, the milk production of such poor-quality animals is low; without a total change in prevailing attitudes toward livestock there is no chance that India will be able to convert this major burden into an advantage. Unrestricted breeding has produced inferior cattle, and even if religious objections were removed overnight, the task of distilling a good-quality herd out of the present livestock population would be an enormous one. The Indian government has begun programs of sterilization, selective breeding, and the expansion of pasture land, but the root of the problem is far from solution.

Regions of the Subcontinent

The Indian subcontinent is a land of immense physiographic variety, but it is possible to recognize several rather clearly defined physical regions. In the most general terms, there are three such regions: the northern mountains, the southern peninsular plateau, and between these two, the belt of river lowlands. In this division, the northern mountains consist of the Baluchistan, Kashmir, Himalayan, and Assam uplands.

The southern plateau is constituted mainly by the huge Deccan (a relative, as we have noted, of the Drakensberg in Africa). And the tier of river lowlands extends from Sind (the lower Indus valley) through the Punjab and middle Ganges to the great delta, and on into Assam's Brahmaputra valley (Map 11-6).

But this set of regions only introduces the varied Indian landscape. Depending on the degree of detail employed, it would be easy to subdivide each of these three into many more specific physiographic subregions. On the other hand, a list of several dozen physical regions would lose its usefulness, and so most geographers who have tried to outline the major regions of India have sought to balance their work between excessive detail and excessive generalization. Thus, in the mountain wall that all but surrounds India and Pakistan, we may recognize the desert ranges of Baluchistan and the Afghanistan border, the snow-capped ranges of the Himalayas, and, in the east, the jungle-clad mountains of the Assam–Burma margin. And of course these northern mountains do not simply rise out of the river valleys below: there is a persistent belt of foothills between the lofty peaks and the moist river basins.

The belt of river lowlands that extends from the Indus to the Brahmaputra also is anything but uniform. This is sometimes called the North Indian Plain, and some geographers consider it to include the Kathiawar Peninsula and part of the Thar Desert as well as the zone of river lowlands. But even without these this is a region full of contrasts, well reflected, incidentally, by Plates 1 through 4. The Indus, which has its source in Tibet and which outflanks the Himalayas to the west in its course to the Arabian Sea, receives its major tributaries from the Punjab ("land of five rivers"). The Punjab contains the divide between the Indus drainage to the west and the Ganges system to the east. It is a transition zone in more than one way: to the west and southwest rainfall falls off markedly (Plate 1), but to the east it increases to where it is among the highest of any part of India. The

Punjab itself is steppe country (Plate 2), flanked by desert to the west and a moist subtropical climate to the east. In the western Punjab, irrigation is necessary; to the east, conditions are wet enough for millet and sorghums. And the Punjab is an ethnic/cultural transition zone as well: it is Moslem-dominated in the west and largely Hindu and Sikh in the east, and on this basis was divided by international boundary between Pakistan and India.

The physiographic region of the Punjab, then, extends into Pakistan as well as India. The region of the lower Indus is characterized by its low precipitation, its desert soils (Plate 3) and its irrigation-based cluster of settlement. This is the heart of West Pakistan, and the farmers grow wheat for food and cotton for sale. Some 70 percent of all cultivated land in West Pakistan is irrigated, most of it by an elaborate system of canals; wheat occupies more than one-third of the cropped land. But in the other direction

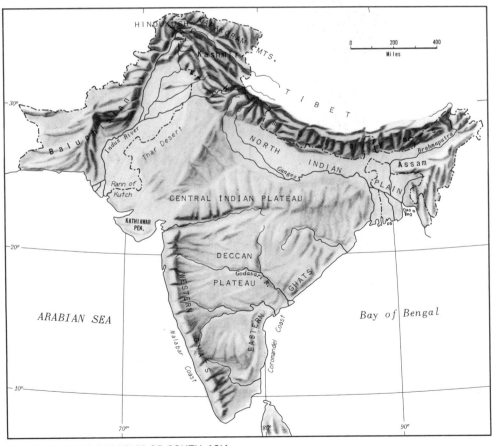

Map 11-6 PHYSIOGRAPHY OF SOUTH ASIA.

Rice is the mainstay for much of India, and it must be planted and harvested by hand, as is done by these women in the paddy fields of a village in Orissa. (Information Service of India)

from the Punjab, in the middle and lower Ganges valley, the situation is very different. This is Hindustan, the heartland of India ever since the decline of the Indus civilization. The region roughly extends from the vicinity of the historic city of Delhi to the delta at the head of the Bay of Bengal; again it extends into Pakistan, this time East Pakistan.

Hindustan is wet country: the precipitation exceeds 100 inches in sizable areas and 40 inches almost everywhere (Plate 1). Deep alluvial soils cover much of the region (Plate 3), and the combination of rainfall, river flow, good soils, and a long growing season make this India's most ·productive agricultural area. As Plate 6 proves, this is also India's largest and most concentrated population core; not only do rural densities in places exceed a thousand people per square mile, but from Delhi to Calcutta lies a series of the country's leading urban centers, connected by the densest portion of its rail and road transport network. In the moister east, rice is the chief food crop; jute is the main commercial product, notably in the delta, in East Pakistan. To the west, toward the Punjab and around the drier margins of the Hindustan region, wheat and such drought-resistant cereals as millet and sorghum are cultivated. This agriculture sustains not much less than half of India's population on not much more than one-fifth of its total land. Indeed, East Pakistan, only about one-sixth as large as the West, contains nearly 70 million people as against less than 60 million in the West. *Seventy* million people on 55,000 square miles of land with one major city counting less than 1 million inhabitants—here is a measure of population density in Asia!

The easternmost extension of the North Indian Plain comprises the Brahmaputra valley in Assam. This valley is much narrower than the Ganges basin, it is very moist, and suffers from frequent flooding, and the river has only limited use for navigation. Up on the higher slopes, tea plantations have been developed, and in the lower sections rice is grown. Assam is just barely connected with the main body of India through a corridor of only a few miles between the southeast corner of Nepal and the northern boundary of East Pakistan, and it remains one of Asia's frontier areas.

Turning now to the peninsular part of India, referred to previously as plateau India, we find once again that there is more variety than first appearances suggest. There are physiographic bases for dividing the plateau into a northern area (the Central Indian Plateau would be an appropriate term) and a southern sector, which is already well known as the Deccan Plateau. The dividing line can be drawn on the basis of the roughness of terrain or on the basis of rock type (much of the Deccan is lava-covered), but the Tapti and Godavaria Rivers—in the west and east respectively—form a rather clear lowland corridor between the two regions. The Deccan (meaning "south") has been tilted to the east, so that the major drainage is eastward; the plateau is marked along much of its margin by a mountainous, steplike descent or *Ghats*. Parallel to the coast in both east and west are the Eastern and Western Ghats; they meet near the southern tip of the subcontinent in the Cardamon Upland, recognized by some as a distinct physiographic region.

There are also cultural reasons for recognizing a break in the peninsular plateau, for here is the historic transition zone between that part of India which came under the direct and effective influences of the Aryans and Moslems, and that region of the country which remained relatively distant and isolated: the Dravidian south. The dividing line between Indo-European and Dravidian languages (Map 11-3) emphasizes this contrast. The hills and mountains of the south functioned as protectors of the pre-Aryans. Today, the plateau region of peninsular India is still relatively sparsely populated. This is not a very productive part of the country, with its latosols (Plate 3), low and variable precipitation, and considerable slope incidence. Wheat, and in the drier parts, sorghum and millet, form the staples.

Peninsular India possesses a coastal lowland zone of varying width. Along the west coast lies the famous Malabar Coast, and along the eastern shore is the Coromandel Coast. These two physiographic regions lie wedged between the interior plateau and the Indian Ocean; the Malabar Coast is the more clearly defined since the Western Ghats are more prominent and

Where the forested slopes of the Ghats give way to the more manageable terrain along the coast, terraces are constructed and irrigated with the ample supplies of water from the adjacent hills. (Authenticated News International)

higher in elevation than the Eastern Ghats. Thickly forested and steep, the Malabar escarpment dominates the west coast and limits its width to an average of less than 50 miles; the Coromandel coastal plain is wider and its interior margins are less pronounced. While not very large in terms of total area, these two regions have had and continue to have great importance in the Indian state. Plate 1 indicates how well-watered the Malabar Coast is; this rainfall supply, and the warm, tropical temperatures, have combined with the coast's fertile soils to create one of India's most productive regions. On the lowland plain, rice is grown; on the adjacent slopes grow spices and tea. Of course this combination of favorable circumstances for intensive agriculture had led to the emergence of southern India's major population concentration (Plate 6), but the Malabar and Coromandel Coasts had other advantages as well. Along these coasts the Europeans made their first contact with India beginning not with the Portuguese, but with the Greeks and Romans. Later these regions became spheres of British influence, and it was during this period that two of India's greatest cities, Bombay and Madras, began their growth.

The Monsoon

The Malabar Coast of India exemplifies a particular kind of climatic regime referred to as the *monsoon* climate. The term apparently has been derived from an old Arabic word meaning *season,* which probably also applied to the prevailing winds during such a period, which were essential to the ocean sails. In English usage the term monsoon now refers to a particular precipitation pattern, one that is especially well defined in southwestern India. The monsoon climatic type is characterized by warm, tropical temperatures and by precipitation totals resembling those found in tropical rain-forest areas—with the difference that there is a marked dry season and a heavy concentration of rainfall in two or three summer months.

The case of Calicut, a town on the Malabar Coast, is an excellent example. There, December, January, February, and March are normally quite dry, none of those months receiving even two inches of rain. April is somewhat wetter, but not much: 3 or 4 inches would not be unusual. May can record as much as 10 inches, most of it likely to fall late in the month, but then the skies open, and June averages nearly 35 inches! July is not far behind with over 30 inches also, but then a rapid decline sets in: August averages around 16 inches and September only 8. Calicut receives slightly more than 120 inches of rain per year, and the two months of June and July account for more than half of it—more than the other 10 months combined.

Such a precipitation regime is not unique to India, or even to Asia. As Plate 2 shows, there are areas of *Am* climate in northern South America, in West Africa, and on the west and east coasts of Southeast Asia. But the regime is most clearly expressed in Asia, and some geographers have even gone so far as to call South, Southeast, and East Asia "Monsoon Asia." That may be a bit misleading, for the areal extent of Asia's monsoon lands is quite limited, as the map shows. But nowhere is the monsoon of greater importance to a larger number of people than it is in India and its neighbors in the Orient.

Any map of the world's climatic regions is to some extent arbitrary and to a considerable degree unsatisfactory. What we call "desert" merges into what we identify as "steppe," and nature certainly never drew the line we put on our maps. Anyone who looks into the question of climatic classification soon discovers that there are about as many different classifications as there are classifiers, each with its own assets and liabilities. Some have opted for simplicity and utility and have based their classifications essentially on monthly temperature and precipitation data, as have the authors of Plate 2. Others have sought to plug in as many variables as can be measured, including, for example, the amount of moisture actually or potentially lost through surface evaporation and transpiration by plants: *evapo-transpiration.* Sometimes, as we saw in our study of Europe, the resultant climatic regions can help us, if not in solving problems of weather mechanics, then at least in raising relevant questions about climate.

Such is the case with the monsoon-type climate. Why does this regime prevail on the Malabar Coast, and especially toward the southern tip of the Indian peninsula? What we must look for is a mechanism in which moisture-bringing winds, which surely must come from the ocean, are held back until something triggers that lets them pour inland, to dump their stored-up water on the shore. When this occurs we can read off the Malabar weather stations' records. But how?

For a long time there has been a glib solution, which can be found in almost any textbook: during the hot summer, the great continental landmasses heat up, the air above them rises, and air from the adjacent oceans is drawn in to replace the rising air at the surface. As this moist air from the oceans hits the coast, it is forced to rise and cool; large amounts of rain result. In winter the cycle is reversed: the continents are cold and so is the overlying air, which flows outward toward the now comparatively warm oceans. Thus no ocean air can flow inland, and the monsoon has to wait for another summer season. But is it really that simple? To the question: Why does this not happen more effectively in the Americas, Africa, and Australia, there is the answer that these continents are somehow "not big enough" to generate the mechanism. Then, when it is pointed out that the Himalayas are rather in the way of any air movement from the Indian Ocean to the Asian interior, the answer is that the heat cell that draws in the ocean air actually forms over northern and north-central India itself. How does that jibe with the argument that the non-Asian continental landmasses are not large enough? They are larger than India! In fact, the whole idea of a continental-scale inward pull during the summer months raises all sorts of unanswerable questions. Why should the Malabar Coast, of all places, have such a pronounced monsoonal climate? How can certain other coastal areas *not* be affected by a monsoon-type clime?

Recently, alternative explanations have begun to appear, one of them relating to the jet stream, an influence on weather and climate long unknown. One thing that has puzzled climatologists about the jet stream is the suddenness with which it starts; surely a build-up of summer heat over Asia would be marked by spring-season incursions of moist air? Now the suggestion has been made that this suddenness has to do with a major shift of the jet stream—that the jet stream moves north of the Himalayas during the summer, and lies south of the great range during the winter. During the winter, it is argued, the jet stream is associated with a high pressure zone which would cover much of the subcontinent if the jet stream were to lie south of the Himalayas. Then, in summer, when the jet stream is diverted to the north of the mountains—an event that could indeed take place quite suddenly—the anticyclone is removed and conditions prevail that would draw in the moist ocean air. Again, with the approaching winter, the jet stream returns to its southern-flank position and the rainy season ends. This, of course, constitutes another possible explanation; it is not proved, nor does it do away with all the problems the monsoon mechanism presents.

The Partition

The Indian subcontinent therefore is a region of physiographic as well as cultural variety. Not until the British government took control did unity come to this realm, and then, of course, the unity was superimposed by an alien power. And neither was Britain's administration capable of consolidating the numerous autonomous and semi-autonomous political entities that formed part of this, London's largest colony. The British were unable to streamline the government; they managed, nevertheless, to make the system with which they found themselves in 1858 work. When they saw the opportunity, they did merge formerly separate states, but this was a slow process. In 1947, when independence came, there were more than two dozen provinces in India and still no less than 562 "princely states."

The modernization of the subcontinent's political framework would have been a big enough problem all by itself, but even before independence it was clear that British India would not survive the coming of sovereignty in one piece. As early as the 1930's the idea of a separate

Pakistan was being pushed by Moslem activists, who circulated pamphlets arguing that India's Moslems were a distinct nation from the Hindus and that a separate state consisting of the Punjab, Kashmir, Sind, Baluchistan, and a section of Afghanistan should be created. The first formal demand for partition was made in 1940, and the idea had the almost universal support of the region's Moslems, as subsequent elections proved. As the colony moved toward independence, a crisis developed: the majority Congress Party would not consider partition, and the minority Moslems refused to participate in any unitary government. But partition was not simply a matter of cutting the Moslem areas off from the main body of the country; true, Moslems were in the majority in what is today Pakistan, but other Moslem clusters were scattered all over the subcontinent (Map 11-7). The boundaries between India and Pakistan would have to be drawn right through transitional areas, and people by the millions would be dislocated.

The Indian Independence Act of 1947 embodied the principle of partition; by then it had become clear to almost everyone that this was the only alternative to chaos. But now the difficult process of boundary definition was to take place, and Moslem claims were ever-widening. There were Moslem demands for all of Bengal and Assam in the east and the whole of the Punjab (as well as Sind and adjacent areas) in the west. There was even a call for an 800-mile corridor to link the two fragments of Pakistan! Ultimately the solution which now shows on the political map was arrived at, though Kashmir has remained an area of dispute between the two countries.

What happened in India was the *superimposition* of a political boundary upon a cultural landscape in which such a boundary had hitherto not existed or functioned. Political geographers consider boundaries in this light, that is, with reference to the stage of development of the cultural landscape at the time boundaries were established. On this basis a genetic classification of boundaries has been developed (as opposed to the morphological classification discussed in the context of the Middle East), whose terminology was proposed by Professor Hartshorne in 1936.[2] Essentially, three classes of boundaries can

[2] R. Hartshorne, "Suggestions of the Terminology of Political Boundaries," abstract, *Annals of the Association of American Geographers,* Vol. 26, March 1936, p. 56.

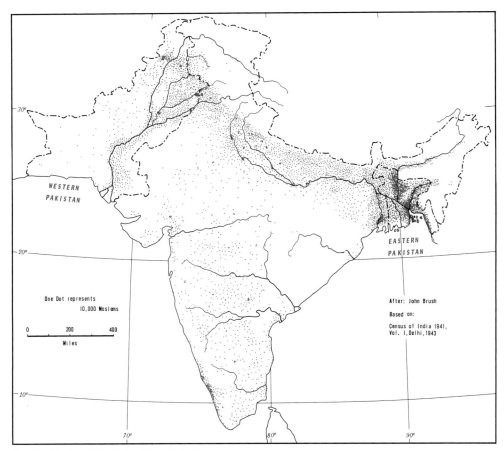

Map 11-7 MOSLEMS IN SOUTH ASIA.

be recognized: those that were established before the main patterns of settlement had evolved, those that were created during the development of the major elements of the cultural landscape, and those that have been delimited after such development.

Boundaries that were defined and delimited before the main elements of the present-day cultural landscape began to develop are identified as *antecedent*. For example, a boundary may be drawn, even geometrically, through a desert area occupied, if at all, only by some nomadic groups of people. Then, however, oil may be discovered in the area through which the boundary lies, and a whole new pattern of settlement emerges. But the boundary was there before the pipelines, the roads, the airports, towns, and other installations: it is antecedent to the cultural landscape. A specific example is the boundary between the United States and Canada, west of the Great Lakes, which was defined when the

only representatives of the present-day majority population were some hunters, traders, and explorers. Now the boundary separates towns, farms, schools; it affects the transport network, interrupts road travel. It is antecedent: it was there before all these things were constructed, and as the cultural landscape emerged, it adjusted to the reality of the boundary.

Some boundaries display a certain degree of conformity to the main elements of the cultural landscape through which they lie. They came about as part of the process of spatial organization in the regions they now fragment. For example, such a boundary might have come about when two state systems found their spheres of influence meeting in a frontier area, and so the boundary is referred to as *subsequent*. Not infrequently this sort of boundary conforms to linguistic, religious, or ethnic breaks or transition zones in the landscape. In the partition of the Indian subcontinent, the newly created bound-

Nevertheless, there are areas in India where the people are reasonably well fed, where the soils are good, where the rewards of increased production are directly evident—but where, despite all this, yields are far below what they might be and where soils are not used to their full capacity. Basin irrigation prevails where perennial irrigation would be possible; canals which could render farmlands more productive have not been dug. The Indian government has tackled these problems with vigor, and through successive Five-Year Plans the productivity of Indian agriculture is being raised. But the problem is a huge one, and despite its fertilizer-distribution projects, its loan program to farmers, its village development projects, its irrigation schemes, and various other forms of encouragement, India's agricultural output continues to lose ground to the demands of a population that is increasing in the late 1960's by 13 million people per year.

Indian agriculture's chief enemy is drought, and there are dams on all the major and many minor rivers to conserve water as much as possible. The farmers must cope with a variable precipitation, a parching, dry winter, and great heat prior to the breaking of the monsoon—a pattern that is broken only in Assam, Bengal, and parts of the west coast. For uncounted centuries men have struggled to bring water to the surface and to distribute it, and on every Indian farm and in every village the center of life is the well or water hole. Along with the large dams on the major rivers there are thousands of smaller, local projects, designed to bring water to the surface or to catch and hold rain water, to channel this water to where it can be used, and to store it in order to make it available during extended droughts. Wells supplement the stream waters; underground water is brought to the surface by ancient methods such as the "Persian wheel," whereby an ox moves a chain of buckets which fill down below and are spilled into a pond or channel at the surface, or by motorized pumps. Together, the great dams, the wells, tanks, ponds, and other storage reservoirs combine to transform the river basins into vast jigsaws of water conservation and distribution systems at every level and on every scale.

But despite the problems faced by the farmers, agriculture must be the basis for economic development in India, for it employs the vast majority of the labor force, produces most of the government's tax revenues, contributes its chief exports by value (jute, 18 percent; tea, 16 percent; cotton fabrics, 7 percent), and generates the main moneys available for investments in other sectors. Not surprisingly, India has channeled most of its energies in the cause of massive improvements in the agricultural sector, but industrialization has not been neglected. The industrial core of India lies in and on the margins of the Chota Nagpur upland; good coking coal lies at Anansol in West Bengal, and India is quite rich in high-grade iron ores, some of which lie in the same area. Large-scale foreign aid has helped the development of the Chota Nagpur industrial heartland, especially the iron and steel industry. Jamshedpur in southeastern Bihar is the center of India's industrial zone (Map 11-9), which is advantageously positioned on the margin of the great Ganges core area and not far from the city of Calcutta, one of the country's great urban hubs.

Industrialization in India, however, is yet in an initial stage, and once again the obstacles are many. Some of these relate to India's attempt to generate rapid development under a part-socialist, part-free enterprise system. Others have to do with the resource base itself: the known coal fields are quite limited and the time is approaching when India may have to begin importing this commodity, always a costly business so far from other world sources. Unlike its Southwest Asian neighbors, India does not have large oil fields (some oil is being produced in Assam), and while there are good hydroelectric power sites in southern as well as northern India, their development is expensive and the rate at which projects can be constructed is necessarily slow. Still, the coal situation being what it is, and the expense of moving coal over great distances to the widely distributed major cities, the Indian government is investing quite heavily in hydroelectric projects, especially those near major cities (Bombay and Madras are served by hydroelectric power) and those where dam construction leads not only to improved power supply but also to new irrigation areas and improved flood control.

In addition to the iron and steel industry, the other leading factory industries in India are the cotton industry, centered on Bombay and Ahmadabad (1.5 million), and the jute milling of Calcutta and its environs. The cotton industry has long been an industrial mainstay in India,

Famine and hunger still afflict India, and are likely to continue to do so unless the population increase can be slowed down substantially. The pictures dramatize the tragedy: the women appealing and weeping for food at a famine-relief handout station, and the baby showing the effects of malnutrition and undernourishment—effects which will be detrimental to him the rest of his life. (Werner Bishof/Magnum—P. Pittet/World Health Organization)

Environmental contrasts in the Indian subcontinent. East Pakistan has vast delta areas where the Ganges and Brahmaputra-Jamuna Rivers reach the Bay of Bengal. During the monsoon season, July to September, large parts of East Pakistan receive as much as 100 inches of rainfall; serious flooding often occurs. The figure right illustrates such a situation along the banks of the Pasur River. The striking contrast with below is emphasized by the ancient Persian wheel, still used to raise water in the drier parts of the subcontinent, in Rajasthan. Here the crops are wheat or dry-climate cereals such as jowar and bajra; in the wet regions of East Pakistan, it is rice. (United Nations)

and it was one of the very few industries to derive some benefit from the new economic order brought by the Europeans. While other small-scale and handcraft industries were destroyed, the local textile industry benefited to some extent from the availability of cheap cotton yarn, cheap because it was mass-produced in the large, power-driven mills. With its own cotton harvest, its large local market, the abundant and cheap availability of labor, and the power supply for the Western Ghat's hydroelectric stations, the large scale cotton industry thrived. Today India, once the victim of colonial mercantilism, outranks Britain itself in textile exports: in recent years its external sales have placed it second in the world (after Japan), with a total value of 7 percent of all the country's exports combined. Although Bombay remains the center of the cotton industry, textile manufacturing takes place in all the major cities, especially Madras and Calcutta, even though the latter is better known for its jute mills. In fact, Calcutta is India's most diversified industrial center, with its railway assembly plants, food processing factories, ship yards, chemical industries, printing and bookbinding establishments, and textile factories. Its port is one of the busiest in the world, handling the products of India's heavy industries (such as pig iron), mines (manganese ore), plantations and farms (tea and jute), and factories (textiles); it also transfers much of India's import trade. One result of the partition of East Pakistan has been the expansion of the jute acreage in Calcutta's hinterland of West Bengal, but despite the political barriers the city's mills have been receiving a large volume of raw jute from Pakistani soil. Nevertheless, there are efforts on both sides of the boundary to reduce this international interdependence: the increases in West Bengal's jute acreage are matched by the construction of jute mills on the Pakistani side of the border. And so political fragmentation once again has led to unnecessary economic duplication—as it has in many other parts of the world.

Population: the Crisis of Numbers

In Chapters 3 and 5 we discussed several aspects of a problem that gnaws at India unendingly: the population problem. If Malthus were alive today, he surely would point to India as an example of the truth of his prophecy about man's inability to keep his food supply rising as

fast as his numbers. India was already quite densely populated before the Europeans came; some demographers have suggested that India's population had already reached its upper limit, under prevailing knowledge of technology and social organization. There was high fertility, but also high mortality. Productive years would witness a rise in the population, but lean years produced famines and epidemics, which offset the increase. One specialist in population problems, Kingsley Davis, has suggested that India's population had remained practically stationary (within the range of fluctuation imposed by gains and losses in years of comparative plenty and famine) for centuries before the Europeans intervened.[7]

The British colonial period terminated this balance, and India's population has been rising at a rapidly increasing rate ever since, and especially during the twentieth century. Late in the nineteenth century the population of the subcontinent was about 250 million. In the 50 years from 1900 to 1950 about 150 million were *added* to that number. Then in the next decade, from 1950 to 1960, it increased by another 100 million. And in the decade from 1960 to 1970 it is estimated that the subcontinent's population will grow by a further 125 million. Thus with a combined population for India and Pakistan of approximately 625 million in 1970, *one-fifth* of all those people were added during the last ten years.

This enormous increase in the rate of population growth did not come about simply because the British improved Indian standards of living. In fact, the living standards of millions of Indians changed very little. But the British did end the incessant, destructive wars that ravaged India; they built the transport network that did so much to improve the distribution of food to areas facing famines; they set up systems of relief supply in stricken areas. Their irrigation works greatly expanded the available cultivable area, and they attacked the rampant diseases, including cholera, smallpox, typhoid, dysentery, and others.

In the process, the death rate was brought down rapidly. Late in the nineteenth century the death rate in India was over 40 per thousand, in 1940 it was near 30, and in the mid-1960's it went below 20 (the world average today is

[7] Kingsley Davis, *The Population of India and Pakistan,* Princeton University Press, Princeton, 1951, p. 27.

Nowhere is the mass of humanity that is India more dramatically represented than on the banks of the Ganges at the time of Hindu pilgrimage. (Marc Riboud/Magnum)

approximately 16). But the birth rate failed to show any sharp decline, and so the excess of births over deaths grew ever larger. Although the reliability of demographic data relating to India is quite low, it is apparent that the birth rate in the late 1960's was still over 40 per thousand. This is not much less that it apparently was in the late nineteenth century, when Davis estimates it to have been about 49; in the 1930's informed guesses put it around 45. The most recent figure published by the United Nations is 41 per thousand.[8] This gives an annual rate of increase of over 2 percent, which must of course be seen in light of the immense population base India already possesses. In the decade of the 1970's India will come close, at this rate, to adding a total population as great as that of the United States today.

What this means is that no matter how hard the Indians struggle to increase their supply of food, they cannot hope to close the gap between demand and supply. Since 1950, indeed, the

[8]United Nations, *Report on the World Social Situation, 1967,* New York, 1969, p. 6.

Indians have managed to raise their farm output by almost one-third—but India is feeding just about one-third more people today than in 1950. Thus, despite all this increased agricultural productivity, the *per capita* production of food has not been raised—in fact, it may have slightly declined.

This situation alarms everyone, and especially the Indians themselves. India in 1952 adopted family planning as national policy as part of the state's first Five-Year Plan (1951–56), and Pakistan followed suit in 1955. But the decision to encourage family planning is one thing; its successful introduction is quite another. Indian policy favored the prevention of conception through education about the rhythm method. Unlike the Japanese, the Indians rejected sterilization and abortion as acceptable methods. The clinics established during the first several years were mostly in the larger cities. The budget for so huge an undertaking was diminutive—less than $1.5 million was initially allocated. Predictably, the impact of the program was very minor, and with their second Five-Year Plan the Indians stepped up their efforts. Now some $10 million was appropriated, and 2000 clinics were planned for the rural areas, where the need is great. But no one is encouraged by the results; India's population is still growing explosively.

One reason for this is obvious: so huge is the problem, so massive the investment necessary to make an impact, that India simply cannot afford to do what is required. India is not a rich country, and its needs are many, and the choice is all too often between some project or scheme that will show quick results and the invisible, long-range benefits of birth control. Another dam, another ten thousand acres of farmland—or another hundred birth control clinics? Another reason lies in the choice of methods. One key to the success of the Japanese (see Chapter 14) has been their adoption of abortion as an acceptable means of limiting families; the thoroughness with which the Japanese tackled their

population problem after World War II was based on this premise. The rhythm method is not particularly effective, and mistakes are easily made. And there simply have not been the funds or systems of distribution to make available birth control devices to the millions of Indian women who might use them. Then there remains the unchanging nature of life for so many of the people in rural India, where for centuries the farmer has struggled in the same way to reap a meager harvest from his plot of soil. He cannot afford new equipment, nor can he spend money on enough fertilizer to raise his production. All he knows is that hard work, manpower, results in greater productivity, and so he is motivated to enlarge his family. This creates more hands to do the work, but it also produces more mouths to be fed. It is a vicious cycle, and so long as the Indian farmer is trapped in it, population-control efforts will be deterred.

PAKISTAN

The Islamic Republic of Pakistan is actually two countries—two countries in two different worlds. Except for the common Moslem goal of partition from India, few basic bonds exist between the two so-called "wings" of this country. In fact, even in the path whereby the Islamic faith arrived, West and East Pakistan differ. In dry-world West Pakistan it arrived over land, carried eastward by invaders from the west and northwest. In monsoon-wet East Pakistan it arrived by the sea, brought to the delta by Arab traders. But in the end the two flanks of the Indian subcontinent joined in their determination to be free of Hindu domination in an Islamic state which would overcome the enormous difficulties imposed by a thousand miles of separation and thus the Republic was born.

Discussions of the Indian subcontinent frequently stress the former unity of the region, especially the unifying effect British colonial administration should have had, and then proceed on various grounds to criticize the partition that took place. There is another way to look at it. In some respects it is amazing that British India did not split into many more countries, once the principle of partition became reality in the Pakistani case. And even more remarkable is the survival of both India and Pakistan, severe centrifugal forces notwithstanding. Had India faced successful secessionist movements the

explanations would have been readily forthcoming: the religious division still existing within the state, the language problem, racial overtones in regional confrontations, political exploitation of regional dissatisfactions (which in Kerala has made the Communist movement very strong), governmental maladministration, and so forth. But the fact is that despite all these problems, and others as well, India in 1969 was still unified. And the case of Pakistan is no less noteworthy. It is easy to point to the Moslem religion and to the bygone days of Mogul glory as the overriding, binding forces of the state of Pakistan, but the fact is that the people of the West and those of the East, in their daily life, have little in common. In early 1969 there were signs that Pakistan's unity was in danger, as a political crisis loomed that threatened to overturn the established order. But it might also be said that for so divided a country to survive nearly a quarter of a century constitutes something of a miracle.

Having made the point of Mogul history and Moslem religion, it is difficult to find any additional area of correspondence or similarity between East and West Pakistan. The two wings of the country lie in different culture realms, with all that implies. West Pakistan lies on the fringes of Southwest Asia, East Pakistan adjoins Southeast Asia. In West Pakistan the problem is drought, and the need is for irrigation. In East Pakistan the problem is floods, and the need is for dikes. West Pakistanis build their houses of adobe and matting; Easterners use grass and thatch. Westerners eat wheat and mutton; East Pakistanis eat rice and fish. Westerners grow cotton for sale, Easterners grow jute.

Even the map itself reveals contrasts between West and East (Map 11-10). With over 300,000 square miles, West Pakistan is nearly six times as large as the East. West Pakistan is bordered by Iran, Afghanistan, and Kashmir as well as India (if the Kashmir partition holds, then Pakistan has a common border with China as well); East Pakistan, on the other hand, is virtually an enclave within India and only a short boundary

with Burma in the southeast technically prevents it from being so. West Pakistan has considerable topographic variety, reaching from the lowland of the Indus to the snowy uplands of the Hindu Kush, but East Pakistan is a flat land except for some low hills in the southeast. Environmentally the contrast is sharper still: the wettest place in West Pakistan is still drier than the driest place in the East. East Pakistan is subject to the monsoon, and some eastern and northern areas get over 200 inches of rain. Dacca, the centrally located capital of the East, receives some 90 inches annually, while the coastal port of Chittagong gets over 110 inches. Karachi, West Pakistan's largest city and the country's former capital, also with a coastal location, records only

Map 11-10 PAKISTAN.

8 inches. That figure is symptomatic of more than half of West Pakistan as a whole, since most of the area south of the latitude of Multan receives less than 10 inches of highly variable precipitation. And there are no dramatic increases toward the higher north: Lahore has 20 inches, and Rawalpindi, in the Himalayan foothills, only 35. Plates 1 and 2 summarize these differences between East and West.

Pakistan is the world's fifth most populous country. With 125 million inhabitants it competes for fifth place (after the United States, fourth with just over 200 million) with Indonesia and Japan, while Brazil also is in the group of states whose populations number around the 100 million mark. Of these 125 million, some 68 million or 55 percent live in East Pakistan, so that the minority reside in the much larger West. This leads to another prominent contrast between the two areas: East Pakistan's average population density is 1220 people per square mile, while that of West Pakistan is only 184. Of course we know that such averages for people are never evenly distributed by the square mile: there are crowded farms and nearly empty desert areas, populous cities and barren mountain slopes. But in East Pakistan rural areas do indeed carry over a thousand people per square mile: urban development is very slight and Dacca and Chittagong, the two major cities, together have less than a million of East Pakistan's nearly 70 million population. While East Pakistan is thus crowded with people almost everywhere, West Pakistan has room for more variety. Baluchistan in the west is very sparsely populated, but in the Punjab there are areas whose densities are comparable to those of East Pakistan. And West Pakistan is more urbanized than the East: Karachi has a population approaching 2.5 million, and the Punjab center of Lahore has 1.5 million. In West Pakistan as a whole, the axis of denser population predictably lies along the Indus artery, with a primary core area in the triangle formed by the Indus, its Sutlej tributary, and the Kashmir border, and a secondary core in Sind, with Karachi as its focus.

Dominance of the West

Although West Pakistan's population is smaller than that of the East, it has since independence been the leading section of the country. West Pakistan, of course, was the historic center of Moslem culture in the subcontinent, and it was also the major base of political activity. Here lay the source of the drive for partition, and here, too, occurred the most violent confrontations with India when the boundary was laid down. In East Pakistan, there were many more misgivings, on both sides, over the principle of partition. Despite its religious heterogeneity, the colonial province of Bengal had seen a Bengali nationalism emerge, one basis of which was the common language spoken throughout the delta and adjacent areas. But Bengal was nevertheless fragmented into East Pakistan and the Indian state of West Bengal, with East Pakistan becoming an adjunct of the distant government in Karachi.

Of the two Pakistani "wings," the West is by far the most strongly Moslem. Some 97 percent of West Pakistan's 57 million people adhere to Islam; less than 2 percent are Hindus. East Pakistan, on the other hand, is much less purely Islamic. Of its 68 million people, some 80 percent are Moslems but fully 20 percent are Hindus who have opted to remain in the Islamic republic. There is something symbolic in this Eastern mixture, for there have been times in the short history of Pakistan when causes intensely pursued in the West received only lukewarm commitment in the East. Indeed, almost from the beginning there were objections in the more populous East to the domination of Pakistani affairs by representatives of the West. Issues such as those involving the irrigation waters of the Punjab, and the struggle with India over Kashmir, which were capable of raising West Pakistani emotions to a fever pitch, in general aroused East Pakistan much less. From very early on there were times when representatives of East Pakistan described the status of their territory as a colony of West Pakistan; there was always resentment over the inadequate allotment of positions in government and the military to East Pakistanis. It was felt that Bengali, the East's major language, took second place to the West's Urdu in the country's affairs. And there were claims that East Pakistan's economy was being exploited for the benefit of the West.

The government repeatedly took steps to overcome the very special difficulties confronting a state so basically fragmented and divided. Bengali was accorded equal status with the Urdu as one of the country's official languages (English remains as an alternate language); in 1959 President Ayub Khan's Basic Democracies System was introduced. By this system, every adult Pakistani was involved in the political system through the election of so-called Basic Democrats (one for every 1000 to 1500 voters), who in turn elect higher officials of the government. The country's National Assembly took to alternating its meetings between the federal capital, Islamabad, and the provincial capital of the East, Dacca. In 1962, the new Constitution omitted the term "Islamic" from the title of the republic, to the satisfaction of the country's 14 million non-Moslems, but in 1963 it was added again by Constitutional Amendment.

Thus the quest for stability and unity in Pakistan is far from over, and the country experienced almost continuous internal political difficulties. This was so despite a major government-sponsored effort in the area of economic development, whereby the investments made in East Pakistan were greatly increased so as to give evidence of the advantages to East Pakistanis of continued unity with the West. In 1969 a crisis loomed which seemed to be the most serious in the country's history, attended by widespread riots against the government, not only in East Pakistan—where order was completely disrupted—but also in sections of West Pakistan, where regional dissatisfaction with the administration also was intense. With the West-East division as strong as it is, it is easy to take insufficient account of other aspects of internal disunity, within the two "wings" themselves. In some respects the problem of fragmentation helped worsen local frictions. For example, the considerable degree of autonomy granted East Pakistan resulted in demands for similar autonomous status from four political divisions within West Pakistan. The political leaders who demanded this used the undeniably special situation of East Pakistan to further their own objectives. And of course the increased effort to integrate East Pakistan economically with the West led to charges of undue expenditures there.

An additional internal problem of long standing has involved the border area of West Pakistan and Afghanistan. Here live Pakistan's Pathans (or Pukhtuns, sometimes Pushtuns), whose ethnic ties lie across the boundary in Afghanistan. For many years there have been demands, supported by the government of Afghanistan, for a separate state of Pathanistan or Pukhtunistan to be carved out of Pakistan's northwest. Such a state would incorporate over 40,000 square miles of land including northern Baluchistan, inhabited by perhaps as many as 9 million people. It would command the vital Khyber Pass, the route through the mountains in this area. Pakistan's government has refused to consider any such secession, but it has repeatedly had to back up its determination to reject it by sending in the armed forces.

Kashmir

Still another area of conflict—again specifically affecting West Pakistan—is that of Kashmir, an area of high mountains bordered by India as well as China. Although known simply as Kashmir, the area actually consists of several political divisions, including the state properly referred to as Jammu and Kashmir (one of the 562 Indian states at the time of independence) and the administrative areas of Gilgit in the northwest and Ladakh (including Baltistan) in the east. Ladakh gained world attention as a result of Chinese incursions there in the early 1960's, but the main conflict between India and Pakistan over the final disposition of the territory has focused on the southwest, where Jammu and Kashmir are located (Map 11-11).

When partition took place in 1947, the existing states of British India were asked to decide whether they would go with India or with Pakistan. In most of the states, this issue was settled by the local authority, but in Kashmir there was an unusual situation. There were about 5 million inhabitants in the territory at that time, nearly half of them concentrated in the so-called Vale of Kashmir (where the capital, Srinagar, is located). Another 45 percent of the people were concentrated in Jammu, which leads down the foothill slopes to the edge of the Punjab. The small remainder of the population is scattered through the mountains, and include Pathans in Gilgit and other parts of the northwest. Of these population groups, the people of the mountain-encircled Vale of Kashmir are almost all Moslems, while the majority of Jammu's population is Hindu. But the important feature of the state of Jammu and Kashmir was that the rulers were Hindu, not Moslem, although the overall population was more than 75 percent Moslem. Thus the rulers were faced with a difficult decision in 1947: to go with Pakistan and thereby exclude themselves from Hindu India, or to go with India and thus incur the wrath of the majority of the people. Hence the

Map 11-11 EMBATTLED KASHMIR.

Maharajah of Kashmir sought to remain outside both Pakistan and India, and to retain the status of an autonomous, separate unit. This decision was followed, after partition of India and Pakistan, by a Moslem uprising against Hindu rule in Kashmir. The Maharajah asked for the help of India, and Pakistan's forces came to the aid of the Moslems. After more than a year's fighting and through the intervention of the United Nations, a cease-fire line was established which left Srinagar, the Vale, and most of Jammu and Kashmir in Indian hands, including also nearly four-fifths of the territory's population. In due course this line began to appear on maps as the final boundary settlement, and Indian governments have even proposed that it be so recognized. But Pakistan has never agreed to such recognition, and Islamabad's position continues to be that no final solution has been reached for the Kashmir problem.

Why should two countries whose interests would be served by peaceful cooperation allow a distant mountainland to trouble their relationship to the point of war? There is no single answer to this question, but there are several areas of concern for both sides. In the first place,

Pakistan is wary of any situation whereby India would control vital irrigation waters needed in Pakistan, and as the map shows, Kashmir is traversed by the Indus River—the country's lifeline. Moreover, other tributary streams of the Indus originate in Kashmir, and in the Punjab Pakistan learned the lessons of dealing with the Indians for water supplies. Second, the situation in Kashmir is analogous to that which led to the partition of the whole subcontinent; Moslems are under Hindu domination. Since the majority of Kashmir's people are Moslems, Pakistan argues that free choice would deliver Kashmir to the Islamic Republic, and a free plebiscite is what the Pakistanis have sought and the Indians have thwarted. Furthermore, Kashmir's connections with Pakistan prior to partition were much stronger than those between Kashmir and India (though India has invested heavily in improving its links to Jammu and Kashmir since the military stalemate). And in addition, Pakistan argues that it needs Kashmir for strategic reasons, in part to cope with the Pathan secession movement which extends into this area.

India, on the other hand, refuses to yield for its own reasons, primarily the nearly 1.5 million

The high mountains of Kashmir form the sources of several of South Asia's most important rivers. Shown here is the area's magnificent scenery, near the rest house at Sonamarg. (Information Service of India)

Hindu residents of the territory (mainly in Jammu), and because it has backed up its rule over the Indian portion of Kashmir by heavy investments in communications, land redistribution programs, and general economic development. And so the latent conflict continues to erode Pakistani-Indian relations, and to a certain degree, intra-Pakistani relations as well. It is not difficult to see why the Kashmir issue, while crucial to West Pakistan, is not viewed in quite the same light in the East, a thousand miles away.

Livelihoods

Contrary to the predictions made at the time Pakistan became a separate state, it has managed to survive economically and has even made a good deal of progress. This is so despite the country's poverty in known mineral resources; apart from natural gas in Baluchistan (which may be related to oil reserves) and chromite, Pakistan has only some small iron deposits which are being used in a small plant at Multan. So the country must make up in agricultural output what it lacks in minerals, and since independence it has made considerable strides. Favored by high prices generated by the Korean War, the economy got an early boost from jute and cotton sales. Then the mid-1950's brought a period of stagnation, made worse by the political conflicts with India. But during the 1960's a considerable expansion of irrigated acreages, land reform, the

improvement of marketing techniques, and the success of the textile industry all contributed to considerable progress.

Certainly there was room for such progress: Pakistan shares with India the familiar low yields per acre for most crops, per capita annual incomes are very low (around $70), and manufacturing is only now emerging as a significant contributor to the economy. In West Pakistan, the low yields for vital crops such as wheat and rice have long been due to the inability of the peasant to buy fertilizers, to the poor quality of seeds, and to inadequate methods of irrigation. In many areas there has not been enough irrigation water; elsewhere waterlogging is a problem. In still other places there is excessive salinity in the soil. In Sind, where large estates farmed by tenant farmers existed before the British began their irrigation programs in the Punjab, yields are kept down by outdated irrigation systems and by the low incentives to landless peasants. But the government has pressed for change through several Five-Year Plans, and in the late 1960's these were bearing fruit. Early in 1969 it was reported that wheat production had risen by 40 percent to 6 million tons and that of rice had increased by 15 percent to 12 million tons, and the Pakistani government announced that it expected the country to be self-sufficient in food by 1970. As a result, loan money was becoming more easily available to the country, and the World Bank announced that it would support the construction of a huge new dam on the Indus River, above Islamabad, to be known as the Tarbela Dam. This dam is part of the modernization program of the Indus plain, a program that is providing perennial irrigation to areas formerly under basin irrigation only, while at the same time making available new sources of water for canals cut off by the political partition. It also constitutes a new source of hydroelectric power in the Mangla Dam already finished on the Jhelum River (1969). Pakistan will need all the soil it can irrigate: the population is increasing at 3 percent annually, one of the highest rates in the world.

Pakistan's virtually new textile industry, based on the country's substantial cotton production, has developed rapidly. It now provides all of the home market, and textiles have quickly risen to become the first contributor to exports with 37 percent by value, followed by jute with 19 and cotton with 17. In 1968 textiles contributed 30 percent, jute 22, and cotton 11. And other industries have also been stimulated; increasingly Pakistan is able to produce at home what formerly it had to import from foreign sources. A small chemical industry has emerged, an automobile assembly plant has opened in Karachi, and at Karachi and Chittagong small steel-producing plants have been built. Thus Pakistan is slowly trying to reduce its dependence upon foreign aid, investing heavily in the agricultural sector and encouraging local industries as much as possible. But there is a long way to go, and a difficult one: there are few natural resources, and the country's agricultural export products, jute and cotton, are both subject to sudden price fluctuations on the world market and to the increasing pressure of competition. Still, the third Five-Year Plan (1965–1970) in early 1969 was meeting its objectives, and if the political crisis does not do severe damage, Pakistan's financial position for a developing country may be quite strong by the end of the decade.

Nevertheless, the country still faces a host of unsolved problems. Literacy, so vital in the nation-building effort, still was only about 25 percent in the mid-1960's, although the current Five-Year-Plan envisages the goal of universal literacy. An urgent need exists for the dissemination of the government-approved program to encourage family planning. In East Pakistan, the livestock situation is similar to that prevailing in so much of India—there are too many un-productive, undernourished cattle who compete with human beings for a precarious food supply. And despite the optimistic tone of government reports relating to economic development, the vast majority of Pakistan's many million of people live in poverty, subsist on an ill-balanced diet, have a short life expectancy, and experience little tangible progress during their lifetime. Of course there are successful farmers in the Punjab, in Sind, and in Bengal. But they are still far outnumbered by those whose life is mainly a struggle for survival.

The Cities

It is understandable that the young state of Pakistan chose Karachi as its first capital. Clearly it was desirable to place the capital city in the Moslem stronghold of the country, the West, but the outstanding center of Islamic culture, Lahore, lay too exposed to the nearby, sensitive Indian boundary.

Lahore, with 1.5 million residents, grew rapidly as a result of the Punjab's partition. In 1950 its population was only about half as large. Founded in the first or second century A.D., Lahore became established as a great Moslem center during the Mogul period. As a place of royal residence the city was adorned with numerous magnificent buildings, including a great fort, several palaces, and mosques which remain to this day monuments of history, with their marvellous stone work and excellent tile and marble embellishments. The site of magnificent gardens and an old university, Lahore also was the focus of a large area in prepartition times, when its connections extended south to Karachi and the sea, north to Peshawar, and east to Delhi and Hindustan. Its Indian hinterland was cut off by the partition, but Lahore has retained its importance as the center of one of Pakistan's major population clusters, as a place of diverse industries (textiles, leather goods, gold and silver lacework), as the unchallenged historic headquarters of Islam in this area, and as an educational center.

Karachi grew even faster than Lahore. After independence it was favored not only as the first capital of Pakistan, but also as the West's only large seaport. Its overseas trade rose markedly; new industries were established; new regional and international interconnections developed between Karachi and East Pakistan and between Karachi and other parts of the world. A flood of Moslem refugees came to the city, presenting serious problems—there simply was not enough housing or food available to cope with the half million immigrants who had arrived by 1950. Nevertheless, Karachi has survived and in its growth has reflected the new Pakistan. As a result of the expansion of cultivated acreages in the Pakistani Punjab and in Sind, Karachi's trade volume has increased greatly, especially in wheat and cotton. Imports of oil and other critical commodities required construction of additional port facilities. Karachi became one of the great modern air terminals of the world.

But Karachi never was the cultural or the emotional focus of the nation. It lies symbolically isolated, along a desert coast, almost like an island; the core of West Pakistan still lies far inland. In 1959, after just over a decade as the federal capital of Pakistan, Karachi surrendered its political functions to an interior city, Rawalpindi. This was a temporary measure: an entirely new government headquarters is being completed a small distance from Rawalpindi, at Islamabad. This new town, as the map shows, lies near the boundary of Kashmir. It confirms not only the internal position of Pakistan's cultural and economic heartland, but also the state's determination to emphasize its presence in the contested north. It is evidence of a sense of security Pakistan did not yet possess when it chose Karachi over exposed Lahore. In this context, Islamabad is a prime example of the principle of the *forward* capital.

CEYLON

Ceylon, the compact, pear-shaped island located off the southern tip of the peninsula of India, is the third independent state to have emerged from the British sphere of influence in South Asia. Sovereign since 1948, Ceylon has had to cope with political as well as economic problems, some of them quite similar to those facing India and Pakistan, and others quite different (Map 11-12).

Good reasons exist for Ceylon's separate independence. This is neither a Hindu nor a Moslem country: the majority—some 70 percent—of its 13 million people are Buddhists. And unlike India or Pakistan, Ceylon is plantation country, a legacy of the European period and still the mainstay of the external economy. Plantations cover over 2.5 million acres of the country's farmlands; rice, the staple food crop, is grown on less than 1.5 million acres, though about one-third of this area is double-cropped. And historically, too, Ceylon differs greatly from the land across the Gulf of Mannar.

The majority of Ceylon's people are not Dravidian, but rather Aryans who link their history to ancient northern India. After the fifth century B.C. they began to migrate to Ceylon, a migration that took several centuries and which brought the advanced culture of the northwestern part of the subcontinent to this southern island. Part of that culture was the Buddhist religion; another part of it was a knowledge of irrigation techniques. Today the descendants of these early invaders, the Sin-

Map 11–12 CEYLON.

India and Ceylon produce tea, an important cash crop for both countries. Shown here is a tea plantation; women are harvesting the tips of the new green shoots. (Camera Press—PIX)

halese, speak a language that is related to the Indo-Aryan language family of northern India, and especially to Bengali.

With their knowledge of irrigation, the early Sinhalese transformed the northern, drier part of the island (see Plates 1 and 2) into a richly productive region, with intensive rice cultivation. They built a splendid capital at Anuradhapura, and Ceylon was a distant outlier of the culture sphere of the Aryans in the Indus-Ganges area. But by the second century A.D. new invaders came on the scene: the Dravidians from southern India. They did not come in large numbers, but nevertheless they constituted a threat, and there was intermittent warfare between the two peoples. In the thirteenth century, the Sinhalese society broke down, and the focus of activity shifted to the moister highlands of the south; ever since, this has been the center of productivity. The Dravidians, however, never came in sufficient numbers to challenge the Sinhalese. They introduced the Hindu way of life, brought the Tamil language to Ceylon, and eventually came to constitute a substantial minority (20 percent) of the country's population. Their number was strengthened substantially during the second half of the nineteenth century, when the British brought many thousands of Tamils from the mainland to Ceylon to work on the plantations that were being laid out. Ceylon has sought the

repatriation of this element in its population, and an agreement to that effect was even signed with India.

Ceylon is not a large island (25,332 square miles), but it has considerable topographic diversity. The upland core lies in the south, where elevations reach over 8000 feet and sizable areas exceed 5000 feet. Steep, densely overgrown slopes lead down to a surrounding lowland, most of which lies below 1000 feet; northern Ceylon is entirely low-lying. The rivers, which are the sources of the irrigation waters for the rice paddies, flow radially from the highland across this lowland rim. As Plate 1 shows, southwestern Ceylon is the wettest part of the country, for here the highlands combine with the summer monsoon to draw over 100 inches of rain from the clouds. The lower north and the lee-side east receive less rainfall, but the 40 to 80 inches recorded here come during the winter monsoon, between November and February, when the winds have turned and come from the northeast.

Since the decline of the Sinhalese empire centered on Anuradhapura, the moist southwestern uplands have been the major areas of productive capacity (the plantations are concentrated here) and the population core as well. Three plantation crops have been successful: coconuts in the hot lowlands, rubber up to about 2000 feet, and in the highlands above, tea, the product for which Ceylon is famous. In fact, tea was not the first plantation crop to be tried in those uplands. In the first half of the nineteenth century the British laid out extensive coffee plantations, with a view to the growing European market for this commodity, but in the late 1860's disease struck a disastrous blow to the industry. Then the British turned to tea, a fortuitous choice: the product continues in most years to account for two-thirds of Ceylon's exports by value.

Ceylon's plantation agriculture is very productive and quite efficient, but the same cannot be said for the island's rice lands. Yields in Ceylon were lower even than in the Indian subcontinent, and among the major reasons was carelessness on the part of farmers who would fail to maintain irrigation canals, fertilize inadequately or not at all, and who generally paid little attention to the crop. Thus, with its rapidly growing population, Ceylon saw its rice imports grow steadily. In recent years it has been necessary to import half the rice consumed in Ceylon, a situation that was detrimental to the general economic situation. Thus the government made it a priority to reconstruct the lowland irrigation systems, to repopulate the lowlands (until its recent eradication, malaria was an obstacle to lowland settlement), and to intensify rice cultivation. The results have been impressive: in 1967 there was a record rice harvest of 55 million bushels (as recently as 1965 it had been 36 million), and it was possible in 1968 to begin reducing rice imports substantially. The government anticipates that only 25 percent of the country's rice needs will have to be imported in 1969.

In a country so heavily agricultural, it is not surprising to find very little industrial development except for factories processing farm and plantation products. Ceylon appears to have very little in the way of mineral resources; graphite is the most valuable mineral export, though gemstones (sapphires, rubies) once figured importantly in overseas commerce. Such industries as have developed, other than those treating agricultural products, depend on Ceylon's relatively small local market. Predictably, these include cement, shoes, textile, paper, china and glassware, and the like. The majority of these establishments are located in the country's major port, largest city, and capital: Colombo (750,000); there are indications that plans to build a small steel mill will go ahead. But Ceylon is an agricultural country; it is in a position yet to close the gap between the demand for staple foods and the local supply capacity. Its chief concern must be to increase productivity of its soil.

THE NORTHERN TIER

Along the northern margins of the Indian subcontinent lies a string of countries, all of them landlocked, all mainly mountainous, and all with a history of stress related to their position between larger and more powerful political spheres of influence (Map 11-13). Afghanistan, we know, was a buffer between Russian and British interests in this part of Asia. Kashmir has become an arena of competition between India and Pakistan. Nepal survives as an independent state. Sikkim and Bhutan are in effect protectorates of India, though with a measure of autonomy. And Tibet has been absorbed by China.

Nepal, Sikkim, and Bhutan lie in a zone of cultural transition and intermittent conflict between India and Chinese Tibet. Nepal, by far the largest territorially (over 54,000 square miles) and in terms of population (14 million), is also the most complex. Into its fertile valleys came Indo-Aryan peoples from the south and Mongoloid peoples from the north, and some groups from Assam and Burma moved in as well. The immense Himalayan mountain barriers obstructed contact and intermixture among these invaders, and tribal and cultural identities have survived through the centuries. This is reflected by Nepal's linguistic diversity. In the southern, India-bordering lowland known as the Terai, and in the adjacent lower Himalayan valleys, Indo-Aryan languages (Hindi and Pahari) are used. In the central and northern, Himalayan part of the country, Tibeto-Burmese languages dominate, the most important being Newari, the language of the Newar people, an old trading group who today occupy the core area of interior Nepal, the Katmandu Valley. With about a dozen mutually unintelligible languages and

464

MODIFIED AFTER: RAND MCNALLY, GOODE'S WORLD ATLAS

Map 11-13 THE NORTHERN TIER.

numerous local dialects, Nepal needs a lingua franca. The official language is Nepali, a derivative of Hindu with a strong Tibeto-Burmese influences—but Nepali is still a minority tongue.

There is far more unity in terms of religion in Nepal: some 85 percent of the people are Hindus, and only about 8 percent are Buddhists. The mountain lands north of the North Indian Plain provided a refuge for Hindus seeking to escape the Moslem domination of the subcontinent; since the Gurkha conquest of 1769, Hinduism has been the religion of Nepal's ruling class. Again, the north has regional identity, for in the northern mountains Lama Buddhism, similar to that of Tibet, prevails.

Thus, Nepal is a physiographic as well as a cultural transition zone, and in the consequent absence of internal ethnic, linguistic, or even religious unity, national cohesion has been difficult to attain. Population concentrations are heavy in and around the capital, Katmandu (150,000) and in the Terai, but the majority of Nepal's inhabitants are scattered throughout the country along the lower valleys of its many rivers. Tenacious opposition to outside interference on the part of the powerful central authority maintained the political status quo, but the second half of the twentieth century has begun to produce change in both the economic and the political spheres. In 1956 the first road linking Katmandu to India was completed, and during the 1960's construction was in progress on a road connecting Katmandu to Lhasa in Chinese Tibet. These developments were related, of course, to the renewed competition of outside forces for the lands of the northern tier. India, China, and other states have made investments in Nepal and have provided various kinds of assistance, and these gestures were not motivated simply by charity. India's concern over events in Nepal has grown since the Chinese takeover of Tibet and especially since the Chinese incursions in Ladakh; China has a yet unsettled boundary disagreement in northern Nepal.

Similar concerns have motivated India in its involvement in the affairs of Nepal's eastern neighbors, the small states of Sikkim (with less than a quarter of a million people) and Bhutan (with a population of less than one million). These two countries straddle the same precarious transition zone as Nepal. But political awareness in both protectorates has brought demands for sovereignty and the ouster of India from the countries' affairs, notwithstanding the absence of any real national framework and despite increasing pressure from the north. The best road from Tibet to India runs through Sikkim, and after the Chinese takeover a stream of refugees have used this route; there have been fears in India that China might find it necessary to involve itself in Sikkim's affairs. In the case of Bhutan, which has long been more effectively tied to Tibet than to India, a boundary dispute looms along the northern border. Although Bhutan's population is dominantly Mongoloid and the Lama Buddhists recognize the formerly Lhasa-based Dalai Lama as their spiritual leader, political events in Tibet have changed Bhutan's formerly northward orientation.

The Indian Sphere

Thus modern India is faced with challenges from both within and without. Internally the great task is to increase the production of food and to slow down the rate of population growth; a second internal challenge lies in the creation of a political system that can withstand the forces of division and promote the stability any developing country so greatly needs. Externally the challenges are several. To the east and west lie the "wings" of Pakistan, areas torn from the economic framework of India; the impact of partition is still being overcome. To the south, the only sensitive issue with Ceylon relates to the repatriation of that country's late-arriving Tamils. But to the north, along the mountain wall, India faces tests of diplomacy and strategy. The Kashmir issue is still latent. Friction with China has occurred within an area that is clearly an Indian sphere of influence (Ladakh), and along a boundary in the northeast known as the McMahon Line in the Northeast Frontier Agency (NEFA). Furthermore, India faces pressure along Bhutan's northern boundary, having committed itself to the defense of the territories of Sikkim and Bhutan when British responsibilities in these areas were terminated in 1949.

Add these problems (on whose solution Indians are by no means of the same opinion) to the already existing internal obstacles to unity and stability, and it is indeed reasonable to speak of the state's survival as the miracle of modern India.

Still the standard geography is the large volume by O. H. K. Spate, *India and Pakistan: A General and Regional Geography,* a new edition of which was published in 1967 by Methuen, in London. Also useful is C. C. Davies, *An Historical Atlas of the Indian Peninsula,* Oxford University Press, published in New York in 1959. On the partition of the Punjab see A. A. Michel, *The Indus River: A Study of the Effects of Partition,* published by Yale University Press in 1967. Partition in Kashmir is discussed by A. Lamb, in *The Kashmir Problem: A Historical Survey,* a Praeger publication of 1967.

On India, the volume by J. H. Hutton, *Caste in India: Its Nature, Function and Origins,* published in a fourth edition by Oxford University Press, New York, in 1963, contains a wealth of detail and insights. In the Van Nostrand Searchlight Series, No. 24 is *India: the Search for Unity, Democracy, and Progress,* by W. C. Neale. Two compendiums of essays are R. R. Platt, et al., *India: A Compendium,* published by the American Geographical Society in New York in 1962, and P. Mason, *India and Ceylon: Unity and Diversity,* published in New York by Oxford University Press in 1967. The Indian population question is discussed in R. O. Whyte, *Land, Livestock and Human Nutrition in India,* a Praeger volume published in 1968. For a sympathetic and easily read volume see P. Griffiths, *Modern India,* published in a third edition by Praeger in 1962.

On Pakistan, in addition to the Spate volume mentioned above, see K. S. Ahmad, *A Geography of Pakistan,* published in New York by Oxford University Press in 1965. The volume by R. D. Campbell, *Pakistan: Emerging Democracy,* is No. 14 in the Van Nostrand Searchlight Series. East Pakistan is covered in detail by N. Ahmad in *An Economic Geography of East Pakistan,* a second edition of which was published by Oxford University Press, New York, in 1965. The whole of the country is dealt with by J. R. Andrus and A. F. Mohammed in *The Economy of Pakistan,* published by Stanford University Press in 1958.

Among the countries of the perimeter, Ceylon is the subject of a volume by S. F. De Silva, *A Regional Geography of Ceylon,* published in 1954 by Apothecaries in Colombo. This book, like that by E. K. Cook entitled *Ceylon: Its Geography, Its Resources and Its People,* published in 1951 in New York by St. Martin's, is rather outdated. On Nepal, P. P. Karan has written *Nepal, a Cultural and Physical Geography,* published by the University of Kentucky Press in 1960. Professor Karan has written a similarly titled volume on Bhutan, published in Lexington in 1967. In collaboration with W. M. Jenkins, Jr., Karan also has produced a volume on *The Himalayan Kingdoms: Bhutan, Sikkim, and Nepal,* No. 13 in the Van Nostrand Searchlight Series.

12

CHINA AND ITS SPHERE

FOR MANY years, until black people began to object to the expression, Europeans and Americans would refer to Africa as the "dark continent." The darkness, Africans argued, had more to do with the minds of ignorant people than with their homeland itself. Anyone who really wanted to lift that veil of ignorance could do so by reading what Westerners and Africans themselves have written about the realm of Ghana and Zimbabwe; those who could afford it might go to find out for themselves what Africa and Africans are like. And thousands of Americans have done just that. Individually, as members of tour groups, as workers for the Peace Corps, as representatives of the government, and in a host of other capacities, they have gone to Africa and come back—always better informed than when they went. The people in America who have recently been to Africa would fill several good-size football stadiums, and through what they have written, lectured, and shown on slides, films, and television programs they have in one way or another contributed to better inform just about everyone in this country about the African continent.

But if all the people in the United States who have recently been to China were to gather, they would not even fill a room. If any major part of the world is "dark" in our minds today, it must be China—and rather than inform ourselves better we seem to have isolated ourselves more and more from a country where 900 million people live. There is in the American image a Darkest China, from where news comes through "the listening post in Hong Kong" or from "reports of the Japanese News Agency." Yet these reports, inaccurately third-hand as they must often be, relate to one-quarter of the entire world's population—or more people than there are in all of North America, South America, and Africa combined. Obviously there are dangers

Map 12-1 CHINA: PHYSIOGRAPHY.

in such a situation. Africa's "darkness" involved misconception and annoyance, but any misjudgments arising from such ignorance affected a still-divided, not particularly populous, and not especially rich or powerful part of the world. Those are not valid excuses, but Africa today is not yet a world power in the making—which China is. And with its revolutionary fervor China has become Washington's prime ideological foe, a situation that calls for an understanding of (if not appreciation for) Chinese feelings and viewpoints, achievements and goals.

This is not to suggest that the Chinese themselves have not contributed to the isolation of their country from foreign (and especially American) contacts and influences. Certainly American governments have done their share, but China's current position is by no means without precedent in its long history.[1] Indeed, we might argue that a certain degree of isolation is an integral part of the Chinese tradition, which has repelled and rejected foreign influences time and again. And as with the Indian subcontinent, China has its own natural barriers to contact with the outside world. To the north lie the great mountain ranges of eastern Siberia and the barren country of Mongolia. To the northwest, beyond Sinkiang, the mountains open—but into the vast Kirgiz Steppe. To the west and southwest lie the Tien Shan, Pamirs, the forbidding Plateau of Tibet, and the Himalayan wall. And to the south there are the mountain slopes and tropical forests of Southeast Asia—the same that mark India's eastern flank with Burma (Map 12-1). But more telling even than these physical barriers to contact and interaction is the factor of distance. China has always been far from the source areas of innovation and technological change. True, it was itself such an area, but its external contributions were comparatively limited as were the outside influences that reached it. China interacted directly with Japan and with Southeast Asia. The Arabs, by comparison, ranged far and wide through the Old World, from Southern Europe to India, Indonesia, and Eastern Africa. Later, when Europe became the center of change, China was farther away by sea than almost any other part of the world, farther even than the distant, sought after Indies.

[1] For a brief description of the adventures of one American writer who sought to enter China in the late 1950's see Edgar Snow's introduction to *The Other Side of the River*, Random House, New York, 1961, pp. 3–12.

Still today, modern communications notwithstanding, China is a distant land. Overland from the heartland of its Eurasian neighbor, the Soviet Union, is a long and tedious journey of several days by rail. Direct overland communications with India are virtually nonexistent. Although a railroad was under construction to Burma in the late 1960's, the only long-existing rail connection to any part of Southeast Asia is the line to Hanoi. And China's long Pacific coastline looks into the world's largest body of water; Japan is China's only nearby, industrialized trading partner.

China in the World Today

And yet, for all its isolation and remoteness, China has in recent years moved to the center stage of world attention. It was Napoleon who long ago remarked that China was asleep, and whoever would awaken the Chinese giant would be sorry. Today China is awake. It was stung by Japanese aggression in the 1930's and 1940's, and after the end of World War II the growing Communist tide took power following a bitter civil war in which the United States supported the losing side. European colonialism, Japanese imperialism, and a Communist ideology combined to stir China into action. The foreigners were ousted, China's old order was rejected and destroyed, and since 1949 the Communist regime has been engaged in a massive effort to remake China in a new image of unity and power.

This effort has taken China from a position of backwardness and weakness to one of considerable strength. China's rise has been dramatic— so dramatic, in fact, that outside observers have probably exaggerated its dimensions at times. China is not yet a third world power, equal to the United States or the Soviet Union. But China is on the move, and by almost any combination of measures it now ranks in third place. China has more than one-quarter of all the people of the world, and they are organized today as never before, to work for the progress of the state. In terms of territory, China is the world's third largest country, and for all its millions of people there still are areas that are virtually empty and mountainous zones that are only partially explored. With initial Soviet aid the Chinese made rapid strides in their program of industrialization and military preparedness. Intensive exploration produced crucial discoveries of minerals and fuels, and in the early 1960's China proved that

MODIFIED AFTER: SCOTT, FORESMAN &CO., ALDINE UNIVERSITY ATLAS

Map 12-2 CHINA: LOCATION MAP.

it could create its own nuclear weapons. Earlier, in the 1950's, China had also proved that it was prepared to stand with its Asian allies against the Americans and their friends. In Korea, which was partitioned into a Communist North and a non-Communist South (see Chapter 14), armed hostilities erupted between the two sides. In this conflict, United Nations forces—mainly United States men and material—assisted the South Koreans, while China joined the North.

The Korean War first raised the spectre of an aggressive, militant China, bent on control over most of Asia. There were dark predictions of a Chinese campaign of expansion, and fears arose that a Sino-Soviet axis would come to threaten all of Eurasia. American policy was guided by a philosophy of "containment," and from Japan to Thailand to Europe the United States created a ring of bases to counter any expansionism the Chinese and Soviets were thought to have in mind. The Korean War ended in a stalemate roughly along the line of partition where the conflict had first begun, but soon the United States was involved again in military operations—this time in Southeast Asia, where Viet Nam, also divided into a Communist North and a non-Communist South, were locked in combat. The Domino Theory became current in American thinking; according to this idea the countries of the Asian perimeter would fall, one by one, to the Communist drive for power which, although carried forward by local movements such as the Viet Cong, was thought ultimately to be of Chinese making. And certainly China was making itself felt. In Viet Nam and Laos as in North Korea, the Communist forces were strengthened by Chinese assistance. Burma was forced to settle a boundary dispute in China's favor, and in India Chinese forces advanced into northern borderlands.

But in all fairness, the great aggression of a hungry China, so vividly forecast by Sinophobes and political alarmists, had still failed to materialize after 20 years of Communism in Peking. The Communist push in Viet Nam, Laos, and Thailand still appeared to be sustained overwhelmingly by indigenous forces (although there can be no doubt of Chinese sympathy and support), but the crises in northeast Burma and northern India turned out after all to be border skirmishes and not, apparently, preludes to large-scale aggressive warfare. And Chinese-oriented Communist parties in Asia did not

To quote the official Chinese caption that accompanied this photograph: "Chairman Mao Tse-tung, the great teacher, the great leader, the great supreme commander and the great helmsman of the Chinese people, and his close comrade-in-arms Lin Piao and Comrade Ho Lung in the company of Comrades Hsieh Fu-chih and Yang Cheng-wu meet the 500,000 revolutionary teachers and students in the Tienanmen Square (Peking)." Note the hands waving the Red Book of Quotations. (Eastfoto)

always fare well either: the Indonesian domino, in any case, is a good deal steadier today than it was in the 1950's and early 1960's. In fact, the whole Sino-Soviet axis has weakened rather than consolidated in recent years. After decades of tutelage by the Russians, the Chinese Communist movement went its own way, and quarrels broke out over ideology between Moscow and Peking. Soviet technicians, who had helped give China its nuclear capability and who had greatly assisted in the country's massive industrialization program, were recalled. The Chinese accused the Soviets of diluting their communism with liberal doses of capitalism, and in 1968 they even charged the Russians with being in league with the United States with the aim of destroying the Chinese nation itself. And now China got a dose of its own medicine as armed conflicts erupted along the boundary with the Soviet Union. So it was that China, awakened to Communism by the inspiration of Russia, took on the mantle of the "purest" of Communist systems and rejected the Soviet version. When anti-Soviet feeling in China reached its peak in the late 1960's, it was not difficult to hear the echo of Napoleon's famous words.

China, after all, is an ocean away from California's beaches, but only a stone's throw from Soviet soil (Map 12-2). It was Russian imperialism that cost China dearly, not American imperialism. It is Soviet territory that China claims on historic grounds as its own, not American soil. And as China's capacity for war grows, it is Russia that will lie in its reach first: even in these modern times of intercontinental missiles an ocean provides a time cushion and therefore constitutes an advantage to the technologically superior foe. That was proved by the Cuban missile crisis during the Kennedy administration: rockets poised to strike the United States from Russia was one thing, but rockets aimed at America from Cuba was something else. China is testing rockets and bombs, but it still does not have the capacity to strike effectively—it does not as yet possess a sufficient arsenal. When the time comes, will the U.S.S.R. accept missiles poised to strike it from Sinkiang any more than the United States did in the case of Cuba?

What is most remarkable about all this, of course, is that after barely two decades of Communist control, China has been able to place itself in a position of serious potential contention for world power, something that hardly seemed possible less than a generation ago. Chinese assistance is going to countries in Asia and America; Chinese technicans are building railroads in Africa. In Albania, China even has a European ally. And at a time when the two greatest powers in the world are attempting to achieve a world in which they can coexist, China looms as a threat to this joint monopoly of ultimate power. The rise of China is indeed the story of the mid-twentieth century.

The Original China

China may have developed in comparative isolation for more than 3000 years, but there is evidence that China's earliest core area, which was positioned about the confluence of the Hwang-Ho (Yellow) and the Wei Rivers, received stimuli from other, distant, and possibly slightly earlier civilizations: the river cultures of Southwest and South Asia. From Mesopotamia and the Indus, techniques of irrigation and metal working, innovations in agriculture, and possibly even the practice of writing reached the Yellow River basin, probably overland along the almost endless route across desert and steppe. The way these early Chinese grew their rice crops gives evidence that they learned from the Mesopotamians; the water buffalo probably came from the Indian subcontinent.

MODIFIED AFTER: A. HERRMANN, AN HISTORICAL ATLAS OF CHINA, ALDINE PUBLISHING COMPANY

Map 12-3 THE SHANG (YIN) DYNASTY.

In fact, archaeologists have dug up remains of very early human occupants in China whom they report to have lived there a half million years ago; other remains indicate the presence of paleolithic man, about 50,000 years back. But China, like all the landmasses of the Northern Hemisphere, was afflicted by the glacial advances of the Pleistocene Ice Age, and there are huge gaps in the human record until after the last advance (the Wisconsin in North America, known as the Wurm in Eurasia). Thus the continuous record in China began about 3000 B.C., and from the earliest times there was evidence of contact with Central and Southwestern Asia— especially in the shape and decoration of the pottery these people made. But quite soon the distinctive Chinese element began to appear, and by the time the record becomes reliable and continuous, shortly after 2000 B.C., Chinese cultural individuality was already strongly established.

The oldest dynasty of which much is known is the Shang Dynasty (sometimes called Yin), which was centered in the Hwang-Wei confluence from perhaps 1900 B.C., to about 1050 B.C. Walled cities were built during the Shang period, and the Bronze Age commenced during Shang rule. The Shang capital was at Anyang, about 70 miles from the Hwang Ho to the north (Map 12-3). As Plate 3 shows, there is a large, crescent-shaped area of alluvial soil in this part of China, and adjacent areas have fertile loess—a combination which, if adequately supplied with water, can support a large and quite dense farming population. Thus, for more than a thousand years after the beginning of the Shang dynasty, northern China was the center of development in this part of Asia, and as agricultural methods improved, better tools became available, and more land was brought into production, the population increased.

Eventually China's southward march began. Agricultural techniques and population numbers combined to press settlement in the obvious direction—southward, where the best opportunities for further expansion lay. During the Ts'in (Ch'in) Dynasty the lands of the Yangtze Kiang were opened up, and settlement spread even as far southward as the lower Hsi Kiang (West River), in what is today the area of Canton (Map 12-4). A pivotal period in the historical geography of China was the tenure of the Han Dynasty (202 B.C. to 220 A.D.). The Han rulers brought unity and stability to China, and they enlarged the Chinese sphere of influence to include Korea, Manchuria, Mongolia, Hsinkiang (Sinkiang), and, in Southeast Asia, Annam (located in what is today Viet Nam) as well. This was done to establish control over the bases of China's constant harassers, the nomads of the surrounding steppes, deserts, and mountains, and to protect (in Sinkiang) the main overland avenue of westward contact between China and the rest of Eurasia. The Han period was a formative one in the evolution of China. Not only was Chinese military power stronger than ever before, but there were changes in the systems of land ownership as the old feudal order broke down and private, individual property was recognized, and the silk trade grew into China's first external commerce. To this day, most Chinese, recognizing that much of what is China first came about during this period, still call themselves the "people of Han."

This description of the Han period, or maps showing China in those days, should not suggest that Han China was a fully integrated, effectively unified political state.[2] Literally hundreds of tribes still occupied the southern hills and mountains, and the control established by the Han rulers was maintained through military garrisons and outposts. Even during the Han Dynasty, much of the south still remained outside the effective national sphere, not to be fully involved in the Chinese state until the T'ang Dynasty, 618–906 A.D. And neither were the centuries following the Han and T'ang Dynasties free from retrogression, disunity, and disorder. After the end of Han rule, in 220 A.D., China reverted to a state of great disarray, with more than a dozen competing states replacing the old authority; this weakness led to the invasion and absorption of some nomadic elements from the north and west. This situation lasted until the Sui Dynasty, the T'ang Dynasty's immediate predecessor, but following the T'ang period there was division once again, and this time it led to the first external conquest, namely that of the Mongols under Genghis Khan. The Mongol authority, which made China a part of a vast empire that extended all the way across Asia to Eastern Europe, lasted less than a century

[2] An excellent series of maps showing the successive Chinese dynasties is incorporated in Albert Herrmann's *An Historical Atlas of China*, Norton Ginsburg (Ed.), Aldine Publishing Company, Chicago, 1966.

Map 12-4 THE CH'IN DYNASTY.

(1280–1368), and had little effect on Chinese culture; in fact, the converse was the case as the Mongols adopted Chinese civilization. The Mongols were succeeded by another great indigenous dynasty, the Ming rulers (1368–1644), and China was once again consolidated to the Great Wall in the north and Annam in the south. Finally, a Manchurian nomadic element, the Manchus, gained control and established a dynasty that lasted into the present century (1644–1911). Unlike the Mongols, the Manchus made use of and built upon the traditional system of Chinese governmental administration and authority. Their empire was never as large as that of the Mongols, but the Chinese sphere of influence nevertheless grew to incorporate not

MODIFIED AFTER: T.R. TREGEAR, A GEOGRAPHY OF CHINA

Map 12-5 GREATEST EXTENT OF THE MANCHU EMPIRE.

only present-day Sinkiang and Mongolia, but also a large part of southeastern Siberia, eastern Kasakhstan, Tibet, Burma, Viet Nam, and adjacent interior territories, and Korea (Map 12-5).[3]

A Century of Convulsion

Of course, the Chinese until very recently would have had a ready answer to any mention of their country's relative isolation: they would have said that it is not China that is far away from the culture cores of the rest of the world, but the rest of the world that is far from China. And they had good reason for saying so. Unlike many other regions of the world which were quickly overcome by the new economic order the colonizing Europeans brought, China was productive and self-sufficient and long withstood the Europeans with a self-assured superiority. There was no market for the British East India Company's rough textiles in a country long used to

[3] For a discussion of the historical geography of China, see T. R. Tregear, *A Geography of China,* Aldine Publishing Company, Chicago, 1965, pp. 45–100.

finely made silks and cottons. There was little interest in the toys and trinkets the Europeans produced in the hope of bartering for Chinese tea and pottery. And even when Europe's sailing ships made way for steam-driven vessels, and newer and better factory-made textiles were offered in trade for China's tea and silk, China continued to reject the European imports which still were, initially at least, too expensive to compete with China's handmade materials and still of poorer quality. Long after India had fallen into the grip of mercantilism and economic imperialism China was able to maintain its established order. This was no surprise to the Chinese; after all, they had held a position of undisputed superiority among the countries of Eastern Asia as long as could be remembered and they had dealt with invaders from land and from the sea.

But the nineteenth century shattered the convictions of China as it proved the superiority of the new Europe—even here. On two fronts, the economic and the political, the European powers destroyed China's invincibility. In the economic sphere, they succeeded in lowering the cost and improving the quality of their manufactured goods, especially textiles, and the handcraft industry of China began to collapse in the face of unbeatable competition. In the political sphere, the demands of the British merchants and the growing British presence in China led to conflicts. In the early part of the nineteenth century, the central issue was the importation into China of opium, a dangerous and addictive intoxicant. The Manchu government had long prohibited the importing of opium, but the foreign merchants brought it in anyway, and they bought off Chinese officials to get their goods into the country. Peking grew worried about this situation, and in 1839 the government sent a special commissioner to the coast, who was empowered to stop the trade. All this he did, and he even got some of the foreigners to sign agreements not to import opium any longer. But other merchants objected to this move by Peking to stamp out the trade, which was a very lucrative one. Thus armed hostilities broke out, and soon the Chinese sustained their first defeats. Between 1839 and 1842 the Chinese fared very badly, and the first *"opium war"* signaled the end of Chinese sovereignty. British forces penetrated up the Yangtze and controlled several areas south of that river; Peking hurriedly sought a peace treaty. As a result, leases and concessions were granted to foreign merchants; the Emperor's Commercial Guild ceased its interference in external trade. Hong Kong Island was ceded to the British, and five ports (Canton and Shanghai among them) were opened to foreign commerce. No longer did the British have to accept a status inferior to the Chinese; henceforth, all negotiations were to be pursued on equal terms.

The treaties of 1842 did not contain any stipulation concerning the opium trade, but despite continued Chinese opposition the foreign merchants were now emboldened to expand it. Opium flooded into China, and the effect of it on Chinese well-being and morale was devastating. Opium is not only an intoxicant: it can do severe damage to the body, and it is very easy to become "hooked." The opposition to the trade on the part of the Chinese government was not based on an unwillingness to grant the European merchants the profits from this thriving business, but on the fear of what the flow of opium might do to the Chinese nation. And so, despite their setbacks, the Chinese continued to struggle against the influx of it. Fifteen years

after the first "opium war," another large-scale conflict erupted between China and its foreign parasites. France was in on this struggle on the side of the British, and this time the cause of opium won out. Not only was the importing of opium from India and elsewhere permitted, but the cultivation of the opium poppy in China itself was legalized. Not until China began to reassert itself in the early 1900's was the scourge of opium defeated again. But by then it had done uncalculable damage.

As the nineteenth century wore on, what remained of China's sovereignty was steadily eroded away. The Germans obtained a lease on Tsingtao in 1898, and in the same year the French acquired a sphere of influence in Kwanchowan (Map 12-2). The Portuguese took Macao, the Russians obtained a lease on Liaotung in Manchuria and railway concessions as well, and even Japan got in the act by annexing the Ryukyu Islands and, more importantly, Formosa (Taiwan) in 1895. After millenia of cultural integrity, economic security, and political continuity, the Chinese world at last lay open to the aggressions of foreigners whose innovative capacities China had denied to the end. But now, ships flying European flags lay in the ports of China's coasts and rivers; the smoke stacks of foreign factories rose above the urban scene of its great cities. Japan was in Korea, which had nominally been a Chinese vassal; the Russians were in Manchuria. The foreign invaders even took to fighting among themselves, as did Japan and Russia in Manchuria in 1904.

Extraterritoriality. A sign of China's weakening during the second half of the nineteenth century was the application on its territory of a European doctrine of international law, extraterritoriality. This principle originated with the French jurist Ayraut (1536–1601), and it denotes a situation in which foreign states or international organizations and their representatives are immune from the jurisdiction of the country in which they are present. This, of course, constitutes an erosion of the sovereignty of the state hosting these foreign elements, but it was nevertheless a situation China was forced to accept. And in few countries anywhere in the world did extraterritoriality go to the extremes it did in China. The best residential areas in China's large cities, for example, were deemed not to be part of China any longer but actually an "extraterritorial" part of some foreign power; hence they were made inaccessible to Chinese. Soon the Chinese found themselves unable to enter their own public parks or buildings without foreign permission to do so. Christian missionaries, fortified with extraterritorial security, commenced the spread of their faith. They had a lot of work to do, for one of the great tragedies in China was the diffusion of opium addiction—achieved through the efforts of their Western (presumably non-Christian) countrymen.

All these events naturally had an enormous impact on the structure of life in China. The weakening of the authority of the Manchu rulers encouraged local viceroys and warlords to rebel. A segment of Chinese society was urging reform along Western lines, arguing that Japan's strength and China's weakness lay in Japan's willingness to Westernize and China's refusal to modernize even the most outdated of its institutions. Bitter opposition grew to the presence of the foreigners in China, and in 1900 a large-scale revolt broke out aimed at all foreign elements. Bands of revolutionaries roamed the cities and the countryside, attacking not only foreigners themselves but also Chinese citizens who had adopted some aspect of Western culture, for example a Christian faith. Known as the Boxer Rebellion (after a loose translation of the Chinese name for these revolutionary groups), the 1900 uprising was put down with a great deal of bloodshed. Meanwhile another revolutionary fervor was gaining strength, aimed this time against the Manchu leadership itself. In 1911 the Emperor's garrisons were massacred all over China, and the outcome was no longer in doubt; the official termination of the centuries-old dynasty, indirectly still another casualty of the foreign intrusion, came in the following year.

One of the many lessons of Chinese history is that the fall of a dynasty is followed by decades of civil strife and disorder—sometimes centuries, as in the aftermath of the fall of the Han dynasty (202 B.C.–221 A.D.). So it was after the fall of the Manchus, despite the unifying enmity toward foreigners which prevailed in China. Of course there were feudal warlords and local chieftains who sought to establish greater autonomy during the period of countrywide instability, but the main problems were deeper than that. Despite the fall of the Manchus and the creation of a republican form of government there was intense division among those who had influence in the affairs of China on many issues.

For nearly 2000 years China had experienced the kind of central authority the Manchus had represented, but now a new form of government had to be organized—almost overnight—to administer a country of hundreds of millions of people. There were differing views on the structure such a government should have. There were also conflicts over the question of how to deal with the continuing foreign presence: the revolution had neither wiped out the outstanding loans nor had it diminished the foreign penetration of China's political and economic affairs. On top of this came World War I, which had unfavorable consequences for China. Germany held some territory on the Shantung Peninsula, including the city of Tsingtao; the Japanese, under the pretext of an Anglo-Japanese alliance, captured the city and proceeded to put pressure on the Chinese state for recognition of Japan's sovereignty there. In this effort the Japanese had the support of Western powers, and at Versailles in 1919 Japan's rights in Shantung were confirmed. This brought about a most significant event in the modern history of China, for when news of the Versailles decision reached the country there was widespread indignation and, on May 4, 1919, a student-led demonstration in Peking against the government's acquiescence with the Versailles treaty. This demonstration spread to other cities and developed into a large-scale boycott of Japanese goods for sale in China, and the Peking government instructed its Versailles delegation not to sign the Shantung properties away to the Japanese. This May 4 Movement, as it is now known, was among the first displays of effective reform in the life of the country; university students and their Western-trained professors were able to sway the country and hence the government. One participant was a young man named Mao Tse-tung. But despite this significant event, the fact is that China emerged from the 1914–1918 War in a state of considerable disarray. By the early 1920's there were two governments—one in Peking and another in the south, in Canton, where the famous Chinese revolutionary Dr. Sun Yat-sen was the central figure. Neither government could pretend to control much of China. Manchuria was in complete chaos, petty states were emerging all over the central part of the country, and the Canton "parliament" controlled only a part of Kwangtung. Yet it was just at this time that the power groups which were ultimately to struggle for supremacy in China were formed. While Sun Yat-sen was trying to form a viable government in Canton, the Chinese Communist party was formed by a group of intellectuals. Some of these intellectuals had been leaders in the May 4 Movement, and in the early 1920's they received assistance from the Communist party of the Soviet Union. Mao Tse-tung was a prominent figure in these events. Initially there was cooperation between this Communist party and the Nationalists of Dr. Sun Yat-sen. The Nationalists were stronger, and they hoped to use the Communists in their anti-foreign (especially anti-British) campaigns. By 1927 the foreigners were on the run: the Nationalist forces entered cities and looted and robbed at will while aliens were evacuated or, failing that, sometimes killed. As the Nationalists continued their drive northward and success was clearly in the offing, the luxury of internal dissension could be afforded. Soon the Nationalists were as busy purging the Communists as they were pursuing foreigners, and the central figure to emerge in this period was that of Chiang Kai-shek. When in 1928 the Nationalists established their capital at Nanking, Chiang was the country's leader.

The Three-Way Struggle: Nationalists, Communists, and Japanese. The first years of the Nanking government's rule brought further armed campaigns against the Communists, who retreated into the difficult mountains of the west and there established permanent bases. This was the so-called *Long March,* a germinal event in the Chinese Communist movement. The Nationalist government was apprehensive about the role of the Soviet Union in this struggle, and relations with the Russians were not good at first—especially since the Russians also continued to hold a concession over the Manchurian Railroad and thus represented a lasting foreign presence. Moreover, Moscow had gained a sphere of influence in Mongolia and seemed to be on the verge of doing the same in Sinkiang. But ultimately it was not the Soviet Union, but Japan which proved to be the most immediate threat to Chinese sovereignty. Through its ownership of ports and railroads in Manchuria, Japan was virtually in control of that region, despite its Chinese-majority population. The Nanking (Nan-ching) government sought to outflank the Japanese by building its own railroads and developing its own ports. In 1931 the first armed hostilities took place as the Japanese decided to protect their

Manchurian sphere of influence. Their objective was to separate Manchuria from China, to prop up a puppet government there, and to treat the territory as a sort of protectorate. In 1932 they achieved this aim by setting up a government under the last of the Manchus and calling the "state" Manchukuo. But even this success was not enough. Slowly they eroded Nanking's authority; in 1934 the Japanese issued a declaration that stipulated that all China was, in effect, a Japanese sphere of influence and that no other foreign power could operate there without the consent of Tokyo. Meanwhile they increased their pressure on Chiang Kai-shek's government by ousting officials from several parts of the country and taking them over (for example, the provinces of Chahar and Ho-pei were thus annexed); but Nanking simply did not have the power to resist. One telling side effect of the Japanese encroachment was the blunting of the anti-Communist effort by Nanking's forces; in the first place there were fears of committing too many forces to this cause while Japan posed such an immediate threat, and in the second, there was a loss of enthusiasm among Chinese for civil strife at a time when a foreign invader stood at the door.

The inevitable war broke out in 1937. There were calls for a suspension of the Nationalist–Communist struggle in the face of the common enemy but after a brief armistice they still continued to fight each other while both were engaged in combating the Japanese. In fact, China was divided into three parts: the areas taken by the Japanese during their quick offensive of 1937–1938, during which they took the ports and much of the east, the Nationalist region centered on Chiang's new capital at Chungking, and the Communist territories in the west. The Chinese offered far stronger resistance than the Japanese had anticipated, despite an effective blockade of the country's ports; even in areas held by the Japanese, security often did not extend much beyond the limits of the larger towns not far away from the railroads and main roads. Even during the war, the Communists were gaining strength and prestige while the Nationalist government was plagued by the problems of wartime administration—and blamed for its failures. The war with Japan ended, we know, not through the successes of Chinese armies but because the Japanese were defeated in the Pacific by the Western powers, principally the United

This photograph, taken in the late 1940's, shows communist forces as they drive forward against their nationalist adversaries. China's new era was about to begin. (Sovfoto)

States. With Japan out of the picture, the Nationalist–Communist struggle was quickly resumed. The United States, hopeful for a stable and friendly government in China, sought to mediate in the conflict; but did so while recognizing the Nationalist faction as the legitimate government. The chances of mediation were impaired by this position, and also by the military aid given to the Nationalists' forces. By 1948 it was clear that Mao Tse-tung's armies would defeat the forces of Chiang Kai-shek, and that the final victory was only a matter of time. Chiang kept moving his capital—back to Canton, scene of the first government, and then to Chungking, where the Nationalists had held out during World War II. But in December, 1949, the remnants of Chiang's forces fled to the island of Taiwan, and the mainland came under its first Communist administration.

Thus little more than a century after the British had taken Hong Kong, and just a half century after Manchu China had been forced to give many of its concessions to the foreign powers, the Chinese were once again masters in their own house. Manchuria was secure, so was Sinkiang, and only Hong Kong itself and a few islands off the southeast coast remained in foreign hands, or in the hands of the Taiwan-based Nationalists. The Communists themselves had not anticipated so rapid and so complete a success.

And in a way, the stage was now set for a return of the old times in China. The foreign element quickly dwindled to even less than had remained through the violence of the civil war. Even the Soviet allies eventually proved unwelcome and hundreds of Russian technicians who had come to assist in China's reconstruction were called or sent home. It was recently estimated that there are perhaps a mere 500 Westerners among China's 900 million people today. But in other ways, the new China was unrecognizable from the old. Mao's government had heeded the lesson of the nineteenth century, that power lay in productivity, and that the most power-productive industries are large scale, heavy industries. From the failing Nationalists the Communists had learned the dangers of ineffective national control. And thus the Peking government of the 1950's set about reorganizing China as it had never been organized before. To catch up with the major powers of the world, China was set to take the Great Leap Forward.

The Northern Focus

The extensive basin of the Hwang Ho and its tributaries is sometimes identified as a region called the North China Plain. For a long period of Chinese history this was the country's core area, politically as well as economically the focus of the developing state. Today, the North China

Irrigation canals and wheat fields near the north bank of the Yellow River. Wheat acreages have reportedly been greatly enlarged, in this area by the reduction of the soil's alkali content, which had kept production dismally low prior to the mid-1950s. (Yang Ping-wan/Eastfoto)

Plain is no longer China's leading economic area, but as the historic heartland of the nation it continues to function as the political center of gravity. Here still lies China's capital, as has been the case with a few interruptions for nearly forty centuries. Since before the Ming Dynasty the headquarters of China's government have been in Peking, and although the Century of Convulsion witnessed the temporary removal of the government's functions from this historic city, the Communist regime quickly reinstated it. Thus the new rulers brought their headquarters back to where the old rulers also had theirs: to Peking, whose name means "northern capital."

Through much of China's history Peking functioned as a forward capital, confirming and strengthening the nation's presence in an area of potential threat. Until China's enemies came across the oceans, the main direction from which adversaries could be expected was the north, beyond the limits of agriculture. There, nomadic peoples moved across the steppe and desert, always capable of striking at the vulnerable farmers in the valley of the Yellow River. Before the emergence of Peking as the national capital, other rulers chose other sites—but almost always in the same general area, close to the northern-

most line of defense. When the Communists took control in 1949, the official capital was the Nationalist choice, Nanking (the nationalist regime had also renamed Peking Peiping, or "northern peace"), but both the original name and function were quickly restored to the city of Peking. And as it turned out, Peking still symbolizes China's concern with its northern margins. Since the Communists took control, at least three areas in the north have been the scenes of competition or conflict: Mongolia, a historic sphere of Chinese influence now in the orbit of the Soviet Union; Korea, where China became involved in an armed conflict with U.S.-dominated United Nations forces; and Manchuria, where in 1969 clashes took place along the river boundary with the Soviet Union.

The North China Plain was the nation's earliest area of sedentary agriculture, irrigation, and urbanization, but it did not long remain the only productive zone. Especially when the use of iron tools, oxen for plowing, and animal fertilizers for higher yields came into general use, the lands to the south, along the Yangtze Kiang, beckoned the northerners. Of course there were other reasons for the attractiveness of the south: here one was farther away from the threat of the nomadic Huns of the steppe. Before long the Yangtze areas began to outproduce the north, and rice was shipped to the northern towns, first by land and river, and later by a more secure canal route built during the Sui Dynasty, the New Pien or Grand Canal. As Professor H. J. Wiens has put it, from the earliest times "the south has been regarded as a granary for the north, and in this manner, the north early became a parasite upon the south."[4] But the Grand Canal and the other improved routes that linked the North China Plain to the southern areas also had important political ramifications, for they helped maintain administrative authority as the People of Han dispersed ever farther over their huge realm.

While northerners migrated southward, those who remained had other concerns, chiefly how to withstand the constant pressure from the northern steppelands, and how to secure the farmlands of the Yellow River against incursions by the destructive nomads, who would turn farmland into pasture and ruin the irrigation works that had been built. For centuries, the 16-inch isohyet was the approximate boundary

[4] Herold J. Wiens, *China's March Toward the Tropics*, Shoe String Press, Hamden, Conn., 1954, p. 170.

A small section of the Great Wall of China. (Rene Burri/Magnum)

between farmer and nomad—it was never a sharp line, of course, but a transition zone tended to lie astride this line. On Plate 1, the 16-inch isohyet shows as a dashed line in a green zone that happens to represent the transition zone rather well; note how quickly the precipitation drops off northward, and how it increases to the south. During the first millenium B.C., various feudal rulers in northern China built stone walls to protect their domains, most of them in this transition zone, to separate the settled population of the moister areas from the nomadic, unstable way of life in the drier North. It was during the first real empire in China, that of Ts'in (Ch'in), when a ruler decided to connect these walls into one Great Wall, and extend it all the way to the coast on the Liaotung Peninsula and westward around the great bend of the Yellow River (Map 12-4). For much of its length, as a comparison of Map 12-4 and Plate 1 will show, the Great Wall lies in the transition zone about the 16-inch isohyet, though to the west it incorporates a great deal of drier land. These drier areas, though inside the Great Wall, often gave trouble to China's central authority, for

they were difficult to control despite their inclusion within the Wall. Still, there were good reasons for building the Great Wall where it was built: there was a profitable silk trade with Rome across inner Asia, and the Chinese wanted to protect the route from plunder: Rome paid in gold. Building the Wall to protect the route meant the incorporation of some steppelands.

For centuries the Great Wall—which was extended still farther westward by the emperors of Han—was, in effect, China's northern boundary, though today the Chinese state extends far beyond it into Inner Mongolia and Manchuria. There was a time when China's feudal rulers built walls not only to keep the "barbarians" out, but also to prevent the loss of their subjects to the nomadic way of life—that is, people on the drier margins who gave up their difficult agricultural existence to turn to pastoralism on the steppes. Even when China's political control extended beyond the Great Wall, such control was not accompanied by stabilization and sedentary farming: the environment was always too much. But recently this has changed. Not only has Manchuria become one of China's major population cores and an industrial heartland as well, but even in Inner Mongolia, the Communist regime is putting every kind of modern technology into an effort to overcome the environmental obstacles and to integrate this area into the national economic sphere.

Thus the north indeed remains a national focus. Here lies the beginning of China, its historic capital, its protective northern barrier. Here the innovations that made the opening of the south possible were first applied. Here was the traditional center of authority, intermittently for three dozen centuries. And here, too, lie some of modern China's greatest challenges and achievements: the dry margins, mushrooming Manchuria, buffer-state Mongolia, sensitive Soviet boundaries, and troubled Korea.

China's Regions

China's total area of 3,691,500 square miles (including Tibet) is only a few thousand square miles larger than that of the United States including Alaska, and thus it is not too difficult

for Americans to gauge the dimensions of the Chinese realm. Latitudinally, too, China lies in the same general range as the contiguous United States, though it extends somewhat farther to both north and south (Map 12-6). Thus a journey from Peking to Canton is of the same order as one from Chicago to Miami; to fly from Philadelphia to Seattle is slightly longer than from Shanghai to Urumchi in Sinkiang. And there are some other similarities between the two countries. In the United States as well as in China, the core area lies in the east of the national territory (Plate 6). Both China and the United States have lengthy east coasts. But on the other hand there is nothing to compare to California—and neither is there a west coast—in China; no subsidiary core areas or large urban areas are developing in the far west. This is so despite the fact that China has well over four times as many people as the United States: China's population core constitutes the greatest human concentration on earth.

One advantage of the similarity in size just

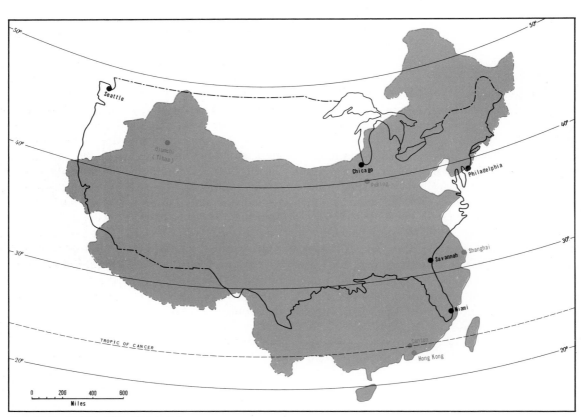

Map 12–6 CHINA AND THE U.S.A.

mentioned is that it is fairly easy to judge the magnitude of China's political and physical regions. Politically China has small provinces in the east and Texas-sized units toward the west. Physiographically, China divides into about the same number of regions as the United States, as seen in Chapter 6.

Political Divisions. Since the Communists assumed control in 1949, China's administrative framework has been changed several times, and undoubtedly it will be changed again. Prior to 1949, China was divided broadly into the eighteen provinces of the east (this area was referred to as China proper), Manchuria, Inner Mongolia, Outer Mongolia (now the buffer state of Mongolia), Sinkiang, and Tibet. Outer Mongolia and Tibet never were fully under the control of pre-Communist China, and Mongolia, following a referendum on the question of independence, in 1950 was recognized by China as a separate and sovereign state. In Tibet, on the other hand, the Communist regime took a different course. In 1951 Chinese forces entered and occupied the country, despite the opposition of the Dalai Lama and the entire Buddhist priesthood in this intensely religious country. For some years there was an uneasy standoff, with China declaring its intention to create in Tibet an Autonomous Region and announcing plans for the closer integration of Tibet with Peking, and the ruling priesthood of Tibet adopting an attitude of aloofness. In 1958 the implementation of Chinese reform programs began to affect the Tibetans to the point of resistance, an outbreak that was decisively countered by China's armed forces. Thousands of refugees, including the Dalai Lama himself, fled by way of the Sikkim Road to India, and China's program of development continued apace.

The Communist regime, thus, found itself faced with the task of organizing the administration of (1) China Proper, (2) Manchuria to the northeast, (3) Inner Mongolia to the north, (4) Sinkiang to the west, and (5) Tibet. The Peking government wanted to create a framework that would ensure direct control at the provincial level in China Proper, while at the same time granting certain measures of autonomy to China's minority nationalities, on the Soviet pattern. Eventually a system was developed involving three levels of administration (Map 12-7):

LEVEL 1
Centrally Governed Municipality
   Peking
   Shanghai
Province
   22 Provinces
Autonomous Region
   Inner Mongolian A.R.
   Sinkiang–Uighur
   Tibet A.R.
   Kuang-si Chuang A.R.
   Ningsia-Hui A.R.

LEVEL 2
Autonomous Chou (A.C.) in Province
Autonomous Chou (A.C.) in Autonomous Region
   For sizable minority nationalities not large enough to be granted
   Regional status

LEVEL 3
Autonomous Hsien (A.H.) or Autonomous Area (A.A.) in Province
Autonomous Hsien (A.H.) or Autonomous Area (A.A.) in Autonomous Region
   For small concentrations of minority peoples

Compared to the system of states in the United States, this seems a very complicated framework, but the Chinese Communist government was determined to accord a certain status to all minority groups large enough to warrant it; previous administrations had failed badly in integrating the state politically and in giving recognition to the country's non-Chinese peoples.

Among Mao Tse-tung's many Famous Sayings is one that goes as follows: "China has a large territory, abundant material resources, and numerous population. Of these three the Hans possess one, namely, they are many in number; [the] minority people possess two, namely they have vast territories and rich resources."[5] In fact, the people of Han constitute about 94 percent of China's population, and thus China comes closer than most countries to being a nation-state. But when there are 900 million people, 6 percent of the total is no small minority; there are over 50 million Chinese subjects who are not

[5]Quoted from Josef Kolmas, "The Minority Nationalities," in Ruth Adams (Ed.), *Contemporary China*, Vintage Books, New York, 1966, p. 51.

Map 12-7 POLITICAL DIVISIONS OF CHINA.

ADMINISTRATIVE SUBDIVISIONS AND BOUNDARIES

1st Level
Centrally Governed Municipality
Province
Autonomous region

2nd Level
Autonomous chou in province
Autonomous chou in autonomous region

3rd Level
Autonomous hsien or autonomous area in province
Autonomous hsien or autonomous c'hi in autonomous region

MODIFIED AFTER: ALDINE PUBLISHING COMPANY, AN HISTORICAL ATLAS OF CHINA.

484

People of Han. These 50 million represent more than 50 separately recognized nationalities, some of them numbering several million, others just a few thousand. The largest minorities are the Chuangs of the Kuangsi-Chuang Autonomous Region, who number 10 million and who are concentrated in their Region and in adjacent Yunnan Province, north of Viet Nam and Laos, and the Uighurs of Sinkiang, of whom there are 6 million along with nearly a million Kazakhs, who live near the boundary with Soviet Kazakhstan. In Inner Mongolia, whose total population of over 12 million is several times that of its independent neighbor, perhaps 20 percent of the people are Mongols. More than a million Koreans live in the northeast, and a similar number of Manchus (who are dispersed all over northeastern China); current estimates for the Tibetan minority suggest that there are under 1.5 million inhabitants in the Tibet Autonomous Region. As this description suggests, there are two major zones of minority nationalities in China, one along the northern sector of the country (including the Uighurs, Kazakhs, Mongols, Manchus, and Koreans), and the second in the south (the Tibetans, Shans, Lao, and Chuang).

Hence China Proper is Han China, made up of the relatively small, eastern, crowded provinces; China of the minorities is the northern, western, and southwestern sphere of larger units with much sparser populations. As Chairman Mao says, these peoples do occupy better than half of China's land area, but whether they also possess—as one might deduce from his statement—an abundance of China's material resources, remains to be seen.

The integration of the minority nationalities of China in the framework of the state has not taken place without opposition. In Tibet and elsewhere, Communist China has been criticized for committing genocide in the process of subduing recalcitrant minorities. Edgar Snow comments:

"In view of the record of thousands of years of Han coexistence with these non-Han peoples, current talk of genocide in China falls wryly from the lips of American officials whose ancestors seized a continent from native Indians and wiped out nations of them scarcely more than a century ago. The march of civilization? No doubt. But let us leave claims to moral superiority in the vestibule when we come to consider the present-day Han people's relations with aboriginal subjects whom their ancestors neglected to massacre."[6]

The fact is that China's minorities have for centuries been undergoing acculturation toward the Han majority, and today the minority nations and tribes can be seen at all stages of the process of sinicization. What has helped this process is that the large majority of these peoples differ from the Chinese not so much in race as in language, way of life, religion (there is a Hui minority of about a million Moslems), and social structure. These, of course, are big obstacles to overcome in the effort to unify, but it does mean that almost no one in China is bedeviled by the uniform of color. Apart from the Turkish element, which is evident in the northwest, most of China's minority peoples are—like the people of Han—of Mongoloid racial stock.

Physiographic Regions. We have already come to know China as a land of rivers and fertile alluvium and loess (Plate 3), temperate to continental-type climates (Plate 2), northern and western dryness (Plate 1), and great mountains. To get a better picture of the spatial arrangement of things in China, both physical and human, it is worth establishing a very general frame of reference. As in the case of India, we can identify several major regions, and then break these down into more detailed subregions. Thus China's four major regions are (Map 12-8): (1) the river basins and highlands of east China and Manchuria; (2) the plateau steppe of Mongolia; (3) the high plateaus and mountains of Tibet; and (4) the desert basins of Sinkiang. Strictly speaking, of course, these are not physiographic *provinces* under the rules with which we became acquainted in Chapter 6; they are broader than that. Similar "regions" in the United States would be, say, the Midwest—which includes several different kinds of landscapes—or the Rocky Mountains, which are diverse enough to be divided into several distinct regions as well.

Region 1, *river basins and highlands of east China and Manchuria,* contains the greater length of the course of China's three most important rivers: the Hwang Ho (Yellow), Yangtze, and Hsi (West), marked as (A), (B), and (C) respectively on the larger map (Map 12-8). All three of these rivers rise on eastern slopes of the Tibet-Yunnan Pla-

[6]Edgar Snow, *The Other Side of the River,* Random House, New York, 1961, p. 596.

Map 12-8 CHINA: PHYSIOGRAPHIC REGIONS.

teau and flow eastward, the Yellow River through the most circuitous and longest course, the West River through the most direct and shortest. The upper courses of the Yellow and Yangtze Rivers lie in two distinct physiographic provinces, but their lower reaches lie in an area which can with justification be identified as a single region, the Eastern Lowland. In human terms this Eastern Lowland is China's most important region, including as it does the North China Plain and the cities of Peking and Tientsin, and the productive lower Yangtze area with Nanking and China's largest city, Shanghai. In every respect this is China's core area, with its greatest population concentration, largest percentage of urbanization (35 percent, as opposed to approximately 20 percent for the country as a whole), enormous agricultural production, growing industrial complexes, and intensive communications networks. China's heartland, as Map 12-8 shows, extends into Manchuria, where the lowland of the Liao River and the city of Mukden (Shenyang) form part of it.

The Hwang Ho, in its upper basin, makes an immense bend and in the process almost encircles one of China's driest areas, the Ordos Desert. Although Plate 1 does not show it, the rainfall here is less than 10 inches, and when it does rain the skies open in torrents that wash away what little soil has formed. And what the rain does not wash into the Yellow River, the gale-force winds of winter blow away: much of the Ordos is a rocky surface where only a few drought-resistant weeds can hold on. But below the Ordos, the Yellow River enters the loess plateau, and here conditions are quite different. Loess is a wind-borne deposit whose origin in this area is related to the nearby deserts (possibly the Ordos) and the Pleistocene glacial epoch, during which the deposits were laid down in a mantle, from a few to 250 feet thick, covering the pre-existing landscape. This loess is quite fertile and, unlike ordinary soil, its fertility does not decrease with depth. Loess, also, has peculiar vertical, columnar structure, which comes about when water percolates downward through root canals and cements the calcium in which the soil is rich. Characteristically, erosion scars in loessic soils are deep and steep-sided, and so are the scars made by human movement: in China's loess plateau, long-traveled roads have been worn as much as 24 feet below the former surface. And since building materials are scarce and the people are

poor, they have used this same quality of vertical strength in the loess to dig cavelike homes into these vertical walls. Unfortunately, part of the loess area in the upper Hwang Ho Basin lies in an active earthquake zone, and while the loess can withstand its own overhead weight, it is vulnerable to earthquakes, which cause great landslides on the ·cliff faces and slopes. In the 1920 quake an estimated quarter of a million people died, many of them buried in their dwellings, and many others in the famine that struck the ravaged area soon afterwards.

The regimen of the Yellow River has always been marked by violent floods and frequent changes in course. Alternately it has drained into the Yellow Sea to the north and to the south of the Shantung Peninsula, with numerous distributaries forming and shifting position over time. Flowing as it does through the loess plateau, the river brings enormous quantities of silt to the North China Plain. There it deposits all this sediment and then proceeds to flow over the new accumulation; for uncounted centuries man has tried through dykes and artificial levees to keep the river's various channels to some extent stable. But as the river brought more alluvium downstream, its level rose, and so did the dykes and levees; soon the river would flow above the surrounding land area. All that awaited a disastrous flood was a season with a particularly high volume of water, enough to overflow and breach the dykes. It has happened dozens of times, and the lives lost directly in these floods or subsequently in the inevitable famine must number in the tens of millions.

The course of China's middle river, the Yangtze Kiang, is usually divided into three basins, of which the westernmost, the Red Basin, constitutes one of China's largest population clusters. The 1969 population of Szechwan province (Ssu-ch'uan) was 94.1 million; about 68 million of these are clustered in the Red Basin, where lies one of the most intensively cultivated areas in the world. Apart from the Chengtu plain, where there is relatively level land, the slopes of the basin's hilly country have been transformed by innumerable terraces, on which rice grows in summer and wheat in winter; apart from other cereals, including corn, such crops as soybeans, sweet potatoes, sugar cane, and a wide range of fruits are grown; on the warmer slopes, tea flourishes. In China it is said that anything that can grow anywhere in the country can be grown in Szechwan, and, in fact, most crops are. Yet the Red Basin lies a thousand miles into the continental interior, where one would expect a rather rigorous climate with winters too cold and a growing season too limited for such variety. What makes Chengtu's climate, 1500 feet above sea level and so far inland, *milder* than that of coastal Shanghai? The whole answer is not known, but part of it probably lies in the protection afforded the Red Basin by the very high mountains that lie to the north and especially the northwest—protection against the bitter winds of interior Asia's cold winters. The Yangtze's middle course begins in the vicinity of Ich'ang, where the river emerges from a series of gorges that limit navigation to Chungking to small motorized vessels. It enters an area of more moderate relief and flows through middle China's lake country—agriculturally one of the most productive parts of the nation. This is the southern part of the Eastern Lowland, and along or near the Yangtze lie the large cities of Wuhan (3.5 million), Nanking (2 million), Hangchow (1.1 million), and China's leading urban center, Shanghai (10 million).

China's southernmost major river is actually called the West (Hsi) River, and neither in length nor in terms of the productivity of the region through which it flows is this river comparable to the Yangtze or the Hwang. In the Hsi Kiang Basin there is much less level land; in the hills and mountains of south China remain many millions of tribal, aboriginal peoples not yet acculturated to the civilization of the people of Han. Hence the western part of the Hsi Kiang Basin is constituted largely by the Kuangsi-Chuang Autonomous Region. To the east lie Kuangtung province, the populous delta, and Canton, the south's largest and most important city (2.2 million).

Between the deltas of the Hsi and Yangtze Rivers lie the Southeast Uplands, a region of rugged relief which for a long time in Chinese history remained a refuge for southern tribes against the encroachment of the people of Han from the north. With its steep slopes and narrow valleys, this region has very limited agricultural possibilities, although in 1961 the government initiated a massive program of dam construction and slope terracing, in which the army was used as labor. For many years this has been one of China's most outward-looking regions, with a considerable emigration (via Kuantung) to the Philippines and Southeast Asia, substantial overseas trade in the leading commercial product, tea, and a seafaring tradition for which the people of Fuchien Province have become famous.

Broadly speaking, the regions discussed here collectively make up what is referred to as China Proper, though it is reasonable, in view of modern developments, to include Manchuria in this concept. Manchuria's Liao-Sungari lowland, while not comparable to the three rivers in some respects (for example, this Manchurian lowland is largely an erosional rather than a depositional basin), is attaining population densities and productivities that are in a class with those of the heartland provinces farther to the south. The Northeast Uplands, comparable in terms of relief to those in the southeast, are not coastal; Manchuria's nearest coastline is silt-plagued Liaotung Gulf, and its most effective direct outlet, Talien-Lüshunk'ou (Dairen-Port Arthur) on the eastern tip of the Liaotung Peninsula.

Region 2, the *Plateau Steppe of Inner Mongolia*, actually constitutes the southern rim of the basin of the Gobi Desert. Near the physiographic boundary line, elevations in a series of ranges reach an average of 5500 feet, but toward the heart of the Gobi the land lies much lower, as low as 2500 feet. As is suggested by Plates 1 and 2, the moistest parts of this region lie on the slopes of those southern, higher hills, though even there the vegetation is only a poor steppe grass with some scrub; some of the mountains are rocky and barren and the whole aspect of the area is one of drought. Not surprisingly, this is an otherwise difficult environment as well: summer temperatures are often searingly hot, while winters are bitterly cold. Vicious winds blow up sand and dust, and altogether this is unlike the image we have of populous, productive, rice-paddy China. Despite the Peking government's efforts to spread sedentary agriculture here through river dams and irrigation projects, as a whole the Plateau Steppe remains one of the country's more sparsely populated areas. Although the physiographic region and the political area of the Inner Mongolian Autonomous Region do not overlap exactly, the population statistics for the latter give an indication of the general situation: the 1969 estimate is about 13 million, an average of just under 30 per square mile in a country where a thousand people per rural square mile is no rarity. Only Sinkiang and Tibet are less populous.

Another face of China: a camel convoy near Chungwei, Ninghsia, just north of the Yellow River not far from the Mongolian Autonomous Region. (Eastfoto)

Region 3, the *high plateaus and mountains of Tibet,* consists of one of the world's greatest assemblages of high, snow-covered mountain ranges and high-elevation plateaus. The southern margins of this region are the great Himalaya Mountains themselves, and in the north lie the Kun Lun Mountains and the Nan Shan. From the Pamir "Knot" in the west, these great mountain chains spread out eastward in a series of vast arcs, which eventually approach each other again and turn southward into the Yunnan Plateau. The *average* elevation in the region as defined on Map 12-8 is probably around 15,000 feet, with the higher mountain ranges standing 10,000 feet above this level and the valleys—where most of the people live—going down to about 5000 feet above sea level. The central plateau of Tibet is desolate and barren, cold, windswept, and treeless; here and there some patches of grass sustain a few animals. In terms of human development, there are two areas of interest. The first of these is the area between the Himalayas in the south and the Trans-Himalayas which lie not far to the north. In this area there are some valleys below 6000 feet, where the climate is milder and cultivation is possible. Here is Tibet's main population cluster; the capital of Lhasa is situated at the inter-section of roads which lead east-west along the valley and northward into Ch'ingai (Tsingai) and China Proper. The Chinese government, since its confirmation of control in Tibet, has made investments to speed up the snails-pace of development that prevailed under the previous administration. The southern valleys contain excellent sites for hydroelectric power projects, and some of these have been put to use in a few light industries. More importantly, the south of Tibet may be an area of considerable mineral wealth, and despite the enormous difficulties of distance and terrain, such minerals may become vitally important to the developing China of the future.

One other part of the Tibetan region is of political economic importance, namely the Tsaidam Basin, in Ch'inghai on the edge of Sinkiang. This basin lies several thousand feet below the surrounding Kun Lun and Nan Mountains, and as such it has always contained a cluster of nomadic Tibetan pastoralists—in fact, there were times when these nomads plundered the trade route to the west, which skirted the Nan Shan to the north. Recently, however, exploration has revealed the presence of oil fields and coal reserves below the surface of the

Tsaidam Basin, and the development of these resources has already begun.

Region 4, the *desert basins of Sinkiang,* comprises two huge, mountain-enclosed basins and several smaller basins. The two largest of Sinkiang's basins are separated by the Tien (Tyan) Shan, a mountain range that stretches across the region from the Kirghiz border in the west to Mongolia in the east. Climatically these are dry areas: the Tarim Basin is in fact a desert (the Turkestan or Takla Makan Desert), while the basin of Dzungaria is steppe country. Both the Dzungaria and the Tarim Basins are areas of internal drainage; that is, the rivers that rise on the adjacent mountain slopes and flow to the basin floor do not continue to the sea. Only in the far north does the Urungu River eventually drain into the Irtysh system, which in turn leads to the Ob River and the Arctic Ocean. The rivers that flow off the adjacent mountains are the chief supply of water for the region's rough gravels that have been washed from the mountains; they disappear below the surface as they reach these coarse deposits. But then they reappear where the gravels thin out, and there, oases have long existed. Along the southern margin of the Takla Makan Desert lie a string of these oases, which at one time formed stations on the long trade route to the west.

Since 1949 the Chinese have made a major effort to develop the agricultural potential of the Tarim area. Although *qanat*-sustained irrigation on the south flank of the Tien Shan is centuries old, a great deal of water always went to waste and land that could have sustained irrigated agriculture was not brought under cultivation. This work has borne fruit: canals and *qanats* were built, oases enlarged, and the acreage of productive farmland has probably quadrupled by now from what it was just 20 years ago. Especially the northern rim of the Tarim Basin, which had long been neglected, has been brought into the sphere of development, and fields of cotton and wheat now attest to the success of the program. And in the Basin of Dzungaria, too, important progress has occurred. Along the northern slopes of the Tien Shan there are opportunities similar to those in the Tarim Basin, and east and west of Urumchi lie newly cultivated acreages of cotton, rice, and wheat. Unlike the barren desert of the Tarim interior, the Dzungaria Basin is moist enough to sustain some grasses, and hence a nomadic pastoral economy based on sheep.

Dzungaria, though it contains perhaps just one-quarter of Sinkiang's more than 7 million inhabitants, has a number of assets of importance to China. First, it has long been the site of strategic east-west routes. Second, the main westward rail link to the Kazakh S.S.R. and Soviet Russia goes from Hsian in China Proper via Yumen in Kansu and Urumchi (Tihua) in Dzungaria. Third, Dzungaria has proved to contain sizable oil fields, notably around Karamai, not far from the Soviet border. Pipelines have been laid all the way to Yumen and Lanchow, where refineries have been built. Thus it is not altogether surprising that the region's capital of Urumchi (250,000) lies in the less populous but strategically important northern part of Sinkiang.

Drainage Patterns. These, then, are China's major physiographic regions, and what stands out in any discussion of the country's physical geography is the pattern of drainage. Whenever China Proper is discussed, reference is made to the three great rivers of the east—but China also gives rise to such major streams as the Indus, which is West Pakistan's lifeline, the Brahmaputra, vital to India's Assam and East Pakistan, and the Mekong, one of Southeast Asia's most important rivers. In fact, rivers flow from China in all directions: north to the Arctic Ocean, east to the Pacific, south to the Indian Ocean, and west to interior Asia. Professor H. J. Wiens identifies this as China's *radial* drainage pattern, using one of several terms employed to distinguish the plan or arrangement which the individual streams collectively form.[7] The opposite of this diverging pattern can be observed in the Tarim Basin, where streams converge into a central depression: this is *centripetal* drainage. In China Proper, where streams flow at several angles, like the branches of a tree, to the Yangtze and Hsi Rivers, the drainage pattern is dendritic. In some places, where rock formations have a parallel orientation and the rivers have been forced to flow parallel to these formations and hence parallel to each other, a *trellis* pattern emerges. In southern and southeastern Tibet, such trellis patterns occur. In the upper reaches of the Yangtze Kiang, rivers which formerly flowed southward into the Hong Kiang (Red River) and the Mekong River have been captured (i.e.,

[7] Herold J. Wiens, "China, Physical Diversity" (Chapter 8), in N. Ginsburg (Ed.), *The Pattern of Asia*, Prentice-Hall, Englewood Cliffs, N.J., 1964, p. 157.

diverted) by the Yangtze's headwaters. Such river capture gives rise to *barbed* drainage patterns, which involve very sharp bends where tributaries join the main stream. Even a map on the scale of an atlas map will display this feature of the Yangtze's upper courses. Finally, there are *rectangular* drainage patterns, in which faults or joints in the rocks control the courses of streams. This situation is usually best seen on large-scale maps, though there are places where rectangular patterns occur over a wide region—for example, on the scoured bedrock of the Canadian Shield. Nothing on that scale appears to exist in China, though part of the drainage pattern in the Southeast Uplands physiographic region may be fault-controlled.

CHINA PROPER

Two regions of China contain the great majority of the country's 900 million people, most of the agricultural land, and almost all the cities and major industries. More even than in the United States, "the east" is China's historic and present heartland, where the civilization of the people of Han had its roots and where the modern core area of the state evolved. In China, "the east" consists of China Proper and Manchuria, and of the two, China Proper still retains its primacy in every way. Let us bring this part of the country into clearer focus.

Humid China

In our discussion of China's physical regions, we defined a set of physiographic provinces defined on several bases—river drainage basins, relief, soils, climate, and so forth. Considering China's enormous agricultural output and the intensity of cultivation almost wherever farming is possible, we might have suggested still another kind of regional division, namely between "humid" China and "arid" China. Geographers often employ this designation to separate the China where farming is possible from the steppes and deserts where pastoralism prevails. Thus China Proper is identified as "humid" China, for it is the China of rice and wheat—but the term is not altogether appropriate. As Plates 1 and 2 show, southern China is subtropical and humid, but northern China Proper is much less moist; virtually the entire course of the Hwang Ho classifies as a steppe region, and northward the climate is continental and quite dry. Eastern China, then, is often called "humid" China, but there is quite a difference between the northeast and the southeast.

China Proper is the China south of the Great Wall, the China of the three rivers. Of the three, the middle river—the Yangtze—and its basin are in almost every respect the most important. From the Red Basin of Szechwan in its upper catchment area to its delta in the Eastern Lowlands (Map 12-8) the Yangtze lies in China's most populous and most productive areas. As a transportation route it is China's most navigable waterway (Map 12-9); ocean-going ships can sail over 600 miles up the river to the Wuhan conurbation (Wuhan is short for Wuchang and Hankow), and boats of up to 1000 tons can even reach Chungking. Several of the Yangtze's tributaries also are navigable, and depending of course on the size of the ships capable of using these various stretches, over 20,000 miles of water transport routes exist in the Yangtze basin. The Yangtze thus is one of China's major transport arteries, and with its tributaries it attracts the trade of a vast area including nearly all of middle China and large parts of the north and south. Funneled down the Yangtze, most of this enormous volume is transferred at Shanghai, whose size reflects the productivity and great population of this hinterland.

Early in China's history, when the Yangtze's basin was being opened up and rice and wheat cultivation began, a canal was built to link this granary to the northern core of old China. Today the Grand Canal no longer serves as a cross-country waterway, for only parts of it remain navigable for junks, but China still has its north-south water route; all along the east coast moves a huge fleet of vessels transporting domestic interregional trade. The bulk of this trade, including industrial as well as agricultural products, is either derived from or distributed to the Yangtze region. International trade, also, goes mainly through Shanghai, which normally handles half the overseas tonnage, with the rest split among China's other ports, including Tientsin on the northern coast and Canton in the south.

Shanghai, just a fishing town until the mid-nineteenth century, rose to prominence as a result of its selection as a treaty port by the British, and ever since then its unparalleled locational advantages have sustained its position as China's leading city in almost every respect. The city lies at one corner of the Yangtze delta, an area of about 20,000 square miles in which live perhaps as many as 50 million people. Some two-thirds of these are farmers who produce food as well as silk and cotton for the city's industries. Thus Shanghai has as its immediate hinterland one of the most densely populated areas in the world, and beyond the delta lies what must be the most populous region in the work to be served by one major outlet. During the nineteenth century and up to the war with Japan, the principal exports to pass through Shanghai were tea and silk; large quantities of cotton textiles and opium were imported. At that time Shanghai's position was undisputed: it handled two-thirds of all of China's external trade. But its fortunes suffered during and after World War II. First the Nationalists blockaded the port (1949) and made

Map 12-9 CHINA PROPER.

bombing raids on it; then the Peking government decided to disperse its industries upcountry, thereby reducing their vulnerability to attack. Meanwhile safer Tientsin had taken over as the leading port. But Shanghai's situational advantages promised a comeback, and it came. In the 1960's the port regained its dominant position, and the industrial complex (textiles, food processing, metals, shipyards, rubber, chemicals) continued to expand.

Shanghai is the largest city in China, but it is by no means the only sizable urban center in middle China. Between Shanghai and Nanking (also on the Yangtze) lie a half-dozen cities of which several have over a half million people. Nanking, Wuhan, and Chungking all have more than 2 million residents. But of course Shanghai's hinterland is predominantly rural, and the cities are mainly service centers for regional parts of this enormous, populous area. The vast majority of the Yangtze basin's more than 300 million people are farmers on cooperatives and collectives which produce the country's staples and cash crops. We have already seen the amazing variety of crops grown in Szechwan; there, as in the Yangtze's lower basin, rice predominates. However, a look at Plates 1 and 2 will suggest why the Szechwan-Yangtze line is more or less the northern margin for rice cultivation. To the north, temperatures go down and rainfall diminishes as well, and even in the irrigated areas wheat rather than rice is the grain crop. In the Yangtze basin, rice is mixed with winter wheat, and in effect this is the transition zone; a line drawn from Hanchung, in Shensi to Singailien in coastal Kiangsu approximates the northern limit of rice cultivation (Map 12-10).

As Map 12-8 shows, the Eastern Lowland merges northward into the lower basin of the Yellow River and the North China Plain. Here spring wheat is grown in the northern areas, while winter wheat and barley are planted in the south; in the spring other crops follow the winter wheat. Millet, sorghum (kaoliang), soybeans, corn, a variety of vegetables, tobacco, and cotton are cultivated in this northern part of China Proper; on the Shantung Peninsula's higher ground, fruit orchards do well. This part of China was marked by very small parcels of land, and the Communist regime has effected a major reorganization of landholding here; at the same time an enormous effort has been made

Water transport always has played an important role in China, with its many rivers and canals. This view is over part of the crowded port of Shanghai. Ocean-going vessels are at quayside (top) as well as being serviced in lighterage (left center); a barge train winds its way among the inevitable armada of junks. (Rene Burri/Magnum)

to control the flood problem that has bedeviled the Hwang Ho basin for uncounted centuries, and to expand the acreage under irrigation.

The North China Plain is one of the world's most overpopulated agricultural areas, with over 1200 people per square mile of cultivated land. Here the ultimate hope of the Peking government lay less in land redistribution than in raising the per-acre yields through improved fertilization, expanded irrigation facilities, and more intensive use of labor. Dam construction on the Hwang Ho has reduced the flood danger, but outside the irrigated areas there is the ever-present problem of rainfall variability and frequent drought. The North China Plain simply never has been able to produce a little surplus to save for future years, and though the spectre of famine has receded, the food situation is always precarious in northern China Proper.

The two major cities of the Yellow River Basin are the historic capital, Peking, and the port city of Tientsin, both positioned near the northern edge of the plain. In common with many of China's harbor sites, that of the river port of Tientsin is not particularly good, but the city is well situated to the northern sector of the Eastern Lowlands, the capital (now with a population of about 7 million), the loess plateau to the west, and Mongolia beyond. Like Shanghai, Tientsin had its modern start as a treaty port, but the city's major growth awaited Communist rule. For many decades it had remained a center for light industry and a flood-prone port, but

after 1949 a new artificial port was constructed and flood canals were dug. More importantly, Tientsin was chosen as a site for major industrial development, and large investments were made in the chemical industry (in which Tientsin now leads China), in heavy industries, especially iron and steel production, in heavy machinery manufacture, and in the textile industry. Today, with a population of over 4 million, Tientsin is the center of the country's third largest industrial complex, after Manchuria's Liaoning Province and its old competitor, Shanghai. Peking, on the other hand, has remained mainly the political, educational, and cultural center of China; although industrial development occurred here also after the Communist takeover, this has not

Map 12–10 AGRICULTURAL REGIONS.

Fishing boats and a harvest of yellow croakers drying in the sun, in the Chusan Archipelago, southeast of Shanghai. China is working hard to expand its fish catch every year. (Yuan Ling/Eastfoto)

been on a scale comparable to Tientsin. The Communist administration did, however, greatly expand the municipal area of Peking, which is not controlled by the province of Hopeh (Hopei) but resorts directly under the central government's authority. In one direction Peking was enlarged all the way to the Great Wall—35 miles to the 'north—so that the "urban" area includes hundreds of thousands of farmers!

The upper basin of the Yellow River—and that of the Wei River—includes the Loess Plateau, an area of winter wheat and millet cultivation whose environmental problems were discussed earlier (p. 487). Beneath the loess lie major coal deposits, and already Taiyuan in Shansi is the site of a large iron and steel complex, machine manufacturing plants, and chemical industries. As the map indicates, the upper Yellow River area is not especially well positioned for the transportation of raw materials, though rail connections do exist to Manchuria's source areas.

Southern China is dominated by the basin of the Hsi Kiang—the West River. While northern China is the land of the ox and even the camel, southern China uses the water buffalo. Northern China grows wheat for food and cotton for sale; southern China grows rice and tea. Northern

China is rather dry and continental in its climate; southern China is subtropical and humid. And the people of northern China are much more clearly Mongoloid in their appearance: northern China has long looked inward, to interior Asia, while southern China has oriented itself outward, to the sea and even to lands beyond. With its very mixed population and multilingual character, southern China carries strong Southeast Asian imprints.

The map suggests that the Hsi Kiang is no Yangtze, and this impression is soon verified. Not only is the West River much shorter, but for a great part of its course it lies in mountainous or hilly terrain. Above Wuchow the valley is less than a half mile wide and, confined as it is, the river is subject to great fluctuations in level. Except for the delta, whose 3000 square miles support a population of some 15 million, there is little here to compare to the lower Yangtze basin. Cut by a large number of distributaries and by levee-protected flood canals, the delta of the Hsi Kiang is southern China's largest area of flat land and the site of its largest population cluster; here, too, lies Canton, the south's leading city.

Subtropical and moist, southern China pro-

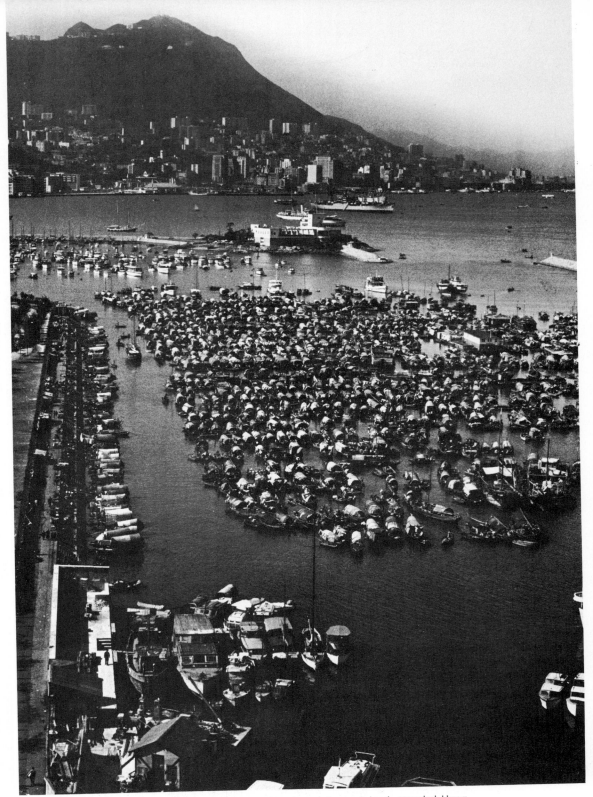

A British prize still held: Hong Kong. Perched on the edge of the mainland, crowded Hong Kong bears witness to another age in Chinese history. This photograph looks across Typhoon Shelter into Hong Kong proper. (Burt Glinn/Magnum)

496

vides a year-round growing season. In the lower areas rice is double cropped: one planting takes place in mid- to late winter with a harvest in late June or shortly thereafter, and a second crop is then planted in the same paddy and harvested in mid-autumn. It is even possible to raise some vegetables or root crops between the rice plantings! As in Szechwan, there are whole areas where the hillslopes have been transformed into a multitude of man-made terraces, and here, too, rice is grown, though the higher areas permit the harvesting of only one crop. But again there is time for a vegetable crop or some other planting before the next rice is put down. Fruits, sugar cane, tea, corn—wherever farming is possible south China is tremendously productive, and the range of produce is almost endless. If only there were more level land—and fewer people: this is one of China's food-deficient regions, always requiring imports of grain.

With its 2.5 million people, Canton is the urban focus for the whole region, and despite its rather narrow valley the West River is navigable for several hundred miles (Map 12-9), so that hinterland connections are rather effective. Canton (or Kwangchow) was the first port city to feel the full impact of Europe and its merchants; the city's fortunes have varied a great deal. Large ships cannot reach it and must dock 12 miles away at Whampoa, the head of the Hsi Kiang's Pearl River Estuary; smaller vessels often bypass it and head for Wuchow, 230 miles upstream. Meanwhile Hong Kong, just over 100 miles away, overshadows Canton in size as well as trade volume. And Canton has faded as an industrial center; there was a time when its factories and production made it a competitor for Shanghai, but in modern times the paucity of natural resources in the West River basin gave caused its industrial development to lag. In current development plans Canton's harbor is undergoing improvement and the port facilities are being expanded (with an eye to the competition of British-owned Hong Kong), but industrial growth is envisaged as relating to the agricultural production of the hinterland: sugar mills, textile factories, fertilizer plants. There is a fuel problem in southern China, which has long been countered by coal imports from northern China, but now the hydroelectric potential of the West River basin, which is good, is being put to use.

Hong Kong

Whatever the future of Canton, undoubtedly it will be bound up with the role of its nearby competitor, the political anomaly of Hong Kong. The British dependency actually consists of three parts: the 32-square-mile island of Hong Kong, a peninsula opposite this island, the Kowloon Peninsula, and a mainland territory of over 350 square miles, the so-called New Territories. Hong Kong island and the Kowloon Peninsula were ceded to Britain in perpetuity in 1841, but the New Territories were leased on a 99-year basis in 1898. With its excellent deep-water harbor, Hong Kong is the major entrepôt of the Western Pacific between Shanghai and Singapore. At various stages of its history the city, by virtue of its British administration, has been the goal of hundreds of thousands of refugees: during the revolutionary chaos of 1911 and thereafter, during the Japanese war, and after 1949. These people have created very crowded conditions in the city, but they also brought with them skills that have been the foundation of an extensive and productive complex of light industries. With 4.5 million inhabitants, Hong Kong obviously cannot feed itself, and despite the often strained relations between Britain and China most of the city's food requirements come from across the Chinese border. Indeed, the mainstay of the dependency's economic life also is trade with China. All this is not to imply a case of generosity on China's part: there are advantages to a foreign port on the mainland, through which China can maintain overseas trade connections which might otherwise not be possible.

Manchuria

Manchuria—home of the Manchus, who were the last to impose a dynastic rule upon China—has become one of China's most vital regions. There was a time, early during the Manchus' rule, when Chinese were barred from entering this area, but after the collapse of the Manchu dynasty in 1911, a massive immigration from northern China has completely submerged the old Manchu culture and has firmly established the Chinese presence here. The Chinese themselves do not even refer to this part of their country any longer; they call it simply the northeast.

Manchuria is bordered on three sides by political boundaries. To the west is Mongolia (and

Chinese Inner Mongolia), to the north and northeast is the Soviet Union, and to the east is North Korea. Close to the point where North Korea, the Soviet Union, and Manchuria meet lies the Soviet port of Vladivostok, near the southern end of what is in effect a southward proruption of the Soviet Far East. As the political map (Map 12-11) shows, the shortest route from Vladivostok to the Soviet heartland is right across Manchuria, via Harbin, China's northernmost large city. This is the railroad in which the Soviets long had an interest, which was finally relinquished by treaty as late as 1950. With the increasing tension between the two countries, the Soviets have come to rely on the route that skirts the Manchurian border, via Khabarovsk—a trip that is 300 miles longer.

Manchuria has still another neighbor that has

Map 12-11 MANCHURIA.

involved itself in its affairs: Japan. Japan, we know, modernized much more rapidly than China during the nineteenth century, and quite early the Japanese saw in Manchuria a nearby source of raw materials and food production, and a developing market for its manufactures. In the war between Russia and Japan (1904–1905), Manchuria was a major prize, and the two foreign powers struggled bitterly for supremacy here. The Japanese made use of the revolutionary and post-revolutionary crisis in China Proper to entrench themselves strongly in Manchuria, and in 1932 they even set up the puppet state of Manchukuo, propping up the last representatives of the Manchu dynasty as their stooges. Manchukuo lasted only briefly, but the Japanese impact in Manchuria was more permanent. They built railroads and factories, opened mines, developed the agricultural possibilities of the region, and generally laid extensive foundations for its large-scale development. Much of this activity was focused on Manchuria's south, in the Liao River valley and its environs; the Russians retained something of a sphere of interest in the north and they, too, busied themselves by opening up the country through railroad building. In the strange economic struggle of the 1920's and the 1930's the Chinese themselves got into the act by building their own railroad from Peking northward toward Shenyang (Mukden).

The aftermath of Japan's defeat in World War II was the incorporation of Manchuria into China, although China was not simply heir to all the facilities the Japanese and the Soviets had built. The Soviets made use of the chaos attending the war's end by taking away most of the locomotives and train cars and a good deal of machinery; there are also claims that the Russians left with many millions of dollars' worth of gold. This, of course, was before the Communist regime took power in China, and after 1949 replaced much of what they had taken. In any event, China did find itself in control of an area with enormous possibilities, a good economic structure, and (for China) a comparatively light population.

The physical and economic layout of Manchuria is such that the areas of greatest productive capacity (and consequently largest population) lie in the south, where they form an extension of the core area of China Proper. The lowland axis formed by the basin of the Liao

River and the Sungari valley extends from the Liaotung Gulf north and then northeastward (Map 12-8). Although the growing season here is a great deal shorter than in most of lowland China Proper, it is still long enough in the Liao valley to permit the cultivation of grains such as spring wheat, kaoliang (sorghum), barley, and corn, and large acreages are planted to soybeans as well. Earlier we commented on the marginal character in terms of precipitation of this part of "humid" China (see Plate 1), but southern Manchuria has some excellent, fertile soils. According to the map of world soil distributions (Plate 3), the Liao basin has an area of the kind of gray-brown podzolic soils that sustain not only nearby Japan's intensively cultivated lands, but also fertile Western Europe and the American Midwest (area 5). Toward the northern margins, though, the problems of drought and cold begin to inhibit agriculture. This is true also of the higher areas that surround the Liao-Sungari lowland axis to the west, north, and east. Though not nearly as high or rugged as the western mountains of China Proper, the elevations, generally up to 5000 feet, are enough in these latitudes and in this continental position to create cold and dry conditions. In timber-poor China, the oak and other hardwood forests on these slopes are of great value.

Driven by hunger, pushed by devastating Hwang Ho floods, dislocated by decades of war, and attracted by the land and the space of Manchuria, millions of Chinese have crossed the Great Wall boundary and have resettled in the northeast. And not all of them have come to be farmers, for Manchuria has become China's industrial heartland, and the northeast's largest city, Shenyang (whose urban area has a population of 5 million) has emerged as a Chinese Pittsburgh. All this is based on Manchuria's large iron and coal deposits, which—like the best farmland—lie in the area of the Liao basin. Shenyang lies in the middle of it: there is coal within a hundred miles to the west and less than 30 miles to the east of the city (at Fushun), and about 60 miles to the southwest, near Anshan, there is iron ore. Since the iron ore is not very pure, the coal is hauled from Fushun (and also from northern China Proper) to the site of the iron reserves; this is the cheapest way to convert these low-grade ores to finished iron and steel. Anshan has thus become China's leading iron and steel producing center, but Shenyang remains the northeast's largest and most diversified industrial city. Machine building and other engineering works in Shenyang supply the entire country with various types of equipment including drills, lathes, cable, and so forth. As the northeast's most productive farming area, Shenyang has other functions as well: it is the leading agricultural-processing center of Manchuria.

Southern Manchuria, however, is not the only area where coalfields and iron ores exist. From near the Korean border (where the iron ore is considerably more pure) to the northeast, where the coal is of best quality, Manchuria has natural resources in abundance, and there is also aluminum ore, molybdenum, lead, zinc, and an important ingredient in the manufacture of steel, limestone. To prove the northward march of development in Manchuria, the railroad crossroads of Harbin has become a city of over 2 million, with large machine manufacturing plants (especially agricultural equipment such as tractors), a wide range of agricultural processing factories, and such industries as leather products, nylons and plastics, and so forth. Not only does Harbin (Ha-Erh-Pin) lie at the convergence of five railroads; it also lies at the head of navigation on the Sungari River, connecting the city to the towns of the far northeast.

In some ways Manchuria resembles Soviet Asia, with its great resources, its rapidly growing cities, its population influx, and its planned development, although there are readily visible differences. The cultivable lowland of the Chinese northeast is no Virgin Lands, and in terms of absolute numbers the population influx into Manchuria has been far greater; the three provinces of Manchuria today probably count over 75 million people. But as in Soviet Asia, Communist doctrine in Manchuria has always been somewhat relaxed; wages have always been rather higher than in China Proper in order to attract the skills the industries needed. In terms of planning, the rebuilt industries of Shenyang, Harbin, and Anshan in some instances are models of everything the Communist regime would like them to be: schools, apartments, recreation facilities, hospitals, and even old-age homes, all part of the huge industrial plants, and having a job means access for the worker to all these. Much the same can be said for Khabarovsk, where the Soviets have used incentives and attractive facilities to help lure the people from the west. But most of all, in Manchuria, the large-scale and efficient agricultural development stands in contrast to the parceled chaos of old China. The northeastern farms have been laid out more recently, they have always been larger, more effectively collectivized, and more quickly mechanized. In many ways Manchuria is the image in which Chinese planners would like to remake all China.

The Search for Food

Manchuria's impressive industrial development, and the drive toward greater industrialization elsewhere in China, still cannot alter the fact that this remains an overwhelmingly agricultural country, in which over two-thirds of the people live directly off the land. This amounts to some 600 million people, who must not only feed themselves but also the nearly 300 million who live and work in cities and towns, factories and mines. We know that China has a total area only slightly larger than the United States, and from the discussion of its physiographic regions we can quickly deduce that most of China—some estimates say as much as 90 percent—is too high, too dry, too steep, or otherwise unsuitable for farming. Even if this figure is way off, so that not 10, but as much as 15 percent of China would be cultivable, then the total acreage of available land would be around 400 million acres, or less than half an acre per person in China.

Statistics on China are not easy to obtain, and all too often one has to rely on old or unreliable sources. What is certain is that China, after 1949, engaged in a massive effort to expand its acreage of cultivated land, to improve yields, to increase the availability and use of fertilizer, to conserve water, irrigate more land, and control floods, and to mechanize farming where feasible. Obviously, under the circumstances, any equal division of land among all the farm families in the country simply was not a realistic proposition; if, in 1949, all the arable land in China had been divided equally among farm families (taking 5 persons as the average size of such families) then the normal farm size would have been about $2\frac{1}{2}$ acres. Today it would be even less. To achieve what the Peking government wanted to achieve, collectivization was imperative. Larger farm units would increase the efficiency of farming, it would make the purchase and distribution of fertilizer more attainable, it would raise yields

The new side of China: teachers and students of the Peking Aeronautical Engineering Institute harvest wheat along with members of the Tungpeiwang Commune near Peking. (Na Yi/China Photo Service)

and generate more capital. Actually, this did not happen right away. In 1960, Chou En-lai told Edgar Snow that China had 275 million acres under cultivation, which was of course more than had been under crops in 1949, but also represented a slight decline during the late 1950's. Mr. Snow comments:

". . . In the euphoria of 1958, official policy encouraged communes to seek increased yields by deep plowing, close planting, more irrigation, more fertilizer, etc., and more double cropping on good land, at the neglect of poorer land. While the government's eight-point charter for agriculture, which included such advice, might be sound enough if applied by experienced farmers, bureaucratic and indiscriminate enforcement was probably costly. Second, some common administrations in 1958 abruptly curtailed or even ended the cultivation of small plots privately owned by the peasants, and these fell fallow. Third, commune diversions of farm labor to nonagricultural projects often caused neglect of available land. Finally, many communes had simply expropriated good land of prosperous peasants for public works without compensation. With the erasure of the last vestiges of private ownership, inducements were lacking to encourage enterprising peasants to combine to exploit new fields."[8]

The progress of the early 1950's, interrupted during the Great Leap Forward, seems again to have picked up during the late 1960's. In the middle years of the 1950's the grain harvest totaled somewhere in the vicinity of an average of 170 million tons; the 1967 harvest, reported to have been one of the best lately, appears to have amounted to 215 million tons. But against these increases stands China's 2 percent population growth rate, which is being maintained despite renewed government-sponsored propaganda for birth control. The increased grain harvest over the decade 1957–1967 is impressive, but it constitutes an annual growth rate of only just over 2 percent, so that the actual gain for China is minimal. As we have noted, every decade China adds a population equivalent to that of the entire United States to its already existing total, and just to keep abreast of this increase requires an enormous effort. Hence the contin-ued need for grain imports from capitalist countries: the gap has not yet been closed.

[8]Edgar Snow, *op. cit.*, p. 81.

China has mobilized its many millions of people in an unprecedented way to speed the country toward its goals. Where there are no bulldozers or trucks, people by the thousands carry rocks and earth in baskets. In this photograph, taken near Nanning, students are spending part of their 'spare' time helping in the construction of a road. (Marc Riboud Magnum)

In a way, collectivization was the road to several of the other objectives the Chinese have for their agricultural industry. More rewarding than mechanization, irrigation, or the expansion of cultivated land was the intensification of labor on the existing land. For uncounted centuries China's farming has been labor intensive; it has been agriculture by hand. Yields were always high where much work was expended on planting, replanting, fertilizing, and weeding. Through the collectives and their cooperative labor, and the communes and their labor brigades, it was envisaged that even more labor could be applied to the farmlands and the yields could be raised even more. This was seen as the most direct route to greater harvests, for it would encompass even those lands where mechanization was not feasible anyway, where water con-

servation was already practiced in the most meticulous way, where fertilizer from livestock (the pig, mainly) already was being applied. With the collectivization program a going concern, China could turn to the other objectives: the building of dams and irrigation works, flood canals, tractor and fertilizer factories, the search for additional farmland (as in the Tarim Basin), and the mechanization of farming where it would really pay off, as in Manchuria's wheat land.

But the fact remains that whatever China does, it will be ever harder pressed to feed its staggeringly large and rapidly growing population. Twenty years ago, when China had but two-thirds of the population it carries today, it seemed inconceivable that 900 million people could be fed here, but it is being done, and with relatively modest imports. Now it seems equally unbelievable that China will have to feed a billion people by 1975; but who is to say where the limits are? If ever they are reached, the world will pay the price for it.

The New China

Having viewed some of the main characteristics of China's territory, let us return now to the China of the past two decades and the efforts of its government and people to reconstruct their homeland. China at mid-century was a country torn by war almost as long as the people could remember. Foreign interference and exploitation and civil strife had been added to the problems China already faced, stemming from its huge population, its rapid population growth, the limited fertile land in the country, and the low levels of output. In the absence of any effective central government, the worst aspects of human nature could freely run their course: feudal warlords held parts of the country, village landlords mercilessly exploited their victims, corruption was rife, the cities had the world's most terrible slums where rats thrived by the millions, children were dying of starvation by the thousands, beggars were everywhere. Of course, in so large a country, there were variations. There were areas where things were not quite so bad, and elsewhere they were worse. But when the new government took control, it faced an enormous task of reconstruction, one at which the Nationalist regime of the 1920's and 1930's had only begun to nibble.

We in America sometimes pretend that the

Communist element which in 1949 gained the reigns of government in China came to power through some sort of fluke—that the people of China really never wanted the Mao-led faction to rule, and that a combination of circumstances (including the war with Japan) prevented the people from confirming their own choice. The fact is that the Nationalists made quick gains after the war and were in addition supported by United States materials; yet they overextended themselves and quite quickly lost the contest. If the people of China had no opportunity to make their choice known, they certainly did not give much support to the "legitimate" faction led by Chiang Kai-shek. And many thousands of members of the Nationalist faction, upon the success of the Communists, committed themselves to the cause of Peking—in the interests of a new and better China, as they saw it. The United States has for many years refused to "recognize" the present government of China's 900 million people, in part on the grounds that the remnant of the Nationalist faction in Taiwan represents the "legitimate" authority. Elsewhere in the world, we have normally taken the criterion of effective control as the basis for recognition.

We also have displayed a tendency to rejoice over any signs that China's government might be failing in its efforts to improve the lot of the people (as proof that the Communist system does not work), and to take pleasure in reports of revolutionary activities or indications of internal disorder, such as the Red Guard affair in the mid-1960's. These are thoughtless reactions. If we do not want to see China forced to seek an aggressive course of territorial expansion, then an alleviation, not an intensification of the threat of hunger is what we should hope for. And indeed, if our quarrel is with the Chinese government more than its millions of peasants, then surely we can wish the Chinese people well in their search for a daily sustenance. As for any hopes for a breakdown of order in China, we are talking here of a breakdown of order among one-quarter of the world's people. Would this really be in our interest? During the 1950's and 1960's there was a breakdown of stability in a country just south of China, with just one two-hundredth of the world's population. Consider what that struggle has cost the United States, where, too, there were people egging one of the factions on.

Whatever our reactions, in 1949 a Communist government initiated a program of reform in China. The central theme of this program, initially, was the reorganization of agriculture, following the doctrines applied earlier in Soviet Russia. But conditions in China were different from those of Russia. In the Soviet Union, there was a distinction—in terms of productivity—between the poor peasants who formed the majority of the inhabitants of thousands of rural villages, and the subsistence farmers; when the Russian Communist government expropriated the landowning minority it thereby acquired most of the productive land. In China, on the other hand, no such clear distinction existed. There were poor peasants, rich peasants, and "middle" peasants; almost every peasant family sold some produce on the market. As in Russia there was a system of tenancy, which the new government was determined to wipe out. But in China the reorganization of agriculture was not simply a matter of expropriating the rich. Proportionally fewer peasants lived a pure subsistence existence; the risks to production of collectivization were therefore greater.

Nevertheless, the reform program was initiated almost immediately, in every village of Han China (the minorities were at first excluded), involving the redistribution of the land of the landlords among all the landless families of the village. This distribution was based on the number of people per family and was done strictly according to these totals; the landlords themselves were included and received their proportionate share of what had been their own land. Only those landlords who had been guilty of the worst excesses were executed; most were absorbed into the new system. Meanwhile the villagers who were not landlords but who did own the land they worked were allowed to keep their properties; hence the reform program did not produce a truly egalitarian situation.

Cooperatives and Collectivization. During the early 1950's, thus, the whole pattern of land ownership in China was being radically changed. In response, there was an increase in agricultural output, but not as great as the authorities had hoped. True, the taxes the new landowners had to pay on their acquired land were only of the order of one-third of what they previously had to pay their landlords, and so they had some money available to buy fertilizer and improve yields. But in the absence of any prospect of large-scale mechanization of agriculture, the Chinese leaders now began to seek a method whereby still greater productivity could be achieved. The year 1952 produced exceptionally good harvests and encouraged the government to press ahead with a collectivization program. This program envisaged the creation of country-wide cooperatives, in which each peasant was encouraged to pool his land with that of his co-villagers. Compensation would be in proportion to the size of his share; the farmer retained the right to withdraw his land from the cooperative should he so desire. In the beginning participation was slow; by early 1955 only about 15 percent of China's farmers were in these new Agricultural Producers' Cooperatives (A.P.C.). But then the government pressed for greater participation, and by coercion and often quite brutal methods compliance was raised to nearly 100 percent by 1956. Meanwhile the cooperatives were being turned into collective farms, in which it was not the share of the farmer's land, but the amount of his labor that determined his returns; as in Soviet collectives the farmer was able to hold on to a small plot of land for private cultivation. Thus by the end of 1956 virtually the whole countryside was organized into social-

ist collectives; the Chinese peasant, who had briefly held some land under the Agrarian Reform Law of 1950, had lost it again to the collectives into which he had been pressed.

Communes. In 1958, the program of collectivization was carried still one step further. In an incredibly short time—less than a year—over 120 million peasant households, most of them already organized in collectives, were reorganized into about 26,500 People's Communes numbering about 20,000 people each. This was not just a paper reorganization: it was a massive modification of China's whole socioeconomic structure. This was to be China's Great Leap Forward from socialism to Communism. Teams of party organizers traveled the country; opposition was harshly put down while the new system was being imposed. In effect, the new Communes were to be the economic, political, as well as social units of the Chinese Communist State, and they were linked to the already existing *hsien* (see p. 483). But most drastic was the way these Communes affected the daily life of the people. The adults, male and female, were organized into a hierarchy of "production teams" with military designations (sections or teams, companies, batallions, brigades). Communal quarters were built, families disrupted through distant work assignments, children put in boarding schools, households were viewed as things of the past. (Later, the Chinese Communist Party rescinded this order and allowed families to stay together.) The private lands of the collective system were abolished, and even the private and personal properties of the peasants were ruled communal, although this rule was soon relaxed. The wage system of the collectives was changed in favor of an arrangement whereby the farmer or worker received free food and clothing plus a small salary—another step toward the Communist idea.

The impact of the Commune system on the face of China was enormous and immediate. Workers by the thousands tackled projects such as irrigation dams, roads, and so forth; fences and hedges between the lands of former collectives were torn down and the acreages consolidated. Villages were leveled, others enlarged. New roads were laid out to serve the new system better. Schools to accommodate the children of parents in the workers' brigades were newly built. Each Commune was given the responsibility for maintaining its own budget, to make capital

Land reclamation in Sinkiang. In another of the Communist Party's programs, previously uncultivated land in the Gobi Desert area of China is being transformed into farmland. (Eastfoto)

Scene at a railroad stop near Lanchow, as modern train meets ancient handcart. The photograph has more than ordinary importance, for it was taken at the time when the Chinese government was pushing the production of iron at small commune-run blast furnaces. Scrap iron was sought everywhere: here, citizens bring whatever they can muster to the trainstop so that it can be taken to the nearest smelter. (Henri Cartier-Bresson/Magnum)

plored. China still has its own resource frontier, and important discoveries (such as the 1966 Taching oil field in the north) are still being made.

Does China have the resources to allow it to become a major world power, a third force more or less equal to the United States and the Soviet Union? This takes us into the field of *power analysis,* an area of political geography where attempts are made to weigh many factors, geographical and nongeographical, to arrive at an index of national power for each state. In this analysis, resources play an important role, but so do such assets of the state as its communication systems, circulation, quality and quantity of the population (health, skills, etc.), spheres of organization such as education and the military, and so forth. In 1960, Professor F. C. German evaluated 19 world states on these and other bases, and according to his analysis China on

a point basis ranked as follows among the first five states:[11]

United States	6459
U.S.S.R.	6321
United Kingdom	1257
China	999
West Germany	666½

It is likely that a 1970 analysis based on the same premises would raise China's rank to third and close the gap between it and the leading states somewhat, though probably not yet substantially. After the rapid push toward industrialization, there are signs that a slowdown occurred—first during the second Five-Year Plan, when investment priorities had to be altered, and then again during the late 1960's, when the

[11] F. Clifford German, "A Tentative Evaluation of World Power," *Journal of Conflict Resolution,* Vol. 4, 1960, pp. 138–144.

Cultural Revolution may have had something to do with the slower rate of growth. Nevertheless, the industrial base China inherited in Manchuria from the Japanese, and the effective (if temporary) assistance program of the Soviet Union, were diligently exploited in the broadening and strengthening of China's industries and power position. In 1964, using uranium and thorium drawn from domestic sources, China exploded its first nuclear bomb; as a nonsignatory to subsequent nonproliferation agreements it has continued these tests throughout the decade, indicating that an arsenal of such weapons is in the making.

Coal and Iron. Of all its important resources, China is best endowed in coal. In total it may have the world's third largest reserve of this fuel, and more importantly most of the known coalfields lie near transport lines and major urban-industrial centers. We have already seen the wide distribution of coalfields in Manchuria, where they occur at Fushun and Fuhsin, east and west of Shenyang respectively, and in the far northeast, where enormous reserves await exploitation. Large mines exist also in the immediate hinterlands of Tsingtao (Ch'ing-tao) (in Shantung) and Tientsin (in Hopeh), sufficiently near the coast to permit the cheap transportation of coal to the huge city of Shanghai. Neither is interior China without its supplies: a large reserve extends across the provinces of Shenhsi and Shanhsi, and here Taiyuan has become a major industrial center. Shanhsi coal also goes to Inner Mongolia, where it is used with local iron ore in the iron and steel plants of Paotou on the Hwang Ho. So dispersed is China's coal that the Chinese government has been able to encourage widespread industrial development, from Kunming in southwestern Yunnan all the way to Harbin in Manchuria, and from Wuhan in centrally positioned Hupeh Province to Paotou in Inner Mongolia. There is a certain security in such industrial dispersal (something we had occasion to refer to with reference to the Soviet Union as well), but like all industrial states China also has distinct core areas where the majority of the heavy industries are located and where most of the resources they require are consumed. China's industrial heartland extends from Manchuria to Shanghai, and all the major centers in it—Shenyang, Anshan, Tientsin, Wuhan, Shanghai—either have coal nearby or can obtain it cheaply by water. In terms of coal, there is

hardly a limit on industrialization in China; the quality is good, and the deposits are mainly near the surface and easily mined.

The same cannot be said for China's iron ores. Although there is a fair quantity of iron ore, much of it in southern Manchuria, the ore often is of rather poor grade, requiring the removal (at considerable expense) of numerous impurities. The highest grade ores appear to lie near the Korean border and near Paotou in Inner Mongolia, and an apparently significant deposit has been found on the island of Hainan. It may be, of course, that additional iron ores will be discovered in China; one recent find was in the Nan Shan on the Chinghai-Kansu border. But the pattern of iron reserves in China is well established: they are scattered, most of mediocre or low quality, costly to convert and, in their impure state, too expensive to move very far. Thus the iron situation in China, insofar as is known, represents something of an obstacle to a large-scale industrialization program, at least in terms of the exceptionally large investments required for the treatment and conversion processes. Nevertheless, iron and steel production in China has increased greatly since 1949 (when steel production was *under* 1 million tons); in 1964 it appears to have been 11 million tons and the plants at Anshan alone were producing 5 million tons. In 1970 production may approach 20 million tons, compared to the 1968 figures for the U.S.A. (127 million tons), West Germany (37) and the United Kingdom (24).

Iron and steel cannot be produced without the addition of other substances, including a variety of alloys, depending on the function the steel is to perform. In these nonferrous metals China is rich. Tungsten occurs at Tayu in the Southeast Uplands, and antimony in Hunan Province; China may have more than half the world's reserves of these metals. A large tin deposit lies in Yunnan, near the Vietnamese border, and there is copper near the Yunnan-Szechwan provincial boundary. Thus several of these nonferrous metals are scattered through the southern hills and mountains of China—not the most fortunate location but preferable still to a dependence on overseas sources. In any case, they need not be moved in large quantities. China also has deposits of aluminum ore as well as lead and zinc—and, as has been noted previously, lots of prospecting still remains to be done in China's distant west.

A view of the No. 1 and No. 2 blast furnaces of the Wuhan steel center. This plant, along with many others in China, was built with Soviet aid and technical assistance during the postwar period of friendship and cooperation. (Eastfoto)

Petroleum. For a long time China depended on imports of petroleum, after 1949 from the Soviet Union, but according to Chinese claims, self-sufficiency has now been achieved. This of course is difficult to verify; China's needs must be rising rapidly and none of the reserves to have been discovered (northwest Sinkiang, the Tsaidam Basin, the old Yumen field, the Szechwan field north of Chengtu) has a particularly fortunate location with reference to the main markets. But again, production has risen rapidly after 1949 (when it was reported as 1.5 million tons); in 1963 it was 5.5 million metric tons and in 1968 it was thought to be some 10 million tons. Oil is also being won from Manchuria's oil shales, found in association with the coal deposits. But by comparison China is much better off with coal as its power source, and even its hydroelectric potential is brighter. On the Hwang Ho, the Yangtze, in Szechwan Province, and in the Southeast Uplands, among other places, dams are being constructed and hydroelectric plants put into operation. Despite China's great opportunities in this sphere—which are far greater than those of the United States, incidentally—it cannot move as fast as it would like in its devel-

opment. Unlike other developing countries, China does not have the World Bank at its disposal; investments in hydroelectric projects are very high and the returns, considering the capacity of the local market, cannot be expected to be very substantial. As in so many other areas, China has to go it alone, and so development is to some extent delayed.

Power Resources. China affords an interesting illustration of a concept contributed by Professor S. B. Jones, relating to the problem of evaluating the actual and potential power of a state. After all, China seems to be full of potential, but much of it is yet unrealized; how can all these different kinds of resources be weighted in any power analysis? Professor Jones suggested that we recognize five groups of actual and potential resources, as follows:[12]

1. *Power resources available immediately.* These include factories and mines producing equipment and minerals for immediate use, the standing army, stockpiled warheads. In China, the aluminum ores (essential in nuclear bomb pro-

[12] S. B. Jones, "The Power Inventory and National Strategy," *World Politics,* Vol. VI, No. 4, 1954, p. 421.

duction), thorium extracting facilities, and uranium plants are all part of this category, as well as the truck factories of Yunnan and the steel mills of Anshan.

2. *Power resources available after activation.* The reserve army (an exceptionally potent force for China, hence American fears of a land war in Asia), mothballed military equipment, factories and mines not operating now but capable of operation in a few weeks in the event of need. China is not likely to have a large number of inactive factories and mines.

3. *Power resources available after conversion.* Factories building tractors can convert to military material; shipyards can start producing naval vessels rather than merchant ships. Here China has a sizable base, which could be turned into a military complex in the event of need. At the time of the Great Leap, the communes were encouraged to build their own "backyard" iron furnaces. The program failed, but chances are that a threat to national security could mobilize the entire effort and direct it toward military purposes.

4. *Power resources available after development.* Included here are mineral deposits known to exist, but not yet exploited. In China, there are coal fields still dormant as more favorably located reserves are being mined; there are iron ores not touched while higher-grade reserves are tapped. All these are available and must count in the analysis of China's potential power.

5. *Hypothetical resources.* These are resources whose existence is conjectured but not yet proved. In China there are still numerous areas where what is known of the geological formations, their age, and rock associations, suggest that minerals may well be found. Part of this hypothetical resource base is likely to be discovered and then moves to category (4). China is unusually strong in category (5), and it is likely that we shall hear continuously of new discoveries of oil, iron ore, alloys, or other mineral resources, as the less hospitable parts of the country are put to modern tests of exploration.

Power. This brings us to the concept of state power itself. Just what is the "power" of a state, and how does such power manifest itself in international affairs? We use terms such as "second-rate" and "fourth-rate" power (North Korea, after it arrested the Pueblo, was described by a Republican Presidential candidate as a "fourth-rate" power) without really defining just

what we mean. Surely China now has "second-rate" power status, but despite its obvious superiority over its Asian perimeter (except the Soviet Union) China has not really asserted itself. It has relinquished a sphere of influence in Outer Mongolia, engaged in half-hearted skirmishes over boundary issues with India, avoided direct involvement in the Viet Nam conflict, allowed North Korea to slip out of its ideological sphere, and permitted Taiwan to remain as a hostile force on its very doorstep. True, Taiwan is in effect a U.S. protectorate, but the fact remains that China, despite its nuclear bombs and its self-proclaimed search for power, has not yet really asserted itself in Eastern Asia.

Professor J. G. Stoessinger has defined power as "the capacity of a nation to use its tangible and intangible resources in such a way as to affect the behavior of other nations."[13] Note that this definition stresses not the mere existence of the necessary resources, but the state's *capacity to use these resources* effectively. This involves organization—efficient organization of the state's many functions. And it is exactly in this area that China seems after 1949 to have failed repeatedly. After the initial successes of land reform, the Great Leap Forward demanded too much too soon and the whole framework broke down. Then the ideological equivalent of the Great Leap, the Cultural Revolution, once again damaged the order that was being achieved. It may be that China in 1970 is once again at a takeoff point, but there can be no doubt that the nation has yet to achieve a position in the world commensurate with the numbers of its people and the range of its resources.

The Railroads and Feng Shui

One aspect of a state's organization is its railroad network, a reflection of its internal circulation and external interconnections. China's first railroad was laid in 1876, a short line from Shanghai westward. What happened to this innovation is in some ways symptomatic of old, rejection-minded China. So intense was the local opposition to the rail line that it seemed unlikely ever to function there, so it was taken up again and shipped to Taiwan.

What was the reason for this bitter opposition to the foreigners' invention? It was the same resistance that had delayed the exploitation of

[13] J. G. Stoessinger, *The Might of Nations,* Random House, New York, 1961, p. 31.

minerals, the building of bridges, the stringing of telegraph wires between towns: *feng shui*. Feng shui is an ancient belief, still held in rural China, that powerful spirits of dragons, tigers, and other beings occupy all natural phenomena, mountains, hills, rivers, the air, trees, and so forth. For fear of incurring their wrath, nothing should be done to nature without consulting these spirits, and this was the function of China's *geomancers*, men who knew the desires of the spirits. Everything must be done according to their grand plan, and while some matters could be adjudicated by the geomancers, others were so obviously contraventions of feng shui that whole communities rose in opposition to them. Not surprisingly something so objectionable as a railroad simply could not be tolerated by feng shui; one did not need the geomancers to know that this was against the will of the spirits. And so, in 1876, that first Shanghai railroad had to be taken away.

Feng shui is not dead by any means. When the Peking regime, in the process of implementing the commune program, leveled burial mounds to make room for farmland and communal facilities, it thereby aroused still greater opposition from those who considered this a violation of feng shui. Burial had always been a matter of geomancy, the higher a person's rank or class, the higher was the level at which he was buried. Burial grounds were an important part of the structure of Chinese communities, and to negate this importance was a major mistake on Peking's part—a mistake, as it happens, the foreigners before 1949 had often made. In the name of feng shui, China's people resisted foreign activities such as mining (an especially onerous practice if not sanctioned by geomancers), the building of roads and railroads, and the erection of telegraph posts and lines. This was not just verbal opposition: these facilities were literally torn down by the angry populace.

Neither is feng shui merely a matter of the most isolated interior areas of China today. Within the past decade the British administration in Hong Kong was prevented from implementing plans to build a road in the colony because of opposition from the area's peasantry— based on feng shui and clearly stated in these terms. Geomancy is still a powerful force in China.

Unquestionably feng shui has contributed to the slowing down of economic development in China, possibly even after 1949; there can be no doubt that the opposition encountered during the late nineteenth century was largely responsible for the virtual standstill in railroad building, for example. Not until the turn of the twentieth century did railroad construction begin in earnest, mostly with foreign capital and loans. Peking was connected to Shanghai and to what is now Wuhan, and when these lines were later extended farther south ferries were employed to carry the trains over the Yangtze. Today the Yangtze is bridged at Wuhan, a remarkable engineering feat, and also at Chungking. Peking was also connected westward with Paotou and northward with Shenyang, then still called Mukden. Railways were also laid from the coalfields of northern China to the ports of Tsingtao and Tientsin, from where the coal was shipped to Shanghai.

The next active phase of railroad construction came in the late 1920's and 1930's, when Japan and Soviet Russia both developed the Manchurian network; by the time the war with Japan broke out Manchuria alone had about half of the total mileage in the whole country—but that total was a mere 6500 miles, a minimal figure for so large and populous a country. To put the comparison another way: in the United States, railroad mileage was about 250 for every 100,000 people shortly after World War II, and in China the figure was 2. Check an atlas dated in the late 1940's and look at the map of world railroad, and the contrast in density between the United States and China is staggering.

The contrast is still there today, but the Chinese now have what they never really had before: a coordinated plan for the development of their transport network (Map 12-12). The last such plan to be drawn up (though never executed) was that of Dr. Sun Yat-sen, but China in this century rarely had the stability and never had the capital to carry out his scheme for connecting the provinces and the resource locations in the country. Peking was not connected to Canton until the 1930's, with the Yangtze still awaiting a bridging; different gauges were being used, and the whole system was haphazard, except in Manchuria.

In 1949 the Communist regime found itself with two areas where a reasonably dense network of railroads existed: Manchuria and the eastern part of northern China Proper. With its limited capacity it was faced with a difficult decision·

Map 12-12 CHINA'S RAILROAD NETWORK.

This is the famed Yangtze River Bridge at Wuhan, which links the northern and southern provinces of China. (Li Chiu-ling/Eastfoto)

to weigh various factors in determining where the greatest need for rail construction lay. This was not just a matter of economics. There was always the possibility of further conflict with invasion-minded Taiwan, and no railroad to the Formosa Strait existed. No connection had been built to North Viet Nam, another area of strategic importance. Then there was the obvious need for railroads linking the Chinese core area to the distant west: such railroads would constitute crucial aids in the effort to consolidate the state and make the Chinese presence felt in an ever-problematic frontier. In the first Five-Year Plan these considerations were put ahead of any intensification of the northern and northeastern networks. From the Shanghai-Canton line railroads were thus laid to Amoy and Foochow on the Formosa Strait, and from Nanning a link was built to connect to Viet Nam's Hanoi line. Almost all of the remaining construction took place in the west: from Paotou to Urumchi, and from Lanchow north to this line and south to Szechwan Province. The great western railroad,

subsequently extended to connect to the Soviet system, is known as the Lunghai Railway, and its completion represents a major Chinese engineering feat: the line crosses shifting sand areas and rugged mountains. Soviet assistance was vital in this immense effort.

During the late 1950's and 1960's attention has been given to the eastern railroads, while plans were laid for further extensions in the west. Double-tracking was achieved all the way from Harbin via Peking to Canton, electrification was completed on the first of several main lines, the Yangtze was bridged (1957) at Wuhan and Chungking, and local areas were successfully encouraged to build their own, often makeshift feeder lines. This renewed attention to eastern railroads was necessary because the rate of industrial growth was beginning to put severe pressure on the available facilities. But the Chinese have not forgotten their plans in the west: railways are to be built to Sufu (Kashgar), China's western-most town in Sinkiang, and into the Tsaidam Basin and through Tibet to Lhasa.

TAIWAN

One hundred miles off the coast of China, opposite Fukien Province and the Southeast Uplands, lies an island with a population one-seventieth as large as China and with a territory about half the size of West Virginia. This is Taiwan, the "other" China, or, as some public officials in the

United States still maintain, the "legitimate" China. After World War II, Taiwan received more than two million refugees from mainland China, among them the remnants of the faction led by Chiang Kai-shek, including the General himself. Assisted by the United States, Taiwan has maintained itself on China's very doorstep, and from time to time there is talk on the island of an invasion and a return of the Nationalist government to its mainland capital.

Mountainous Taiwan is one of a huge belt of islands and archipelagos that stretches along Asia's eastern coasts from the Kurlis north of Japan to Indonesia south of the Philippines. Unlike many of the other islands, Formosa (as the Portuguese called the island when they tried to colonize it) is quite compact and has rather smooth coastlines; no good natural harbors exist. In common with several of its northern and southern neighbors, the island's topography has a linear, north-south orientation and a high, mountainous backbone. In Taiwan this backbone lies in the eastern half of the island (Map 12-13); elevations in places exceed 10,000 feet. Eastward from this forested mountain backbone, the land drops very rapidly to the coast and there

is very little space for settlement and agriculture, but westward there is an adjacent belt of lower hills and, facing the Formosa Strait, a substantial coastal plain. The overwhelming majority of Taiwan's 14 million inhabitants live in this western zone, near the northern end of which the capital, Taipei, is positioned.

There have been times when Taiwan was under the control of rulers based on the mainland, but history shows that its present, separate status is really nothing new. Known to the Chinese since at least the Tang Dynasty, it was not occupied until the 1400's. For a time there was intermittent Chinese interest in Taiwan, and then the Portuguese arrived. The 1600's witnessed a period of conflict among the Europeans, including the Hollanders and the Spaniards, but in 1661 a Chinese general landed with his mainland army and ousted the foreign invaders. This was during the decline of the Ming Dynasty and the rise of the Manchus, and, much like the Nationalists of the 1940's, hundreds of thousands of Chinese refugees fled to Taiwan rather than face Manchu domination. But before 1700 the island fell to the Manchu victors and became administratively a part of Fukien Province. By then the indigenous population, of Malayan stock, was already far outnumbered and in the process of retreating into the hills and mountains.

Taiwan did not escape the fate of China itself during the second half of the nineteenth century. In 1895 it fell to Japan as a prize in the war of 1894–1895, and for the next half century it was under foreign rule. The Japanese saw some of the same possibilities in Formosa as they saw in Manchuria: the island could be a source of food and raw materials, and, if developed effectively, a market for Japanese products. Thus Japan engaged in a prodigious development program in Taiwan, involving road and railroad construction, irrigation projects, hydroelectric schemes, mines (mainly for coal), and factories. The whole island was transformed; farm yields rose rapidly as the area of cultivated land was expanded and better farming methods were introduced; in the sphere of education the Japanese attacked the illiteracy problem and oriented the entire system toward their homeland.

Japanese rule ended with Japan's collapse (1945) in World War II, and briefly the island was to return to Chinese control, but before long

MODIFIED AFTER: FOCUS, A. G. S.

Map 12-13 TAIWAN.

it became the last stronghold of the Nationalists and once again its mainland connections were severed. A large influx of refugee Nationalists arrived, and they were fortunate: here was one of the few parts of China where they could have found a well-functioning economy, productive farmlands, the beginnings of industry, good communications—and the capacity to absorb an immigrant population of some two million or more. True, American assistance made a major difference. But Taiwan had much to offer.

Nationalist "China." From 1945 until 1949, Taiwan was officially part of China, torn as it was by the civil war. This period was one of quite ruthless and generally unpopular government on the island, which had to do with the degree to which the Japanese had transformed its life and culture, and the large residue of sympathy they had left behind. But 1949 brought a new phase with the installation of Chiang Kai-shek's regime in Taipei and the arrival of some two million immigrants, who constituted about 25 percent of the total population in 1950. In some ways the problems Chiang's regime faced were similar to those of the new Peking government. The Japanese had achieved much, but the war had brought destruction; as on the mainland there was a need for land reform and increased agricultural yields. While American assistance helped reconstruct Taiwan's transport network and industrial plants, the Taipei government initiated a program whereby the farm tenancy system was attacked. In 1949, most Taiwanese farmers still were tenant farmers who paid rent amounting to as much as 70 percent of the annual crop. By law, the maximum allowed rent was reduced to 37.5 percent; the land of large landowners was bought by the government in exchange for stock holdings in large government-owned corporations, and this land was sold to farmers at low interest rates. Through incentives, seed improvement, additional irrigation layouts, more fertilizers, and new double-crop rotations it has been possible to double the yields per acre of Taiwan's major crops.

These efforts to raise the volume of the grain harvest are familiar to anyone who has studied the Peking regime's attempts to do the same on a much larger scale. Taiwan lies astride the Tropic of Cancer and thus its lowland climate is comparable to that in the Hsi Kiang basin; rice is the leading staple and about two-thirds of the harvest is from double-cropped land.

Wheat and sweet potatoes also are important staples; sugar cane in the lower areas and tea higher up are grown for cash. But in at least one respect Taiwan faces problems even more serious than China itself: the rate of population growth is an almost unequalled 3.4 percent, so that annual harvests must increase spectacularly just to keep pace. And of course Taiwan's available land is very restricted, so that on the coastal plain rural population densities rank among the highest in the world. In the best farming areas they are reported to exceed 3400 people per square mile.

Notwithstanding Japan's big effort to promote industrialization on Taiwan, the country remains mainly agricultural; domestic resources (except sources of power) are very limited and Taiwan is cut off from its most natural supplier of external raw materials. Along the western flank of the main mountain chain there are coal deposits, and numerous streams on this well-watered isle provide ample opportunities for hydroelectric power development (see Plates 1 and 2). There is even petroleum and natural gas. But the mineral wealth is not great; there are no large iron ores, for example. Hence the food processing industries rank first in Taiwan, and include pineapple canning, flour milling, and tea processing; the textile industry, which ranks second, gets its raw material in the form of cotton from the United States. The mountains' forests serve as a good local base for the thriving paper and pulp industry, which also uses the waste from sugar cane processing plants, and local limestone and silica-rich sand sustain a cement industry. But ferrous and nonferrous minerals must be imported: the small aluminum industry, for example, gets its bauxite from Indonesia.

Taiwan remains in effect an American protectorate, a solid bastion in support of the American presence in eastern Asia. A military alliance was signed in 1954, during the Eisenhower administration, with Chiang Kai-shek. But neither Peking nor Taipei consider the domestic struggle closed: Peking awaits the "liberation" of this last of China's provinces, while Taipei continues to talk of the "liberation" of the mainland. Once again the United States is directly involved on one side in a latent intra-Asian conflict; in the eyes of many Asians, America is allied with a replica of the Old China against the rising power of the New China.

ADDITIONAL READING

China, like India, is covered in detail in several geographies that deal with Asia as a whole or with all of non-Soviet Asia. Among the standard works in this group are J. E. Spencer's *Asia, East By South, a Cultural Geography,* a Wiley publication of 1954 and highly recommended, N. S. Ginsburg, et al., *The Pattern of Asia,* published by Prentice-Hall, Englewood Cliffs, N.J., in 1954, and G. B. Cressey's *Asia's Lands and Peoples,* a third edition of which was published by McGraw-Hill in 1963.

On China specifically, a good source is K. Buchanan, *The Chinese People and the Chinese Earth,* published by Bell in London in 1966. A basic geography is T. R. Tregear's *A Geography of China,* published by Aldine in Chicago in 1965. Also see T. Shabad, *China's Changing Map: A Political and Economic Geography of The Chinese People's Republic,* a Praeger publication of 1956. The book by G. B. Cressey, *Land of the 500 Million: A Geography of China,* was brought out by McGraw-Hill in New York in 1955 (when China may have had a mere 500 million people). Of more recent vintage are O. E. Clubb's *Twentieth Century China,* published by Columbia University Press in 1964, and A. Donnithorne, *China's Economic System,* published by Praeger in 1967, as well as V. P. Petrov, *China: Emerging World Power,* in the Van Nostrand Searchlight Series, in 1967. Another volume focused on China in this series is that by Chiao-Min Hsieh, *China: Ageless Land and Countless People,* Van Nostrand, 1967.

Historical perspectives are obtained by consulting Albert Herrmann's *Historical Atlas of China,* edited by N. Ginsburg and published by Aldine, Chicago, on 1966, and through the book by K. S. Latourette, *The Chinese: Their History and Culture,* the fourth edition of which was published in 1964 by Macmillan, in New York. Also see O. and E. Lattimore, *China, a Short History,* published in New York by Norton in 1947.

Despite its alleged shortcomings, Edgar Snow's *Other Side of the River—Red China Today,* published in 1962 by Random House in New York, remains one of the most challenging and informative works currently available. Also see the collection of essays edited by Ruth Adams, *Contemporary China,* a Vintage paperback published by Random House in 1966. To put the discussions of China's power in a politico-geographic perspective, see J. G. Stoessinger, *The Might of Nations,* another Random House publication of 1961.

On the interaction between China and its Asian neighbors, see A. Lamb, *The China-India Border: The Origins of the Disputed Boundaries,* Oxford University Press, 1964, published in New York; O. Lattimore, *Pivot of Asia: Sinkiang and the Inner Asian Frontiers of China and Russia,* published in Boston by Little Brown in 1950; and H. J. Wiens, *China's March Toward the Tropics,* published by Shoe String Press, Hamden, in 1954; G. S. Murphy, *Soviet Mongolia: A Study of the Oldest Political Satellite,* published by the University of California Press, Berkeley, in 1966; and C. P. Fitzgerald's *Chinese View of Their Place in the World,* published by Oxford University Press in New York in 1964.

13

SOUTHEAST ASIA: BETWEEN THE GIANTS

SOUTHEAST Asia, Asia's southeastern corner of peninsulas and islands, is bounded by India to the northwest and by China to the northeast; its western coasts are washed by the Indian Ocean and to the east stretches the vast Pacific. From all these directions, Southeast Asia has been penetrated by outside forces. From India came traders and from China came settlers. Across the Indian Ocean came the Arabs to engage in commerce and the Europeans in pursuit of empires. And from across the Pacific came the Americans. Southeast Asia has been the scene of countless contests for power and primacy; the competitors have come from near and far. Like Eastern Europe, Southeast Asia is a region of great cultural diversity; again like Eastern Europe, it is also a shatter belt.

Southeast Asia is nowhere nearly as well-defined a culture realm as India or China. It does not have a single, dominant core area of indigenous development. Culturally it is diverse in every respect: there are numerous ethnic and linguistic groups, various religions, and different economies. Even spatially the region's discontinuity is obvious: it consists of a peninsular mainland section and hundreds of islands which form the archipelagos of Indonesia and the Philippines. The mainland area, which is smaller than India, is divided into no less that eight political entities (Map 13-1), and the islands, in addition to Indonesia and the Philippines, contain two of the world's remaining dependencies, namely Portuguese Timor (the eastern part of that island) and the British-protected Sultanate of Brunei, in Sarawak on Kalimantan. This political fragmentation not only underscores the cultural complexity of the region, but also relates to the history of European colonial intervention. Except Thailand, none of the states on the map of Southeast Asia today was independent when World War II came to an end.

Map 13-1 SOUTHEAST ASIA: POLITICAL UNITS.

520

Population Numbers. After the population totals and densities to which we became accustomed in India and China, and figures for the countries of Southeast Asia almost seem to suggest a sparseness of people (Table 13-1).

Table 13-1   Data on Southeast Asia

State	Area (square miles)	Population (1970 Estimate)
Burma	261,788	26,600,000
Thailand	198,455	34,700,000
Laos	91,428	2,900,000
Cambodia	69,898	7,000,000
North Viet Nam	61,293	20,500,000
South Viet Nam	65,987	16,900,000
Malaysia	128,298	10,900,000
Singapore	224	2,070,000
Indonesia	735,268	117,000,000
Philippines	115,830	37,200,000
Brunei	2,226	126,000
Portuguese Timor	5,763	569,000

Laos, quite a large country, has a population of under 3 million and an average density per square mile of 32, which is even less than desert-dominated Iran. In fact, much of the interior of mainland Southeast Asia has population densities comparable to those of savanna Africa, Soviet Central Asia, and the highland rim of South America's Amazon Basin (Plate 6). And even the coastal areas of this region have fewer people than those elsewhere in Southern and Eastern Asia. As Plate 6 shows, there is nothing in Southeast Asia to compare to the immense human agglomerations of the Ganges Basin, the Malabar Coast, or China's Yangtze and lower Hwang Ho Basins. The whole pattern is different: Southeast Asia's few dense population clusters are relatively small and lie separated from each other by areas of much sparser occupance.

Why the difference? When there is such population pressure and such land shortage in adjacent regions, why has Southeast Asia not been flooded by waves of immigrants? Several factors have combined to inhibit large-scale invasions. In the first place, the overland routes into Southeast Asia are not open and unobstructed. In discussing the Indian subcontinent, we noted the barrier effect of the densely forested hills and mountains that coincide with the border between Indian Assam and western Burma. North of Burma lies forbidding Tibet, and northeast of Burma and north of Laos is the high Yunnan

Plateau. Transit is easier into North Viet Nam from southeastern China, and along this avenue considerable contact and migration has occurred. Neither is contact within Southeast Asia itself helped by the rugged, somewhat parallel ridges which hinder easy east-west communications between the fertile valleys within the region (Map 13-2). Second, and the population map reflects this, Southeast Asia is not an area of limitless agricultural possibilities and opportunities. Much of the region is covered by dense, tropical monsoon forest and rain forest, parts of an ecological complex whose effect on human settlement we have observed in other parts of the world as well (Plate 4). Except for certain local areas, the soils of mainland Southeast Asia are excessively leached by the generally heavy rains (Plate 1). In the areas of monsoonal and savanna regime (the latter prevails over much of the interior), there is a dry season, but we know the effect of savanna conditions upon agriculture (Plate 2). High evapotranspiration rates, long droughts, hard-baked soils, high runoff, and erosional problems add up to anything but a peasant's paradise in these parts of Southeast Asia.

The Exceptions. Where and how do those people who have made Southeast Asia their habitat make a living? Or, to put the question differently, under what conditions have those scattered clusters of dense population developed that are revealed by the population map? Natural barriers and historic circumstances combined to delay the arrival in Southeast Asia of those groups that are now dominant in the northern part of the region. And those areas that have proved to be most suitable for the development of productive agriculture also came into use at a late stage—in fact, this was done by the European colonial powers, mostly within the past century and a half. Even today, Southeast Asia is less urbanized and industrialized than India and China, and the major population agglomerations (apart from the cities) are centered in the limited areas of really good agricultural land. Three kinds of areas can be identified. First, there are the valleys and deltas of Southeast Asia's major rivers, where alluvial soils have been formed. Four major rivers stand out. In Burma, the Irawaddy rises near the border with Tibet and creates a delta in the Bay of Bengal. In Thailand, the Chao Phraya traverses the length of the country and flows into the Gulf of Siam.

Map 13-2 SOUTHEAST ASIA: PHYSIOGRAPHY.

In South Viet Nam lies the extensive delta of the great Mekong River, which rises in the high mountains on the Chinghai–Tibet border in interior China and crosses the whole peninsula. And in North Viet Nam lies the Red River, whose lowland probably is the most densely settled area in Southeast Asia (the Tongkin Plain). Each of these four river basins contains one of mainland Southeast Asia's major population clusters, which are in effect the core areas of those four countries.

Southeast Asia is a tropical region. It lies almost entirely within the tropics, and its peninsulas and islands are interdigitated with warm, equatorial ocean waters. As Plate 2 indicates, the region is dominated by the three varieties of A climate—the monsoon, the savanna, and the rain forest. Only in northern Burma and in northern Laos and North Viet Nam do subtropical conditions occur; elsewhere the only amelioration is provided by elevation (the H, gray-colored areas on Plate 2 reflect this). The prevalence of latosolic soils is reflected by Plate 3; note the exceptions in the Irawaddy, Chao Phraya, and Mekong areas. And the forest-dominated vegetative situation is shown by Plate 4. Southeast Asia, however, is known not only for its heat and humidity, but also for its volcanic mountains—at least in the archipelagos. Mount Krakatoa, in the Sunda Strait between Djawa (Java) and Sumatera (Sumatra),[1] erupted with such force in 1883 that the whole mountain blew away, sending heavy layers of ash to cover forests and hillslopes over hundreds of square miles around, causing ocean waves that killed perhaps as many as 40,000 people in coastal villages, and blowing dust into the skies that darkened the daylight of Djakarta and Bandung, a hundred miles away. Not all of the islands of Southeast Asia have soils that are derived from volcanic lava, nor do all lavas necessarily produce fertile soils. But in certain parts of the islands the conditions are all right for the formation of deep, dark, rich volcanic soils, especially in much of Djawa. The population map indicates what this fertility means: the island of Djawa is one of the world's most densely populated and most inten-

sively cultivated areas of the world. On Djawa's productive land live well over 70 million people—nearly two-thirds of the inhabitants of all the islands of Indonesia and more than a quarter of the population of the whole Southeast Asian realm. It is hardly necessary to identify Djawa as Indonesia's core area. But there are other areas where volcanic soils support concentrations of people: in Sumatera's center and north, in Sulawesi (Celebes), and in a few localities in the Philippines.

Thus the major population concentrations seem to be based on the large mainland areas of alluvial soil and the islands' volcanic soils. In the Philippines and in Kalimantan (Borneo), smaller clusters of people depend on the more confined alluvial deposits; in the intervening zones, shifting, subsistence cultivation is the rule. But another look at Plate 6 indicates that one area remains to be accounted for: the belt of comparatively dense population that extends along the western coast of the Malay Peninsula, apparently unrelated to either alluvial or volcanic soils. This represents a third basis for population agglomeration in Southeast Asia, namely the plantation economy. Actually, plantations were introduced by the European colonizers throughout most of insular Southeast Asia, but nowhere did they so transform the whole economic geography as here in Malaya. Rubber trees were planted on tens of thousands of acres, and Malaya became the world's leading exporter of this product. Fortunately, as it happens, there is a belt of tin deposits along the western edge of the spine of the Malay Peninsula, and so the transport system that was developed could serve both the rubber industry and the mines. In terms of soil quality, certain plantation crops (especially the tree crops) do not depend so much on rich soils, so that the problem of competition with food crops is lessened. Undoubtedly Malaysia would not have developed so populous a core area without its plantation economy.

Where there are no alluvial soils, no volcanic soils, and no profitable plantations, as in the rain forest areas and in the higher areas where there is more slope and not very much level land, far fewer people manage to make a living. Here the practice of shifting, subsistence cultivation prevails, augmented sometimes by hunting, fishing, and the gathering of wild nuts, berries, and the like. This practice is similar to that we observed in South America and Equatorial Africa, and

[1] As in Africa, names and spellings have changed with independence in Asia. In this chapter, the modern spellings will be used except when reference is made to the colonial period. Indonesia's four main islands are Djawa, Sumatera, Kalimantan, and Sulawesi. The Dutch called them Java, Sumatra, Borneo, and Celebes.

There are ancient conflicts in Southeast Asia between the peoples of the mountains and the peoples of the lowlands. In Viet Nam, for example, the Montagnards have long lived in relative isolation, and they are now caught in the struggle for that country. This photograph shows part of a Montagnard village in South Viet Nam. (Fujihira/Monkmeyer)

it is capable of sustaining a rather sparse population: land that has once been cleared for cultivation must be left alone for years to regenerate. Here in the forests and uplands of Southeast Asia the nonsedentary cultivators still live in considerable isolation from the peoples in the core areas; the forests are dense and difficult to penetrate, distances are great, and the strong relief adds to the obstacles in the establishment of surface communications. Over a major part of their combined length, the political boundaries of Southeast Asia lie through these rather sparsely peopled, inland areas. Here are the roots of some of the region's political troubles: these interior, unstable zones never have been effectively integrated in the states of which they are part. The peoples who live in the forested hills may not be well-disposed toward those who occupy the respective core areas, and so these frontierlike inner reaches of Southeast Asia, with their protective isolation, distance from the seats of power, and dense forest cover, are fertile

ground for revolutionary activities if not for farm crops.

When the shifting cultivators girdle-burn the forest, they plant a variety of crops. Some they will harvest very soon, but others they will simply leave, to collect perhaps two years later when and if the need arises. As elsewhere in wet, tropical areas, the root and ground crops are tried most often. Cassava, yams, squash, and melons grow readily for a season or two where the forest has been partially cleared, and patches of corn, rice, millets, taro, and beans are also grown. This kind of fire-field agriculture is still prevalent over most of the total area of Southeast Asia, although most of the total population is concentrated in the areas of greater productive capacity.

This brings us back to the question of why the majority of Southeast Asia's people are clustered so strongly in a few prominent core areas. Having explained this on the basis of the fertility of alluvial and volcanic soils, we are left to wonder why these clusters are not larger—why Southeast Asia's agricultural frontier has not expanded more deeply into those zones where conditions, though not ideal, are still good enough to support productive agriculture. Of course Southeast Asia does not have the population pressure of either China or India, but there are beckoning opportunities for the expansion of rice production (and ready markets nearby), and sizable areas could carry many more people than they do. This concerns several thresholds: the index of population numbers and pressure is not great enough yet; the available local capital cannot yet be allocated to the large-scale forest clearing, swamp draining, and construction of the necessary irrigation works; and there has to be a determination of the people themselves, or their government, to push outward into these untried areas.

Indochina[2]

Although there are barriers to immigration from the north, and although the environment of Southeast Asia presents limitations to human settlement over large areas, the region has nevertheless been invaded by outsiders for many centuries. As we reported earlier, these invasions involved smaller numbers than we might have expected in view of the very large and dense

[2]The term *Indochina* is used here as a cultural name for the whole region; the French adopted it to identify their colonial possessions in the eastern mainland.

populations of the adjacent Indian and Chinese realms, but the overwhelming majority of the people today are of Mongoloid stock. The strongest ethnic affinities are with the people of southern China; anthropologists suggest a differentiation between the peoples of the northern mainland, such as Viet Nam's Annamese (Vietnamese), Thailand's Thai, and Burma's Burmese (Burmans) and those of the Malay Peninsula and the Indonesian and Philippine islands, who are identified as Malay. In any case, the peoples who moved southward into Southeast Asia displaced still earlier inhabitants, and pockets of these aboriginal peoples still exist in isolated parts of Southeast Asia. Some were Pygmy Negritos. Djawa is one of those places in the world where evidence exists of very early human habitation (Pithecanthropus).

Mainland Southeast Asia is appropriately called *Indochina*, suggesting the two main Asian influences that have affected the region over the past 2000 years. While most of the migrants moved southward from southern China, often pushed by the imperialism of the people of Han, the Indians were the first to give much of Southeast Asia a strong cultural imprint. By the beginning of the Christian era, Indian trading ships were plying the coasts, and communities of Indian immigrants from Bengal, Orissa, and the Coromandel Coast were being established on the Malay Peninsula, in the Mekong Delta, on Djawa, and on Kalimantan. These settlers created numerous small states, imposed on the indigenous population, and they introduced the Indian faiths of Buddhism and Hinduism. A French geographer, Professor C. Robequain, describes what happened on Sumatera during the first several centuries of Indian contact:

". . . The first Indo Malay kingdom had its center in the Palembang District, which, however, is very poor in archeological remains. Sanscrit inscriptions and Chinese and Arab writings describe this kingdom of Sri Vijaya as a rich and powerful state. The territory under its rule had shifting boundaries, but between the 8th and 12th centuries extended not only over the Minangkabau and Batak districts of Sumatra, but also went far beyond the island to reach Cambodia, Siam [Thailand], and Ceylon, to cover the greater part of Java and the coastal districts of Borneo, and thence to stretch out through Banjermasin and Brunei to the Philip-

pines. . . . [The] capital was a center of Buddhist teaching and missionary activity to which men came from very great distances to study the sacred volumes."[3]

During the period from about 800 to 120 A.D., while the Indian influence in Southeast Asia was at its zenith, many great temples were built, some of which still survive to bear witness to this phase of history. Angkor Wat, capital of the Khmers, in present Cambodia, was built during this phase. Among the most impressive structures is the Stupa of Borobudur, in central Java. Elsewhere, numerous temples built later still carry the old Indian form, often mixed with more recent Chinese influences.

The Hindu and Buddhist Indians were followed by the Moslem Indians—Arabs as well as Moslem Indian traders and priests contributed to the conversion of millions of people to Islam, though once again the Indian factor was significant. While the Arabs themselves participated in the spread of Islam, the majority of its protagonists came from northwestern India; Gujeratis outnumbered Arabs in many of the trading settlements on Southeast Asia's coasts. The conversions in this area took place as they did elsewhere, often by the threat of the sword. But the Islam that came to Southeast Asia was a Hinduized version of the faith, and ever since that time (the twelfth and thirteenth centuries) Southeast Asia's Moslems set up many sultanates of varying sizes; Brunei on Kalimantan is a last remnant of this kind that still has political identity today, and 9 of the 11 states of Malaya (west Malaysia) also are remnants at the subnational level. Political control, trade domination, and missionary zeal combined to convert about 80 percent of Indonesia's people to the Islam faith; even the Philippines, where the Spaniards quickly arrived to blunt the spearhead of Moslem penetration, is still 5 percent Moslem, virtually all in Mindanao.

This, then, represents the *Indo* part of Indochina: the Buddhist and Hindu faiths (Ceylonese merchants played a big role in the introduction of the former), Indian architecture, art, especially sculpture, writing and literature, and social structure. But the Chinese role in Southeast Asia has been even greater. Apart from the fact that upheavals within China itself contributed to the

[3]Charles Robequain, *Malaya, Indonesia, Borneo, and the Philippines,* (Transl. by E. D. Laborde), Longmans, London, 1954, pp. 66, 67.

A view over part of the Borobudor, an eleventh-century Buddhist shrine near Jogjakarta, central Java. (Burt Glinn/Magnum)

intermittent southward push of sinicized peoples, there were Chinese traders, pilgrims, sailors, fishermen, and others who sailed from southeast China to the coasts of Malaya and the islands and settled there. Invasions and immigrations continued into modern times, and the relationships between the recent Chinese arrivals and the earlier settlers of Southeast Asia have often been strained.

Thailand affords an example of the sequence of events: in the sixth century A.D. an Indian-influenced state existed, with its capital not far from Chiengmai in the very north of the country. In Cambodia, to the southeast, the great Khmer civilization was developing into one of the most impressive early Southeast Asian states. Intermittently there were infusions of people from the north, but in A.D. 1250 the Mongol–Chinese forces of Kublai Khan expelled the Lao–Thai people from their base in Chinese Yunnan. Pushing into the Chao Phraya basin, the Lao–Thais destroyed the existing state and created their own kingdom. Their capital, Ayutthaya (just 50 miles north of Bangkok), lay not far from the mouth of the Chao Phraya river system

and for the first time the center of power was near the coast rather than on the flanks of the hills of the north. About 1350 the first king of Siam ruled over most of the country which for centuries was to be called by that name; the change to Thailand was made only in 1939. Having established themselves, the Thais now attacked the Khmer state in Cambodia and destroyed it; conflict with Cambodia continued off and on for four centuries. Northward, into the hills beyond Chiengmai, and northeastward, into present-day Laos, the Thai influence was spread. Luang Prabang and Vientiane, the twin Laotian capitals, long were Thai dependencies. Meanwhile the Siamese themselves faced invasions from Burma and European encroachments as well—there was hardly a time when there was peace in the region. In 1767, during a period of weakness in the Siamese state, the capital city of Ayutthaya, famed for its riches and consequently beseiged many times, finally fell to the Burmese and was destroyed. It was after this phase, during the second half of the eighteenth century, when Bangkok was chosen as the new capital and the state managed to resurrect itself.

Unlike the rest of Southeast Asia, Thailand was able to retain its identity as an independent state throughout most of the colonial period, although the French precipitated a boundary dispute in the east and the British sought to draw the country tightly into their Burmese–Malayan commercial sphere. It was during this period when the modern Chinese community of Thailand began to arrive, to engage in retail trade and in the newly stimulated international commerce, in industry, and in mining. Although Chinese immigration has been strongly discouraged and restricted since 1947, these people still make up more than 12 percent of the country's current population of nearly 35 million. Many have intermarried with the Thai. In some other countries of Southeast Asia the percentage of Chinese is even greater: Singapore, the city state at the southern tip of the Malay Peninsula, has a population of 2 million that is 80 percent Chinese; the state of Malaysia, which includes the southern part of the Malay Peninsula excluding Singapore and Sabah and Sarawak on Kalimantan, has a population of nearly 11 million, of whom more than one-third are Chinese. Elsewhere the percentages are lower but still substantial: North Viet Nam is 14 percent Chinese; South Viet Nam, 8 percent; Cambodia, 6 percent; and although Indonesia is only about 3 percent Chinese, this 3 percent out of a total of nearly 120 million people constitutes one of Southeast Asia's larger minorities. The Philippines are less than 1 percent Chinese. The map of the distribution of Chinese in Southeast Asia (Map 13-3) shows that they are concentrated mainly in those areas where the big cities and the commercial opportunities are located. For instance, although they make up just 12 percent of the population of Thailand, they are nearly half the population of the capital, Bangkok, and not long ago came close to making up two-thirds of that city's population.

The Chinese benefited enormously from the arrival of the Europeans, who stimulated the agriculture, trade, and industries, and in which they found opportunities they did not have at home. They tapped the plantation trees, found jobs on the docks and in the mines, cleared the bush, and transported goods in their sampans. They brought with them skills that proved to be very useful, and as tailors, shoemakers, blacksmiths, and fishermen they did well. They proved to be astute businessmen, and soon they domi-

The cities of Southeast Asia all have Chinese sectors, where the Chinese whose families migrated into this region are now concentrated. The Chinese immigrants have been successful in several spheres of economic life. This photo was taken in the Chinese market of Bangkok, Thailand. (Tiers/Monkmeyer)

nated not only the region's retail trade, but held prominent positions in banking, industry, and shipping. Thus their importance has always been far out of proportion to the size of their Southeast Asia minorities. The Europeans used them for their own designs but found them stubborn competitors at times, and the British from 1928 to 1937 maintained a quota system to restrict their influx into Malaya. The Americans tried to exclude Chinese from any immigration into the Philippines. The Hollanders had trouble with the Chinese in Borneo (Kalimantan), where during the nineteenth century Chinese had come to dominate the mining industry; with their customary solidarity, the Chinese had formed miners' associations which were on their way to becoming small, self-governing republics. Eventually, after the European withdrawal, Southeast Asia's independent states were left to come to

Map 13-3 CHINESE IN SOUTHEAST ASIA.

Main Areas of Chinese Concentrations

Other Areas of Chinese Concentration

TRIBES IN CHINA

H Hokkien   K Kwongsai
C Cantonese   U Hokchiu
Hakka   A Hokchia
T Teochiu   E Henghua
S Hainanese

terms with their Chinese minorities, not an easy task. The Chinese were always quite aloof, and formed their own, separate societies within the cities and towns of Southeast Asia; they kept their culture and language alive by maintaining social clubs, schools, and even residential suburbs which in practice if not by law were Chinese in character. There was a time when they were the middlemen between European and Southeast Asian, and when the hostility of the local people was directed to the white man as well as the Chinese. But with the withdrawal of the Europeans, the Chinese became the chief object of such antagonism, which was strong because of the Chinese position as money lenders, bankers, and trade monopolists. In addition there is the spectre of an imagined or real Chinese political imperialism along Southeast Asia's northern flanks, and thus the Chinese presence in the region contains many difficulties for the young states there. In the case of Malaysia, a federation that consisted initially of the Malay Peninsula, Singapore, Sarawak, and Sabah (Kalimantan's north), and numerous intervening islands, the centrifugal forces related to the Chinese population were too much: its greatest city, the Chinese-dominated famed port of Singapore, split off from the state and became a separate, sovereign entity. In Indonesia, the political crisis of 1965–1966, which touched off massive reprisals against members and sympathizers of the Indonesian Communist Party, saw many Chinese killed—Western observers estimate as many as 18,000 may have perished.

The *china* in Indochina, thus, represents a wide range of conditions: the source of most of the old invasions was in southern China, and Chinese territorial consolidation provided the impetus for successive immigrations. The Mongoloid racial strain carried southward from East Asia mixed with the earlier Malay stock to produce a transition from Chinese-like people in the northern mainland to dark-skinned, Malay types in the distant Indonesian east. Although Indian cultural influences remained strong, Chinese modes of dress, plastic arts, types of boats, and other cultural attributes were adopted in Southeast Asia. During the past century, and especially during the last half century, a renewed Chinese immigration brought Chinese skills and energies which propelled their minorities to positions of comparative wealth and power in this region.

The Europeans

If there is something wrong with the term Indochina, it is that it does not take account of the third major influence that has penetrated Southeast Asia: that of the European colonial powers. If, as we suggested, there are obstacles in the way of the southward movement of people overland from Eastern Asia, there were few obstructions to face the Europeans when they came from the west by sea. Indeed, the Strait of Malacca forms a funnel for shipping that comes eastward from the Indian subcontinent; Singapore is seen by practically every vessel that turns toward the Philippines, China, and Japan. European ships made their appearance first around 1500, when the Portuguese found the way to the Indies. Until the nineteenth century, the Sunda Strait was also an important passageway. One of the early theaters of competition among European powers was the Philippine archipelago, which during the sixteenth century was contested between the Portuguese and the Spaniards. The Iberian powers had the advantage of being the first European forces to arrive in this part of the world, but their empires did not last very long. The Spaniards managed to maintain their sphere of influence in the Philippines, but the Portuguese were ousted from all their possessions except the island of Timor. This was the work of the Hollanders, who had taken over the supremacy of the oceans.

The Hollanders took control of the "spice islands" through their Dutch East India Company, and the wealth that was extracted from what is today Indonesia brought the Netherlands its Golden Age. From the mid-seventeenth to the late eighteenth century, Holland could develop its East Indian sphere of influence almost without challenge, for the British and French were mainly occupied with the Indian subcontinent. By playing the princes of Indonesia's states off against each other in the search for economic concessions and political influence, by placing the Chinese in positions of responsibility, and by imposing systems of forced labor in areas directly under its control, the company had a ruinous effect on Indonesian society. Java became the focus of Dutch administration, and from its capital at Batavia (now Djakarta) the company extended its sphere of influence into Sumatra (Sumatera), Borneo (Kalimantan), Celebes (Sulawesi), and the smaller islands of the East

Singapore is one of the world's busiest ports. This photograph overlooks the South Boat Quay on the Singapore River, and shows the tall buildings of the waterfront in the background. (Fujihira/Monkmeyer)

Indies. This was not accomplished overnight: the struggle for territorial control was carried on long after the company had yielded its administration to the Netherlands government. Northern Sumatera was not subdued until early in the present century.

These territorial gains notwithstanding, the initial objectives of the European colonizers in Southeast Asia were the development and exploitation of the region's resources more than the mere acquisition of territory. The fact is that the Dutch could go about expanding their sphere without much obstruction from powerful rivals, and frequently the exploitation of an area's resources had to be preceded by the subjugation of the local political authority. But the British did not become interested in Southeast Asia until the nineteenth century, and then, at first, primarily to protect the sea routes to China. The

French appeared still later, and on the eastern coasts of Southeast Asia's mainland they hoped to create a beachhead from where overland routes might be established into China Proper, outflanking the British treaty ports.

As elsewhere in the colonial world, the pace of change accelerated greatly during the nineteenth century. Singapore—it hardly seems possible to conceive of a Southeast Asia without Singapore—was founded in 1819. Burma, adjacent to British India, first lost territory to the British in the early 1820's; gradually the British sphere in Burma and Malaya expanded. On the opposite side of the Indochinese peninsula, the French established themselves in the Mekong Delta. By treaty and by conquest the European spheres of influence grew. Between the two lay embattled Siam (Thailand), facing encroachment from both sides. Eventually Thailand lost its Laotian provinces to the French, areas in the northwest to British Burma, and four border territories in the south to British Malaya. But the two colonial powers recognized the advantages of a buffer state between their respective spheres of influence, and in 1896 Britain and France signed a convention guaranteeing the sovereignty of the Siamese state. At the same time some of the remaining boundary problems in Southeast Asia were disposed of, a process whereby the ultimate territories of the region's future states were allocated. The Mekong River was taken to be Thailand's eastern boundary with Laos although the French did obtain one area west of the river; Thailand's northern boundaries were fixed along the crestlines of several ranges of hills and mountains, including the Luang Prabang Range in the north (Laos) and the Tanen Range (Burma). The western borders with Burma also were delimited in part along crestlines, including those of the Dawna and Blaukraung Ranges (Map 13-4).

Colonial Frameworks. As in Africa, the colonial political structure that emerged in Southeast Asia during the nineteenth century had the effect both of throwing diverse peoples together in larger political units while at the same time dividing people with strong ethnic and cultural unity. The Thais, Cambodians, Malays, and the Shans of northern Thailand and eastern Burma all experienced the latter at the hands of the British, as did the Annamese of Indochina at the hands of the French. The peoples unified by the Dutch in their East Indies are numerous and varied; at least 25 distinct languages are spoken in this insular realm now comprising the state of Indonesia. And, again as in Africa, the internal divisions within the respective colonial realms had lasting significance, for eventually the individual fragments were to emerge as separate states. Thus the French organized their Indochinese dependency into five units—four protectorates (Cambodia, Laos, Tongkin, and Annam) and the colony of Cochin China in the Mekong Delta, with Saigon as its focus. Eventually Cambodia and Laos emerged as independent states, but Viet Nam (consisting of Tongkin, Annam, and Cochin China) faced fragmentation.

The British held two major entities in Southeast Asia—Burma and Malaya, in addition to a number of islands in the South China Sea and the northern sector of Kalimantan. In Burma they created an entity incorporating many different peoples, but from 1886 to 1937 Burma's administration was linked to that of India. Burma's lowland core area is the domain of the majority of Burmese (or Burmans), but the surrounding areas are occupied by other peoples. The Karens live in the Irrawaddy Delta and the eastern margins, the Shans inhabit the eastern plateau that adjoins Thailand, the Kachins live on the Chinese border in the north, and the Chins are in the highlands along the Indian border in the west. Still other groups also form part of Burma's complex population, and this complexity was compounded still further during the period of India-connected administration, when more than a million Indians entered the country. These Indians came as shopkeepers, moneylenders, and commercial agents, and their presence intensified Burmese resentment against British policies. The national division of Burma was brought sharply into focus during World War II, when the lowland Burmese welcomed the Japanese intrusion, while the peoples of the surrounding hill country, who had seen less of British maladministration and who had little sympathy for their Burmese countrymen anyway, generally remained pro-British. The independence of Burma has led to a retreat of perhaps more than half of the Indian population, but on the other hand the Chinese presence in the country has been considerably strengthened. An estimated three-quarters of a million Chinese were in Burma in 1969.

In Malaya, the British also developed a complicated framework of different-status territories. Several scattered areas on the mainland as well as islands in combination made up the colony of Straits Settlement (including Singapore, Malacca, and some islands in the Indian Ocean and off the Sarawak coast); the rest of the dependency was organized into nine indirectly ruled protectorates. This concept of a far-flung political unit materialized into the Federation of Malaysia when independence was achieved, but the centrifugal forces were strong: early during its existence the Federation lost its major city, Singapore, through secession.

Exploitation and Commerce. The political colonial consolidation of Southeast Asia was accompanied by accelerated economic development in several spheres. The introduction of plantations completely revolutionized the economies of such areas as Malaya, eastern Sumatra, and Java. In

Map 13-4 FORMER COLONIAL SPHERES IN SOUTHEAST ASIA.

Java, the Hollanders introduced a concept known as the Culture System, which in effect embodied a variety of forced-crop and forced-labor practices carried on by other colonial powers as well, notably the Portuguese. Called the *Cultuur Stelsel,* the system required the growing of stipulated crops on predetermined acreages; Indonesians who had no land to cultivate were forced to make themselves available as laborers for the state. As Professor Spencer has put it, the system turned Java into one great farm estate.[4] Undeniably, it stimulated production: sugar, coffee, and indigo were the leading exports and they brought in increasing revenues as the system was made to work. Crops such as tea, tobacco, manioc, and the oil palm were introduced or their strains were improved. But the price was high: the Javanese resented the principle, harsh and cruel methods were used to sustain the system—to such an extent that there was an outcry about it in the Netherlands. With the Dutch and the Chinese in control of all phases of the trade based on the products of the forced croplands, including collection, treatment, packaging, and despatch, the local people got little or no experience in this aspect of their island's economy. There was little incentive, other than fear, to cultivate what was required. And lands that had been producing rice and other staples were turned over to these cash crops, without (initially) satisfactory arrangements for adequate replacements of the food supplies thus lost.

The Culture System was officially abandoned in 1870 (although the forced growing of coffee continued until 1917), and the Dutch now sought to apply some of the lessons they had learned through its operation. A new and more liberal colonial policy evolved, whose success is reflected by the sustained production of export crops in Java: there was no general retreat to the cultivation of subsistence crops, as might have been expected. Now, well into the final century of supremacy in Indonesia, the Hollanders encouraged private enterprise, and a great influx of Netherlands immigrants raised their total from under 100,000 to a quarter of a million. But the new system could not erase completely the liabilities of the old. The economic well-being of Java and Sumatra notwithstanding, and despite the rapid growth of the

[4] J. E. Spencer, *Asia, East by South,* Wiley, New York, 1954, p. 273

A worker harvesting rubber in a plantation in Malaysia. (Forbert/Monkmeyer)

Indies' population (indigenous as well as Dutch), the Indonesian drive toward self-determination grew in intensity throughout the first half of the twentieth century.

The Hollanders were in Indonesia, and the British in Burma and Malaya, to seek raw materials, produced cheaply by the colonized population under European management, in exchange for the manufactured products their home countries could supply. To this end, commercial agriculture was stimulated, to raise cultivated acreages, to improve yields of cash as well as subsistence crops. On Sumatra's eastern coastal plain, this was accomplished through the introduction of the rubber tree and the oil palm. In the Philippines, coconuts and sugar provided the main cash exports. In southern Viet Nam, the French colonizers introduced rubber plantations, as well as estates for tea, pepper, tobacco, and coffee. But the potential of Southeast Asia was not confined to its agricultural production. This region is by no means poor in mineral resources; before the Europeans arrived the Chinese were already exploiting most of the deposits later mined by the colonial powers.

The great revenues have been derived from a belt of tin deposits that extends through Southeast Asia from beyond its northern margins in China through Thailand (which mines and exports a small quantity every year) into the Malay Peninsula and on into northwest Indonesia. Both the British in Malaya and the Hollanders in Indonesia benefited enormously from these deposits; in Malaya the juxtaposition of the tin mines and the rubber plantations created a very productive core area, and in the Netherlands East Indies, the islands of Bangka and Billiton (Beliton) contained a major source of wealth in tin. Map 13-1 shows how fortuitous were the locations of these tin fields: they could hardly have been nearer the main ocean routes to the west, nor could they be positioned much closer to Singapore, the obvious collecting and smelting station. Today, tin is still the second export by value (after rubber) of Malaysia, though Indonesia's production is not what it used to be while the Hollanders were in charge.

In Indonesia, another of Southeast Asia's underground resources has taken the lead: petroleum, also first made profitable by the Dutch. Petroleum occurs throughout much of the Indonesian archipelago, in Sumatera, Kalimantan, Djawa, and West Irian. It is also brought to the surface in Burma, and it figures in the export trade of Malaysia through a small field in Sarawak; the main fields are in Brunei. Altogether the production from these reserves is less than 5 percent of the world total, whereas Southeast Asia exports some 60 percent of the world's tin. But the petroleum is nevertheless of major importance, for it remains the largest yield of this commodity over an area that extends from Southwest Asia to North America.

Southeast Asia's tin and petroleum are exported to foreign markets, and herein lies the story of this region's mineral base. Still today, after the countries have become independent, the extraction of their mineral resources and their shipment elsewhere for treatment and use is standard procedure. It is true of the bauxite of Malaya and Indonesia, the iron ore of the Philippines, the tungsten of Thailand, the zinc of Burma, and several other, less significant deposits worked on a small scale. It is the story of much of Black Africa all over again: there are minerals, but there is not the balanced association of resources that can promote local industrialization. Good-quality coal, for example, appears to exist only in North Viet Nam. And even if cheap coal were available to the whole region, there are still the other obstacles to industrialization: capital is scarce, markets are small (except Indonesia's) and rather poor, technology and skills are inadequate. And so Southeast Asia, the most productive tropical realm in the world, has yet been unable to extract itself from the cycle of poverty.

Rice. The period of European colonization saw major developments in rice production. Until the Europeans came, the rice produced in Southeast Asia was almost all consumed by the families of the people on whose land it was grown. The Europeans built and expanded cities, they laid out the transport routes to link these cities to the rice-growing areas, so that the people in the cities would be supplied with their necessities. Rice cultivation now became a major cash proposition in those areas. And the Europeans also found and developed new areas for additional rice production.

These new areas, which have played so large a part in the transformation of this region, were (1) the delta of the Irrawaddy River, in the hinterland of Rangoon in Burma, (2) the delta of the Chao Phraya, below Bangkok in Thailand, and (3) the delta of the Mekong River, south of Saigon in Viet Nam. The British were responsible for bringing the Irrawaddy and Chao Phraya deltas into production, and the French developed the Mekong area. British, French, Chinese, and Indian capital were involved in this significant operation; European planning made possible the draining of the waterlogged delta fields; and Indian and Chinese labor were used in the preparation of the new rice-growing areas. And not only Southeast Asia's growing cities benefited. The rice also went to the plantation labor of other parts of the region, and eventually large exports began to flow to the food-deficient countries outside of Southeast Asia.

Spatial Arrangements

Viewed in the perspective of its lengthy history, Southeast Asia emerged from the colonial period with a wholly new character. The many small states and principalities overpowered by the Europeans had been welded into less than a dozen countries. The population was growing at an increasing rate: that of the island of Djawa had grown to nearly ten times what it was just a century previous. Towns had been founded and some had developed into major cities, the capitals of the new countries (Map 13-5). Populous core areas had become the centers of gravity for the young states, although generally population totals and densities fortunately were still far below what was common in Southern and Eastern Asia. Political boundaries ran mostly through the region's emptier areas, though several states—especially Thailand—seemed to have grounds for irredentist attitudes toward neighbors. Chinese minorities held positions of economic power, inherited from their advancement during the European period. The region produced and exported most of the world's rubber, tin, and copra, three-quarters of the rice, and more than a third of the sugar. Irrigation, flood control, new transport methods, and large-scale processing plants made possible cultivation, harvesting, packaging, and exporting of rice in unprecedented quantities, notably from the Irrawaddy, Chao Phraya, and Mekong valleys and deltas. Burma, Thailand, and Viet Nam were Southeast Asia's great rice exporters, although Burma's production had suffered as a result of World War II and that of Viet Nam's declined for reasons with which we are all too familiar. For other parts of Asia, Southeast Asia looked like a rice bowl: the rice exports went to such areas of chronic deficiency as India, Pakistan, Japan, and Hong Kong, as well as to Malaysia, Indonesia, and the Philippine Islands.

Territorial Morphology. It is difficult to look at the political map the Europeans left behind in Southeast Asia (Map 13-1) without being struck by the several different spatial forms (that is, shapes) these states display. Indonesia is dispersed over a number of large and small islands, North and South Viet Nam extend narrowly along the east coast of Indochina, Burma and Thailand share the narrow, northern part of the Malay Peninsula. This aspect of the state—its shape—is a very important variable that can have a crucial effect on its cohesion and political viability. A state that consists of several separate parts, located far away from each other, obviously has to cope with problems of national cohesion not afflicting a state that has a contiguous territory. We saw what problems Pakistan's division has caused: we know the advantages Uruguay derives from its compactness.

Political geographers identify five major shape categories on the world map, and four of them happen to be illustrated by the states of Southeast Asia; hence the region should be considered in this context. Consider, for example, the country of Cambodia. This country's territory is quite nearly round in shape; in fact, it looks a great deal like Uruguay, and as it happens, it is almost the same size (69,000 square miles against Uruguay's 72,000). Cambodia is a *compact* state, which means that points on its boundary lie at about the same distance from the geometric center, near Kompong Thom on the Sen River. Theoretically, a compact state encloses a maximum of territory within a minimum of boundary, and this may hold advantages, for boundaries often still lie in sensitive areas. In the case of Cambodia, the common boundary with Viet Nam is less than half as long as that of adjacent Laos; the Cambodians have sought to secure their border areas against intrusions, while the Laotian boundary is violated as though it did not exist. It is noteworthy that the initial United States penetration of Cambodia was related to Viet Cong advantages gained through the use of minor protrusions of Cambodian territory into South Viet Nam. Compact countries like Cambodia, moreover, do not face the problems, confronting some other states, of integrating faraway islands, lengthy peninsulas, or distant extensions of territory in the national framework. Effective national control, something Southeast Asian states have yet to achieve over all their domains, is most easily attained in a compact area, and the Cambodians have perhaps come closer to it than any other Southeast Asian nation. Of course, Cambodia has some other advantages. Some 85 percent of its 7 million people are Cambodian; 8 percent are Vietnamese and 16 percent are Chinese. For Southeast Asia, this represents a high degree of unity. In addition, the rice-producing Cambodian core area lies not on the coast, but in the country's interior, along the Mekong River, where also the capital of Phnom

536

Map 13-5 CORE AREAS OF SOUTHEAST ASIA.

Penh (500,000) is located. With the unifying factors of the Khmer language (and the legacy of French) and the Buddhist religion—as well as the charismatic leadership of Prince Norodom Sihanouk—Cambodia's compactness played its role in withstanding the war threat. But after a decade on the fringe of conflict, Cambodia in 1970 finally fell victim to a struggle in which the forces were too strong and the stakes too high to be matched by the country's national cohesiveness.

By contrast, the states of Malaysia, Indonesia, and the Philippines are inherently far less fortunate. These are *fragmented* states, meaning exactly what the term implies, namely states whose national territory consists of two or more individual parts, separated by foreign territory or by international waters. Indonesia and the Philippines represent one of the three subtypes of this category, in that their areas lie entirely on islands. Malaysia is a second subtype, since it lies partly on a continental mainland and partly on islands. The third subtype is not present in Southeast Asia: in this case the fragmented state lies entirely on a continental mainland, and of course Pakistan is the prime example.

Fragmented states must cope with problems of internal circulation and contact, and often with the friction of distance. Indonesia's government, based on Djawa, has had to put down secessionist uprisings in Sumatera, Sulawesi, and in the Moluccas (between Sulawesi and West Irian). The Luzon-centered government of the Philippines has had to combat rebels on Mindoro Island and elsewhere, who took advantage of the insular character of the national territory. The Huk movement focuses in central Luzon, near Manila itself. Far-flung Malaysia was forced to yield to the centrifugal forces inherent in its ethnic complexity and its spatial and functional structure, and permitted the microstate of Singapore to secede. Even in the choice of a capital the fragmented state has difficulty: of all the separate parts of the country, one must be chosen to become the seat of government, the national headquarters. The choice may bring resentment elsewhere in the state, as it did in Pakistan. In Southeast Asia's fragmented states, there could not have been much doubt. Djawa contains some 60 percent of Indonesia's nearly 120 million people, and the choice of Djakarta (the Hollanders' Batavia) as the capital could hardly be disputed. Nevertheless, the peoples of

the outer islands have shown resentment against Javanese domination in Indonesia's affairs. In the Philippines, Luzon is the most populous island and Manila without doubt the country's primate city. In Malaysia, sparsely peopled Sarawak and Sabah can hardly compete with the mainland core area, and naturally Kuala Lumpur, the preindependence headquarters, continued as the new state's capital city.

Still another spatial form is represented by Burma and Thailand. The main territories of these two states, where their core areas are located, are essentially compact—but to the south they share sections of the Malay Peninsula. These peninsular portions are long and narrow, and states that have such extensions leading away from the main body of territory are referred to as *prorupt* states. Obviously Thailand is the best example: its proruption extends nearly 600 miles southward from the vicinity of Bangkok! Where the Thailand–Burma boundary runs along the peninsula, the Thai proruption is in places less than 20 miles wide. Naturally such proruptions can be troublesome, especially when they are as lengthy as this. In the whole state of Thailand, no area lies as far from the core or from Bangkok as the southern extreme of its very tenuous proruption. But at least Thailand's railroad system extends all the way to the Malaysian border; in the case of Burma, not only does the railroad stop more than 300 miles short of the end of the proruption, but in 1968 there was not even a permanent road over its southernmost 150 miles.

In Burma, the spatial structure of the state is compounded by a shift in the core area, a shift that has taken place during colonial times. Prior to the colonial period, the focus of the embryo state was in the so-called Dry Zone, between the Arakan Mountains and the Shan Plateau. Mandalay, now with about a quarter of a million people, was the urban node. Then the British developed the rice potential of the Irrawaddy delta, and Rangoon became the new capital city. The old and new core areas are connected by the Irrawaddy in its function as a water route, but the center of gravity in modern Burma lies in the south.

Proruptions do not always lie on peninsulas. Some landlocked states are prorupt: Afghanistan's Wakhan Corridor is in this class. Other states reach the open ocean via a narrow corridor that is in effect a proruption, as for example the

Congo (Kinshasa) in Africa, which has a narrow extension of territory from the capital to the port of Matadi. And the Congo has a landward proruption too: the famous Katanga. The Congo illustrates a problem neither Burma nor Thailand faces, for there, much of the productive capacity of the state is concentrated not in the main body of territory, but on or near those proruptions. What would seem on a blank map to be the heart of the Congo actually presents great liabilities of distance and cost and comparatively few assets of production or potential. At least Burma's and Thailand's core areas lie in the largest areal zones.

The fourth type of state shape represented in Southeast Asia is the *elongated* or attenuated territory. By this is meant that a state is at least six times as long as its average width. Familiar examples are those of Chile in South America, Norway and Italy in Europe, and Malawi in Africa. Elongation presents obvious and recurrent problems: even if the core area lies in the middle of the state, as in Chile, the distant areas in either direction may not be effectively connected and integrated in the state system. In Norway the core area lies near one end of the country, with the result that the opposite perimeter takes on the characteristics of a frontier. In Italy, the contrasts between north and south pervade all aspects of life, and they reflect the respective exposures of these two areas to different mainstreams of European–Mediterranean change. If a state is elongated and at the same time possesses more than one core area, there will be strong divisive stresses on it. This has been the case in Viet Nam, one of the three political entities into which former French Indochina was divided.

French Indochina incorporated three major ethnic groups: the Laotians, the Cambodians, and the Vietnamese. Even before independence, the Laotians and the Cambodians possessed their own political areas (which later were to become the states of Laos and Cambodia), and this left a 1200-mile belt of territory, averaging under 150 miles in width, facing the South China Sea. This was the domain of the Annamese, extending all the way from the Chinese border to the Mekong delta. Administratively, the French divided this elongated stretch of land into three units: Cochin China in the south, Annam in the middle, and Tonkin in the north. The capitals of these areas became familiar to us during the Viet Nam War; Saigon was the focus of Cochin China and the headquarters of all Indochina; Hué, the ancient city, was the center of Annam; and Hanoi was the capital of Tonkin. Cochin China and Tonkin both were incipient core areas, for they lay astride the populous and productive deltas of the Mekong and Red Rivers, respectively.

During 1940 and 1941, the Japanese entered and occupied Indochina, and during their nearly four years of occupation, the concept of a united Viet Nam emerged. The Japanese did not discourage the rising Vietnamese nationalism, especially when it became apparent that the tide of war was turning against them; they would rather have an Annamese government in Viet Nam than a French one. So it was that during the period of Japanese authority a strong coalition of pro-independence movements arose in Indochina, the Viet Minh League. In 1945, when the Japanese surrendered, this organization seized control and proclaimed the Republic of Viet Nam a sovereign state under the leadership of Ho Chi Minh. For some time after the Japanese defeat, the Chinese actually occupied Tonkin and northern Annam, but theirs was a sympathetic involvement as well, and the Viet Minh grew in strength during that time. Meanwhile the French had proposed that Viet Nam, Laos, and Cambodia should join as Associated States in the fourth-republic concept of the French Union, but this plan was rejected everywhere in Indochina. As the French reestablished themselves where they had always been strongest—in Saigon and the Mekong Delta of Cochin China—they started negotiations with the Viet Minh. These talks soon broke down, and in 1946 a full-scale war broke out between the French and the Viet Minh. France's base lay in Cochin China, and the Viet Minh had their greatest strength in Tonkin, a thousand miles away. Viet Nam's pronounced elongation favored the Viet Minh and their numerous nationalist, Communist, and revolutionary sympathizers. After an enormously costly eight-year war it was in the north, not far from the Chinese border, where the crucial battle was fought (at Dien Bien Phu) and lost by the French.

Following the request for a truce by the French, a conference was held in July, 1954, in Geneva. There, the ribbonlike country of Viet Nam was cut in half at about the 17th parallel—that is, in Annam—and the north became the Viet Minh's domain while the south came under the administration of a government in Saigon. This was to be a temporary arrangement, and elections were to be held within the following two years to create a national government representing all of Viet Nam's people. Those elections were never held. North Viet Nam, adjacent to and allied with China, became a Communist state; South Viet Nam, after a period of disorganization worsened by the arrival of about a million refugees from the Communist north, eventually got an authoritarian regime shored up by the United States. Ever since, this government has had to cope with a movement, which, like the old Viet Minh League, consists of Communist, nationalist, and other revolutionary elements. Supported by North Viet Nam with men and material and aided indirectly by China, this Viet Cong movement has wrested large parts of the country from the control of Saigon's unstable regime; during the 1960's the United States entered the contest with an army that eventually numbered more than a half million men.

The Insurgent State. The national revolution in Viet Nam, though magnified into an international conflict of major proportions, is symptomatic of what appears to be happening with increasing frequency in modern times: armed insurrections aimed at the overthrow of national governments. In part this is due, no doubt, to the weakness of so many governments that have taken over at short notice the administration of former colonies. In many cases the good relationships between these new governments and their former colonial masters constitutes an irritant to large groups of the population, and if conditions permit it, an antigovernment movement may arise. But these insurrections are not all of this kind: the Chinese Communist insurgency was the final chapter in a long history of civil conflict; in Indonesia it was the returning colonial (Dutch) government itself that faced a successful revolution. In Cuba the national revolution led by Castro ousted a government that was quite typical of the norm for Middle and South America. In Greece after World War II there was a serious threat from Communist guerilla forces which for some time held sizable parts of the mountainous interior on the Albanian and Yugoslav borders.

The struggle in South Viet Nam began slowly and at first was a minor rebellion, but quite soon it evolved into a full-scale war. Political scientists recognize stages in such a growing conflict: the beginning is marked by "contention," the ensuing stage is one of "equilibrium," during which the government's forces are successfully withstood in certain critical areas, and eventually a final stage of "counter-offensive" or regular war occurs. This took place in Viet Nam from 1946 to 1954, when the French lost their war against the Viet Minh; it occurred also in Cuba in the 1950's, when the Castro revolution was carried from the protective mountain forests into the countryside and eventually even into the cities. In South Viet Nam the final stage would no doubt have come to pass were it not for the involvement of the United States, but even the U.S. and South Vietnamese forces combined could not prevent the Viet Cong from controlling sizable parts of the country and from threatening the rest.

Professor R. W. McColl has recently translated these military stages of national revolution into territorial terms, suggesting that what eventually happens is that an *insurgent* state is progressively created within the boundaries of the threatened state.[5] Correlating the spatial and military evolution of the insurgent state, he suggests that first, a guerilla area is established; this may be small, and protected by mountainous terrain, dense vegetation, or some other physical attribute. During this stage of "contention" or "mobile war," the guerilla forces consolidate themselves, and then strike out suddenly at some provincial capital, road, or bridge—in any case, an objective whose destruction has important psychological consequences. Eventually the sphere of influence and control of the guerilla base grows, and true core areas emerge (using this term in a different context than was done in Chapter 2), whose most important quality is the quick access they provide to important targets, such as major transport routes as sizable urban places. Core areas may now also emerge in other parts of the country, and these insurgent cores begin to take on stable and permanent characteristics: boundaries or marked transition zones separate them from the government-held domain, and internally organizational steps are taken to administer most efficiently the growing guerilla sphere. Now the movement is on the verge of achieving

[5] R. W. McColl, "The Insurgent State: Territorial Bases of Revolution," *Annals of the Association of American Geographers,* Vol. 59, No. 4, December, 1969, pp. 613–631.

The insurgent state often emerges in forested, protective terrain where counter-insurgency operations are difficult. This Viet Nam combat photograph shows whole areas of dense forest that have been destroyed to open the countryside and render concealment difficult. (UPI)

the insurgent state, which can be identified when the "equilibrium" stage of revolutionary war comes about. The government forces are held in check and a sort of intrastate frontier emerges, where the insurgents compete with the government for the support and eventual control of the people and their land. Within the insurgent state, the revolutionary elements can afford to develop a more complex administration, and at this stage the insurgent regime normally establishes a politico-territorial framework incorporating not only the insurgent state's areas, but the whole country. This map can be used to political advantage, for example by promising privileges for certain government-controlled areas; as Professor McColl points out, the Viet Cong in Viet Nam promised autonomy for the *Montagnards,* a tribal people of the western mountains, while at the same time guaranteeing unity to the rice farmers of the Mekong Delta. The insurgent state permits the revolutionaries to add political persuasion to

their military activities: in their larger sphere they are now able to demonstrate successful reform programs, while the hard-pressed government is at the same time shown to be incapable of protecting the villages and towns still under its authority.

In the decade of the 1960's South Viet Nam has been a laboratory for the testing of this concept of the insurgent state (Map 13-6). All the conditions were right: an authoritarian yet ineffective regime without general popular support, numerous areas where natural protection could be found for insurgent groups, easy guerilla access to vulnerable villages and towns, quick psychological rewards for successful strikes against governmental authority and installations. In addition, the insurgent Viet Cong had the assistance of North Viet Nam, whose men and material flowed into the insurgent core areas via Laos. Although it can be argued that most of South Viet Nam's productive capacity and

Map 13-6 THE INSURGENT STATE.

population still remain in Saigon/American hands, the map shows an undeniable reality: in 1969 there were three Viet Nams, namely the North, the South, and the Viet Cong's insurgent state.

Water and Land

In a region of peninsulas and islands such as Southeast Asia, the surrounding and intervening waters are of extraordinary significance. They may afford more effective means of contact than the land itself; as trade and migration routes they sustain internal as well as external circulation. On the other hand, the waters between the individual islands of a fragmented state also function as a divisive force. There are literally thousands of islands in Southeast Asia, and while some are productive and in effective maritime contact with other parts of the region and the world, many of these islands and islets are comparatively isolated. In this respect, of course,

Southeast Asia is not unique in the world. In previous chapters we discussed the peninsular character of Western Europe and the insular nature of Caribbean America. In all these areas, peoples and governments have an awareness of the historic role of the seas. And everywhere, states that have coasts have extended their sovereignty over some of their adjacent waters. This is not a new principle: in Europe, the concept that a state should own some coastal waters is centuries old. Initially these coastal claims—on a world scale—were quite modest. But as time went on, some states began to extend their territorial waters farther and farther out, and other countries followed suit. Since World War II, the situation has been complicated still further by the sometimes vast demands of newly independent, decolonized states. Various United Nations-sponsored conferences, called to seek a generally acceptable consensus among states as to just how wide a sovereign sea should be under

Map 13-7 TERRITORIAL WATERS: CLAIMS IN SOUTHEAST ASIA.

542

international law, have failed to produce such an agreement.

In the course of time, numerous justifications have been brought up by the governments of states seeking to consolidate maritime claims. Among these have been the protection of domestic fishing fleets into coastal waters, the maintenance of neutrality and immunity during wartime conditions, the right to apprehend smugglers in their approach to the coast, and the prevention of water pollution by waste-disposing vessels. In recent years the continental shelf has attracted the interest of coastal states, for below the shallow coastal waters lie potentially valuable mineral resources; already, oil is pumped from fields located offshore. In places the continental shelf extends dozens of miles out beyond the coast, and of course the best way for a state to guarantee its primacy there is to claim the overlying water (the epicontinental sea) and all that exists below it.

Certain maritime claims of coastal states are sanctioned under international law, and disputes that have arisen over violations of maritime boundaries have been fishing disputes between the United Kingdom and Iceland and between the United Kingdom and Norway. But the fact is that the power of international law and jurisdiction remains very limited and depends, in effect, on the willingness of the states involved to subject themselves to such judgment. Even though there are rules concerning the delimitation of maritime boundaries, states actually can claim what they wish—so long as they have the power to back up their demands. And there are even some legal loopholes. For example, although international conferences have stipulated the maximum width of bays and estuaries that may legally be closed to international use, states can circumvent these restrictions by laying claim to adjacent waters under the justification that they have special, long-term national significance. Such waters, then, are recognized as *historic waters*. This was the chief basis for Indonesia's announcement, late in 1957, that it would claim as national territory not only all waters within 12 miles of its island areas, but also all waters between the far-flung islands of the whole archipelago. It had the effect of including the entire Java Sea, Celebes Sea, and Banda Sea in Indonesia, and it meant that sections of some of the world's major shipping lanes now lay within Indonesian territorial waters (Map 13-7),

for instance that connecting Singapore to Australia. Early in 1958 the government of the Netherlands, then still in control of West New Guinea, tested Indonesia's capacity to sustain this huge claim by announcing that it would not recognize the measure. Several factors lay behind Indonesia's move, and the obstruction of Dutch shipping was only one of them: the late 1950's was a difficult time for the Djawa-based government in its effort to hold the archipelago together as a single state, and there were fears that outside powers might assist secessionist movements. Certainly the implied expansionism fits well in the cycle theory model (Chapter 10).

A Frontier. There is no doubt about it: the seas and oceans and the subsoil below them represent a last frontier, and one with great productive potential. States are pushing outward into this last unclaimed part of the world's surface, and the open high seas are subject now to steady encroachment. Not satisfied with total sovereignty within their territorial seas, states are even demanding certain rights in waters beyond their maritime boundaries, in the so-called contiguous zone (which lies immediately adjacent to the territorial sea) and even beyond this ill-defined belt, in the zone of diffusion. All these legalistic terms in fact represent devices by which states are slowly extending their spheres of influence farther into the seas and oceans, and the lack of universally accepted rules and limits underscores the unstable state of affairs. This situation is well illustrated in Southeast Asia, where in the late 1960's neighboring states still claimed territorial waters of different widths. The three colonial powers which ruled in this realm all claimed three miles as the width of their territorial sea, but in 1969 only Malaysia still adhered to this practice. South Viet Nam claimed 12 miles, while North Viet Nam's claim, though apparently never clearly stated, probably is similar, following the demands of China and the Soviet Union. Cambodia's claim is 5 miles, and only one other country in the world (Uruguay) has adopted this width. Thailand has long claimed 6 miles; Burma and the Philippines have not announced permanent claims. The contrast between Indonesia and the Philippines in this context is noteworthy, for what became a major issue early in Indonesia's existence as a sovereign state held little apparent interest in the other archipelago. Nor can this be readily explained on the basis of Indonesia's designs on the Dutch-held

Intensive rice paddying in the Philippines. (George Holton/Photo Researchers)

section of New Guinea: the Philippine government has also coveted external territory, notably the Sabah area of northern Kalimantan, held by Britain and now by Malaysia.

A further problem relating to the territorial seas of the world's states is just how the maritime boundaries are to be delimited and, if need be, demarcated (that is, visibly and permanently outlined by markers, as is done on land). This is not quite so simple a matter as might at first appear. If a state claims, say, 12 miles as the width of its territorial sea, is this 12 miles to be measured from a series of baselines along its coast? And if so, how long may these baselines be? How large are the bays and estuaries that may be cut off by baselines? Thailand, for example, might with a single baseline cut off the whole northern end of the Gulf of Siam; Burma could do the same with the Gulf of Martaban. Various international conferences have addressed themselves to these problems, but no final and permanent solutions have yet been found. Baselines have grown progressively longer, and states desiring to do so can always cut off estuaries and bays on the grounds that they are historic or strategic waters. As for the method of delimita-

tion, a method has come into general use whereby the maritime boundary is delimited through a series of arcs, drawn with a protractor; the radius is the width claimed and the resultant boundary is a coalescence of these arcs based on the coastline. This method makes triangulation from the sea quite easy, and even when the boundary is unmarked (undemarcated) its violation is not difficult to avoid. These aspects may seem to be less relevant in Southeast Asia than in smaller, narrower, fish-rich Western Europe— but the strategic implications of Eastern Asia's maritime boundaries have recently been emphasized by several serious international incidents.

However, there is more to the territorial sea issue than fishing rights. As Man's technological abilities have improved, the mineral potential of the continental shelf has come within his reach. Oil is already taken in great quantities from undersea reserves, but other minerals also can be expected to lie in the subsoil. As any atlas map of Southeast Asia shows, the region has very extensive, shallow continental shelves. The whole sea zone surrounded by the southern mainland, Sumatera, Djawa, and Kalimantan (including the entire Java Sea) is in effect a continental shelf, lying at less than 500 feet below sea level. This means that the underlying rocks are not far beyond man's grasp, and they may contain rich minerals. No country with adjacent continental shelf areas would want to give up its actual or potential rights there; it might be giving up a great advantage. Southeast Asia's continental shelf is one of the largest such areas in the world, and it may well become an arena of exploration and competition.

The Philippines

Several justifications exist for a separate view of the Philippines in a discussion of Southeast Asia. After Indonesia, this is Southeast Asia's most populous state, but few of the generalizations that can be made about the region apply without qualification to the Philippines. Although the peoples of the Philippine islands have ethnic affinities with those of the Asian mainland—Mongoloid peoples in successive waves displaced the older hunters and gatherers, the Negritos—their culture history has been quite distinct from that of the rest of the East and Southeast Asian realm. The Philippines' location relative to the mainstreams of change in this part of the world has had much to do with this. It is thought that the oldest inhabitants may have come to the islands via land bridges that may have existed during the Pleistocene glacial times, but the Mongoloid mainlanders later came across the South China Sea. The Philippines shared with much of the rest of Southeast Asia the early period of Indian cultural influence, but its impact was greatest in the southern and southwestern islands and diminished to the north. The subsequent Chinese impact, on the other hand, was greatest in Luzon in the north. Again, the impact of Islam was somewhat delayed by the position of the Philippines well to the east of mainland Southeast Asia and to the north of Indonesia. It was delayed sufficiently to have allowed the Spanish invasion of the sixteenth century to overwhelm the Moslem beachheads within a comparatively short time. The contrast is expressed today by Indonesia, which remains a Moslem state, while in the Philippines more than 80 percent of the people now adhere to Catholicism with another 10 percent belonging to other Christian denominations; the 5 percent Moslems are predictably concentrated in southern Mindanao and in the Sulu Archipelago, near Malaysia's Sabah and Indonesia's Kalimantan.

These are the Moros (from Moors), who have for centuries resisted full integration in the political life of the Philippines.

Out of the Philippine melting pot, where Mongoloid–Malay, Arab, Persian, Chinese, Japanese, Spanish, and American elements have met and mixed, has emerged the distinctive culture of the Filipino. It is not a homogeneous or a unified culture, but in Southeast Asia it is in many ways unique. One example of its absorptive qualities lies in the way the Chinese infusion has been accommodated: the "pure" Chinese minority numbers a mere 1 percent of the total population (far less than normal for Southeast Asian countries), but in fact a much larger portion of the Philippine population carries a marked Chinese ethnic imprint. What has happened is that the Chinese have intermarried, producing a sort of Chinese-mestizo element—constituting more than 10 percent of the entire population. In another cultural sphere, the country's ethnic mixture and variety is paralleled by its great linguistic diversity. Nearly 90 Malay languages, large and small, are spoken by the less than 40 million people in the Philippines; about 1 percent still use Spanish and approximately 40 percent of the population is able to use English. Upon independence the largest of the Malay languages, Tagalog or Filipino, was adopted as the country's official language, and its general use is being promoted through the educational system. English is learned as a second language and remains the chief *lingua franca*.

The widespread use of English in the Philippines, of course, results from a half century of American rule and influence, beginning in 1898, when the islands were ceded to the United States by Spain under the terms of the Treaty of Paris, following the Spanish-American War. The United States took over a country in open revolt against its former colonial master (the Philippines had declared their independence from Spain on June 12, 1898) and proceeded to destroy the Filipino independence struggle, now directed against the new foreign rulers. It is a measure of the subsequent success of United States administration in the Philippines that this was the only dependency in Southeast Asia which during World War II sided against the Japanese in favor of the colonial power. United States rule had its good and bad features, but the Americans did initiate reforms that were long overdue, and they were already in the process of negotiating a future independence for the Philippines when the war intervened.

Spain in Asia. The reforms begun by the United States in the Philippines, and continued by the Filipinos themselves after independence in 1946, were designed to eliminate the worst aspects of the Spanish legacy in this country—conditions that are familiar to anyone who has knowledge of the Spanish colonial period in Middle and South America. Manila, ever since its founding in 1571 the primate city of the Philippines, was the colonial capital from where political and religious power, working hand in hand, radiated outward until only the most isolated parts of the archipelago remained unaffected. At lower levels of administration the existing indigenous political units were maintained, but provincial control was facilitated through the creation of *econo-miendas,* huge estates of good farmland given as rewards to loyal representatives of church and state. The quick acceptance of Catholicism and its diffusion through the islands helped consolidate this system; exploitation of land and labor (often by force) were the joint objectives of priests and political rulers alike. The Spaniards and the small Hispanicized Filipino elite thrived on the labor of the many thousands of now-landless peasants; they left the bulk of the trading to the Chinese–Filipino minorities. Only in the mountains, where Spanish rule failed to reach, and in the south, where the Moros succeeded in keeping the Spanish off balance through raids and guerilla-style attacks, did the system fail to take hold.

When, after three centuries of such exploitation, Spanish colonial policy showed signs of change during the nineteenth century, it was too late—and however well intentioned, the Spaniards were trapped by their own errors. Crops from the Americas were introduced (tobacco became a lucrative product), and a belated effort was made to integrate the Philippine economy with that of Spanish-influenced America. But what the Philippines needed most the Spaniards could not provide: land reform. As everywhere else in the colonial world, the main issue between the colonizers and the colonized in the Philippines was land, agricultural land, and as elsewhere, the Spanish colonizers found that what had been easy to give away was almost invariably impossible to retrieve. Long after the Spaniards had lost their Philippine dependency the Americans, with a much freer hand, found the same still to be true—and even today the Filipino government, after a quarter of a century of sovereignty, still faces the same issue.

The Land. During World War II, guerilla bands based in the mountains continuously harassed the Japanese occupation forces. After the war, the Filipino government was faced with a continuation of the guerillas' campaign, directed now against the large landowners and the laws that permitted the gross inequities in land ownership to continue. Especially in Luzon and Mindoro Islands, these revolutionaries—the Hukbalahaps or Huks—came near to establishing an insurgent state within the Philippine Republic. But with American assistance, the government, through vigorous pursuit and well-propagandized land redistribution, managed to cope with this outbreak, and eventually the Manila regime prevailed. Land resettlement was especially effective in Mindanao, where farmland was available and where even former revolutionaries were induced to settle instead of fight. But land redistribution is a slow process and its slowness is a constant irritant; in the late 1960's the Huks, now a political as well as an armed revolutionary force, were again engaged in attacks on the government's forces and installations while intimidating and even murdering noncompliant politicians.

As Plate 6 shows, the Philippines' population, concentrated where the good farmlands lie in the plains, is densest in three general areas: the south–central and northwestern part of Luzon (Manila lies at the southern end of this zone), the southeastern proruption of Luzon, and the islands between Luzon and Mindanao, the Visayan Islands. The Philippine archipelago is reputed to consist of over seven thousand mostly mountainous islands, of which Luzon and Mindanao are the two largest, accounting for nearly two-thirds of the total area. In Luzon the farmlands, producing rice and sugar cane, lie on alluvial soils, but in extreme southeastern Luzon and in the Visayan Islands there are good volcanic soils. Copra is the most productive export of the agriculture-dominated Philippines; the forests yield valuable timber, while sugar ranks next—but most Filipino farmers are busy growing the subsistence crops, rice (in which the country achieved self-sufficiency in 1968) and corn. As in the other Southeast Asian countries,

Southeast Asia's extensive continental shelves have already yielded oil; more is likely to be found in the future. This offshore rig symbolizes the importance of this resource in this region. (Arabian-American Oil Co.)

there is considerable range of supporting food crops. Along with other countries in the region, the Philippine state does not yet face a major overpopulation problem (although in parts of Luzon and the Visayan Islands the threshold is being reached). And when it comes to more intensive farming, few peoples in the world could provide a better example than northern Luzon's Igorots, who have transformed the hillslopes in their domain into impeccably terraced, irrigated rice paddies.

Thus the Philippines provide both rule and exception in Southeast Asia. A fragmented, island country, it must cope with internal revolutionary forces and intermittent Irredentist encroachment from outside (Indonesia has at times expressed support for Mindanao's Moslems), but

it is no stranger to expansionist demands, as evidenced by Manila's claim to Sabah. Culturally it is in several ways unique; economically it shares with the rest of Southeast Asia the liability of imbalanced and painfully slow development. And politically the Philippine Republic is an unusual phenomenon: like the other colonial territories it has been bequeathed a spiritual and material colonial legacy, but unlike the norm the Filipino government has quite consistently aligned itself with the Western powers, even to the point of anti-Communist involvement in the Korean conflict and membership in SEATO. It is symbolic that the Philippines should stand at the very edge of Southeast Asia.

ADDITIONAL READING

Although geographers sometimes disagree on what exactly constitutes South East Asia, a number have written textbooks about the region as they conceive it. These include E. H. G. Dobby, whose *Southeast Asia* is now in a seventh edition, published by London University Press in 1960. Another major work is C. A. Fisher's *Southeast Asia: a Social, Economic and Political Geography,* published in New York by Dutton in 1964, and one of the standards is the book written by C. Robequain and translated from the French by E. D. Laborde, *Malaya, Indonesia, Borneo and the Philippines; a Geographical, Economic, and Political Description of Malaya, the East Indies, and the Philippines,* a second edition of which was published by Longmans in London, in 1958. Other basic works are G. Hunter's *South-east Asia: Race, Culture, and Nation,* printed by Oxford University Press in New York in 1966, and T. G. McGee, *The Southeast Asian City: A Social Geography,* a Praeger, New York, publication of 1967. Also see the useful *Atlas of South-East Asia,* published by St. Martin's, New York, in 1964.

On the Chinese in South East Asia, see V. Purcell, *The Chinese in Southeast Asia,* a second edition of which was published by Oxford University Press in New York in 1965, and L. E. Williams, *The Future of the Overseas Chinese in Southeast Asia,* published by McGraw-Hill in New York in 1966.

Individual countries also have been discussed. On Burma, see J. L. Christian, *Modern Burma, a Survey of Political and Economic Development,* an older book (it was published by the University of California Press, Berkeley, in 1942) but still useful. On Thailand, see one of the Country Survey Series of the Human Relations Area Files, W. Blanchard's *Thailand: Its People, Its Society, Its Culture,* published in 1958. On Viet Nam see C. H. Schaaf and R. H. Fifield, *The Lower Mekong,* No. 12 in the Van Nostrand Searchlight Series. On Cambodia, another H.R.A.F. volume, by D. J. Steinberg, *Cambodia: Its People, Its Society, Its Culture,* was published in 1959 and is still full of interest. From the same publisher and in the same series is a volume edited by F. M. LeBar and A. Suddard on *Laos,* brought out in 1960. On Malaysia, see F. C. Cole, *Peoples of Malaysia,* again an older book (1945), but still valuable. It was published by Van Nostrand, Princeton. Also on Malaysia, there is a good volume by Jin-Bee Ooi, *Land, People, and Economy in Malaya,* published by Longmans in London in 1963. On Indonesia, see the volume by B. and J. Higgins, *Indonesia: The Crisis of the Millstones,* No. 10 in the Van Nostrand Searchlight Series. And on the Philippines there are several good works. One of the best is an older volume, by J. E. Spencer, *Land and People in the Philippines; Geographic Problems in Rural Economy,* published by the University of California Press in Berkeley, in 1952. Later, Professor Spencer collaborated with F. L. Wernstedt in *The Philippine Island World: a Physical, Cultural, and Regional Geography,* published by the University of California Press at Berkeley and Los Angeles, in 1967. Also see R. E. Huke, *Shadows on the Land: An Economic Geography of the Philippines,* published in Manila by Bookmark in 1963.

On the question of colonial policies see J. S. Furnivall, *Colonial Policy and Practice, a Comparative Study of Burma and Netherlands India,* published by New York University Press in 1956. On spatial morphology and seaward national boundaries, see Chapters 2, 10, and 11 in H. J. de Blij, *Systematic Political Geography,* a Wiley publication of 1967. Also see, for a wealth of information about the origins of maritime boundary delimitation, L. M. Alexander's *Offshore Geography of Northwestern Europe; the Political and Economic Problems of Delimitation and Control,* published by Rand McNally in 1963. And among many writings on Viet Nam, see the book of readings edited by W. Fishel, *Vietnam: Anatomy of a Conflict,* published by Peacock, Itasca, 1968.

14

PRODIGIOUS JAPAN: THE AFTERMATH OF EMPIRE

In the non-Western, non-European world there is no country quite like Japan. All of a sudden the familiar phrases—underdevelopment, stagnation, subsistence—no longer apply. Japan is an industrial giant, an urban society, a political power, a vigorous nation. No city in the world is without Japanese cars in its streets; few photography stores are without Japanese cameras; many laboratories use Japanese optical equipment. From tape recorders to television sets, from ocean-going ships to children's toys, Japanese manufactures flood the world's markets. The Japanese seem to combine the precision skills of the Swiss with the massive industrial power of prewar Germany, the forward-looking designs of the Swedes with the innovativeness of the Americans. How have the Japanese done it—does Japan have the kind of resources and raw materials that helped boost Britain into its early, revolutionary industrial prominence? Are there locational advantages in Japan's position off the eastern coast of the Eurasian landmass, just as Britain benefited from its situation off the western coast? Could Japan's rise to power and prominence have been predicted, say, a century ago?

In the mid-nineteenth century, Japan hardly seemed destined to become Asia's leading power. For 300 years the country had been closed to outside influences; Japanese society was stagnant and tradition-bound. Very early during the colonial period, European merchants and missionaries were tolerated and even welcomed, but as the European presence on Pacific shores grew greater, the Japanese began to shut their doors. By the end of the sixteenth century Japan's overlords, fearful of Europe's imperialism, decided to expel all foreign traders and missionaries. Christianity (especially Catholicism) had gained a considerable foothold in Japan, but it

Map 14-1 THE JAPANESE COLONIAL EMPIRE.

Legend:
- 1875–1937
- World War II

Labels on map: HAWAII, COOK ISL., SAMOA, TONGA, NEW CALEDONIA, MARSHALL ISL., MARIANA ISL., CAROLINE ISL., BONIN ISL., KURIL ISL., SAKHALIN, RYUKYU ISL., TAIWAN, KOREA, MANCHURIA

Scale: Miles 0 400 800

was now viewed as a prelude to colonial conquest, and in the first part of the seventeenth century it was practically stamped out in a massive, bloody crusade. Determined not to share the fate of the Philippines, which had by then fallen to Spain, the Japanese permitted only the most minimal contact with the Europeans. Thus a few Dutch traders, confined to a little island near the city of Nagasaki, were for many decades the sole representatives of the European world. And so Japan retreated into a long period of isolation that lasted past the middle of the nineteenth century.

Japan could maintain its aloofness when other areas were falling victim to the colonial tide because of a timely strengthening of its central authority, because of its insular character, and because of its distant position, far to the north along East Asia's difficult coasts. There were other factors, of course—Japan's isolation was far less splendid than that of China, whose silk and tea and skillfully made wares always attracted hopeful traders. But in the eighteenth and nineteenth centuries the modernizing influences brought by Europe to the rest of the world passed Japan as they did China, and when Japan finally came face to face with the steel ships and the firepower of the United States as well as Britain and France, it was no match for its enemies, old or new.

In the 1850's it was the United States which first showed the Japanese to what extent the balance of power had shifted. American naval units sailed into Japanese harbors and extracted trade agreements through a show of strength; soon the British, French, and Hollanders were also on the scene seeking similar treaties. When there was resistance in local areas to these new associations, the Europeans and Americans quickly demonstrated their superiority by shelling parts of the Japanese coast. By the late 1860's no doubt was left that Japan's lengthy isolation had come to an end.

The New Japan. In the century that has passed since 1868, Japan emerged from its near-colonial status to become one of the world's major powers, overtaking a number of its old European adversaries in the process. The year 1868 is an important one in Japanese history, for it marks the revolt that overthrew the old rulers and brought to power a group of reformers whose objective was the introduction of Western

ways in Japan. The supreme authority officially rested with the emperor, whose role during the previous, militaristic period had been pushed into the background. Thus the 1868 rebellion came to be known as the Meiji Restoration—the return of enlightened rule. But despite the divine character ascribed to the imperial family, Japan in fact was ruled by the revolutionary leadership, a small number of powerful men whose chief objective was to modernize the country as rapidly as possible, to make Japan a competitor, not a colonial prize.

The Japanese success is written on the map: even before the turn of the twentieth century a Japanese empire was in the making and the country was ready to defeat encroaching Russia. Japan took Taiwan from China (1895), occupied Korea (1910), and established a sphere of influence in Manchuria. Various archipelagos in the Pacific Ocean were acquired by annexation, conquest, or mandate (Map 14-1). In the early 1930's Manchuria was finally conquered, and Tokyo began calling for an East Asia Co-Prosperity Sphere, which would combine, under Japanese leadership of course, all of China, Southeast Asia, and numerous Pacific island territories. By the late 1930's deep penetrations had been made into China, but Japan got its big chance as a result of World War II. Unable to defend their distant Asian possessions while engaged by Germany at home, the colonial powers offered little resistance to Japan's takeover in Indochina, Burma, Malaya, and Indonesia; neither was the United States able to protect the Philippines. Japan's surprise attack on Pearl Harbor, a severe blow to the United States military installations there, was a major success—and is still admired in Japanese literature and folklore. The Japanese war machine was evidence of just how far Japan's projected modernization had gone: airplanes, tanks, warships, guns, and ammunition all were produced by Japanese industries and they were a match for their western counterparts. Japan, we know, was ultimately defeated by superior American power (two cities were wiped out by atomic bombs), but this defeat coupled with the loss of its prewar as well as wartime empire has failed to destroy the country's progress. It has rebounded with such vigor that its overall economic growth rate currently is the highest in the world.

One aftermath of empire: the ruins of Hiroshima shortly after the city was destroyed by an American atomic bomb. The bomb may have saved lives by shortening the war, but here, children, pregnant women, and elderly folks were among tens of thousands who perished in minutes in a hell of fire. It broke the back of the Japanese war effort and it marked the end of its half century of imperialism. (Shunkidi Kikuchi/Magnum)

The Island Arc

Japan is an archipelago, one of a long series of island groups that extend along the whole East Asian and Australian coast, from Sakhalin and the Kuril Islands in the north, through Taiwan, the Philippines, Indonesia, New Guinea, New Caledonia and the New Hebrides to New Zealand in the south. These are the great island arcs of the western Pacific, and with their numerous volcanic mountains and frequently severe earthquakes they do justice to the term "Ring of Fire." Here the Earth's crust is very unstable, and the island arcs themselves are manifestations of this instability. Their great length and arcuate shape suggest that a major mountain chain is being built which is just rising out of the sea, a system that will in time rival the Andes and Rocky Mountains in its dimensions. This may be exactly what is happening, for the land portions of the island arcs, such as Japan, represent the uppermost parts of giant folded structures in the crust—the sial—beneath the sea. Even where these folds are not high enough to emerge above sea level, they still persist, as for example between

southern Japan and Taiwan (where only the small Ryukyu Islands rise above the water) and between New Guinea and New Zealand, where there are huge undersea mountain ranges. Adjacent to the island arcs lie the world's deepest ocean troughs, much deeper than Mount Everest is high. Immediately east of Japan lies the Japan Trench, nearly 35,000 feet deep; east of the Philippines is the Philippine or Mindanao Trench, which reaches nearly 38,000 feet below sea level. These trenches, which tend to lie east of the island arcs, provide additional support for the idea that island arcs are giant folds, for they seem to represent the downward sections of the same folds of which the islands themselves are the upper parts.

Japan, then, seems precariously poised at the edge of its eastern trough. Certainly whatever processes are going on in the crust affect the people who live on the islands: probably no area in the world has seen more destructive or more intense earthquakes than Eastern Asia's archipelagos. *Tsunamis* or seismic sea waves, caused by undersea earthquakes, volcanic eruptions, or landslides, slam into the Japanese coast with walls of water sometimes over 50 feet high. But we know that people do not hesitate to take up residence in the most dangerous places, on the very slopes of volcanoes known to be likely to erupt, in areas below sea level where only a dike stands between life and death, in lands below river levees which have been broken time and again during floods. Where there are good soils, men will go to farm them, whatever the risks. An so earthquakeprone Japan is no exception, although it is like living on a giant geologic escalator.

Access. This leads to the question of who the first people were that made the decision to seek a home in the Japanese islands, and how they managed to reach the archipelago. A look at the map suggests some likely routes. The peninsula of Korea juts out of the Asian mainland to within 120 miles of southern Japan, and the waters of the Korea Strait are shallow—so shallow that a land bridge probably connected Japan to Korea during periods of low sea level in the Pleistocene glaciation. Another possible route lies through Sakhalin to the north, which virtually connects Japan's northern island, Hokkaido, to the northeast Asian mainland. Still a third available route extends from Kamchatka via the Kuril Islands, also to Hokkaido. Finally, it is likely that some influences reached Japan via the sea, from the south along the east coast of China.

Probably the most effective route has been the Korean one, but it is reasonable, in view of Japan's population makeup, to look for alternate avenues as well. The Mongoloid people who today form the vast majority of Japan's population were preceded by other arrivals—including a significant Caucasoid element! These early people, the *Ainu,* ("hairy-ones") probably came to Japan some 5000 years ago. Some thousands of Ainu have survived the onslaught of the later invaders, mainly in Hokkaido, but their continuing presence has not made the search for their origins much easier. Physically they simply do not resemble any East Asian peoples, although there are traces of intermixture; their skin color and the color of their eyes are not the same. Their "complexion, in short, resembles that of a tanned European," writes Professor G. P. Murdock, but their language provides even less of a clue—it is unrelated to any known tongue.[1]

[1]George P. Murdock, *Our Primitive Contemporaries,* Macmillan, New York, 1934, p. 164.

In view of their physical appearance it is not unreasonable, perhaps, to ascribe to the Ainu a northern migration route into Japan, where they carried on—and still practice—a hunting and fishing economy, supplemented now by a little cultivation. The Ainu always have been great fishermen, and although they now do a little farming along with their fishing, their prime source of food remains the sea. With Eskimo-like inventiveness they fish with spear and net, trap and harpoon; they even go after seals, walruses, and whales in their rather flimsy boats.

The relationships between the Ainu and the succeeding Neolithic cultures in Japan are uncertain. The record only begins to become clear during the first millenium B.C., when the Jolon culture and its successor, the Yayoi (both Neolithic cultures) were superseded by a large influx of people from Eastern Asia, who brought with them knowledge of rice cultivation, the use of bronze, and new tools and techniques. The new invaders came across the Korean Peninsula and made southern Japan their base, and eventually they began to push the older settlers northward while their own sphere expanded. The first eight centuries A.D. constituted a vital period in Japan's development, for the settled realm grew rapidly, a new social organization was achieved, and a political structure emerged under the authority of a succession of emperors. A great deal of borrowing from Chinese culture took place, and the introduction of iron tools and weapons gave the central power a stronger hand over the people. Buddhism appeared in the sixth century, and the Chinese script was adopted. After the eighth century, Japanese history was a matter of record.

Migration

The movement of peoples into Japan, and their subsequent spread throughout the archipelago, is an example of a process that has occurred as long as man has existed: migration. In Europe we discussed the great Volkerwanderung, in the Americas the spread of the Indian peoples is still not completely understood, and in Africa linguistic clues are used to reconstruct the Bantu invasion. People migrate—but what *causes* them to migrate? What brought the ancestors of the Ainu to Japan, and what drove the migrants of two thousand years ago to cross the Korea Strait to settle on Kyushu Island?

Migration as a geographic concept is by no

means a simple matter. "To seek a new way of life" is the basis for only one kind of migration: people also migrate to be able to maintain the way of life they already have. The Boers in Southern Africa, for example, moved away from the Cape of Good Hope because their way of life was being threatened by British cultural superimposition. But hundreds of thousands of Europeans moved voluntarily to colonial areas in America, Africa, Asia, and Australia because of opportunities (especially the ownership of land) there that did not exist at home. There is also the problem of scale: even local, valley-and-mountain transhumance is after all a form of migration, although, like the movement of the Masai and their cattle, it is a *cyclic* or seasonal form of migration.

One way to gain an insight into the migration process is to classify examples of it on the basis primarily of inducement. Thus *natural* or ecological causes, including long-term environmental changes and shorter-term climatic cycles, have caused the displacement of millions of people throughout human history and still induce more localized migrations today. *Cultural* causes take several forms: movement away from an actual or potential threat (a push effect) or toward a more attractive cultural environment (a pull effect). Clearly, different types of cultural inducements can be recognized; one noteworthy example is that of the annual pilgrimage, made by thousands of worshippers, to the holy city of Islam, Mecca.[2] In the fourteenth and fifteenth centuries masses of people made the pilgrimage from the savanna kingdoms of West Africa—it is said that 50,000 to 60,000 people marched the long route to Mecca in some years. Many thousands of them never returned, settling elsewhere on the way. In the Sudan today, more than a million people can trace their ancestry to Nigeria! *Economic* causes also are varied, including the great voluntary migrations from Europe to the overseas colonies, but also the enforced movement of African slaves to the New World. Indentured Asian labor came to both Africa and America, mostly through British manipulation. Chinese settlers moved from the Canton area into many parts of Southeast Asia. Finally, there are movements that involve vari-

ous kinds of inducements, without a clearly defined, primary focus. Everywhere in the world people are moving to the cities, attracted by the (sometimes more imagined than real) economic opportunities there, and by the image of city life. Sometimes the seeming lack of progress in the rural areas induces the farmer to abandon his efforts on the land to seek a city job; thus many variables, natural, cultural, as well as economic come into play. Still, there is no doubt that a rural-to-urban migration exists today, on a world scale, and that it is one of the chief characteristics of current human movement.[3]

The migration of mainlanders to Japan, as we have seen, went on intermittently for centuries. It is difficult, so far back in history, to pinpoint the causes that led the Ainu and the later Mongoloid peoples to enter Japan. Perhaps the ancestors of the Ainu sought for environmental reasons to move southward, and found the Japanese islands a more attractive and productive domain for their hunting and fishing economy than the mainland coast. Possibly the expansion of Han China led the people of Korea and nearby areas on the Asian mainland to move away and eventually across the Korea Strait; what is certain is that during the Han and Tang periods in China Proper, a great deal of cultural borrowing took place in Japan. The early Japanese were eager to continue to learn from the more advanced culture of China, and students and priests continuously traveled to the mainland for Chinese tutelage. In the seventh century there were political upheavals in Korea, which brought a large number of artisans and priests to Japan: steadily numbers and the skills of the Japanese people increased. The Chinese calendar was adopted, and Chinese practices of land allocation were also employed. In China, land was divided by the state among families according to their size; in return for the use of irrigation systems in the North China Plain each family had to pay the state a certain proportional percentage of its rice income. This was one of those Chinese practices which, when introduced in Japan, turned out to be less than ideal for the island country: Japan's mountainous landscape and its fragmented farmlands are a very different scene for the extensive flatlands of the North China Plain.

---

[2] For a fascinating discussion of human movement in this part of the world, and an analysis of the many factors involved, see R. Mansell Prothero, *Migrants and Malaria*, Longmans, London, 1965.

[3] A more complete discussion of the concept of migration occurs in J. E. Spencer and W. L. Thomas, Jr., *Cultural Geography*, Wiley, New York, 1969, pp. 194–196.

And so, quite early on, the Japanese began to modify their essentially Chinese heritage to suit their own domain. Until the eighth century there had been no central capital, but since China had an impressive primate city, Japan should have one as well, and thus the first capital, Nara, was built along the lines of the T'ang capital of China—strictly in accord with the principles of geomancy. The choice of the site, on the Yamoto Plain, suggests how the focus of Japanese settlement was shifting: first it had been on Kyushu, opposite Korea, but now it was pushing northeastward into Honshu. In the process, the conflict with the Ainu grew in intensity, and it was a costly struggle for the Japanese, for at this stage the Ainu offered bitter opposition and the Japanese army was still poorly organized and led. It was a war that had to be carried on during a difficult formative stage in Japan's development, when the central authority was having trouble with dissident clans, when the Buddhist priests threatened to take too much power, and when the peasants grew restive over the conscription of sons for the army. It was probably the Buddhist issue that led, twice during the eighth century, to shifts in the capital city; in 794 it was finally moved to Kyoto, where it remained until the Meiji Restoration.

Japanese Landscapes

To assess the migration of the Japanese within Japan itself, from the original base on Kyushu eastward and northward into Hokkaido and briefly even into Sakhalin, the character of the islands is naturally relevant. Japan consists of one large and three smaller islands, strung in an arc along the east Asian coast. Two of the three smaller islands—Kyushu and Shikoku—lie at the southern end of Honshu, the main island (Map 14-2). Southern Honshu, Shikoku, and Kyushu enclose a body of water known as Setonaikai—the Inland Sea. On its shores lay Japan's first core area; by the third or fourth century A.D. the settlers who had first come to Kyushu from Korea spread all around this Inland Sea, and before long the eastern end of it emerged as Japan's most productive and populous area. Not surprisingly, it was here, where Osaka and Kobe are now located, that the first capital was built, and the later headquarters of Kyoto lie nearby in the interior. The Inland Sea has been referred to as a Japanese Mediterranean, with its many smaller islands and its mountainous coasts, and

the comparison serves to emphasize how little flat agricultural land Japan has—even here, in its historic coreland.

This scarceness of flat land is the dominant impression Japan conveys. Everywhere there are mountains and steep slopes, and everywhere the Japanese have built the most meticulously constructed terraces against hillsides that seem surely too steep for farming, to eke out every potential half-acre of soil. All along the shores of the Inland Sea this scene is repeated time and again: small coastal lowlands, all of them intensively cultivated, rise rapidly to the slopes of the mountains that overlook the Setonaikai. Settlements are placed so as to use a minimum of cultivable land, and everywhere the terraces rise to hundreds of feet above the coastal areas. Obviously those lowland areas that are somewhat larger than usual have always been of extraordinary importance, especially the Kinki District, which includes the Yamato Plain and the old capital of Nara, and now contains the great Kobe–Osaka conurbation and Kyoto. In modern times Japan's largest lowland area, the Kanto Plain, has emerged as the leading agricultural as well as urban-industrial zone, but even this region occupies only 2500 square miles—a mere 1.8 percent of Japan's 142,726 square miles of territory. And here, the farming areas compete for space with the huge Tokyo-Yokohama urban area.

Thus Japan's territory is not only fragmented by water; it is fragmented by its many mountains and mountain ranges as well, so that farming areas and cities alike are perched on or near the margins of the land, separated from each other not only by the mountainous backbones of the islands' main ranges, but also by spurs leading from these mountains to the sea. Eighty percent of Japan is mountainous or hilly, too steep or otherwise unsuitable for cultivation. Yet the country must support a population in excess of 100 million, just about double that of that other island country off the Eurasian coast, the United Kingdom. Small wonder that the Japanese have tried to bring every cultivable acre into production, and practice multiple cropping and intercropping techniques to raise their farm output to the maximum. And, in contrast to India, the Japanese make very effective use of the available nightsoil and manure, which with typical efficiency is distributed from the cities to the farmlands where—as the rural traveler in Japan will

Map 14-2 JAPAN: HISTORICAL GEOGRAPHY.

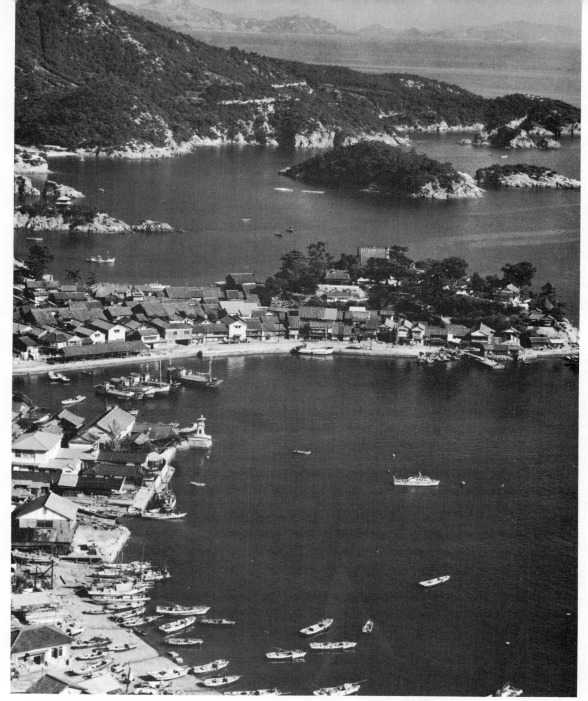

On the shores of the Inland Sea. The drowned coastline here is reminiscent of some Mediterranean shores. Villages cluster on the coast; only the roughest, steepest slopes have been left unused by the farmers. (Japan National Tourist Organization)

North and south in Japan. The photo above shows the cultivated slopes of Shikoku Island; the photo below is a reminder that Japan has a cold northern sector, and that winter scenes are part of northern Japan's environment. (Burt Glinn/Magnum—Japan National Tourist Organization)

know—its odor is an inescapable part of life.

Japan is mountainous country, but fortunately most mountainous areas are clothed in forests, which still cover more than half of its total territory. As Plate 4 shows, these forests range from the so-called temperate rainforest (2 on the Plate) in the south to the needleleaf trees (8) in the north, on Hokkaido. The southern forests could also be described as broadleaf evergreen trees; the northern forests are pine. Between these two areas lie a broad zone of deciduous forest, much of it mixed with needleleaf trees, but despite this rich forest cover Japan does not have all the types of wood it needs, and in addition to local exploitation, wood is also imported. The importance of Japan's forests is clear when it is seen against a background of comparative poverty in fuels: the country has very small petroleum reserves and although there is coal, the best seams have long been worked out and good coking coal is in short supply. So wood is all-important—as a fuel as well as a building material. Japanese houses are built mainly of wood, people burn charcoal in their fireplaces, and the paper and furniture industries demand great quantities of it as well. Where the forests are easily accessible they have been cut over, and as is so often the case, provisions for planting are inadequate. The Japanese, of course, are well aware of the importance of the forest cover as it prevents erosion, conserves moisture, and protects streams and valleys where hydroelectric plants (a crucial alternative for any fuel-poor country) are established. But so great is the pressure on the forest resource that even in this conservation-conscious country the danger of overexploitation is real.

Climate. As Plate 2 indicates, Japan lies in an east coast latitude which in North America would place its northern island, Hokkaido, opposite Maine, while Kyushu would lie off the coast of South Carolina and Georgia. In North America, this would mean that only Hokkaido would have a "snow" or continental climate, while the rest of the country would lie under the relatively warm temperature climate identified as *Cfa* on Plate 2 (extrapolate the dashed line off southern Maine). But in Japan the colder *Dfb* clime extends farther south than that, deep into Honshu. As the boundary line on the map suggests, Japan's high mountains have something to do with this: interior Honshu is colder than the coastal zones. This also tells us something Plate 2 does not reveal, namely that maritime conditions affect Japan's climate to make it, if not as mild as eastern North America, certainly much milder than mainland Asia at the same latitudes. The waters off Japan are warmed by the famous Kuroshio Current (in the east) and the less well known Tsushima Current (in the Sea of Japan, between the islands and the mainland). These waters help ameliorate the cold winds that blow off the Asian mainland during the winter months, so that Japan's coastal areas are somewhat warmer than they would be if this was simply a part of mainland Asia. On the other hand there are cold countercurrents such as the Okhotsk in Japan's northern waters, and Hokkaido's winters are long, snowy, and cold for farming. The coldest month in Tokyo (January) averages 37 degrees Fahrenheit, and even at Nemuro, halfway up the east coast of Hokkaido, only three months of the year register below freezing and none of these averages below 20 degrees. In the south, the growing season is long (240-300 days) and summers are hot. If Japan is thwarted by its mountainous topography, at least it has an ally in its climate.

This is true also of its precipitation. Plate 1 reveals that Japan is quite moist: the three southern islands get more than any adjacent Asian mainland area. Again, what the map does not show is that this rain comes more reliably than it does in much of China and India. Variability is much lower, the seasonal distribution is favorable, and the vicious droughts and violent river floods that strike other parts of Asia do not afflict Japan. Southern Honshu, Shikoku, and much of Kyushu receive over 80 inches of rainfall annually; Tokyo gets about 60 inches, and the Hokkaido station at Nemuro, one of Japan's driest zones, still records nearly 40 inches.

Of course, all cannot be perfect. Japan is infamous for its destructive, death-dealing typhoons. These are the hurricanes of the western Pacific, so named after the Chinese (*tai-fung*, meaning big wind), and this is one of the two world areas of greatest frequency for these storms. (The other is the western Atlantic and Caribbean Sea). The mechanism that causes these storms to form is still imperfectly understood, but the answer appears to lie in studies of the upper air. Typhoons are huge cyclonic storms whose convergent winds rotate at a high velocity around a low-pressure center. The diameter of the whole system may exceed 500 miles; vertically it extends from 40,000 to 50,000 feet. Air is drawn at speeds upward of 100 miles an hour toward the "eye," around which it is then corkscrewed upward at a high rate, producing intense rainfall and thick, dark clouds, with associated thunderstorms. The entire mechanism moves slowly in a path that usually begins with a westward direction, but then it tends to turn toward the north and eventually to the northeast—a giant curve that may take many days to complete. In the process, the storm builds up greater and greater strength, and the damage is done by those surface winds rushing toward the center. In Japan, the relatively flimsy houses are demolished by the thousands, waters are whipped into raging attack, and few typhoons pass over the south without a loss of life.

Just how a typhoon or hurricane begins is still being worked out, but it appears that the process has a chance of developing when a surface low-pressure area happens to coincide with an upper-air anticyclone. In other words, the upper-air anticyclone is so positioned that it lies precisely on top of the surface low-pressure center. Now a pumping system is created: the air at the surface converges into the cyclonic center (as it was already doing, by the nature of a cyclonic low-pressure center; see Chapter 2), and as it is fed upward, it is rapidly dispersed by the process of divergence inherent in the anticyclone. Thus the combination of surface convergence and upper-air divergence, produced when the two circulation cells are juxtaposed, may create the beginnings of a typhoon. Although the typhoon does not move forward fast, and its impact on a land area can usually be forecast, history is full of examples of occasions when the unexpected happened—a change in course, a rapid buildup of an apparently weak storm, a huge wall of water, whipped up by the winds, smashing suddenly into the coastal towns. This is a price southern and central Japan have to pay for their otherwise favored locations; hardly a year goes by without typhoon-caused destruction of life and property. In 1959, Typhoon Vera on September 26 and 27 killed 4500 people in Honshu. In 1967, nearly 350 died on July 9 during Typhoon Billie.

To return to our theme of internal Japanese migration, it is understandable that the south remained for a long time the main base of settlement. Not only did southern Japan lie nearest to the culture hearth from which ideas and innovations were constantly borrowed, but southern Japan, topographically and climatically, was the most livable and productive part of the islands, insofar as they were known. For several centuries the Japanese sphere remained confined to the area south of the narrow "neck" of Honshu, where Lake Biwa lies (Map 14-2), spread about the Inland Sea and focused on its eastern margins. This was not all due to the favorable southern environment, of course. Japan was badly divided, and rich and powerful families struggled with each other, mostly through force of arms, to gain greater influence and control. But during the seventh and eighth centuries the northward push gained momentum. Here lay a frontier of expansion where new spheres of control could be created, and conscript armies fought a desultory war against the tenacious but retreating Ainu. Driven off the Honshu fishing coasts, the Ainu fled northward, out of the immediate reach of the encroaching southerners, or into the montane forests in the northern quarter of Honshu Island. Eventually they lost the whole Honshu coast and only Hokkaido remained, but Hokkaido's highlands are barren and cold, and soon the coasts would be threatened. The lot of the Ainu was sealed, and it was only a matter of time, by the eleventh century,

until their last, localized resistance was broken.

The conquest of Honshu did not lead to a mass migration by Japanese southerners to the newly opened lands of the north, although the Kanto Plain was an early goal for migrants and Tokyo was founded in the twelfth century. But people who settled in the north often took up the fishing of the Ainu (and some intermarried with those who survived the conquest) and abandoned the agriculture they had known in the south. There was no rush to the northern frontier. But then, in the thirteenth century, there suddenly seemed good reason to leave the south: Japan was facing what almost all of Asia faced, namely the might of the Mongol armies of Kublai Khan. From their base in Korea the Mongols twice launched armadas against Kyushu and southern Japan, first in 1274 and again in 1281. The attack of 1274 was on the verge of success when a typhoon destroyed most of the Mongol fleet, and Japan's eternal enemy was an ally in this time of crisis. The invasion of 1281 found the Japanese better prepared, and they managed to hold off the Mongols until, incredibly, another summer typhoon annihilated their enemy's fleets. Small wonder the Japanese have a famous word for their "divine wind"—the *kamikaze*.

Meanwhile, the sociopolitical structure of Japan was jelling into something resembling Europe's feudal condition. The long internal struggles and the general defense against the Mongol invaders had helped bring into the foreground the most powerful and successful military leaders: these *shoguns* and their vassals, the *samurai*, represented a new and larger-scale order. Actually, order is not quite the right term, because the competition for absolute supremacy went on. But eventually the shogunate based in the Kanto Plain gained more or less complete superiority over Japan, and early in the seventeenth century the Tokugawa family-clan obtained control of this shogunate. The productivity of this, the largest lowland agricultural area of Japan, overshadowed that of the Inland Sea fringes; the old fishing town of Edo grew into a metropolis which midway through the eighteenth century had over a million inhabitants and may at that time have been the largest city in the world. During the Tokugawa Shogunate, which lasted until the Meiji Restoration of 1869, Japan went into its long period of isolation, which, we know, involved the rejection of the initial European contributions, the virtual elimination of trade with Europe, and the closing of all doors to external influences. After Japan's confused formative cultural period from about 900 to 1400 A.D., this was the time for the country's political, cultural, economic, and demographic stabilization. Unified for the first time, Japan attained economic self-sufficiency with a population that had leveled off at nearly 30 million—not an average maintained by periodic disasters offsetting periods of rapid growth, but a truly stable figure. In view of what was happening to populations elsewhere in the world during this period, this was a remarkable achievement.

Thus the landscape of Japan—the cultural landscape especially—was changing, but the changes, unlike those in India or Southeast Asia, were indigenously Japanese and not the result of the imposition of some foreign culture. Thirty million people in the mid-nineteenth century was a large number of people for an area the size of Japan, and even then most of the alluvial lowlands had been transformed into rice fields and the terraces were climbing the adjacent hillslopes. The cities, and cities of impressive size they were, carried one cultural imprint: a Japanese imprint. Roads were connecting the widespread settlements, the forests were being cut increasingly, more and more boats sailed the Inland Sea and the coastal waters as internal trade and fishing grew. Japan was a world in itself, and it emerged from its isolationist period with the overt liability of military weakness, but also with an inherent strength derived from its national cohesion and cultural unity.

Modern Japan

With the Meiji Restoration, just a century ago, Japan began its phenomenal growth that brought it an empire and involved it in a disastrous war, and which eventually saw the country achieve an industrial capacity unmatched not only in the non-Western world, but in much of the Western world as well. As if to symbolize their rejection of the old, isolationist position of the country, the new leaders moved the capital from Kyoto (long known as Saikyo or Western Capital) to Tokyo (or Eastern Capital), the new name for the city of Edo in the Kanto Plain. Kyoto had been a landlocked capital; Tokyo lay on the Pacific coast, on an estuary that was to become one of the busiest harbors

in the world. The new Japan looked to the sea as an avenue to power and empire, and it looked to industrial Europe for the lessons that would help it achieve those ends. While China suffered as conservatives and modernizers argued the merits of an open-door policy to Westernization, Japan's leaders committed themselves—with spectacular results.

Japan's success might lead to the assumption that the country's raw material base was sufficiently rich and diverse to support the same kind of industrialization that had characterized England. Britain's industrial rise, after all, was accomplished largely because the country possessed a combination of resources that could sustain an industrial revolution. But Japan is not nearly so well off. Although Japan has coal deposits in Kyushu, Hokkaido, and Honshu, there is little good coking coal. The most accessible coal seams were soon worked out, and before long the cost of coal mining was rising steadily. There was a time when Japan was a net coal exporter, when coal was shipped to the coastal cities of China, but today Japan has to import this commodity. Nevertheless, Japan did have enough coal, fortuitously located near its coasts, to provide a vital stimulus to industry during the late nineteenth century. Both in northern Kyushu and in Hokkaido, the coal deposits lie so near the coast that transport is not a problem, and by the nature of Japan's landscape the major cities and industrial areas also lie on or near the coast. Thus coal could be provided quite cheaply and efficiently to every locale of incipient industrial development, and hence the fringes of the Inland Sea were among the first areas to benefit from this.

As for the other main ingredient of major industry—iron ore—Japan has only a tiny domestic supply, and one of small, scattered deposits of variable but generally low quality. In this respect there is simply nothing to compare to what Britain had at its immediate disposal, and neither is Japan rich in the various necessary ferroalloys (zinc, lead, tin, etc.), except for some copper. The Japanese extract what they can, including ores that are so expensive to mine and refine that it is hardly worth the effort, in view of import alternatives, and they carefully assemble and use all available scrap iron. But still, iron ore ranks high on each year's import list, and Japan buys iron all over the world, even as far away as Swaziland in southern Africa.

Tokyo, with nearly 12 million people, is Japan's modern capital and the world's largest city. The view overlooks downtown Tokyo and the central railroad station. (Consulate General of Japan, New York)

Neither does Japan have sizable petroleum reserves. More of this fuel is needed every year, and today Japan already pays about twice as much for its annual petroleum purchases as it does for its iron ores. The few barrels of oil that are produced at home are hardly worth mentioning, and the hydroelectric power that is derived from the country's many favorable sites is not enough to fill its needs. This in itself is a measure of the amount of electricity Japan consumes: with its high mountain backbone, its ample rainfall, deep valleys, and urban markets clustered on the mountains' flanks, all conditions are right for efficient hydroelectric power production and use. But although nearly two-thirds of Japan's electricity comes from these dams, the demand rises faster than does their production, and the day seems far off when petroleum and coal imports can be reduced.

This, then, answers one of the questions we posed in the first paragraph of this chapter. Japan really has nothing to compare to what Britain had during its industrial transformation. In fact, after what has just been said, it would seem that we are dealing with just another of those many countries whose underdevelopment can be attributed largely to a paucity of resources, and whose future is therefore deemed to depend on agriculture. Certainly it would have been difficult in 1869 to predict that the enthusiastic leaders of the Meiji Restoration would really have much success with their plan for the Westernization and strengthening of their country. What, then, gave Japan its opportunity? Obviously foreign trade must have played a great part in it. Was Japan's location the key to its fortune?

The Foundations. It is tempting to turn immediately to Japan's external connections, and to call upon factors of location to account for Japan's great industrial growth. The Japanese built an empire, much of it right across the narrow Korea Strait, and this empire contained many of the resources Japan lacked at home. But before we rush into such a deduction, let us consider what Japan's domestic economy was like in the mid-nineteenth century, for, as we shall see, what Japan achieved was in the first place based upon its internal human—and to a large extent natural—resources.

In the 1860's manufacturing—light manufacturing of the handcraft type—was already widespread in Japan. In home industries and community workshops they produced silk and cotton-textile manufactures, porcelain, wood products, and metal goods. At that time the small metalliferous deposits of the country served their purpose: they were enough to supply these local industries. Power came from human arms and legs and from wheels driven by water; the chief source of fuel was charcoal. Thus there was industry, and there was a labor force, and there were known manufacturing skills.

The planners who took over the country's guidance after 1868 realized that this was but an inadequate base for industrial modernization, but that it might nevertheless generate some of the capital needed for this process. Thus the community and home workshops were integrated into larger units, thermal and hydroelectric power began to be made available to replace older forms of power and energy, and for the first time Japanese goods—still of the light manufacture variety, of course—began to beat Western producers on the world's markets. Meanwhile the Japanese continued to resist any infusion of Western capital, which might have accelerated the industrialization process but would have cost Japan its economic autonomy. Instead, the farmers faced higher taxes, and the money thus earned by the state was poured into the industrialization effort.

Now Japan's layout and topography contrib-uted to the modernization process. Coal could be mined and—in the relatively small quantities then needed—transported almost wherever it was needed. Nearly everywhere, a hydroelectric site was nearby, and soon electric power was available virtually anywhere, in the cities as well as the populated countryside. Japan was still managing on what it possessed at home, and the factories and workshops multiplied. Soon the beginnings of a chemical industry made their appearance, and government subsidies and support led to the establishment of the first heavy industry. The decade of the 1890's was a period of great progress, stimulated by the war against China and Russia. Nothing served to emphasize the need for industrial diversification as did the war, and Japan's victories vindicated the ruthlessness with which industrial objectives were sometimes pursued.

Area Organization

Japan is not a very large country, but it was already quite populous when its modern economic revolution began: more than 30 million people were crowded in its confined living space. Since then, Japan's population has more than tripled, and its economic complex has grown enormously. With a million people Tokyo may have been the world's largest city in the 1800's, but today the capital is more than ten times as large and it has merged with the urban area of Yokohama. The whole conurbation is still the largest of its kind, although—if it matters—it can be argued that New York remains the biggest single metropolis in the world. That Tokyo should be in New York's league at all is the amazing thing, and a reflection of the phenomenal development of the country.

With such growth, space in Japan has been at a premium, and urban and regional planners have shared with farmers the problem of how to make every square foot count. It is often said that the most densely populated rural areas in Japan are more crowded than any other similar areas in the world, but the cities, towns, and villages must also vie for the country's limited livable terrain. While Japan's industries were still of the light, handcraft variety the many workshop-type establishments were located in cities as well as villages. The small-scale smelting of iron could be carried on in many places; charcoal was widely available, and textile manufacture could be done in many places. But then the age of the factory arrived, and the rules of industrial location went into effect. We discussed some of these rules earlier in this book: the advantages of treating heavy or bulky raw materials at the spot where they are mined to save transport costs, the transport of lighter and more easily moved raw materials to the areas where heavier, more costly-to-move resources are found. In Britain during the industrial revolution this led to the rapid growth of some towns into industrial cities, the decline of others, and the founding of still other industrial centers. In Japan, on the other hand, urban-industrial growth was straitjacketed: apart from northern Kyushu's coal no great reserves of raw materials were found, and it soon was clear that these would have to be imported from elsewhere. Internally, the distribution of coal was mainly by water; externally, the iron ore and other requirements of course came by sea. Hence the coastal cities, where also the labor forces were concentrated, became the sites of the modern factories and industrial plants.

The modern industrial growth of Japan, thus, tended to confirm the development that was already taking place—in any case, Japan did not have the resources to merit any major reorientation. But obviously some cities had advantages over others in terms of their facilities or their position with reference to the source areas of the imported raw materials. Now began the kind of differentiation that has marked regional organization everywhere in the world. Professor A. K. Philbrick in 1957 published an article about this principle, which he called Areal Functional Organization.[4] Briefly, this idea is based on five precepts. First, human activity has focus, that is, it is concentrated in some locale—a farm, or factory, or store. Second, such "focal" activity

[4]Allen K. Philbrick, "Principles of Areal Functional Organization in Regional Human Geography," *Economic Geography,* vol. 33, No. 4, October 1957, pp. 299–336.

Yokohama is eastern Japan's largest port. The extensive urban sprawl and large port facilities are shown in this photograph. (Consulate General of Japan, New York)

is carried on in certain particular places. Obviously no two establishments can occupy exactly the same spot on the earth's surface, and so every one of them has an absolute location, which can be measured by latitude and longitude calculations, but more relevantly, every establishment has a location *relative* to other establishments and activities. Obviously, no human activity is carried on in complete isolation, and thus the third precept is that interconnections develop among the various establishments. Farmers send crops to markets, they buy equipment at service centers. Mining companies buy gasoline from oil firms, lumber from sawmills; they send ores to refining plants. Thus a system of interconnections emerges, which grows more complex as human capacities and demands expand. With their spatial character, these systems form units of area organization. Philbrick's fourth principle is that the evolution of these regions of human organization is the product of man's "creative imagination" as he applies his total cultural experience as well as his technical know-how when he decides how to organize and reorganize his living space. And finally, it is possible to recognize levels of development in area organization, a sort of ranking on the basis of type, extent, and intensity of exchange. To quote Professor Philbrick: "The progression of area units from individual establishments to world regions and the world as a whole are formed into a hierarchy of regions of human organization."[5]

The broadest ranking, of course, is into subsistence, transitional, and exchange-type organization, and Japan's area organization obviously reflects the last of these. But within each type of organization, and especially within the complex exchange-type unit, individual places can also be ordered or ranked on the basis of the number and kinds of activities and interconnections they generate. A map of Japan showing its resources, cities and towns, and surface communications, can tell us a great deal about the kind of economy the country has. It looks just like maps of other parts of the world where exchange-type organization has developed—cities and towns ranging from the largest to the smallest hamlets, a true network of railroads and roads connecting these places, and productive agricultural areas near and between

the urban centers. In Japan's case, Map 14-3 shows us something else: it reflects the country's external orientation, its dependence on foreign trade. All the primary and secondary regions lie on the coast: of all the cities of any size, only Kyoto lies in the interior, and it lies at the head of the Kinki Plain. If we were to deduce that Kyoto does not match Tokyo, Yokohama, Nagoya, or Kobe–Osaka in terms of industrial development, that would be correct: the old capital remains mainly a center of light manufacture. Actually, Kyoto's ancient character has been preserved deliberately, and large-scale industries have been discouraged. With its old temples and shrines, magnificent gardens, and its many workshop and cottage industries, Kyoto remains a link with Japan's pre-modern past.

As Map 14-3 shows, Japan's leading primary region of urbanization and industry, along with very productive agriculture, is the Kanto Plain. The Kanto Plain contains not only Tokyo and

Made in Japan: the New Tokaido Line between Tokyo and Osaka. The electric trains run at speeds up the 125 miles per hour. Mount Fujiyama is in the distance. (Japanese National Railways)

[5]For a brief and clear statement of the idea see A. K. Philbrick, *This Human World,* Wiley, New York, 1963, pp. 12–36. The quotation is from p. 32.

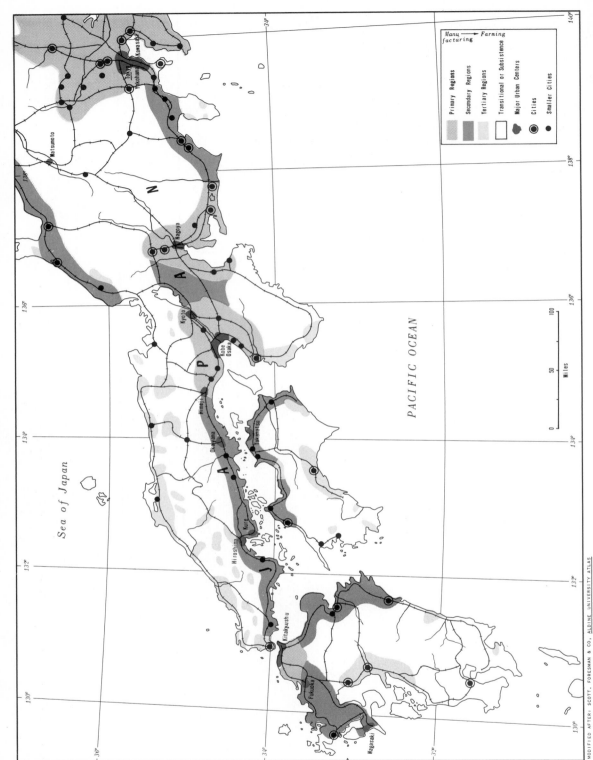

Map 14-3 JAPAN: AREA ORGANIZATION.

MODIFIED AFTER: SCOTT, FORESMAN & CO., ALDINE UNIVERSITY ATLAS

566

Yokohama, but also Japan's most extensive low-land farming area. This giant cluster of cities and farms forms the eastern end of the country's elongated, fragmented core area (Map 14-4), and apart from its flatness it has other advantages: its fine natural harbor at Yokohama, its relatively mild and moist climate, and its central location with reference to the country as a whole. It benefited also from Tokyo's selection as the capital, at a time when Japan embarked on its planned economic growth. Many industries and businesses chose Tokyo as their headquarters in view of the advantages of proximity to the government's decision makers.

MODIFIED AFTER: RAND MCNALLY, GOODE'S WORLD ATLAS

Map 14–4 JAPAN: CORE AREA.

The Tokyo–Yokohama conurbation has become Japan's leading manufacturing center, with between one-fifth and one-quarter of the country's output. But the raw materials for all this industry come from far away: Tokyo is the chief steel producer in Japan, using iron ores from the Philippines, Malaya, India, and even Africa. Although electric power comes from the mountain valleys to the northwest, and coal is shipped in from Hokkaido, coal also comes from Australia and America, and petroleum from Southwest Asia and Indonesia. As for the supply of food to the huge population, the Kanto Plain cannot produce nearly enough, and imports come from Canada, the United States, and Australia, as well as from some other areas in Japan itself. Thus Tokyo depends completely on its external trade relations for food, for raw materials, and for markets for its wide variety of products, ranging from children's toys to steel, and from the finest optical equipment to ocean-going ships.

The second-ranking urban-agricultural region in Japan's core area is the Kobe–Osaka–Kyoto triangle, at the eastern end of the Inland Sea. The position of the Kobe–Osaka conurbation with the old Manchurian empire of Japan may have been strong: the facts confirm it. Osaka was the major Japanese base for the exploitation of Manchuria and trade with China, and it suffered when the empire was destroyed and the trade connections with China were not regained after World War II. Kobe, with Yokohama Japan's chief shipbuilding center, has remained the country's busiest port, adding the traffic of the Inland Sea to its overseas connections. Among these connections has for many decades been the import of cotton for Osaka's textile factories: with its large skilled labor force and its long history of productive settlement, Osaka until very recently was Japan's first textile producer—before as well as after the Meiji Restoration. Kyoto, as we noted, remains much as it was before Japan's great leap forward; with nearly 2 million people it nevertheless remains a city of small workshop-type industries.

The Kobe–Osaka–Kyoto region, like the Kanto Plain, is an important farming area (the area is known as the Kinki District) as well as an industrial complex that rivals Tokyo–Yokohama. Rice, of course, is the most intensively grown crop here in the warm, moist lowlands, but the Yamato Plain (one of five fault basins in the Kinki District) is not nearly as extensive as that of Kanto. Here, then, is another cluster of people that requires a large infusion of foodstuffs every year.

The Kanto Plain and the Kinki District, as Map 14-3 suggests, are the two leading primary regions within Japan's core. But between them lies the Nagoya or Nobi Plain, focused on the textile center of Nagoya (2 million), the city that recently ousted Osaka from first place among Japan's textile producers. The map indicates some of the Nagoya area's advantages and liabilities. The Nobi Plain is larger than that of the Kinki District, and thus its agricultural productiveness is greater—though not as great as that of the Kanto Plain. But Nagoya has neither Tokyo's centrality nor Osaka's position on the Inland Sea; its connections to Tokyo are by the tenuous Sen-en Coastal Strip, the Tokaido Road (Map 14-4). Its westward connections are somewhat better, and there are signs that the Nagoya area is merging with the Kobe–Osaka region. Still another disadvantage of Nagoya lies in its port, which is not nearly as good as Tokyo's Yokohama or Osaka's Kobe, and has been plagued by silting problems.

Westward from the three regions just discussed extends the Inland Sea, along whose shores Japan's core area is developing. The most impressive growth has occurred around the western entry to the Sea, where Kitakyushu (a conurbation of five north Kyushu cities) constitutes a fourth Japanese manufacturing center. Honshu and Kyushu are connected by road and railway tunnels, but the northern Kyushu area does not have any equivalent on the Honshu side of the Strait of Shimonoseki. The Kitakyushu conurbation includes the town of Yawata, site of northwest Kyushu's large coal mines, and it is on the basis of this coal that the first steel plant in Japan was built there—a complex which for many years was Japan's largest. The advantages of transportation here at the western end of the Inland Sea are obvious. No place in Japan is better situated to do business with Korea and China, and when relations with China are normalized this area should be quick to reflect the results. Elsewhere on the Inland Sea coast, the Hiroshima–Kure urban area has a manufacturing base that includes heavy industry, and on the coast of the Korea Strait, Fukuoka and

Japanese-made automobiles being loaded onto a Japanese-built freighter. In 1967 the country's automobile industry became the world's second largest, after that of the U. S. A. Exports increase annually (Japanese vehicles have taken a substantial share of the American small-car market), and domestic consumption rises rapidly as well. (Consulate General of Japan, New York)

Nagasaki are the principal centers, Fukuoka as an industrial town, and Nagasaki as a center of large shipyards.

The map of Japan's area organization, thus, shows the country's core area to be constituted largely by four individual primary regions—each of them primary because each duplicates to some degree the contents of the others. Each of the four areas contains iron and steel plants, each is served by one major port, each lies in or near a large and productive farm area. What the map does not show is that each has its own overseas, external connections for the acquisition of raw materials and the sale of its products. These connections may be stronger than those among the four core sections within Japan itself; only in the case of Kyushu and its coal have domestic raw materials played much of a role in shaping

nature and location of heavy manufacturing. In the shaping of the country's areal functional organization, therefore, more than just the contents of Japan itself is involved. In this respect Japan is not unique—all countries that have the exchange-type organizations must to some degree adjust their spatial form and structure to the external interconnections that are required for progress. But it would be difficult to find a country in which this is more true than in Japan.

Japanese Agriculture

After what has been said about agriculture in India, in China, and in Southeast Asia—what can be said about farming in Japan that is not merely a repetition of evidence pointing to its high intensity, great productivity, and large rural population density? The fact is that Japan's industrialization so occupies the center stage that it is easy to forget that considerable achievements have been made in the field of agriculture. Japan's planners, no less interested today in closing the food gap than in expanding industries, have created an extensive network of experiment stations and information services to diffuse as rapidly as possible information to farmers that is useful in the raising of crop yields. But although this program has been very successful, Japan faces the unalterable reality of its stubborn topography: there simply is not the necessary land to farm. Less than one-fifth of the country's total area is in cultivation, and although there still may be some land that can be brought into production it is so mountainous or cold that its contribution to the total harvest would be minimal anyway. Japanese agriculture may resemble Asian agriculture in general, but nowhere else do so many people depend on so little land. Japan's population density overall is over 700 per square mile to begin with, but when the actual cultivable land is brought into the picture it turns out that more than 4000 people depend on the average square mile of farmland. Thus Japan's *physiological density* may be the highest in the world. For an idea of the situation compared to other regions and countries: Asia as a whole has about 500 persons per square mile of cultivated land, Europe 220, and the United States 260. Egypt, one of the most highly concentrated populations in the world, still has a physiological density of under 3000. Thus Japan in these terms is almost in a class by itself.[6]

In other ways, too, Japanese agriculture stands apart from farming elsewhere in "monsoon" Asia. It is by far the most efficient in all of Asia, and, on its scale, in the entire world, given the existing conditions of soil, slope, and climate. Previously we commented on the way in which yields could be increased in India, and how additional land could be brought into cultivation in the Philippines, how irrigation could be improved or expanded, and better care taken of the fields and paddies. This commentary would hardly apply to Japan, whose rice yields are the highest in the world: over 80 bushels per acre, while China has under 60, the United States just over 50, and South and Southeast Asia are used

[6]For a summary of the concept of physiological density, other examples, and some comparisons to other kinds of applications of the density concept, see Howard F. Gregor, *Environment and Academic Life,* Van Nostrand, Princeton, 1963, pp. 252–254.

to 20 to 30 bushels. In only a few places are higher yields than those of Japan recorded, and then because of special circumstances, as in southern China, where multiple cropping can raise the amount of rice *annually* taken off an acre of soil above that harvested in a single harvest of Japan, and in a few places in Mediterranean Europe, where rice is grown on soils far better than those of Japan.

Population Control. Among Japan's many remarkable achievements is its low rate of population increase, which currently stands at about 1 percent per year. This has been an achievement of the past 25 years; prior to and during World War II the Japanese encouraged population expansion to support imperialist designs. But when the war ended, Japan found itself without an empire to which an overflow population could emigrate, with an influx of people ousted from the former colonies, and with a continuing high birth rate at home. In 1947 this birth rate still was as high as 34.3 per thousand, a figure that is familiar to Asian and Latin American countries. Then, in 1948, the government took action. It passed the so-called Eugenic Protection Act, whereby abortions for "social, medical, and economic reasons" were legalized. Also made legal was the sale of contraceptives. Family planning offices were set up throughout the country, but it was the high number of abortions rather than the quick adoption of contraceptives that brought the birth rate down dramatically—so dramatically, in fact, that the Japanese government became afraid for the well-being of the nation, and warned that contraception was the preferred method for limiting family size. By 1958, just 10 years after the Eugenic Protection Act was passed, Japan's birth rate was 18.0, and an estimated 2 million abortions per year were taking place. Meanwhile, the death rate, which in 1948 still stood at 14.2, was down to 7.5 by 1958, so that the rate of increase in the late 1950's was just about what it is today—just 10 per thousand. It is not difficult to see what would have happened to Japan had the birth rate remained near its postwar levels (as it has in so many Asian and Latin American countries) and had the death rate declined as it did. Japan's population explosion would have spelled trouble; no conceivable rate of industrial growth or agricultural expansion could have coped with such a circumstance. The Japanese were aware of what faced them, and with characteristic thoroughness they decided to tackle the problem with massive programs and saturation methods of diffusion. What they managed to do stands as an example to the world.

Japan's population is now well over 100 million, having tripled over the past century, and although the annual growth rate of 1 percent is much lower than those of other Asian countries, Japan no longer enjoys its one-time luxury of self-sufficiency in food. But it is remarkable that the country is somehow able to produce more than three-quarters of its annual needs—enough to feed over 75 million people. It is achieved through turning over more than 90 percent of all farmland to food crops (and 53 percent of this is under rice), through diligent irrigation (more than half of Japan's agriculture is irrigated), through painstaking terracing, multiple cropping wherever possible (south of the 37th parallel, including the major farm districts), intercropping, intensive fertilization (animal manures, night soil, and chemical fertilizers are bought by the farmers in great volume) and transplanting—that is, the use of seed-beds to raise the rice plants while another crop matures on the paddy. Japan's second and third crops, wheat and barley, occupy less than one quarter of the farmland. Wheat is used as a winter crop with rice in the warmer parts of the country, and barley is grown in the north as a summer crop. Sweet and white potatoes, as the major root crops, occupy less than 10 percent of the soil. Still less is given over to cash crops, such as tea, grown for local consumption, and mulberry, the plant whose leaves are used to feed the silkworms, the basis of Japan's once-rich silk industry. Of course, there are the vegetable gardens to be expected around the large cities. But Japan's major agricultural effort goes into raising the production of its staple—rice.

As an urbanizing country, Japan is in a class with the United States and Western Europe, in that some 64 percent of the people live in cities and towns (the corresponding figure for the United States is 72 percent, the United Kingdom, 81 percent). Some sources suggest that nearly three-quarters of Japan's population are now concentrated in cities and towns of 20,000 and over, but this seems a high estimate; nevertheless, the percentage of farmers in Japan's population has declined steadily during most of the past hundred years. In the mid-nineteenth century some 80 percent of the Japanese people were farming; today the total number of farmers on the land is not much larger, although the population has more than tripled. But even this is a large number of farmers for so small an agricultural area, and not surprisingly, farms in Japan are quite small—not much over 2 acres on the average. And even these small farms are subject to subdivision as a farmer leaves his land divided among his sons upon his death. This is not Japanese practice, but it was introduced as part of a reform program imposed after World War II. Previously, the Japanese custom was to leave the land to the oldest son. If this program was not altogether desirable, from the point of view of land fragmentation, it did have good results in another area. As was the case in pre-Communist China and in so many other parts of Asia, Japan before its World War II defeat had a very high rate of tenancy in agriculture, with landlords owning great tracts of soil which were farmed by tenants in return for a harvest-rent which went as high as 70 percent of the total yield. Two out of three Japanese farmers were renters, and the system demanded reform. This was also imposed during the American occupation, with the result that today, nearly all the land cultivated is farmed by its owners.

The Frontier. Is there an agricultural frontier in Japan as in China, where further development can yet take place? Discussions of Japanese agriculture tend naturally to focus upon its achievements and its severe limitations, but it is worth recording that other views also exist. Here, for example, in what Professor J. E. Spencer wrote in 1954:

"It is normal to suggest that much of Japan is too steep in slope to permit a significant increase in the amount of agricultural land beyond that cultivated in recent centuries. It is also a common generalization that in all Japan today there are only a few hundred thousand acres of reserve land that can be turned into productive farms, and that most of this is marginal land for which the costs rival the potential return. This is a traditional view, and it is the traditional attitude of Japanese leaders, both in and out of government. It fails to take into account the most efficient developments in mountain agriculture elsewhere in the world. There is ample ground to suggest that, if a realistic program were conceived beyond the traditional oriental pattern, looking to tree and bush crops, upland animal economy, industrial crop production, the extension of mountain and arctic-margin varieties of land use, there are many upland areas of relatively gentle slope that could be brought into effective land use.

"What is needed is a fresh outlook upon land use, a program of the world study of agriculture, which could promote something of the same kind of revolution that the Japanese leaders brought to the manufacturing economy of Japan. One is duly impressed by the facts that in Japan today there is very little wastage of lowland arable lands, and that on the existing farms there is no small patch of ground wasted. It is also impressive that in the industrial expansion of Japan a minimum of farmland was diverted from agricultural productivity, in contrast to the situation in parts of the United States where urban land developments of the last century have often driven the farmer off the very best crop lands. Nevertheless, a large area of uplands is still unused, and it does not seem true that there is no opportunity for the expansion of Japanese agriculture. . . ."[7]

Apart from Japan's uplands, the other area that comes to mind when the Japanese frontier is discussed is that cold, sparsely peopled northern island, Hokkaido. With a growing season of less than five months, bitter winter colds, heavy snows, and limited farmable land, Hokkaido's agricultural possibilities are obviously rather restricted. Nevertheless, it is possible that the comments made by Professor Spencer apply as much to Hokkaido as to any part of Japan. There are 2 million acres of land which can be improved by fertilizers, the clearing away of stones, and the kind of meticulous care the Japanese normally give their soil. There are crops that are adapted to the harsh climate Hokkaido presents. And there are minerals in Hokkaido; a larger population almost certainly could be supported in Japan's far north.

An Alternative: Fishing

With its restricted arable area, Japan naturally seeks to produce the maximum calories per cultivated acre, and this, combined with the cultural attachment involved, naturally led to the choice of rice as the country's staple. Rice, wheat, barley, and potatoes make for a very starchy diet, and what the Japanese need more than an extension of farmland to produce yet more starch

[7] Joseph E. Spencer, *Asia, East by South,* Wiley, New York, 1954, pp. 389–390.

is protein. Happily they have the opportunity to get it, and not by trade, but by their own harvesting of the oceans' fishing grounds. With customary thoroughness the Japanese have developed a fishing industry which on all accounts is larger than that of the United States or any of the long-time fishing nations of northwestern Europe, and now supplies the people with a second staple after rice.

Although mention of the Japanese fishing industry brings to mind fleets of ships and trawlers scouring the oceans and seas distant from Japan, in search of salmon, whales, tuna, and herring, the fact is that most of the catch comes from waters within a few dozen miles of Japan itself. With the warm Kuroshio and Tsushima currents meeting colder water off Japan's coasts, a rich fish fauna exists, including sardines, herring, tuna, and mackerel in the warmer waters, and cod, halibut, and salmon in the cooler northern seas. Along Japan's coasts there are about 4000 fishing villages, and tens of thousands of small boats ply the waters offshore to bring home catches that are distributed locally. Many of the fishermen divide their time between farming and fishing, just as so many Japanese farmers still increase their income by some seasonal or part-time light manufacturing. Japanese farmers also raise fresh-water fish in artificial ponds and in flooded rice paddy fields, and they are experimenting with the cultivation of algae for its food potential.

Apart from the wealth of fish in Japan's waters, the industry has other advantages. The coastal orientation of Japan's population, and hence the proximity of the markets to the supply points, facilitates both the fishing itself and the quick distribution of the catch. Japan's long coastlines are indented in many places to create good harbors, and offshore, the continental shelf is quite shallow and wide, despite the deep trench that lies off the east coast of the archipelago. And with their empire and their shipbuilding industry, it is not surprising that Japanese fishing boats soon ranged far and wide. That advantage—the empire—has vanished: Japan once held not only the Kuril Islands, part of Sakhalin, and Korea, but also the rich fishing waters around these areas. They were lost as a result of World War II, and this was a blow to the large-scale sector of the industry. But the Japanese have simply increased the tonnage of their fleets and widened the search to all the world's oceans, and they

Japan's fishing industry is enormous; its fleets sail all the world's waters in search of their catch. The photo, taken in Yaizu, shows a local catch and processing plant. (Rene Burri/Magnum)

now bring in more than one-quarter of the combined annual catch of all the fishing countries in the world.[8] Fish, in fact, figures in the Japanese list of exports, as high-priced canned salmon and crab, as well as tuna. The fishing industry is another of Japan's amazing success stories.

Prospects

So remarkable has been the Japanese capacity for adaptation and innovation that it seems unwise to suggest that difficult times might yet lie ahead for this energetic country. After 1868, practically every obstacle has been overcome; the country staved off any colonial threat, defeated its mainland enemies and created an empire of its own, lost that empire and saw enormous destruction of its cities and industries, and yet regained its old momentum and more: in the 1950's and 1960's Japan's economy grew faster than the economy of any country had ever grown. This is all the more remarkable in view of the fact that Japan's economy had been so tightly integrated with the various parts of the empire, to which it sold its products at high

[8]For a description of Japan's effort in this sphere see G. A. Borgstrom, *Japan's World Success in Fishing,* Fishing News Ltd., New York, 1964.

prices and from which it imported raw materials—including three-quarters of its coal and more than half its iron—at very low cost. Japan had made heavy investments in the mines and factories of its Chinese, Manchurian, Korean, and Taiwanese colonies, and it depended on them for most of its food imports. Of course Japan received considerable American aid after the war, but the question was whether the country could establish new external connections to replace the old, lucrative ones, and whether in the long term, its economy could survive on the basis of foreign trade.

The answer to the first part of this question is yes: with its twin assets of low-cost production (against Western competition) and large skilled labor force (against Asian competition) Japan has in fact exceeded its old standards. The domestic market, never much of a factor because its purchasing power was very limited, has become much wealthier and is buying more of Japan's production. But it also requires more imports, and the only way Japan can pay for these imports is by selling its own products overseas. Thus there are always risks to face: raw material imports can become more expensive, eventually perhaps prohibitively so, and as time goes on, alternatives may dwindle. Japan's major trade partner is the United States, from where vital imports are drawn, including cotton and scrap iron; other Western, "developed" countries such as Canada, Australia, Britain, and Germany are high on the list. These countries are glad to see Japan buy raw materials, but they are concerned over the competition Japanese products create against their own. And so Japan faces another risk: that the success of it products will eventually mean that countries will erect barriers to their sale, or that states will join in economic unions (such as the European Common Market), to protect themselves against Japanese goods. So far, Japan has not suffered and the country thrives. But its economic base, however fast its current growth, is nevertheless a precarious one.

Japanese leaders not infrequently voice the opinion that their country's greatest hope for the future lies in the immense market of China. Major trade with China would greatly reduce the impact of one of Japan's greatest enemies, distance. But when and if China becomes Japan's trading partner once again, the terms will be very different from those that prevailed during the 1930's. Still, China is proof that Japan's alternatives are not yet exhausted, and it could be that when the time comes that rising costs and closed markets in America and Europe slow the Japanese economy down, Japan will once again have an answer to it problems. If so, the Japanese can be trusted to make the fullest possible use of it, in their unparalleled tradition of adaptation and innovativeness.

KOREA

Peninsular Korea, Asia's bridge to Japan and Japan's bridge to Asia, has had a turbulent history. For uncounted centuries it has been a marchland, a pawn in the struggles among the three powerful countries that surround it (Map 14-5). Korea has been a dependency of China and a colony of Japan; when it was freed of Japan's oppressive rule, in 1945, it was divided for administrative purposes by the victorious powers. This division gave North Korea beyond the 38th parallel to the Russian forces, and South Korea to the American forces. In effect Korea traded one master for two others. The country never was reunited, as North Korea fell under the Communist ideological sphere and South Korea, with massive American aid, became part of Asia's non-Communist perimeter. Once again it was the will of external powers that prevailed over the desires of the Korean people themselves. In 1950 North Korea sought to reunite the country by force, and invaded the south across the 38th parallel, an attack which drew an American-led United Nations response. This was the beginning of a devastating conflict in which North Korea's forces pushed far to the south, only to be driven back into their own half of Korea nearly to the Manchurian border, upon which Chinese armies entered the war to drive the United Nations forces southward again. Eventually a cease-fire was arranged in mid-1953, but not before the people and land of Korea had been ravaged in a way that was unprecedented even in its violent past.

Regional Associations. Should Korea be viewed in relation to Japan, as is done in this chapter? It is difficult to justify this: Korea's main cultural heritage comes from China, not Japan. Neither is there any strong friendship between South Korea and Japan, for the Koreans remember well the exploitative and harsh rule the Japanese imposed on them during the colonial period. On the other hand, the division between North and South Korea is such that more than two-thirds of the combined population of 45 million live in South Korea, and South Korea's orientation for a quarter of a century now has been away from China and toward the West. Moreover, Korea's modern economic infrastructure—even that of the North—was laid out by the Japanese, who first developed the mines and built factories, and who brought their efficient agricultural methods to the mountainous landscape of the peninsula. And for comparative purposes, Korea in some ways resembles Japan. Although it has not enjoyed Japan's insular security, Korea is nevertheless a quite clearly demarcated region, whose boundaries in the north are reinforced by mountainous terrain and the deep gorges of the Yalu and Tumen Rivers (Map 14-6). In total area and population the Koreas are in Japan's range: with somewhat more than half Japan's territory, the two Korean states have slightly less than half of Japan's population. And in terms of topography the situation is quite similar to that in Japan: Korea is a mountainous country of which only about

Map 14-5   KOREA AND ITS NEIGHBORS.

Map 14-6 KOREA: AREA ORGANIZATION.

Legend:
- Railroad Network
- Rugged Mountain Areas
- Industrial Areas
- Main Rice Producing Areas
- Secondary Rice Producing Areas

one-fifth is flat enough to be arable.

But the fact is that Korea really cannot be satisfactorily included in any larger Asian region. It is situated, alone, in the middle of three of the world's major powers: its three neighbors are respectively the most populous country in the world, the largest territorial state in the world, and the greatest industrial power in Asia.

Korean Origins

The Koreans share with the Japanese and the Chinese a central Asian origin, but they are today a distinctive people with their own culture and language. This is so despite the fact that from the very beginning they have had to cope with Chinese influence. Prior to its unification in the seventh century A.D., there were several Korean kingdoms and numerous Chinese settlements in Korea; during the Han dynasty Korea was a Chinese dependency. Later the Mongols took Korea with a campaign of great destruction, and the last great Chinese dynasty, that of the Manchus, also ruled Korea. Intermittently there was autonomy in Korea; the Koreans see the period of the Koryo Kingdom (918–1392 A.D.) as their formative period, and they trace the name of their country to this ancient state.

Korea, like Japan, wanted to maintain a policy of isolation and closure against European colonialism, but the Koreans did not go about this in the same way as Japan. By the time Japan had switched from isolationism to aggressive expansionism, Korea was still a feudal country; the weakening Manchu period was followed not by internal unity and progress, but by political conflicts and economic stagnation. During the wars between Japan and China in the 1890's and Japan and Russia in the early 1900's, Japan's influence in Korea grew rapidly, and in 1910 Japan formally annexed the country.

The Prize. In Korea, Japan acquired a productive colony. It is one of the tragic truths of Korea's tortured history that the country has been held back more by its misfortunes in being caught up in internal and external struggles than by its own economic limitations. Korea is inherently not a poor country, and the role it soon played in the growing Japanese economy attests to its productive capacity. Korea when the Japanese entered it was mainly an agricultural country, and had a subsistence-type of economy. The Japanese managed to double Korea's farm output, mainly of rice, and they siphoned half of this production off for consumption in Japan itself. These increases were not simply the result of information and encouragement given to the Korean farmers. Rather, it was achieved through large-scale land alienation by Japanese companies, an increase in tenancy, and the creation of large marketing boards to funnel the commercial rice production toward the home consumers. One result of this system was that Korea, after the end of World War II, was in dire need of sweeping land reform—which took place in North Korea on the Chinese model, but was slow in coming in the South.

But the Japanese contribution in other spheres of the economy was even more revolutionary. Korea has what Japan lacks: sizable iron ore deposits, and much high-quality coal. Large, though lowgrade, iron ores are concentrated in northeastern Korea, near Musas (low-grade coal also lies here), while high-grade anthracite comes from the northwest. Coal also lies at Yongwol in the South. In addition, Korea offered Japan a wide range of other necessary minerals, such as tungsten, copper, lead, zinc, and manganese ore. The exploitation of all these resources increased especially during the 1930's, after the takeover of Manchuria and while Japan was arming itself for wider war. In the process, the Japanese made major investments, in refining plants built to reduce the bulk of the ore prior to transportation to Japan's industrial complexes, in hydroelectric projects (especially in the North) to supply the power necessary for the processing of the raw materials, in railroads to transport these materials to the coasts, and in port facilities, highways, public health, education, and a host of other fields. Some industrialization even took place, but the Japanese held the management positions and took the profits. Despite all these developments the average Korean was probably no better off after three decades of development-oriented Japanese rule than before. We could replace the names of Japan and Korea with those of some European colonial power and some African dependency, and it would prove once again that exploitive colonialism need not be a Western monopoly. Perhaps Japanese colonialism was even more repressive and extractive than that of most European powers; in any case, the Koreans remember the period, now nearly three decades behind them, with intense bitterness. The Japanese certainly paid little heed to the needs of Korea: they cut the forests and

Two Koreas: the Demarcation Line along the River Han. This is the "bridge of no return," seen from the South Korean side. (M. Siverstone/Magnum)

hauled away the lumber without the necessary reforestation programs, thereby contributing to local fuel shortages, to increased erosion problems, and to the loss of water that must be conserved. They built one of the world's leading fishing industries in Korea, but they sent more than 70 percent of the catch to Japan. They trained few Koreans as administrators, technicians, or teachers. They took away much, but they invested almost exclusively in those areas where the returns for Japan demanded it.

Still, in 1945 Korea at least found itself with a modern economic base, and it had something to show for its long period of subjugation—something upon which the country could have built. But the political disaster of partition was followed by the calamity of the Korean War, in which Korea lost much of what the Japanese had left behind. It is estimated, for example, that between 80 and 90 percent of all industries in Seoul, the largest Korean city, were destroyed. Port facilities, hydroelectric stations, railroad bridges, farms—the destruction was wholesale. It set Korea decades back, reduced productive farmlands to subsistence, dislocated millions of people and ripped apart the fabric of society. And when Korea began to rebuild, there was still the spectre of political division, a division which denies Korea its opportunity to benefit to the fullest from its own regional complementarities.

The Two Koreas

The fragmentation of Korea into two political units happen, to a considerable extent, to coincide with the regional division any physical geographer might suggest as an initial breakdown of the country. As Plate 1 shows, South Korea is moister than the North; Plate 2 emphasizes the temperate maritime climatic conditions that prevail over most of the South as opposed to the more continental, "snow" character of the North. Plate 3 shows much of Korea as a whole to lie under mountain soils, with the most extensive belt of gray-brown podzolic soils in a belt that lies largely in South Korea. Plate 4 proves South Korea to possess a sizable area of broadleaf evergreen trees like that of southern Japan; the remainder of the country has a deciduous forest as its natural vegetation. And Plate 6 shows that the majority of Korea's 45 million people live in a zone in the western part of the country, a zone that widens southward so that the great majority of the population finds itself in South Korea, in a triangle roughly to the west of a line drawn from Seoul to Pusan.

Thus a great number of contrasts exist between North and South: the North is continental, the South peninsular; the North is more mountainous than the South; the North can grow only one crop annually and depends on wheat and millet while in much of the South multiple cropping is possible and the staple is rice; the North has significantly fewer people than the South, but the North has a large food deficit while the South comes close to feeding itself (it has had surpluses in the past). But perhaps the most striking contrast lies in the distribution of Korea's raw materials for industry. North Korea has always produced vastly more coal and iron ore than the South, and the overwhelming majority of all other Korean mineral production also comes from the North. Similarly, North Korea has maintained its great lead in hydroelectric power development, an advantage that was initiated by the Japanese and has been maintained. In recent years several discoveries of coal and iron have been made in South Korea, but the overall balance relating to the bases for heavy industry remain strongly in favor of North Korea.

Although it is practically impossible to obtain data concerning North Korea's external trade, one thing is certain: the superimposed political boundary (can there be a better example?) between the two Koreas is virtually watertight and in effect no trade passes across it. North Korea's trade connections are with China and U.S.S.R.; those of South Korea are with the United States, Japan, and West Germany. Yet it must be clear from what has just been said that North Korea, the focus of heavy industry in Korea as a whole,

could do a great deal of business with South Korea, the more agriculturally productive region of the country. The two Koreas are interdependent in so many ways; even within the industrial sector itself there are complementarities. While North Korea specializes in heavy manufacturing, light industries (cotton textiles, food processing) still prevail in the South, although heavy industries are now developing, especially around Pusan, since the discovery of iron ores and of Samchok's coal (Map 14-6). North Korea's chemical industries produce fertilizer; South Korea needs it. South Korea long exported food to what is today North Korea. North Korea has electric power aplenty; the transmission lines that used to carry it to the South were cut soon after the postwar division of the country. South Korea has the largest part of the domestic market as well as the largest cities, the old capital of Seoul (4 million) and the southern metropolis of Pusan (1.5 million). Pyongyang, the Northern headquarters, is just approaching 1 million. Seoul is the country's primate city, whose location in the waist of the peninsula, midway between the industrialized northern and agriculturally productive southern regions of Korea, has been an advantage. Pusan is closest to Japan and grew in phases during the twentieth century, first under Japanese stimulus and later as the chief American entry point. Pyongyang was also

Pyongyang, the capital of North Korea, has grown substantially since it became the headquarters of a separate state. This view is over a new district of the city, built after the division of Korea during the Communist regime. (Novosti from Sovfoto)

developed by the Japanese, and it lies at the center of Korea's leading industrial region, which is merging with the mining-industrial area of the northwest (Map 14-6).

The Koreans, like the Vietnamese, are one people, with common ways of life, religious beliefs, historic and emotional ties, and with a common language. In the entire twentieth century the Koreans have barely known what self-determination in their own country would be like, but they have not forgotten their aspirations. After its seemingly endless suffering through conflicts and wars, most of them precipitated by the "national" interests of other states, Korea today, divided as it is, once again contains the ingredients for trouble. What the two countries are building for themselves today, to make the best of their separation, may be destroyed again tomorrow. Korean nationalism cannot now overcome the will of the external powers in whose interests the country's fragmentation continues. But either side, emboldened by the presence or promises of its allies, acting under the pretext of a provocation, or aggressively confident of its own strength, can once again plunge the peninsula into still another devastating conflict. So, for that matter, can the great powers, conscious of "spheres of influence." Meanwhile Korea suffers, a country of considerable potential which has never been allowed, in modern times, to prove to itself and the world what it could have achieved as a unified sovereign state.

ADDITIONAL READING

Standard works on Japan include G. T. Trewartha's *Japan: A Geography,* published by The University of Wisconsin Press in Madison in 1965, and E. A. Ackerman, *Japan's Natural Resources and Their Relation to Japan's Economic Future,* which, though published in 1953 by the University of Chicago Press, proves that a dated book can remain current and provocative. Also see P. Dempster, *Japan Advances: A Geographical Study,* published in New York by Barnes and Noble in 1967, and H. Borton, *Japan's Modern Century,* published in New York by Ronald Press in 1955. The volume by E. O. Reischauer, *The United States and Japan,* a Compass Book of Viking Press, New York, published in 1965, sees Japan through the eyes of one who knows it better than most other. Of interest also is R. B. Hall's *Japan: Industrial Power of Asia,* No. 11 in the Van Nostrand Searchlight Series, published in 1963.

On some specialized topics, see I. B. Taeuber, *Population of Japan,* a 1958 publication of Princeton University Press which traces population patterns from the twelfth century to the mid-twentieth, and T. C. Smith, *The Agrarian Origins of Modern Japan,* published by Stanford University Press, 1959. On the land reform imposed after World War II by the United States, see R. P. Dore, *Land Reform in Japan,* published in New York by Oxford University Press in 1959.

On Korea, the standard work remains S. McCune, *Korea's Heritage: A Regional and Social Geography,* published in Rutland by Tuttle in 1956. Also see McCune's *Korea: Land of Broken Calm,* published by Van Nostrand, Princeton, in 1966.

Migration is discussed in detail in several sections of J. E. Spencer and W. L. Thomas, Jr., *Cultural Geography,* published by Wiley in 1969; also see the paperback by R. M. Prothero, *Migrants and Malaria,* published in London by Longmans in 1965 and which, though focused on Africa, contains a wealth of general information on the migration process. The Ainu are the subject of an essay in G. P. Murdock, *Our Primitive Contemporaries,* published by Macmillan in London in 1934 and reprinted many times subsequently. On areal functional organization, see the textbook by A. K. Philbrick, *This Human World,* published by Wiley in 1963.

15

AUSTRALIA AND THE ISLANDS

CONCEPTS AND IDEAS:

Biogeography
Territoriality
Peneplanation-Pediplanation
Endoreic-Exoreic Drainage
Artesian Water
Introduced Capital
Eugenic Population Policies
Remoteness

Beyond the islands and archipelagos of Southeast Asia, in the vast expanse of water of the Pacific and Indian Oceans, lies the island continent of Australia. Smaller even than Europe, Australia might never have come to be recognized as an individual continent and, with New Zealand, as a distinct culture realm. If Australia were populated by peoples of Malayan stock with ways and standards of living resembling those of Indonesia and mainland Southeast Asia, and had a comparable history, then it is quite possible that it would be viewed today only as an exceptionally large island sector of the Asian continent. But just as Europe merits recognition as a continental realm despite the fact that it is merely a peninsula of "Eurasia," so Australia has achieved identity as the island realm of "Australasia." Australia and New Zealand are European, white man's outposts in an Asian Pacific world, as unlike Indonesia as Britain and America are unlike India.

While Australia and New Zealand are culturally similar, and together constitute a culture region, they are quite different in terms of their geologic origins and surface terrain. Australia is mainly a "shield" area (see Chapter 2), but New Zealand belongs to that great belt of island arcs that fringes the western Pacific Ocean (see Chapter 14). Much of Australia is flat-lying and relatively low in elevation; Australia's eastern mountains only average between 2000 and 3000 feet. The highest point in all of Australia is somewhat over 7000 feet. In New Zealand, on the other hand, there is the familiar island-arc topography complete with high mountains, rugged terrain, volcanoes, and relatively little low-lying, flat land.

Although Australia was spawned by Europe, and its people and economy are Western in every way, Australia as a continental realm is a far cry from the crowded, productive, populous

Map 15-1 THE AUSTRALASIAN LAND BRIDGE.

582

European world. Apart from the fact that its area of 2,971,000 square miles is only three-quarters of that of Europe (and smaller than the Soviet Union, Canada, China, the United States, or Brazil), its population is really diminutive: the 1970 estimate is 12.5 million. This gives an average population density of less than 4.5, and suggests that Australia is a virtual population vacuum on the very edge of teeming Asia. But the physiological density is actually quite high. So much of Australia is arid or semiarid that only about 8 percent of it is agriculturally productive, and much of this moister part of the continent (mainly the east and north, Plate 1) is too rugged for farming. By some calculations, only 1 percent of Australia's total area is prime land for intensive agriculture; even if this figure were doubled the physiological density would exceed 210 per square mile—not an awfully high index but far more representative of the situation than the 4.5 per square mile for Australia as a whole. And it suggests that Australia, while capable of absorbing many more people, is no feasible outlet for Asia's huge overpopulation. Australia's *total* population is only about half the annual *increase* of China's alone.

The First Australians

Australia apparently has been sparsely peopled throughout its history. When the Europeans first made contact with Australia there were less than a half million indigenous Australians on the continent—most estimates suggest between 300,000 and 350,000, and some place the total was even lower. As with the Bushmen of Southern Africa and the Ainu of Japan, anthropologists do not agree on the origins of Australia's black peoples. Some have suggested that the early Australians, like the Ainu, had a Caucasoid ancestry. Another theory held that the indigenous Australians showed evidence of at least three racial infusions: a first Negrito invasion, a later Caucasoid immigration, and finally the "true" Australoids arrived, whose ancestral ties lay westward, in southern India and Ceylon. Still another idea holds that the aboriginal Australians are but one stock among several that descended from a Malaysian–Indonesian source, and of which various strains moved into India and Ceylon, into Eastern Asia, into New Guinea, and into Australia and the Pacific islands.

Whatever the real sequence of events was, it is likely that the first Australians reached the

The first Australians: these four men were photographed at Kuranda in Queensland, in front of their dwelling. (American Museum of Natural History)

continent via the landbridge that even today almost connects New Guinea to Australia's Cape York Peninsula (Map 15-1). It is of course possible that canoes or rafts carried some of the people across the water, but the peopling of Australia, like that of Japan, was probably facilitated by low sea levels during glacial times. Certainly the map of the distribution of Australia's pre-European population (Map 15-2B) indicates such a northern approach, and it suggests that people moved southward along both coasts. Eastern Australia was more densely peopled than the west, and it may be that this was due not only to the southward migration from Cape York, but also in some measure to the arrival on eastern coasts of late immigrants from Melanesia. In Tasmania lived a Stone Age people who perhaps were remnants of a people pushed southward by these invasions; whatever their history, they have been completely wiped out by their successors, Asian and European.

The indigenous population, like Africa's Bushmen, subsided on hunting and gathering. In their life, everything depended on the availability of water, and the water hole was the critical element in their existence. In their semi-nomadic search for food and water they constantly traversed their domain and learned to

MODIFIED AFTER: MCGRAW HILL BOOK CO., MANS'S DOMAIN

Map 15-2 AUSTRALIA: THEMATIC MAPS.

know it with amazing accuracy; practically every description of Australia's original peoples remarks on their incredibly detailed knowledge of the terrain and, importantly, its boundaries.[1] These boundaries were necessary, for the few hundred thousand black Australians were fragmented into several hundred tribes and clans using numerous different and mutually unintelligible languages. Here is another manifestation of the early development of Man's sense of territoriality, or, as Robert Ardrey has called it, his territorial imperative. The original Australians did not have much—they built no permanent dwellings, knew no agriculture, had no pottery, made no weapons beyond spear and stick. They suffered greatly during droughts. But they knew boundaries and neutral zones as we do today. Ardrey said it about the African Bushmen, but he might have said it about Australia's hunters as well: they "may wound an animal in the . . . desert; but, famished though they may be, they will not follow it if the animal crosses the neutral zone into the next band's territory. . . ."[2]

Remoteness. Australia's isolated position helped the Australian indigenous population to survive in this form for a very long time. In Southeast Asia, the older peoples long before the coming of the Europeans were already in full retreat before the expansion of more advanced peoples, seeking refuge in forests and mountains as did the Ainu of Japan and the Negritos of the Philippines. But even the arrival of the Europeans in Australia was delayed centuries after they had begun to penetrate and occupy other Pacific and Indian coastlands. This was in part due to the lack of evidence that Australia could produce desirable goods in trade, or gold and silver, so that the colonial governments put a low priority on Australian coastal exploration. On an individual basis there were fears of the dangerous coasts and the winds that might drive ships against it. Briefly the Dutch East India Company from its base in Batavia sent some of its captains toward Australia's shores, but Captain Tasman's reports of his landings on Van Diemen's Island off Southeastern Australia and New Zealand in 1642 were such that the company lost interest in the whole area. It was not until the journeys of James Cook, the British

captain, in the 1770's that Australia finally entered the European orbit permanently. It is a reflection of the island-continent's isolation that when Cook visited the east coast in 1770, he was the first European to do so. It was then that Australia's better lands came to Europe's attention, and specifically, the attention of the British. Cook returned in 1772 and again in 1776, and by 1779 the British could see a use for the new Australian territory—a use which, ironically, was once again related to its remoteness.

A Penal Settlement. Among the uses Europe's colonial powers had for some of their overseas possessions was their hospitality to people convicted for some offense and sentenced to deportation—or "transportation" as it was called. Such "transportation" had long been bringing British convicts to American shores, but in the late 1770's this traffic was impeded by the Revolutionary War. British jails soon were overcrowded, and judges continued to sentence violators to be deported; this was a time when Britain was undergoing its own economic and social revolution, and the offenders were numerous. An alternative to America for purposes of "transportation" simply had to be found, and it was not long before Australia was suggested. From the British viewpoint, the far side of Australia, the east coast, was an ideal place for a penal colony. It lay several thousand miles away from the nearest British colony and would hardly be a threat; its environment appeared to be such that the convicted deportees might be able to farm and hunt. In 1786 an order was signed making the corner of Australia known as New South Wales such a penal colony. Within two years, the first party of convicts arrived at what is today the harbor of Sydney and began to try to make a living from scratch.

For those of us who have learned of the horrible treatment of black slaves transported to America, it is revealing to see that European prisoners sentenced to deportation were not any better off, despite the fact that the offense of many of them was simply their inability to pay their debts. The story of the second despatch of deportees to Australia gives an idea: more than a thousand prisoners were crammed aboard a too-small boat, and 270 died on the way and were thrown overboard. Of those who arrived at Sydney alive, nearly 500 were sick, and another 50 died within a few days. What the colony needed was equipment and healthy workers;

[1] See, for example, W. Lloyd Warner, *A Black Civilization* (*A study of an Australian Tribe*), Harper, New York, 1958.
[2] Robert Ardrey, *The Territorial Imperative*, Dell, New York, 1966, p. 251.

what it got was hardly any tools and ill and weakened people. The first white Australians hardly seemed the vanguard of a strong and prosperous nation.

Continental Contents

To say the very least, Australia is an unusual continent, and not just because the beginnings of white settlement came so late and had such a bad start. In physical and biological terms, too, Australia has its own peculiar character. Australia's zoogeography is quite unlike that of Eastern and Southeastern Asia, something that can be attributed to the same quality of remoteness and isolation that so strongly influenced the human course of events. Many animals that long ago were exterminated in other parts of the world still thrive in Australia, protected from destructive invasions. Thus Australia is the world's last major refuge of the marsupials (animals that have a pouch on the abdomen which contains the teats and serves to carry the young,) including the kangaroo, koala, and wombat, and it is the home of the only two remaining egg-laying mammals, the platypus and the Australian anteater. Map 15-1 shows the westward limit of the marsupial fauna, and it also shows some of Asia's more famous animals (the elephant, tiger, and rhinocerous) that have failed to negotiate the Indonesian Archipelago. The elephant made it no farther than Sumatra, the rhino did not reach beyond Java, and the tiger got little closer to Australia.

This suggests that there is a transition zone in the Indonesian Archipelago, where Asia's fauna gives way to that of Australia. In the mid-nineteenth century the famous zoologist Alfred Russell Wallace, a colleague of Darwin, made an intensive study of what is today Indonesia and concluded that the strongest break was along a line between Borneo and Celebes (Kalimantan and Sulawesi) and between Bali and Lombok. This became known as "Wallace's Line," a concept that has aroused debate ever since. As the fauna of the archipelago became better known, other lines were suggested, notably that of another zoologist, Weber (see Map 15-1), and these alternatives served to emphasize that the Indonesian islands really constitute a broad transition zone between Asian and Australian mainland faunas.

Even in the field of phytogeography—plant geography—Australia has its particular character. The eucalyptus is native to Australia and represented by about 600 species, including huge trees that may exceed 300 feet in height. Certain kinds of acacia trees and conifers are unique to Australia, and the continent's vegetation matches the animal life in terms of its singular nature. The Dutch botanist Van Steenis argued that Australian-type vegetation takes over quite far to the west, between Java and Sumatra (Djawa and Sumatera) (Map 15-1). But others have suggested that any line separating Australian from non-Australian vegetative assemblies should lie through the Arafura Sea and the Torres Strait—thus separating Australia even from New Guinea. Once again it is clear that a transition zone of a sort exists in the islands between Asia and Australia.

The Surface. As if to match its individuality in the plant and animal spheres, Australia's physiography also is distinctive. Though on the very edge of the Pacific Ocean, and with an eastern mountain chain that seems clearly to be a continuation of the Pacific "Ring of Fire," Australia has not a single active volcano. Neither are those eastern mountains very high; Australia seems least affected by the Alpine mountain building that also created the Alps and the Himalayas. The highest mountain on the whole continent is Mt. Kosciusko (7316 feet), and this is not high enough to produce permanent snow: Australia is the only continent without it. Again, only about 7 percent of Australia's nearly 3 million square miles lie at elevations exceeding 2000 feet, making this the lowest in altitude of all the continents.

Australia's physiography is dominated by three main regions: (1) the western "shield," a mostly flat surface averaging between 1000 and 2000 feet in elevation but with some local hills and depressions, (2) the eastern mountains, and between these (3) a belt of large basins extending from the Carpentaria Basin in the north through the Great Artesian Basin in the continent's mid-section to the Murray River Basin in the south (Map 15-2C). Along the coast lie a number of separate lowlands, including the Nullarbor Plain, the Perth Lowland, and the smaller lowland areas of the east coast, where the first white settlers were put ashore to forge an existence. This physiography is reflected by Plates 1 to 5 in the rear of this volume, for in the latitude of Australia there is a strong correlation between elevation and rainfall. Plate 1 shows that the western shield of Australia ranks with the world's driest regions; if fact, nearly half of the continent gets an average of less than 12 inches of rainfall per year, and this minimal total is coupled with great unreliability. The moister areas lie against the slopes of the eastern mountains and in the tropical north, along with a small corner in the southwest. Plate 2 shows how narrow Australia's more humid rimland really is: the whole heart of the continent is occupied by the desert and its marginal steppe. Northern Australia lies under an essentially moisture-deficient savanna-type climate, and in the southwest, around Perth, and in the south, around Adelaide, there are areas of Mediterranean-type climate (*Cs* on the map)—not exactly known for excessive precipitation, and where a short, wet winter follows a hot, dry spring and summer. Only along the eastern coast, from the south in the vicinity of Melbourne to the base of the Cape York Peninsula in the north (and including Tasmania), does the climatic classification reflect better moisture-temperature conditions. The overlap between these areas and the leading zone of agriculture is proved by Fig. 15-4A. The soil map (Plate 3) again repeats the familiar pattern: a huge area of desert soils in west and central Australia, surrounded by a belt of prairie soils; latosolic soils in the tropical north, mountain soils in the eastern highlands, and podzolic-latosolic soils in the lower Pacific coastlands.

Pediplanation. In 1889, Professor W. M. Davis proposed the term *peneplain* to identify the world's extensive, nearly flat surfaces such as that of Western Australia. Such a plain, he argued, would form near the end of a long period of erosion, during which the agents of weathering and erosion—river water, the chemical breakdown of rocks, mass wasting etc.—were at work, constantly wearing the landscape down. Toward the end of such a cycle of erosion, mountains and hills would be worn down to a very low relief, with very gentle, broad slopes and wide valley floors. Since most of the material weathered loose had to be carried toward the sea, the plain could never be completely flat; thus Davis coined the term *pene*plain—almost a plain. Later, some geographers began to spell it pene*plane*, suggesting that it was the geometrically abso-

lutely flat *plane* that was almost achieved by nature.

The concept of peneplanation has occupied physical geographers ever since Davis' definition of it. Peneplains can be formed and then buried—under lavas, under glacial tills, or under a marine invasion of the land. Buried peneplains were recognized in many parts of the world, and soon attempts began to be made to correlate these various surfaces. The peneplain that is presently forming in Western and Central Australia is only the most recent of at least five peneplanation cycles. As long as nothing disturbs the cycle—a major uplift, or a tilting, or a volcanic eruption, the land surface continues to be worn down. Some areas of exceptionally hard rock, or areas that lie remote from rivers and streams, may stand up longer than the rest of the region, and they may be left as remnants. Such remnants stand out as islands on the otherwise flat plain—as *inselbergs,* to use one term that has been applied to them, or as *monadnocks.* In Australia, there are mesalike erosional remnants of this kind in the Barkly Tableland (b, Map 15-2B), and more sizable mountains in the Arnhem, Kimberley, and Hamersley Plateaus.

The concept of peneplanation has had its detractors. Professor Davis got into one of the most celebrated debates in the history of geography with a German physical geographer, Penck, who held that the idea was based on erroneous principles. Peneplanation presupposes that the slopes of a mountain wear downward, that is, that the mountain becomes lower and lower as the forces of weathering and erosion attack it. Its slopes will, therefore, become convex, until finally the whole peneplain consists of convex interstream tracts. Penck argued that he saw many concave rather than convex slopes, and proposed that the slopes of a mountain wear *backward* rather than downward. If, for example, we started with a mesa-shaped mountain made of homogeneous rock material, Davis would argue that it would become progressively rounded; Penck proposed that its sides would retreat until only a small butte was left—and finally only the slight slope around its foot, the *pediment.* Recently Professor L. C. King took up Penck's side of the debate and suggested that the whole idea of peneplanation be abandoned in favor of *pediplanation,* that is, the formation of erosion surfaces as a result of the retreat of

An *inselberg* rising out of the flat, planed surface of interior Australia. Ayers Rock lies in the southwestern part of the Northern Territory, and it has withstood the forces that beveled the surrounding pediplane. (Australian News and Information Bureau)

slopes. He further sought to correlate such erosional pediplanes in many parts of the world, from Brazil to Africa, and from India to Australia.[3]

For this kind of research, Australia is a veritable laboratory. Its older surfaces have not been so badly disturbed by violent earth movements that they can no longer be recognized; for a long period in its recent geologic history, the continent has been relatively quiet. Still it has mountains and deserts, each with their own physiographic problems; it has different kinds of coastlines; it has unusual drainage patterns. To take these drainage systems as an example: some of Australia's rivers, like the Murray, reach the sea, but others, like the Finke Macumba (which drains the Macdonnell Ranges in the middle of the continent), never make it to any coast. The Finke Macumba River drains into a so-called lake, a playa named Lake Eyre, which lies several dozen feet below sea level in the heart of Australia. As any atlas map shows, numerous Australian rivers and streams lead into such basins of internal drainage, and here is an example of *endoreic* drainage systems (as opposed to *exoreic* systems, which allow the water to flow into the sea). And many Australian rivers are intermittent or emphemeral in their flow. Even the continent's two largest streams, the Darling (over 1700 miles long) and the Murray (1600 miles), have on some occasions failed to flow during an entire year. The Darling once failed for 18 consecutive months, and the Murray lay dry in 1830, 1912, and 1922.

Artesian Water. As Map 15-2C shows, the east-central part of Australia is occupied by a physiographic region called the Great Artesian Basin. This is reputedly the largest of its kind in the world, and artesian water also occurs south of the Great Artesian Basin in the Murray River Basin, and north of it in the Carpentaria Basin. The supply of water by artesian wells demands several conditions (Map 15-3). First, there must be an aquifer, a layer of rock that is both porous and permeable and which thus provides a large underground storage reservoir for water. This aquifer must be continuous—if a fault cuts it off the whole system breaks down. Additionally, the aquifer layer must be bent or warped upward as shown in Map 15-3, so that its edge is exposed at a higher level and in an area where rainfall provides a substantial supply of water for seepage. Above the aquifer there must be a rock layer of low permeability that prevents ground water flow—an aquiclude.

Essentially what happens is that water received as rainfall or snow in the mountains of

[3]L. C. King, "Canons of Landscape Evolution," *Bulletin of the Geological Society of America,* Vol. LXIV, No. 7, 1958, pp. 721–753. Also see Dr. King's *Morphology of the Earth,* Oliver & Boyd, Edinburgh, 1962.

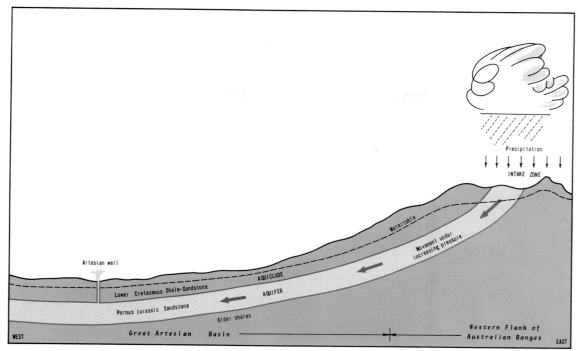

Map 15-3 THE ARTESIAN BASIN.

Eastern Australia flows downward into the aquifer layer; in doing so, it exerts pressure on the water already in the lower parts of this aquifer. When a well is drilled through the aquiclude, the water under pressure may be pushed upward to the surface. Sometimes artesian springs are created through natural processes, for example a small fault or crack in the aquiclude which permits the aquifer's water to escape. If this description suggests that the water in the deepest parts of the aquifer might well heat up under such pressure, that is an accurate impression: in several places in interior Australia the water comes to the surface warm and in a very few locales it is actually boiling.

At first sight, this natural distribution system of the rainfall that comes to the mountains of eastern Australia would seem to open up all sorts of possibilities for agriculture under irrigation in the east-central interior of the continent—but another look at Map 15-4A reveals that the Great Artesian Basin is grazing country, not cropland. Many artesian wells, even where water at first spouted out of the ground, have been known to dry up, and with fearsome frequency. Of the 2000 or so boreholes drilled in Queensland during the first half of this century, two-thirds had

dried up by 1950! in fact, pumping is normal over much of the area today; the salinity and unreliability of much of the artesian supply combine to make this an unsuitable basis for irrigation.

Australian Frontier

The inmates of Australia's first penal colony at Sydney Cove were not free to move away as they wished, nor could they have—even though Eastern Australia proved to be one of the continent's most livable areas, this was no horn of plenty. There was a good harbor (one of the best in the world, in fact), there was building stone, timber, and good fresh water, and there was some cultivable land. But the British did not give the prisoner-colonists the needed equipment to start a self-sufficient colony. Deliberately, and understandably, the settlement was kept to a large extent dependent on British-provided supplies. This dependence was perpetuated also by the constant arrival of new deportees, who made the establishment of additional colonies necessary. At Hobart on Van Diemen's Island (now Tasmania), at Brisbane, and at Newcastle, about 80 miles north of Sydney; convicts were put ashore. By the time the practice ceased, in the

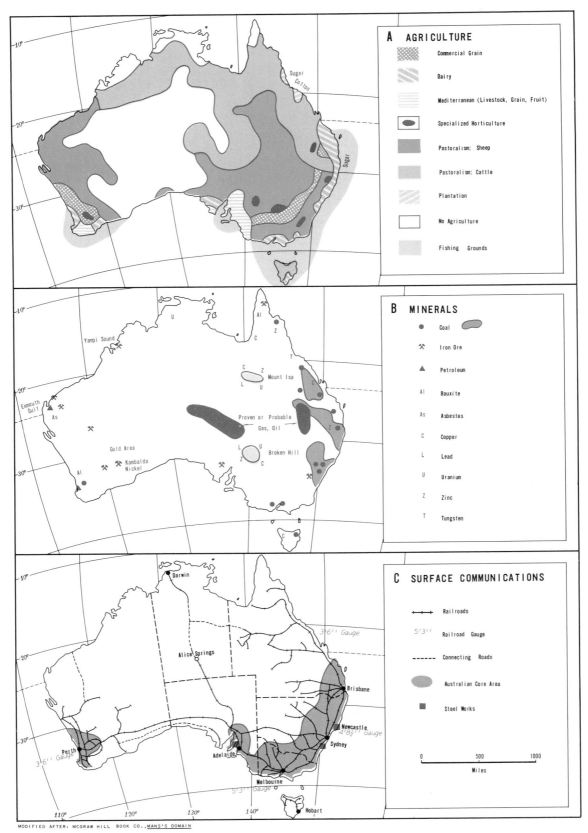

A AGRICULTURE

- Commercial Grain
- Dairy
- Mediterranean (Livestock, Grain, Fruit)
- Specialized Horticulture
- Pastoralism: Sheep
- Pastoralism: Cattle
- Plantation
- No Agriculture
- Fishing Grounds

Sugar
Cotton
Sugar

B MINERALS

- Coal
- Iron Ore
- Petroleum
- Al Bauxite
- As Asbestos
- C Copper
- L Lead
- U Uranium
- Z Zinc
- T Tungsten

Yampi Sound
Al
C
Z
T
Exmouth Gulf
As
C Z
Mount Isa
L U
Proven or Probable Gas, Oil
Gold Area
Kambalda Nickel
Z
C
U
Broken Hill
Al
Z
C

C SURFACE COMMUNICATIONS

- Railroads
- 5'3'' Railroad Gauge
- Connecting Roads
- Australian Core Area
- Steel Works

Darwin
3'6'' Gauge
Alice Springs
Brisbane
Newcastle
4'8½'' Gauge
Sydney
Perth
3'6'' Gauge
Adelaide
Melbourne
5'3'' Gauge
Hobart

0   500   1000
Miles

110° 120° 130° 140°

MODIFIED AFTER: MCGRAW HILL BOOK CO., MANS'S DOMAIN

Map 15-4 AUSTRALIA: THEMATIC MAPS.

The first European settlement in Australia has grown into a city of 2.5 million. The homes in the foreground, between the wharves and the viaduct, are Sydney's oldest and represent Australia's oldest residential area. They stand in sharp contrast to the tall skyscrapers fronting Sydney Cove in the background. (Australian News and Information Bureau)

mid-nineteenth century, around 165,000 deported offenders had been sent to Australia.

The effect of British policy relating to the Australian settlements was to discourage expansion, although inevitably individual explorers and small groups of free Britons soon began to explore the coast beyond the settlements and the mountains inland. In any case, free settlements began to develop parallel to the penal colonies, as prisoners who finished serving their sentences chose to remain in Australia rather than return home. With the officers in charge of the prison colony, and personnel retired from British government employ who decided to stay, the free settlements eventually began to outstrip the penal colonies. There were no constraints on these free settlers to investigate the hinterland of their home bases, and of course the new Australians wanted to seek profit. Soon it was known that behind the mountains overlooking Sydney lay a vast grassland pasture, and one obvious source of income lay in the provision of meat and wool. Now the push into the interior began: a road was completed in 1813 to the first town to be founded in the frontier, Bathurst; and

explorers ranged far and wide looking for good pastures. Wherever they went, the herdsmen followed, driving their growing flocks before them. Breeding animals were brought to Australia from Britain, wool went to the English mills and markets, first in 1821 and in rapidly increasing quantities after that year. The decade of the 1820's was one of great settlement expansion, and the ranchers who laid claim to the lands they had occupied were usually confirmed by the government in their right to own it.

However, not all the new settlements of Australia came about as a result of the continent's opportunities for pastoralism. Australia was still a colonial frontier, and British supremacy, despite Cook's formal annexation of "New Holland" in 1770, was still not beyond possible challenge. Thus, in 1828, the British government decided to found a settlement in the southwestern corner of Australia, for strategic as well as economic reasons; the following year a party of British settlers, attracted by generous land grants, arrived at the site of what is today the city of Perth. Compared to the vigorous expansion of the east, the Perth colonization scheme was a dismal failure, and this was one of the causes for change in the whole liberal land-grant program that had been in operation. Now it was decided that land should be sold at a minimum rate per acre, and the income thus derived should be used to support the immigration of more free settlers, to bolster Australia's population and to relieve Britain of its comparative surplus.

For a time, the whole issue of land and its ownership, superimposed on the problems of the relationship among the growing Australian colonies and the relationships of Australia's settlements with the London government, produced a difficult and potentially even violent time for Australia. Australians wanted the "transportation" system stopped (it was, on the mainland in 1840 and in Tasmania in 1853) and they wanted a greater voice in local government. When the early 1840's brought bad rains and poor harvests of wool, the frustration of the herdsmen was focused in a militant Pastoral Association (reminiscent of southern Africa's Diggers' Republic of the late 1860's), which demanded an end to the compulsory land purchase system and promised armed rebellion if its aims were ignored.

Gold. Pastoralism had catapulted Australia into the commercial age, had opened the frontier, and for half a century had dominated the continent's economic life when, in 1851, rich gold fields were discovered to exist in New South Wales and Victoria. At that time, there were perhaps a half million white Australians, whose wool exports accounted for over half of Britain's wool imports, and whose political power and aspirations were based on the influence this productivity gave them in Britain and the needs they saw in their pastoral frontier-land. In no way—politically, socially, or materially—was Australia prepared for what followed the news of the 1851 gold finds. True, gold had been known to exist prior to 1851, but only when some Australian diggers returned from California's fields to prove the lucrative character of the Australian ores did the rush begin. During the decade from 1851 to 1861, the population of the colony of Victoria increased more than sevenfold, from just over 75,000 to well over a half million; yet Melbourne, the colony's leading town, was just 16 years old when the rush began and had none of the amenities needed to serve so large a number of people. New South Wales, which in those days still included all of what

is now Queensland and the Northern Territory, saw its population rise to over 350,000, and its pastoral economy rudely disturbed by the rush of diggers to Bathurst. Overall, Australia's population nearly tripled in the 1850's: gone was the need for subsidized immigration in all areas but the west and northeast (Queensland).

The new wealth of Australia brought problems of accommodation for the new settlers, and inevitably it brought on some of the disorder and political conflict that also marked the gold rushes in North America and Africa. But it also brought great advantages. A new prosperity came to every Australian colony; in the 1850's the continent produced 40 percent of the world's gold. It brought successful searches for other minerals (Tasmania, with its tin and copper, was an early beneficiary). It also gave impetus to the political progress of the colonies. In the late 1850's selfgovernment came to all of them except Western Australia; Queensland was severed from New South Wales as a separate colony, and by 1861 the now-familiar framework of boundaries (Map 15-2A) had been created, although the Northern Territory remained under Sydney's administration as part of New South Wales for another two years.

Australia came out of its mid-nineteenth century gold rush with new strength and new direction. Before long the pastoral industry regained its former position as the most productive industry, but now it had serious competitors. In response to the vastly greater market, agriculture had made enormous strides and continued to expand as more cropland came into cultivation and modern, mechanized techniques of cultivation developed. And of course the mining industry itself was not confined to gold alone: other minerals also proved lucrative. In 1883 the silver, lead, and zinc deposits at Broken Hill (near the New South Wales–South Australia border) were found, and from time to time the other deposits marked on Map 15-4B were located. Additionally, the need for improved transportation brought the accelerated expansion of the railroad network. Here is a good example of the motor-effect of the mining period of the 1850's: by 1860 there still were only about 200 miles of railroad in Australia, but by 1870 there were nearly 1000; by 1880 the total exceeded 3500; in 1890 it was close to 10,000, and by the turn of the century, 13,000 miles of railroad had been laid by the colonies. Unfortunately, each colony was making its decisions concerning extent and direction of its network independently, so that the Australian system was not planned as a functioning whole. On the contrary: the rivalry between the colonies focused at the port cities, and the railroad systems resulted from deliberate policy designed to capture large hinterlands for these ports. Worse: the colonies decided on different gauges, so that the system could not be effectively interconnected. This problem has been tackled during the twentieth century through the single-gauge modernization of the network; the railroad system is shown on Map 15-4C, and reflects the eastern focus of development in Australia.

The Federal State

As the nineteenth century progressed the colonies of Australia found themselves competing with each other and developing, if not in isolation, then at least less in concert than was surely desirable. As in the case of the South American colonies and their relations with Spain, there was better contact between the individual Australian colonies and Britain than there was among the colonies themselves. From their similar beginnings, their common cultural heritage, their shared frontier expansion, and their joint mineral rush one might conclude that the eventual unification of Australia was an easy matter, but such was not the case. The colonies had competed with each other for a long time, and in a variety of spheres. For example, Victoria and New South Wales were locked in a battle over the issue of protective tariffs (wanted by Victoria's manufacturers) versus free trade, preferred by New South Wales. In a way Australia was six ocean ports with their respective hinterlands, each with an administration devoted to their individual interests. When the discussions leading to eventual federation began, the full extent of division became clear, and when the vote was taken, it was far from unanimous. Nevertheless, on January 1, 1901, the seven colonies of Australia became the seven states of the independent Commonwealth of Australia, a federal union.

Although there was a large measure of compromise involved in the seven colonies' acquiescence to federation (for example, Victoria secured some of its tariff demands, but much below what it wanted), the main impetus toward federal union was the states' clear and numerous areas of mutual interest. The rise of Japan as a major power in the Pacific had not gone unnoticed in Australia, nor was there much disagreement over the eugenic population and any potential nonwhite immigrants. While the aboriginal population was being driven before the white invasion and was partially absorbed through sporadic intermarriage or destroyed by force, there were demands at an early stage that other nonwhites should be prevented from immigrating. These demands were not completely adhered to, for after 1840 and especially during the decade of the 1850's nearly 50,000 Chinese entered Australia, mainly Victoria. Beginning in 1854 there were occasional anti-Chinese riots in Australia, and by 1886 every colony had passed laws discriminating against Chinese. Still another nonwhite immigration came from the South Sea Islands, from where indentured laborers were brought to the Queensland coast to work in the sugar plantations. This traffic, full of abuses and shameful conditions of work and living, began in the early 1860's and brought some 60,000 so-called Kanaka laborers from the Solomon Islands and the New Hebrides (and some from New Guinea) to Australia. But once again there was opposition, both from white workers and from humanitarian groups, and the practice tapered off. Shortly after federation arrangements were made to repatriate all but a few of the Kanaka laborers.

Thus Australia's States agreed, in one form or another, on a white-Australia policy, and even today there is only a smattering of nonwhites on the continent, apart from the 75,000 or so pure or half-blooded indigenous people. The Chinese now number less than half of what they did a century ago (20,382 according to the latest census): here is what a faculty member at an Australian university recently said about it:

"The primary motive in the formulation of exclusionist regulations was fear of the economic efficiency of the Chinese. This fear was hidden beneath ostentatious pretensions to cultural and racial superiority; it was bolstered by the knowledge of the great reservoirs of humanity waiting, it seemed patently obvious, to come streaming after their precursors into the South Seas if given half an opportunity. In framing the regulation of Chinese immigration as a blanket ban on virtually all non-Europeans, the Australians undoubtedly erred in terms of the politics of the mid-twentieth century. Still it would be absolutely wrong to disguise the gains made from the policy: lack of racial strife within the country, an inestimable boon; and freedom from fears of internal sedition such as are rife in all the countries of Southeast Asia."

"In the short and medium terms White Australia has brought benefits. In the long run it contains great dangers in a world in which political hostilities arising from the consciousness of racial difference tend to increase rather than diminish. For half a century Australia has maintained what has been a policy of immigration based on a "wooden horse" philosophy. She has been unwilling to trust or test in any substantial way the strength and absorptive powers of her own society; her policy has been defeatist, almost pusillanimous, stemming from fear rather than from confidence. Hence the rigidity of the policy, which will permit the merest trickle of affluent non-Europeans to gain citizenship, and then only if their skills are unprocurable from other sources."[4]

The Capital. If the Australian states could agree on a white-Australia policy, the selection of a capital city for the new federation was quite another matter. None of the states was especially eager to grant the great advantage of the capital's many functions to one of its competitors. Shortly before the implementation of federation, the premier of New South Wales managed to obtain from the other members a concession whereby the federal capital would be on his state's territory—though it would not be Sydney. In 1908 the site of Canberra was chosen, and a Chicago architect won the international competition for the best development plan. Ground was broken in 1912, and in 1927 the Commonwealth Parliament moved from its temporary quarters in Melbourne to the new national center. Here, then, is an example of an introduced capital (see Chapters 3 and 11). Today Canberra's size reflects its still essentially administrative functions (125,000).

Space and the Australian Economy

Geographers, when they discuss the pattern of development in Australia, often talk about "occupied" and "empty" Australia, and point out that occupied Australia lies in a discontinuous rim around the continent's empty heart. On Plate 1, the 20-inch isohyet can be seen to correspond quite closely to the break between the occupied and empty zones as reflected by Map 15-4A and C. Some Australian geographers, notably Professor Griffith Taylor, have taken a determinist view of Australian settlement, which, in view of the particular features of the Australian environment and economy, is not surprising.[5] Australia's European occupants seemed constantly to be limited in their frontier expansion by the constraints imposed by drought. The "drought gradient" from the comparatively moist east coast into the desert interior seemed to impose inescapable economic straitjackets, and the "discomfort" of the northern tropical areas seemed to have much to do with the southern and southeastern clustering of the white population. In 1917 Taylor published a "habitability map" of Australia whose main elements appeared to be confirmed by the population distribution of 1937.

Today we know that exaggerated determinism involving absolute, environment-imposed limitations on human settlement and achievement cannot stand the light of scientific enquiry, but on the other hand it is obvious that the Australian habitat presented obstacles as well as opportunities to the new invaders. For decades Australia was almost exclusively pastoral country, and ranching carried the economy. The Murray-Darling basin was an area of woodland and grass whose pastures were greatly improved when the trees were cleared; this early became the center of gravity of the sheep industry, and still today nearly half of Australia's estimated 175 million sheep are in New South Wales. As Map 15-4A shows, this is the heart of the great eastern crescent formed by the sheep industry, which extends now from South Australia into Queensland. What the map does not show is that the density of sheep decreases toward the drier west; the optimum zone for sheep raising lies where the rainfall annually is between 15 and 25 inches and where average temperatures are between 55 and 70 degrees Fahrenheit. As the herdsmen drove their sheep deeper into the Australian frontier, the problem of water supply loomed ever greater, and this is what led to the drilling of those wells that proved the existence of the Great Artesian Basin. Limitations of water provision, of course, have contributed to the comparatively low numbers of sheep in the

[4]A. James Rose, *Dilemmas Down Under,* Van Nostrand, Princeton, 1966, pp. 85–86.

[5]T. G. Taylor, *Australia: A Study of Warm Environments and Their Effect on British Settlement,* (6th ed.), Methuen, London, 1951.

Western Australian pastoral zone.

What Map 15-4A does show is that where the country becomes too hot and moist cattle take over. The beef cattle industry cannot compare to the sheep industry in terms of numbers of animals (about 20 million) or value in the exports of the country (about one-third of that of wool), but cattle have put to productive use what otherwise might have been empty Australia. As it is, the ranches are huge; in some areas the carrying capacity of the land is one animal per square mile! About three-quarters of the cattle are in the northern and northwestern sections of the cattle zone outlined on the map; they formerly were driven in long treks to the coastal railheads serving the (mainly Queensland coastal) stockyards, where the meat was prepared and refrigerated for export. Today, processing is decentralized, and cattle are trucked and railed to the meat production plants. The cattle areas that extend into southern Queensland are tributary to the local market (which takes about four-fifths of the annual production), mainly in New South Wales and Victoria.

Australian opportunity and British demands for wool (increased because of steam-power in the English textile factories) and meat (always in short supply in Britain) combined to give the pastoral industries first place in the continent's economic picture, a position briefly relinquished only during the famed gold rushes. In 1810 there were 34,000 sheep in Australia, in 1820 over 150,000, by mid-century more than 12 million. In the following decades there were fluctuations which tell the story of drought in Australia: around 1890 the herd reached 100 million, but a decade of rain deficiency reduced this total by nearly 50 percent, and the 1890 total was not reached again until the late 1920's. A quarter of the herd was lost during the 1942–1945 drought, when once again the sheep population fell sharply, from 125 million to 96 million. Since 1950 the herd has grown at an unprecedented pace, to exceed 164 million according to the 1967 count, the latest official figure that is available.

Apart from the water crises, Australia's ranchers have faced another, man-made problem. One great advantage the pastoral industry in Australia enjoyed, and which was especially important at the beginning, was the absence from the

The protective effect of a rabbit fence in Australia is displayed dramatically by this photograph. The men are standing in the guarded pasture, while the land to the left has been utterly denuded by the rabbit menace. The contrast can be seen to stretch as far as the eye can see, even to the vegetation on the hillslope in the distant background. (Australian News and Information Bureau)

continent of carnivores capable of seriously damaging the livestock—lions, tigers, leopards, and the like. It is thus a supreme irony that man himself imported the animals that damaged the industry—not carnivores, but rabbits. In 1863 some rabbits were let loose near Melbourne and, enjoying the same advantage (the absence of natural enemies), multiplied and spread over the entire continent. By the hundreds of millions they ate the grasses, drank the limited available water, and consumed as much each year, after they had spread to all parts of Australia, as a hundred million sheep could have. Worse, their entry into the ecological balance provided the impetus for an increase in the dingo population. The dingoes are Australia's wild dogs, probable descendants of dogs brought into the continent by the indigenous peoples. They were never numerous enough to be a real threat to the sheep industry—until the millions of rabbits presented themselves as easy prey and the dingo thrived and multiplied. Now the dingo did become a threat, and year by year larger numbers of sheep were destroyed by these wild dogs, until the annual kill ran into the thousands. Several methods of control have been attempted, including many miles of fences, and biological control also has been tried through the introduction of mixomatosis, a disease that killed rabbits by the millions.

Farming. Refrigeration boosted the beef industry: irrigation and mechanization did it for agriculture. Commercial crop farming centers on the production of wheat, a commodity of which Australia is one of the world's leading exporters and which brings in nearly half the amount of income derived from wool. When the penal colony at Sydney was struggling to survive, wheat was the main grain crop tried there, but the soils were not very good and the outlook was not bright. Eventually, however, other parts of Australia, especially near Adelaide in South Australia and in the colony of Citoria, proved to be good wheat country, and the farmers moved in to plant the fields. Soon there was trouble between the ranchers, for whom this was good land too, and the farmers—a conflict that was not resolved until after the whole Australian economic scene had been transformed by the gold rush of the 1850's and its attendant population influx. Eventually lands were set aside for agriculture, and before long Australia was self-sufficient in wheat.

Wheat production is concentrated in a broad belt that extends from the vicinity of Adelaide into Victoria and New South Wales (along the rim of the Murray Basin) and even into Queensland, where a separate area is shown on Map 15-4A in the hinterland of Brisbane. It also exists in an area behind Perth in Western Australia. Although these zones are mapped as "commercial grain farming," a unique rotation system exists here, whereby sheep and wheat share the land. Under this mixed crop and livestock farming the sheep use the cultivated pasture for several years, and it is then plowed for wheat sowing; after the wheat harvest the soil is rested again. This system, plus the innovations in mechanized equipment the Australians themselves have made, have created a highly lucrative industry. Australian yields per acre are still low by American standards, but the output per man is about twice what it is in the United States.

As Map 15-4A shows, dairying, with cultivated pastures, takes place in what Plate 1 shows to be areas of high precipitation. The location of the dairy zones suggests also their development in response to Australia's comparatively large urban markets. Even these humid areas of Australia are occasionally afflicted by damaging droughts, but the industry is normally capable of providing more than the local market demands. It is, in fact, the most important rural industry measured by the number of people who find work in it.

It is natural that irrigation should be attempted in a country as dry as Australia. Unfortunately, the opportunities for irrigation-assisted agriculture are quite limited: the major potential lies in the basin of the Murray River, shared by the states of Victoria (which leads in irrigated acreage) and New South Wales. Rice, in the Wakool and Murrumbidgee valleys, both tributary to the Murray, grapes and citrus fruits, are among the irrigated crops in the Murray River Basin; on the east coast, in Queensland and northern New South Wales, the cultivation of sugar cane is partly under irrigation. Seasonal irrigation, also, occurs in northern Queensland, on the Kimberley Plateau. The sugar areas of Queensland (Map 15-4A) were the scene of the much-debated Kanaka labor indenture. When the Kanaka laborers were repatriated, white men were left to do the work, albeit at much higher wages. The Australian government provided the necessary tariff protection,

Mechanized harvesting of wheat in New South Wales. The two harvesters are Australian-made. (Australian News and Information Bureau)

and the white workers of Queensland proceeded to prove that talk of the white man's incapacity to do manual labor under a tropical sun is nonsense.

Mining. Everything seems to conspire to keep Australia's empty heart unoccupied; in many places in the world the discovery of valuable mineral resources has drawn settlement deep into the desert, as in the Atacama in Chile, in the Sahara in Mauritania and Algeria, in the Tsaidam in China. But in Australia most of the known and valuable mineral resources lie along the edge of the great shield, confirming still more the continent's peripheral development (Map 15-4B).

Australia is known for wool and wheat, but it is the gold that brought waves of immigrants to the continent, in different spurts (the greatest in the 1850's), and the mining industry and the people it attracted stimulated the agricultural, pastoral, and manufacturing sectors of the economy. To Victoria and New South Wales they came in the 1850's, to Queensland in the 1870's, and to Western Australia in the 1890's, all in search of wealth and all in need of goods and services, food and supplies. Many failed in mining and turned to other economic pursuits; others joined the search for minerals still awaiting discovery. And in the process they proved that in Australia there is a great deal more than gold. Discoveries are still being made. Long believed poor in petroleum and natural gas (oil always ranks high on the import list for this urbanized, industrialized society), Australia seems now to have sizable supplies of both fuels

after all, if recent finds on the margins of the Great Artesian Basin are large enough to warrant exploitation; additional reserves have been located on the continental shelf between the mainland and Tasmania. Still more recently a deposit of nickel was found near the center of Western Australia's gold mining, Kalgoorlie. For nearly 80 years men have been mining and searching in this area, and yet the nickel discovered recently at Kambalda may constitute the world's largest known reserve of this important alloy. The Australian frontier still holds its secrets and surprises.

As Map 15-4B suggests, Australia's mineral deposits are scattered and very varied. The country is fortunate in being well endowed with coal deposits when its water power prospects are so minimal (except in Tasmania and the Southeast) and its petroleum so long in being located. The chief coal fields, as the map indicates, lie in the east, notably around Sydney—north at Newcastle, west at Lithgow, and south at Bulli. In the hinterland of Brisbane and about 125 miles inland from Rockhampton (Queensland), bituminous coal in mined. From the New South Wales field coal is sent by coastal shipping to areas that are coal deficient. But coal is widely distributed: even Tasmania and Western Australia have some production and can thus keep their import necessities down. There is no doubt that coal is Australia's more important mineral asset: it is used for the production of 90 percent of the electricity the country consumes, for the railroads, factories, artificial gas production, and a host of other purposes.

One of these purposes is the manufacture of iron and steel. Australia has iron ores which have recently been proved to be of major proportions, attracting United States and Japanese investments for their exploitation. These can keep the country self-sufficient for decades to come, and the quality of the ores in several places is very high. The Middleback Range deposit not far from Adelaide, and only 30 miles inland from the port of Whyalla, has long been the country's chief producing zone. The Yampi Sound deposits in the northwest are less favorably located for the east coast steel industry, but they have stimulated iron production in Western Australia, using local coal.

The most famous Australian mining district undoubtedly is that of Broken Hill, which has neither gold, coal, nor iron. Discovered in the aftermath of the gold rush, Broken Hill became one of the world's leading lead and zinc pro-ducing areas (in the beginning it was important also for silver), and the enormous income derived from the export sales of these minerals provided much of the capital for Australia's industrialization. In Queensland, the Mount Isa ore body yields a similar mineral association, and in both areas uranium also has been mined. But these products are only a small part of the total Australian inventory. Tasmania's copper, the large, recently discovered bauxite (aluminum ore) deposit in the York Peninsula, Queensland's tungsten, and Western Australia's asbestos only represent the range of the continent's minerals; there are but a few nonmetallic resources in which the country is really deficient. And the search has only just begun. Recent discoveries of iron and petroleum promise a bright future, and many possibilities still exist, in the opinion of geologists. Australians view their country as an emerging industrial power, and its list of deficiencies is only

Australia has been fortunate in its endowment of coal; there is enough to permit substantial exports to countries that need to buy this commodity. This stockpile, from the Moura field, lies at Gladstone Harbour awaiting transportation to Japan. (Australian News and Information Bureau)

With its peripheral development, Australia depends heavily on coastwise transportation for the internal exchange of bulk commodities. Several of the vessels in this photograph of the Victoria Docks at Melbourne are engaged in this coastal trade. (Australian News and Information Bureau)

very short compared to what we have become used to in much of Australasia. Australia is in a fortunate position.

Manufacturing. Once Australia ceased to be primarily a source of raw materials for Britain, its industrial progress was rapid. Its membership in the Commonwealth, tariff protection for local industries seeking to develop, its good resource base, and, importantly, the effects of the two World Wars, have all helped speed Australia's industrialization. For a long time Australia was to Britain what colonies always were: a supplier of needed resources and a ready market for British-manufactured products. But World War I, at a crucial stage, cut these trades connections—and Australia for the first time was largely on its own. Now it had to find ways to make some of the consumer goods it had been getting from Britain, and by the time the war was over Australian industries had made a great deal of progress. An iron and steel industry was begun with a plant at Newcastle, N.S.W. (on the local coal field), during World War I; in 1928 another works was opened at Port Kembla, near Wollongong, south of Sydney. In 1941, in response to the threat of Japan and in an effort to decentralize the important industry, a blast furnace was opened at Whyalla, the iron ore port on the Eyre Peninsula. Wartime also pushed the develop-

ment of the food-processing industries, and today Australian manufacturing is varied, producing not only machinery and equipment made of locally manufactured steel, but also textiles and clothing, chemicals, foods, tobacco, wines, and paper—among many other items.

Not surprisingly, the state capitals of Australia became the country's major industrial centers. Rural Australia, despite its productivity, is not densely populated, and Australia is a highly urbanized country, with well over 80 percent of the people living in cities and towns. The capitals, at the foci of the railroad networks of the individual states, were also the main ports for overseas as well as coastwide shipping; they had not only the best amenities for industry— concentrated labor force, adequate power supply, water, access to government—but they were at the same time the major markets. The process of self-perpetuation to which we referred in the context of some of Europe's larger cities is well illustrated here. Additionally, the cities and manufacturing industries have absorbed the vast majority of Australia's recent European immigrants. The Australian government pursues a vigorous policy of encouraging still-selective immigration; the majority of the new Australians come from Britain, but numerous Italians, Hollanders, Germans, and Greeks also

Map 15-5 NEW ZEALAND IN THE PACIFIC.

seek new homes in Australia. In recent years about 200,000 immigrants have annually come to Australia, and most of them have settled in the cities and found work in industry. The advantages to Australia are obvious: a strengthening of the country's still-small population, an expansion of the skills and numbers of the labor force, an increase in the size of the local market, and a growing consolidation of the Australian presence on the large island-continent, from a strategic viewpoint.

Australia's cities, thus, are large for a country with only 12.5 million people: Sydney leads with 2.7 million, Melbourne is second with 2.4 million, and Brisbane and Adelaide are approaching 1 million. Perth, Western Australia's capital, has over 600,000 inhabitants, and Hobart, the capital of Tasmania, has nearly 200,000 residents. Canberra, the federal capital, actually is the smallest of Australia's seven largest cities with an estimated 125,000 people in 1970; the smallest Australian town with capital functions still is Darwin, with under 25,000. Thus the seven largest cities account for some 7 million of Australia's 12.5 million population. This again underscores the commercial character of the Australian economy, and the importance of manufacturing in it. According to the latest available figures, industry now contributes over 40 percent of the Gross Domestic Product (the figure for agriculture, by comparison, is 14 percent).

Despite all this evidence for Australia's industrial revolution, the fact remains that the continent's domestic market is still quite small, and foreign markets are expensive distances away. Notwithstanding all its valuable resources and its undoubted industrial progress, Australia still imports large quantities of consumer goods—many of which could be manufactured at home. Certainly Australians can afford to pay for these imports; the people enjoy one of the highest

standards of living in the world. But the situation reflects the limitations a small local market and the liability of distance to world markets combine to impose on Australian manufacturing. Of course the case of Japan immediately comes to mind—is not Japan just as far away from the very markets where its products have had such a success? And Australia has so many of the raw materials Japan lacks! Explanations for the difference sometimes refer to the Australian preference for the leisurely life. More likely it is not a question of preference, but opportunity. Japan has a hundred million mouths to feed; it cannot feed them and must buy much of its food. It must trade with its best products and compete on world markets; it must constantly search for more efficient production methods to achieve lower prices than the competition. Australia's industries are not especially efficient. It is estimated that the industrial worker in Australia produces about half as much as his American equivalent per unit of time. Australian industry is not oriented to world competition; on the limited scale imposed by the local market it cannot achieve the volume and efficiency of output that would be required. Instead, Australian industries are protected by tariff walls from the competition of just such producers as Japan and the United States.

As Australia's population grows, so its economy will diversify further and the restrictions it faces will slowly relax. But Australia is choosy about its immigrants, and with so urbanized a population its natural increase in numbers is obviously not very great. It does not need to speed the pace of change; barring the kind of interference brought to mind by a look at the population map of Asia, it can afford to plan for generations of steady development rather than Great Leaps Forward.

NEW ZEALAND

There is a province in the Netherlands named Zeeland—land of the sea. On December 13, 1642, the Dutch sea Captain A. J. Tasman, who had headed east into the Pacific after stopping at Van Diemen's Island, saw the great mountains of a large island rising out of the ocean. After weeks of endless water here, finally, was land— Nieuw Zeeland. It was an appropriate name (as Map 15-5 shows), and although the spelling has

changed, it has survived through the centuries. New Zealand, like the Dutch province, consists of islands, large and small. The two large ones are South Island, 58,000 square miles, whose high mountains Tasman first saw, and North Island, 44,300 square miles, separated by what became known later as Cook Strait. Among the small islands the most substantial are Stewart Island, off New Zealand's south coast, and the Chatham

Islands, 450 miles east of South Island in the southwestern Pacific. Together these smaller islands amount to just over 1000 square miles. Tasman's party found New Zealand to be occupied by a people of apparent Polynesian ancestry, the Maori. The story goes that the first attempt by the Dutch party to set foot on New Zealand soil was not exactly fortuitous: the Maori did not take well to the white men and in a brief but bitter skirmish several Dutchmen were killed. Tasman's men retreated, and for more than 125 years New Zealand's shores were not visited again by any Europeans.

The Maori. It is thought that the Maori came to New Zealand, perhaps from Southern or Southeastern Asia, via the southern islands of Polynesia, principally Tonga (Map 15-5). They were skilled boatsmen, and may first have

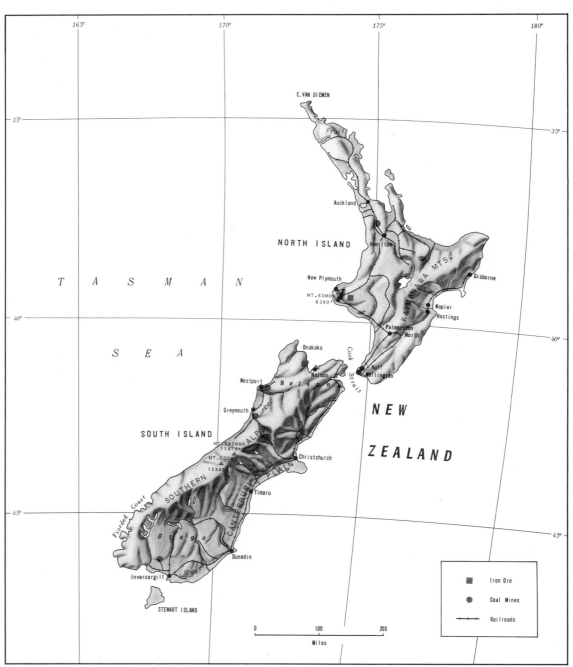

Map 15-6 NEW ZEALAND.

reached New Zealand in the tenth century A.D., although their largest migration to the islands apparently occurred in the fourteenth century, from Tahiti. In any case, the Maori found an aboriginal population already living in New Zealand, subsisting on hunting and gathering. Before long these ancient people had been driven southward until they were, for all practical purposes, exterminated; there was some absorption into Maori society as well. Today not a single representative of New Zealand's oldest inhabitants survives.

The Maori not only were good seafarers: they were excellent settlers on land as well, showing great adaptability to the new environment. Although Plate 2 does not extend into the Pacific Ocean, it does show New Zealand to be a cool and wet environment; the islands from where the Maori came were tropical. They needed more clothing, and more substantial houses than they used to have; they had to get used to new crops. But the Maori proved to be good builders, and inventive farmers. They found locally growing root crops and introduced those they had domesticated. They fished the waters of New Zealand's many streams and fjords. They made implements of wood and soft stone. However, the Maori were not a unified nation; they probably numbered about a quarter of a million when the Europeans finally came to settle, but they were divided into many tribal groups.[6]

Like so many other pre-European peoples, the Maori suffered greatly at the hands of the white man. European settlement, so long delayed after Tasman's first look at New Zealand, finally began after Cook's voyages of the 1770's. For over a half century settlers trickled in, missionaries, whalers, sealers, traders. Initially the relationships with the Maori were quite good; there were occasional quarrels but nothing to presage the violent wars that were to follow. Then, in 1840, the New Zealand Company received a charter to issue land and encourage immigration. Britain formally annexed the islands, and land alienation and social injustice followed the familiar colonial pattern. The Maori fought back, bravely but ultimately ineffectively. For a time it appeared that they were surely destined for extinction, especially during the destructive decade of the Maori War, 1860-1872. But subse-

[6]For a readable account of the Pacific Islands and New Zealand see O. W. Freeman (Ed.), *Geography of the Pacific*, Wiley, New York, 1951, especially pp. 423–459.

A Maori girl works on her sculpture. Art in New Zealand owes much to the Polynesian traditions introduced by the Maori people—not only in the visual arts but also in music. (New Zealand Information Service)

quently a combination of more enlightened policy and the Maori's famed adaptability to new circumstances reversed the situation. Today the Maori component in New Zealand's population of nearly 3 million is approaching 200,000, more than half of it of mixed Maori–European ancestry. Many of the latter have found a place in the white man's society of present-day New Zealand, although in economic terms the Maori as a whole remain a depressed class in the country. They do not have enough land left, in their reserve areas, to pursue their traditional way of life, which involved systems of communal ownership. Neither have they had the full opportunities of the white man in education, health facilities, housing, and so forth. But the Maori are a resilient people, and their position is steadily improving in a society that can afford to be racially tolerant.

The Land

Certainly New Zealand's land was worth fighting for. In sharp contrast to Australia, the islands are mainly mountainous or hilly, with several mountains rising far above Australia's highest peak. South Island has a mountainous, snow-capped backbone appropriately called the Southern Alps (Map 15-6), and the mountains

Wool is a major source of revenue for New Zealand, and the lower slopes of the country's mountainous topography afford excellent pastures. Here Merinos are gathered for clipping at Glentanner Station, Canterbury. The flock here numbers some 9000 sheep, and in the summer months they graze all along the slopes of the Ben Chau Range in the background—right up to the ice and snow. (New Zealand Consulate)

Tasman first saw were Mts. Cook (12,349 feet) and Tasman (11,475 feet). Smaller North Island has proportionately more land under low relief, but it too has a set of central ranges and high mountains, including Mt. Raupehu (9175 feet); in the west lies Mt. Egmont (8260 feet), on whose lower slopes lie the pastures for New Zealand's chief dairying district.

New Zealand, unlike Australia, is part of the Pacific "Ring of Fire," and in addition to its volcanic mountains and soils, geysers, and hot springs, the islands also get their share of earthquakes. The nearest deep ocean trench, however, lies off the Kermadec Islands rather than New Zealand itself; the islands' geology suggest that this is an ancient land area, not simply a recent overfold created by lateral movement (see Chapter 14). In any event, whereas Australia's

land is largely low in elevation and low in terms of relief, New Zealand's is on the average quite high and its relief is mostly rugged. Thus the most promising areas must be the lower-lying slopes and lowland fringes on both islands. On North Island, the largest urban area, Auckland (700,000) lies on a comparatively low-lying peninsula, and on South Island, the largest lowland is the Canterbury Plain, centered on Christchurch (200,000). What makes these lower areas so attractive, apart from their availability as cropland, is their magnificent pastures. Such is the range of soils and pasture plants that both summer and winter grazing can be carried on, and a wide variety of vegetables, cereals, and fruits can be produced. About half of all New Zealand is pastureland, and much of the farming is done to supplement the pastoral industry, to provide fodder when needed. But in the drier areas of South Island wheat is grown, truck farming exists around the cities, and fruit orchards of apples, pears, and grapes are widely distributed from Otago in the south to Nelson in the north.

Thus New Zealand's pastoral economy is based on sheep and cattle, wool, meat, and dairy products. As in Australia, sheep greatly outnumber cattle (over 60 million against under 8 million in 1967), and wool normally ranks highest in terms of value among the exports. There is a danger in such a situation, as evidenced by the 1967 reduction of world market prices for wool and the consequent economic crisis in New Zealand (unemployment, unknown since 1930, reared its head for the first time in more than a generation), but in modern times New Zealand has been among the world's leaders in terms of per capita trade; its standard of living is very high. Meat (mutton and beef) and dairy products (principally butter) combine with wool to provide nearly 75 percent of the islands' export revenues.

Patterns. Despite their contrasts in size, shape, physiography, and history, New Zealand and Australia have a great deal in common. Apart from their joint British heritage, they share a pastoral economy, a small local market, the problem of great distances to world markets, and a desire to nevertheless stimulate and develop domestic manufacturing. New Zealand does not have Australia's quantities of coal. There is bituminous coal at Greymouth and Westport on South Island, but difficult access has impeded development. Subbituminous coal is available in the northern one-third of the North Island and is used with coastal iron for the production of steel. Brown coal also is mined at Waikato on North Island and in Otago in the far south. On the other hand, New Zealand has a considerable hydroelectric power potential, being developed to serve the cities and industries. In short, New Zealand is not rich in industrial raw materials, but it does have the basics for a modest industrial base.

The fairly high degree of urbanization in New Zealand indicates another similarity to Australia: the high employment in the city-based industries. About two-thirds of all New Zealanders live in cities and towns, where the industries are mainly still those that treat and package the products of the pastoral-agricultural indus-

tries. But although New Zealand is of course behind Australia in its industrial development, the country's manufactures are becoming more diverse all the time. Textiles and clothing, wood products, fertilizers, cement and building materials, metal products, and some machinery are now made in New Zealand.

Spatially, New Zealand shares with Australia its pattern of peripheral development, imposed not by deserts but by high, rugged mountains. The country's major cities, Auckland and the capital of Wellington (with its satellite of Hutt, 300,000) on North Island and Christchurch and Dunedin (150,000) on South Island all are located on the coast, and the whole railroad and road system (Map 15-6) is peripheral in its orientation. This is true to a greater extent on South Island than in the north, for the Southern Alps form New Zealand's most formidable barrier to surface contact. Cook Strait, on the other hand, has been a unifier rather than a divider; it is only 16 miles wide at its narrowest point and Wellington is the regional focus for an area that encompasses both the southern part of North Island and the northern region of South Island. In its particular position and with its special functions it has deterred the development of any insular sectionalism.

On a broader, regional scale, the cultural unity of Australia and New Zealand and the growing economic complementarity of the two countries suggest that a common market would serve both well. Australia's industrialization and New Zealand's agricultural productivity provide opportunities for exchange and a lowering of barriers to the free flow of goods. The concept of an *Anzac* common market has been championed by enlightened leaders in both countries, but as always there have been opponents in Australia as well as New Zealand to any scheme that would in some way dilute national sovereignty. In 1969 the common market proposals for the two countries were still being negotiated.

Considerable justification exists, then, for the identification of Australia and New Zealand together as one of the world's major regions—perhaps "culture realm" is a bit much in view of the very small combined population of the two countries. But whether viewed as the eastern extremities of Australasia or as portions of the Pacific realm, Australia and New Zealand merit distinction on many grounds: ethnic, cultural, economic, and political, to mention but

Cattle have great importance in the economic life of both New Zealand and Australia. But the pastures are very different: the lush slopes and grasses of New Zealand (above) can sustain a higher concentration per unit area of the livestock than the grasslands of northern Australia (below). (New Zealand Consulate and Australian News and Information Bureau)

a few. So sharp are the contrasts with their neighbors, so distant the sources of Australian society, that it is as though a portion of the Western world had been cut adrift to fend for itself beyond the Far East. It is, of course, in the context of their location that Australia and New Zealand hold their greatest interest for geographers. If New Zealand lay wedged between Switzerland and Germany it would without a second thought be viewed as part of the European core area. If Australia were part of North Africa's Mediterranean fringe, its European connections would be easily conceived. But the two countries lie halfway around the globe from Europe, and nearly halfway from America. In the tense Asian–Pacific world, will they be given time and chance to gain the strength they may require for their survival?

ADDITIONAL READING

The most recent geography on Australia to appear seems destined to become the standard work on the continent. This is O. H. K. Spate's *Australia*, published in New York by Praeger in 1968. This is a primary source. A much older work that still retains its flavor and interest is T. Griffith Taylor's *Australia, a Study of Warm Environments and Their Effect on British Settlement*, a volume that went through numerous editions by Methuen in London. Another Methuen publication is the regional work by K. B. Cumberland, *Southwest Pacific; a Geography of Australia, New Zealand and Their Pacific Island Neighbourhoods,* now in its second (1958) edition. Also see K. W. Robinson, *Australia, New Zealand and the Southwest Pacific,* published in New York by London House and Maxwell in 1962. Another general discussion is A. J. Rose, *Dilemmas Down Under, Australia and the Southwest Pacific,* No. 29 in the Van Nostrand Searchlight Series, published in 1966.

On Australia's indigenous people see the famous study by W. Lloyd Warner, *A Black Civilization,* a Harper Torchbook of 1964, published in New York. On the white man's dispersal, see T. M. Perry, *Australia's First Frontier: The Spread of Settlement in New South Wales 1788–1829,* published in New York by Cambridge University Press in 1965. Also see the monograph by D. W. Meinig, *On the Margins of the Good Earth: The South Australian Wheat Frontier, 1869–1884,* published by Rand McNally in 1962. Another useful volume is A. L. McLeod's *Pattern of Australian Culture,* published in 1963 by Cornell University Press.

On New Zealand, a standard work is K. B. Cumberland and J. W. Fox, *New Zealand, a Regional View,* published in a second edition in 1963 by Whitcombe and Tombs, Christchurch. G. J. R. Linge and R. M. Frazer have published an *Atlas of New Zealand Geography* (Reed, Wellington, 1966). R. F. Watters has edited a volume entitled *Land and Society in New Zealand: Essays in Historical Geography,* published by Reed in Wellington in 1965.

On the Pacific, see the somewhat outdated but still interesting book of essays edited by O. W. Freeman, *Geography of the Pacific,* a 1951 Wiley publication. Also see H. J. Wiens, *Pacific Island Bastions of the United States,* No. 4 in the Van Nostrand Searchlight Series, published in 1962.

Good discussions of the pediplanation-peneplanation controversy appear in the volume entitled *Principles of Geomorphology* by W. D. Thornbury, which was published in a second edition by Wiley in 1968. This volume also contains discussions of drainage patterns and structures, and the stratigraphic aspects of artesian water. R. Ardrey's speculative volume *The Territorial Imperative,* published by Dell, New York, in 1966, is worthwhile reading for any geographer. On biogeography, see P. J. Darlington, *Zoogeography: the Geographical Distribution of Animals,* published by Wiley in 1957, and R. D. Good, *The Geography of Flowering Plants,* published by Longmans in London in 1947. And Alfred Russell Wallace's classic *The Geographical Distribution of Animals* has been reprinted by Hafner Publishing Co., New York, in 1962; it was first produced in 1876.

Author Index

Subject Index

Aarhus, population, 100
Abadan, 388, 418
Abidjan, 342, 352
Aborigines (Australia), 583, 585, 593
Abstractions, 1
Abyssinia, see Ethiopia
Acapulco, 247
Accessibility, 197, 201-204, 226, 228-229
Accra, 352
Acculturation, 4
Achaean league, 22
Ackerman, E.A., 393
Acre (Brazil), 302
Adana, 415
Addis Ababa, 32, 402, 403
Adelaide, 587, 596, 598, 601
Aden, Gulf of, 410
Adriatic Sea, 30, 44, 108, 111, 112, 125, 126
  coastal precipitation, 49, 50
Aegean Sea, 8, 15, 20, 44, 105, 126, 127, 377
Aegean-Marmara lowlands, 415
Afan and Issa, Territory of the, 403
Afghanistan, 17, 89, 131, 137, 373, 405, 417, 439, 446, 455
  agriculture, 421
  as a buffer zone, 115, 463
  climate, 421
  ethnic groups, 420, 421
  and India, 430
  landlocked location, 420, 421
  language, 420
  Pakistan, conflicts with, 421, 457
  political history, 421
  prorupt shape, 537
  religion, 420
  topography, 420, 421, 434
Africa, 53, 102, 103, 131, 176, 177, 434, 521, 531, 538, 583
  agriculture, subsistence, 326, 327
  ancient, 20
  and Arabs, 469
  aridity, 326
  boundaries, fixing of, 335
    superimposed, 441
  boundary morphology, 387
  and China (Mainland), 472
  cities, ancient, 330
  culture hearth, western, 328-332
  culture in Middle America, 257,

258
  culture in South America, 278, 279, 283, 285, 295, 296, 308, 310
  culture in United States, 173
  cycle theory, 412, 413
  "Dark continent," 17, 467
  Europe, contact with, 245, 332, 333
  European colonial policies, 338, 339
    Belgian, 342, 343
    British, 339-342
    French, 342
    Portuguese, 343-345
  European competition for, 335, 336
  gold, 592
  herding, 328
  history, precolonial, 323, 324
  independence, problems of, 340
  irridentism, 118
  and Japan, 568
  languages, 325, 326
  long lot farming, 230
  and Middle America, 251
  migrations, 20, 553, 554
  pediplanation, 588
  periodic markets, 13, 51
  physiography, Continental Drift Theory, 319-323
    elevation, 318
    and Land Hemisphere, 318, 319
    mountains, 315, 316
    rift valleys, 316, 317
    rivers, 317, 318
  post-war political changes, 336, 337, 348
  precolonial outside contacts, 332
  prehistoric man, 323
  as a region, 17
  regions of, 345
    Central, 346
    East Africa, 346, 354, 356, 357-359
    Equatorial Africa, 346, 360, 361
    Southern Africa, 361-370
    West Africa, 346, 347, 350-354
  Sahara Desert transition zone, 398
  slave trade, 181, 192, 332, 333
  South African colonization, 333-335

states, ancient, 330-332
    precolonial, 332
  territoriality concept, 583, 585
  topography, 41
  trade, trans-Saharan, 328, 329
  tribes, 324, 325, 332
  as part of World-Island, 155
  see also North Africa and Southwest Asia
Afrikaners, see Boers, the
Agglomeration, 204, 205
Agra, 446
Agrarian Reform Law, Chile (1967), 307
Agriculture, Agrarian Reform Law, Chile (1967), 307
  China (1950), 504
Agricultural Producers' Cooperatives (China), 504
  agriculture-based cities, 332
  analogue area analysis, 153, 155
  ancient Middle Eastern, 20
  and Areal Functional Organizations, 564, 565
  basin irrigation, 392
  chinampa irrigation, 242
  collectivization of, 151-153, 155, 500-504
  communal ownership, 242, 244, 262, 263
  communes, 504
  cooperatives, 503, 504
  crop-livestock rotation, 596
  Culture System (Southeast Asia), 533
  decline of, 209
  development, ancient, 327, 377
  diffusion of innovations, 378
  economiendas, 546
  the fazenda, 312
  "forced crop" program, Africa, 344
  functional specialization, 56
  the hacienda, 244, 251, 252, 262
  individual ownership, 12, 178, 179, 181, 190, 262, 263
  and industrial revolution, 70
  Isolated State concept, 31-33, 37
  Kibbutzim, Israel, 406
  marginal land, use of, 571, 572
  mechanization of, 12
  methods, Japan, 571

Cairo, founding of, 390
Islamic empire, 385
and Israel, 405
in Kenya, 356
in Maghreb, 395
Middle East refugees, 405, 408
in Saudi Arabia, 410
and Southeast Asia, 519, 525
in Southwestern Europe, 26, 27, 29
in Spain, 102
trade, ancient, 469
in Uganda, 358
"Arab World," 3, 10, 11, 17, 395
languages, 375
Aragon, kingdom of, 29
Aragura Sea, 586
Arakan Mountains, 537
Aral Sea, 137, 385
Aramayo family, 302
Ararat, Mount, 415
Aravalli Range, 446, 447
Arawak Indians, 236, 237
Archimedes, 390
Ardennes, 84
Ardrey, Robert, 585
Areal Functional Organization, 564, 565
Area of Decision, the, 157
Argentina, 56, 115, 271, 272, 276, 295, 296, 306
agriculture, 283, 305
area, 303
boundary disputes, 281
Brazil, war with, 282
climate, 283
foreign investments in, 305
and Incas, 273
independence of, 280
industrialization, 305
the Pampa, 303, 305
petroleum, 305
population, 283, 303
population distribution, 305
railroads, 303
regions of, 303
size of, 259
urbanization, 285, 287-289, 291, 305
Arica, 302
Aristotle, 21
Arizona, 212, 223
Arkansas, 188, 192, 222, 224
Arkansas River, 183, 204
Armenia, 149, 150, 413, 415
Arnhem Plateau, 587
Artesian wells, 588, 589
Artic Circle, 98
Artic lands, 132, 136
Artic Ocean, 15, 38, 47, 96, 131, 143, 188, 201, 490
Artificial capitals, 447
Aruba, 252
oil refining, 295
Aryans, the, and Ceylon, 461, 462
in India, 425, 426, 437
in Nepal, 463
Ashanti empire, 333, 352
Ashkelon, 406
Asia, 23
migrations, 20, 26

population density, physiological, 570
see also North Africa and Southwest Asia; Southeast Asia; and Southern Asia
Asians, in Africa, 17
in Kenya, 356
in South Africa, 364, 367
in Tanganyika, 356
in Uganda, 358
Askum Kingdom, 401
Asoka, 427
Assab, 403
Assam, 423, 431, 434, 436, 439, 443, 445, 449, 463, 490, 521
Assam Hills Division (India), 445
"Assimilation" (France), 338
Assyria, 407
and ancient Egypt, 383
and Persians, 384
rise of, 381
Astrakhan, 160, 161
Asuncion, 287, 288, 303
Aswan High Dam, 393
Atacama Desert, 281, 283, 306, 307, 597
Ataturk (Mustafa Kemal Pasha), 413, 415, 416
Athens, 21, 45, 46
focality, 105
as permanent capital, 447
population, 104
as primate city, 77
Atlanta, 45, 207
Atlantic Ocean, 3, 5, 34, 67, 73, 96-99, 136
and climate, 48
land bridges, 235
typhoons, 559
Atlas Mountains, 43, 315
climate, effect on, 395
trade across, 398
Auckland, 604, 605
Augelli, J.P., 251, 283, 285
Augsburg, 80
Augusta, 185
Aurangzeb, 430
Australasia, 581, 605
Australia, 11, 15, 64, 66, 86, 543
aborigines, 583, 585, 593
and Africa, 335
agriculture, 583, 592, 596
plantation, 593, 596, 597
area, 583
artesian wells, 588, 589, 594
cattle, 595, 596
China, trade with, 505
climate, 583, 587, 594, 595
coal, 597
colonies, competition, 593
self-government, 592
Continental Drift Theory, 320-323
convict "transportation," 585, 586, 592
dairying, 596
dingoes, 596
drainage patterns, 588
drought areas, 594-596
European exploration, 585
European orientation, 593

federation, 593
the frontier, 115
geology, 588
gold rush, 592, 595-597
grasslands, 303
history, 583, 585
immigration, 597, 599, 601
indentured labor, 593, 596
industrialization, 597-599, 601
iron and steel, 598, 599
irrigation, 596
isolation of, 585
Japan, trade with, 574
and Japan, 568
compared, 601
land purchase system, 591, 592
minerals, 592, 597, 598
New Zealand, cultural unity with, 581, 605, 607
and New Zealand, compared, 605
pastoralism, 589
pediplanation concept, 587, 588
penal colony, British, 585, 586, 589, 591, 596
peneplanation concept, 587
Perth colonization scheme, 591, 592
petroleum, 597
physiography, regions, 586, 587
phytogeography, 586
population, 280, 583, 601
aboriginal, 593
eugenic policies, 593, 594, 599, 601
Chinese, 593
urban, 599, 601
population density, 583
physiological, 583
population distribution, 280, 594
precipitation, 587
rabbit problem, 596
railroads, 592, 593
as a region, 17
settlement of, 591, 592
sheep, 591, 592, 594-596
size of, 581
soils, 587
standard of living, 601
territoriality concept, 585
topography, 41, 581, 586, 587, 603
urban centers, 601
zoogeography, 586
Australoid aborigines, 427, 583, 585, 593
Austria, 30, 36, 57, 58, 72, 88-99, 100, 109, 113, 123
area, 7
eastward orientation, 92
economic potential, 92, 93
industry, 88, 92, 93
irridentism, 118
petroleum, 50
political history, 92, 115
population, 8
Austrian Alps, the, 112, 123
Austro-Hungarian Empire, 92, 112, 113, 121, 123
Autonomous Oblasts, 149, 150
Autonomous Soviet Socialist Republics, 149

railroads, 513
situational assets, 497-498
Soviet occupation, 499, 513
topography, 499
and Trans-Siberian Railway, 145
Manchuria and east China, as a region, 485, 487-488
Manchurian Railroad, 478
Mandalay, 537
Mandate territory, definition of, 339-340
Mangla Dam, 460
Manhattan, 227, 289
Manila, 297, 537, 546-547
Manitoba, 188, 201, 211
Mannar, Gulf of, 461
Manufacturing industries, *see* Secondary industries
Maori peoples, 602-603
Mao Tse-tung, 477-479, 483, 485, 503, 507-508
Maracaibo, 288, 295
Maracaibo, Lake, 295
Marathi language, 445
Mariinsk Canals, 160
Maritime boundaries, 541, 543-544
Maritime Provinces (Canada), 185, 196, 233
Maritsa River, 128
Marmara, Sea of, 413, 415
Marne River, 76
Marrakech, 398
Marseilles, 21, 51, 73, 75, 100
Marshall Plan, 57
Martaban, Gulf of, 544
Martinique, 252, 255, 258
Maryland, 177, 179, 181, 185, 190, 210
Marx, Karl, 149
Masai, the, 326, 328, 356, 554
Mashad, 418
Massawa, 403
Massif Central, 77-78
Matadi, 343, 361, 538
Mato Grosso, 80, 271, 309-310
Mau Mau revolt, 340, 356
Mauritania, 330, 347
    Arabic language, 375
    boundary disputes, 388
    minerals, 597
    Moroccan claims on, 398, 413
    pastoralism, 352
    population, 347, 352
Mauryan state, 427, 430-431, 446
May 4 Movement (China), 477-478
Maya empire, 236-237, 239-240, 242, 249, 269, 276
Mayan language, 239
Mbeya, 354
Mecca, 331, 385, 400, 411, 554
Mecklenburg, 32
Medellín (Colombia), 287-288, 296
Medina, 385, 411
Mediterranean Europe, 44-46, 72, 93, 141, 143
    agriculture, 50, 105, 107
    climate, 49-50, 101
    colonialism, 102-103
    Greek and Roman heritage, 100-101
    hydroelectric power, 101
    industry, 106-109

land erosion in, 45-46, 101
languages, 101
mineral resources, 101
political history, 101-103
population, 103-104
precipitation, 101
as a region, 61, 100-109
rice, 570
and Scandinavia, compared, 103-104
site and situation, 100
soils, 50
standard of living, 104
topography, 101
Mediterranean Sea, 3, 5, 17, 20-21, 23, 25, 30, 38, 47-48, 73, 126, 373
    and Israel, 405
    land bridges, 235
    and Roman Empire, 384
Megalopolis, 225-229, 288
Meiji Restoration, 551, 555, 560-561, 563, 568
Meinig, D.W., 157
Mekong River, 490, 523, 525, 530-531, 534-535, 538 540
Melanesia, 583
Melbourne, 587, 592, 594, 596, 601
Memphis-Chattanooga railroad, 195
Mendoza, 305
Menelik II, Emperor (Ethiopia), 402-403
Mercantilism, 30-31, 63
Meroë, 330
Merrimac River, 185
Meseta Central (Spain), 238, 101, 104
Mesoamerica, 237-242, 246, 249
    Aztec empire, 244
Mesolithic period, 19-20
Mesopotamia, 272, 377-379, 381-385, 409, 417-418, 425, 472
Mestizos, 236, 258, 264
Mestizo-Transitional culture sphere (South America), 285, 295
metamorphic rock, 200-201
Metz, 52
Meuse River, 40, 74, 87
Mexica empire, *see* Aztec empire
Mexican-Americans, 173, 222
Mexican Empire, 258
Mexican Plateau, 261
Mexican Revolution (1910), 262
Mexico, 8, 15, 176, 235-237, 241, 249, 265
    agriculture, 261-263
    area, 252
    climate, 262
    distinctiveness of, 258-259, 262
    ejidos, 262-263
    Indian culture in, 262
    Indian villages, 247
    industrialization, 263
    land reform, 262-263
    land tenure, 244
    Mainland-Rimland concept, 251
    minerals, 201, 263
    and oscillation principle, 29
    petroleum, 263
    physiography, 238-239
    population, 252, 259

Population characteristics, 259, 261
population, Indian, 236
Revolution (1910), 262
U.S. boundary, 258-259
urbanization, 261-262
Mexico City, 238, 247, 261-264
Mexico, Gulf of, 176, 179, 185, 188, 199, 245, 247
Mexico, Valley of, 238, 241-242, 247
Miami (Florida), 45, 482
Miami River, 183
Michigan, 40, 172-173, 183, 187, 195, 201
Middleback Mountain Range, 598
Midlands, the (Britain), 77, 106
Middle America, 30, 173, 249, 258-259, 261-263, 295, 324, 377, 539
    agriculture, 251-252, 255, 262, 266-267
        development of, 377
    Aztec empire, 237, 240-242, 244-245
    banana revolution, 266-267
    boundary disputes, 258
    Carib and Arawak Indians, 236-237
    Caribbean America, 249, 255-258
    Ciboney Indians, 236
    countries of, 136, 235
    cultural diversity, 252, 254, 263-264
    culture hearth, 15
    ejidos, 252
    ethnic diversity, 236
    European powers in, 249, 251-252, 254
    finca system, 266
    forest dwellers, 249
    Greater Antilles, 252, 254-255
    the *hacienda,* 251-252, 262
    Indian villages, 247
    as land bridge, 235
    land erosion, 246
    land reform, 242, 244
    language, 258
    Lesser Antilles, 252, 254-255
    mainland republics, 252, 258, 263-267
    Mainland-Rimland concept, 249, 251-252
    Maya empire, 236-237, 239-240, 242
    Mesoamerica, 237-244
    minerals, 255
    Nahuatl Indians, 241
    periodic markets, 353
    plantations, 251-252
    political divisions, 252
    population, Indian, 236
    population explosion, 255
    physiography, 238-239
    as a region, 15
    shape of, 235
    social problems, 265-266
    Spanish-Indian culture contact, 245-247, 262
    Spanish colonialism, 247, 249
    Tarascan Indians, 241, 245
    Toltec Indians, 239-240
    tourism, 256

in Taiwan, 516
topography, 40
Netherlands Antilles, 251, 255
Nevada, 201, 417
Newar peoples, 463
Newari language, 463
New Brunswick (Canada), 185, 188-189, 196
New Caledonia, 552
Newcastle (Australia), 589, 597, 599
Newcastle (Britain), 34, 64
New Delhi, 445-446
New Economic Policy, the (Soviet Union), 151
New England, 177-178, 181, 185, 187, 190, 192
Newfoundland, 46, 179, 185, 196, 201
New France, 176, 196
New Granada, 276, 279-280, 287
New Guinea, 552, 583, 586, 593
New Hampshire, 177-178
New Hebrides, 552, 593
New Jersey, 177, 183, 226
"New Holland," 591
New Orleans, 176, 207-208
New Pien see Grand Canal
Newport (Wales), 67
New South Wales, 585, 592-597
New Spain, 176
New Territories (Hong Kong), 497
New York City, 45, 173, 185, 206, 216, 219, 225, 288-289
and megalopolis, 225
as primate city, 202, 208
and Tokyo, compared, 563
New York state, 177, 183, 187, 202
New Zealand, 11, 17, 47, 64, 552, 581, 585, 601-605, 607
Nhambiquara Indians, 310
Niagara Falls, 201
Nicaragua, 239, 235-236, 249, 258, 263-265
Nicaragua, Lake, 265
Niger, 347, 352, 398
Niger River, 318, 321-322, 330, 350, 352
Niger-Congo languages, 325
Nigeria, 17, 56, 318, 323-324, 326, 330-331, 339, 346, 554
agriculture, 352
art, ancient, 332
and Cameroun, 361
civil war, 351
economy, 350
foreign trade, 347
independence, problems of, 338, 340-341
and Islam, 375
periodic markets, 353
population, 347, 350, 352
Nikopol, 137, 162
Niksic, 125
Nile River, 21, 317-318, 321-322, 330, 383, 425
flood stages, 379
and modern Egypt, 390
population, 347, 373, 375
rock beds, 381
and Sudan, 398, 401
technology, ancient, 377

Nilo-Hamitic peoples, 358
Nilotic peoples, 326, 356, 358
Nizhnil, 163
Nkrumah, Kwame, 347
Nobi Plain, see Nagoya
Nodality, 205
Norden, 93; see Nordic Europe
Nordic Europe, as a region, 61, 93-94, 96-100, 109
Norfolk, 195
Norilsk, 139, 161
Normandy, 13
Normans, the, 13, 27, 63
North Africa, 20-23, 30, 43
North Africa and Southwest Asia, as a region, 17, 390, 392-393, 403, 405-410, 415-417, 420-421
agriculture, ancient, 377
as "Arab world," 375
boundary disputes, 388
boundary morphology, 386-388
Bronze Age, 377
civiliations, ancient, 375, 377-379, 381-384
climate, 373, 375, 382, 395
cycle theory, 411-413
diffusion of innovations, 377-378
as "dry world," 373, 375
ecological trilogy concept, 418, 419
Islamic empire, 385
languages, 375
as "Middle East," 375
mineral deposits, 397
modern States of, 386
Ottoman Empire, 385-386
petroleum territories, 388-389
physiography, 402
population distribution, 373
religions, 375
Roman Empire in, 384
Sahara Desert, economic functions, 398
source area, ancient, 373
Sultanates and Sheikdoms, 411
technology, ancient, 377
topography, 395
urban dominance concept, 419
as "World of Islam," 375
North America, 40-41, 178
accessibility, 201-204
agriculture, 188, 201, 209-210
biculturalism, Canadian, 195-196
Black America, 180-181, 183, 192, 195-197, 214-216, 219
British convicts in, 585
British culture in, 195
circulation, 230-231, 233
climate, 46-47, 49, 186-188
colonies, Dutch, 177
colonies, English, 176-177
colonies, pre-Revolutionary, 178
complementarity, 174-176
Continental Drift theory, 322
core region, 202-209, 213-214, 225, 231-233
environmental determinism, 174-176
European colonialism, 176
explorations of, 176
the frontier, 115

grasslands, 188
indentured servants, 177-178
Indians, 174-176, 179, 190-191, 222-223
industrialization, 197
and Interactance Hypothesis, 229-230
megalopolis, 225-228
migrations, 223-224
minerals, 197-201
New England settlement, 177
and oscillation principle, 29
physiographic provinces, 185-190, 197-199
plantation economy, 177-178, 180-181
plural societies, 172-173
population distribution, 171-172
poverty, 219, 221-223
as a region, 15
regional separatism, 190-193, 195-197
regions, crop, 210-213
regions, physiographic, 136
settlement patterns, 178-179
single-owner farms, 178
slavery, 177-178
standard of living, 171
topography, 137, 177
transportation networks, 54
"triangular trade," 181
urbanization, 197, 204-205, 208-209, 213-214, 228-229
westward expansion, 183
westward movement, 187, 189-190
Northampton, 64
North Atlantic Drift, 48
North Atlantic Treaty Organization, 58, 98, 415
North Cape (Norway), 93
North Carolina, 177, 181, 187, 190, 192, 210, 213
North China Plain, 479-480, 487, 493, 554
North Dakota, 211, 289
Northeast Frontier Agency (NEFA) (India), 465
Northeast Uplands (China), 488
Northern Ireland, see Ulster
Northern Italian Merchant League, 29
Northern Oriental (Congo), 343
Northern Territory (Australia), 592
North European Lowland, 34, 39-42, 46-49, 57, 63, 73, 77-78, 94, 112, 119, 137
North German Plain, 80
North Indian Plain (India), 435-436, 465
North Island (New Zealand), 601, 604-605
North Sea, 3, 11, 29, 38, 44, 47-48, 67, 73, 75, 86, 119
North-South Canal, Europe, 73
Northwest Territory (United States), 190, 195
Norway, 3, 58, 63, 93-94, 96-101, 105
area, 7
coastal claims, 543
elongated shape, 538
fishing, 50, 97

MEAN ANNUAL PRECIPITATION OF THE WORLD

Simplified and modified from Plate 3, World Climatology,
Volume I, The Times Atlas, Editor John Bartholomew,
The Times Publishing Company, Ltd., London, 1958

Inches	0	4	12	20	40	80	200	above 200
Centimeters	0	10	30	50	100	200	500	above 500

Isohyets labeled in inches (approximate), centimeters in parentheses.

MILES

| 0 | 1,000 | 2,000 | 3,000 |

(True distances on mid – meridians and parallels 0 to 40)

Based on Goode Base Map, copyright by the University of Chicago;
used by permission of the University of Chicago Press. Prepared by
J. P. Tremblay, cartographer, for John Wiley and Sons, Inc., Publishers.

TROPIC OF CANCER

TROPIC OF CAPRICORN

Plate 1
MEAN ANNUAL PRECIPITATION
OF THE WORLD

KÖPPEN-GEIGER SYSTEM OF CLIMATE CLASSIFICATION
After R. Geiger and W. Pohl (1953)

Key to letter code designating climate regions:

FIRST LETTER

A C D   Sufficient heat and precipitation for growth of high-trunked trees.

A *Tropical climates.* All monthly mean temperatures over 64.4° F (18° C).

B *Dry climates.* Boundaries determined by formula using mean annual temperature and mean annual precipitation (see graphs).

C *Warm temperate climates.* Mean temperature of coldest month: 64.4° F (18°C) down to 26.6° F (−3° C).

D *Snow climates.* Warmest month mean over 50° F (10° C) Coldest month mean under 26.6° F (−3° C).

E *Ice climates.* Warmest month mean under 50° F (10° C)

SECOND LETTER

S Steppe climate.   ⎫
W Desert climate. ⎬ Boundaries determined by formulas (See graphs).

f Sufficient precipitation in all months.

m Rainforest despite a dry season (i.e., monsoon cycle).

s Dry season in summer of the respective hemisphere.

w Dry season in winter of the respective hemisphere.

THIRD LETTER

a Warmest month mean over 71.6 °F (22°C).

b Warmest month mean under 71.6°F (22°C). At least 4 months have means over 50°F (10°C).

c Fewer than 4 months with means over 50°F (10°C).

d Same as c, but coldest month mean under −36.4°F (−38°C).

h Dry and hot. Mean annual temperature over 64.4 °F (18°C).

k Dry and cold. Mean annual temperature under 64.4 °F (18°C).

H Highland climates.

TROPIC OF CANCER

TROPIC OF CAPRICORN

MILES

0   1,000   2,000   3,000

(True distances on mid-meridians and parallels 0 to 40)
Based on Goode Base Map, copyright by the University of Chicago; used by permission of the University of Chicago Press. Prepared by J. P. Tremblay, cartographer, for John Wiley and Sons, Inc., Publishers.

Plate 2
CLIMATES OF THE WORLD

EQUATOR

BOUNDARIES OF DRY CLIMATES

BW
Desert

BS
Steppe

$R = 22 (T-19.5)$

$R = 44 (T-19.5)$

A, C, or D
(Humid climates)

Precipitation evenly distributed throughout year

BW
Desert

BS
Steppe

$R = 22 (T-7)$

$R = 44 (T-7)$

A, C, or D
(Humid climates)

Precipitation concentrated
in summer (high-sun) season

BW
Desert

BS
Steppe

$R = 22 (T-32)$

$R = 44 (T-32)$

A, C, or D
(Humid climates)

Precipitation concentrated
in winter (low-sun) season

R = MEAN ANNUAL PRECIPITATION, INCHES

GREAT SOIL GROUPS OF THE WORLD

Simplified and modified from a world map of soils by
A. C. Orvedal in Goode's World Atlas 11th edition,
1960, Edward B. Espenshade, Jr. Editor.
Used by permission of Rand McNally & Company,
Chicago, Illinois.

Color Significance: Podzolization-blues. Laterization-reds. Calcification-browns.
Salinization-yellows. Gleization-greens.

SOILS OF PLAINS AND HILLS

1 Alluvial

2 Tundra
(including lithosol and marsh)

3 Arctic brown forest

4 Podzol and weakly podzolized

5 Gray-brown podzolic

6 Red-yellow podzolic-latosolic

7 Degraded chernozem.

8 Prairie and chernozem

9 Chestnut and brown

10 Reddish prairie, reddish chestnut,
and reddish brown

11 Sierozem and desert
(including lithosol and sand)

12 Red desert
(including lithosol and sand)

13 Black and dark gray soils of
wet-dry tropics

14 Terra rossa, brown forest,
and rendzina

15 Latosolic soils of wet-dry tropics

16 Latosolic soils of continuously
humid tropics

SOILS OF MOUNTAINS
(including some hill areas with steep slopes)

17 Mountain soils of tundra zone
with lithosols.

18 Mountain soils of podzolic soil zones
with lithosols, including islands of
tundra and alpine meadow soils at
high elevations.

19 Mountain soils of chernozem,
chestnut, reddish chestnut, brown
and reddish brown soil zones with
lithosols, including islands of podzol,
alpine meadow or tundra soils at high
altitudes.

20 Mountain soils of sierozem, desert, and red desert
soil zones with lithosols, including islands of brown,
reddish brown, chestnut, reddish chestnut, or
chernozem soils at high elevations.

21 Mountain soils of latosolic soil zones with lithosols,
including islands of podzolic and alpine meadow
soils at high elevations.

22 Mountain soils of desert and latosolic soil zones
with lithosols, including reddish brown, reddish
chestnut and black soils of tropics at intermediate
elevations and islands of alpine meadow soils at high elevatic

23 Mountain soils of brown forest, terra rossa and rend
soil regions with lithosols, including podzolic and alp
meadow soils at high elevations.

Plate 3
GREAT SOIL GROUPS
OF THE WORLD

EQUATOR

TROPIC OF CAPRICORN

MILES
0   1,000   3,000

(True distances on mid-meridians and parallels 0 to 40)
Based on Goode Base Map, copyright by the University of Chicago;
used by permission of the University of Chicago Press. Prepared by
J. P. Tremblay, cartographer, for John Wiley and Sons, Inc., Publishers.

NATURAL VEGETATION REGIONS
OF THE WORLD

Simplified and modified from a world vegetation map by
H. Brockmann Jerosch, 1951, showing vegetation-classes
according to the classification of Eduard Rübel.

	Map classes	Equivalent formation classes
1	A. EQUATORIAL & TROPICAL RAINFOREST	1. Equatorial rainforest 2. Tropical rainforest
2	B. TEMPERATE RAINFOREST (LAUREL FOREST)	4. Temperate rainforest (Laurel forest)
4	C. RAINGREEN FOREST, WOODLAND, SCRUB, & SAVANNA	3. Monsoon forest (Tropical deciduous forest) 8. Savanna-woodland 9. Thornbush & Tropical scrub 10. Savanna
3	D. EVERGREEN-HARDWOOD FOREST	7. Evergreen-hardwood forest (Sclerophyll forest)
7	E. SUMMER-GREEN DECIDUOUS FOREST	5. Summer-green deciduous forest
8	F. NEEDLELEAF FOREST	6. Needleleaf forest
5	G. STEPPE & PRAIRIE GRASSLANDS	14. Prairie 15. Steppe
6	H. DRY DESERT & SEMI-DESERT	17. Dry desert 11. Semi-desert
9	I. TUNDRA (ARCTIC & ALPINE)	16. Grassy tundra 17. Cold woodland 18. Arctic fell-field
10	J. ICE-CAPS & GLACIERS	
	Highland areas with sufficiently high elevation to have alpine tundra.	

TROPIC OF CANCER

MILES

0   1,000   2,000   3,000

(True distances on mid–meridians and parallels 0 to 40)

Based on Goode Base Map, copyright by the University of Chicago;
used by permission of the University of Chicago Press. Prepared by
J. P. Tremblay, cartographer, for John Wiley and Sons, Inc., Publishers.

Plate 4
NATURAL VEGETATION REGIONS
OF THE WORLD

LANDFORMS OF THE WORLD
by Richard E. Murphy, Ph.D.

ALPINE SYSTEM	CALEDONIAN AND HERCYNIAN (OR APPALACHIAN) REMNANTS	GONDWANA SHIELDS	LAURASIAN SHIELDS
AM	CM	GM	LM
AW	CW	GW	LW
AT	CT	GT	LT
AH	CH	GH	LH
AD	CD	GD	LD
AP	CP	GP	LP

RIFTED SHIELD AREAS	SEDIMENTARY COVERS OUTSIDE SHIELD EXPOSURES	ISOLATED VOLCANIC AREAS
RM	SM	VM
RW	SW	
RT	ST	VH
RH	SH	
RD	SD	
RP	SP	

M–Mountains
W–Widely spaced mountains
T–High tablelands
H–Hills and low tablelands
D–Depressions or basins
P–Plains

.... i-Ice caps at present
- - - - - w-Wisconsin or Würm glaciated areas
– – – g-Pre-Wisconsin, pre-Würm and undifferentiated Pleistocene glaciated areas
h-Humid landform areas
d-Dry or arid landform areas
———— Division between humid and dry landform areas

Continental shelf

– – – Undersea axial connections of the Alpine system

Major oceanic rift and fault lines

TROPIC OF CANCER

ANTARCTIC CIRCLE

Plate 5
LANDFORMS OF THE WORLD

CHi

CIRCLE
Mw

SPw

SPg

SPh

SHh

AMg

SPd

ADd

AMh

SHd

GHh

AMh

GHh

SHd

EQUATOR

70°

80°

90°

100°

110°

120°

10°

GMh

AMh

GHd

SPd

CHh

TROPIC OF CAPRICORN

GHd

SMh

AMh

AMg

60°

50°

40°

30°

20°

10°

0°

10°

20°

30°

40°

50°

60°

70°

10° 20° 30° 40° 50° 60° 70°

100° 110° 120° 130° 140° 150° 160° 170°

GTi

GTi

Map supplement number 9,
Annals of the Association of
American Geographers
Volume 58, number 1, March 1968.

Based on Goode's base map,
copyright by the University of Chicago;
used by permission of the University of
Chicago Press.
Prepared by J. P. Tremblay, cartographer,
for John Wiley and Sons, Incorporated,
Publishers.

0   500   MILES 1000   1500

(True distances on mid-meridians
and parallels 0 to 40)

WORLD POPULATION DISTRIBUTION

INHABITANTS

Per square kilometer		Per square mile
Under 2		Under 2
1 to 9		2 to 24
10 to 24		25 to 59
25 to 49		60 to 124
50 to 100		125 to 250
Over 100		Over 250

○ Cities with 1,000,000 to 2,500,000 population

● Cities with over 2,500,000 population

```
0          1,000          2,000
            MILES
```

True distances on mid-meridians and parallels 0 to 40

TROPIC OF CANCER

TROPIC OF CAPRICORN

WORLD
POPULATION DISTRIBUTION

TIC CIRCLE

Leningrad

Gorki
Moscow

mbur
Berlin
Warsaw
Prague
Kiev
Kharkov
Munich
Vienna
Budapest
Milan
Bucharest
Rome
Istanbul
Naples

Novosibirsk

Tashkent

Harbin
Shenyang
Peking
Tientsin
Taiyuan
Lushun Tsien
Sian
Chengtu
Chungking
Wuhan

Pyongyang
Seoul
Fusan
Kyoto
Tsingtao
Kitakyushu
Nanking
Shanghai

Tokyo
Yokohama
agoya

Tehran

Baghdad

Alexandria

Cairo

Lahore

Delhi
Kanpur
Ahmedabad
Calcutta

Karachi

Bombay

Hyderabad

Madras

Bangkok

Canton

Taipei

Manila

EQUATOR

Saigon

Singapore

Djakarta
Surabaja

NTARCTIC CIRCLE

Sydney

Melbourne

Based on Goode Base Map,
copyright by the University of Chicago;
used by permission of the University of Chicago Press.
Prepared by J. P. Tremblay, cartographer,
for John Wiley and Sons, Inc., Publishers.

WORLD POLITICAL DIVISIONS

WORLD POLITICAL DIVISIONS

Based on Goode Base Map, copyright by the University of Chicago; used by permission of the University of Chicago Press.

Prepared by J. P. Tremblay, cartographer, for John Wiley and Sons, Inc., Publishers.